PL V.

UNIVERSITY OF MAINE

RAYMOND H. FOGLER LIBRARY

DENSE Z-PINCHES

Illustration on the inside front cover

Martinus van Marum's capacitor bank was built around 1790 by John Cuthbertson in Amsterdam. At the time, the cost of materials was 2,200 Dutch guilders, maybe enough to purchase a nice house (but now US $ 1,000). The bank's footprint is 3.3 m square. Its 100 capacitors ("Leyden Jars") are 4 mm thick Bohemian glass bottles about 300 mm in diameter and 600 mm high. Their area is 0.5 m,^{2} covered inside and out by tin foil. A single jar's capacitance is around 5 nF, the bank's capacitance 0.5 μF. The permissible voltage might have been up to 60 kV, in which case the bank's charge is 30 mC and the energy is 1 kJ. The bank's energy density is about 160 J/m,^{3} and its specific cost 1$/J.

The picture does not show explicitly where to put the exploding wires. The likely place is in the center, where 1 m long rods connect the capacitors to the small sphere. Each rod's inductance may be about 2 μH, for a total of 0.5 μH. A characteristic discharge time is then 0.5 μs and a peak current is 60 kA.

The remaining one quarter of this bank is on display at Teylers Museum, Spaarne 16, Haarlem, the Netherlands (ph: 23 531 9010, fax: 23 534 2004) (Drawing courtesy Teylers Museum. Reproduced with permission).

Illustration on the inside back cover

Sandia National Laboratory's PBFA-Z generator is today's most powerful successor to van Marum's device. Two hundred years later the world's largest pulse power machine cost tens of millions in materials and untold additional millions in expertise. For a scale, try to find the person (Waldo?) close to the machine's center.

Various papers in these proceedings discuss PBFA-Z's parameters and physics results. Its capacitors are still about the same size, but now their capacitance is 250 times larger, 1.3 μF, and they are charged higher (up to 85 kV, sometimes even 100 kV): an upgrade to 3 μF capacitors is planned. A substantial more complicated connection to the load now has its inductance is reduced also 250 times to 10 nH. The peak current is more than 250 times larger, 20 MA, and 10 times shorter, 0.1 μs. Ten years ago, the nominally 14 MJ marx was built for 40 M$, almost the same nominal price per Joule (Photograph courtesy of Dr. R.B. Spielman, Sandia National Laboratory. Reproduced with permission).

DENSE Z-PINCHES

Fourth International Conference

Vancouver, Canada May 1997

EDITORS
Nino R. Pereira
Berkeley Research Associates

Jack Davis
Peter E. Pulsifer
Naval Research Laboratory

AIP CONFERENCE PROCEEDINGS 409

American Institute of Physics   Woodbury, New York

Authorization to photocopy items for internal or personal use, beyond the free copying permitted under the 1978 U.S. Copyright Law (see statement below), is granted by the American Institute of Physics for users registered with the Copyright Clearance Center (CCC) Transactional Reporting Service, provided that the base fee of $10.00 per copy is paid directly to CCC, 222 Rosewood Drive, Danvers, MA 01923. For those organizations that have been granted a photocopy license by CCC, a separate system of payment has been arranged. The fee code for users of the Transactional Reporting Service is: 1-56396-610-7/ 97/$10.00.

© 1997 American Institute of Physics

Individual readers of this volume and nonprofit libraries, acting for them, are permitted to make fair use of the material in it, such as copying an article for use in teaching or research. Permission is granted to quote from this volume in scientific work with the customary acknowledgment of the source. To reprint a figure, table, or other excerpt requires the consent of one of the original authors and notification to AIP. Republication or systematic or multiple reproduction of any material in this volume is permitted only under license from AIP. Address inquiries to Office of Rights and Permissions, 500 Sunnyside Boulevard, Woodbury, NY 11797-2999; phone: 516-576-2268; fax: 516-576-2499; e-mail: rights@aip.org.

L.C. Catalog Card No. 97-76959
ISBN 1-56396-610-7
ISSN 0094-243X
DOE CONF- 9705120

Printed in the United States of America

CONTENTS

Preface xv

FUSION AND ALTERNATE CONCEPTS

The Quest for a Z-Pinch Based Fusion Energy Source--
A Historical Perspective. 3
    J. Sethian

The Role of Z-Pinches and Related Configurations in Magnetized
Target Fusion 11
    I. R. Lindemuth

Computational Modeling of Wall-Supported Dense Z-Pinches 17
    P. Sheehey, R. A. Gerwin, R. Kirkpatrick, I. R. Lindemuth, R. Moses,
    and F. Wysocki

Driver Coupling to Quasistatic Z-Pinches 21
    G. Decker, W. Kies, and S. Stein

An Overview of the DZP Project at Imperial College 27
    M. G. Haines

UCI Staged Z-Pinch Facility. 39
    F. J. Wessel, B. Moosman, N. Rostoker, Y. Song, and A. Van Drie,
    P. Ney and H. U. Rahman

Comparative Studies on a Gas Embedded Compressional Z-pinch
in H_2 and D_2 47
    L. Soto, H. Chuaqui, R. Saavedra, M. Favre, E. Wyndham,
    M. Skowronek, P. Romeas, R. Aliaga-Rossel, and I. Mitchell

On the Radiative Collapse Phenomenon. 51
    P. Choi and C. Dumitrescu-Zoita

A Novel Cryogenic Fibre Maker for Continuous Extrusion. 55
    R. Aliaga-Rossel and J. Bayley

Neutron Production in the MAGPIE Generator Using CD_2 Fibres ... 61
    R. Aliaga-Rossel, I. H. Mitchell, and H. Schmidt

Modeling of the Neutron Yield in a Z-pinch 67
    E. O. Baronova and V. V. Vikhrev

Hot Spots In Fiber Pinches. 71
    J. P. Chittenden, R. Aliaga-Rossel, S. V. Lebedev, I. H. Mitchell,
    A. R. Bell, F. N. Beg, A. Lorenz, A. E. Dangor, M. G. Haines,
    and G. Decker

Spatial and Temporal Evolution of Neutron Emission From
a Small Plasma Focus 75
    S. L. Yap, S. P. Moo, C. S. Wong, P. Choi, and C. Dumitrescu-Zoita

Coronal Plasma Behavior in C and D_2 Fibres on the MAGPIE Generator 79
    S. V. Lebedev, R. Aliaga-Rossel, J. P. Chittenden, A. E. Dangor,
    M. G. Haines, J. F. Worley, and R. Saavedra

Steady State of Z-Pinch Taking into Account the Absorption
of Own Radiation 83
    A. Muravich

**Influence of the Hall Effect on a Neck Development
in the Z-Pinch Discharges** 87
    V. V. Vikhrev and O. Z. Zabaidullin
**Modeling of Z-Pinch Processes for Studying Initial Conditions
to Get Inertial Confinement Fusion** 89
    V. V. Vikhrev, A. V. Dobryakov, and O. Z. Zabaidullin
The MAGO System 93
    S. F. Garanin

SUPERPOWER GENERATORS

PBFA Z: a 60-TW/5-MJ Z-Pinch Driver 101
    R. B. Spielman, C. Deeney, G. A. Chandler, M. R. Douglas, D. L. Fehl,
    M. K. Matzen, D. H. McDaniel, T. J. Nash, J. L. Porter, T. W. L. Sanford,
    J. F. Seamen, W. A. Stygar, K. W. Struve, S. P. Breeze, J. S. McGurn,
    J. A. Torres, D. M. Zagar, T. L. Gilliland, D. O. Jobe, J. L. McKenney,
    R. C. Mock, M. Vargas, T. Wagoner, and D. L. Peterson
**Recent ACE 4 Z-Pinch Experiments: Long Implosion Time Argon
Loads, Uniform Fill Versus Annular Shell Distributions
and the Rayleigh-Taylor Instability Problem** 119
    P. Coleman, J. Rauch, W. Rix, J. Thompson and R. Wilson
Z-Pinch Implosion for ICF Physics Study on ANGARA-5-1 125
    A. V. Branitsky, M. V. Fedulov, E. V. Grabovsky, S. L. Nedoseev,
    G. M. Olejinik, V. P. Smirnov, and S. V. Zakharov
**Radious Scaling of X-Radiation from Gas-Puff Implosions
on an Inductive Driver** 135
    D. Mosher, S. J. Stephanakis, J. P. Apruzese, D. C. Black, J. R. Boller,
    R. J. Commisso, M. C. Myers, G. G. Peterson, B. V. Weber, and F. C. Young
**Wire Array Implosion Experiments on the Inductive Storage Generators
GIT-4 and GIT-8.** 141
    R. B. Baksht, I. M. Datsko, A. A. Kim, V. A. Kokshenev, B. M. Kovalchuk,
    S. V. Loginov, A. Yu. Labetsky, A. V. Fedunin, A. G. Russkikh,
    and A. V. Shishlov
**Comparison of Ribbon and Wire Array Load Characteristics
on Double Eagle** 145
    E. J. Yadlowsky, R. C. Hazelton, J. J. Moschella, B. H. Failor,
    P. D. LePell, C. A. Coverdale, J. P. Apruzese, K. G. Whitney, J. W. Thornhill,
    and J. Davis
**The Experimental Investigations of Imploding Plasma as a Source
of Hard X-Ray** 149
    Yu. L. Bakshaev, A. V. Bartov, P. I. Blinov, A. S. Chernenko,
    S. A. Dan'ko, Yu. M. Gorbulin, Yu. G. Kalinin, V. D. Korolev,
    V. I. Mizhiritskii, L. I. Rudakov, A. Yu. Shashkov, and S. A. Shibaev
The Dense Z-Pinch Program at the University of Nevada, Reno 153
    B. S. Bauer, V. L. Kantsyrev, F. Winterberg, A. S. Shlyaptseva,
    R. C. Mancini, H. Li, and A. Oxner

High Velocity Implosions on PBFA Z 157
  J. S. DeGroot, C. Deeney, T. W. L. Sanford, R. B. Spielman,
  K. G. Estabrook, J. H. Hammer, D. D. Ryutov, and A. Toor
Characteristics of X-Ray Emission in a Composite Pinch 161
  C. Dumitrescu-Zoita, P. Choi, A. Chuvatin, and B. Etlicher
**Improvement of X Rays Power of Gas Puff Z'Pinch: Dependence
on the Initial Density Profile**... 165
  P. Grua, J. M. Sajer, and A. Sevastianov
Plasma Focus Current Shell Implosion onto Foam Liner 169
  L. Karpiński, M. Scholz, W. Stepniewski, A. Szydlowski, A. V. Branitsky,
  M. V. Fedulov, S. F. Medovschikov, S. L. Nedoseev, V. P. Smirnov,
  and M. V. Zurin
Dynamic Hohlraum Experiments on SATURN 175
  T. J. Nash, M. S. Derzon, G. Allshouse, C. Deeney, J. F. Seamen,
  J. McGurn, D. O. Jobe, T. L. Gilliland, J. J. MacFarlane, P. Wang,
  and D. L. Peterson
Increasing A Z-Pinch's Hard X-Ray Yield 183
  L. I. Rudakov
Conversion Efficiency of Heat Flux into Hard X-ray Radiation.............. 187
  L. I. Rudakov, A. N. Starostin, I. I. Yakunin, A. B. Kukushkin,
  and V. S. Lisitsa
Decade Quad Load Performance... 193
  J. W. Thornhill, F. L. Cochran, J. Davis, J. P. Apruzese, and K. G. Whitney

RADIATION HYDRODYNAMICS

Application of 2-D Simulations to Hollow Z-Pinch Implosions.............. 201
  D. L. Peterson, R. L. Bowers, J. H. Brownell, C. Lund, W. Matuska,
  K. McLenithan, H. Oona, C. Deeney, M. S. Derzon, R. B. Spielman,
  T. J. Nash, G. A. Chandler, R. C. Mock, T. W. L. Sanford, M. K. Matzen,
  and N. F. Roderick
Implosion Dynamics of a Radiative Composite Z-Pinch..................... 211
  R. Benattar, P. Ney, A. Nikitin, S. V. Zakharov, A. A. Otochin,
  A. N. Starostin, A. E. Stepanov, V. K. Roerich, A. F. Nikiforov,
  V. G. Novikov, A. D. Solomyannaya, V. A. Gasilov, and A. Yu. Krukovskii
Diffusion in Multicomponent Liners 215
  V. I. Oreshkin
Compressional Pinch Simulations and Experiments....................... 219
  N. G. Kassapakis, H. M. Davies, and M. G. Haines
Magnetohydrodynamic Simulation of Capillary Plasmas................... 225
  N. A. Bobrova, S. V. Bulanov, D. Farina, R. Pozzoli, T. L. Razinkova,
  and P. V. Sasorov
**Comparison of Simulation and Experimental Results for A Gas Puff
Nozzle on Ambiorix** 229
  J-N. Barnier, J-M. Chevalier, and B. Dubroca
Pinhole Images and XRD Signals from Numerical Simulations 233
  R. Benattar and A. Nikitin

The Electric Resistance and Electron Viscosity of Z-Pinch Plasma 237
    A. A. Esaulov and P. V. Sasorov

Numerical Analysis of MHD Instability Suppression in a Double Gas Puff.. 243
    I. V. Glazyrin, O. V. Diyankov, N. G. Karlykhanov, and S. V. Koshelev

Radiation MHD Modeling of a Proposed Dynamic Hohlraum 247
    J. H. Hammer, J. S. DeGroot, M. Tabak, A. Toor, and G. B. Zimmerman

Modeling a Dense Z-Pinch Plasma With a Cold, Dense Core Using a 2D 2-Temperature MHD Code.................................... 253
    G. V. Ivanenkov, A. R. Mingaleev, S. A. Pikuz, V. M. Romanova, T. A. Shelkovenko, and W. Stepniewski

UCI Staged Z Pinch... 259
    P. Ney, H. U. Rahman, N. Rostoker, and F. J. Wessel

2D MHD Simulations of a Rayleigh–Taylor Instability in Z-Pinches 265
    D. Zdravkovic, A. R. Bell, and M. Coppins

Simulations of Radiatively-Driven Implosions on the PBFA-Z Facility 271
    J. B. Aubrey, R. L. Bowers, and D. L. Peterson

AXSTRAN: A Non-LTE Radiation Transport and Ionization Dynamics Code in Axisymmetric Two-Dimensional Geometry 277
    Y. K. Chong, T. Kammash, and J. Davis

A Detailed Postprocess Analysis of an Argon Gas Puff Z-Pinch Plasma Using SPEC2D 283
    Y. K. Chong, T. Kammash, and J. Davis

SPECIAL PINCHES (AND LASER SCHEMES)

Compact Plasma Focus Devices: Flexible Laboratory Sources for Applications.. 291
    R. Lebert, A. Engel, K. Bergmann, O. Treichel, C. Gavrilescu, and W. Neff

Control of X-ray Spectrum Emitted from a Gas-puff Z-pinch 299
    K. Takasugi, T. Miyamoto, K. Tatsumi, and T. Igusa

Stimulated VUV Radiation from Z-Pinch Necks........................... 303
    K. N. Koshelev, P. S. Antsiferov, L. A. Dorokhin, and Yu. V. Sidelnikov

Influence of Preionization on Dynamics of a Gas Puff Implosion............. 307
    A. G. Russkikh, R. B. Baksht, A. Yu. Labetsky, and A. V. Shishlov

High Density Plasmoid Acceleration by Phased Implosion of Capillary Z-Pinch and its Application to Hyper-Velocity Projectile Acceleration 311
    K. Horioka, M. Nakajima, T. Aizawa, and M. Tsuchida

Carbon Fiber Z-Pinch with Current Prepulse.............................. 315
    A. Lorenz, F. N. Beg, J. Ruiz-Camacho, J. F. Worley, and A. E. Dangor

Enhanced K Shell X-ray Yield from Over-massed Targets 319
    A. Fisher, R. W. Clark, J. Davis, and J. Giuliani, Jr.

Density, Temperature and Size of a Plasma Produced in Single and Double Shell Liner Implosions.. 323
    S. A. Chaikovsky and S. A. Sorokin

About Plasma Points' Generation in Z-Pinch 329
  V. I. Afonin, A. V. Potapov, V. P. Lazarchuk, V. M. Murugov,
  and A. V. Senik

Z-pinch Discharges-Bare and Plastic Coated Copper Wires 333
  F. N. Beg, J. Ruiz-Camacho, and A. E. Dangor

X-ray Emission from a Small 2 kJ Plasma Focus 339
  F. N. Beg, I. Ross, and A. E. Dangor

**Enhanced Propagation Rate of Magnetic Field in Plasmas Due
to the Hall Effect: Analytic Solutions in Electron MHD** 345
  K. V. Cherepanov and A. B. Kukushkin

Capillary X-Ray Laser Research 349
  J. P. Chittenden, M. Michaelis, S. N. Bland, M. D. Eaton,
  and J. F. Worley

**Experimental Studies on a Pulsed Hollow Cathode Capillary
Discharge** 353
  P. Choi, M. Favre, C. Dumitrescu-Zoita, J. Moreno, H. Chuaqui,
  and E. Wyndham

Observations of Vacuum Spark Dynamics from its X-Ray Emission 357
  H. Chuaqui, M. Favre, R. Saavedra, E. Wyndham, L. Soto, P. Choi,
  and C. Dumitrescu-Zoita

**Interaction of Plasma Jets Produced from Pinch Plasma
with Neutral Atoms in Order to Achieve an Effective Charge
Exchange Table Top X-laser** 361
  A. Engel, R. Lebert, K. N. Koshelev, Yu. V. Sidelnikov, S. S. Churilov,
  C. Gavrilescu, and W. Neff

Transition from Column to Micropinch Regime in Z-Pinches 367
  A. Engel, R. Lebert, K. N. Koshelev, Yu. V. Sidelnikov, C. Gavrilescu,
  and W. Neff

**Hotspot Features in a Small Plasma Focus Operating in H_2-Ar
and H_2-N_2 Mixtures** 373
  M. Favre, P. Choi, C. Dumitrescu-Zoita, P. Silva, H. Chuaqui,
  and E. Wyndham

**Large-Scale Spheromak-Like Magnetic Configuration (SLMC)
in High-Current Discharges: Self-Formation and Self-Compression
of the SLMC in Plasma Focus Experiments** 377
  A. B. Kukushkin, V. A. Rantsev-Kartinov, and A. R. Terentiev

**Short-Scale Mixing of the Plasma and Magnetic Field, and Magnetic
Flux Ropes in Plasma Focus Experiments** 381
  A. B. Kukushkin, V. A. Rantsev-Kartinov, A. R. Terentiev,
  and K. V. Cherepanov

Studies on a Small Modified Plasma Focus Opening Switch 385
  W. S. Leong, C. S. Wong, P. Choi, and S. P. Moo

Spectral Investigations of Micropinches in a Plasma Focus 389
  M. H. Liu, X. Feng, and S. Lee

**Effect of Anode End Structures on Plasma Pinching and Neutron
Yield in a Plasma Focus** 393
  M.-F. Lu

Evolution of the Filamentary Current Sheath in a Plasma Focus Device 397
    M.-F. Lu
Steady State of Elliptic Z-Pinches 401
    T. Miyamoto
Laser Probing of Fibre Z-Pinch Plasmas 407
    J. Ruiz-Camacho, F. N. Beg, and A. E. Dangor
Capillary and Transient Inversion Table-Top X-ray Lasers.................. 413
    V. N. Shlyaptsev, J. J. Rocca, P. V. Nickles, M. P. Kalachnikov,
    W. Sandner, A. L. Osterheld, and D. C. Eder
Preliminary Results on a Pulsed Capillary Discharge 417
    C. S. Wong, P. Choi, and T. Serguei
Study of Pulsed Soft X-ray Source Employing a Gas-puff Z-pinch
Plasma Device For Lithography Applications 423
    G. X. Zhang, X. M. Guo, C. M. Luo, S. Lee, and X. Feng

SPECTROSCOPY

Studies of X-pinch Plasma Fine Structure Using High Resolution
Optical and Imaging Spectroscopy Methods 429
    S. A. Pikuz, T. A. Shelkovenko, V. M. Romanova, G. S. Sarkisov,
    D. A. Hammer, and D. F. Acton
High Resolution Monochromatic X-ray Imaging System Based
on Spherically Bent Crystals.. 437
    Y. Aglitskiy, T. Lehecka, S. Obenschain, S. Bodner, C. Pawley, K. Gerber,
    J. Sethian, C. M. Brown, J. Seely, U. Feldman, and G. Holland
Study of X-ray Polarization and E-beams Generation during Hot-Spots
Formation in PF-Discharges.. 443
    L. Jakubowski, M. Sadowski, E. O. Baronova, and V. V. Vikhrev
Carbon Fiber X-Ray Lasing 449
    P. Kubeš and J. Kravárik
Spatially Resolved Thomson Scattering on a Gas-liner Pinch................ 455
    Th. Wrubel, I. Ahmad, S. Büscher, and H.-J. Kunze
Measurement of Gas Distributions from PRS Nozzles 459
    B. V. Weber, G. G. Peterson, S. J. Stephanakis, R. J. Commisso,
    and A. Fisher
Electron Beam Measurements in Carbon Fibres in MAGPIE................ 463
    R. Aliaga-Rossel, I. H. Mitchell, J. P. Chittenden, A. E. Dangor,
    M. G. Haines, and A. Robledo
SBS Pulse Compression Applied to a Commercial Q-Switch
Nd-YAG Laser 467
    R. Aliaga-Rossel, J. Bayley, A. Mamin, and Y. Nizienko
Optical Multi-Slit and X-Ray Measurements From Carbon
and Deuterium Pinches 471
    R. Aliaga-Rossel, S. V. Lebedev, J. P. Chittenden, A. E. Dangor,
    and M. G. Haines
Polarized X-rays from a Z-pinch Plasma 475
    E. O. Baronova

**Suprathermal Electron Diagnostics, Based on X-ray Line Radiation
Polarization Measurements** 483
    E. O. Baronova
**Influence of the Transient Plasma Dynamics on the Scaling
of the K-shell Line Emission in Pinch Plasma** 487
    K. Bergmann, R. Lebert, and W. Neff
XRAYFIL, an Analysis Code for Non-Dispersive X-ray Diagnostics 491
    C. Dumitrescu-Zoita and P. Choi
**Nonlocality of Radiative Transfer in Continuous Spectra
and Bremsstrahlung Radiation Transport in Hot Dense Plasmas** 495
    V. V. Ivanov and A. B. Kukushkin
**New EUV and X-ray Optical Instrumentation for Hot Plasma Imaging,
Polarimetry, and Spectroscopy, Using Glass Capillary Converters
and Multilayer Mirrors** 499
    V. L. Kantsyrev, B. S. Bauer, A. S. Shlyaptseva, R. F. Bruch,
    and R. A. Phaneuf
Diagnostics of Small Carbon Fiber XUV-Pulse 503
    J. Kravárik and P. Kubeš
Diagnostics of the Thick Carbon Fibre Z-pinch 507
    H. M. Davies, A. Lorenz, J. Kravárik, and P. Kubeš
Investigation of the Linear Z-Pinch Plasma by Means of Laser Scattering 513
    V. A. Rantsev-Kartinov and E. E. Trofimovich
Faraday Rotation Measurements in MAGPIE Generator 517
    M. Tatarakis, R. Aliaga-Rossel, A. E. Dangor, and M. G. Haines
Monochromatic X-ray Backlighting for Application to PBFA-Z 523
    S. A. Pikuz, T. A. Shelkovenko, D. A. Hammer, and D. F. Acton
Diagnostic of Energetic Electrons in Dense Z-pinch Plasmas 527
    A. S. Shlyaptseva and R. C. Mancini

STABILITY

A Review of the Stability of the Z-pinch 533
    M. Coppins
Stabilized Z-pinch Loads with Tailored Density Profiles 549
    A. L. Velikovich, F. L. Cochran, and J. Davis
Stability and K-shell Radiation of Z-Pinches 555
    R. B. Baksht, A. V. Fedunin, A. Yu. Labetsky, V. I. Oreshkin,
    A. G. Russkikh, and A. V. Shishlov
**Variation of High-Power Aluminum-Wire Array Z-Pinch Dynamics
with Wire Number, Load Mass, and Array Radius** 561
    T. W. L. Sanford, R. C. Mock, B. M. Marder, T. J. Nash, R. B. Spielman,
    D. L. Peterson, N. F. Roderick, J. H. Hammer, J. S. DeGroot, D. Mosher,
    K. G. Whitney, and J. P. Apruzese
Instabilities in Z-pinch and Liner Systems 575
    P. V. Sasorov, A. A. Esaulov, and S. L. Nedoseev
Metallic Wire Pinch Instability 579
    P. L. Auer and D. D. Ryutov

Recent Progress on Large Larmor Radius Theory 585
    M. Coppins, T. D. Arber, P. G. F. Russell, and J. Scheffel
Resistive Stability of the Z-pinch Revisited 589
    M. Coppins and I. D. Culverwell
K-shell Radiation Power and Yield From Double Shell Plasma Liner Implosions 593
    S. A. Sorokin and S. A. Chaikovsky
Double Shell Liner Implosions 597
    S. A. Sorokin and S. A. Chaikovsky
Analysis of Magnetic Interlayer Staged PRS Loads 601
    R. E. Terry and R. W. Clark
Effect of Enhanced Thermal Dissipation on the Rayleigh–Taylor Instability in Emulsion-Like Media 607
    A. Toor and D. D. Ryutov
Electron Beam Generation in the Turbulent Plasma of Z-pinch Discharges 611
    V. V. Vikhrev and E. O. Baronova
Resistance of Z-Pinch Current Sheath 615
    V. V. Vikhrev

E-Mail List of All First Authors 621
Author Index 625

Preface

These Proceedings contain virtually all the papers presented at the 4th International Conference on Dense Z-Pinches, held from May 28 to May 31, 1997, in Vancouver, BC, Canada. The Conference brought together some 129 delegates from 13 countries, with largest contingents from the USA, Russia, UK and France. Most groups active in z-pinch research are represented, with the latest developments in continuing research programs. The results represent a quantum increase in knowledge of the z-pinch, thanks to new and more comprehensive diagnostics, better computer models, and ever more powerful pulse generators. The most prominent example of the latter is the PBFA-Z machine at Sandia National Laboratories, whose record-shattering radiation output is being studied at Sandia and collaborating laboratories elsewhere.

There was welcome continuity with earlier conferences in this series. The substantial presence from the Imperial College group at this Conference is particularly appropriate because it is now 4 years after the grand opening of their MAGPIE facility, on the occasion of the 3rd Z-pinch Conference in 1993 at Imperial College, London, UK. As shown here, MAGPIE has proved to be quite useful in exploring z-pinch stability and issues relating to fusion. Opening the present conference was a review of prospects for thermonuclear fusion given by J. Sethian, the Chairman of the 1st Dense Z-Pinch Conference (Alexandria, VA, USA, 1984), which focused on fusion.

The Conference was especially honored to have Dr. R. S. Pease, FRS, as its after-dinner speaker. Prominent in nuclear research, Dr. Pease was an early motivator of z-pinch activity, and is well known for his work on pinch equilibrium leading to the derivation of the "Pease-Braginskii current." This fundamental concept was developed simultaneously by S. I. Braginskii, whom he met for the first time at the 3rd Dense Z-Pinch Conference. Dr. Pease's seminal article appeared in 1959 in the Proceedings of the Physical Society of London, immediately following that of J.D. Lawson defining the "Lawson criterion" for break-even thermonuclear fusion.

In the course of his remarks, Dr. Pease denied his paternity of the field by quoting an early observation of z-pinching, around 1900 in an Australian lightning rod. Interestingly, L. Rudakov then brought forward an even earlier ancestor, the late-1700's Dutch scientist Martinus van Marum. His capacitor bank was powerful enough to vaporize wires of different metals, up to bismuth. These Proceedings begin with a drawing of van Marum's bank (in the Teylers Museum, Haarlem, the Netherlands, courtesy J. Kistemaker), and end with a photograph of its most powerful modern counterpart, PBFA-Z (at SNL, Albuquerque, NM, USA, courtesy R. Spielman).

In view of the significant and ongoing funding reductions in defense-motivated z-pinch research since the last conference, the large attendance at this conference was a show of vitality that pleasantly surprised the Conference Organizers. The present number of attendees (129) exceeded the 117 in London at the 3rd Conference (AIP Proceedings 299) and the 107 at the 2nd Conference (Laguna Beach, 1989, AIP Proceedings 195). One interesting factor was a sizable contingent (36) of Russian scientists, many with contracts to collaborate with colleagues elsewhere, and others (13) based abroad full-time. Another factor is increasing interest in radiating Z-pinches in France, and continuing but smaller programs in many other countries.

One modern aspect of the present Conference was its extensive use of electronic communications and the Internet. A World-Wide-Web page was established at the Naval Research Laboratory and mirrored in Europe by Peter Röwekamp at the Heinrich-Heine University. Most abstracts were submitted electronically, in a variety of formats, and those arriving on paper were scanned. All abstracts were converted to TeX format for printing and to HTML for posting on the Internet. This allowed submission deadlines to be kept flexible and all attendees, regardless of country of origin, to have access to the most current information. This procedure will likely become routine for future conferences, but presently it is still challenging because of the lack of standardization in word processing, e-mail, and WWW interface formats.

To encourage continued electronic communication among z-pinch researchers, these Proceedings include attendees' e-mail addresses and a list of web pages on Z-pinches. Date and venue for a possible 5th Dense Z-Pinch Conference have not yet been determined, but should be discussed on the Conference Web page by the active Z-pinch community, starting in about two years. The address is http://wwwppd.nrl.navy.mil/zpinch/announcement.html. Your contribution is welcome.

Finally, we note that J. Davis (NRL) and C. Deeney (SNL) are editing a supplementary Special Issue on Z-Pinches for the IEEE Transactions On Plasma Science, with expected publication date mid-1998. Some abbreviated papers here will appear in full in the Special Issue, which in this sense can be considered as a companion volume to these Proceedings.

It is a pleasure to acknowledge financial support for the Conference by the US Defense Special Weapons Agency and the Naval Research Laboratory, who have been at the forefront of Z-pinch research in the US over the last two decades, and by the US Department of Energy's Sandia and Los Alamos National Laboratories, who are presently pushing the Z-pinch envelope.

N. R. Pereira
Berkeley Research Associates
J. Davis
P. E. Pulsifer
Naval Research Laboratory

FUSION AND ALTERNATE CONCEPTS

The quest for a z-pinch based fusion energy source--a historical perspective

John Sethian
Plasma Physics Division
Naval Research Laboratory
Washington DC 20375

Abstract. Ever since 1958, when Oscar Anderson observed copious neutrons emanating from a "magnetically self-constricted column of deuterium plasma," scientists have attempted to develop the simple linear pinch into a fusion power source. After all, simple calculations show that if one can pass a current of slightly less than 2 million amperes through a stable D-T plasma, then one could achieve not just thermonuclear break-even, but thermonuclear *gain*. Moreover, several reactor studies have shown that a simple linear pinch could be the basis for a very attractive fusion system. The problem is, of course, that the seemingly simple act of passing 2 MA through a stable pinch has proven to be quite difficult to accomplish. The pinch tends to disrupt due to instabilities, either by the m=0 (sausage) or m=1(kink) modes. Curtailing the growth of these instabilities has been the primary thrust of z-pinch fusion research, and over the years a wide variety of formation techniques have been tried. The early pinches were driven by relatively slow capacitive discharges and were formed by imploding a plasma column. The advent of fast pulsed power technology brought on a whole new repertoire of formation techniques, including: fast implosions, laser or field-enhanced breakdown in a uniform volume of gas, a discharge inside a small capillary, a frozen deuterium fiber isolated by vacuum, and staged concepts in which one pinch implodes upon another. And although none of these have yet to be successful, some have come tantalizingly close. This paper will review the history of this four-decade long quest for fusion power.

INTRODUCTION

The linear z-pinch, in which a current both confines and heats a plasma column, has been considered as the basis for a very attractive fusion system since the earliest days of controlled fusion research. Because the current is large and the column diameter small, the magnetic field is high enough to confine a plasma at very high density. Simple 1-d models show that if one could pass 4 MA through a 20 µm radius, solid density pinch, it could produce up to 8.6 MJ of fusion energy, burning up to 14% of the D-T fuel in the process (1). All this assumes, of course, that the pinch can be made to be stable, and therein lies the challenge. Attempts to achieve stability have been based on two fundamental approaches: One is to form the pinch through a snowplow implosion of a gas load, the other is to start it at a small diameter and high density and keep it that way as the current is increased. In the early experiments the current was driven with low voltage capacitor banks. In later experiments the pinch was formed with high voltage pulsed power drivers.

Z-PINCH PHYSICS

In a seminal paper published in 1958, Anderson *et al* (2) describe most of the phenomenological behavior that would be observed in pinches over the next four decades: namely that the pinch is subject to rapid growth of the m=0 (sausage-mode) instability. Anderson and his team used a 20 kV capacitor bank to implode a linear plasma column in about 1.5 μsec. They observed 10^8 neutrons per pulse emanating along the length of the compressed D-D plasma column. This was a very impressive number for a laboratory experiment in 1958. However, using voltage, current, and neutron diagnostics Anderson concluded that the neutrons were not produced by a Maxwellian plasma, as originally hoped, but by deuterons accelerated in the large electric fields produced by the m=0 instability. Thus they concluded the neutrons were not of "thermonuclear origin", and that the pinch would not be appropriate for a fusion reactor.

Others had a different interpretation to Anderson's results, and that was that if the pinch could be driven fast enough, it would implode to fusion conditions faster than it would take the instability to grow. The two most prominent experiments of the day were the Columbus-II machine at Los Alamos (3) which used a circular array of conventional capacitors in a very low inductance arrangement, and the SuperFast Pinch at Space Technology Laboratories (4) which mounted the pinch in the center between two parallel circular plates. In effect the pinch was the load of a radial transmission line. The SuperFast could drive currents up to 300 kA at a rate of 20 kA/nsec. These are parameters worthy of a modern pulsed power device. Unfortunately pinches formed with either machine still exhibited the classic m=0 instability. As a result the pinch was abandoned in favor of increasingly more complex schemes. These included adding an axial magnetic field, adding a central core, closing the pinch on itself to form a toroid, and eventually adding a sheared magnetic field. While all showed the promise of enhancing the plasma stability, the beautiful simplicity of the linear pinch was lost.

In 1964, E. Smars at the Royal Institute in Stockholm tried a different tack. He formed the pinch at a small diameter inside a uniform volume of hydrogen at 1 atmosphere (5). This later came to be known as the gas embedded pinch. By starting the pinch at a small diameter, Smars hoped to avoid having to dissipate the energy associated with the radial motion of an implosion pinch. He also expected the surrounding gas to impeded any unstable motion. Smars used a 1 A vortex stabilized DC arc to initiate the pinch, and then drove 100 kA through the plasma at a rate of 0.1 kA/nsec. In a beautiful series of Kerr-cell photos, Smars observed the pinch to undergo an m=1 (or kink) instability, in exact accordance with ideal MHD plasma theory. He also observed the pinch accreting into the surrounding gas...I.E. the line density increased and the temperature decreased. As a result the pinch was abandoned again as a candidate for a fusion power source.

The next phase in z-pinch development was spurred by the advent of modern pulsed power generators. These were capable of generating voltages that were factors of ten above what had been achieved before..up to 1000 kV vs. 20 kV, and thus enabled very large currents (100-1000 kA), with fast rise times (1-15 kA/nsec) to be driven through inductive loads such as z-pinches. One of the first experiments was carried out at Imperial College, in which a 600 kV pulse forming line was coupled to a conventional implosion pinch (6). The pinch was formed inside a quartz tube 1 cm in diameter. The current through the pinch rose to 100 kA with a risetime of 1 kA/nsec. The plasma contracted from a radius of 5 mm to about 0.2 mm in about 20 nsec. Electron temperatures as high as 1 keV were reported, with a peak density of 10^{18}/cm^3. However there was no evidence that the ions themselves were getting hot. Others used even larger machines to drive an implosion pinch. For example, the Angara 5-1 device formed the pinch at currents up to 3 MA (7). While these experiments generated an impressive number of neutrons (10^{12} per pulse), it was clear from anisotropy measurements that these were generated through the m=0 instability, and were, in fact, far less than would be generated if they were coming from a uniformly stable pinch carrying all the current. As a result, the energetics were not favorable for fusion energy. In retrospect, both these experiments probably duplicated the Anderson experiment, albeit on a much faster time scale.

Pulsed power drivers were also used with the gas embedded pinch. A simple formula, dubbed the Haines-Hammel curve (8), gives a prescription for the current waveform required to keep the pinch at a constant radius. The curve is calculated by balancing the internal energy of the pinch with the energy gained by Ohmic heating and that lost by radiation, and assumes the pinch is always in Bennett equilibrium (9). The current at which the Ohmic heating exactly balances the radiation loss is called the Pease-Braginskii current (10, 11), and is around 1.5 MA for a hydrogen pinch. The Haines-Hammel curve calls for the current rate of rise (dI/dt) to be infinite at t=0, and only after the first few nsec does dI/dt fall within the capabilities of mortal pulsed power generators. While the requirements are mitigated somewhat in that it takes some time for the plasma to form, an unavoidable consequence is that the pinch will always undergo some expansion in the early stages.

One of the first experiments to test this theory was carried out at Los Alamos. Currents up to 150 kA, with dI/dt of 5 kA/nsec, were driven through a gas embedded pinch formed in 15 psi of hydrogen (12). The pinch was initiated with a pulsed ruby laser. The pinch attained an initial density as high as $2 \times 10^{20}/cm^3$, but it also exhibited the behavior observed earlier by Smars: m=1 unstable and accretion into the surrounding gas. Experiments at NRL at pressures up to 1500 psi exhibited the same behavior (13). To try to combat this behavior, a follow-on set of experiments at NRL formed the pinch inside small quartz capillaries (100 μm I.D.) filled with D_2 (13). The capillary quenched both the accretion and the m=1, and the plasma did heat up rapidly, as evidenced by the sudden onset of 1 keV x-rays. However the x-ray signal dropped within 10 nsec after formation. The NRL group found that the inner surface of the quartz turned into a conductor and shunted all the current away from the pinch. This was later shown to be quite a nice technique for rapidly cooling a plasma and making an x-ray laser (14), but the experimentalists, having fusion on their minds, did not recognize this potential.

Faced with these difficulties, the solution seemed to be to form the pinch at a very small diameter, away from any walls or surrounding gas, and at very high density. The only way to achieve this was to form the pinch from a fiber of solid deuterium surrounded by vacuum. This formation technique was first proposed in simultaneous articles by Pereira *et al* (15) and Hammel, Scudder, and Shalchter (16) in the proceedings of the first conference in this series. As deuterium freezes at 19 °K, this requires a cryogenic fiber extruder. However the difficulties associated with this were overcome (17) and experiments were performed first at Los Alamos at currents up to 250 kA (18) and soon after at NRL at currents up to 620 kA (19). In both cases the pinch remained stable for many radial Alfvén transit times. The NRL experiments were particularly intriguing, as the pinch was stable for as many as 100 radial transit times, and more importantly, seemed to be stable as long as the current was rising. In these experiments the end of the stable period was marked by a rapid expansion of the plasma, as recorded by a visible light streak camera, and the simultaneous onset of neutrons and x-rays.

These results sparked a flurry of theoretical activity. Most took the course of modifying ideal MHD to include more realistic conditions. The first step was to consider models with arbitrary current distributions (20, 21). These predicted somewhat lower growth rates, but never predicted stability when ideal MHD predicted stability. The second step was to add resistivity. This work was carried out at NRL, with a 2-D resistive MHD code (22), and at Imperial College, using a linearized initial value code (23). In both cases it was found that the pinch could be stabilized as long as the Lundquist number, $S = 4\pi\sigma a v_A$, was less than a critical value $S^* = 50 \, (ka)^{-0.8}$. Here σ is the conductivity, a is the pinch radius, v_A is the Alfvén speed and k is the wavenumber. Unfortunately, S went up to 240 in the experiments, leading to the conclusion that resistivity alone could not account for the observed stability. Attempts to include viscosity (24) met with the same fate: some stable regimes were predicted, but not within the experimental parameters. In none of these cases could the models show a relation between stability and the rising current. Concurrent with this analytical work, 2-D simulations of the pinch were performed by Los Alamos(25). These simulations attempted to model the entire history of the pinch, starting with a frozen solid fiber. They produced the rather discouraging results that the enhanced stability was simply due to the fact that a frozen core of the fiber lasts for a very long time.

Despite these theoretical results, the experiments were so promising that NRL and Los Alamos went ahead and built new generators anyway. These were to be capable of driving Mega-Ampere currents through the an inductive, single fiber, pinch. Both were built on extremely modest budgets, and the pulsed power suffered accordingly. When the preliminary experiments were finally performed they were not instantly encouraging (26, 27), and the funding environment was no longer conducive for alternative fusion concepts. As a result both programs were deleted. Nevertheless, Imperial College has forged ahead and built a high current driver called MAGPIE, and results from these experiments will be presented in these proceedings (28).

As of this writing, the consensus opinion is that it is not possible to produce a simple linear pinch for fusion energy applications because of its intractable tendency to go m=0 unstable. However more complicated z-pinch schemes have been proposed in which the pinch is the "target" for a liner, driven either by lasers (29), an imploding theta pinch (30), or explosive generators (31). The latter two are being actively pursued and are reported in the proceedings of this conference. Whether or not they are viable scientifically only research will tell.

Z-PINCH REACTOR STUDIES

The potential advantages of the z-pinch had not gone unnoticed by the reactor designers. The most comprehensive burn and energy balance calculations were carried out in a Level III reactor study by Hagenson *et al* at Los Alamos (32). They assumed they had a stable, gas embedded pinch carrying a current of 1.4 MA. With an energy input of only 140 kJ, the pinch produced a fusion output of 4.4 MJ. I.E. a "gain" of 30. They did not, however, address the engineering issues.

The engineering issues were addressed in four other studies. Hartmann (33) envisioned a gas puff pinch surrounded by high pressure helium and a liquid Lithium "waterfall". Robson (34), following a suggestion from McCorkle (35) considered a capillary pinch in which the "capillary" was in reality a vortex of liquid lithium. This design looked very much like a pressurized water fission reactor. Bolton at Imperial College (36) again assumed a gas embedded pinch and a liquid lithium vortex. Robson (37) considered a reactor based on the fiber pinch in which the fiber was injected in vacuum to bridge a gap between two liquid lithium jet electrodes.

These studies revealed three key issues for a dense z-pinch based reactor. First, the energy per pulse in a z-pinch is relatively small, so the system must be run at a fairly high rep-rate (60 Hz), and the pulsed power system has to have an efficiency of around 70%. Second, the pinch electrodes carry a large amount of current, on the order of a few MA, and electrode erosion is a serious issue. Third, the high voltage insulator must hold off several MV in a steady environment of 14 MeV neutrons. All of these issues were recognized by the reactor designers, but none offered satisfying solutions.

LINKING FUSION RESEARCH WITH FUSION ENERGY

It is important to realize the significance of these reactor studies-- they point out engineering issues that are just as important as the physics issues. As such they need to be addressed by any credible research program. While as scientists we tend to concentrate on the physics problems, we must keep in mind that our objective is to make a viable electrical energy source. Put more bluntly, what we are trying to do is boil water. And there are many existing technologies that boil water just fine. The only way fusion will be workable as a source of electricity is if it is economically competitive. And that means the entire concept-- *including the reactor,* must be competitive. For example, the reactor studies for the dense z-pinch have given some surprising results: the "simple" linear z-pinch has problems that are not easily solved, if at all. So if we want to add complications such as imploding liners or external magnetic fields, we need to clearly think about how such a reactor would look, and use this to define and direct the research program. We cannot ignore the reactor issues and assume they can be solved by someone else at a later time. Such thinking leads to lack of credibility, and eventually, a decrease in research support. I submit that the reactor issues should be addressed at the outset, and that they should be given equal weights in judging any fusion scheme.

ACKNOWLEDGMENTS

The author wishes to acknowledge stimulating discussions with A.E. Robson. This work was sponsored by the US Department of Energy.

REFERENCES

1. Robson, A.E., *Proceedings Second International Conference on Dense Z-Pinches*, (1989) 362-375.
2. Anderson, O.A., Baker, W.R., Colgate, S.A., Ise, J., and Pyle, R.V., *Physical Review*, **110**, 24-70 (1958).
3. Hagerman, D.C., and Mather, J.W., *Nature*, **181**, 226 (1957).
4. Clauser, M.V., and Weibel, E.S., *Proceedings Second U.N. Conference on Peaceful uses of Atomic Energy*, **32**, 161 (1958).
5. Smars, E.A. *Arkivf. Fysik* **19**, 97 (1965).
6. Choi, P., Dangor, A.E., Folkierski, A.E., Kahan,E., Potter, D.E., Slade, P.D., and Webb, S.J. *Proceedings Seventh International Conference on Plasma Physics and Controlled Thermonuclear Fusion Research*, (1978) 69.
7. Batyunin, A.V., Bulatov, A.N., Vikharev, V.D., Volkov, G.S., Zaitsev, V.I., Zakharov, S.V., Komarov, S.A., Nedoseev, S.L., Nikandrov, L.B., Oleinik, G.M., Smirnov, V.P., Trofimov, S.V., Utyugov, E.G., Fedulov, M.V., Frolov, I.N., Ya.Tsarfin, V., Sov J. Plasma Physics **16**, 597 (1991).
8. Haines, M.G., *Proc. Phys. Soc.* **76**, 250 (1960). and Hammel, J.E., "An Ohmically Heated High Density Z-Pinch" Los Alamos National Laboratory Report, # LA-6203-MS, Jan, 1976. Also see Ref 12.
9. Bennett, W.H., *Phys. Rev.* **45**, 890 (1934).
10. Pease, R.S. *Proc. Phys. Soc.* B**70**, 11 (1957).
11. Braginskii, S.I., *Sov.Phys. JETP*, **6**, 494 (1958).
12. Hammel, J.E., Scudder, D.W., and Shlachter, J.S. *Nucl. Instrum and Methods* **207**, 161-168 (1983).
13. Sethian, J.D., Gerber, K.A., Desilva, A.W., and Robson, A.E., *Proceedings of the Fourth International Conference on Megagauss Magnetic Field Generation and Related Topics*, (1986) 131-1
14. Rocca, J.J., Shlyaptsev, V., Tomasel, F.G., Cortazar, O.D., Hartshorn, D., and Chilla, J.L.A., *Phys.Rev. Lett.* **73**, 2192 (1994).
15. Pereira, N.R., Rostoker, N., Riordan, J., and Gersten, M., *Proceedings of the First International Conference on Dense Z-Pinches for Fusion*, (1984) 71.
16. Hammel, J.E., Scudder, D.W., and Shlachter, J.S., *Proceedings of the First International Conference on Dense Z-Pinches for Fusion*, (1984) 13.
17. Sethian, J.D., Gerber, K.A., and Sy, M.O., *Rev. Sci. Instrum.* **58**, 56 (1987).
18. Scudder, D.W., *Bull. Am. Phys. Soc.* **30**, 1408 (1985).
19. Sethian, J.D., Robson, A.E., Gerber, K.A., and DeSilva, A.W., *Phys. Rev. Lett.* **59**, 892 (1987); Ibid 1790.
20. Culverwell, I.D.,and Coppins, M., *Plasma Phys. and Contr. Fusion*, **31**, 1443 (1989).
21. Coppins, M., *Plasma Phys. and Contr. Fusion*, **30**, 201 (1989).
22. Cochran, F.L., and Robson, A.E., *Phys. Fluids B*, **2**, 123-128 (1990).
23. Culverwell, I.D., and Coppins, M., *Phys. Fluids B*, **2**, 129 (1990).
24. Cochran, F.L.,and Robson, A.E., *Proceedings of the Workshop on the Physics of Alternative Magnetic Confinement Schemes*, (1990) 395.
25. Lindemuth, I.R., *Phys. Rev. Lett.* **65**, 179 (1990).
26. Sethian, J.D., Robson, A.E., Gerber, K.A., and DeSilva, A.W. *Proceedings of the Workshop on the Physics of Alternative Magnetic Confinement Schemes*, (1990) 511.

27. Scudder, D.W., Shlachter, J.S., Hammel, J.E., Venneri, F., Chrien, R, Loveberg, R., and Riley, R. *Proceedings of the Workshop on the Physics of Alternative Magnetic Confinement Schemes*, (1990) 519.
28. Haines, M.G. Proceedings, this conference.
29. Parks, P. *Proceedings of the Workshop on the Physics of Alternative Magnetic Confinement Schemes*, (1990) 1083.
30. Wessel, F.J. Proceedings, this conference.
31. Lindemuth, I.R. Proceedings, this conference.
32. Hagenson, R.L., Tai, A.S., Krakowski, R.L., Moses, R.W., *Nucl. Fusion*, **21**, 1351 (1981).
33. Hartmann, C.W., Carlson, G., Hoffman, M., Werner, R., and Cheng, D.Y., *Nucl. Fusion*, **17**, 909 (1977).
34. Robson, A.E. *Proceedings of the First International Conference on Dense Z-Pinches for Fusion*, (1984) 2.
35. McCorkle, R.A. *Nouvo Clemente*, **77B**, 31 (1983).
36. Bolton, H.R. Choi, P., Dangor, A.E., Haines, M.G., Javedi, M. Peerless, S.J., Power, A. Robson, A.E., *Proceedings Eleventh International Conference on Plasma Physics and Controlled Thermonuclear Fusion Research*, (1986) 367.

The Role of Z-Pinches and Related Configurations in Magnetized Target Fusion

Irvin R. Lindemuth

Los Alamos National Laboratory, Los Alamos, New Mexico

Abstract. Magnetically driven z-pinch liners coupled with z-pinch plasma formation schemes may make the achievement of controlled fusion conditions in the laboratory possible in a shorter time frame and at much lower cost than with any other approach.

As discussed in these proceedings (1), z-pinches were one of the earliest candidates to heat a deuterium-tritium (D-T) plasma to thermonuclear conditions. Unfortunately, violent instabilities prevented the early approaches from achieving the goal of controlled fusion, and fusion in the laboratory has been an elusive goal for more than four decades. As many different approaches have been tried and abandoned, controlled fusion research has evolved into two mainline approaches, magnetic confinement (MFE, or magnetic fusion energy), as now embodied primarily in tokamaks, and inertial confinement fusion (ICF), as now embodied primarily in laser driven targets. With proponents proposing two multi-billion, multi-year next-generation facilities (ITER, or International Toroidal Experimental Reactor, for MFE, and NIF, or National Ignition Facility, for ICF), the conventional approaches have reached a funding crossroads.

Although many approaches have been tried in the four decades or so of fusion research, the plasma conditions that must be achieved have remained essentially unchanged. In general, these conditions are stated as an ion temperature, T_i, greater than 4 keV and either a number-density/confinement-time product, $n\tau$, greater than 10^{14} s/cm^3 (MFE) or an areal density, ρR, greater than 0.4 g/cm^2. It is the inability of any approach to obtain all of the required conditions simultaneously that has frustrated fusion researchers, and even after forty years of increasing knowledge about the behavior of plasmas, there remains no guarantee that next-generation machines, after a major capital investment, will finally achieve the long sought goal.

To achieve the required temperature, all approaches must involve one or several heating processes to overcome plasma cooling processes that increase rapidly with temperature. At the risk of oversimplification, this fundamental challenge can be written as

$$\frac{dT_i}{dt} = \text{heating processes - radiation losses - thermal conduction losses}.$$

Tokamaks use such processes as microwave heating and neutral beam injection to provide the plasma heating, whereas ICF relies entirely on compressional (pdV) heating. In principle, at least, magnetic confinement schemes eliminate the last term, since the plasma is not in contact with its cold surroundings. Nevertheless, such losses are not eliminated, and such things as "anomalous transport" are encountered in tokamaks. Even though such losses are not totally eliminated, it is hoped that they can be held at a sufficiently low level that confinement times exceeding 1 second will be achieved.

In contrast, in ICF, electron thermal conduction is a dominant loss mechanism. The thermal loss rate is so high that confinement times are limited to less than, say, 1 ns. Concurrent with the nine or ten orders of magnitude difference in confinement times between MFE and ICF is a nine or ten orders of magnitude difference in densities, with typical tokamak densities being $10^{14}/cm^3$ and ICF densities near $10^{24}/cm^3$. The high density of ICF means that the radiation loss rate, as well as the thermal conduction rate, is many orders of magnitude higher than can be tolerated in MFE.

The compressional heating rate in a target implosion is proportional to the target pusher's implosion velocity. To overcome the losses, the implosion velocity of a fusion target containing unmagnetized fuel must exceed 30 cm/µs. Furthermore, prior to the main heating phase, the fuel must experience carefully timed shocks to raise the fuel "adiabat" from the initial, ambient temperature value so that fusion conditions can be reached at an achievable target final radius. Nevertheless, NIF ignition targets require the final radius to be smaller than 1/30 of the initial radius, a formidable challenge requiring an extremely symmetric implosion. Initial fuel densities must therefore be greater than $10^{19}/cm^3$.

Various computational models (e.g., 2), including the LASNEX code on which the US ICF program is based, predict that fusion can be achieved under much less stringent implosion conditions if the compressed fuel is preheated and magnetized prior to implosion. Although the possible benefit of a magnetic field in a fusion target was recognized in the 40's by Fermi at Los Alamos and at approximately the same time by Sakharov in the former Soviet Union, it is only in light of recent advancements in plasma formation techniques, implosion system drivers, plasma diagnostics, and large-scale numerical simulation capabilities that the prospects for fusion ignition using this approach can be evaluated.

The use of magnetized fuel within a fusion target is now known as Magnetized Target Fusion in the US and as MAGO (Magnitnoye Obzhatiye, or magnetic compression) in Russia (3). In contrast to direct, hydrodynamic compression of initially ambient-temperature fuel (e.g., ICF), MTF involves two steps: (a) formation of a warm (e.g., 100 eV or higher), magnetized (e.g., 100 kG), wall-confined plasma of intermediate density (e.g., $10^{18}/cm^3$) within a fusion target prior to implosion; (b) subsequent quasi-adiabatic compression and heating of the plasma by imploding the confining wall, or pusher. In many ways, MTF can be considered a marriage between the more mature MFE and ICF approaches, and this marriage potentially eliminates some of the hurdles encountered in the other approaches. When compared to ICF, MTF requires lower implosion velocity (perhaps as low as 1 cm/µs), lower initial density, significantly lower radial convergence (e.g., 1/10 of initial radius), and larger targets (e.g., 1-10 cm), all of which lead to substantially reduced driver intensity, power, and symmetry requirements. When compared to MFE, MTF does not require a vacuum separating the plasma from the wall, and, in fact, complete magnetic confinement, even if possible, may not be desirable. The higher density of MTF and much shorter confinement times should make magnetized plasma formation a much less difficult step than in MFE.

The substantially lower driver requirements and implosion velocity of MTF make z-pinch magnetically driven liners, magnetically imploded by existing modern pulsed power electrical current sources, a leading candidate for the target pusher of an MTF system. Although the most elementary liner z-pinch is a simple cylinder (e.g., 4), a z-pinch current drive can provide quasi-spherical implosion of a liner (5). The attractiveness of z-pinch liners stems from the fact that existing magnetic flux compression generators, and possibly existing and near-term pulsed power machines, appear to exceed the energy necessary to implode a suitable plasma chamber and demonstrate fusion ignition. For example, a Russian Disk Explosive Magnetic Generator (DEMG) has delivered a 100 MA current pulse to an imploding liner, which achieved a kinetic energy of more than 20 MJ at a velocity greater than 0.7 cm/ms (6).

Whereas the fundamental driver technology for MTF exists, a plasma formation scheme for MTF's pre-implosion plasma cannot readily be identified from "off-the-shelf" approaches. The cryogenic fiber z-pinch, once pursued as a fusion source (1), is perhaps the simplest candidate. As predicted computationally at the Second Dense Z-Pinch conference (7), and subsequently confirmed by second generation experiments (8), previous attempts to drive fibers directly to fusion conditions encountered the "classical," explosive m=0 instabilities even though initial data suggested some "anomalous stability."

Although the direct Ohmically heated fiber z-pinch had to be abandoned as a fusion concept, recent two-dimensional MHD computations predict that the exploding plasma will settle into a Kadomtsev-stable wall confined plasma when it contacts its confining walls (9). With proper choice of initial fiber diameter, driving current, and outer wall radius, the late-time, wall-confined plasma may have the parameters required for MTF's pre-implosion plasma. Fiber z-pinch experiments are underway at Los Alamos to confirm or refute these new computational predictions.

The Russian MAGO plasma formation scheme is a leading candidate for MTF's pre-implosion plasma because detailed two-dimensional computations, which match preliminary and yet incomplete data, predict near ideal pre-implosion conditions (10). The unique MAGO chamber combines features of coaxial guns, dense plasma foci, inverse z-pinches, and hard-core z-pinches. However, it differs from all of these in that, during the initial dynamic phase, two distinct current paths drive two distinct, but interacting, magnetically driven shocks. The two-dimensional computations indicate that the late-time state of the MAGO plasma is a quiescent, Kadomtsev-stable hard-core z-pinch configuration with density, temperature, and magnetization appropriate for subsequent implosion. Although the early, dynamic interaction of the two shock-heated, magnetized plasmas is not of primary interest from an MTF perspective, a small fraction (e.g., 5 %) of the plasma is briefly heated to temperature above 1 keV and emits a pulse of 10^{13} fusion neutrons, the highest number of reactions ever achieved in a Los Alamos fusion experiment.

Experimental diagnosis of the plasma and liner conditions in an MTF system looms as a major challenge to MTF. Our experience to date suggests that MTF will require a strong synergism between experimental and computational endeavors, since MTF success will be determined not simply by liner and plasma stability or lack thereof, but by a detailed understanding of the long-term, non-linear evolution of instabilities including, perhaps, the evolution of one unstable configuration into a new, and stable, configuration as computations predict for the fiber pinches (9). Our success to date in predicting and interpreting z-pinch experimental data through detailed multi-dimensional computational modeling (7-11) is encouraging.

Because the economy in the Soviet Union could not sustain the long-term development of high-energy capacitor bank technology such as developed in the US, Russian scientists use the much cheaper magnetic flux compression techniques (the US has also pursued similar technology, but at a much lower level) for both liner implosion and plasma formation. Russia has developed "one-shot" devices which can produce higher magnetic fields, higher electrical currents, and higher electrical energies than any US "off-the-shelf" technology. Magnetic flux compression generators offer to the fusion scientist a low-cost way to do scientific experimentation which, today, cannot be performed in any other manner. For MTF in a fusion energy context, of course, high-explosive energy sources would be replaced by a capital intensive non-explosive facility, but this investment would be made only after the MTF physics had been unquestionably demonstrated so that only engineering uncertainties remained.

Existing magnetic flux compression generators and existing pulsed power systems and near-term systems being developed in other contexts (e.g., the Los Alamos Atlas facility) mean that the cost of research in MTF is devoted primarily to plasma physics, not to large capital investments in driver technology or voluminous plasma containment vessels. Hence, Z-pinches and related configurations in an MTF context hold out the prospect of achieving controlled fusion in a shorter time frame at substantially lower cost than any other known path.

REFERENCES

1. Sethian, J., "The Quest for a Z-Pinch Based Fusion Power Source--An Historical Perspective," these *Proceedings*.
2. Lindemuth, I., and Kirkpatrick, R., *Nuclear Fusion* 24, 263 (1983).
3. Lindemuth, I., et al., "Magnetic-Compression/Magnetized-Target-Fusion (MAGO/MTF): A Marriage of Inertial and Magnetic Confinement," to be published in *Proceedings of 16th IAEA International Conference on Plasma Physics and Controlled Nuclear Fusion*, Montreal, Canada, October 7-11, 1996.
4. Atchison, W., "Examination of Instability Growth in Solid Liner Surfaces Using Comparisons of Two Dimensional MHD Calculations and Measured Data," these *Proceedings*.
5. Degnan, J., et al, *Phys. Rev. Lett.* 74, 98 (1995).
6. Chernyshev, V., et al., Reinovsky, R., et al., Clark, D., et al., Faehl, R., et al, series of papers to be published in of *Proceedings 11th IEEE International Pulsed Power Conference*, Baltimore, Maryland, June 29-July 2, 1997.
7. Lindemuth, I., "Solid Fiber Z-Pinches: 'Cold-Start' Computations," in *Proceedings of Second International Conference on Dense Z-Pinches*, Laguna Beach, California, April 26-28, 1989, p. 327.
8. Sheehey, P., et al., *Phys. Fluids B* 4, 3698 (1992).
9. Sheehey, P., et al., "Computational Modeling of Wall-Supported Dense Z-Pinches," these *Proceedings*.
10. Lindemuth, I., et al., *Phys. Rev. Lett.* 75, 1953 (1995).
11. Lindemuth, I., et al., "Computational Modeling of the Laser Initiated Z-Pinch," in *Proceedings of the First International Conference on Dense Z-Pinches for Fusion*, Alexandria, Virginia, March 29-20, 1983, p. 46.

Computational Modeling of Wall-Supported Dense Z-Pinches

Peter Sheehey, Richard A. Gerwin, Ronald Kirkpatrick,
Irvin Lindemuth, Ronald Moses, and Frederick Wysocki

Los Alamos National Laboratory, Los Alamos, NM 87545 USA

Abstract. In our previous computational modeling of deuterium-fiber-initiated Z-pinches intended for ohmic self-heating to fusion conditions, instability-driven expansion caused densities to drop far below those desired for fusion applications; such behavior has been observed on experiments such as Los Alamos' HDZP-II. A new application for deuterium-fiber-initiated Z-pinches is Magnetized Target Fusion (MTF), in which a preheated and magnetized target plasma is hydrodynamically compressed, by a separately driven liner, to fusion conditions. Although the conditions necessary for a suitable target plasma--density $O(10^{18}$ cm$^{-3})$, temperature $O(100$ eV), magnetic field $O(100$ kG)--are less extreme than those required for the previous ohmically heated fusion scheme, the plasma must remain magnetically insulated and clean long enough to be compressed by the imploding liner to fusion conditions, e.g., several microseconds. A fiber-initiated Z-pinch in a 2-cm-radius, 2-cm long conducting liner has been built at Los Alamos to investigate its suitability as an MTF target plasma. Two-dimensional magnetohydrodynamic modeling of this experiment shows early instability similar to that seen on HDZP-II; however, when plasma finds support and stabilization at the outer radial wall, a relatively stable profile forms and persists. Comparison of experimental results and computations, and computational inclusion of additional experimental details is being done. Analytic and computational investigation is also being done on possible instability-driven cooling of the plasma by Benard-like convective cells adjacent to the cold wall.

INTRODUCTION

Fast-current-rise Z-pinches initiated from frozen deuterium fibers appeared to show anomalous stability at current peaks up to 600 kA, leading to hopes that fusion conditions could be reached in machines designed for peak currents in excess of 1 MA. However, pinches produced on high-current machines, such as Los Alamos' HDZP-II, showed instability-driven plasma column expansion, dropping densities far below those desired for fusion applications. We developed a detailed two-dimensional computational magnetohydrodynamic (MHD) model for such discharges, which showed good agreement to experimental results (1,2). A new application for such dense Z-pinches is as a preheated and magnetized tar-

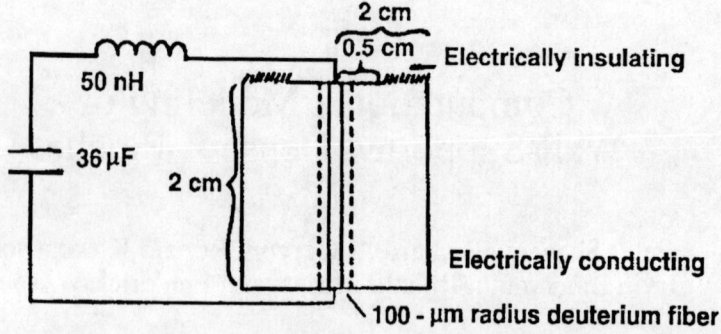

FIGURE 1. LANL Z-pinch target plasma: a 2-cm radius, 2-cm long, electrically conducting plasma chamber, containing a deuterium-fiber-initiated Z-pinch, driven by a capacitor bank (200 kJ, 100 kV, 2.2 μs risetime).

get plasma for subsequent hydrodynamic compression, by a separately driven liner, to fusion conditions, in a scheme known as "Magnetized Target Fusion."

Magnetized Target Fusion (MTF) is an approach to controlled fusion that is intermediate between magnetic confinement and inertial confinement fusion (ICF) in time and density scales (3,4). MTF uses a pusher-confined, magnetized, preheated plasma fuel within a fusion target. The magnetic field suppresses losses by electron thermal conduction in the fuel during the target implosion heating process. Reduced losses permit near-adiabatic compression of the fuel to ignition temperatures, even at low (e.g., 1 cm/μs) implosion velocities. An MTF system requires two elements: (1) a target implosion system (2) a means of preheating and magnetizing the thermonuclear fuel prior to implosion. An optimal driver source for MTF may be relatively inexpensive electrical pulsed power. This could utilize either fixed pulsed-power facilities, such as Los Alamos' Pegasus or Atlas, or explosive-flux-compression generators, such as Los Alamos' Procyon or the 200-MJ-class disk flux compression generators developed by the All-Russian Scientific Research Institute of Experimental Physics (VNIIEF) (5,6). Such energy-rich sources might allow a demonstration of fusion ignition via MTF, without a major capital investment in driver technology; evaluation of such implosion systems is ongoing at Los Alamos.

Optimal target plasma conditions for MTF are temperature O(100 eV), density $O(10^{18} cm^{-3})$, and magnetic fields O(100 kG). Sufficient plasma lifetime to allow implosion is necessary (on the order of several microseconds), and plasma-wall interactions must not lead to excessive introduction of impurities, which could result in rapid cooling of the plasma. A deuterium-fiber-initiated Z-pinch might well produce an acceptably hot, dense, magnetized target plasma for subsequent MTF compression. Using the same computational tool--a version of Lindemuth's MHRDR code (7,8)--with which the Los Alamos HDZP-I and HDZP-II fiber Z-pinches were modeled (obtaining excellent agreement with experiment (1,2,8)), a fiber Z-pinch target plasma experiment has been designed and modeled. This experiment has been built and is now operating at Los Alamos.

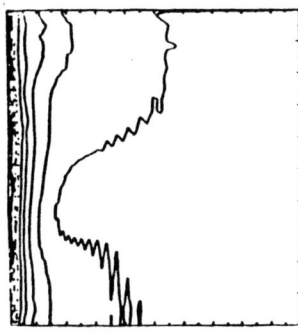 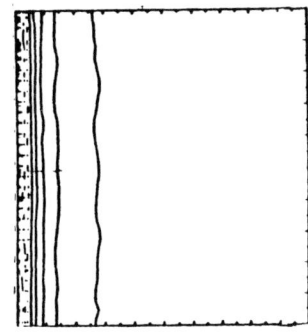

FIGURE 2. Computed axial current contours (r-z 2 cm by 2 cm): (left) early (1.1 μsec) unstable, expanding phase; (right) later (2.4 μsec) stable, wall-supported phase.

FIBER Z-PINCH TARGET PLASMA EXPERIMENT

The Z-pinch target plasma experiment (Figure 1) will be driven by the Colt capacitor bank at Los Alamos (200 kJ, 100 kV, up to 2 MA with a 2.2 μsec risetime), which is considerably lower voltage and slower than the original Los Alamos HDZP experiments. In addition, the plasma will be contained inside a 2-cm-radius conducting wall; the HDZP experiments were in a chamber with tens-of-centimeter distant walls.

Detailed two-dimensional MHD modeling of such an experiment predicts early behavior similar to the HDZP experiments: the fiber-initiated plasma becomes unstable and expands explosively. However, when the plasma finds support and stabilization at the conducting wall, it appears to settle into a dense, hot, Kadomtsev-stable state, capable of carrying megamp-plus currents in a few-mm-radius column, over several microseconds (Figure 2). The pinch varies from temperatures near 1 eV, and densities near 10^{17} cm^{-3} at the 2-cm-radial wall, to near 1 keV and 10^{20} cm^{-3} at r=0. To the extent that such an experiment lives up to these predictions and remains free of contaminants, it would certainly be an acceptable MTF target plasma, and would be of considerable research interest, even without MTF liner-on-plasma compression.

Of course, such predictions must be verified experimentally. Such problems as insulator flashover and wall-plasma interactions must be investigated. Three-dimensional effects will have more time to develop than in the HDZP experiments, although the relatively close conducting wall should have some stabilizing effect. Since these are issues critical to many MTF liner/plasma schemes, experimental investigations on such a device can be extremely useful parts of an MTF research program.

As an MTF target plasma, such a dense Z-pinch has advantages and disadvantages when compared to other possible target plasmas, such as compact toroids or the Russian-originated "MAGO" plasma (9). MAGO plasmas, in their late-time relatively steady state, are calculated to be much more uniform in temperature, density, and field, and to load the electrodes with smaller current per unit area. Hence a more nearly adiabatic compression of such plasmas by wall implosion might be possible; on the other hand, dense hot plasma may spend more time in contact with the walls, which could lead to deleterious wall-plasma interactions. Compact toroids have the potential advantage of being electrically detached from the electrodes after formation, avoiding current-driven electrode material influx; but it remains to be seen whether or not the necessary plasma densities can be achieved in a compact toroid. Detailed experimental and computational analysis will have to be continued to sort out these differences and evaluate the best target plasma for MTF. The Colt Z-pinch facility will not only allow the development of diagnostics and detailed computational models for evaluating the Z-pinch as an MTF target plasma, but should be adaptable to other configurations if they appear advantageous.

Dense magnetized plasmas in contact with a cold wall (fixed or imploding) may form Benard-like convective cooling cells. The MHD code being used for these Z-pinch simulations has been employed to study such convective processes (10,11). Semi-analytic models are being developed in conjunction with computational work to predict the occurrence and consequences of such phenomena.

CONCLUSIONS

Deuterium-fiber-initiated Z-pinches may provide a suitably hot, dense, magnetized target plasma for compression inside a heavy metallic liner to fusion conditions in a Magnetized Target Fusion (MTF) scheme. Modeling of such a target plasma inside a 2-cm-radius conducting liner predicts that, after initial explosive instability-driven expansion, the plasma may find a state stabilized by wall support of the plasma and magnetic field. An experiment to investigate these plasmas, and allow development of target plasmas for MTF, is beginning operation at Los Alamos National Laboratory.

REFERENCES

1. P. T. Sheehey and I. R. Lindemuth, *Phys. Plasmas* **4**, 146 (1997).
2. P. T. Sheehey et al., *Phys. Fluids B* **4**, 3698 (1992).
3. I. R. Lindemuth and R. C. Kirkpatrick, *Nuclear Fusion* **23**, 263 (1983).
4. V. N. Mokhov et al., *Sov. Phys. Dokl.* **24**, 557 (1979).
5. V. K. Chernyshev et al., in *Megagauss Fields and Pulsed Power Systems*, ed. V . Titov and G. Shvetsov, p. 347, Nova Science Publishers, New York (1990).
6. A. I. Pavlovskii et al., ibid., p. 327.
7. I. R. Lindemuth, University of California Lawrence Livermore Laboratory Report No. UCRL-52492 (1979).
8. P. T. Sheehey, Los Alamos National Laboratory Report LA-12724-T (1994).
9. I. R. Lindemuth, V. K. Chernyshev, V. N. Mokhov et al., *Phys. Rev. Lett.* **75**, 1953 (1995).
10. I. R. Lindemuth and T. R. Jarboe, *Nuclear Fusion* **18**, 929 (1978).
11. I. R. Lindemuth et al., *Phys. Fluids* **21**, 1723 (1978).

DRIVER COUPLING TO QUASISTATIC Z-PINCHES

G. Decker, W. Kies, S. Stein

Institut für Experimentalphysik, Heinrich-Heine-Universität Düsseldorf
Universitätsstraße 1, 40225 Düsseldorf, FRG

Abstract

Quasistatic Z-pinches formed "on axis" and magnetically confined by a current layer larger than a critical current (> 1.5 MA for deuterium) are supposed to undergo radiative collapse providing high energy density thermal plasmas. Z-pinches created from solid fibers do not well couple to the necessarily high power drivers due to lacking initial conductivity and compressibility resulting in fast global plasma expansion and current leaks. Therefore experiments starting from plasmajets ($n_e \leq 10^{23}$ m^{-3}, $T_e \approx 1$ eV) have been performed using two different drivers, namely the terawatt pulseline KALIF (2 MV, 900 kA) and the fast condenser bank SPEED 1 (200 kV, 900 kA) in order to investigate driver - load coupling with different initial power conditions. The main results of this study are: (i) plasmajets show much better initial coupling than fiber experiments, (ii) there is a critical limit of the reduced electrical field (E/n $\approx 10^{-16}$ Vm^2) above which fast plasma erosion and decoupling takes place preventing pinch formation and (iii) plasma loads need high initial densities ($n_e \geq 10^{24}$ m^{-3}) in order to well couple to terawatt drivers providing pinch electric fields above 10^7 V/m.

A conductive, dense and narrow plasma column is a necessary condition for an efficient energy transfer from the driver to the load. Whether or not this is sufficient to induce radiative collapse still remains to be experimentally demonstrated.

Introduction

Z-pinches have been studied for several decades in different configurations. The linear dynamical Z-pinch has been found prone to violent instabilities producing non-thermal plasma components and runaway particles instead of a magnetically well confined thermal pinch plasma. Z-pinches whose plasma radii change slowly compared with the Alfvén velocity (quasistatic) seem to offer the possibility of creating thermal plasmas with extra high density via radiative collapse. Deuterium pinches of a few cm length need terawatt pulse power for radiative collapse. Though high power pinches created on axis have been experimentally found to be much longer stable than ideal magnetohydrodynamic theory predicts, high power drivers do not well couple to that load. Driver - load coupling here refers to energy transfer to the load, the pinch plasma. This energy reads

$$W_p(t) = \frac{1}{2}\int_0^t L_{dot} I^2 dt' + \int_0^t RI^2 dt',$$

where L_{dot} + R is the load impedance, the sum of the inductance derivative and the plasma resistance. Thus energy can be only transferred to that load as long as (i) its impedance remains positive, (ii) the current is concentrated to the pinch, and (iii) the current carriers are confined to the pinch. Especially pinches created from solid fibers tend to escape coupling to the driver by surface plasma expansion ($L_{dot} < 0$) and electron runaway during fiber ablation. This seems mainly due to lacking initial conductivity and compressibility of fibers. Therefore we have tried to improve the driver-load coupling by starting from a plasma created in the auxiliary plasma focus device DAVID and injected through a nozzle into the vacuum feed of a driver where the Z-pinch is to be formed. The plasmajet experiments have been performed at the TW pulseline KALIF (2 MV, 900 MA) and the fast condenser bank SPEED 1 (200 kV, 900 kA).

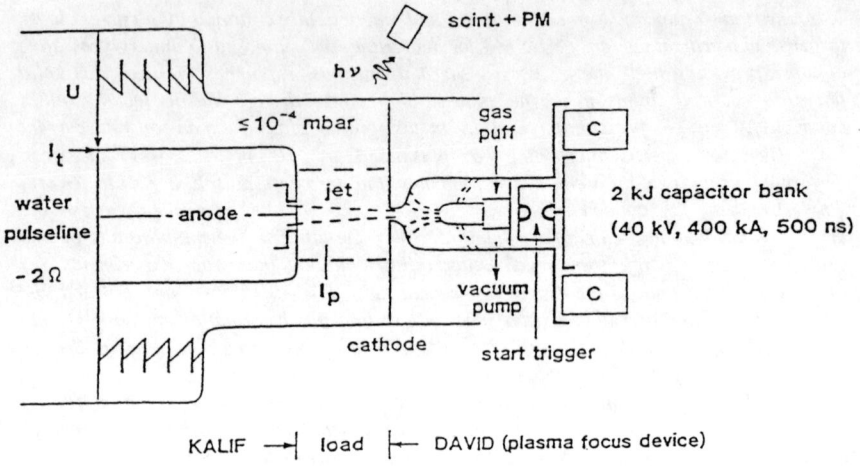

Figure 1: Plasma focus device DAVID connected to the vacuum feed of KALIF

KALIF experiments

Figure 1 shows the experimental set-up of the plasmajet DAVID connected to the vacuum feed of the water pulseline KALIF (2 MV, 900 kA). The vacuum feed consists of a water-vacuum interface and a variable anode-cathode gap. Gap widths were 19 mm and 35 mm, the jet radius expands from 4 mm (i.e. the nozzle radius) up to 15 mm at the anode. The jet plasma parameters ($n_e \approx 10^{22}$ m^{-3}, $T_e = 0.5 - 1$ eV) have been estimated by spectroscopic means.

The quality of the coupling of KALIF to the jet plasma can be judged from the signals voltage U, total current I_t, pinch current I_p (measured at a position about 12 cm from the symmetry axis), and hard X-rays occurring when runaway electrons hit the anode surface indicating bad coupling to the load or decoupling of the load by plasma erosion. The typical signals from a discharge with a gap width of 35 mm (figure 2, left) show a perfect coupling of KALIF to the jet plasma for the first 60 ns: (i) negligible breakdown delay, (ii) no leakage current ($I_p = I_t$), and (iii) no detected runaway electrons (no X-ray signal). Then a strong disruption takes place.

Experiments with a gap width of 19 mm show decoupling already at 40 ns after discharge ignition and at significantly lower values of voltage and current. This discharge behavior is not caused by a possible difference in line density with a different gap width. Evidence is given by figure 2 that shows on the right the electrical signals for a discharge with enhanced jet line density obtained by a larger time delay between jet injection and ignition of the KALIF discharge. The delayed pinch current rise indicates a current flow starting at outer regions with respect to the pinch current probe that is then moving towards the symmetry axis. Hard X-rays occur even before the current carrying region reaches the position of the pinch current probe. Too high values of the reduced electric field ($E/n > 10^{-15}$ Vm^2) cause a plasma erosion which is so fast that sheath and pinch formation is prevented. The collision mean free path is so large that electrons gain enough energy to run away the more so the smaller the gap width is.

This sudden turn from well coupled discharges into a diode-like behavior resembles fiber discharges where disruptions take place below a critical line density of 10^{19} m^{-1} when the current exceeds 300 kA.

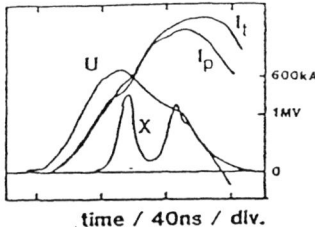 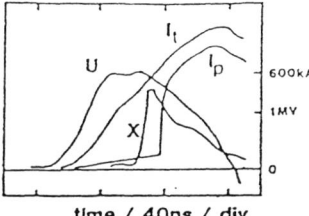

time / 40ns / div.   time / 40ns / div.

Figure 2: KALIF discharges, gap width 35 mm; 'normal' (left), enhanced line density (right)

SPEED 1 experiments

The SPEED 1 experiments (200 kV, 900 kA) were performed to investigate if driver-load decoupling can be avoided with lower values of the reduced electric field. The set-up of the vacuum feed at SPEED 1 is similar to that at KALIF shown in figure 1. Experiments have been carried out with a gap width of 15 mm and nozzle radii of 6 mm and 15 mm. Figure 3 shows on the left the electrical signals of a discharge with a nozzle radius of 6 mm ($n_e = 10^{22} - 10^{23}$ m^{-3}). About 250 ns after ignition a strong plasma compression takes place that leads to a pinch indicated by the voltage overswing, the I_{dot} dip, and the emission of hard X-rays and neutrons. X-rays and neutrons are measured with a scintillator-photomultiplier combination with the pulses separated by time of flight. Unfortunately, a strong competition between pinching and disruptions is characteristic for this type of discharge. The time integrated X-ray pinhole picture of a typical discharge (figure 4, left, $\lambda < 2$ nm) shows local axial compressions and also strong radiation in front of and from the anode surface due to fast electrons. Even with an electric field reduced by about one order of magnitude compared with the KALIF experiments driver-load decoupling could not be completely suppressed.

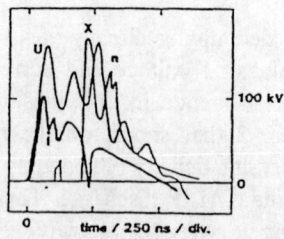 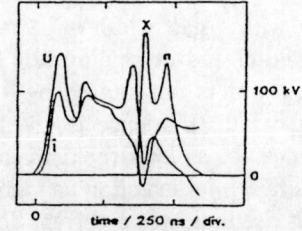

Figure 3: SPEED 1 discharges, nozzle radius 6 mm (left) and 15 mm (right)

A nozzle radius of 15 mm leads to an enhanced density of about $n_e = 10^{23}$ - 10^{24} m^{-3} and therefore to a lower value of the reduced electric field, i.e. $E/n \approx 10^{-17}$ Vm^2. A sequence of optical framing pictures, given in figure 5, shows a good driver-load coupling that results in sheath formation and pinching. From the electrical signals it can be seen that the maximum compression takes place at the current maximum (figure 3, right). No hard X-rays are detected except for the time of pinch decay and neutrons are produced during pinch life. The time integrated X-ray-pinhole picture (figure 4, right) also indicates that a homogeneous axial pinch develops. Unfortunately, only a few percent of the discharges show such efficient plasma compression. In most cases efficient pinching is prevented by current paths which occur in outer radial positions caused by the extended jet plasma.

However, all discharges that fulfilled the condition $E/n < 10^{-16}$ Vm^2 did not lead to driver-load decoupling.

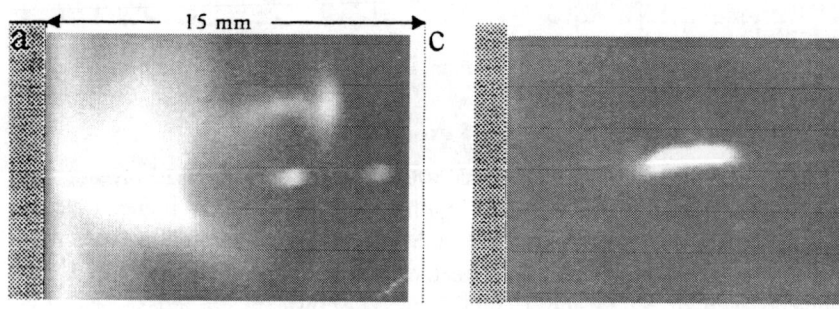

Figure 4: time integrated X-ray pinhole pictures ($\lambda < 2$ nm) from SPEED 1 discharges, nozzle radius 6 mm (left) and nozzle radius 15 mm (right)

Conclusion

Due to initial conductivity and compressibility jet plasmas provide perfect initial driver-load coupling and therefore much better start conditions than fibers. However, in the SPEED 1 experiments it has been shown that for efficient plasma compression a maximum value of the reduced electric field of $E/n = 10^{-16}$ Vm^2 should not be exceeded. All KALIF experiments are characterized by strong plasma erosion preventing pinch formation because the upper limit of the reduced

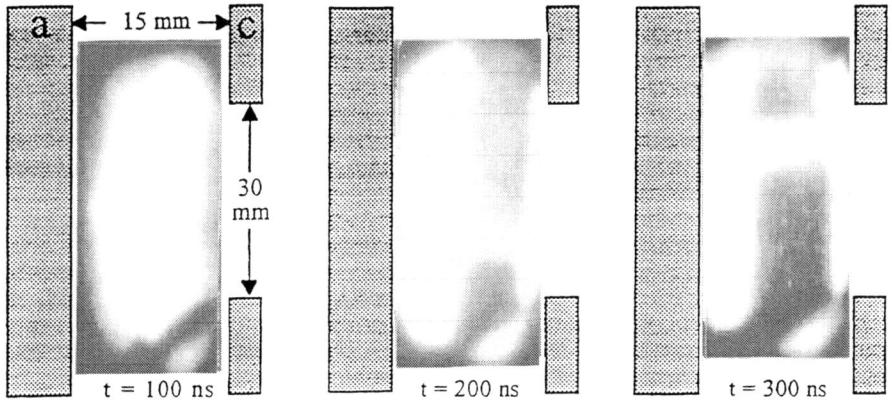

Figure 5: framing pictures (t_{exp} = 5 ns) for a SPEED 1 discharge with a nozzle radius of 15 mm

electric field was exceeded by almost two orders of magnitude. Since terawatt drivers are necessary to force radiative collapse in linear deuterium Z-pinches high electric fields are inevitable. This means that the jet density must be increased to about $n_e = 10^{24}$ m^{-3} for gap widths of a few cm. Simultaneously a good concentration of the jet to the axis is necessary in order to avoid current leaks and to prevent dynamical plasma compression. This could be achieved using a tailored pulseline prepulse for jet plasma compression so that the main pulse hits a well prepared pinch plasma channel.

Acknowledgements

Support by the KALIF crew and EC is gratefully acknowledged.

References

F. C. Young et al., *J. Appl. Phys.* **48**, 3642 (1977)
J. D. Sethian et al., *Phys. Rev. Lett.* **59**, 892 (1987)
M. G. Haines, *AIP Conf. Proc.* **195**, 203 (1989)
W. Kies et al., *J. Appl. Phys.* **70**, 7261 (1991)
S. Stein et al., *AIP Conf. Proc.* **299**, 509 (1994)

AN OVERVIEW OF THE DZP PROJECT AT IMPERIAL COLLEGE

M. G. Haines

Blackett Laboratory, Imperial College, London SW7 2BZ, UK

Abstract

The main long term objectives of the DZP Project at Imperial College are to achieve radiative collapse from cryogenic hydrogen fibres, to study Z-pinch conditions relevant to controlled fusion, and to study the basic plasma physics processes.
  Initial experiments have been conducted on carbon fibres using both the IMP (250 kA) and MAGPIE (1.5 MA) generators. Recently cryogenic deuterium fibres have also been employed. A wide range of diagnostics has been developed including short pulse laser interferometry, schlieren and Faraday rotation, as well as optical and X-ray streak and framing pictures. We have made a special study of the transient (~ 1 ns) intense X-ray bright spots and their bifurcation, and found good agreement with a 2-D MHD simulation with ionisation and recombination included. The magnetic field and current distribution have been measured in the coronal plasma showing that over 90% of the circuit current is flowing in the pinch. A current prepulse has led to a much more uniform and more intense X-ray emission during the main current pulse.
  Schemes are being explored for non-linear saturation of instabilities using sheared axial flow, large ion Larmor radius and axial magnetic field.

M. G. Haines, m.haines@ic.ac.uk, +44-171-594 7656 (f. 7658)

I. INTRODUCTION

The dense Z-pinch offers a compact and direct method of achieving very high energy density plasmas of relevance to controlled fusion[1] and to very intense X-ray sources[2]. The radiation loss by bremsstrahlung can lead to the possibility of radiative collapse when the current exceeds the Pease-Braginskii value of approximately 1MA[3,4], which might lead to compression in excess of 10^4 x solid[5,6].

However none of these objectives can be achieved unless the Z-pinch is sufficiently stable and formed to give a plasma of sufficient density and temperature.

At Imperial College we undertake a broad programme of research into Z-pinches that encompasses experiment, simulation and theory, together with the development of suitable diagnostics. The team is listed in Appendix A, and without their work this report would not be possible; approximately 30% of those listed undertake theory and simulation. I should also stress the importance of collaboration within the international Z-pinch community, and within the frameworks of the USDOE, INTAS and the EU as well as various other national institutions, where there are several schemes that allow worthwhile collaboration.

The DZP Project was initiated in 1989 with the funding of the MAGPIE generator by the Science and Engineering Research Council (now EPSRC) with the objectives of studying (i) radiative collapse of a hydrogen Z-pinch to achieve a plasma at a density of ~ 10^4 x solid, (ii) controlled fusion conditions at a current of 10^6A, at a number density of about 5×10^{21}cm^{-3} and confinement time about 100ns, (iii) the basic properties of the Z-pinch, especially pinch formation and stability, and (iv) application to X-ray sources and an X-ray laser.

During the last International Conference on Dense Z-pinches in 1993 the MAGPIE generator was officially opened by Dr. R. S. Pease and Prof. S. I. Braginsky, who independently predicted the critical current at which bremsstrahlung balances Joule heating. The generator is of high impedance (1.25Ω) and with maximum parameters 2.4MV, 1.9MA and a current risetime of 150ns[7]. It was built to schedule, to cost and to specification, and within a few months was fully tested with the magnetically insulated transfer line. A suite of diagnostics including optical and X-ray streak and framing cameras was implemented, and in particular a SBS compressed laser pulse ($\lambda = 0.53\mu m$, 300ps) for simultaneous schlieren, interferometry and polarimetry was developed[8,9,10]. A smaller generator, IMP (800kV, 250kA, 100ns) has since been built for small scale experiments and development work. At this conference there are papers on the effect of current prepulse on fibre Z-pinches[11] and on a compressional pinch using this generator[12]. We also have an extensive programme on the development of X-ray backlighter sources such as the micro plasma focus (optimised for various gases) [13], X-pinch and capillary X-ray laser[14]. Two techniques have been developed for the manufacture of cryogenic fibres[15] and in this paper an overview of the main results of carbon and deuterium fibre Z-pinch will be presented.

The Imperial College Plasma Physics Group has an almost equal effort in theory and simulation in all its research fields, not least in Z-pinches. Through Prof. A. R. Bell we have 1-D, 2-D and 3-D MHD codes. Of particular interest are results from the 2-D resistive MHD code with ionisation, recombination and lower hybrid turbulence (for anomalous transport) included. This has allowed good comparison with experiment of the phenomena of bright spot formation and bifurcation[16]. In both 2-D and 3-D (ideal) we have demonstrated that instabilities can saturate at small amplitude if there is sufficient sheared axial plasma flow present in the equilibrium. The linear stability of the Hall fluid[17] and resistive models[18] have recently been revisited, while we have shown large reductions in growth rate with large ion Larmor radius (LLR) effects present. A Vlasov hybrid code follows the non-linear development of the $m = 0$ mode under LLR conditions [19].

In this overview the principal experimental and theoretical milestones will be presented.

II. CARBON FIBRE EXPERIMENTS
II.i Optical and X-ray emission

The current rises to a peak in 200ns, but by this time a "disruption" has occurred, typically for 33μm diameter at 160 to 180ns depending on the applied voltage. Results at about half voltage (45kV charge) of 1ns exposure gated optical images show that the pinch appears to be relatively stable until 110ns, but then develops dominantly $m = 0$ MHD instability leading to disruption and break-up into plasma islands. Time integrated X-ray pinhole pictures for a series of discharges show the presence of bright emission spots of less than 50μm size mostly on the axis. However, as we will see, there is only a little correlation between the two phenomena. An axial optical streak photograph (fig.1) reveals that emission occurs at random positions close to the Z-axis and at random times from ~ 2ns from the start of the current. Furthermore each spot with a typical velocity of 1 - 2 x 10^5m/s. After 50ns this activity gives way to more slowly growing large scale $m = 0$ structures.

Two-dimensional MHD simulations with ionisation, recombination and lower hybrid drift turbulence (to limit the electron drift velocity in the low density corona) have been carried out, and can describe the observed phenomena.

Mass density, temperature and Z effective profile in the r-z plane are shown in fig.2 at (a) 26ns and (b) 28ns. The bifurcation process is clearly demonstrated and its understanding can be helped by plotting $J_r B_\theta$ (the axial component of $\underline{J} \times \underline{B}$). Whilst $J_z B_\theta$ provides the $m = 0$ necking, the bifurcation is driven substantially by $J_r B_\theta$, and leads to a more rapid ionisation of the fibre. A comparison (fig.3) can be made between a sequence of gated soft X-ray (MCP) images and a calculation from the 2-D simulation of the expected X-ray emission, and shows the bifurcation over a 6ns period.

II.ii Coronal plasma

There is a low density coronal plasma formed at early times which expands radially with a velocity of up to 4 x 10^4m.s^{-1}. We have diagnosed this using a 300ps 0.53µm wavelength laser pulse. The experimental layout is shown in fig.4, to indicate how simultaneous shadowgraphy, interferometry and Faraday rotation can be obtained. The interferometer has a self-referencing layout. Figure 5 is a series of shadowgrams of 33µm diameter carbon fibres taken at different times (but on separate discharges). $m = 0$ structures are apparent by 10ns and have grown non-linear by 14ns. By 70ns there are discrete density islands and the low density outer corona reveals narrow density minima extending radially.

A simultaneous polarogram and interferogram at 52ns is shown in ref.10. We note that the polarogram has a brighter image above as compared to below the Z-axis. From detailed Abel inversion techniques the density, magnetic field and current could be calculated at this time at various axial positions. Within experimental error of ± 15% about 90% of the current measured externally is being carried by the fibre pinch.

We have also compared experimental schlieren results with computer images generated by suitable line integral calculations postprocessing the 2-D simulations.

II.iii Effect of prepulse

Two current prepulse experiments have been undertaken. In the first[11] a separate pulsed power supply could discharge up to 7kA in a damped ascillator discharge with a variable time delay before the main current. Ref.11 shows axial X-ray streak and framing pictures during the subsequent main current in a 7µm diameter carbon fibre. It is clear that a coronal plasma is formed surrounding the fibre during the prepulse. The main current flows in this coronal plasma at first and pinches this plasma (as in a snowplough model) onto the fibre. In this way the main current is by then at a larger value as it transfers to the fibre, and a very much more intense X-ray emission occurs at this time; but it is not simultaneous along Z, but zippers from cathode to anode over a 10ns period, as can be seen in both the axial streak and framing pictures. An interesting effect of a prepulse is a delay in time of the onset of instabilities.

By putting a resistor in parallel with one pulse forming line switch, it was possible to have a prepulse in MAGPIE, although not so controllable. Figure 6 shows a 33 μm carbon fibre with laser probing (interferogram and schlieren) at -56ns (prior to main current) and then an optical streak. Here a single region of plasma which in the prepulse phase had the greatest $m = 0$ instability activity, formed a region of greatest emission which bifurcated in both directions (seen also in the 2ns gated soft X-ray images) ionising the fibre. MHD activity followed with a disruption (accompanied by hard X-ray emission from the anode) at a later time of 120ns. The improvement in performance on MAGPIE with prepulse was not considered sufficient for its continuance.

III. CRYOGENIC DEUTERIUM FIBRE EXPERIMENTS

Cryogenic deuterium fibres of about 80μm diameter have been fired in MAGPIE with a ~ 50-60kV charging voltage to give maximum currents in the 0.85 to 1.05MA range. Figure 7 shows a sequence of shadowgrams together with a gated 2ns X-ray MCP image at 44ns. An axial optical streak of the first 200ns is shown in fig.8, indicating transient hot spots that bifurcate axially at ~ $10^5 ms^{-1}$, followed at 50ns by a slower evolution of mainly $m = 0$ MHD modes. Thus we conclude that deuterium fibres behave in a very similar way to carbon fibres.

IV. DISRUPTIONS

At about 120ns a hard X-ray burst lasting 20-30ns is detected by scintillators. From our earlier work on CD_2 fibres[20] we know that the neutrons are produced at this time and are of beam-plasma origin. In the case of deuterium fibres the yield is about 5×10^9. There is a marked drop in the circuit dI/dt just prior to the hard X-ray pulse; the plasma at this time is in a series of density islands with a tenuous plasma in between.

The electron beam energy has been inferred from four cross-filtered scintillator-photomultiplier combinations and for 33μm carbon and 25μm aluminium fibres reveal an energy up to 4.8 and 1.7MeV respectively at peak X-ray intensity. The emission continues for some 100ns at an energy in the 1MeV range. Scintillators at 2.5, 6.8 and 16.8m can distinguish the neutron emission, and for CD_2 fibres indicate counterstreaming ion beams; the neutrons having energy up to 5MeV. Two mechanisms are thought to be at work: the transient necking off of the pinched plasma to form density islands could produce an energetic singular ion beam with off-axis oppositely moving guiding centre ions carrying the equal and opposite momentum[21]; electrons will similarly be energised but their guiding centre motion off-axis will give a slow drift to the anode. The low density plasma between the density islands will probably have an anomalous resistivity, and the high voltage across this region will also permit a runaway diode action.

V. STABILITY THEORY

The ideal MHD model of stability strictly only applies when resistivity, viscosity, finite ion Larmor radius, pressure anisotropy and the Hall effect can be neglected. The dimensionless parameters, eg. the Lundquist number, depend on the functions I^4a and N for a pinch pressure balance. (Here I is the current, a the pinch radius and N the ion line density.) Therefore a universal diagram can be constructed[22] in which experiments can be described by a point or a time-dependent trajectory in $I^4a - N$ space.

At Imperial College we have explored the linear stability for most regimes, and have shown that a marked reduction in growth rate can be achieved with large ion Larmor radius (LLR). Using two independent Vlasov models, one an initial value code, the other based on a variational technique, the growth rate of the fastest growing mode has been calculated for $m = 0$[23] and $m = 1$[24]. Reductions of 70% and 80% are found at a ratio of ion Larmor radius to pinch radius of 0.1. This corresponds to N of about $10^{19} m^{-1}$, which coincidentally is the value needed for fusion conditions or radiative collapse. Under these conditions a non-linear Vlasov ion, fluid electron code shows no saturation of the growth[19], but the result is known to be very sensitive to the plasma-vacuum boundary condition where, in the current model, the electron drift velocity exceeds that allowed by lower hybrid turbulence. Further research is required here.

However for LLR conditions to apply the ions must be collisionless over an ion Larmor period ie. $\omega_i \tau_i > 1$. In fibre experiments with a cold start $\omega_i \tau_i$ is less than one during the heating phase, and though resistive effects can lead to slower modes, it would be necessary for the plasma to spend some time in an unstable regime.

Many ideas are currently being explored to overcome this problem. A compressional pinch from a uniform gas-fill pinches to a hot plasma[12], and in principle it could be tailored to meet the $\omega_i \tau_i > 1$ and $a_i/a = 0.1$ conditions which would be useful for testing the theory at low density.

We have shown how a surrounding coronal plasma around a fibre can delay the onset of instabilities. This could be taken to an extreme limit by collapsing a wire or fibre array[2] onto a central fibre. The transfer of current to the central fibre is an uncertain phenomenon that requires study.

We have made a study of the effect of sheared axial flow ie. $v_z(r)$ where v_z is the axial component of velocity[25]. The growth rate reduces with amplitude of axial velocity which has a radial parabolic profile to provide the shear. For the conditions of $ka = 10$ stabilisation occurs for a shear Mach number of 4. There are several experiments being planned to explore this stabilisation technique. There is in general a residual growth rate, but non-linearly there appears to be saturation of the modes at an acceptably low amplitude. Fig.9 illustrates this for 2-D ($m = 0$) and for 3-D ($m = 1$) simulations[26].

VI. SUMMARY

We have made a detailed study of the dense Z-pinch formed from a bare fibre of carbon, CD_2 or deuterium, with and without current prepulse. There is good agreement between 2-D MHD simulations including ionisation effects and experimental observations of transient bright spot formation and bifurcation. Whilst we have shown a 70-80% reduction of MHD growth rates when the ion Larmor radius is one tenth of the pinch radius, no saturation of the modes non-linearly has yet been found. However the presence of sheared axial flow can in theory lead to reduced growth rates and saturation of the instability at an acceptable value. Clearly we are entering an exciting phase of research when the potential of the dense Z-pinch will be put to the test.

REFERENCES

[1]   M. G. Haines, *J. Phys. D; Appl. Phys.* **11**, 1708 (1978)
[2]   T. W. L. Sanford et al., *Phys. Rev. Lett.* **77**, 5063 (1996)
[3]   R. S. Pease, *Proc. Roy. Soc.* **70**, 11 (1957)
[4]   S. I. Braginskii, *Zh. Eksp. Teor. Fiz.* **33**, 645 (1957);
      Sov. Phys. JETP **6**, 494 (1958)
[5]   M. G. Haines, *Plasma Phys. & Contr. Fusion* **31**, 759 (1989)
[6]   J. P. Chittenden & M. G. Haines, *Phys. Fluids* **B2**, 1889 (1990)
[7]   I. H. Mitchell et al., *Rev. Sci. Instrum.* **67**, 1533 (1996)
[8]   R. Aliaga Rossel et al., these proceedings, C1 P2
[9]   R. Aliaga Rossel et al., these proceedings, C1 P3
[10] M. Tatarakis et al., these proceedings, C1 P16
[11] A. Lorenz et al., these proceedings, B2 O6
[12] N. Kassapakis et al., these proceedings, B1 O7
[13] F. N. Beg et al., these proceedings, B2 P2
[14] J. P. Chittenden et al., these proceedings, B2 P4
[15] R. Aliaga Rossel & J. M. Bayley, these proceedings, A1 P1
[16] J. P. Chittenden et al., these proceedings, B2 P4
[17] D. F. Howell, these proceedings, Ce P6
[18] M. Coppins & I. D. Culverwell, these proceedings, C2 P3
[19] T. D. Arber, *Phys. Rev. Lett.* **77**, 1766 (1996)
[20] R. Aliaga Rossel et al., these proceedings, A1 P2
      A. Robledo et al., *Phys. Plasmas* **4**, 490 (1997)
[21] M. G. Haines, *Nucl. Instru. Methods* **207**, 179 (1983)
[22] M. G. Haines & M. Coppins, *Phys. Rev. Lett.* **66**, 1462 (1991)
[23] T. D. Arber, M. Coppins & J. Scheffel, *Phys. Rev. Lett.* **72**, 2399 (1994)
[24] T. D. Arber, P. Russell, M. Coppins & J. Scheffel
      Phys. Rev. Lett. **74**, 2698 (1995)
[25] T. D. Arber & D. F Howell, *Phys. Plasmas* **3**, 554 (1996)
[26] J. P. Chittenden & S. Lucek, private communication.

APPENDIX A

DZP TEAM (1997)

	Raul Aliaga Rossel		*(Stephen Lucek)*
§	*Tony Arber*	†	*Max Michaelis*
§	*James Bayley*	§	*Ian Mitchell*
	Farhat Beg	†	*Stavros Moustaizis*
	(Tony Bell)	*	*Alan Newton*
	Jeremy Chittenden		*Sanjay Pattni*
	Eugene Clark	*	*Nicol Peacock*
	Michael Coppins	†	*Ian Ross*
	Bucker Dangor		*José Ruíz Camacho*
	Huw Davies	†	*Jan Scheffel*
†	*Gernot Decker*	†	*Hellmut Schmidt*
	David Zdravkovic		*Michael Tatarakis*
	Malcolm Haines		*John Worley*
	David Howell		*Technicians:*
	Nicholas Kassapakis		*Jack Beckwith*
	Serguei Lebedev		*Brian Fantini*
	Axel Lorenz		*Alan Finch*

( )	*Part time on DZP*	§	*Former member,*
†	*Visitor*		*still contributing*
*	*Consultant*		

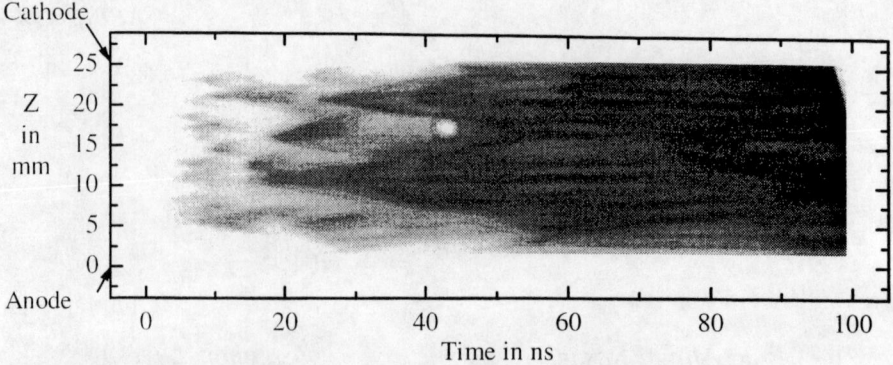

Figure 1. An optical axial streak photograph showing axial bifurcation of brightly emitting spots from 1 to 50ns.

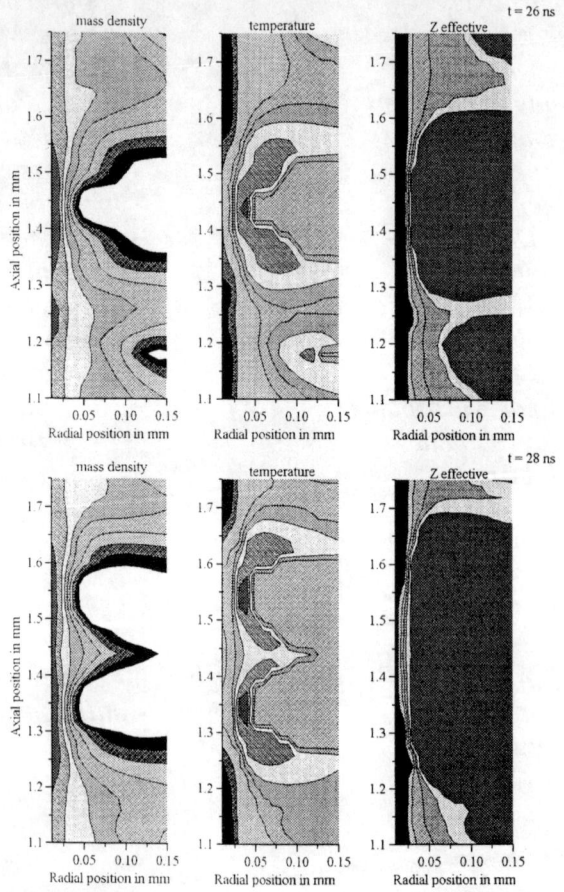

Figure 2. A 2-D MHD simulation of an m = 0 ionising MHD instability. Mass density, temperature and ion Z effective contours are shown as (a) 26ns and (b) 28ns.

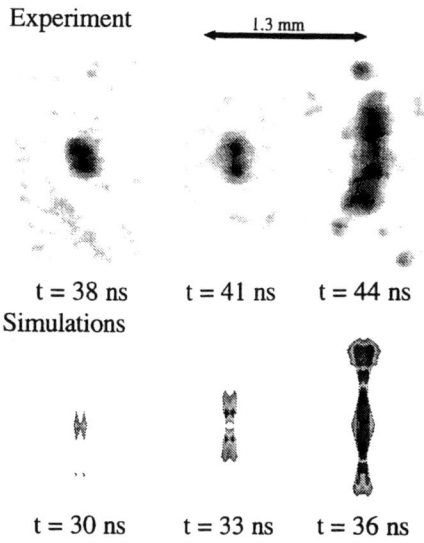

Figure 3. A sequence of three gated soft X-ray (MCP) images, 2ns exposure taken through 10μm beryllium filter and a 50μm diameter pinhole. These are compared to a simulation of similar conditions.

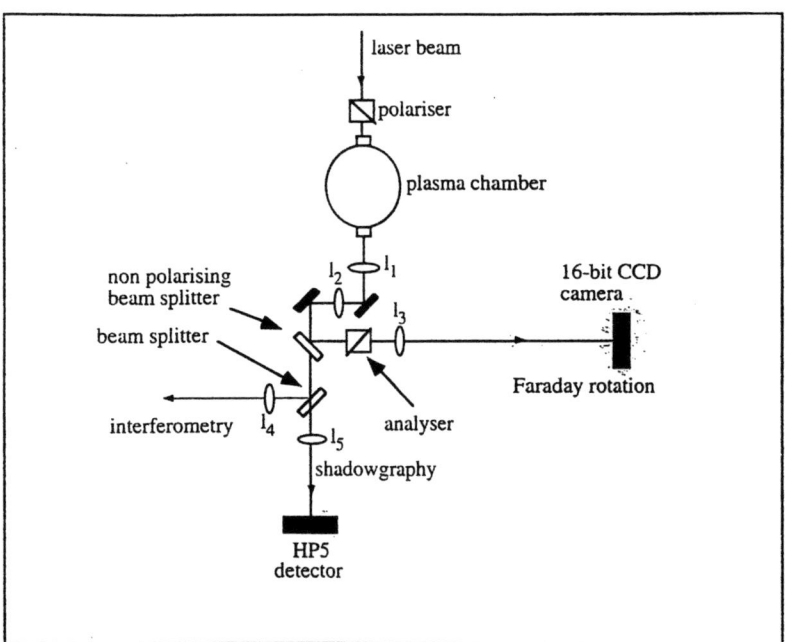

Figure 4. Experimental layout of the simultaneous laser probing for interferometry, shadowgraphy and Faraday rotation.

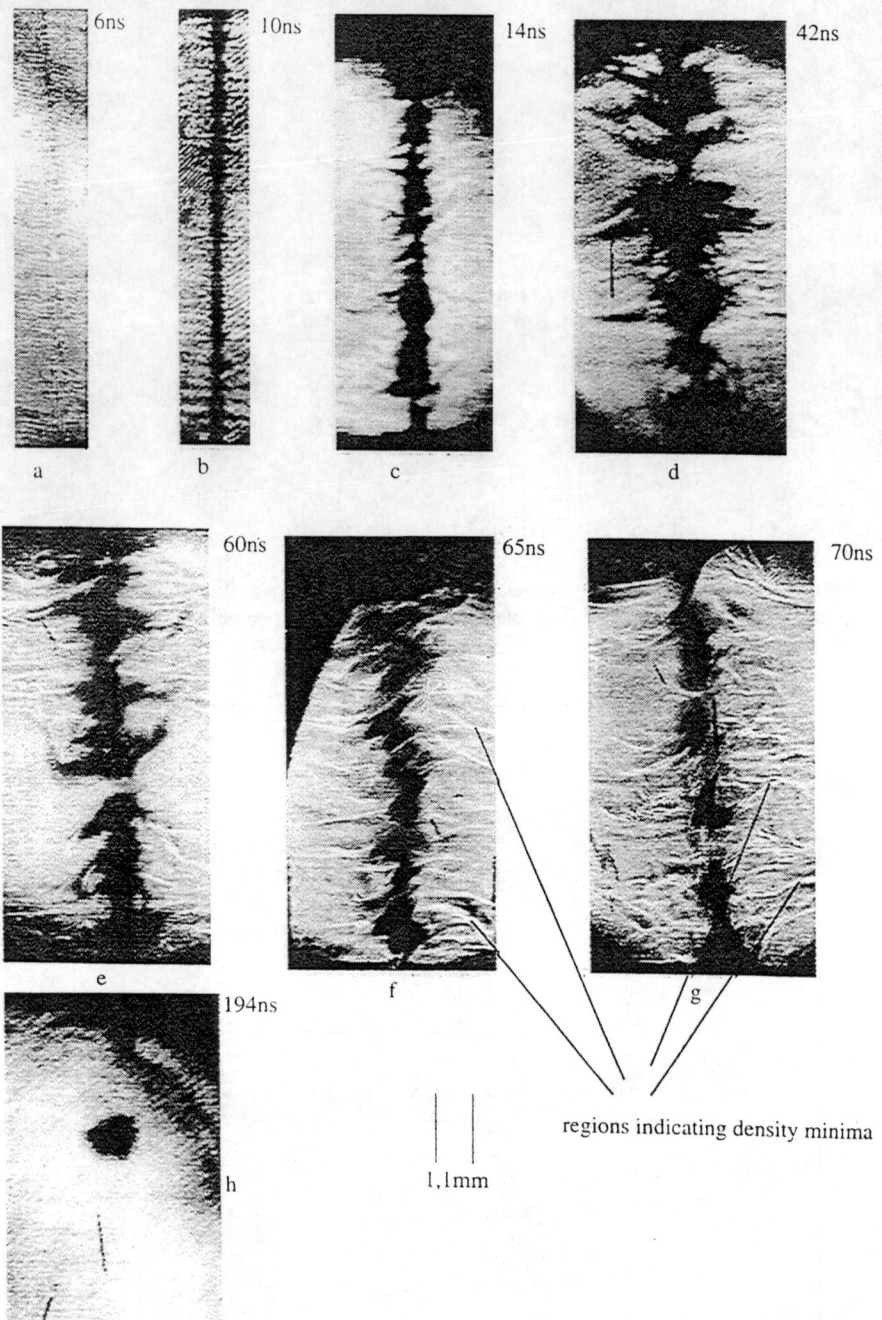

Figure 5. A series of schlieren shadowgrams taken in different shots for 33um diameter carbon and 60kV charging voltage (peak current 1.1MA).

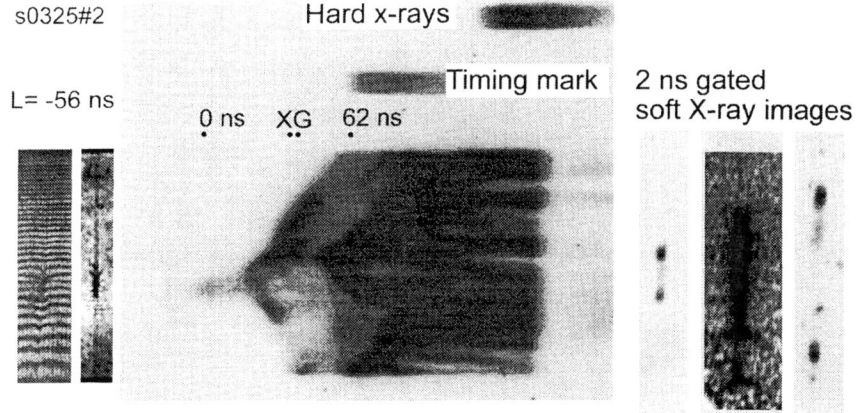

laser probing optical streak soft x-rays

Figure 6. The effect of a prepulse current on a 33µm carbon fibre on the MAGPIE generator. An interferogram and Schlieren image are shown of the prepulse coronal discharge at 56ns before the main pulse. An optical axial streak of the main discharge shows brighter emission in a region of MHD unstable activity of the prepulse discharge. The ensuing bifurcation is shown both in the streak and in the three soft X-ray images. A hard X-ray pulse indicates the start of a disruption.

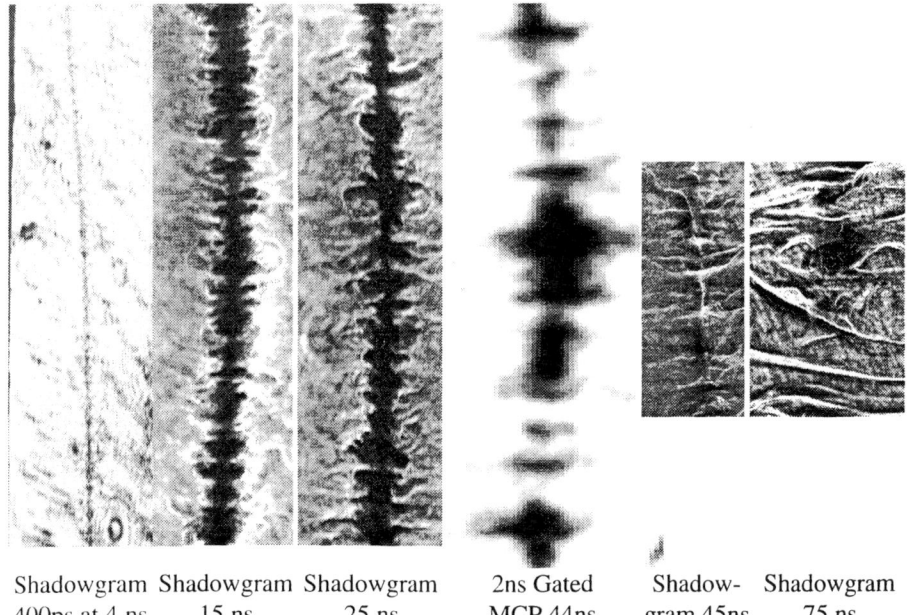

Shadowgram Shadowgram Shadowgram 2ns Gated Shadow- Shadowgram
400ps at 4 ns 15 ns 25 ns MCP 44ns gram 45ns 75 ns

Figure 7. 80µm diameter deuterium fibre results are shown in five shadowgrams together with a gated soft X-ray MCP.

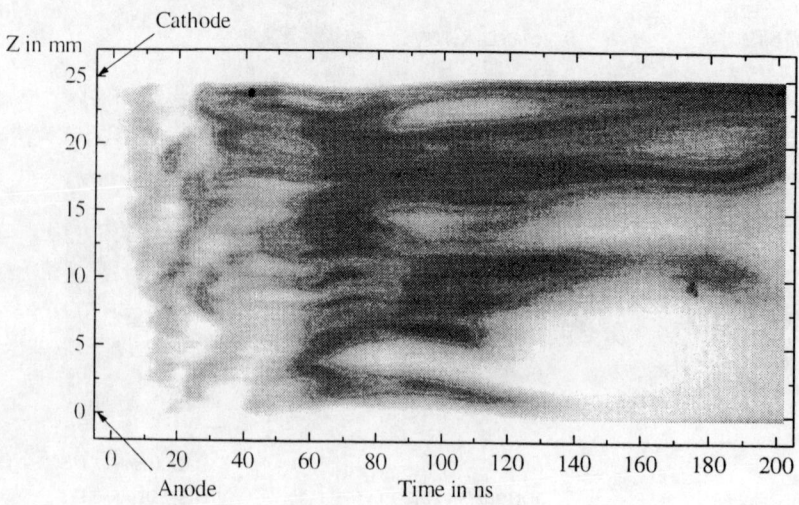

Figure 8. An axial optical streak of a 80μm diameter deuterium fibre shows the same sequence of bifurcating bright spots, a quiescent phase, followed by a disruption with density islands.

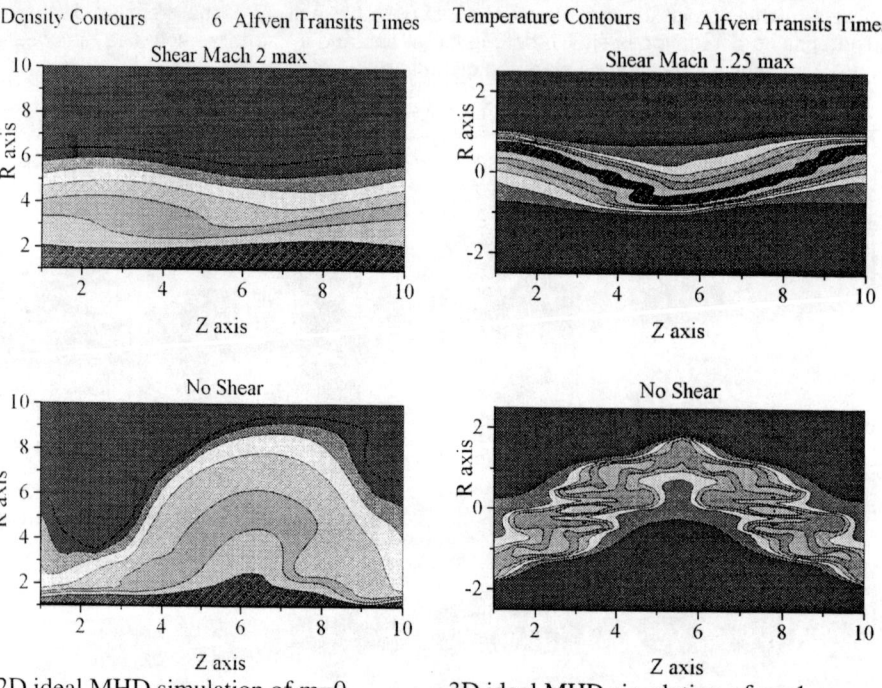

Figure 9. (Left) 2-D ideal MHD simulation showing density contours of m = 0 nonlinear development for a shear Mach number of 2 (above) and with no shear (below) at 6 Alfvén transit times. (Right) 3-D ideal MHD simulation showing temperature contours of m = 1 for a shear Mach number of 1.25 (above) and no shear (below).

UCI Staged Z Pinch Facility

F. J. Wessel, B. Moosman, N. Rostoker, Y. Song, and A. Van Drie

University of California, Department of Physics and Astronomy
Irvine, CA 92697-4575

and

P. Ney and H. U. Rahman,

University of California
Institute of Geophysics and Planetary Physics
Riverside, CA 92521

Abstract. The Staged Z Pinch couples energy to a target plasma, dynamically in stages. The present UCI experiment provides stable, multishell z pinches at 1.2 MA and 1 μs implosion time. Test-stand studies of an exploded- fiber target indicate that the fiber core is not ionized, due to current channeling in the high conductivity ablated plasma.

INTRODUCTION

In the Staged Z Pinch a plasma liner implodes onto a fiber-target plasma. The fiber is initially preionized by a current pulse that turns it into a uniform, high-density plasma before the z-pinch liner implodes; the current pulse also injects an azimuthal-magnetic field (B_θ) inside the pinch, in addition to an axial-magnetic field provided by external Helmholtz field coils. The magnetic fields compress differentially, providing dynamic shear, a means to inductively amplify current in the target, and confinement of fusion-reaction products. The unique feature of the Staged Z Pinch is the near-term prospect for thermonuclear break-even in a laboratory device[1] (http://mainpinch.ps.uci.edu).

RESULTS

The UCI Staged Z-Pinch Facility[2] is schematically shown in Fig. 1a. Two capacitor banks, consisting of ten, 2.5 μF, 50 kV capacitors each, are mounted on the top of the facility and attached to the ends of a plate-transmission line; peak energy - 50 kJ. The banks are switched by two pairs of Maxwell railgap switches. The plate-transmission line is fabricated from 6.4-mm thick aluminum plates and insulated by 1.8-mm thick mylar film. The plate line is 1.25-m long (between banks) and 2-m wide. The polypropelene vacuum interface provides an 0.11-m long vacuum path-length between H.V. electrodes; the anode-cathode (pinch) gap is 15 mm. The total-circuit inductance is 28 nH and the short-circuit current is 2 MA with a 1.8-μs quarter-period risetime. Vacuum is maintained by a 0.254-m diameter cryopump and a 0.15-m diameter liquid-nitrogen- baffled oil-diffusion pump; base vacuum is $1 \times 10^{-4} Pa$.

The pinch-discharge load-region is illustrated in Fig. 1b. The pinch is a multi-shell configuration, consisting of a 0.10-m diameter low-density hydrocarbon plasma imploding onto a 0.04-m diameter gas annulus and target plasma. Two Helmholtz field coils inject a DC, axial-magnetic field in the discharge-load region prior to energizing the pinch ($B_z \leq 0.05$ T).

The gas-shell load mass is injected by a two-stage, high-pressure gas valve ($\mathcal{P} \leq 6.8 \times 10^6$ Pa) and annular-nozzle (0.05-m diameter, 200-μm throat) that are integrated into the cathode electrode. The gas-injection system was characterized by piezo-electric probes to measure the flow-rate. The profile was imaged using a electron-beam scattering diagnostic to generate a time-resolved r-z profile of the gas-jet.[2] The nominal flow characteristics for argon are: risetime - 60-80 μs, gas-exit velocity - 800 m/sec, Mach number - 4.5, and mass-per-unit-length - $\sim 10^{-6}$ kg/m, 100 μs after injection.

The low-density plasma, injected outside the annular-gas shell, effects a rapid-transfer of current to-, and preionization of-, the gas shell near peak current. Current coupling occurs by a "dynamic-snowplow" of the low-density hydro-carbon plasma onto the load-gas shell.[3] The plasma is injected by semi-rigid, coaxial-cable "flashover" guns with nominal parameters, $T_e \sim 1$ eV and $n_e \sim 10^{22}$ m^{-3}.

Z-pinch diagnostics include: current and voltage probes, Ross-filtered P-I-N diode array, multi-channel x-ray pinhole camera, x-ray crystal spectrometer (KAP and LiF crystals), optical-framing and laser-backlit streak-camera imaging, liquid-scintillator neutron time-of-flight detectors, and neutron foil-activation (analyzed by a 0.20 m^2 active area EG&G drifted Germanium detector). An aluminum K-shell x-ray backlighter[4] and N_2 laser-interferometer[5] are used to characterize exploded-fiber targets on a separate test stand.

Pinch experiments involve a bank energy of 35 kJ. Figure 2 displays current, voltage, and x-ray PIN waveforms for a multi-shell neon pinch, and no target

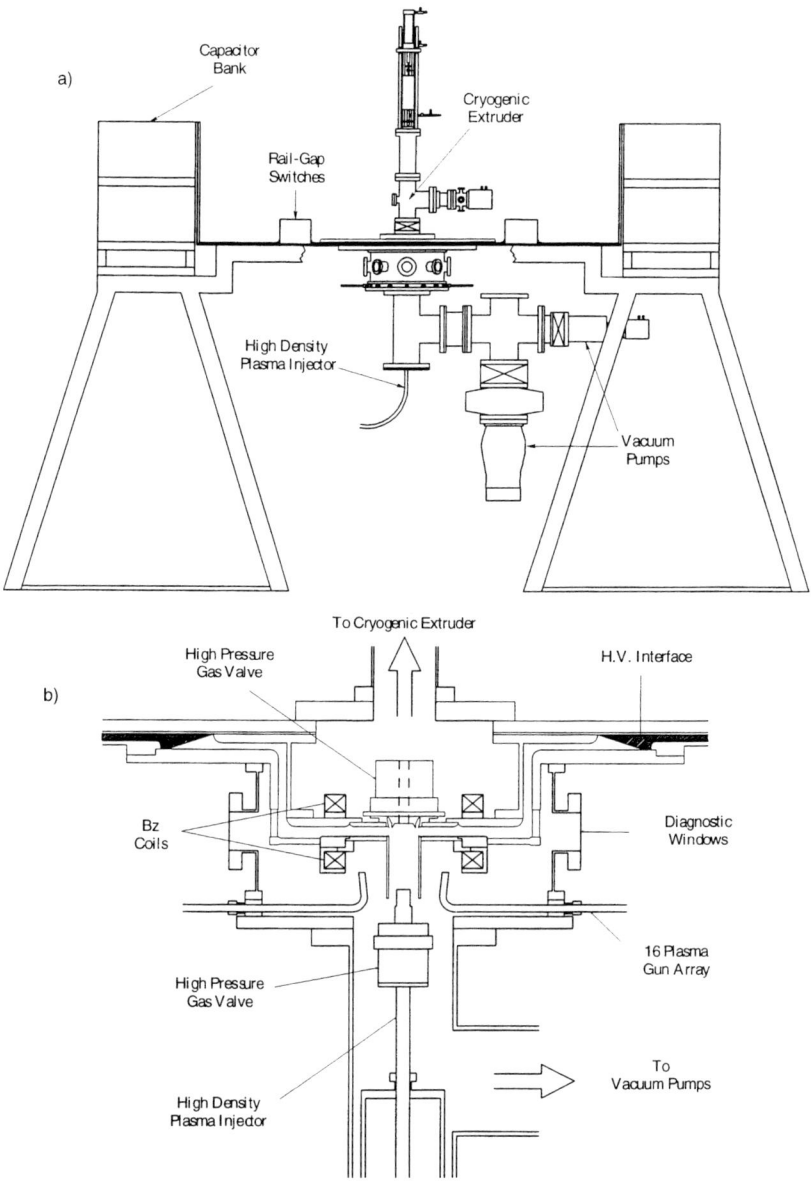

Figure 1: a) Staged Z-Pinch Facility and b) discharge-load region.

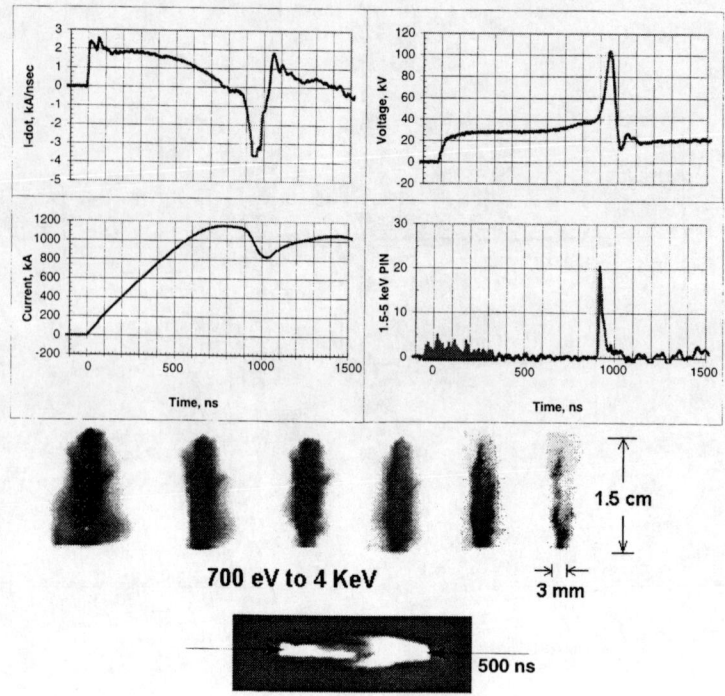

Figure 2: dI/dt, I, V, 1.5-5 keV PIN, 0.7-4 keV pinhole, streak images for Ne.

plasma. At approximately 750 ns there is a slight dip in the I-dot signal, indicating contact between the low-density plasma and the neon-gas shell as the current attains a peak value of 1.2 MA. The neon-gas shell implodes during the next 200 ns, generating 30 ns wide, broad-band (0.5-5 keV) x-ray pulses; only the 1.5-5 keV signal is shown here. Our experiments primarily involve neon, which provides a simpler-emission spectra and is easier to image than Kr for the parameters of our machine.

Time-integrated, x-ray pinhole images are displayed at the bottom of the Figure 2, where the average filtered-photon energy of the image increases from left-to-right in the range, 0.7-4 keV. Of particular significance is the axially-uniform, and slightly-hollow neon pinch at higher-filtration energy, due to the presence of a trapped, axial-magnetic field. Optical streak images of the z pinch confirm that the pinch remains hollow during most of the implosion. From these data we estimate the following peak-implosion parameters: 0.3-0.4 m/μs implosion velocity, 10^{26} m^{-3} density, and 500 eV temperature, with a K-shell yield of the order of a kJ.

A modified version of the LLNL TRAC2 code was used to compute waveforms[6] for the preceding load configuration and lumped-circuit parameters; the results are displayed in Figure 3, where the outer shell is assumed to be, 1.66×10^{23} m^{-3} density and 1 eV temperature, and the neon-plasma shell, 2.5×10^{23} m^{-3} density and 0.2 eV temperature. The close match between the computed waveforms and the data instills confidence in the predictions of the TRAC2 code.

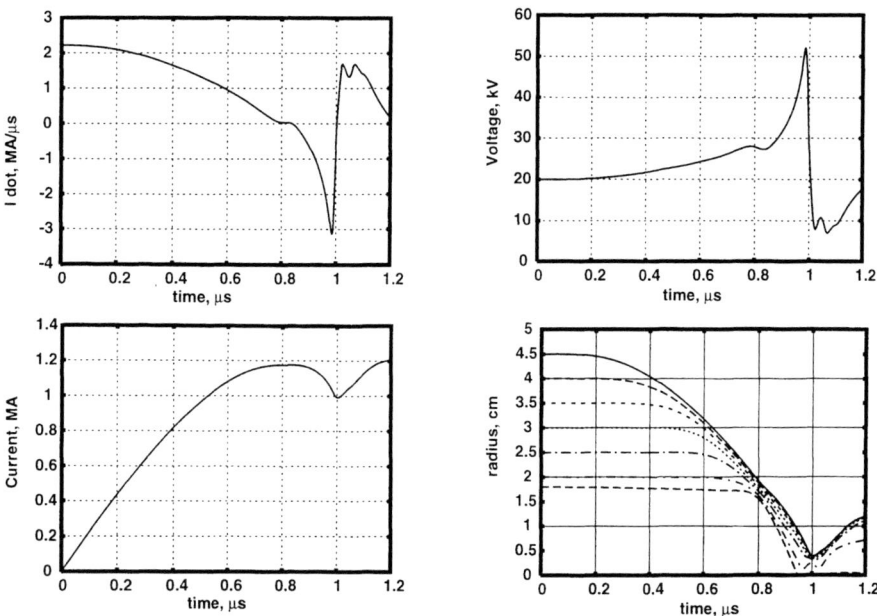

Figure 3: 1-D calculations for the actual parameters of Fig. 2.

The cryogenic-fiber extruder[7] utilizes a closed-cycle, liquid-helium refrigerator and has been used to successfully extrude $100 - \mu$m diameter, 0.38-m long fibers of hydrogen, deuterium and neon; length-limited by the vacuum-chamber. Freely-hanging fibers sublimate in vacuum for up to 20 minutes. A fiber-guiding system was also tested to co-axially position the fiber within a 0.002-m diameter, inside the pinch.

Preionization of a $100-\mu$m diameter D_2 cryogenic fiber requires a coupled-input energy of the order of one Joule, and precise timing to limit the radial expansion of the fiber prior to pinch initiation. On a separate vacuum-test stand we are evaluating fiber-preionization techniques using 25- to 100-μm diameter dielectric (polyethylene) and metallic fibers. The preionization pulse is provided by a compact Marx generator that has an adjustable output in the range: $E_0 < 2$ kJ, $I_{peak} = 15\text{-}45$ kA and $V_0 = 20\text{-}40$ kV with a $\tau_{1/4} = 0.65$ μs. The fiber expansion is time- and space-resolved using a 337 nm, 0.5 ns, TEA N_2 laser interferometer and a 30-ns pulsewidth, 0.5 J, aluminum K-shell, x-ray backlighter.

Figure 4 compares images for 25- and 100-μm diameter copper fibers energized by a 12 kA, 20 kV pulse. The input-energy was $E_{input} \sim 10$ MJ/kg for both fibers, (i.e., $¿E_{vaporization}$ for the solid[8]) and was adjusted by varying the image time relative to the discharge pulse. The electric-pulse timescale is sufficiently long to fully diffuse to the core of the fiber. The nitrogen laser is highly attenuated in the shadowgram and interferogram images for both fibers. Whereas, in the radiograph images the core of the fiber is absent for the 25-μm diameter fiber, but not for the 100-μm diameter fiber.

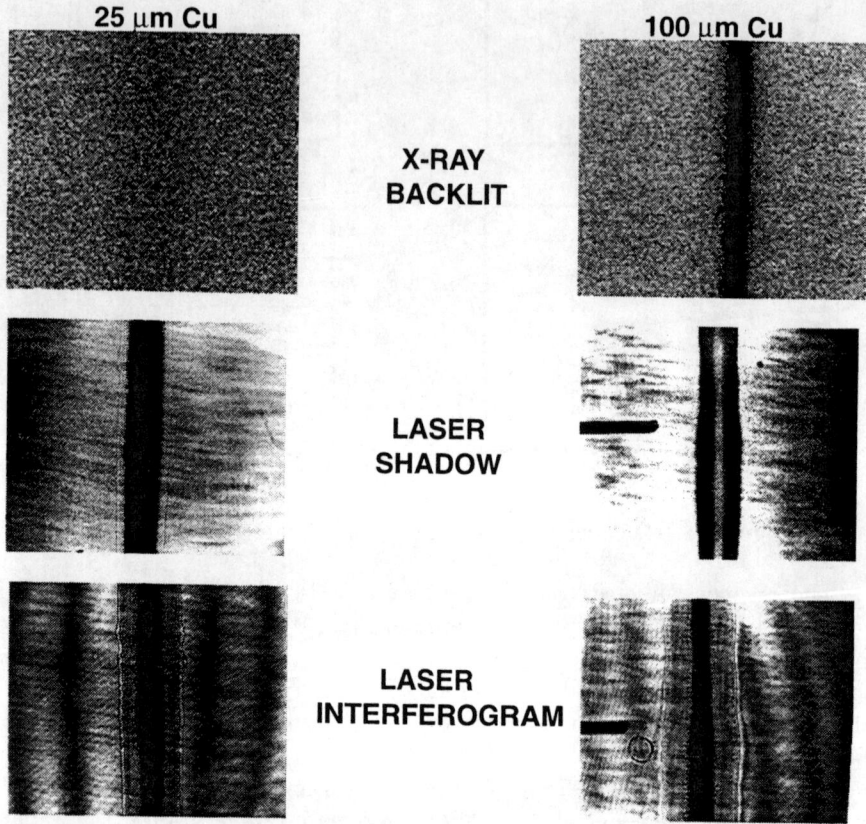

Figure 4: Laser shadow, interferogram, and x-ray radiograph images for a 25 μm diameter and 100 μm diameter copper fiber; $E_{input} \sim 10$ MJ/kg.

Thus, energy coupling to the fiber core is not observed, due to current channeling in the high-conductivity plasma that is ablated from the fiber during energy input. Varying the initial-fiber conductivity, using copper, iron, and aluminum-coated dielectric fibers, causes more energy to be channeled outside the fiber core for lower conductivity. Pulsed-energy coupling to a cryogenic fiber that is sublimating in vacuum is expected to be even more difficult and alternate approaches are under investigation.

ACKNOWLEDGMENTS

Thanks to Drs. E. Ruden, J. Degnan, and J. Bailey for their assistance with various equipment. Supported by U. S. DoE, OFES and in part by LANL.

References

1. Rahman, H.U., Wessel, F.J., Rostoker, N., Phys. Rev. Lett **74**, p. 714(1995).

2. Wessel, F.J., Bystritskii, V.M., Moosman, B., Rostoker, N., Song, Y. Tierney, T., Van Drie, A., Ney, P. and Rahman, H.U., 10th Pulsed Power Conference, Albuq., NM 1995, p. 112(1995).

3. Goldberg, S.M., Velikovich, A.L., AIP Conf. Proc. 299, Dense Z Pinches, p. 42(1994).

4. Moosman, B., Ph.D. thesis, University of California Irvine, May 1997.

5. Moosman, B., Bystritskii, V.M., Boswell, C., Wessel, F.J., Rev. Sci. Inst. **67**, p. 1(1995).

6. Rahman, H.U., Ney, P., Moosman, B., Rostoker, N., Song, Y., Van Drie, A., and Wessel, F.J., these proceedings.

7. Rahman, H.U., Ruden, E.L., Strohmaier, K.D., Wessel F.J., and Yur, G., Rev. of Sci. Inst. 67, p. 3533(1996).

8. Knoepfel, H., Pulsed High Magnetic Fields, North Holland Pub. Co. London, 1970.

Comparative Studies on a Gas Embedded Compressional Z-pinch in H_2 and D_2

L. Soto[1], H. Chuaqui[2], R. Saavedra[2], M. Favre[2], E. Wyndham[2], M. Skowronek[3], P. Romeas[3], R. Aliaga-Rossel[4], and I. Mitchell[4]

[1]*Comisión Chilena de Energía Nuclear, Casilla 188-D, Santiago, Chile*
[2]*Pontificia Universidad Católica de Chile, Casilla 306, Santiago, Chile*
[3]*Université Pierre et Marie Curie, Paris, France*
[4]*Blackett Laboratory, Imperial College, London, U. K.*

The present work deals with a comparative study in a compressional gas embedded Z-pinch in H_2 and in D_2 are presented. The use of H_2 and D_2 allows discharges with the same electrical properties, but different dynamics. Pressures of 1/3, 1/6 and 3/70 *atm* were used to carry out the experiments. The pinch is initiated by a focused laser pulse, which is coaxial with a cylindrical DC microdischarge. This configuration results in a double column pinch at early times, which as current rises, coalesces into a single column becoming a gas embedded *compressional* Z-pinch. The maximum electron density achieved on axis is greater than twice the expected value from the filling pressure, in contrast with a traditional gas embedded pinch. The expansion rate is reduced to a third of the observed value for the single channel laser initiated gas embedded pinch. This observation is consistent with a central current channel in the composite pinch. The experimental results, electron density at the centre and lower expansion rate, confirm the high degree of compression achievable with the composite preionization scheme.

INTRODUCTION

A gas embedded Z-pinch configuration which addresses the density and stability issues has been reported (1). In this configuration a central conduction channel, together with a coaxial current path is initially established. Compression up to twice the expected value from filling pressure is observed, as well as, enhanced stability. It has been conjectured that the stability observed can be explained by resistive effects and finite Larmor radius effects(2,3). In the present work studies in the gas embedded compressional Z-pinch covering a wider pressure range than previously reported in H_2 are presented, as well as, preliminary results in D_2. The comparison between discharges in H_2 and D_2 provides some insight into the relevance of the dynamics in the plasma evolution.

EXPERIMENTAL DETAILS

The experiments were carried out on GEPOPU, a generator capable of delivering currents up to 200 kA to a 1.5 Ω impedance load for 120 ns. The value of dI/dt of the current ramp was approximately $2 \cdot 10^{12}$ A/s. A DC microdischarge is established between two stainless steel conical hollow electrodes with 2 mm diameter, separated by 10 mm. A few nanoseconds before the application of the main voltage from the driver, a pulsed Nd-YAG laser (20 ns, 200 mJ at 1.06 µm) is focused through the anode onto the cathode. As the main voltage is applied, the laser generated plasma acts as an electron source to provide preionization on a central plasma column. Initially this conduction channel is highly resistive. The combined preionization scheme produces two parallel concentric conductive paths.

A Nd-YAG frequency doubled 6 ns laser pulse was used for optical diagnostics to obtain simultaneous single shot image-plane holographic interferometry and holographic shadowgraphy. Visible streak camera provides radial and axial plasma motion. Single frame image converter camera are used to corroborate results obtained with the streak camera. From these diagnostics the electron density profile, $n_e(r)$, the electron line density, N, and the external Z-pinch radius, a, are obtained with good temporal and spatial resolution. The total current, $I(t)$, and the external voltage, $V(t)$, were also measured with a Rogowskii coil and a capacitive divider. Filling pressures of 1/3 1/6 and 3/70 atm were used in H_2 and D_2, although not all diagnostics were employed for all conditions.

RESULTS AND DISCUSSION

Results obtained at 1/3 atm from single frame interferometry indicate that the axial electron density in H_2 becomes larger than $4 \times 10^{25} m^{-3}$ at 75 ns after the current start, being of the same order for discharges in D_2 after 85 ns. From the corresponding shadowgram it can be seen that the refractivity is high at the pinch centre, which is in agreement with the fact that no fringes are visible near the axis of the discharge. The line density in H_2 is of the order of 5×10^{19} m^{-1}, which is significantly higher than reported previously (1), and 2.8×10^{19} in D_2. The Bennett temperature can thus be estimated at only 40 eV in H_2 and 90 eV in D_2, which is consistent with the negative results obtained with soft X-ray diagnostics used to measure temperature. If the observed trend of negligible growth in line density is maintained at higher currents, the Bennett temperature for the maximum current obtained in GEPOPU, 180 kA, would be 75 eV in H_2 and 180 eV in D_2.

From axial streak photographs it is observed that the initial laser spark at the centre has independent dynamics, which is consistent with a metallic laser spark in expansion. The spark remains near the electrode. These observations agree with

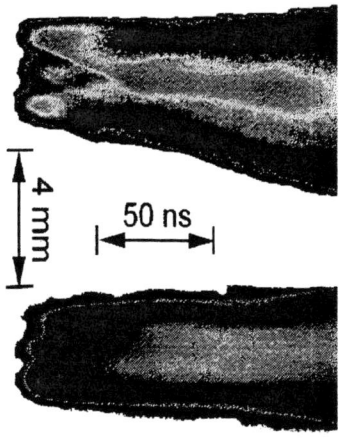
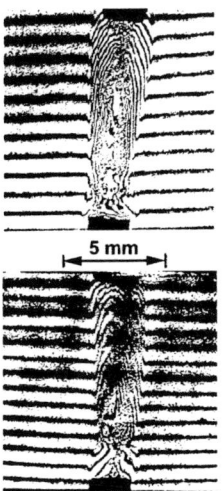

Figure 1. Radial streak photographs and interferograms are shown for H_2 (top) and D_2 (bottom) for 1/3 *atm*. In the streak photographs the different expansion rates for the two cases are observed, having separate coaxial initial current channels coalescing at later times. The H_2 interferogram was obtained at 75 *ns*, whereas the D_2 one was at 85 *ns*.

previous inteferometric results(1) which show that the maximum spark size is 2 *mm* at 50 *ns*. From radial streak photographs, as shown in figure 4, an initial 10 *ns* fast expansion phase is observed, followed by about 50 *ns* expansion of the central channel of 2×10^4 *m/s*. The expansion of the internal wall of the annular plasma is of the same order, whereas the external wall has a slower expanding velocity of 6×10^3 *m/s*. The reduced expanding rate of the external wall compared to the central channel is presumably due to the magnetic field generated by the central current channel. The fact that the laser initiation does make a difference is an indirect evidence that there is an important fraction of the current flowing near the axis. At the end of this phase the two initial plasmas coalesce into one with an apparent internal structure, a brighter central region which is maintained during the following 110 *ns*. The corresponding expansion rates for D_2 at 1/3 *atm* are ~ 1.2×10^4 *m/s* and ~4×10^3 *m/s*, lower than the values obtained for H_2. The expansion rate is reduced slightly less than expected from the mass difference. For the lowest pressure studied no instability has been observed, the density being correspondingly lower.

The relevant parameters regarding stability properties are the Larmor radius over the pinch radius a_i/a, the Lundquist number S, and the Alfvén time τ_A. For the two cases considered, H_2 and D_2 at 1/3 *atm*, the corresponding values are 0.08 and 0.15; 40 and 150; 20 and 18 *ns*. From the values of S it is apparent that the discharges are resistive. The discharges last for many Alfvén times maintaining a stable pinch. The a_i/a ratios observed are consistent with recent theoretical studies which indicate the region of minimum instability for pinch discharges is in the neighbourhood of 0.2 (4,5).

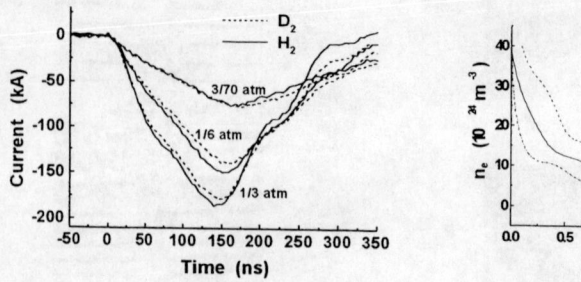
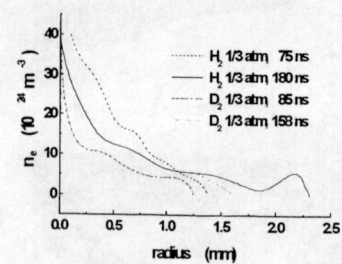

Figure 2. On the left hand side, current traces for H_2 and D_2 at 1/3, 1/6 and 3/70 *atm* are given. The right hand side graph shows the electron density obtained from interferograms. On the 158 *ns* case for D_2 the fringes are lost for radii smaller than 1.25 mm. The driving voltage is the same for all conditions.

Further work is required for discharges in D_2, a complete set of electron density and line density are required to do a full analysis of the influence of a higher mass gas. Future work being considered includes experiments with smaller initial radius for the annulus, lower background pressure and up to 1.2 *MA* peak current in the LLAMPÜDKEÑ generator(6).

ACKNOWLEDGEMENTS

This work has been partially funded by FONDECYT, grants 1950048, 1960555 and 2960008, Andes Foundation grant C-12776, ECOS/CONICYT programme and Imperial College (MAGPIE project). R. Saavedra holds a CONICYT scholarship.

REFERENCES

1. L. Soto, H. Chuaqui, M. Favre, and E. Wyndham, *Phys Rev. Lett.* **72**, pp 2891-2894(1994).
2. H. Chuaqui, L. Soto, M. Favre, E. S. Wyndham, and M. Skowronek, "Parameter Space Comparison with Universal Diagram for Z-Pinch Stability", in Proc. III International Conference of Dense Z-pinches, London (1993), pp 27-33.
3. L. Soto, H. Chuaqui, M. Favre and E. Wyndham., "Gas embedded compressional Z-pinch experiments", in Proc. Int. Conf. on Plasma Physics, 1994, **V.1**, Contributed Papers, pp 216-219 (Iguazu Falls, Brazil)
4. T. D. Arber, M. Coppins and J. Scheffel, , *Phys. Rev. Lett.* **72**, pp 2399-2402(1994).
5. T. D. Arber, P. G. F. Russel, M. Coppins and J. Scheffel, , *Phys. Rev. Lett.* **74**, pp 2698-2701(1995).
6. H. Chuaqui, E. Wyndham, C. Friedli and M. Favre, to be published in *Laser and Particle Beams*.

On the Radiative Collapse Phenomenon

P. Choi and C. Dumitrescu-Zoita

*Laboratoire de Physique des Milieux Ionisés, Ecole Polytechnique,
91128 Palaiseau, France*

Abstract For a pinch under equilibrium, the Pease-Braginskii current limit sets the ceiling when the ohmic heating rate is balanced by the bremsstrahlung loss rate. Under a dynamical condition, Shearer showed that, when the pinch current exceeds the Pease-Braginskii limit, the difference in ohmic heating and bremsstrahlung loss rate could lead to the phenomenon of radiative contraction, producing a plasma structure of ultra high energy density. The work was extended by Haines and Robson independently to describe the conditions for radiative collapse. These analytical results show that under certain circumstances, a high temperature pinch can be self-compressed to an extremely high density, limited only by photon self-absorption or electron degeneracy. In this paper, we present an analysis on the radiative collapse formalism including the effect of an external circuit. It is shown that the degree of contraction is governed by the intrinsic rate of energy delivery to the system. Two expressions are derived which define the ultimate pinch dimension with the plasma parameters and that of the electrical circuit.

I. RADIATIVE CONTRACTION

In the first paper to consider radiative contraction, Shearer examined the situation of power balance in a Z-pinch with constant line density.[1] For a pinch column in thermal equilibrium, the gain or loss of energy will result in the expansion or contraction of the column. The resulting PdV work done on the column is given by the balance of ohmic power input, P_Ω, and the radiation power loss, P_B. This leads to an equation of motion which, in terms of the Pease-Braginskii limiting current I_{PB}, [2,3] is

$$\frac{1}{a}\frac{da}{dt} = C_1 \frac{P_B}{I^2}(\frac{P_\Omega}{P_B} - 1) = \frac{C_2}{\tau_R}(\frac{I_{PB}^2}{I^2} - 1) \quad (1)$$

where $\tau_R = 3NkT/P_B$ is the radiation time constant and a the radius of the column. From this, it can be seen that the pinch column will contract when $I > I_{PB}$ and it was suggested that the pinch radius will approach zero until the contraction is terminated by additional mechanisms, like re-absorption of the plasma radiation at high density or the onset of anomalous resistivity due to microinstabilities. Both of these processes will lead to an increase in I_{PB} and terminate the contraction.

I.1 Radiative Collapse

Haines extended the picture proposed by Shearer to include the rate of change of internal energy in the pinch column. By assuming a constant ln Λ and a linearly rising current, it was shown analytically in a 0-D model for an isothermal hydrogenic plasma, that the pinch radius would be reduced to zero when $I=\sqrt{3}\ I_{PB}$.[4] The same result was obtained independently by Robson [5] and was considered as the condition when radiative collapse is complete.

The effect of an external pulse generator was considered by Haines and the 0-D model solved numerically. It was concluded that finite voltage imposed by the external circuit would prevent the total collapse. Robson considered the compression of hydrogen and helium plasma with a suitable external generator at constant voltage, where the initial rate of rise of current is prescribed.[6] The effects of self-absorption and electron degeneracy was included and it was found from a 0-D calculation that the maximum density achieved is increased by the addition of a series inductance next to the load.

I.2 Limiting current

The concept of the limiting current I_{PB} rests on balancing the ohmic heating rate to the bremsstrahlung radiation loss rate. The assumption of pressure balance allows the Bennett relation to be used to substitute current for temperature, leading to $I_{PB}=0.433*(\ln \Lambda)^{1/2}$ MA, for a hydrogenic pinch column with uniform temperature and uniform current density.[5] This limiting current is in fact not applicable in non-steady state conditions, where the assumption of a Bennett temperature is no longer valid. In fact, the limit of current in a pinch column exists only if the incremental increase in current is converted to incremental increase in temperature. This is not the case in dynamical pinches.

2. ENERGY LIMIT

In the original radiative collapse formalism, the energy available to drive the compression was not considered explicitly. When the current exceeds the value of I_{PB} the energy loss rate through radiation begins to exceed the energy input rate through ohmic heating. The loss in internal energy will result in a contraction of the column radius, driven by the magnetic pressure. To continue the contraction, energy has to be supplied to the pinch column to partially compensate for the energy loss through radiation, in order to maintain the high radiation loss rate necessary to continue the contraction mechanism. On the other hand, the magnetic pressure in the system has to remain higher than the internal pressure of the pinch column for the contraction to take place. These conditions are contained implicitly in Eqt.(1) which requires $I > I_{PB}$ at all time of the contraction.

To examine the energy requirement to drive the compression, we define a radiative compression scale time, τ_{RC}. Only energy which can be supplied to the plasma column within this time could be of use to maintain the contraction. The energy delivery rate is naturally controlled by the speed of light in the local energy storage medium. The way that energy is distributed in this local storage medium therefore governs the overall compression limit. The limiting condition is dictated by the power supplied equal to the power dissipated.

We can consider two limiting cases. In the first case, there is negligible inductance between the plasma column and the pulsed power generator and the bulk of the energy available for radiative contraction is delivered by the generator itself. This is a typical condition encountered in high current pulse forming line driven system. In the second case, the inductance around the load region dominates and the generator is effectively decoupled from the plasma column. This could the situation when the pinch is formed in a high voltage, high inductance system as that in an IES generator.

2.1 High current low inductance system

In a pulse forming line driven system, we can define a length of the pulse line from where energy can be extracted during the radiative compression, given by $\Delta = c\tau_{RC}/\varepsilon_r^{1/2}$, where c is the speed of light and ε_r the dielectric constant of the pulse line medium. Consider the case where the pulse forming line is a coaxial line of impedance Z. The inductance from which the store energy can be extracted to drive the radiative compression is then

$$L_F = \frac{\mu_o}{2\pi}\Delta Z \frac{\sqrt{\varepsilon_r}}{60} = \tau_{RC} Z \qquad (2)$$

after some algebra and substituting for Δ. During the radiative compression the power supplied to the load is $L_F I^2/2\tau_{RC}$ which is equal to $ZI^2/2$ using Eqt.(2). The power dissipated is limited to the ohmic input given by I^2R. Equating the power supplied to the power dissipated, we obtain the limiting value of $R=Z/2$. From the plasma resistance

$$R = \frac{\eta\delta}{\pi a^2} \qquad (3)$$

where δ is the length of the contracted plasma column with resistivity η, we arrived at an expression for the limiting radius a of compression

$$a^2 = \frac{2\delta\alpha Z_A \ln \Lambda}{\pi Z T^{\frac{3}{2}}} \qquad (4)$$

using the resistivity transverse to a magnetic field, $\eta = \alpha Z_A \ln\Lambda/T^{3/2}$ where Z_A is the effective charge of the plasma. What is interesting is that neither the time scale of the compression, nor the voltage of the driver appears in the expression. The impedance of the external driver Z is the only controlling factor external to the plasma column. Fig.1 shows the value of the limiting radius as a function of electron temperature, for a 1 cm long, $Z_A=1$ plasma at driver impedances $Z = 0.25$, 1 & 5 ohm. The dependence on electron temperature and density enters through the coulomb logarithm $\ln \Lambda$ but the variation with density is small.

2.2 High voltage high inductance system

The situation where the inductance around the load region dominates can be examined by considering a pinch column of length h, between two parallel circular electrodes. In this case, energy can be extracted during radiative compression in a zone within a radius b along the electrodes, given by $b = c\tau_{RC}$ with an associated inductance

$$L_R = \frac{\mu_o}{2\pi} h \ln(\frac{b}{r_0}) \qquad (5)$$

where r_0 is the initial radius of the pinch before the start of the radiative compression. Using previous arguments for the radiative compression, the power supplied to the load is $L_R I^2/2\tau_{RC}$ and the power dissipated is given by I^2R. Equating the two gives us the limiting value of $R=L_R/2\tau_{RC}$. Using Eqt.(3) with the length of the plasma set to h, and the resistivity transverse to the magnetic field, we obtain a limiting radius given by

$$a^2 = \frac{4\tau_{RC} \alpha Z_A \ln\Lambda}{\mu_0 T^{\frac{3}{2}} \ln(\frac{c\tau_{RC}}{r_0})} \qquad (6)$$

Unlike the pulse line driven system, the limiting radius is now independent of the length of the discharge but is dependent on the compression scale time. The factor $\ln(c\tau_{RC}/r_0)$ is a slow varying function and changes from 6.9 to 13.8 for $(c\tau_{RC}/r_0)$ of 10^3 to 10^6. Taking this value as 10, Fig.2 shows the limiting radius obtained from Eqt.(6) as a function of the electron temperature, for three values of the scale time τ_{RC}=.1, .5 and 1 ns.

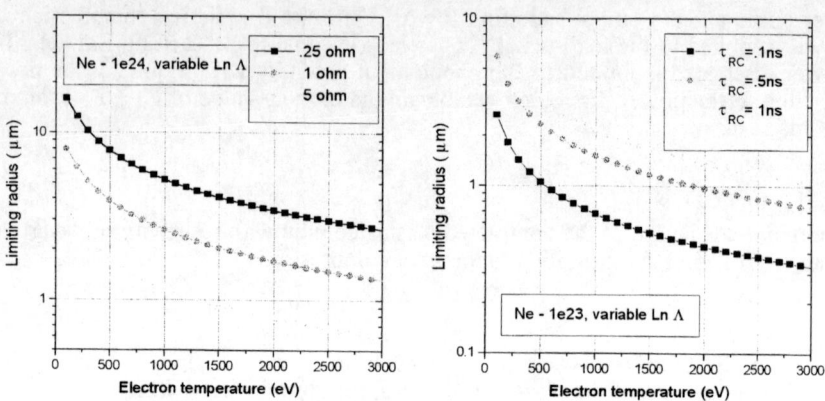

Figure 1 Limiting radius of contraction vs T_e for driver impedance Z=.25, 1 & 5 ohm.

Figure 2 Limiting radius of contraction vs T_e for different radiative compression time

2.3 Conclusion

For both driving conditions considered here, the effect of a finite rate of energy supply is to limit the degree of compression. The pulse line driven system is the more common approach in driving high current Z-pinch discharge. For typical TW class generators the impedance Z is 2 ohm or less and the limiting radius obtained here is larger than the sub-micron size anticipated from earlier work. The results here show that a high local inductance approach should be adopted if a higher degree of compression is to be attained.

REFERENCES

[1] Shearer, J.W., Phys. Fluids, 19, 1426 (1976).
[2] Pease, R.S., Proc. Phys. Soc. London Ser. B 70, 11 (1957).
[3] Braginskii, S.I., Sov. Phys. JETP 6, 494 (1958).
[4] Haines, M.G., Plasma Phys. Control. Fus. 31, 759 (1989).
[5] Robson, A.E., Nucl. Fusion, 28, 2171 (1988).
[6] Robson, A.E., Phys. Fluids B 1, 1834 (1989).

A Novel Cryogenic Fibre Maker for Continuous Extrusion

R. Aliaga-Rossel[a], J. Bayley

*The Blackett Laboratory, Imperial College
London, U. K.*

Abstract. The results of a cryogenics fibre maker which extrudes fibres continuously are presented. The fibre maker is based on a simple concept of differential temperature. Two reservoirs are connected in cascade and are kept at different temperatures. The first reservoir is connected to an external gas line supply (the gas that will made the fibre) and is used to liquefy the gas. The second reservoir is colder that the first and the liquid that comes from the first reservoir is frozen and later is used to form the fibre. The pressure of external gas supply in the first reservoir is used to extrude the fibre. The system is cooled by a two stage closed cycle refrigerator, which uses liquid helium as a working fluid. The nozzles used to extrude the fibre are made of stainless steel capillary with diameters between 50 μm and 250 μm, with a length of 2 mm. The use of a system with two independent temperatures, permits to control the extrusion rate of the fibres and to produce the fibres continuously. Using this system, hydrogen, deuterium, nitrogen and argon fibres of various diameters were extruded.

INTRODUCTION

The primary interest in producing cryogenic fibres was their use as a load in the MAGPIE[1] (Mega Ampere Generator for Plasma Implosion Experiments) generator. This generator is capable of driving over 1 MA into high inductance load such as solid fibres with diameters between 10 to 300 μm. The discharge chamber of the generator is located on top of an MITL (Magnetically Insulated Transmission Line). The separation between anode and cathode is 20 mm. The cryogenic nature of these fibre requires them to be manufactured *in situ*. The fibre maker must be located at least 100 mm away from the electrodes due to mechanical constraints and the risk of damage caused by the discharge. For a correct operation of the MITL the pressure in the load chamber must be less than 10^{-4} mbar. The charging of the generator may take several minutes, therefore the fibre must be stable after it has been placed between the electrodes. As the fibre touches the lower electrode (cathode) it evaporates continuously which sets the condition of continuous extrusion with a controlled speed. For safety reasons personnel have to be at least 5 m away from the target area, so the fibre maker has to be operated by remote control. With these restrictions in mind, the fibre should be 100 mm long, 10 to 300 μm diameter, mechanically stable, not increase the pressure of the load chamber above 10^{-4} mbar and be produced continuously under remote control.

(The pressure requirement was not fulfilled in the MAGPIE chamber, due mainly to the high impedance section of the MITL, where a turbo pump is located.)

PREVIOUS STUDIES

Although little work has been done on the production of solid fibre from high Z gases, controlled thermonuclear fusion programs have stimulated research into forming solid pellets and fibres of hydrogen and deuterium. Deuterium fibres have been used in Z-pinch experiments where very small diameters are necessary in order to have the optimum line density in the plasma. Sethian[2] *et al* (NRL) and Griolly[3] *et al* (LANL) were able to produce fibres of deuterium with sizes between 40 and 300 μm using a different extrusion method. The gas (deuterium) is slowly bled into a cold cylindrical reservoir (which is cooled by liquid helium) where the gas desublimates onto the walls as snow. A ram is then driven into the reservoir to compact the snow. After the compression, the temperature of the reservoir is raised until the fibre is extruded through a nozzle (an aperture or capillary) of a given diameter.

CONTINUOUS EXTRUSION

The use of a ram to force the snow through a nozzle is not very suitable for automation and the solid material that bypass the piston can be larger (Sethian reports that only 8% of the solid formed in the extrusion chamber is actually extruded). For a nozzle of 100 μm diameter and a close fitting piston of 7 mm diameter with a gap of 20 μm, the area between the piston and the bore is about 100 time bigger than the nozzle, therefore such a degree of loss is to be expected. In time, the piston and the bore wear each other away and the losses past the piston increase. If one is using expensive gases (or a radioactive material) it is desirable to use the smallest possible quantity. The great pressure used for the extrusion of the fibre (about 100 bar) requires a strong mechanical structure in the fibre maker, which increases the size and the mass of the apparatus and therefore the cooling power. To overcome these problems a different approach was used, in which the fibre is extruded at high temperature using a gas-liquid system with two reservoirs, in a scheme similar to the one used by Jarboe[4].

THE FIBRE MAKER

A very important aspect of the design is the nozzle through which the material is extruded. The ratio of length to diameter (aspect ratio) must be sufficiently large to extrude a straight and well directed fibre, and also sufficiently short to keep a low extruding force. The first attempt was to use pinholes from an electron microscope as nozzles, which have sizes between 5 to 300 μm, but the small aspect ratio in such pinholes results in curved and unstable fibres. For this reason 2 mm long stainless steel capillaries soldered into a copper disk of 22 mm diameter were used. The copper disk was fixed to the lower reservoir by six M3 SS bolts with a seal of 1 mm diameter indium wire. The smallest diameter of capillary available

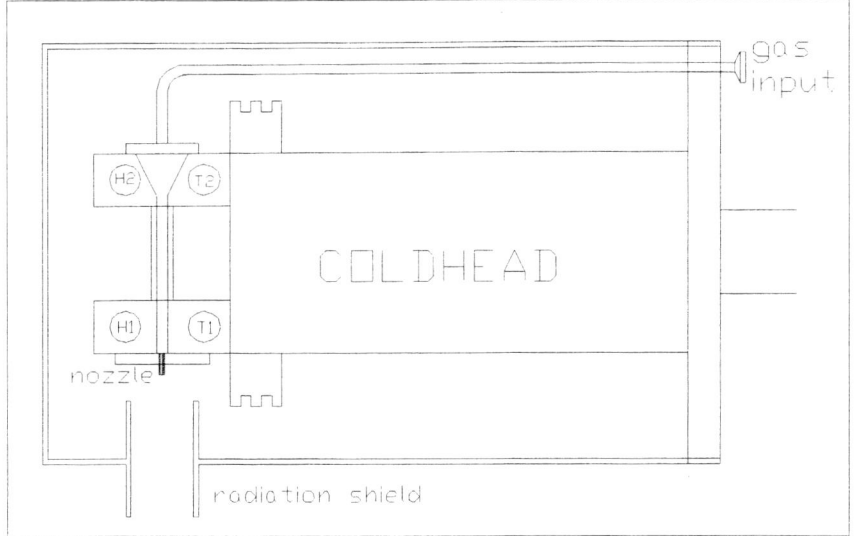

FIGURE 1. Scheme of the fibre maker.

was 50 μm bore. Due to the high temperature of the extrusion, it was possible to use a closed cycle two stage refrigerator (Edward 6/30). The cooling power is only a few hundred mW at 10 K and exceeds 3 W at 15 K, depending of the thermal losses of the coldhead. The cryo-refrigerator serves two purposes: it condenses the gas and cryopump the vacuum chamber into which the fibre is extruded. This second function is very important because the vacuum is lost if there is insufficient cooling power to freeze out the gas which evaporates from the surface of the fibre and from its contact point with the lower electrode. This in turn, produces a loss of thermal insulation in the cryostat and thermal runaway. The fibre maker is composed of two reservoirs made of oxygen free copper blocks (dimensions 40 x 25 x 8 mm). A SS (stainless steel) tube of 5 mm diameter with 3 mm bore and 10 mm long, connects both blocks. The volume of the upper and lower reservoirs are 400 mm^3 and 70 mm^3 respectively. The gas is supplied by a SS capillary 1.25 mm diameter with one end connected to the upper reservoir and the other end to a gas cylinder. Each reservoir has a RhFe temperature sensor (T_1 and T_2) and a 80 W heater (H_1 and H_2). Both sensors are connected to a couples of temperatures controllers, with 0.1 K resolution. The sustainable difference of temperature between both reservoirs depends on the thermal conductivity between the reservoirs and the cold head and also of the conductivity of the SS tube that joins both reservoirs. This is a critical part of the design because that difference in temperature will set what type of gases is possible to use. Figure 1 shows the scheme of the fibre maker. The two reservoirs are bolted to the cold head by two M4 SS bolts. The outside of the cryostat is surrounding by a radiation shield (a copper cylinder 1 mm thick) which is fitted to the first stage of the cold head. This radiation shield was cooled as low as 30 K. The cryostat was coupled to the vacuum chamber with a MW40 flange.

OPERATING CONDITIONS

To extrude fibres, the fibre maker and the vacuum chamber were evacuated to a pressure less than 10^{-4} mbar. The lower reservoir was cooled down to a temperature much lower than the frozen point of the gas used in the extrusion, while the upper reservoir was kept at a temperature near the melting point of the gas. Gas at 2 bar from a 75 cm^3 plenum was let in the feeder capillary until the nozzle was blocked with the frozen gas. More gas is fed until the pressure ceases to fall indicating that the upper reservoir was full of liquid, then the external gas supply line is connected permanently. It is important not to cool the upper

TABLE 1. Temperature of the upper and lower reservoir for the extrusion of fibres of different gases.

Filling gas	T_1 (K)	T_2 (K)
H_2	19	29
D_2	23	33
N_2	65	71
Ar	84	88

reservoir below the freezing point, otherwise the gas may freeze in the supply line and the system will be isolated from the external extrusion pressure and from the gas supply. To characterise the operational mode of the fibre maker the nozzle tip was monitored with a video camera. To start the extrusion, the temperature of the lower reservoir was raised until solid material was seen to flow out through the nozzle. The temperature could then be lower to the operating value and continuous extrusion ensured. By connecting the feed to a large gas cylinder, extrusion of fibres of nitrogen was obtain for over an hour without external intervention. The rate of extrusion could be controlled by changing either the temperature of the lower reservoir and the pressure of the external line gas supply. A typical value of the extrusion rate was 1 mm/s, which is just enough to keep the vacuum under 10^{-4} mbar in the test vacuum chamber (using a turbo pump of 340 l/s). Table 1 shows the extrusion temperatures for different gases when a external pressure of 2 bar was applied.

SUMMARY

A new cryogenic fibre maker which works continuously with different gases, has been presented. New features include: the use of closed cycle refrigerator to reduce running cost, the capability of having fibres under continuous extrusion and the use of most of the gas in producing fibres (no losses are involve in the process, apart of the evaporation of the fibre). The fibre are extruded by gas pressure following a controlled gas-liquid transition. The disadvantages of this scheme are: a) the requirement of a very low pressure to avoid thermal losses (10^{-5} mbar), b) a high pumping rate (340 l/s) which is necessary to get rid of the solid that is evaporated and c) the maximum fibre diameter that the system can extrude is about 300 µm. Modification of the pumping scheme are under way, in order to be able to use this fibre maker in the MAGPIE generator.

REFERENCES

1. Mitchell, I. H., Bayley, J. M., Chittenden, J. P., Worley, J. F., Dangor, A. E. and Haines, M. G., Rev. Sci. Instrum. **75** 1533 (1996).
2. Sethian, J. D. and Gerber, K. A., Rev. Sci. Instrum. **58** 536 (1987).
3. Grilly, E. R., Hammer, J. E., Rodrigues, D. J., Scudder, D. W. and Shlachter, J. S, Rev. Sci. Instrum. **56** 1885 (1985).
4. Jarboe, T. R. and Baker, W. R., Rev. Sci. Instrum. **45** 431 (1974).

[a] Electronic mail: r.aliaga-rossel@ic.ac.uk

Neutron Production in the MAGPIE Generator Using CD_2 Fibres

R. Aliaga-Rossel[a], I. H. Mitchell

The Blackett Laboratory, Imperial College
London, U. K.

H. Schmidt

Plasmaforschung Institut, University of Stuttgart
Stuttgart, Germany

Abstract: A series of experiments have been carried out on MAGPIE (Mega Ampere Generator for Plasma Implosion Experiments) using CD_2 fibre loads. Diameters between 50 µm and 200 µm were used. The generator was operated at a voltage of 1.4 MV, peak current of 1.1 MA, 150 ns rise time and with a stored energy of 215 kJ. The effect of reversing the polarity of the generator on the neutron emission was studied. An average anisotropy in the neutron emission of 1.2 (end-on to side-on) was obtained in the case where the end-on measurement was taken with a detector located behind the anode. When the initial polarity was reversed (the end-on detector, now being behind the cathode) the anisotropy increased to 1.7. This is a novel observation and requires a theoretical model that explains the acceleration of ions in both directions. Correlation between the fibres diameters and the neutron yield will be discussed. The neutron energy was measured using the time of flight technique. Three plastic scintillators were located at different positions in relation to the fibre. Neutron energies up to 4.8 MeV were measured at 45 degrees from the pinch axis in the anode direction. This, together with the detection of hard x-ray emission, indicates a beam-target reaction as the mechanism responsible for the neutron production. A deuteron beam energy of 2.6 MeV was inferred from these results. This is consistent with the electron beam energies inferred from the hard x-ray emission in a similar experiment.

INTRODUCTION

One of the main interests in the study of z-pinch has been the neutron emission when deuterium is used as a load. In linear z-pinches a neutron yield of 8×10^8 was obtained in the late fifties [1]. In order to avoid the leak of current along the walls of the tube that contain the plasma, a new configuration called plasma focus (PF) was designed. In the Poseidon plasma focus, a neutron yield of 2.5×10^{11} was reported [2]. In PF devices a saturation in neutron yield is observed due to effects which impede the current delivered by the generator to couple completely to the pinch plasma. To obtain a higher density in plasma it was developed the fibre pinch. First experiments carried out in Gamble II (50 kJ, 1 MV, 1.5 Ω, 50 ns) with deuterated polyethylene fibres (100 µm diameter), gave a neutron yield of

10^{10} [3]. A similar neutron yield using fibres between 15 μm and 150 μm in diameter was obtained in KALIF, where the neutron yield appears to be isotropic [4]. We present new results obtained on MAGPIE [5] for neutron emission where the neutron yield, anisotropy and high values in energy support the essential role played by beam target processes.

THE EXPERIMENT

The experiments were carried out on the MAGPIE generator [5]. It was operated at 60% and 80% of maximum charge (1.4 MV and 1.9 MV, peak currents 1 MA and 1.4 MA respectively), rise time 150 ns. Figure 1 shows typical current traces for a 13 nH load. Magpie has 4 Marx units, each of them with 24 capacitors charged up to 100 kV. A complete set of diagnostics was used, which include an x-ray streak camera, four frames x-ray framing camera, an optical streak and an optical framing camera, hard x-ray detectors, an x-ray pinhole camera as well as current and voltage monitors. A Nd-YAG laser (400 ps @ 532 nm) [6] was used for interferometry, Faraday rotation and schlieren diagnostics. The set up of the optical and x-ray diagnostics is shown in Figure 2. Time of flight (TOF) measurements of neutrons were carried out using plastic scintillators NE102A coupled to photomultipliers. Three of such detectors were located at : Scintillator 1 at 2.5 m @ 175^0, Scintillator 2 at 6.8 m@90^0 and Scintillator 3 at 16.8 m @ 135^0. The angular position is given in respect to the pinch axis in the anode-cathode direction. Fibres of CD2 , 2 cm long with diameters between 50 μm and 200 μm were used during the experiments. In normal operation the central electrode of the MITL is negative (cathode) while the external part is grounded (anode). Currents of 1.4 MA and 1 MA were used with normal and reverse polarity. Time integrated neutron measurements were taken with silver activation counters (side-on and end-on) and bubble detectors. A collimated filtered PIN diode was used to look at the hard x-ray emission from the anode.

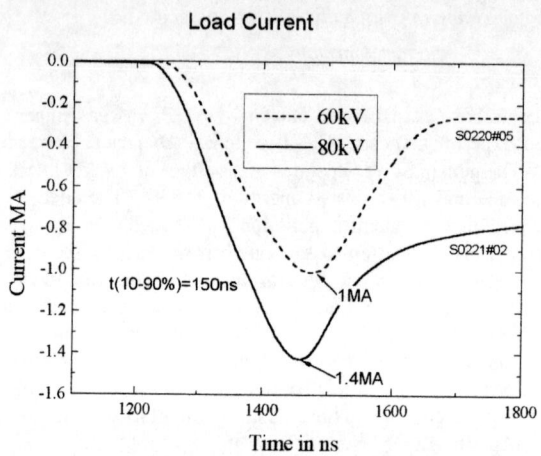

FIGURE 1: Current traces of Magpie at different charging voltages.

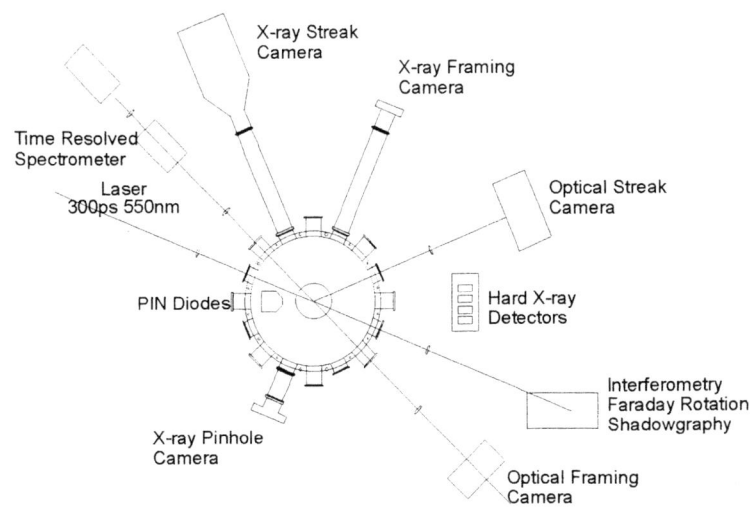

FIGURE 2. Top view of the diagnostics set-up

RESULTS

Figures 3 shows the signals detected in the three scintillators described previously. This particular shot (S0921#01) was taken with a fibre of 200 μm and at a voltage of 80% of full charge (1.9 MV). The signal in the PIN diode shows the emission of hard x-ray with MeV energies [7] consistent with the signal detected in scintillator number 1(filtered with 10 cm lead). This scintillator is too close to the pinch and does not provide information about neutron energy (neutron pulse is covered by hard x-ray signal) but it is useful in the timing of the other scintillators. Scintillator 2, located at 6.8 m, shows clearly a neutron pulse of 3.3 MV (the first hard x-ray pulse is used as timing mark). Scintillator 3, located at 16.8 m, shows a fast neutron pulse of 4.6 MeV. The 2.45 MeV component (if existing) is immersed in the pulse, to which also scattered neutrons contribute on the low energy tail of the signal. In this shot (S0921#01) the time integrated measurements of the neutron emission shows a yield of 7.3×10^8 end-on (anode side) and 6.7×10^8 side on. The anisotropy in total yield is then 1.09. The anisotropy in energy is 1.5. In the shots taken with normal polarity and at 1.4 MV the anisotropy in the total neutron yield varies between 1 and 1.7 (with fibres with diameters between 50 μm and 100 μm). Figure 4 shows anisotropy in the total neutron yield (end-on/side-on) as well as the neutron yield for the 1.4 MV shots.

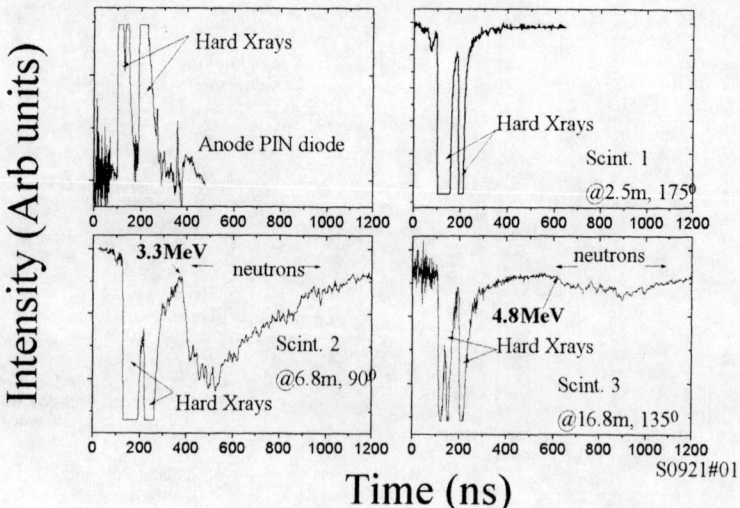

FIGURE 3. Signals from PIN diode and scintillators.

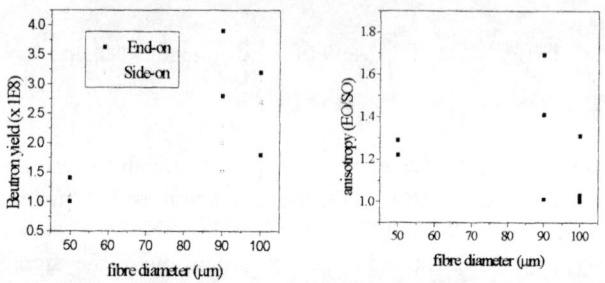

FIGURE 4. Anisotropy and neutron yield for 60 kV.

DISCUSSION

The PIN diode and the scintillators show two large hard x-ray pulses at around 100 ns and 190 ns after the current starts, which is consistent with the signal detected in the scintillator number 1, shielded with 10 cm of lead. The energy associated to these pulses are in the MeV range which is greater than the anode-cathode voltage. It is more likely that this high voltage is due to a sudden increase in the resistance of the plasma and not to dL/dt produced by a m=0 instability [7]. Assuming that the neutrons are emitted in a single burst at the time of the first hard x-ray pulse and using the time of flight information, an energy of 3.3 MeV is detected at 90^0 (scintillator 2) and an energy of 4.6 MeV was measured at

135^0 (scintillator 3). The neutron yield is higher in fibres of 90 µm diameters where also a higher anisotropy is measured (Figure 4). Shots carried out with reverse polarity show an increase in the anisotropy of the neutron energy from 1.2 to 1.7, which support the existence of a deuteron beam with an energy of 2.6 MeV.

Using the data from fibres of 90 µm diameters, the scaling for the neutron yield can be described as follows: a) on average the neutron yield scale as I^5 ;b) taking upper values neutron yield scales as I^6 and c) in side-on measurements, the scaling is proportional to I^4 and I^5 for the average and upper values, respectively.

Considering a beam target model, a deuteron beam with an energy of about 2 MeV is required to explain the presence of a 4.8 MeV neutron pulse emitted in the backward direction (from cathode to anode, which in our case is considered normal polarity). Also the presence of deuteron beams travelling both from anode to cathode and from cathode to anode have to be explained. This is a novel observation and requires a theoretical model that explains the acceleration of ions in both directions; indeed in a quasi-neutral plasma momentum must be conserved [8]. Further angular resolved neutron TOF as well as yield measurements should be performed to clarify ion acceleration mechanisms leading to the observed unexpected neutron emission characteristics of the fibre pinch.

REFERENCES

1. Anderson, O. A., Baker, W. R., Colgate, S. A, Ise, J. and Pyle, R. V., Physical Review **110**, 1375-1387 (1958).
2. Herold, H., Jerzykiewicz, A., Sadowski, M. and Schmidt, H. Nuclear Fusion **8** 1255-1269 (1989).
3. Young, F. C., Stephanakis, S. J. and Mosher, D., Journal Of Applied Physics **48**, 3642 -3650 (1977).
4. Kies, W., Decker, G., Malzig, M., van Calker, C., Westheide, J., Ziethen, G., Bachmann, H., Baumung, K., Rusch, D., Ratajczak, W. Stoltz, O and Bayley, J. M., J. Appl. Phys. **70** 7261-7272 (1991).
5. Mitchell, I. H., Bayley, J. M., Chittenden, J. P., Worley, J. F., Dangor, A. E., Haines, M. H. and Choi, P., Rev. Sci. Instrum. **57**, 1533-1541 (1996).
6. Aliaga-Rossel, R., Bayley, J., Mamin, A. and Nizienko, Y., " SBS pulse compression applied to a commercial Q-switch Nd-YAG laser", this proceeding.
7. Robledo, A., Mitchell, I. H., Aliaga-Rossel, R., Chittenden, J. P., Dangor, A. E., and Haines, M. G., Plasma Physics, **4** 490-492 (1997).
8. Haines, M. G., Nuclear Instruments and Methods **207** 179-185 (1983).

[a] Electronic mail: r.aliaga-rossel@ic.ac.uk

Modeling of the Neutron Yield in a Z-pinch.

Elena O. Baronova and Victor V. Vikhrev

*Nuclear Fusion Institute, RRC Kurchatov Institute,
Moscow 123182, Russia*

Abstract. A simple model for the hot, dense plasma in a z-pinch neck can show how to optimize the neutron yield as function of experimental parameters. The results agree with data from deuterium filled plasma focus discharges.

1. Introduction

It has long[1] been known that a Z-pinch in deuterium creates a significant number of neutrons. Neutrons come from nuclear reactions in the hot, dense plasma that forms best when the peak of the discharge current coincides with the plasma sheath implosion. This paper uses a model[2] for the plasma in the z-pinch neck to optimize the neutron yield for variations of pinch parameters under the experimenter's control, e.g., the capacitor bank's initial voltage and the initial gas pressure.

2. Theoretical model

We modify our zero-dimensional model[2] for the formation of a neck in a z-pinch discharge for use in a plasma focus by defining two stages. Initially, in the first (formation) stage, a neutral gas coexists with a weakly ionized plasma inside a cylindrical current sheath that accelerates the plasma as a magnetic piston. When the inward motion exceeds the sound velocity a shock front forms that ionizes the gas. The neutral gas in part compresses inside the shock front, and in part flows through the shock front as plasma. If the plasma density n in between shock front and current sheath is constant, then

$$n = N/\pi(R^2 - R_s^2), \qquad (1)$$

where N is the number of particles per unit length of the plasma sheath, R is the current sheath radius and R_s is shock wave radius. N increases as the shock wave ionizes the gas, and decreases through outflow through the ends of plasma sheath. This is described by:

$$\frac{dN}{dt} = -2\pi R_s n_0 \frac{dR_s}{dt} - \frac{N}{\tau}; \qquad (2)$$

n_0 is the initial gas density, and dR_s/dt is the velocity of shock wave front. The characteristic time of plasma outflow τ depends on the height of the plasma column h as

$$\tau = \tau_h = h/2V_z \qquad (3)$$

where V_z is the velocity of plasma in z direction. In turn this is given by:

$$\frac{dNV_z}{dt} = \frac{4NT}{hm_i} - \frac{NV_z}{\tau}, \qquad (4)$$

where m_i is the ion mass and T is the plasma temperature in the current sheath.

Plasma pressure and plasma outflow determine the current sheath's velocity,

$$\frac{dNV}{dt} = \frac{2\pi R}{m_i}\left[2nT\left(1-\frac{R_s}{R}\right) - \frac{H^2}{8\pi}\right] - \frac{NV}{\tau}, \qquad (5)$$

Here $H = 2I/cR$ is the strength of magnetic field and V is the velocity of the current sheath, or the magnetic piston. We will assume that all the plasma inside the current sheath moves with velocity V. This simple representation is reasonable because of the experimental fact that current sheath thickness is constant in time.

The velocity of shock wave front may be approximately determined under the condition that the pressure of the gas, running against the shock wave front, is equal to plasma pressure inside the sheath:

$$V_s = (2nT/m_i n^0)^{1/2} \qquad (6)$$

where n is plasma density in the sheath, and n_0 is the initial density of the gas.

The total energy (thermal and kinetic) of the plasma sheath is:

$$W = 3NT + m_i NV^2/2 \qquad (7)$$

and the energy balance is expressed by

$$\frac{dW}{dt} = -2\pi RV\frac{H^2}{8\pi} - \frac{W+2NT}{\tau} + Q_J. \qquad (8)$$

The first term on the right is the energy change due to magnetic field pressure, the second term describes energy loss owing to plasma outflow from the zone of compression, and Q_J is Joule heating.

The circuit voltage including pinch inductance and resistive losses is

$$U_0 - \frac{1}{C}\int_0^t I dt' = \frac{d}{dt}(L_0 + L_p)I + \frac{jh}{\sigma}, \qquad (9)$$

where C is the bank's capacitance, L_p the inductance of the plasma, j the current density, and σ the plasma conductivity.

In the first stage of the model the current is distributed uniformly within the current sheath. Then $j = I/\pi(R^2 - R_s^2)$. In the second stage, once the shock wave has hit the axis and R_s has vanished the current density is $I/\pi R^2$. The second stage of the discharge assumes that the plasma created by the shock wave's convergence on axis can flow out in the axial direction, similar to what is assumed in a simple model for the final stage of a z-pinch[2]. Besides a characteristic time of plasma outflow in the radial direction τ_r from the MHD destruction of plasma column, $\tau_r = \alpha R/v_s$, there is now a second time scale for axial outflow, $1/\alpha = 0.1$ is a factor that accounts for plasma confinement in the radial direction. Then

$$1/\tau = 1/\tau_h + 1/\tau_r \qquad (10)$$

Anomalous resistivity is taken into account as:
$$\sigma = \sigma_C /(1 + v_{eff}\, \tau_{ei}), \tag{11}$$
where σ_C is the Coulomb conductivity and τ_{ei} is the electron to ion collision time. The effective collision frequency is
$$v_{eff} = (v_d/v_s)^2 (\omega_{ci}\, \omega_{ce})^{1/2}, \tag{12}$$
where v_d is the electron current velocity, and ω_{ci} and ω_{ce} are the ion and electron cyclotron frequency at the plasma column's boundary.

In the discharge's second stage the current distribution is assumed to be non-uniform, which is accounted for by increasing the effective drift velocity four-fold, a factor suggested by radial scaling of quantities such as the ion velocity, which can be proportional to pinch radius ($v_r = r/R*V$). Then (5) and (7) become

$$\frac{dNV}{dt} = \frac{4\pi R}{m_i}\left[2nT - \frac{H^2}{8\pi}\right] - \frac{NV}{\tau} - \frac{NV}{\tau_r}, \tag{13}$$

$$W = 3NT + m_i NV^2/4, \tag{14}$$

The last term in equation (13) accounts for the fact that the ions leave the plasma column more rapidly because of their larger radial velocity. This term makes nonphysical oscillations of the plasma column weaker because of significant radial plasma outflow. In the second stage equations (2,4,8,9) remain the same.

Optimization is done with respect to parameters that can be varied by the experimenter such as the bank energy, the initial pressure, the initial radius, and the column height (which is self-generated, but can be limited by choosing the anode-cathode distance).

Neutron production is $dY_{dd}/dt \sim Vn^2 <\sigma v>$, where V is plasma volume and $<\sigma v>$ the reaction rate for the dd reaction averaged over the (Maxwellian) ion distribution. It is a strong function of plasma temperature. The total neutron yield follows from integration of the neutron production.

3. Results

Figure 1 shows a typical result for the optimized neutron yield as function of the height of the plasma column and of the inductance for the discharge's electrical circuit, with the other parameters as given. The maximum yield is about 4.10^{10} neutrons per shot, more or less what is observed in the experiments. Further results include the following:
1. The capacitor bank's energy is converted most efficiently to plasma thermal energy if the initial radius of plasma sheath is

$$R_{opt} = \sqrt{\frac{230 W_0}{U_0 P^{1/2}}}$$

where P is expressed in Torr, the capacitor bank energy W in Joule, the initial voltage U_0 in Volts, and the radius R in cm.
2. The circuit inductance L_o should be comparable to the plasma column's inductance L_p. Then the current peaks at maximum compression.

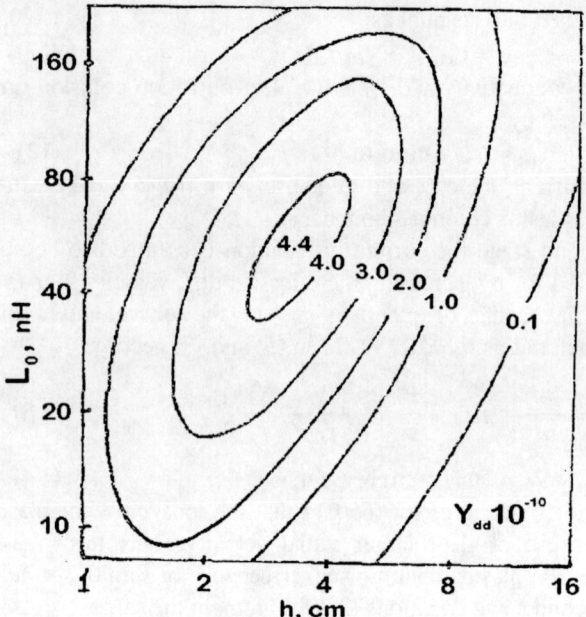

Figure 1. Neutron yield versus L_0 and h, when P= 1 Torr, U_0= 18 kV.

3. Neutron yields are maximized for column height $h_{opt} \sim 10 W^{1/2} P^{-1/3}$, circuit inductance $L_o \sim 30\ h_{opt}$ (with L_0 is expressed in nH or cm, h_{opt} is expressed in cm, and P in Torr).

4. When R_{opt}, L_o, and h_{opt} are optimized the neutron yield scales as
$$Y_n \sim W_0^{1.8} P^{0.6}$$
with W_0 is in Joule and P in Torr. Although this scaling suggests that you can get higher yields by increasing the pressure, excessive pressure gives asymmetric discharges for which the analysis no longer applies.

These theoretical results were obtained for plasma focus devices filled with deuterium without impurities, when anomalous heating and radiative collapse effects are negligible. This is true for $n\tau > 10^{15} T^{1/2}$, where $n\tau$ is in s/cm^3, and T in eV[2]. Good agreement of the modeling with the experimental results obtained in a Filippov type device suggests that the model can be used for optimization purposes.

References

1. Artsimovich, L. A. et al., *Atomic energy*, **3**, 84 (1956).
2. Vikhrev V.V. *Sov. J. Plasma Phys.* **3**, 539, (1977).
3. Vikhrev V.V. and Ananin, S.I. *Fiz. Plasmy* **7**, (1981) 494. (*Sov. J. Plasma Phys*, **7**, 1981)

Hot Spots In Fiber Pinches

J.P. Chittenden, R. Aliaga Rossel, S.V. Lebedev, I.H. Mitchell, A.R. Bell,
F.N. Beg, A. Lorenz, A.E. Dangor, M.G. Haines, and G. Decker[1]

Imperial College, London SW7 2BZ, UK
[1]*Heinrich-Heine-Universität, Düsseldorf, Germany*

Abstract. Results are presented on the behavior of "bright spots" in carbon and deuterium fiber Z-pinches using the MAGPIE generator at the 1 MA current level. The experimental diagnosis was carried out using optical and X-ray framing and streak images, along with Schlieren and shadowgraphy laser probing. After a short (~4 ns) duration formation phase, these bright spots exhibit highly dynamic behavior. Bifurcation of the bright spots gives rise to rapid axial motion at $1\text{-}3 \times 10^5$ m/sec. The post-bifurcation bright spots persist for up to 40 ns. The important features of bright spot evolution can be reproduced using a 2-D MHD code incorporating LTE ionization dynamics and cold start conditions. Construction of "artificial" diagnostic images from the simulation data allows direct comparison to experiment. From the close agreement between experiment and 2-D simulation we infer that the observed bright spot behavior can be explained entirely in terms of the non-linear evolution of the m=0 instability.

INTRODUCTION

Both the optical and X-ray emissions from dense Z-pinch plasmas are dominated by short-duration bursts from highly localized regions known variously bright-spots, X-ray hot-spots, micro-pinches, beads and pinch spots. In fiber Z-pinches or exploding wire experiments, the X-ray yield from these short-lived, localized bright regions can exceed the total yield from the background plasma by orders of magnitude. The results of this paper describe the dynamic evolution of bright-spots during and more importantly after their formation. The essential features of the bright-spot phase can reproduced using a 2D MHD code incorporating LTE ionization dynamics and cold start conditions [1]. The code results are used to construct artificial diagnostic images which are compared to the experimental images at a number of points throughout the paper. The ability of a 2D MHD code to reproduce the observed phenomena implies that they are a product of the non-linear evolution of the m=0 instability.

EXPERIMENTAL OBSERVATIONS

The results presented here are all taken from a series of shots into 33μm diameter carbon fibers and approximately 100μm deuterium fibers using the Imperial College MAGPIE generator. MAGPIE consists of four 2.4 MV, Hermes III type Marx banks storing up to 336KJ and is capable of producing a current pulse up to 1.8MA with a 10-90% rise-time of 150ns. The results presented here are for 60% of full charge producing a 1.05 MA peak current.

The first 50ns of the discharge is dominated by rapid m=0 instability evolution in a low density coronal plasma [2] and hot-spot formation in the higher density plasma near the axis [3]. The rapidity of coronal instability evolution is shown by the sequence of Schlieren images (and their artificial counterparts) in Figure 1.

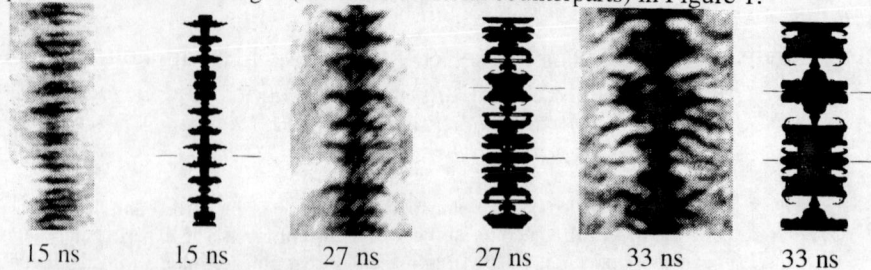

15 ns   15 ns   27 ns   27 ns   33 ns   33 ns

Figure 1. Comparison of experimental and artificial Schlieren images.

Axial streak photographs of the optical emission from the plasma (see Figure 2) show bifurcation of bright-spots into two spots moving axially at equal and opposite speed away from the formation point at speeds of $\pm 1\text{-}3 \times 10^5$ ms^{-1}. X-ray streaks show

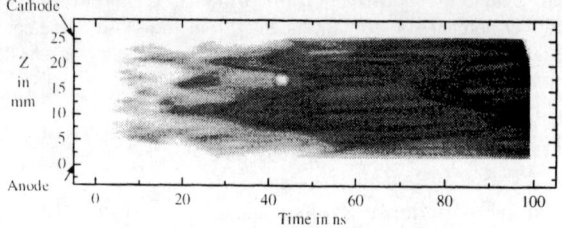

Figure 2. Axial Optical Streak of the first 100ns

similar behavior. The post-bifurcation bright-spots live for up to 40ns in the optical streak photographs and are often destroyed only by collision with another bright-spot. This dynamic behavior may explain discrepancies between measurements of bright-spot temperature using different techniques. Time integrated measurements are dominated by the high intensity formation phase of bright-spot evolution whereas time resolved measurements are dominated by the longer lived post-bifurcation phase.

Figure 3 shows enlarged regions from a sequence of three gated micro-channel plate (MCP) soft X-ray images (from the same shot) showing bifurcation of a bright-spot and subsequent axial motion at 10^5 ms^{-1}, together with the simulated equivalents. 10µm beryllium filters were used with 175µm spatial resolution, 2ns gating duration and 3ns frame separation. Gated soft X-ray images using a variety of filters show emission from the region between bifurcated bright-spots implying the presence of plasma not participating in bifurcation. If the bright-spots are formed by the neck of the m=0 instability then this neck is not penetrating to the axis before bifurcation occurs.

At approximately 50ns, the plasma appears to undergo a transition from a dynamic phase of rapidly evolving instability, to a quiescent phase (lasting 50-150ns) where the plasma remains inhomogeneous, but evolves at a much slower pace (see Figure 2). The transition time corresponds to the point at which the entire length of the plasma has been traversed by either one or two of the many bifurcating fronts. A possible explanation for this transition is that the bulk of the fiber material on axis is ionized by the bifurcating fronts, thereby increasing the plasma line density, decreasing the temperature and hence increasing the time scale for the evolution of

Experiment

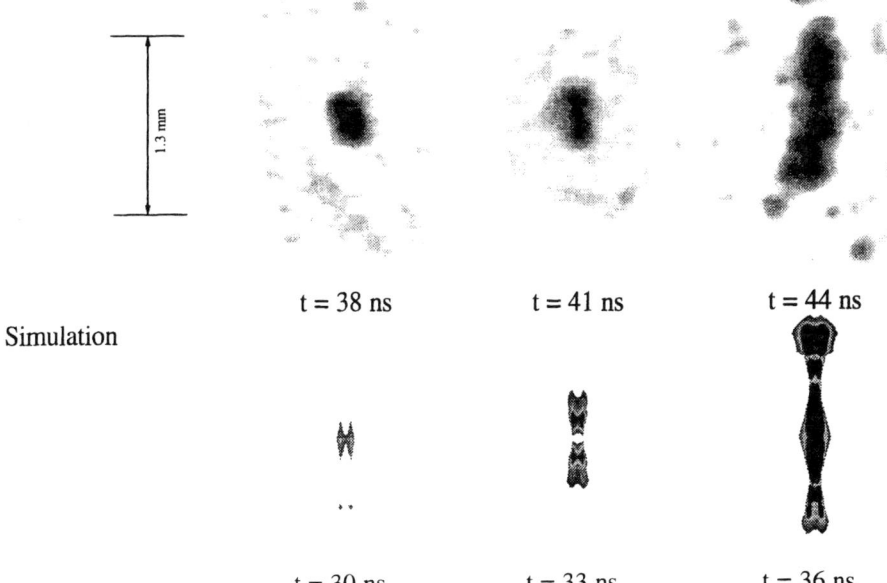

Simulation

Figure 3. Sequence of gated X-ray images showing bifurcation of a bright-spot

the MHD instability. The inhomogeneities seeded by the rapid instability in the bright-spot phase eventually lead to disruption of the plasma column and the formation of electron beams with energies much greater than the applied anode-cathode voltage [4].

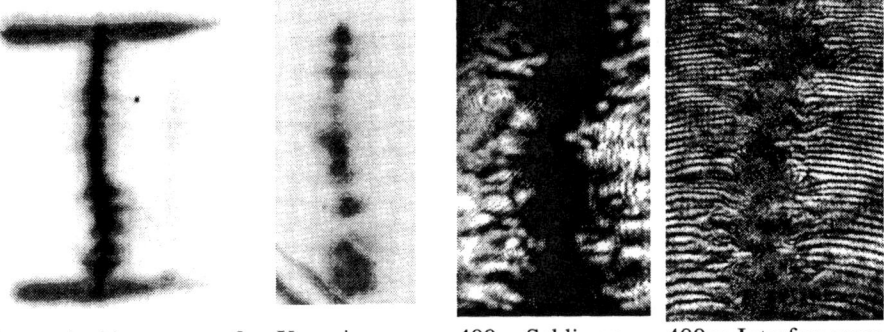

1ns optical image   2ns X-ray image   400ps Schlieren   400ps Interferogram

Figure 4. Simultaneous diagnostics at 70ns for deuterium.

Figure 4 shows simultaneous gated optical and X-ray images along with Schlieren and interferometry images (using a 400ps, 532 nm laser pulse) at 70ns into the discharge. Details on the deuterium fiber results are in reference 5. The dynamic behavior of bright-spots in deuterium fiber Z-pinches is similar to that in carbon.

2D MHD SIMULATIONS

The experiment was modeled using the MH2D code (originally developed by Bell [6]) which solves the equations of resistive MHD in 2D cylindrical (r,z) geometry on an Eulerian grid, and has recently been extended to model a single temperature admixture of electrons, ion species of different charge and neutral gaseous atoms with LTE ionization dynamics and cold start conditions. We use a grid of 200x200 cells, each 20μm long and 10μm wide. The initial fiber is represented as a uniform density neutral gas of the same line density, with radius 50 μm, temperature 0.01 eV and with a seed fraction (10^{-6}) of singly ionized ions. A 1% random density perturbation is applied. Phase transitions from the solid state to the gaseous state are not included. Figure 5 shows a mass density contour plot from 30ns into the simulation. The corona exhibits large amplitude, long wavelength m=0 behavior, yet the predominantly neutral core remains virtually unperturbed. The rapid early instability evolution is due to low initial ionization levels causing high temperatures and growth rates in the corona. At z=2mm, the instability has penetrated almost to the axis of the pinch and a high density, high temperature, X-ray bright-spot begins to form. At this stage in the simulation the neck of the instability does not penetrate to the axis and is stopped by the residual unionized core. Bifurcation of the bright-spot is then caused by an axial $j_r B_\theta$ force which is if opposite sign on either side of the neck of the instability. In the simulation, axial motion of the bifurcating, off-axis regions of high temperature and bright emission is observed at $\pm 5 \times 10^4$ ms^{-1}.

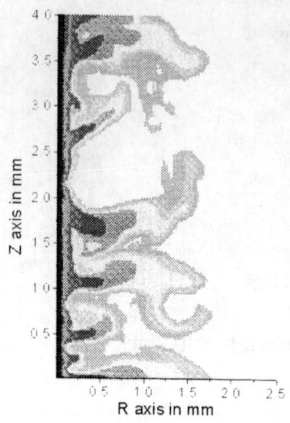

Figure 5. Carbon Fiber Simulation at 30 ns

REFERENCES

1. J. P. Chittenden, R. Aliaga Rossel, S. V. Lebedev, I. H. Mitchell, M. Tatarakis, A. R. Bell and M. G. Haines, submitted to Physics of Plasmas.
2. S.V. Lebedev, R. Aliaga-Rossel, J. P. Chittenden, A. E. Dangor, M.G.Haines and J.F. Worley, in these proceedings.
3. J.P. Chittenden, I.H. Mitchell, R. Aliaga-Rossel, J.M. Bayley, F.N. Beg, A. Lorenz, M.G. Haines and G. Decker, accepted for publication in Physics of Plasmas
4. A. Robledo, I. H. Mitchell, R. Aliaga-Rossel, J.P. Chittenden, A.E. Dangor and M.G. Haines, Physics of Plasmas, **4**, 490 (1997).
5. R. Aliaga-Rossel, S.V. Lebedev, J.P. Chittenden, A.E. Dangor and M.G.Haines, in these proceedings.
6. A.R. Bell, Physics of Plasmas, **1**, 1643 (1994).

SPATIAL AND TEMPORAL EVOLUTION OF NEUTRON EMISSION FROM A SMALL PLASMA FOCUS

S.L.Yap, S.P. Moo, C.S. Wong, *P. Choi and *C. Dumitrescu-Zoita

Plasma Research Laboratory, Physics Department, University of Malaya, 50603 Kuala Lumpur, Malaysia.

* *Laboratoire de Physique des Milieux Ionises, Ecole Polytechnique, 91128 Palaiseau, France.*

Abstract

Neutron emission of a low energy 3.3 kJ (15 kV) plasma focus operated in pure deuterium and in deuterium-argon admixtures is studied. Both the time resolved and time integrated techniques are employed simultaneously. The time resolved measurements are made using six plastic scintillator-photomultiplier detectors placed at different angular positions. Two sets of Indium foil activation counters are used for the time integrated measurements. In general, two periods of neutron emission are observed with the second period emission being more energetic. Thus at least two different mechanisms may be involved in the neutron production.

INTRODUCTION

This paper reports some preliminary results of the spatial and temporal evolution of neutron emission from a 3.3kJ, 15kV plasma focus device at the University of Malaya [1]. Pure deuterium and deuterium doped with different amount of argon are used in the study. Since the mass density of the filling gas controls the sheath dynamics, a constant mass density of the gas mixture which is equivalent to a filling of deuterium at 6 mbar is used.

THE EXPERIMENT

Indium activation counters are used to measure neutron yield. The detectors are placed 45cm from the focus, one end on and the other side on. Each of the activation counters consist of an Indium foil wrapped round a thin-wall cylindrical GM tube and placed inside a paraffin moderator.

The time profiles of neutron emission are obtained by using six scintillator-photomultiplier assemblies placed in lead shields at different positions (table 1). Each of the detectors consists of a cylindrical NE102A scintillator (diameter : 5cm, length :10cm) wrapped with a fluorescent fibre and optically coupled to a normal plastic optical fibre [2]. The optical signals are carried by this optical fibre to the photomultiplier tube, which is placed inside a screen room to diminish the pick up of electromagnetic noise of the discharge. Electrical signals from the photomultiplier tubes are recorded by digital oscilloscopes with a sampling rate of 1Gs/s.

Table 1 : Detector Positions

Ch 1	Ch 2	Ch 3	Ch 4	Ch 5	Ch 6
15°	50°	90°	90°	0°	25°
41cm	50cm	101cm	50cm	80cm	47cm

RESULTS AND DISCUSSION

Neutron Yield

Total yield measurements are shown in the Figure 1 for argon doping range from 0% to 70%. Each point in the graph is the average of 12 shots . The neutron yield is highest at argon doping of 30%-40%. Figure 1 also shows the ratio of end on yield to the side on yield. In general the anisotropy is about 10% to 30%.

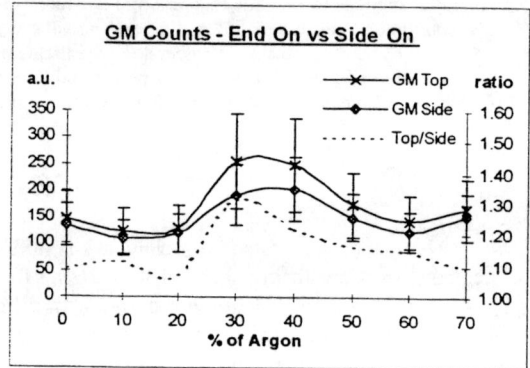

Figure 1 : Neutron Yield

Time Profile

Figure 2 shows typical neutron time profiles for a pure deuterium discharge. The profiles are essentially Maxwellian with a halfwidth of 30ns to 40ns. In discharges where multiple pinches occur, multiple peaks are seen, each with a duration of 60ns to 100ns.

Figure 3 shows typical neutron time profiles for a 30% argon doped deuterium discharge. At the end on direction, the neutron pulse is broader at the nearer position

(Channel 1 - 41cm) than at the further position (Channel 5 - 80cm). At the side on direction, the single peak observed in the nearer detector (Channel 4 - 50cm) evolved into two peaks at the further position (Channel 3 - 101cm).

Figure 2 : Neutron Signals Of A Pure Deuterium Discharge

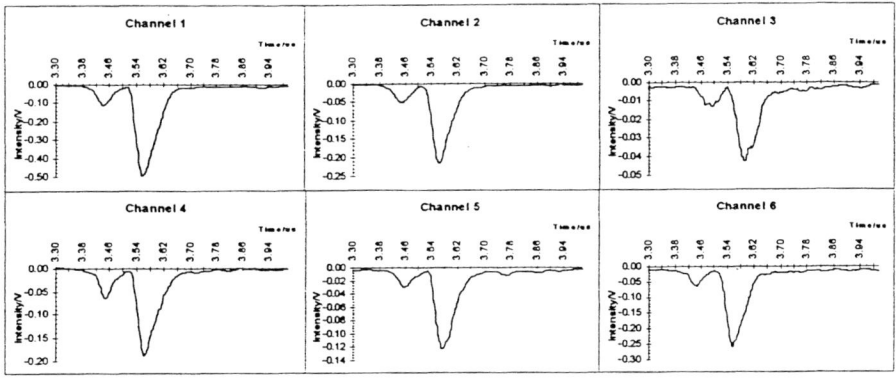

Figure 3 : Neutron Signals of A 30% Argon Doping Deuterium Discharge

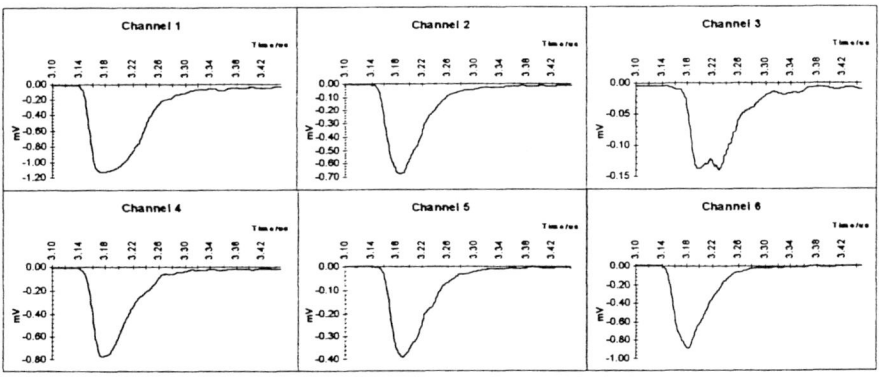

This seemingly complicated results can be understood if the neutrons are emitted in two periods as shown in figure 4. In the end on direction, the neutron emitted later are more energetic so that the neutron profile becomes slimmer a at further position due to the fact that the second faster neutron pulse has caught up slightly with the slower first pulse. In the side on direction, the second pulse is less energetic and as a result, the two pulses are further separated in time at a further distant away. Further work is necessary to substantiate this interpretation.

Figure 4 : Double Pulse Fitting (To Show The Complexity Of The Neutron Signals)

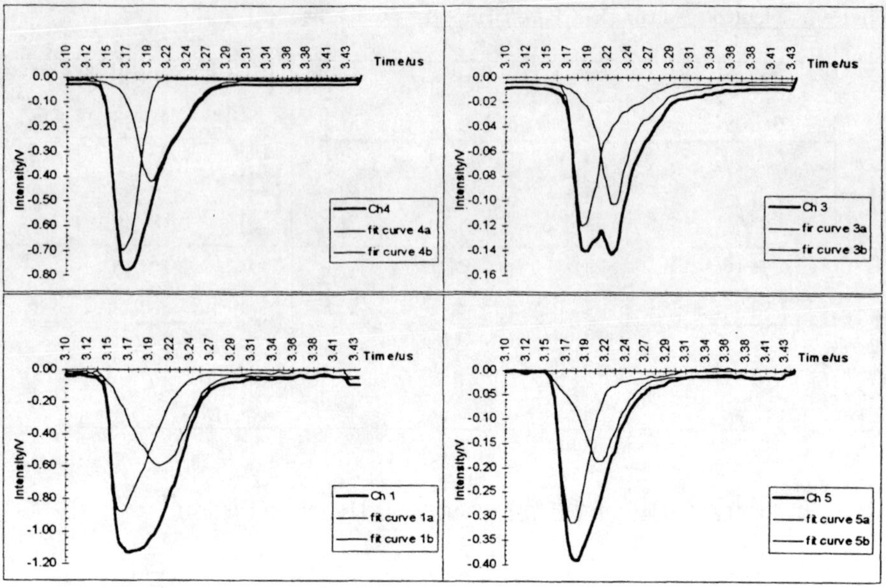

ACKNOWLEDGMENT

This project is supported by IRPA Program under project 02-02-03-0115. Choi P.'s participation in this project is partially supported by an ICTP Visiting Scholar Program. We acknowledge the technical assistance of Mr. Jasbir Singh.

REFERENCES

[1] Lee S. et. al. Am. J. Phys. **56**,62 (1988).
[2] R.F. Aliaga Rossel, Dynamics of the pinch formation in a medium energy plasma focus, Ph.D. Thesis, The Blackett Labotarory, Imperial College (Feb. 1993).

Coronal plasma behaviour in C and D_2 fibres on the MAGPIE generator

S.V. Lebedev, R. Aliaga-Rossel, J. P. Chittenden, A. E. Dangor, M.G.Haines and J.F. Worley

The Blackett Laboratory, Imperial CollegeLondon SW7 2BZ, U. K.

and R. Saavedra

Facultad de Fisica, Pontificia Universidad Catolica, Santiago, Chile

A series of fibre pinch experiments has been carried out on MAGPIE generator (1.8 MA, 150 ns) to study the coronal plasma. The analysis of schlieren photographs, axial streak images and gated x-ray pictures allows the evaluation the radial and axial motion of the corona plasma. Radial expansion velocity of $5.5 \cdot 10^6$ cm/s for carbon fibre and $3.6 \cdot 10^6$ cm/s for carbon with current prepulse was measured. Axial wavelengths of dominant instabilities in the corona were $\lambda_z = 0.05 - 0.2$ cm corresponding to $ka \sim 10 - 20$. Comparisons of the results obtained with carbon fibres with and without current prepulse (30 kA, 200 ns) and deuterium fibres are presented.

Experimental set-up and diagnostics

The experiments were carried out on the MAGPIE generator [1] which was operated at peak current 1 MA with a rise-time of 150 ns. In a series of experiments a shunt resistor across the spark gap in one of the four pulse forming lines was installed, which enabled operation with current prepulse. The choice of the resistor value allowed the duration of prepulse current to be varied in the range 100 - 200 ns.

In most experiments carbon fibres with diameter 33 µm and length 2.3 cm were used. A set of experiments with cryogenic deuterium fibres with diameters ~ 100 µm and thick carbon fibres (diameter 300 µm) were also done.

To study the dynamics of the corona plasma, an extensive set of optical diagnostic techniques was used. A Nd-YAG laser (532 nm) with SBS pulse compression (0.4 ns) was used for schlieren (sensitivity $4 \cdot 10^{-4}$ radian) and a self-referencing interferometer. The laser pulse was split and delayed to obtain two schlieren photographs separated by 3 ns.

Axial optical streak photography was performed. In some shots a set of slits displaced in radial direction was used to investigate the radial expansion of the emitting plasma (see [2]). Gated 4-frame (2 ns gate with 9 ns separation) x-ray camera with different pinholes and filters was used.

Experimental results and discussion

Figure 1 shows a waveform of the current through the pinch, where time t = 0 corresponds to the start of the main current. The prepulse current reached 35 kA in about 200 ns. The amplitude of the main current with and without prepulse were

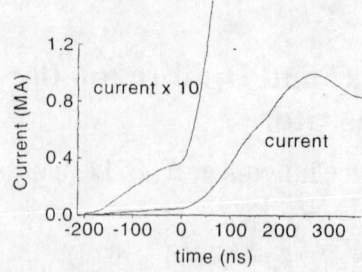

Figure 1. Current in prepulse regime.

essentially the same. The most significant difference between the two regimes was the delay time between voltage pulse and the current start (fibre breakdown). For shots without prepulse for both carbon and deuterium fibres, the main current starts with a delay of 20 - 30 ns. The voltage at the cathode at this moment is about 200 kV. For shots with prepulse the main current starts practically without any delay. It is interesting to note that with the prepulse, when the voltage rise is slow (~1 µs), breakdown occurs at much less voltage of about 20 kV.

Figure 2 shows a set of representative schlieren pictures at different times. Instabilities with m=0 appear earlier than 10 ns from the current start. The instability pattern for both regimes does not change significantly between two schlieren frames separated by 3 ns. This shows that instability growth time is more than 3 ns after at least 20 ns after the current start. Differences between the two regimes are observed only during early times (up to ~ 80 ns), when the size of the coronal plasma is smaller with the prepulse. Later, after about 120 ns there are practically no differences in coronal plasma behaviour. Variations from shot to shot become more significant. From the schlieren pictures it is possible to measure

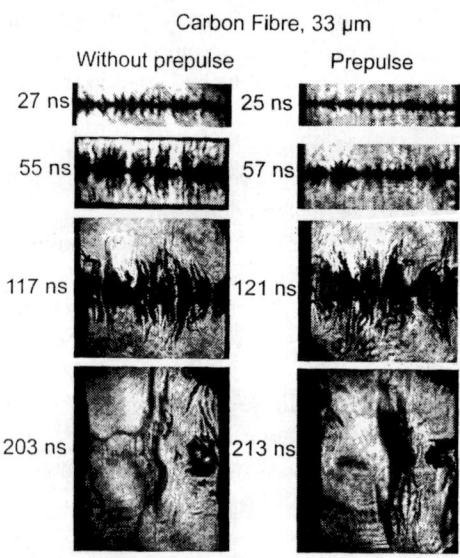

Figure 2. Schlieren images for different moments of time. The pinch length is 2.3 cm, cathode at the left.

the size (diameter) of perturbations and the wavelength. These measurements are shown at figure 3, each point representing one shot, the error bars being the standard deviation of the measured value along the length of the fibre. The diameter of the perturbations increases linearly in time. The radial velocity is $(3.6\pm0.4)\cdot10^6$ cm/s with prepulse. Without prepulse the velocity is slightly larger $((5.5\pm1.5)\cdot10^6$ cm/s). The wavelength of the perturbations is slowly increasing with time from 0.05 cm to about 0.2 cm in 100 ns and is about the same for both regimes. Also shown in the figure 3 is the calculated $ka = \pi D/\lambda_z$, where a is the radius of perturbations. With the prepulse the value of ka is about 10 while for shots without the

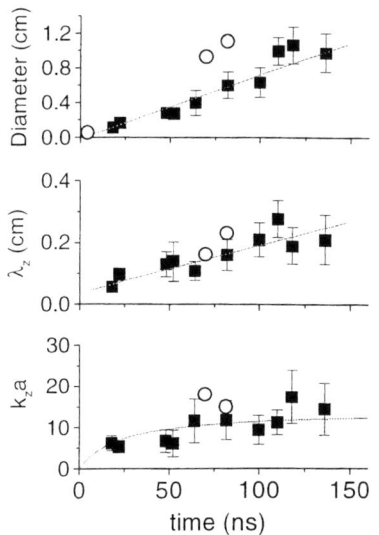

Figure 3. Diameter of corona, instability wavelengths and ka versus time for carbon fibre with prepulse and D_2 fibre (o) shots.

prepulse ka is up to 20. The difference is mainly due to the faster expansion of the corona for the latter regime.

Another value of the velocity of plasma expansion in the corona is measured from schlieren pictures obtained in the same shot with 3 ns time separation. This velocity agrees with the average value obtained from figure 3. However, the velocity varies along the length of the pinch, the maximum values being $2 \cdot 10^7$ cm/s. It is also possible to observe an inward motion in the neck regions with velocities $\sim 10^7$ cm/s.

The local Alfven velocity (V_A) for the corona plasma can be calculated if we assume that the integral $\int (\nabla n_e) dl \sim n_e$, which gives from the sensitivity of the schlieren system estimation of $n_e = 3 \cdot 10^{18}$ cm^{-3}, and assume that all current is flowing inside the corona plasma [3]. This gives $V_A = 2 \cdot 10^7$ cm/s for totally stripped carbon, what is 2 to 4 times higher than the measured average velocity of the coronal plasma expansion.

From the two frames of schlieren pictures it is possible also to obtain information about the axial motion of plasma and compare with data from other diagnostics shown in figure 4. The pictures were taken at about 30 ns after the current start. This time coincides with the moment when the streak picture shows axial motion of the emitting region (see [4]) with a velocity of about $2 \cdot 10^7$ cm/s. This velocity would give displacement of 0.6 mm between the two frames, which is larger than

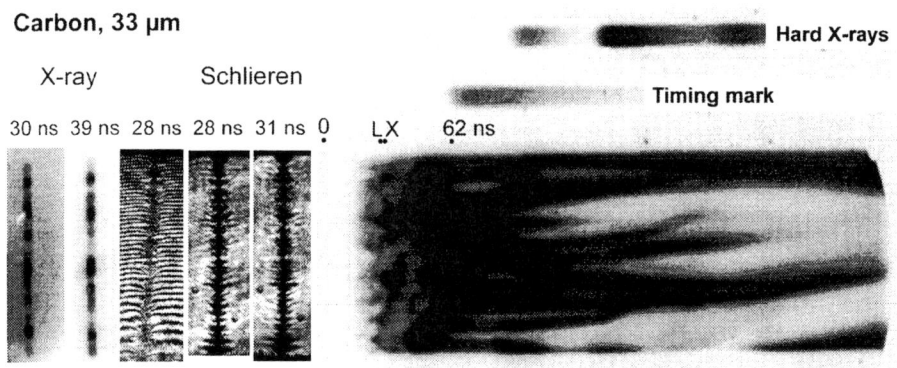

Figure 4. Schlieren, interferometer and soft x-ray images for the shot without prepulse. At the streak picture time of laser pulse (L) and first x-ray frame (X) are marked. Cathode is at the bottom.

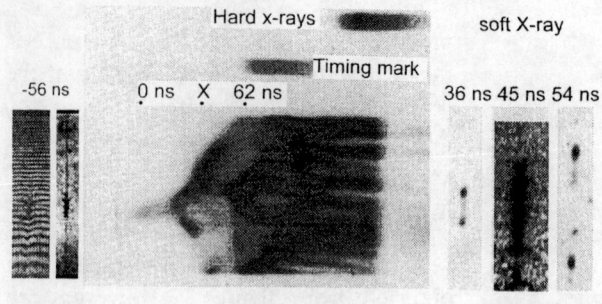

Figure 5. Shlieren and interferometer pictures during prepulse (56 ns before the start of the main current), axial streak and gated soft x-ray images obtained in the same shot.

the wavelength of the perturbations at this moment, but is not observed in the schlieren images. The typical spatial scale of light modulation in optical streak images is 2 to 4 times larger than the wavelength of perturbations in the coronal plasma (from schlieren data) The x-ray gated images show similar spatial scale as the optical streak. This is because x-ray and optical emission originate from the central dense part of the pinch, while the schlieren images show the corona plasma with much less density.

During the prepulse, the interferometry and schlieren show that the plasma was formed along the whole length of the fibre with diameter of ~0.5 mm and line density $N_e \sim 5 \cdot 10^{17}$ m^{-1}. This corresponds to 0.5% of the initial line density of the carbon fibre with 33 μm diameter used (assuming Z=1). This plasma is not homogeneous along the fibre (figure 5). One or more regions with well developed instability are usually observed. Optical emission recorded in the streak pictures shows that during the main current "zippering" starts from these places. The axial velocity of these emitting fronts for this shot is $3 \cdot 10^7$ cm/s. Gated x-ray pictures obtained in the same shot show two bright spots which move in opposite axial directions with the same velocity as the front of light emission.

Summary

Expansion velocity and evolution of instabilities wavelength in corona plasma were measured in carbon fibre pinches. With the prepulse, coronal plasma expansion early in the discharge is slower. Optical and x-ray emission from the core plasma show different wavelengths and axial velocities than observed in corona by laser probing.

References

1. I.H.Mitchell, J.M.Bayley, J.P.Chittenden, J.F.Worley, A.E.Dangor, and M.G.Haines, Rev. Sci. Instrum. 67, 1533 (1966)
2. R.Aliaga-Rossel, S.V.Lebedev, J.P.Chittenden, A.E.Dangor, and M.G.Haines, this conference.
3. M. Tatarakis, R.Aliaga-Rossel, A.E.Dangor, M.G.Haines, Phys. Plasmas 1997 (submitted)
4. J.P Chittenden, I.H. Mitchell, R. Aliaga-Rossel, J.M. Bayley, F.N. Beg, A. Lorenz, M.G. Haines, G. Decker, Phys. Plasmas 1997 (to be published).

Steady state of Z-pinch taking into account the absorption of own radiation.

Alexander Muravich

Atomic Energy Research Institute, College of Science and Technology,
Nihon University, Kanda-Surugadai, Chiyoda-ku, Tokyo 101, Japan

Abstract. The steady state of z-pinch is calculated with due regard for radiation transport and absorption effects. Thermal conductivity effects are also taken into account. Assuming axial and azimuthal symmetry, pressure and energy equations are composed as self-consistent integral-differential equations for the temperature and density radial distributions. In the limit of small pinch radius it is shown some influence of radiation transport on temperature profile in comparison with the case when absorption is not taken into account.

I. INTRODUCTION.

The steady state of an isolated Z-pinch, in which Joule heating balances by bremsstrahlung radiation losses, is established at the specific current (so called, Pease-Braginskii current) and has been studied by many authors [1]-[7]. In most of such a studies absorption of own radiation assumed to be negligible because high temperature plasma (some keV and more) is of considerable interest. However, it is not small for small radius and exert an influence on characteristics of pinch, as shown by extensive literature on radiative collapse problem (for example [5]-[8]). In above papers calculation of absorption is zero-dimensional in order to simplify simulation of pinch dynamics. More complicated 2-D and 3-D codes as a rule takes into account many effects and their results are sometimes difficult for interpretation. In this paper we calculate the steady state Z-pinch in the frames of relatively simple 1-D model, including thermal conduction, bremsstrahlung radiation losses and inverse bremsstrahlung effect.

II. THE MODEL.

Present calculations has been provided for the fully ionized hydrogen plasma. The basic assumptions of the model is cylindrical symmetry, so all quantities are varies only in the radial direction. We neglect the viscous effects too. Plasma assumes to be far from any kind of degeneration. Electron and ion temperature are taken equal: $T_e = T_i \equiv T$. We will note the temperature and density on the axis accordingly as T_0 and n_0 and dimensionless temperature and density distributions as $g(r)=T(r)/T_0$ and $f(r)=n(r)/n_0$. Expressions for the thermal conductivity coefficients given by Braginskii [9] are used. The current density j is given by

$$j = \frac{ET^{3/2}}{\alpha \bar{\eta} \log \Lambda} \qquad (1)$$

where E is the axial electric field, $\bar{\eta}=65.3\Omega \cdot m \cdot K^{3/2}$ is the Spitzer [10] conductivity without magnetic field, and α is a factor, depending on electrons Hall parameter.

Pressure balance and energy balance equations are:

$$\nabla p = j \times B \qquad (2)$$

$$P_{RAD} = P_{HF} + P_J + P_{ABS} \qquad (3)$$

where P_{RAD}, P_{HF} and P_J are radiated power, heat flux and joule heating respectively. Absorbed power is expressed by

$$P_{ABS}(\bar{r_0}) = \frac{1}{\pi}\int_0^\infty \frac{P_{Rv}(r_0)}{I_{Pv}(r_0)} dv \int_0^{\pi/2} d\theta \int_0^\pi d\phi \int_0^{b_1(r_0,\phi)} dR \times P_{Rv}(R,\phi) \exp\left(-\frac{1}{\cos\theta}\int_0^R \frac{P_{Rv}(l)}{I_{Pv}(l)} dl\right) \qquad (4),$$

where $P_{Rv}(r)$ is spectral radiation power and $I_{Pv}(r)$ is a Plank distribution function. The full power radiated and full power absorbed by pinch unit length are accordingly:

$$P_R = 2\pi \int_0^a r P_{RAD}(r) dr \qquad (5)$$

and

$$P_A = 2\pi \int_0^a r P_{ABS}(r) dr \qquad (6)$$

III. RESULTS AND DISCUSSION.

System of eqs. (2),(3) and (4) was integrated numerically. We choose temperature and density on the axis as two free parameters in this system, so the rest of the values and radial distributions were calculated from solutions of the above mentioned equations as a functions of T_0 and n_0. But as it is more valuable to express the results in terms of integrated parameters such as full current I_0, line density N and pinch radius a, such an expression has obtained by implicit conversion. Value $A = P_A / P_R$, is adopted as a

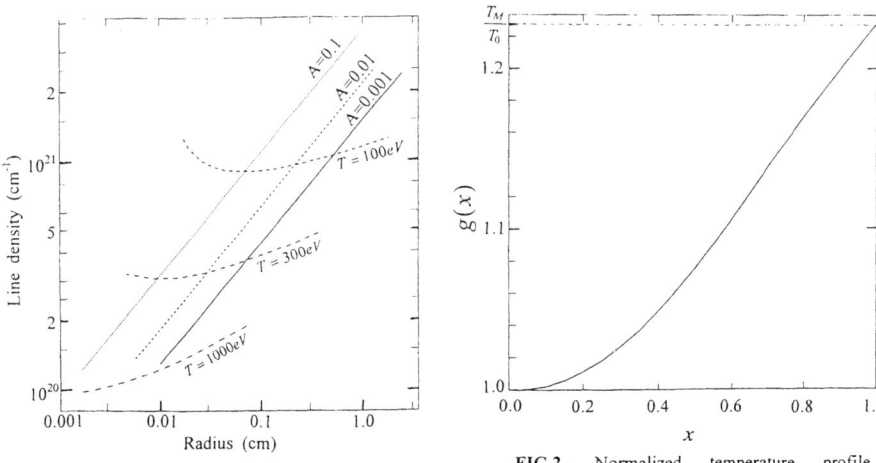

FIG.1. Lines of equal temperature and equal absorption parameter in the plane (pinch radius - line density).

FIG.2. Normalized temperature profile $g(x)=T/T_0$ versus normalized radius $x=r/a$ ($N=3 \times 10^{21}$cm^{-1}, a=0.0326 cm, A=0.71).

measure of absorption of pinch own radiation. For the case of uniform temperature A has an univalent correspondence with $\gamma = a / l_{ph}$, where l_{ph} is mean free pass of photon with thermal energy in the axis of pinch and a is a pinch radius. The maximum temperature T_M is reaches on the outer layer of pinch. We have adopted parameter $\delta = (T_M - T_0)/T_0$ as a measure of temperature non-uniformity.

In Fig.1 lines of equal temperature and absorption degree in coordinate system (radius-line density) are shown. This confirms the fact that it is hardly possible to obtain significant absorption for the hydrogen at high temperatures experimentally, because too small radius of pinch is required. For instance, if $T > 1000$eV (or $N < 10^{20}$cm^{-1}) for the $r = 10 \mu$m we have $A < 0.1$. The example of temperature distribution is plotted in the Fig.2. Calculated dependence of current on radius (Fig.3) is similar with zero-dimensional one, given by Robson ([5], p.1836, Fig.2). It follows from the fact, that zero-dimensional model has a good validity for the range of small line densities because homogeneity of temperature is very high (for $N = 3 \cdot 10^{20}$ cm^{-1} $\delta_{max} \approx 0.012$, Fig.4). Even for the case of considerably large line density $N = 3 \cdot 10^{21}$cm^{-1}, δ is not more than 0.23 (Fig.2,4). In Fig.4 δ is plotted versus A for the given line densities ($3 \cdot 10^{21}, 10^{21}$ and $3 \cdot 10^{20}$cm^{-1}). Calculations without radiation transport account shows that α increases with decreasing of thermal conductivity [11]. This is in agreement with growth of α for the small values of A because thermal conductivity decreases with increasing of A for the given line density. The interesting result, that at $A \approx 0.7$ (it corresponds to $\gamma \approx 3$) $\delta(A)$ has

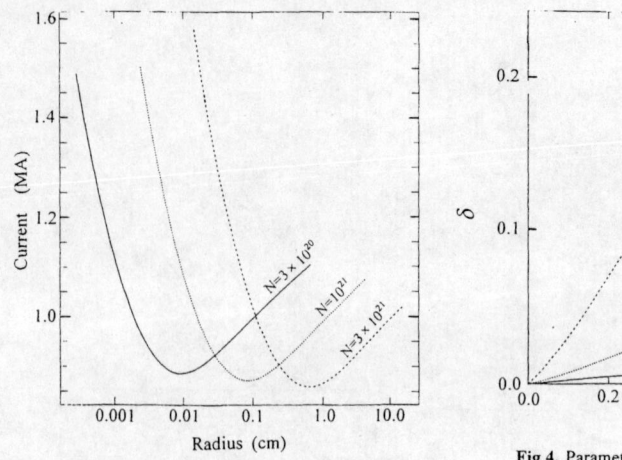

FIG.3. Total current versus pinch radius for different line densities.

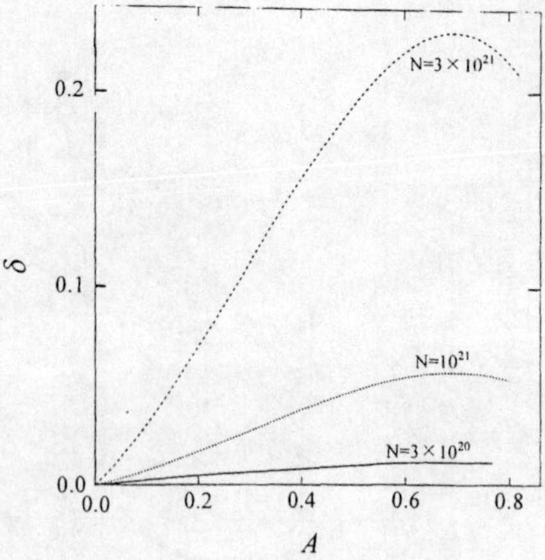

Fig.4. Parameter of temperature non-uniformity $\delta = T_M / T_0 - 1$ versus absorption parameter A for different line densities.

maximum for all line densities. It means that at this characteristic value of A the radiation transport becomes prevail over thermal conductance in the process of heat transfer.

ACKNOWLEDGMENTS

The author wishes to express his gratitude to Prof. T. Miyamoto for many helpful comments and discussions.

REFERENCES

1. R. S. Pease, Proc. Phys. Soc., B70 (1957) 11.
2. S. I. Braginskii, Sov. Phys. JETP, 6 (1958) 494.
3. D. W. Sccuder, Phys. Fluids 26 (1983) 1330.
4. N. R. Pereira, Phys. Fluids B 2(3) (1990) 677
5. A. E. Robson, Phys. Fluids B 1(9) (1989) 1834
6. J. L. Giuliani, Jr, AIP Conf. Proc. 195 (New York, 1989), p124.
7. J. P. Chittenden, A. J. Power and M. G. Haines, Plasma Phys. Contr. Fusion Vol.31(11) (1989) p.1822
8. K. N. Koshelev, N. R. Pereira, J.Appl.Phys. 69(10) (1991) R.21.
9. S. I. Braginskii, "Transport Processes in Plasma" in
Reviews of Plasma Physics (New York: Consultants Bureau, 1965), Vol.1, p.205.
10. L. Spitzer, Jr., Physics of Fully Ionized Gases, New York: Interscience, 1962.
11. P. V. Sasorov (private communication, 1997).

Influence of the Hall Effect on a Neck Development in the Z-pinch Discharges

Victor Vikhrev and Oleg Zabaidullin

*Nuclear Fusion Institute, RRC Kurchatov Institute,
Moscow 123182, Russia*

Abstract. The detailed analysis of the z-pinch neck formation is presented. The two-dimensional MHD modeling is carring out with special attention to the magnetic field transfer by the electron component (the so called Hall effect).

Shown, that the neck development may be nonsimmetrical with respect to the current direction, when the linear density N becomes less than 10^{17} cm^{-1}. There are many experimental confirmations of this fact at z-pinch with the current I < 100 kA. Also, it was found the increasing penetration of the magnetic field into the plasma due to the Hall effect influence when $N < 10^{16}$ cm^{-1}.

Results of the two-dimensional modeling of a neck formation with account of the Hall effect (transition of a magnetic field by electrons) is presented. When the initial number of particles in a cross-section $N < 5 \cdot 10^{17}$ cm^{-1} the neck has asymmetrical elongated shape in the directions to the electrodes.

Previous two dimensional modeling of Z-pinch column sausage instability development within the frame of one-fluid MHD [1] shows only symmetrical neck shape, because MHD equations are not sensitive to polarity of the electrodes. On the contrary, numerous experimental date show asymmetric shape of the neck with respect to a current direction. This fact becomes understandable after including into the numerical model the Hall effect - transition of the magnetic field by the current electrons. Since the current electrons flow in the main direction from a cathode to an anode, the task itself starts to contain implicitly some chosen direction of the magnetic field transition. Often, it's explain the asymmetrical shape of the neck mostly elongated towards to the anode.

The present 2D modeling of the sausage instability development is carring out within the frame of MHD equations with account of the Hall effect, assuming cylindrical geometry. The numerical equations are published in [2]. Initial conditions are shown in the Fig.1a. Uniform plasma column with some density inhomogeneity (the cross-section is shown in the left part of Fig.1a) is surrounded by the magnetic field (in the right part). The grey scale (from white to black) corresponds to a value variation in 1.4 times. Initial number of particles in the cross-section N is $3,2 \cdot 10^{16}$ cm^{-1}. In the Figs.1b-1f the different stages of the sausage instability development are shown. Influence of the Hall effect leads to the transition of the magnetic field from the upper (cathode) part of the neck to the lower (anode) part. Therefore, the magnetic field energy near the cathode part decreases and to the same value increases near the anode part. For this reason, near the cathode part of the neck the plasma compresses slower than in the anode part. Finally, this behavior of the magnetic field leads to the asymmetrical shape of the neck.

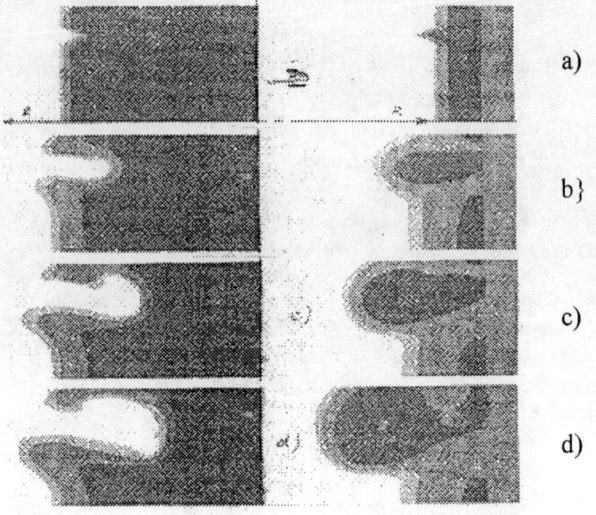

Fig.1. Different stages of the sausage instability development for $N = pr^2n = 3.2 \cdot 10^{17}$ cm^{-1} in the hydrogen plasma.

The theoretical explanation of the Hall effect influence is following. As it was shown in [3] due to the Hall effect the electromagnetic energy flux across any column cross-section is given by expression

$$S = \frac{c}{4\pi} \int_0^R [E \times H] 2\pi r dr \qquad (1)$$

Here R is the plasma column radius and H is the strength of the magnetic field. If the skin layer depth is less than R, then the flux

$$S = \frac{cH_0^3 R}{24\pi^2 en_e} = \frac{I_0^3}{3\pi^2 en_e R^2 c^2}, \qquad (2)$$

where I_0 is a value the current, passing through the column. The flux S depends on the product nR^2. In the case of the density dependence $n = 1/R^{2-\alpha}$ with positive α, then the flux $(nR^2) = 1/R^\alpha$ has maximum value S_{max} in the narrowest place of the neck. Therefore, the magnetic field energy change in the cathode part of the neck $\Delta S_C = S(nR^2) - S_{max}$ is negative. On the contrary, the magnetic field energy change in the anode part is positive $\Delta S_A = S_{max} - S(nR^2)$. But the total energy change in the volume of the neck stays the same $DS = S_{max} = \Delta S_C + \Delta S_A = 0$.

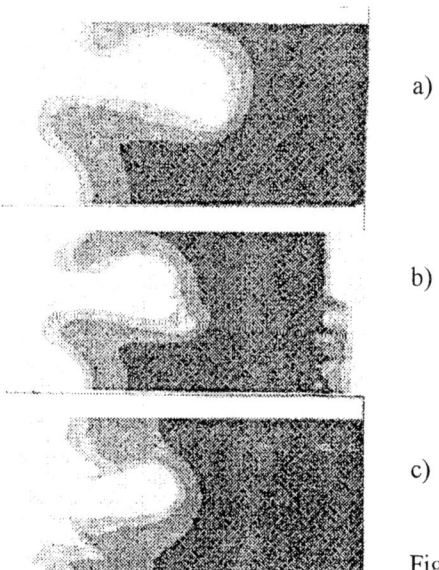

Fig.2. The shape of neck for the final stage of modeling a) $N = 3.2 \cdot 10^{17}$ cm^{-1}, b) $N = 7 \cdot 10^{16}$ cm^{-1}, c) $N = 1,7 \cdot 10^{16}$ cm^{-1}.

Therefore, the Hall effect results to the redistribution of the magnetic field in the neck, but not to it decreasing.

The next important result was obtained for the decreased number of particles in the cross-section $N = 1.7 \cdot 10^{16}$ cm^{-1}. The shape of neck at the final stage of modeling is shown in the Fig.2. The most important difference between Fig.2 and Fig.1 is that for the decreased number of particles the neck is less developed. This is due to the fast magnetic field penetration into the neck plasma with some efficient diffusion coefficient, greater than classical one.

References.
1. Vikhrev V.V., Rozanova G.A., Ivanov V.V., *Sov. J. Plasma Physics*, **15**, 44 (1989).
2. Vikhrev V.V., Terent'ev A.R., Zabaidullin O.Z., *Plasma Phys. Rep.*, **21,** 20 (1995).
3. Vikhrev V.V., Zabaidullin O.Z., *Plasma Phys. Reports.*, **20**, 867 (1994).

Modelling of Z-pinch Processes for Studding Initial Conditions to Get Inertial Confinement Fusion

Victor V. Vikhrev, Alexey V. Dobryakov and Oleg Z. Zabaidullin

Nuclear Fusion Institute, RRC Kurchatov Institute, Moscow 123182

Abstract. Z-pinch is the system with a rather small confinement time. For this reason it is necessary to create in it a plasma of high density in order to get the $(n\tau,T)$ parameters required for the Inertial Confinement Fusion. Shown, that it is more easy to get the required plasma parameters for the thermonuclear fusion in conditions of the radiative collapse, than in conditions of the adiabatic compression. Influence of the initial conditions on the radiative collapse is also discussed.

At the present moment, there is a growing interest to realizing of the Inertial Confinement Fusion (ICF) in a pulsed Z-pinched systems. The approach, discussed in the works [1]-[2], is substantially simpler than those of alternative pulsed schemes (e.g. the laser and the particle beam driven ICF). Instead of employing the high technologies, one can use the advantage of a neck formation in the current-carrying plasma column. During the neck formation, the magnetic field energy concentrates in a small neck cavern volume. As the result, the density of the magnetic field energy can reach here the extremely high values (10^{10} - 10^{15} J/cm3). This energy constriction is rather enough for creation a plasma of the thermonuclear parameters.

Two dimensional dynamics of the neck plasma is numerically studied in [1] on the Euler grid. At the initial moment of time it is supposed an existence of a hot dense plasma (n=1026 cm-3, T=2 keV), a short-wave disturbance on the plasma surface and the current value I=10 MA. Results of the modelling show that the neck plasma can ignite the combustion wave along the D-T fibber under the conditions of the radiative collapse. The work [2] is devoted to the complete (from the beginning to the end) one-and-half dimensional modelling of the neck plasma dynamics in the Lagrangian co-ordinates.

To simulate the neck plasma dynamics it is sufficient to use a one-and-half-dimensional (1.5 dimensional) code. The essential difference of the present 1.5 code from an one dimensional is the plasma escaping from the neck along the z-axis, describing by means of some analytical formulas. The 1.5-dimensional code makes possible to calculate the dynamics of the imploding plasma over the entire range of variations of its parameters (from $n=10^{22}$ cm^{-3}, T=1 eV to $n=10^{27}$ cm^{-3}, T=10 kV).

Due to the short confinement time τ, it is necessary to obtain high dense plasma n to get the product $n\tau$ required for ICF. Using of the thermonuclear fuel (e.g., a DT condensed compound) provides the most convenient opportunity to start from a high dense plasma. Nevertheless, to reach the required density it is necessary to realise conditions when the contribution of radiation losses to Z-pinch dynamics will be significant (the so called radiative collapse). During the modelling [2] the effect of the thermonuclear heat release on the neck dynamics mainly depends on the deceleration of the α-particles in the plasma. The equation of energy balance per unit length of neck is:

$$3\frac{dNT}{dt} = 2\pi RV \frac{H^2}{8\pi} - 5\frac{NT}{\tau} - Q_{rad} + Q_J + Q_\alpha, \quad (1)$$

where the first term on the right-hand side represents the work of the magnetic field; the second term represents the energy losses due to the particles escape through the neck ends; the third represents the radiation losses; the fourth represents the Joule heating; and the fifth represents the thermonuclear heat realise, and V is the velocity of the neck boundary.

When the radiation energy losses is the main term of the right-side (1) the pinch implosion occurs in the radiative collapse regime. Compare the volume losses due to the radiation from an optically thin plasma with the volume losses due to the particles escape, $5nT/\tau$. Here, τ is typical time of a particle escape from the constriction region, $\tau = h/v_s$. The equilibrium between these losses gives the plasma density

$$n = 6 \cdot 10^{26} \frac{T^3}{Z_{eff}^2 I^2}, \quad (2)$$

Hereafter, the density n, the temperature T, and the current I are expressed in cm-3, eV, and A, respectively.

For an optically thick plasma (the «trapped radiation» case), the radiation losses, i.e. radiated energy per unit length of the plasma column is

$$Q_{rad} = 2\pi R \sigma T^4 \quad (3)$$

where σ is the Stefan-Boltzman constant, R-radius of the neck, T. Equilibrium of Qrad with the particles losses from the neck ends, $5NT/\tau$, gives the following expression for the plasma density

$$n = 2 \cdot 10^{17} T^{5/2} \quad (4)$$

where n cm^{-3}, T eV. Therefore, the initial condition for calculation burning wave along the Z-pinch is possible to take from (4).

In the (n,T) diagram (plotted in Fig.1 in logarithmic scale), the expressions (3) and (4) are straight lines with slightly different slopes. They bound the domain of radiative collapse. The location of line 1, representing by (1), depends on the current and the effective charge of ions. Line 2, representing by (4), is independent of any of these parameters. If, after the arrival of the shock wave at the axis and the establishment of a Bennett equilibrium, the system goes to the line 2, the dynamics will steadily follow this

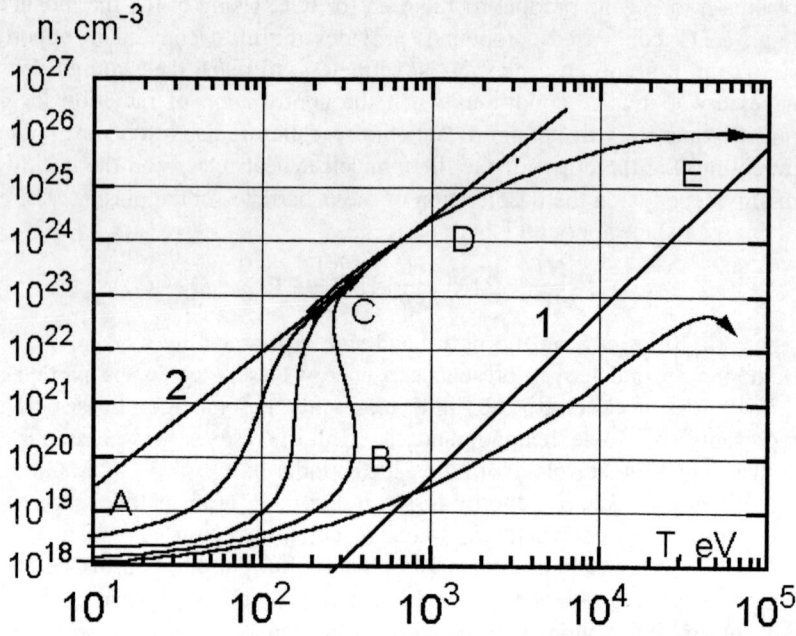

FIGURE 1. Scheme of the pinch-implosion dynamics in the (n,T)-plane. Domains of the radiative collapse and the thermonuclear heat release, as well the region corresponding to the trapped and the untrapped radiation region are shown schematically. ABCD is the dynamics curve (AB correspond to heating by a shock wave; BCD corresponds to the radiative collapse; and DE corresponds to expansion); (1) is the boundary of the radiative collapse domain with untapped radiation; (2) is the same as 1 but with trapped radiation.

line. On the contrary, the system cannot follow the line 1, the dynamics here is unstable. Actually, if equilibrium established in the vicinity of line 1, below it, the consequent evolution will follow the adiabatic curve $n \sim T^{3/2}$. Since the line 1 (formula (2)) is steeper than the adiabatic one, the latter cannot cross line 1 during the temperature increase. If the equilibrium established above the line 1, the radiative collapse will take place. The specific feature of the radiative collapse is that the increase of density is more sharper, than of the temperature. During the transition from the line 1 to the line 2, the magnetic confinement may temporally fail, and restores gradually when the dynamics approaches the line 2. If, due to some fluctuation, the system goes out above the line 2, the dynamics will follow an adiabatic curve and soon will return to the line 2 (because the latter is stepper than the adiabatic curve). Line 2 has a remarkable feature. After the reaching the line 2, the system «forgets» its initial conditions and follows it through all the states.

Results

1) In the frame of the modern high-voltage pulsed power technology, it is substantially easier to get ignition of a thermonuclear reaction in the regime of the radiative collapse rather than during the adiabatic compression.
2) Provided that the compression dynamics is stable, the temperature dependence from the density of an optically thick plasma has the following invariant form, regardless of it's initial conditions: $n = 2 \times 10^{17} T^{5/2}$, where n cm-3, T eV.
3) The radiative collapse makes it possible to convert all the nuclear fusion energy into the radiation.

1. Vikhrev V.V. and Rozanova G.A., *Phys. Plasma Rep.* **19**, 40 (1993).
2. Vikhrev V.V, Dobriakov AV. and Zabaidullin O.Z., *Phys. Plasma Rep.* **22**, 97 (1996).
3. Bilbao L. and Linhart J., Fizika plasmy (*Phys. Plasma Rep*) **22**, 503 (1996).
4. Sheebey P., Hammel J.E., Lindemuth LR., et al, *Phys Fluids B*, **4**, 3698 (1992),
5. Haines M.G., Bailey J., Baldock., et al. *Nucl. Fusion Suppl.*, **2**, 565 (1988).

The MAGO System

S. F. Garanin

*All-Russian Research Institute of Experimental Physics
607190, Sarov, Nizhny Novgorod Region, Russia*

Abstract. This paper reviews theoretical and experimental investigations for the magnetohydrodynamic implosion concept, MAGO. It consists of a pre-heated magnetized DT-plasma with subsequent adiabatic compression by a liner imploded with a magnetic field. Pre-heating is performed in a special plasma chamber wherein magnetized plasma is accelerated in an annular nozzle up to velocities ~100 cm/μs and heated by collisionless shock waves. Subsequent plasma compression up to ignition characteristics can be obtained using explosive magnetic generators with energy 100 to 500 MJ. The paper also discusses MAGO's plasma chamber and the physical effects essential for its operation. Due to its cylindrical symmetry with only a toroidal magnetic field component and the plasma described by of magnetohydrodynamics, the MAGO system and its problems are similar to those related systems, such as Z-pinches, plasma accelerators and liner systems.

INTRODUCTION

The MAGO (MAGnitnoye Obzhatiye or magnetic compression) or MTF (Magnetized Target Fusion) system is an alternative to the main approaches for achieving controlled thermonuclear fusion (CTF), namely magnetic fusion energy (MFE) and inertial confinement fusion, (ICF). MAGO consists of a thermonuclear target located at the center of a cylindrical chamber, which is compressing by a liner that is imploded by a magnetic field. An important advantage of this system is that fusion ignition experiments can be done without expensive stationary energy sources: relatively cheap explosive magnetocumulative generators (EMGs) are appropriate.

To use this approach, it is necessary to combine a hot magnetized plasma generation system and a highly energetic compression system. The All-Russian Research Institute of Experimental Physics (VNIIEF) has implemented a new technique of hot magnetized plasma generation with a special plasma chamber MAGO [1,2]. In experiments with the MAGO chamber a thermonuclear plasma was obtained with neutron yield up to 4×10^{13} per shot. It is computationally shown that ignition can be achieved within this system

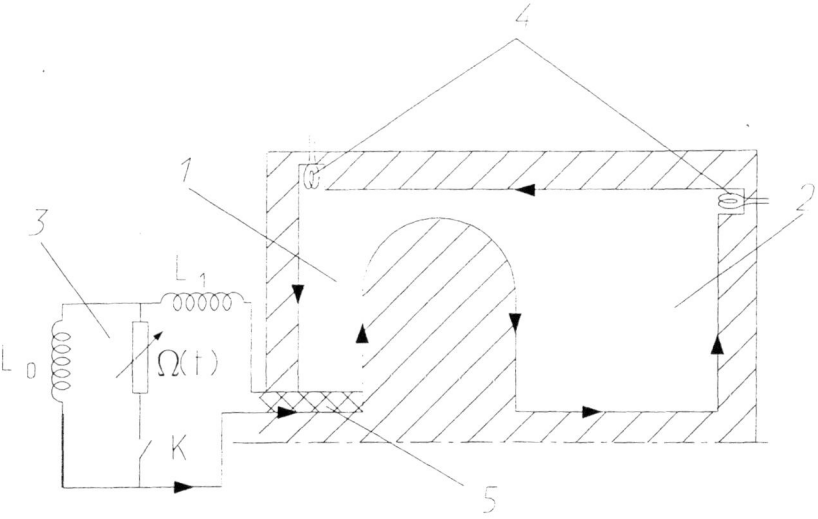

FIGURE 1. Plasma chamber: (1), (2) first and second sections; (3) equivalent scheme of EMG with closing switch K and opening switch $\Omega(t)$; (4) inductive probes; (5) insulator.

at an energy from 100 to 500 MJ, which can be supplied using VNIIEF's disk EMG. This can be done at realistic implosion symmetry, i. e. MAGO sidesteps the main ICF ignition problem - stringent requirements on the implosion symmetry.

PLASMA CHAMBER MAGO

The MAGO chamber consists of two toroidal sections connected by a narrow annular nozzle (Fig. 1) that is filled with low-density DT gas.

By the time of the start of the operation an initial toroidal magnetic field has been created in the chamber (the direction of the current is shown with arrows). When switching on the main current source (after the start of the opening switch operation in Fig. 1, the switch K is closed) the accompanying high electric field leads to a gas discharge that results in the initial magnetic field freezing into the generated plasma. Driven by the source magnetic field the plasma begins to move through the nozzle from the first chamber section to the second.

If the Alfven velocity in the plasma is sufficiently high and the nozzle is sufficiently narrow, then the total pressure will have time to become uniform in each chamber, even though there is still a pressure difference between the chambers. The plasma states at the nozzle input in chamber 1 and at the output in chamber 2 after complete deceleration are related by conservation of total enthalpy of each plasma element at the flow and the condition of magnetic field being frozen in plasma,

$$w_2 = \frac{H_1^2}{4\pi\rho_1}(1 - \frac{H_2}{H_1}),$$

where H_1 and ρ_1 are the magnetic field and plasma density at the nozzle input, and H_2 and w_2 are the magnetic field and plasma enthalpy in section 2. For $H_2 \ll H_1$ the main part of magnetic energy is converted to plasma thermal energy.

COMPUTATION METHODS

The plasma in the MAGO chamber at low temperatures satisfies the criteria for the validity of the MHD description. As for the plasma heated up to keV temperatures, MHD is valid for the electron component and but not for the ions. However, this fact does not severely affect the flow dynamics and may be important only for neutron diagnostics of plasma. Thus, 1D and 2D calculations in the MHD approximation can be considered as quite a good approach to description of plasma in the MAGO chamber.

In the experiments the system parameters varied in the following ranges: preliminary current $1 \div 3$ MA, final current at the chamber input $3 \div 9$ MA with a rise time $1 \div 3$ μs, initial DT pressure $1 \div 20$ torr, chamber radius $6 \div 10$ cm, the chamber length $9 \div 20$ cm, and nozzle width $0.5 \div 2$ cm. According to the calculations, for these parameters the maximum plasma velocity at the nozzle output is $(0.5 \div 2) \cdot 10^8$ cm/μs, the maximum ion temperature $2 \div 10$ keV, the maximum electron temperature $0.5 \div 2$ keV, the DT neutron yield $10^{11} \div 10^{13}$, and the neutron pulse duration $0.5 \div 2$ μs.

The approximations used in the MHD calculations do not take into account properly some effects which, however, can play an important role:
1. Initially plasma is generated in the chamber due to gas breakdown in the magnetic field. It is quite hard to describe the breakdown phenomena with the MHD approximation due to important role of various kinetic processes.
2. Plasma contamination with impurities from the walls and insulator.
3. MHD instabilities. Experimental data show that here the situation resembles that in the Z-pinch where the main instability is the $m = 0$ instability. Azimuthally asymmetric instabilities may be essential in regions of tangential discontinuities [4] or high velocity gradients.
4. Kinetic phenomena arising near electrodes. When plasma flows along the electrodes, plasma heating due to friction on the electrode is possible. The situation analysis is complicated with the Hall effect resulting in plasma

rarefaction near the anode and compression near the cathode that leads to asymmetry with respect to the electrode polarity.

5. Collisionless shock waves. As the plasma downstream from the shock front is strongly magnetized the shock wave is collisionless, i. e. the effective heating occurs not due to particle collisions, but due to flow chaotization.

SOME EXPERIMENTAL RESULTS

The main body of experimental results is in reasonable agreement with theoretical predictions and confirms the MAGO chamber's principle of operation.

The pinhole measurements show that the neutron generation region and therefore the hot plasma occupy a large volume that is located in the middle part of the second section, with a characteristic diameter and length at least half the corresponding sizes of the second section.

To get an insight into the energetics level pertaining to the MAGO chamber experiments and characteristic values, we give the parameters of the MAGO-2 experiment conducted in Los Alamos in October 1994 [3]: the preliminary chamber feed was with a 2.7 MA current. By the time of opening the switch the EMG inductance was 36 nH and its current 17 MA, corresponding to a stored energy 5 MJ. The plasma chamber (see Fig. 1) was 10 cm in radius, the first section width was 2.5 cm, the second section width was 8 cm, the minimum nozzle width was 1.2 cm, and the initial pressure of DT gas was 10 torr. The maximum current flowing in the chamber at the main fast stage was 7.7 MA and its maximum derivative 3.8 MA/μs. Neutron measurements indicated the production of 10^{13} DT neutrons with a peak reaction rate of $3 \cdot 10^{13}$ neutrons per μs.

MAGO SYSTEM PLASMA COMPRESSION

Paper [5] discusses some techniques for adiabatic plasma compression in the MAGO chamber and relevant problems. According to [2,3,5], if warm plasma lives for $\sim 10^{-5}$ s, ignition can be reached in the MAGO system by using a compression system with $\sim$20 MJ in liner energy and $\sim$1 cm/μs in liner velocity. The recent joint VNIIEF/LANL experiment HEL-1 obtained a liner with comparable parameters, namely energy $\sim$25 MJ and velocity $\sim$0.8 cm/μs.

REFERENCES

1. Buyko, A. M., Garanin, S. F., Gubkov, E. V. et. al., *VANT. Ser. Metodiki i programmy chislennogo resheniya zadach matematicheskoi fiziki* No. 3(14), 30–32 (1983).

2. Buyko, A. M., Chernyshev, V. K., Demidov, V. A. et. al., "Investigations of thermonuclear magnetized plasma generation in the magnetic implosion system MAGO," in *Proceedings of the IX IEEE International Pulsed Power Conference*, 1993, pp. 156–162.
3. Lindemuth, I. R., Reinovsky, R. E., Chernyshev, V. K., Mokhov, V. N. et. al. *Phys. Rev. Lett.* **75**, 1953–1956 (1995).
4. Garanin, S. F., and Kuznetsov, S. D., *Plasma Physics Reports* **22**, 674–678 (1996). Translated from *Fizika Plazmy* **22**, 743–746 (1996).
5. Buyko, A. M., Garanin, S. F., Mokhov, V. N., and Yakubov, V. B., "On possibility of low-dense magnetized DT-plasma ignition threshold achievement in MAGO system," presented at the Spring Workshop on Basic Science Using Pulsed Power, Santa Barbara, California, April 5-7, 1995.

SUPERPOWER GENERATORS

PBFA Z: A 60-TW/5-MJ Z-Pinch Driver*

R. B. Spielman, C. Deeney, G. A. Chandler, M. R. Douglas, D. L. Fehl,
M. K. Matzen, D. H. McDaniel, T. J. Nash, J. L. Porter, T. W. L. Sanford,
J. F. Seamen, W. A. Stygar, K. W. Struve,[†] S. P. Breeze, J. S. McGurn, J. A.
Torres, D. M. Zagar, T. L. Gilliland, D. O. Jobe, J. L. McKenney, R. C.
Mock, M. Vargas, and T. Wagoner

Sandia National Laboratories, Albuquerque, NM 87185

D. L. Peterson

Los Alamos National Laboratories, Los Alamos, NM 94545

ABSTRACT

PBFA Z, a new 60-TW/5-MJ electrical accelerator located at Sandia National Laboratories, is now the world's most powerful z-pinch driver. PBFA Z stores 11.4 MJ in its 36 Marx generators, couples 5 MJ into a 60-TW/105-ns FWHM pulse to the 120-mΩ water transmission lines, and delivers 3.0 MJ and 50 TW of electrical energy to the z-pinch load. Depending on load parameters, we attain peak load currents of 16-20 MA with a current rise time of ~ 105 ns with wire-array z-pinch loads.

We have extended the x-ray performance of tungsten wire-array z pinches from earlier Saturn experiments. Using a 2-cm-radius, 2-cm-long tungsten wire array with 240, 7.5-μm diameter wires (4.1-mg mass), we achieved an x-ray power of 210 TW and an x-ray energy of 1.9 MJ. Preliminary spectral measurements suggest a mostly optically-thick, Planckian-like radiator below 1000 eV. Data indicate ~ 100 kJ of x rays radiated above 1000 eV. An intense z-pinch x-ray source with an overall coupling efficiency greater than 15% has been demonstrated.

INTRODUCTION

Progress in the field of applying z pinches to ICF and to creating high energy photons has long been limited by the lack of sufficiently powerful drivers. Moreover, z pinches are complex and they are difficult to model in an integrated quantitative fashion. Previous experiments on the Saturn facility[1-6] have yielded remarkably high x-ray powers and excellent pinch quality over a wide range of array parameters. These surprising results were the result of better azimuthal wire array uniformity, overall improved load quality, and general load design improvements. These improvements moved the z-pinch performance much closer to that predicted by idealized one- and two-dimensional calculations. However, flexibility in load design was limited by the pinch masses that could be imploded. Sandia's new accelerator, PBFA Z, delivers a peak current of 20 MA with a current

---

* This work supported by the U.S. Department of Energy under Contract DE-AC04-94AL85000.
†Present address: Mission Research Corporation, Albuquerque, NM 87106

rise time of 105 ns to wire-array z-pinch loads. This new capability allows us to build multi-mg-mass z pinches. Consequently, we have fielded tungsten wire arrays with up to 240 wires. These highly symmetric arrays have allowed us to obtain total x-ray powers of 200 TW and x-ray energies of 1.9 MJ with x-ray pulses as short as 5-ns FWHM. Simultaneously with our new experimental capability we have developed quantitative modeling tools.

This paper presents the first data from tungsten wire-array z-pinch experiments on PBFA Z and compares these data with two-dimensional radiation-magneto-hydrodynamic calculations. Calculations suggest that the dynamics of uniform wire array z pinches can be modeled assuming that the implosion is dominated by the magneto Rayleigh-Taylor instability.

Fig. 1 A schematic of the PBFA-Z accelerator showing the Marx generators, the water pulse forming section, the insulator stack, and the diagnostic line-of-sight.

THE PBFA Z ACCELERATOR

The experiments described herein were conducted on the Sandia National Laboratories PBFA-Z accelerator.[7,8] The original PBFA-II accelerator[9] at Sandia National Laboratories was originally used by the light-ion-beam ICF Program. We made major modifications to the accelerator in order to optimize coupling to magnetically-imploded loads, typically z pinches. The facility, in operation since September 1996, has been renamed PBFA Z. PBFA Z represents the latest step in z-pinch drivers at Sandia National Laboratories. Like Saturn,[10] PBFA Z is based on classical water-dielectric pulse-forming technology. (See Figure 1.) PBFA Z stores 11.4 MJ in its 36 Marx generators, couples 5 MJ in a 60-TW/105-ns pulse into 36, constant-impedance 4.32-Ω water transmission lines. (See Fig. 2.) PBFA Z, in its normal operating mode, delivers up to 3.0 MJ and 50 TW of electrical energy to four, separate vacuum insulator stacks. Total currents up to 20 MA are measured at the insulator stack. The constant, 120-mΩ impedance driver allows us to model the electrical circuit and load coupling much more accurately than on Saturn.

Current is fed via four magnetically-insulated vacuum transmission lines (MITLs) and a vacuum convolute to the z-pinch load. These vacuum transmission lines (Shown in Fig. 3.) operate far above the electron self-emission threshold and must run in the self-magnetically-insulated regime. The current from the four independent, conical-disk MITLs is combined together in a double post-hole vacuum convolute and then fed to the z-pinch load through a 5-cm-long radial-disk MITL. (See Fig. 4.) Depending on the load geometry, power flow configuration, initial inductance (typically 12 nH), and the implosion time (100-150 ns), we reach peak load currents of 16-19 MA with a time-to-peak current of 105 ns. We can successfully model the behavior of PBFA Z using electrical circuit codes to accuracies of ~ 5%. PBFA Z presently operates with a shot rate of one per day.

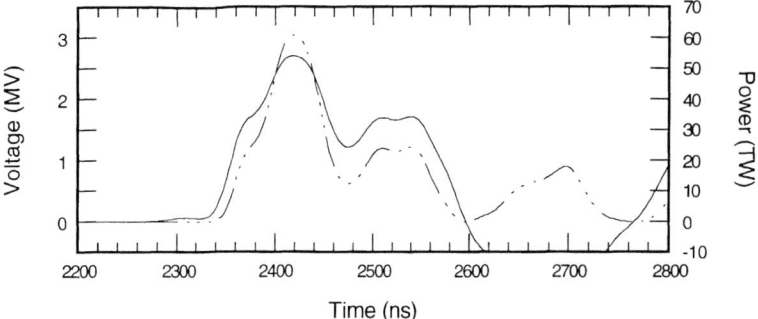

Fig. 2 The forward-going voltage (solid line) and power pulse in the water transmission lines for Shot 51.

Z-PINCH LOAD HARDWARE

The initial condition of the z pinches is determined by a highly-symmetrical cylindrical wire array. These arrays are composed of 120-300 tungsten wires with a 1- to 2-cm length and with array diameters ranging between 17.5 and 40 mm. The diameters of the individual tungsten wires are typically 7.5-15 μm and result in masses of 4.1-10 mg. Figure 5 shows a picture of such a wire array. The locational tolerance of an individual wire is ±75 μm azimuthally and radially. The wire diameter and mass is measured via an SEM and a high accuracy mass scale, respectively. The annular z-pinch plasma is formed by the heating and vaporization of the wire array. PBFA Z has a current prepulse that can be described by a linear ramp 50-ns long reaching a peak current of 200-300 kA. The main current pulse completes the vaporization and ionization process in the next few ns. Data from Saturn experiments suggest that details of this wire initiation process play a significant role in the final pinch quality.[3,6]

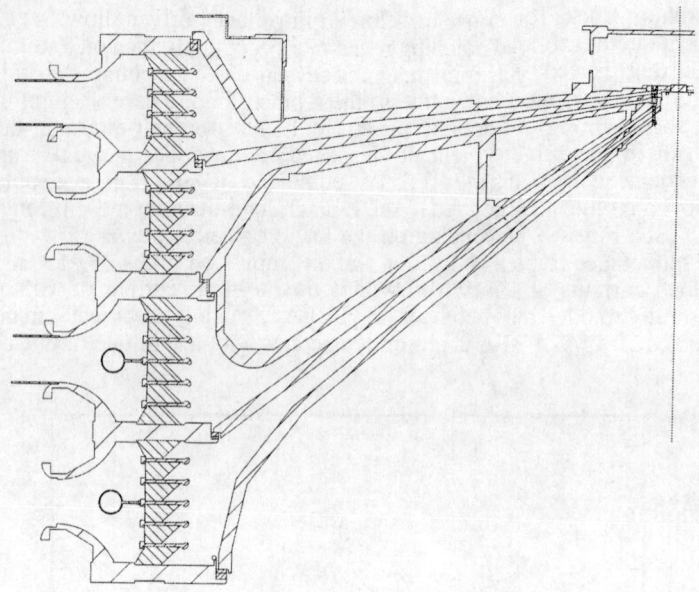

Fig. 3 A drawing of the PBFA-Z insulator stack and MITLs is shown. Note, only half of a cylindrically-symmetric section is displayed. The diameter of the insulator stack is nearly 4 m.

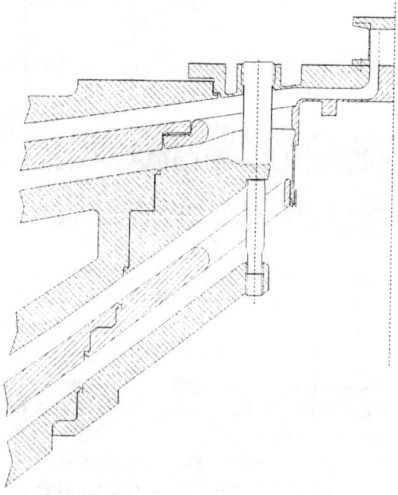

Fig. 4 A drawing of the double post-hole vacuum convolute is shown. Four MITLs are combined into a single radial-disk MITL leading to the load.

Fig. 5 A photograph of a 4-cm-diameter tungsten wire array. The array has 240, 7.5-µm-diameter wires.

X-RAY DIAGNOSTICS & DATA ACQUISITION

We used a variety of x-ray diagnostics for these experiments. Electrical diagnostics such as x-ray diodes (XRDs),[11] resistive bolometers,[12] diamond photoconducting detectors (PCDs)[13] are fielded for all shots. We also fielded time-resolved and time-integrated x-ray spectrographs and a time-resolved x-ray pinhole camera. Figure 6 shows the primary diagnostic line-of-sight, LOS 5/6, used for these experiments. LOS 5/6 is oriented 78° off the axis of the z pinch. The 18- to 25-m length of LOS 5/6 is dictated by the extremely high source flux. Indeed, even the relatively insensitive bolometers are placed 20 m from the x-ray source. LOS 5/6 contained all of the diagnostics but the x-ray pinhole camera.

Electrical signals are recorded on fiber-optically controlled data acquisition systems located adjacent to the detectors. This ensures excellent signal quality because we can use very short, high-bandwidth cables. We use Tektronix TDS684A/B digitizers for detectors that require 1-GHz bandwidth and Tektronix TDS640 digitizers for 500-MHz bandwidth needs. The digitizers are placed in a double-walled shielded enclosure made for Sandia National Laboratories by Lindgren and Assoc. (Again, see Fig. 6.) Power is fed into the shielded enclosure through LC filters and used to power a battery-operated uninterruptible power supply (UPS) to minimize electrical noise. All signal attenuators are made by Barth Electronics (Boulder City, NV) and have bandwidths greater than 5 GHz. Bias voltage is applied to XRDs and PCDs using Sandia-designed DC insertion units. These insertion units have > 2-GHz bandwidth, have a stable capacitance as a function of applied voltage, and have a long RC time constant (~ 2 µs).

Fig. 6 A photograph of the primary line-of-sight on PBFA Z. The XRD detectors in the foreground are 25 m from the x-ray source.

We provide a summary of the x-ray detectors and their applications. The five-channel XRD arrays are used to measure the x-ray flux in the energy range of 100 eV to 1.5 keV. In particular, these detectors are used to determine the color and brightness temperatures of nearly Planckian radiators. A standard filter set is given in Table 1. Typically, the XRD data is unfolded to provide an x-ray spectrum. This unfold can give the peak x-ray power and the total energy radiated from the source. The XRD photocathodes are individually calibrated on the National Synchrotron Light Source at the Brookhaven National Laboratory. An array of six, 1-mm-thick, 3-mm-long diamond PCDs are used to measure the x rays in the 1-10-keV portion of the spectrum. A filter set for the PCDs is given in Table 2. The PCDs are calibrated on a pulsed laser calibration source at Sandia

National Laboratories. We use an array of three, identical, unfiltered thin-film nickel bolometers to measure the total x-ray fluence from the source. These detectors are sensitive to x rays from 5 eV to 1500 eV and have a nearly flat spectral response over that range. A relative spectral-response function of a 1-μm-thick nickel bolometer element is given in Fig. 7. These bolometers have a time response of ~ 2 ns and provide a second, independent, measure of the x-ray energy and power (compared with an unfold of the XRD data). Bolometers are not "calibrated" but have a sensitivity based on the intrinsic properties of nickel and the physical dimensions of the element. The time-resolved x-ray pinhole camera has 100-ps time gates and 2-ns interframe times. The camera has a magnification of 1.73 and uses 50-120-μm diameter pinholes. We have used either 25.4-μm-thick beryllium filters or 5.5-μm-thick aluminized-Kimfol filters on the camera. Images are recorded on Kodak RAR2484 film and digitized with a Perkin-Elmer digitizer.

All of the diagnostic lines-of-sight are independently pumped with turbo-molecular pumps and have optical alignment systems. Most of the diagnostics have pneumatically-operated fast-closing valves to protect the diagnostics from debris from the source.

Detector ID	Photocathode Material	Filter Material	Thickness (μm)
XRDKM	Vitreous carbon	Kimfol (Lexan)	4.75
XRDV	Vitreous carbon	Vanadium	1.00
XRDZN	Vitreous carbon	Zinc + CH	0.8 + 0.5
XRDBE	Vitreous carbon	Beryllium + CH	8.0 + 1.0
XRDBEV	Vitreous carbon	Beryllium + Vanadium	10.0 + 0.8

Table 1 A list of the XRDs and filters used on PBFA Z.

Detector ID	PCD Material	Filter Material	Thickness (μm)
PCD215	Diamond	Beryllium + CH	8.0 + 1.0
PCD212	Diamond	Kapton	8.33
PCD211	Diamond	Kapton	25.4
PCD232	Diamond	Kapton	84.66
PCD213	Diamond	Kapton	254
PCD233	Diamond	Kapton	1016

Table 2 A list of the PCDs and filters used on PBFA Z.

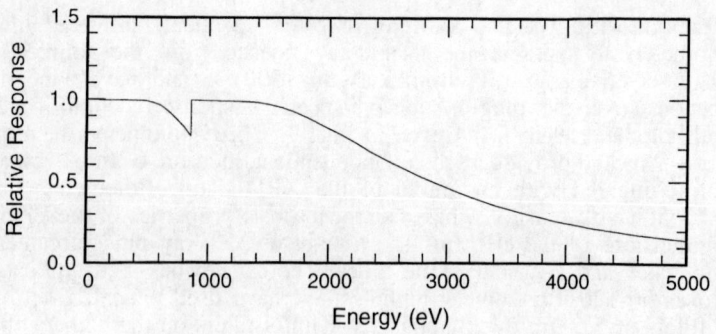

Fig. 7 The relative spectral response of a 1-μm-thick nickel bolometer.

EXPERIMENTS

We conducted a number of experiments on PBFA Z in which the load parameters were varied according to the applications we are pursuing i.e. driving vacuum and internal hohlraums. To drive vacuum hohlraums, high power and high energy sources, consistent with the smallest area return can geometry possible, are needed to drive internal or dynamic hohlraums. High velocity, uniform plasma shells are required. We describe four series of experiments: one in which the load mass was varied while the diameter was fixed at 40 mm and the length was held at 2 cm to optimize the energy delivery, a second experiment where the array diameter was varied between 17.5 mm and 40 mm, with a constant 2-cm pinch length, and the implosion time held constant at ~ 105 ns, a third experiment where the pinch length was varied between 1 cm and 2 cm with the implosion time held to 105 ns and the array diameter held to 30 mm to reduce inductances and hohlraum can volumes, and a fourth where the anode/cathode gap at the base of the wire array was varied between 2 mm and 5 mm while the load diameter was held constant at 30 mm and the load mass was fixed at 5.5 mg again to reduce inductance.

The baseline load design for PBFA Z was a 40-mm-diameter, 2-cm-long tungsten wire array. This was the first load used on PBFA Z. Using this load we varied the wire diameter (total load mass) and wire number to optimize the electrical coupling to the accelerator and the x-ray output. The x-ray energy and power was found to increase as the implosion time was decreased from 150 ns to the 105 ns that is optimal for coupling driver energy to the load. Figure 8 shows the x-ray energy and power as a function of implosion time defined as the time of peak x-ray power. The optimization in energy radiated is directly proportional to the energy coupled into the load and the available magnetic energy near the pinch at the time of stagnation. The voltage (and power) at the insulator typically reverses at 130 ns or so. Implosions later than this time are less efficient because energy is flowing out of the inductor represented by the vacuum MITL. As the implosion time moves earlier than 130 ns then the implosions are occurring before the maximum energy delivery to the load region. As the implosion time is decreased in Fig. 8, the increase in x-ray power is due to the increasing energy in the pinch coupled with a decreasing radiation pulse widths. The decreasing pulsewidths are primarily due to higher implosion velocities and smaller wire sizes. The impact of individual wire diameter on the initial condition of the z pinch is important but it is difficult to quantify and will be studied in the future. The data in Fig. 8 for the 120 wire arrays (solids) show a peak power of 170 TW at an implosion time of 120 ns. The data points represented by the open symbols are the result of doubling the number of wires in the array from 120 to 240. This shot, Shot 26, gave a power of ~ 230 TW and a radiated energy of 2.6 MJ. These shots were performed without apertures to limit the field of view to only the pinch. The very high yields obtained were an indication that there may have been some re-radiation from the return current hardware and the cathode surfaces that was contributing to our measured energies and powers.

Shot 26 was repeated on Shot 51 and 52. These shots, like Shot 26, had a 40-mm-diameter, 2-cm-long tungsten wire array with 240 wires. Shot 51, however, had a 8-mm x 15-mm aperture located 15 cm from the source to limit the field of view to just the pinch. Shot 51 generated a peak x-ray power of 210±20 TW and an energy of 1.9±.2 MJ. The total x-ray pulse width for this shot was 5.5-ns FWHM. These data are seen in Fig. 9. As can be seen from this figure, the peak power measured with and without (Shot 52) an aperture are not very different. However, the late time emissions are much higher without the aperture which is consistent with our hypothesis of can re-radiation. Also on Shot 51, the water pulse-forming switches had been adjusted so there was shorter current risetimes which led to a shorter implosion time. The x-ray powers and energies generated on Shots 26, 51, and 52 represent x-ray power and energy records for any laboratory experiment with any source. Only nuclear explosives and z pinches driven explosively-generated electrical power sources generate more x-ray energy and nuclear explosives produce more x-ray power. The powers and energies shown herein are

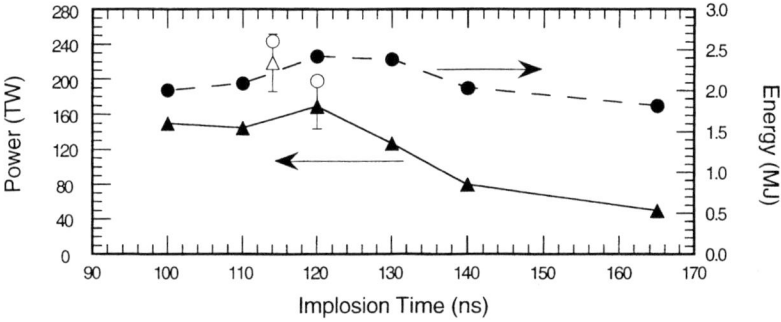

Fig. 8 The x-ray power and energy is plotted as a function of implosion time. The closed triangles are the power from 120 wire arrays and the closed circles are the energy. The open symbols are the energy and power from an array with 240 wires.

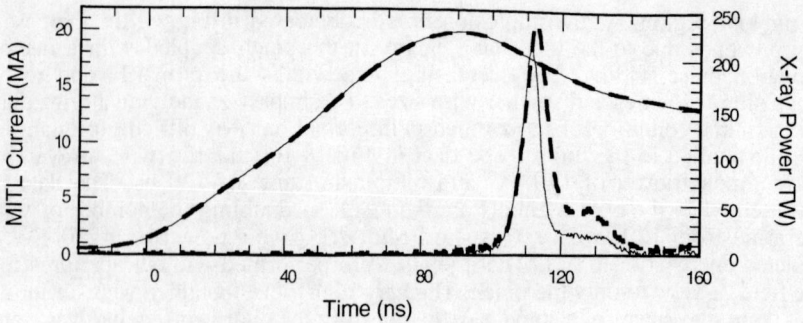

Fig. 9 The currents and x-ray powers from two, 40-mm-diam. arrays with 240, 7.5-μm wires, Shot 51 (solid lines) with an aperture and Shot 52 (dashed lines) without an aperture are compared.

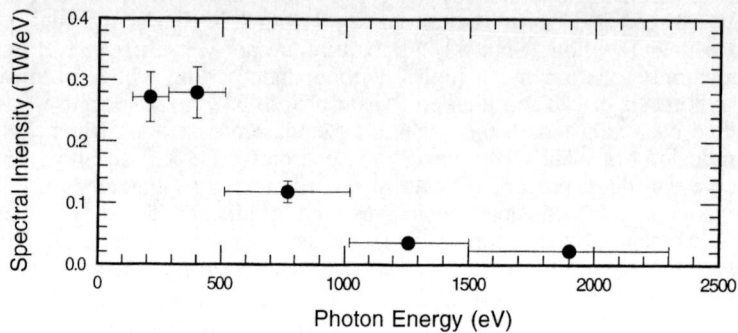

Fig. 10 A spectral unfold generated from XRD data is shown. X-error bars show the energy bin width.

the result of an XRD unfold and are checked against bolometry. An example of an XRD unfold is shown in Fig. 10. These XRD unfolded spectra show that emissions are relatively Planckian up to 1 keV. Above 1 keV, the emissions are above the blackbody curve for the equivalent total pinch brightness temperature.

Although the x-ray powers from PBFA Z are astounding, there is probably still room for improvement. The soft-filtered time-resolved x-ray pinhole camera data (an example is shown in Figure 11) indicate that these tungsten pinches are wider and less uniform than those measured on Saturn. This is not entirely unexpected as the implosion times and initial array radii are twice those of Saturn. Typically, the diameters of the x-ray emissions above 200 eV are between 1.5 and 2.0 mm; on

Saturn, we measured 1-mm diameters. As seen in Figure 11, the emission are axially non-uniform which is probably consistent with the evolution of Rayleigh-Taylor instabilities. Applying RT mitigation techniques should reduce the pulsewidths and result in tighter pinches and would have a major impact in driving internal hohlraums. Another unusual phenomena is the presence of absorption features which may be due to blow-off from diagnostic slots in the return current can. Future experiments will look at minimizing diagnostic hole closure.

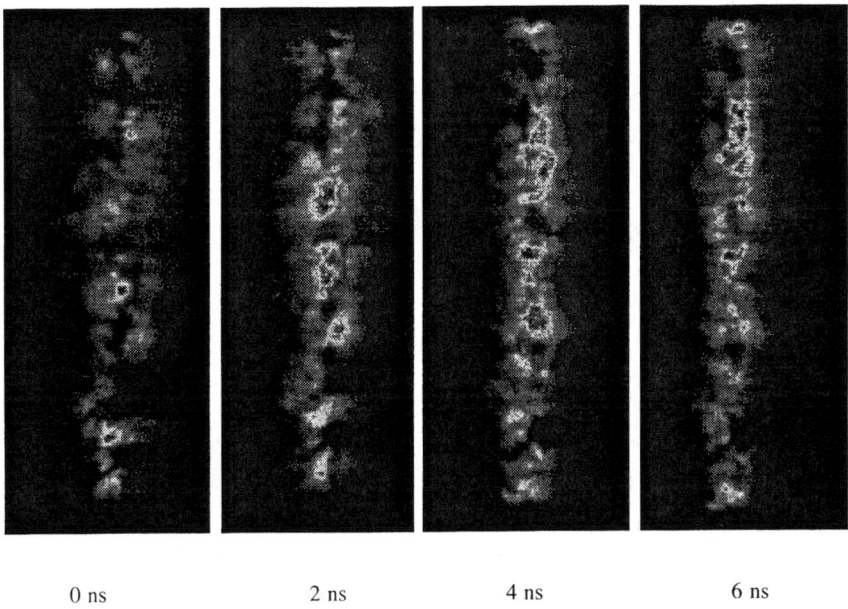

0 ns   2 ns   4 ns   6 ns

Fig. 11   A sequence of time-resolved x-ray pinhole pictures is shown. The peak x-ray power occurred at 4 ns.

After we determined the optimal implosion time for energy coupling to the z-pinch load we conducted a scan of array diameter to see if we could further optimize the power and to explore the power flow issues of driving small diameter arrays for vacuum hohlraums. The mass of the load was varied by changing the wire diameter and, hence, the mass in order to hold the implosion time constant. The interwire spacing of the array was also held nearly constant to remove the effect of changing azimuthal symmetry from the data. The length of the arrays was held constant at 2 cm. Figure 12 shows the total x-ray power as a function of wire array diameter. The data show that the power generated by the pinch decreases rapidly as the array diameter is decreased. By a diameter of 20 mm the power had fallen from a peak of

200 TW to 70 TW. Some of this decrease was expected due to a reduction in implosion velocity. In addition, a reduction in total coupled energy was expected. This would also be reflected as reduction in radiated power. Unfortunately, we observed a larger reduction in x-ray yield and power than expected. This appears to be due to an increase in electrical losses in the vacuum convolute. Our conclusion from these shots is that we need to decrease the inductance of the smaller diameter loads to reduce the electrical losses if we want to improve the x-ray production efficiency.

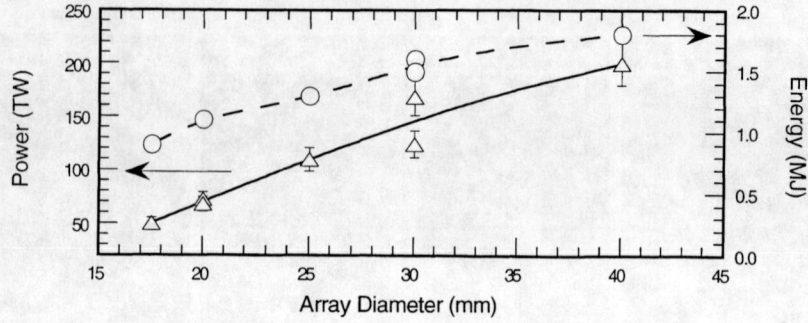

Fig. 12 The x-ray power (triangles) and energy are plotted as a function of array diameter.

One way of reducing the load inductance and, potentially, increasing the x-ray energy and power per unit length of the pinch is to decrease the pinch length. We conducted shots in which the standard 30-mm-diameter, 2-cm-long tungsten wire-array load was fielded with 1.5- and 1.0-cm lengths. We expected to see a 30% increase in the energy per unit length but a decrease in total x-ray energy and power from the 2-cm long pinches on the basis of the changes in kinetic energy per unit length. The slight increase in energy per unit length would be due to the lower initial inductance of the shorter load giving a higher peak current and allowing us to have a 30% heavier mass per unit length and hold the implosion time constant. What we found was encouraging. Figure 13 shows the x-ray power per unit length for these shots. These data show that the energy per unit length on the shortest length increased nearly a factor of two over the x-ray energy from a 2-cm long pinch. This increase cannot be explained in terms of the radial kinetic energy delivered to the pinch.

One possible explanation for the increase being larger than predicted from kinetic energy arguments is that the pinch extracts the energy from the magnetic field surrounding it. A 1-cm long pinch not only reaches a slightly higher peak current (19 MA vs. 18 MA) but the dL/dt dip in the current at stagnation (13.1 MA vs. 9.8 MA, est.) is less because the dynamic change in inductance is 1/2 as large. This argument implies that while there is 1/2 the final inductance around a 1-cm long pinch (vs. A 2-cm-long pinch) there is 1.34X more current at stagnation. The

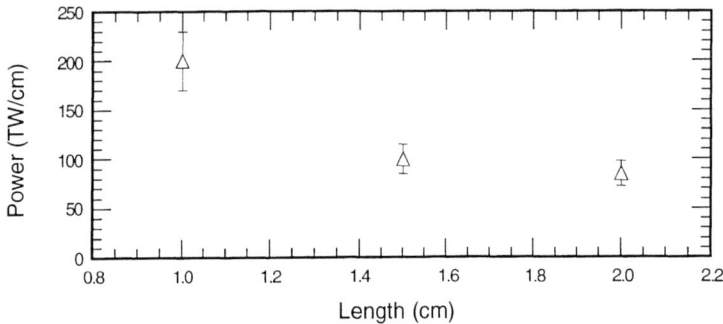

Fig. 13 The measured power per unit length versus the pinch length for a 30-mm diameter arrays with near constant implosion times of 103±2 ns. The number of 10 μm wires were increased (180, 200, 220) as the pinch length was decreased.

amount of energy in the magnetic field at stagnation is nearly the same! Energy from compressional pdV work, ohmic heating, and turbulence enters the pinch and is radiated as x rays. These arguments are consistent with the facts that the total radiated energy is always much larger than the radial kinetic energy and that the 2-D calculations discussed in the next section show that the **J x B** forces acting on the on-axis plasma are important in heating the plasma.

Another way to reduce the inductance of the load is to reduce the vacuum gap between the anode and the cathode in the radial portion of the MITL nearest the load. The minimum gap is set by plasma closure and realistic mechanical assembly tolerances. However, it is not clear that plasma can easily close the gap on PBFA Z because of the huge magnetic fields. At peak current for a typical 30-mm diameter load the magnetic field on the anode near the pinch is > 250 T! We conducted a series of shots decreasing the gap from the standard 5 mm to 3 mm, then to 2.5 mm, and finally to 2 mm with the standard 180, 10-μm W wires. Figure 14 shows the power generated and the implosion time as a function of radial A-K gap dimension. The decreasing implosion time is a strong indication that the current was increasing as the gap decreased i.e. a 10% increase in load current between the 5- and 2-mm AK gaps. This is consistent with the inductance decreasing and also implies that there is no gap closure. The fact that the radiated power stays constant also implies that the energy delivery is good even with the small gaps. This lack of closure is remarkable because the electric field at the base of the z-pinch load reaches several MV/cm. For these shots the z-pinch mass was not optimized to take advantage of the slightly reduced inductance.

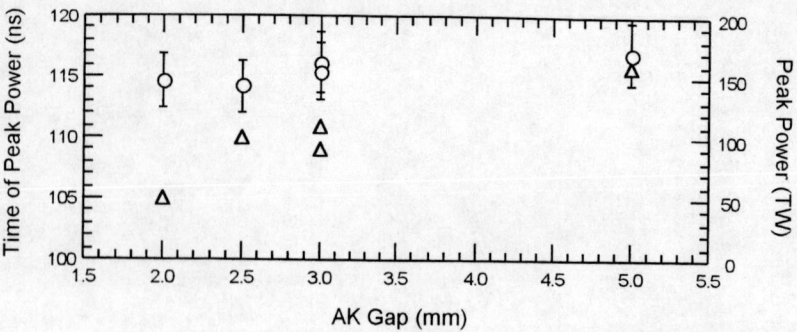

Fig. 14 The implosion time (open triangles) and the peak power (open circles) are plotted as a function of the anode/cathode gap at the base of the array.

THEORY AND MODELING

This section describes the modeling done for experiments on PBFA Z. First, we describe circuit modeling conducted with the Sandia Screamer code.[14] In Screamer we model the electrical circuit of PBFA Z from the Marx generators to the load. Screamer is much preferred over a simple 0-D circuit code because it does a much better job treating the details of the setup of magnetic insulation and self-consistent losses in the vacuum convolute. Screamer will cause the power flow to react to the dynamic impedance of a wide variety of z-pinch loads. For example, a highly inductive load will see less current because of increased power flow losses. Thus, a single pulsed power setup can model a wide range of loads. We have found Screamer to be remarkably accurate. Figure 15 compares the current measured at the insulator stack and in the MITLs with Screamer calculations for the currents at the same locations. The adjustable parameter in Screamer is the effective loss in the vacuum convolute. We determine the value of the loss impedance (a so-called Z-flow impedance) by forcing Screamer to implode a well known mass in the measured time. Once this is done Screamer can be used to model a wide range of load parameters. These circuit model calculations have facilitated the ability to maintain accurate timing of our fast x-ray diagnostics.

Calculations to model the performance of the z pinch itself are much more difficult, however, with PBFA Z we have moved into higher density, higher opacity regimes so making the use of MHD and simple radiation models more appropriate. We are using two, 2-D Eulerian radiation-magneto-hydrodynamics codes (RMHD) with a simple 3-T plasma treatment, Sesame EOS and opacities, and radiation diffusion to model the stability and radiative performance of tungsten; these are the MACH2 code and a LANL Eulerian code.[15] The use of such a simple radiation model is acceptable because of the optically-thick, high-Z nature of these tungsten pinches. We find that only two adjustable parameters are necessary to get good agreement with data: the zoning used to determine the Eulerian mesh (actually

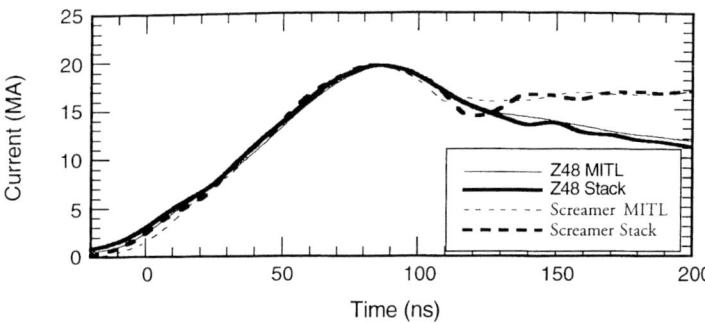

Fig. 15 The circuit model predicted (dashed) and the measured (solid) stack (heavy) and MITL (light) currents on PBFA Z.

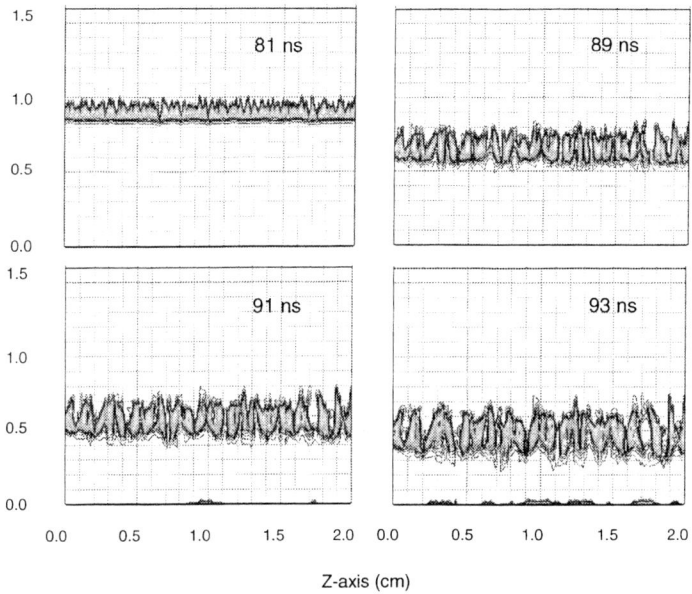

Fig. 16 The RMHD calculation shows the development of the Rayleigh-Taylor instabilities as the tungsten wire array implodes. Each plot represents the implosion at a different time as shown for a 30-mm diameter array implosion (Shot 48) and they correspond to the same times as used in Figure 17(b).

determined by the computing time limits) and the initial random density perturbation.[15] Figure 16 shows plasma isodensity contours for a 30-mm-diameter wire-array z-pinch implosion when a 2.5% initial density perturbation is used. The growth of short wavelength (< 1 mm) RT instabilities are evident as the shell implodes before the shell reaches the axis, the bubbles begin to break and send some mass to the axis. From Figure 17, these features can be seen to correlate with the onset of x-ray emissions. The peak x-ray emission correlates with the arrival on-axis of all the shell mass. Presently, we are using the MACH2 code to develop RT mitigation techniques to improve the radiation outputs of these pinches.

The 2-D, 3-T calculations have given us good agreement with the measured z-pinch performance. Figure 17(a) shows a calculation in which a 4-cm-diameter, 2-cm-long tungsten wire array is compared against PBFA-Z data. In this case, only the perturbation level is adjusted to get agreement with the observed pulse width. The powers and pulse shapes agree well but the small time shift maybe due to inaccuracies in measuring the load current. Figure 17(b) shows a similar comparison for a 30-mm diameter wire array (as shown in Fig. 16). The agreement was very good and assumed the same random density perturbation as the 40-mm case in Figure 17(a).

Since the calculations appear to model the dynamics and radiation energetics of the pinch it is instructive to look at the overall energetics based on these calculations. In Reference 15, calculations show clearly that **J x B** work is a key component in heating the pinch plasma. They show that the radial kinetic energy in the ions accounts for only half of the final radiated energy. The remaining energy is **J x B** work that appears as pdV work and continued radial and axial acceleration. There is also a small amount of ohmic heating.

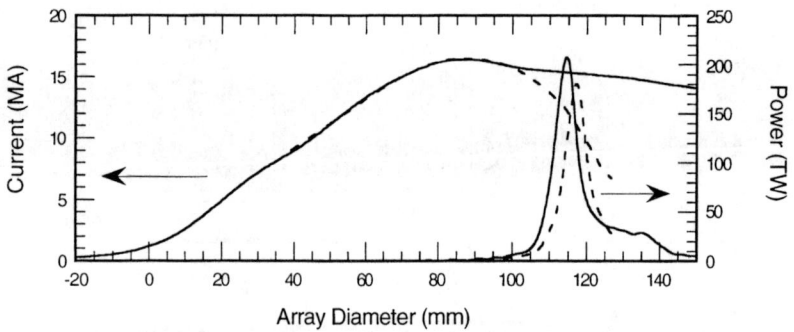

Fig. 17(a) The calculated (dashed lines) current and x-ray power are compared with the data (solid lines) from Shot 26. The perturbation level was lowered to 2.5% to match the data.

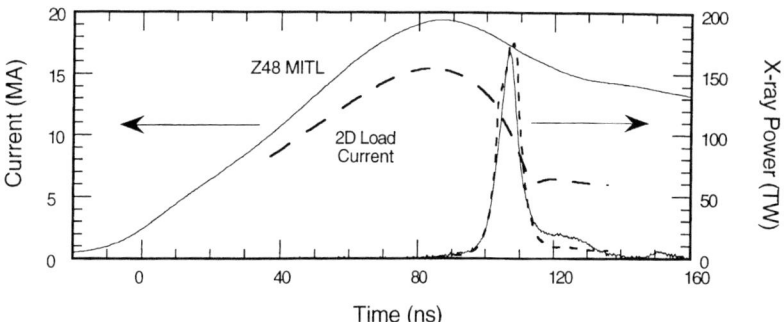

Fig. 17(b) The calculated (dashed lines) current and x-ray power are compared with the data (solid lines) from Shot 48; a 30-mm-diameter, 180-wire array. The perturbation level was 2.5% to match the data. To match the implosion time, the load current (not measured on the shot) was calculated to be 15.5 MA.

SUMMARY AND CONCLUSIONS

PBFA Z is now fully operational at the 20-MA level for z-pinch experiments. We have demonstrated greater than 200 TW of x-ray power and 1.9 MJ of x-ray energy from tungsten wire-array z pinches. We delivered up to 19 MA to a 40-mm-diameter z-pinch load with ~ 5% current losses in the MITLs. Data show significant levels of Rayleigh-Taylor instabilities in the implosions. We have been able to model the dynamics and the x-ray performance of tungsten wire-array z pinches to $\pm 10\%$ accuracies with 2-D RMHD codes.

PBFA Z will be available for basic science applications as well as for use within the DOE and DoD communities starting 1 October 1997. We anticipate continuous improvements in power flow, load performance, and diagnostics over the next year of operation. Up to date information on the progress on PBFA Z can be obtained from our web site at http://plasma.opp.sandia.gov.

ACKNOWLEDGEMENTS

We wish to thank the PBFA-Z operations crew for their help with all aspects of the experiments. The invaluable pulsed power expertise of Pulse Sciences, Inc. is acknowledged.

REFERENCES

[1] R. B. Spielman, *et al.*, in Dense Z Pinches, edited by N. R. Pereira, J. Davis, and N. Rostoker, AIP Conf. Proc. #195 (AIP, New York, 1989), p. 3.

[2] R. B. Spielman, *et al.*, in *Dense Z Pinches*, edited by M. Haines and A. Knight, AIP Conf. Proc. #299 (AIP, New York, 1994), p. 404.

[3] T. W. L. Sanford, *et al.*, Phys. Rev. Lett. **77**, 5063 (1996).

[4] T. W. L. Sanford, *et al.*, Rev. Sci. Instrum. **68**, 852 (1997).

[5] C. Deeney, *et al.*, Rev. Sci. Instrum. **68**, 653 (1997).

[6] C. Deeney, *et al.*, submitted to Phys. Rev. E (1996).

[7] R. B. Spielman, *et al.*, *Proc. of the Ninth IEEE Pulsed Power Conf.*, Albuquerque, NM, 1995, p. 396.

[8] R. B. Spielman, *et al.*, Proc. of the *11th Int. Conf. On Particle Beams*, edited by K. Jungwirth and J. Ullschmied, Prague, Czech Republic, 1996, p. 150.

[9] B. N. Turman, *et al.*, *Proc. of the Fifth IEEE Pulsed Power Conf.*, Arlington, VA 1985, p. 155.

[10] D. D. Bloomquist, *et al.*, *Proc. of the Sixth IEEE Pulsed Power Conf.*, Arlington, VA edited by P. J. Turchi and B. H. Bernstein (IEEE, New York, 1987), p. 310.

[11] G. A. Chandler, *et al.*, Rev. Sci. Instrum. 63, 4828 (1992).

[12] L. P. Mix, *et al.*, in Low Energy X-Ray Diagnostics, edited by B. L. Henke, AIP Conf. Proc. #75 (AIP New York, 1981), p. 25.

[13] R. B. Spielman, *et al.*, Rev. Sci. Instrum. **68**, 782 (1997).

[14] M. L. Kiefer, K. L. Fugelso, K. W. Struve, M. M. Widner, "SCREAMER, A Pulsed Power Design Tool", 25 August 1995 (Sandia Internal Document).

[15] D. L. Peterson, *et al.*, Phys. Plasmas **3**, 368 (1996) and in this proceeding.

Recent ACE 4 Z-Pinch Experiments: Long Implosion Time Argon Loads, Uniform Fill Versus Annular Shell Distributions and the Rayleigh-Taylor Instability Problem

P. Coleman, J. Rauch, W. Rix, J. Thompson and R. Wilson

Maxwell Technologies, San Diego, California 92123

Abstract. Hammer (1996) and Velikovich (1996) have discussed ways to mitigate the growth of the magneto-Rayleigh-Taylor (MRT) instability in z-pinch (PRS) implosions. They predict that initial mass distributions more complex than a simple annular shell will reduce instability development. Sanford (1996) reported experimental data showing a benefit for a uniform mass distribution compared to a shell; those tests used "conventional" load radii of 2.25 and 1.25 cm respectively, and implosion times under 100 ns.

However, the instability problem is expected to grow exponentially as the implosion time, or alternatively the initial radius, increases. Thus we made a comparison of a uniform fill load with a shell but at larger radii, 3.6 and 2.5 cm respectively, and at implosion times well above 100 ns. We see nearly a factor of 10X improvement in peak K-shell power and 2X increase in K-shell yield for the uniform mass load. Hence it appears that suitable tailoring of the imploding mass distribution can significantly limit the instability growth.

For our tests, the ACE 4 inductive energy store (4 MJ) machine was used with a plasma opening switch (POS) to drive the PRS loads (Figure 1). The vacuum inductance between the POS and PRS was only a few nanohenries compared to a system inductance of over 100 nH. For unrelated reasons, on these experiments ACE 4 was operated at reduced current, a peak of 2 ¾ MA into the load. This estimated load current is derived from the observed voltages and currents upstream of the POS and from load parameters: mass, size and implosion times. A separate analysis of the data (Coleman 1997) indicates that the drive current to the pinch was effectively a constant current unlike most PRS machines. Because of the more nearly constant acceleration of the load with constant current drive, instability growth might be expected to be even worse in this case. This may have been a more severe test of instability mitigation techniques.

A standard annular gas puff nozzle produced the 2.5 cm radius shell load; pinch length was 3.3 cm. We generated the 3.6 cm radius uniform fill by puffing the argon into a small volume inside the cathode. The flow into the pinch region was defined by an aperture (Figure 2); pinch length was 3.6 cm. We designed both

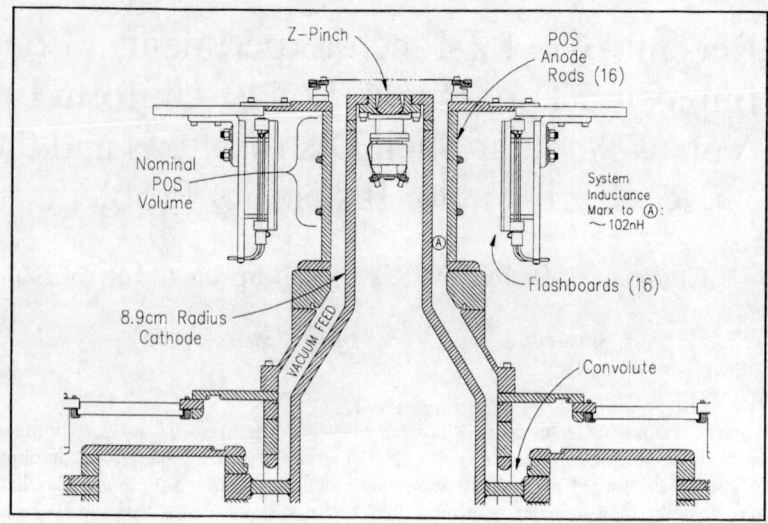

FIGURE 1. Section view of ACE 4 with POS driver and PRS load region.

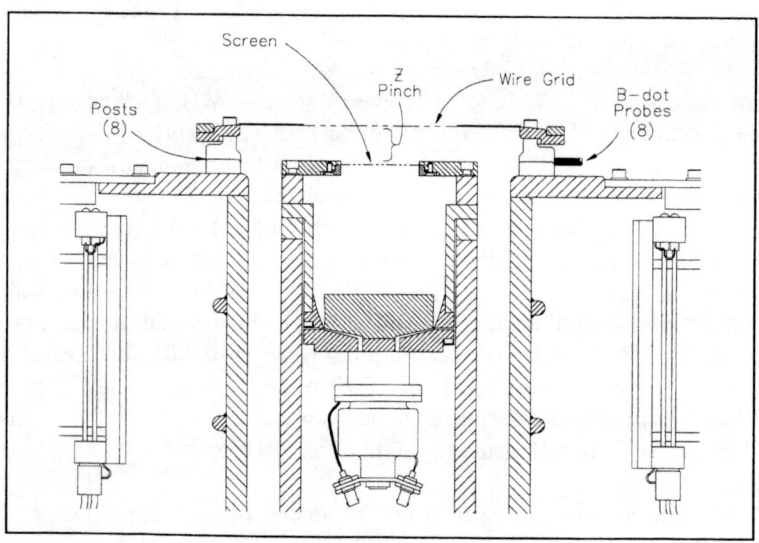

FIGURE 2. 7 cm Uniform fill: Load region.

mass distributions using the gas dynamics code RZ-DELTA. Weber (1997) verified the gas flows using a precision interferometer.

Figure 3 shows observed electrical parameters - current into the POS and voltage across the load - for a typical case of the uniform fill load. (Low levels of plasma that escape from the POS compromise current probes near the load.) Qualitatively similar waveforms are seen for the shell load. But the larger initial inductance of the shell load is reflected in a higher voltage (~ 650 kV) at the time (~800 ns) when current transfers from the POS to the PRS. On the other hand, the lower quality of the pinch for the shell load is evident in its lower voltage (~ 630 kV) at pinch time (~930 ns). Figure 4 compares optical framing images for the two load types. Instability is not strongly evident for the shell load except perhaps for the final image before the pinch.

Figure 5 shows K-shell yield per unit length versus implosion time for the two load types. The model curve is for a simple snowplow prediction of the solid fill. The predicted curve for the shell has a similar shape but a peak value almost twice as high, 2.3 kJ/cm. The observed optimum yield for the uniform fill was comparable to the simple uniform fill model. The observed yield for the shell was 2X lower than for the solid fill and 4 times lower than predicted. Figure 6 compares implosion time versus initial mass load. Simple kinematic models that assume no mass loss during the implosion would predict that the two loads should have had very similar implosion times for a given initial mass. The data clearly show a much early implosion for the shell. To first order, the average implosion velocity for the two loads is the same, a surprising result that is however consistent with recently reported 2-D models (Douglas 1997).

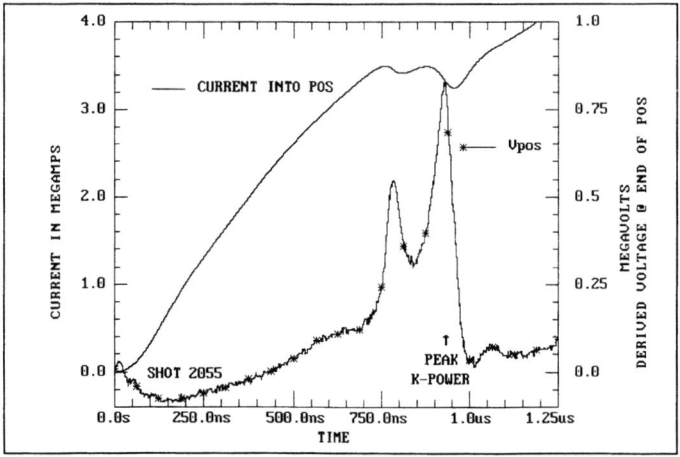

FIGURE 3. Sample electrical diagnostics: Uniform fill load.

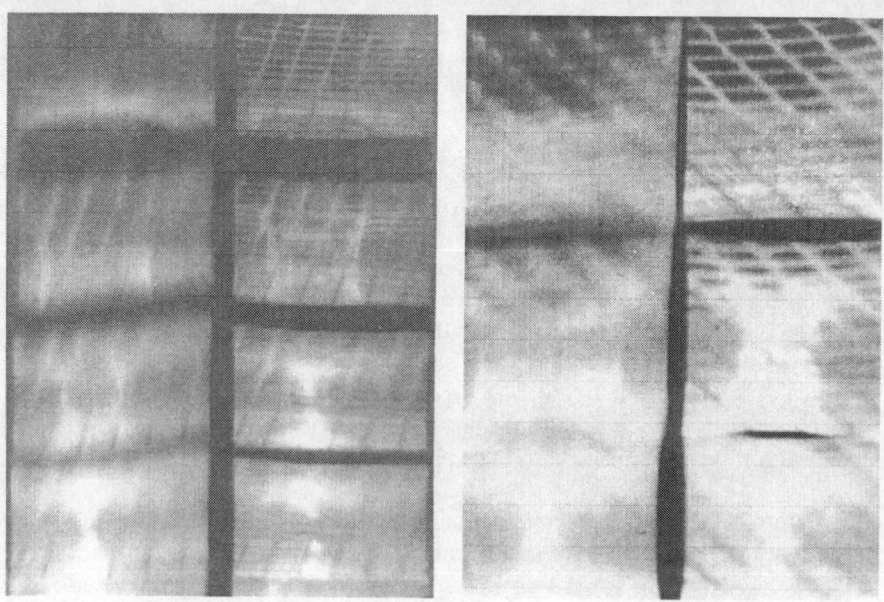

FIGURE 4. Optical framing images, 21 ns between frames. The left image set is for shell load, Shot 2180. The right image set is for the uniform fill load, Shot 2155.

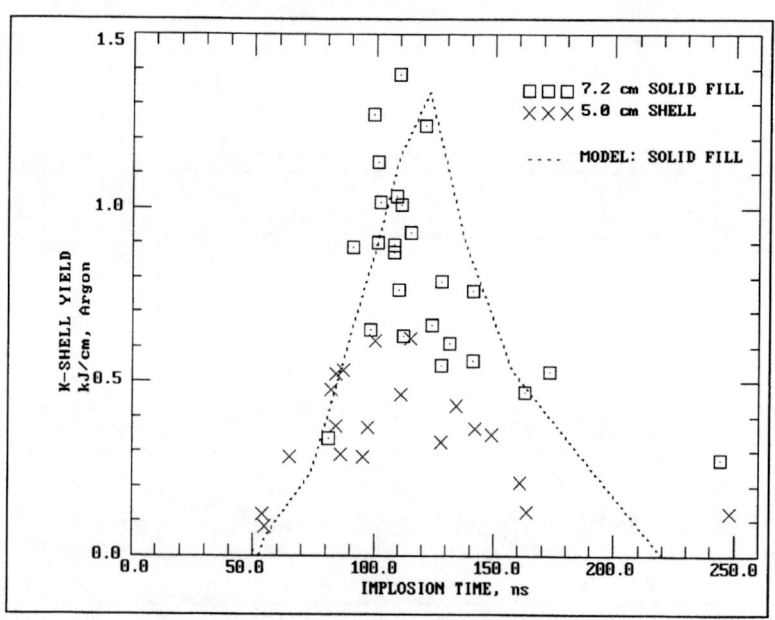

FIGURE 5. K-shell yield vs. Implosion time for solid fill and shell loads. The model is for a simple snowplow with yield derived from the empirical model of Thornhill et. al. (1996).

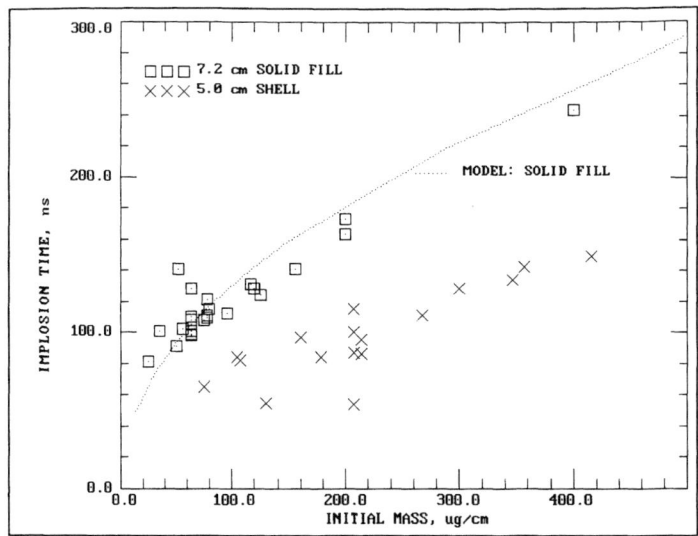

FIGURE 6. Implosion time vs. Initial mass for solid fill and shell loads. The model is a simple kinematic snowplow for the solid fill load. The corresponding slug model for the shell load is only slightly different from the model curve shown.

In summary, by all measures the large radius (3.6 cm) uniform fill load performs better than the 2.5 cm radius annular shell load. K-shell power (>500 GW) is up to 10 times higher, pulse widths (<10 ns) are 2 times smaller, K-shell yield (>1kJ/cm) is 2 times higher, and the K-shell images are tighter and straighter. The performance of the uniform fill is close to that predicted by 0-D, 1-D models that have been benchmarked with smaller radius loads. The implication is that the MRT instability growth is limited in the uniform fill load so that most of the initial mass participates in the implosion.

REFERENCES

1. Coleman, P.L., J.Rauch, W.Rix and J.Thompson, Paper 6B04, *IEEE International Conference on Plasma Science Abstracts*, May 1997, p. 272.
2. Douglas, M., C.Deeney and N.Roderick, Paper 2P59, *IEEE International Conference on Plasma Science Abstracts*, May 1997, p. 183.
3. Hammer, J.H., et.al. Phys. Plasmas **3**, pp. 2063–2069, (May 1996).
4. Sanford, T.W.L., et.al. Paper 4HP14, *IEEE International Conference on Plasma Science Abstracts*, June 1996, p. 251.
5. Thornhill, J.W., K.G.Whitney, J.Davis and J.P.Apruzese, *J.Ap.Phys.* **80**, 710 (1996).
6. Velikovich, A.L., F.L.Cochran and J.Davis, *Phys.Rev.Let.* **77**, 853–856, (July 25, 1996).
7. Weber, B.V., et.al., *Fourth International Conference on Dense Z-Pinches* (this volume), 1997.

Z-PINCH IMPLOSION FOR ICF PHYSICS STUDY ON ANGARA-5-1

A.V. Branitsky, M. V. Fedulov, E.V. Grabovsky, S. L. Nedoseev,
G. M. Olejinik, V. P. Smirnov, S.V. Zakharov

Troitsk Institute for Innovations and Thermonuclear Investigations, TRINITI,
142092, Troitsk, Moscow region, Russia

ABSTRACT

Recent development of soft X-ray sources based on super-fast Z-pinch implosion has demonstrated the great promise of pulsed power for ICF physics study. The main direction of the "Angara-5-1" program is oriented toward using the double liner scheme to confine radiation inside a cavity in order to enhance its intensity significantly. Collision of the external liner shell onto the inner leads to radiation penetration through the inner liner and a decrease in the radiation pulse duration to 3 - 5 ns. Testing this scheme on "Angara-5-1" with current 3.5 - 5 MA demonstrates a flux intensity up to 3 TW/cm^2. In spite of the fact that results of the experiment and a 1D-simulation are close, there are many issues with external liner stability during current sheath formation and implosion.

Recent experimental efforts on Angara-5-1 focused on the "cold start" problem and on the Rayleigh-Taylor instability for thick (gas-puff) and thin (doped foam) shells. Preionization makes the liner's plasma more homogeneous. The method also works in a plasma focus, according to the first results of a joint Polish-Russian experiment. A high current convolute increases ANGARA-5's load current from about 4 MA to 5.8 MA, which moves the radiation temperature toward the region of interest.. We also outline a new approach to a generator intended to produce tens of MA, ANGARA-5-2.

DOUBLE LINER SCHEME TO CONFINE RADIATION INSIDE A CAVITY FOR ICF

ANGARA-5 is an 8-module pulsed power facility with maximum power 9 TW, pulse rise time 90 ns, and peak load current 4 MA. Since 1988, its principal research is developing power sharpening techniques for soft x-ray radiation generated by conversion of kinetic energy of an imploding cylindrical shell (liner), and to the application of intense soft x-ray pulse for target studies.

We are discussing here cylindrical shells, even though the most effective way for energy concentration could be a magnetic implosion scheme in three dimensions, a spherical shell with conical electrodes. This scheme's problems include production of light mass distributed on shaped shells and their enhanced instability during 3-D compression under azimuthal (2-D) magnetic field pressure.

The Double Liner (DL) concept was proposed in Ref. 1. It permits to convert liner kinetic energy to radiation with a pulse that is significantly shorter than the generator pulse. Theoretical and experimental studies of the Double Liner with ANGARA-5 were carried out and first results were reported in Ref. 2, 3.

DOUBLE LINER CONCEPT

The Double Liner is a cascade system with two coaxial liners doped by materials with high atomic number Z>> 1 (Ref.1). After acceleration by magnetic pressure the external liner collides with the internal one. Thermal X-ray radiation generated by a high-velocity shock wave ($V = 4$-$5\cdot 10^7$ cm/s) penetrates into the internal liner cavity and irradiates a hohlraum-like target. The external liner gives energy confinement and at the same time hinders radiation escape to the outside, thereby increasing the radiation intensity in the hot cavity. The inner liner serves both to stagnate the imploding outer plasma, to convert plasma kinetic energy into radiation, and hydrodynamically to isolate the target from the imploding plasma before its ignition.

Converting liner kinetic energy into radiation and realizing energy confinement is best done with a liner made from high atomic number materials, Z>> 1. As shown in Ref. 4, the thermal pressure P of the liner's multicharged plasma is much less than the magnetic pressure, P<< $B^2/8\pi$, because of high radiation losses. Therefore, during liner implosion its thickness is compressed into a skin layer with scale length $\delta = (c^2 t/2\pi\sigma)^{1/2}$. The liner kinetic energy is determined mainly by the current amplitude (I) and the convergence ratio (R/r, where R - initial radius and r - final one). It may be estimated as:

$$mV^2/2 \propto \alpha I^2 h \cdot \ln(R/r) \qquad (1)$$

where m, V - outer liner mass and velocity, h - its length, α–coefficient depending on particular current pulse shape.

The external liner acceleration is accompanied by radiation of energy dissipated in the liner plasma. As a result, the internal liner sublimates from irradiation and disintegrates with the sound velocity. During collision, the external liner's deceleration and conversion of its kinetic energy into radiation occurs with characteristic time

$$\tau = \delta / V.$$
(2)
Efficient conversion of liner kinetic energy into radiation happens when the collision excites a strongly radiative shock wave in the internal liner plasma, while the external liner is decelerated without shock. According to a MHD model of strongly radiative plasma (Ref. 1, 4) such a regime takes place when the Alfven velocity in the external liner plasma is comparable to the liner velocity.

A strongly radiative shock wave is generated in the plasma of the second liner. The kinetic energy conversion into radiation in the shock wave propagating through a multiply- charged plasma is a result of a chain on sequential events. Due to the ion viscosity mechanism, the kinetic energy of a directed motion transfers into a thermal energy of ions. Plasma electrons heated in ion-electron collisions lose their energy mainly through excitation of multiply-charged ions.

A high efficiency of the kinetic energy conversion to radiation and target irradiation may be achieved with an appropriate choice of materials and liner parameters (Ref. 1, 3). If the effective conversion conditions are met, the following expression can be obtained for the thermal x-ray radiation intensity on the target (Ref. 1):

$$W = 1.33 \cdot (1+3 \, \mu\kappa/4) \, \rho D^3 \qquad (3)$$

where κ is the mean mass radiation absorption coefficient in the external liner plasma. The coefficient ($\mu\kappa$) in (3) describes partial energy confinement in the cavity by the external cylindrical shell. Equation (3) has important consequences. Due to the increase of the external liner's optimum mass μ with peak current I, $\mu \propto I^2$ and the kinetic energy flux's increase as I^2, the thermal radiation intensity at $\mu\kappa \gg 1$ increases with peak current as

$$W \propto I^4. \qquad (4)$$

This equation implies that the target irradiation efficiency increases rapidly with an increase in peak current. The temperature on the inside of the hohlraum must be higher than the temperature on the outer (visible) surface. Besides enough current and liner mass magnification, to get into the regime (4) it is necessary to select the outer liner's material properly for energy confinement in the cavity. Spectral calculation in temperature range of 300 eV have their best results with neodymium, xenon and similar elements.

The liner kinetic energy increases with increasing convergence ratio R/r. Unfortunately, MHD instabilities restrict the convergence ratio to R/r < 10: this limits the kinetic energy flux. The external liner instability during acceleration affects the radiation pulse duration through the width of the first cascade liner, and it also affects energy confinement in the cavity due to discontinuities in the external liner's optical thickness.

Among various types of plasma instabilities for current driven liners with multipy charged plasmas are:
- an ionization instability in the early discharge stage (Ref. 7)
- a thermal instability at the initial stage of plasma current heating (Ref. 5), continuing into
- a thermal-radiating instability during radiation-heating equilibrium (Ref. 6)
- a non-isothermal instability after the first shock wave (Ref. 6, 8)
- Rayleigh-Taylor like MHD instabilities as modified by multicharged plasma
- anomalous resistance of liner plasma due to different types of micro instabilities

The influence of instabilities and anomalous resistance decreases with increasing current (at fixed convergence ratio) because the plasma becomes unmagnetized and the electron drift velocity decreases for higher liner densities.

Numerical simulations predict that for a generator with current amplitude I = 15 MA and voltage pulse rise time 100 ns a liner with mass of 2.8 mg and radius R = 16.5 mm is accelerated up to velocity $V = 5 \cdot 10^7$ cm/s. As a result of its collision with an internal liner with mass 4.5 mg at radius r = 2 mm, the cavity of inner liner is filled by thermal radiation with an intensity on target W = 500 TW/cm^2, and FWHM pulse duration τ = 3.7 ns. Calculations show that this is sufficient for target ignition.

DOUBLE LINER COLLISION EXPERIMENT

In experiments on ANGARA-5, the outer liner hollow gas puff with Mach number 6 was produced by a supersonic ring nozzle with diameter of 34 mm and ring gap of 2 mm. Xenon or neon were used as working gases. A jet specific mass was 0.1-0.2 mg/cm, its height between cathode and anode grid - 1 cm. The internal liner with 4 mm diameter, was prepared from a foam with average density of 10 mg/cm^3, doped by Mo metallic powder with a grain diameter less than 1 μm.

Optical and X-ray cameras show a typical liner velocity of $4\text{-}6 \cdot 10^7$ cm/s at the moment of collision. At this time the radiation intensity from the outer surface increases up to 1.5 TW/cm^2 with rise time of 3 ns. During the next 5 ns the outer surface temperature drops rapidly, suggestive of radiative (non-hydrodynamical) cooling of plasma. The radiation yield during the liners' collision is about 7-10 kJ. The cavity internal surface radiation intensity reaches 3 TW/cm^2 with a rise time of 3-5 ns. The radiation intensity in the internal cavity remains on the same level or drops slightly while it decreases outside.

This intensity difference between inside and outside demonstrates the radiation screening effect by the outer liner. The radiation's spectral characteristics at collision differ from a Plankian at the measured intensity level. The color temperature of the outside radiation is close the 100-300 eV reported earlier (Ref. 2, 3). Comparisons of simulations with experimental results permit to explain this difference by the liner collision's local character due to initial axial and azimuthal non-uniformity in the gas puff and instabilities. In fact, the radiation intensity rise time measured locally is 2 ns. Laser shadow and streak camera pictures show the zipper effect and the theoretically expected instabilities.

After collision the liners implode further together due to inertia and magnetic field pressure. A z-pinch forms from the stagnated plasma and a second high power radiation pulse occurs with rise time 5-10 ns, total radiation yield about 50 kJ, and power 3 TW.

MORE EFFICIENT CONVOLUTE INTEGRATED WITH LOAD

The most recent experiments on Angara-5-1 have been directed toward achieving a more efficient convolute integrated with load development. Now a disk shaped output vacuum feed connects 8 conical MITLs to a small final collector close to the load. As a result, the convolute inductance decreases and the total size of the magnetic nulls decreases tenfold. The load current increases from 4.5 MA to 5.5 MA. Due to the closer coupling between MITLs and the load the stability of the current to jitter between the Angara units increases significantly.

The new design of a load for the Double Liner scheme is based on a self sustained foam liner as the outer shell. Advantages of such a design include a more controllable density distribution and the opportunity to suppress zippering and Hall instability. An outer liner with specific mass of 200 µg/cm and 22 mm diameter was produced from 2-5 mg/cm^3 doped foam. Experiments without foam preionization have shown a pronounced filamentation at the initial stage of current sheath formation. To avoid it, a tenuous 10 µg/cm Xe gas corona with outer diameter of 50 mm has been produced by a supersonic nozzle. Due to ionization of Xe gas and plasma implosion onto a foam liner, the uniformity of produced current sheath improved significantly.

However, fine-scale laser shadow and streak pictures revealed the existence of numerous small size filaments in the foam loads. From a re-analysis of our previous experimental data on gas-puff liners we conclude that also the gas loads could have been "fractured" before implosion due to the effect of the "cold start." This experimental basis leads to the hypothesis that the nonlinear phase of the current filamentation is the reason of early liner fracturing, which in turn is the principal cause of subsequent liner noncompact implosion: instabilities of magnetically accelerated plasmas could be a contributing factor.

STUDY OF "COLD START" PROBLEM

Almost all Z-pinch or fast liner implosions on (multi) terawatt pulsed power generators begin with a plasma in the anode-cathode gap that is created in a "cold start", by electric breakdown of a gas, solid or foam. Current filamentation due to thermal instabilities of plasma conductivity is a specific feature of this breakdown, being independent on initial homogeneity of plasma producing matter. Because of breakdown's random nature, filaments are introduced into the liner bulk. The azimuthal magnetic field structure is irregular, which can force azimuthal inhomogeneities of even an initially homogeneous liner plasma precursor. This current contraction is provided by plasma electric conductivity: effects of magnetic pressure are initially small. Eventually magnetic forces become significant with increasing current. So, the initial current filament could evolve under effect of two alternative factors: expansion due to Joule heating, and compression due to magnetic pressure. Especially for powerful drivers the current rate of rise may be too high to allow the filaments to expand, to flow together, and to produce an azimuthally homogeneous liner plasma; the magnetic field around the separate current filament could confine them. Being responsible for liner azimuthal inhomogeneities, the filaments could enhance axial inhomogeneities of the initial liner plasma, which are especially dangerous as starting perturbations for the Rayleigh-Taylor and MHD instabilities. Moreover, cold start effects can destroy the liner early in the implosion. From an energetic standpoint the magnetic energy of the current in filament is large compared to the homogeneous current. Therefore, the filamentary current's magnetic structure could evolve to a rough axial and azimuthal structure. Consequently, the "cold start" current filamentation can fracture the liner matter before the instabilities of magnetically accelerating plasma appear. It seems that the most effective way to overcome "cold start" effects is to apply a powerful currentless preionization at about the 0.1 TW level. For example, an ion beam could be created for the purpose, at the expense of a rather complicated technique.

FOAM LINER PREIONIZATION WITH PLASMA FOCUS CURRENT SHELL (joint Russian - Polish experiment on PF-1000 plasma focus)

We consider the Plasma Focus current shell as a possible preionizer for the multiterawatt driver. The high current plasma shell, being accelerated during some microseconds in a plasma focus accelerator, could deliver its kinetic, thermal and magnetic energy to a liner positioned in the shell focus region in rather short period, ~ 100 ns, providing appropriate initial conditions for fast implosion. We use a microheterogeneous solid foam as the plasma producing substance, because our foam liner technology allows us to produce liners with different sizes, shaping and radiative dopants. Homogeneity of plasma liner produced was the main problem of interest.

To produce a plasma liner the PF-1000 current shell imploded onto an agar foam liner, positioned on the top of inner electrode of the plasma focus' coaxial accelerator. The foam liners have diameter 20 mm and length 15 mm with 20 µg/mm or diameter 5.4 mm and length 15-20 mm with 25-28 µg/mm. The operating parameters of the PF-1000 were capacitance C = 1 mF, charging voltage V_{max} = 25 kV, hydrogen pressure ~5 Torr, current I_{max}~1 MA, current rise time 4-5 µs. The details of the experiment are published in our joint report. The initial results of our joint experiment are encouraging:

1. The low mass plasma focus current shell interacting with a higher mass foam liner has produced rather homogeneous foam plasma liner. We could not see filaments in the liner plasma, and the foam liner did not expand dramatically during the process.
2. The foam plasma produced has temperature T ≥ ~20 eV and, consequently, liner electric conductivity is sufficient to start an effective implosion driven by a 3-5 MA current with $dI/dt \geq 5.10^{13}$ A/s.

3. Combining a plasma focus with a multiterawatt driver and their synchronized operation are possible in principle. We can't couple PF-1000 and "Angara-5-1" to test preionization effects of ≥1 MA current, available on PF-1000. We believe this current level will be good for drivers that are more powerful than "Angara-5-1." Consequently, testing of this preionization method is planned on "Angara-5-1" at the 100 k level.

A NEW APPROACH TO MULTITERAWATT GENERATORS: THE "ANGARA-5-2" SCHEME

TRINITI has initiated a project to build a new multiterawatt machine to produce 10 - 20 MJ pulses of soft x-rays, the "Angara-5-2" generator. The generator is to have electric pulse power 500 - 1000 TW. TRINITI has a unique complex of 3 pulsed electric generators with inductive store and commutators, originally built to power the magnets of the T-14 tokamak. The complex is situated in special buildings in the vicinity of "Angara-5-1." It is proposed as the primary energy source for "Angara-5-2." New systems as intermediate power amplifiers, reactor chamber and control system are to be created. Preliminary evaluations show that the machine's total price is 5-8 times lower when using the available inductive energy source compared to using a capacitor bank with the same stored energy.

Parameters of proposed X-ray generator are:

X-ray pulse energy - 10 -20 MJ with pulse duration - 10 ns;
Method of X-ray generation - plasma shell current implosion;
Load current amplitude - 50 MA;
Load current pulse duration through the imploding load - 100 ns.

The scheme of the necessary electric power multiplication is rather complicated, and it is impossible to go into the details here. A mechanical energy of 3 GJ is stored in three generators TKD-200. The generators supply the primary windings of 32-sectional inductive store TIN-900 with a current of 150 kA for 6 seconds, delivering 900 MJ. Then the TIN-900 is disconnected from the generators and crowbarred by a mechanical switch. The store secondary winding has 32 sections as well, with each section generating a current of ~1 MA. Using a symmetric scheme of triple current doubling the energy transfers into transforming inductive store IN-2, having 16 groups of commutation, and its primary winding is crowbarred as well. The energy transfer into IN-2 secondary winding, coupled with vacuum inductive store, occurs during ~100 μs. Then, 50 synchronized explosive opening switches provide fast current transfer (2 μs) into the last stage of current sharpening.

Two kinds of the last stage are being analyzed now. The former one is a current sharpening system, based on plasma opening switches to produce 50 MA, 100 ns current pulse in the load. The physics of this POS is absent now. So, the second kind is being analyzed, with water insulating pulse forming lines.

SUMMARY

As a result of investigations on ANGARA-5, the general features of implosion and strike of liners are studied. The effect of energy confinement by the outer liner is observed. High radiation power sharpening inside the cylindrical cavity with pulse rise time duration of 3-5 ns and kinetic energy conversion efficiency of 30-50% was obtained in the Double Liner scheme. These values are comparable with code calculations provided that the initial gas puff density distribution and liner instabilities are taken into account. The development of technologies for foam liner production allows the use of liners with appropriate parameters. Experiments carried out on ANGARA-5 and calculations permit us to consider the Double Liner as a high intensity soft X-ray source for ICF target ignition.

The results discussed above have demonstrated a promising perspective of liner implosion to study indirect drive pellet physics as well as a pellet ignition. The problem of external liner compact implosion become the critical point in our approach toward increasing the main specific parameters of a dynamic hohlraum. Our current problem is to overcome "cold start" effects by liner preionization.

The Angara-5-2 generator is a new project at TRINITI. The generator is to have electric pulse power 500 - 1000 TW and a soft x-ray output of 10 to 20 MJ. The availability of 3 pulsed electric generators built for powering the magnetic system of the T-14 tokamak reduces the machine's cost manyfold.

REFERENCES

1. Zakhorov S.V., Smirnov V. P. et. al., Collision of Current Driven Cylindrical Liners. Kurchatov Institute of Atomic Energy. Moscow. Preprint 4587/6, 1988.
2. Smirnov V. P., Grabovskii E.V., Zaitsev V. I., Zakharov S.V. et. al., Progress in Investigations on a Dense Plasma Compression on ANGARA-5-1. Proc. of BEAMS'90. World Scientific, 1991. V.1, p.61 (I.07)
3. Zakharov S.V., Smirnov V. P., Tsarfin V. Ya. ANGARA-5 High Intensity Soft X Ray Source with Imploding Liner Cascade for Inertial Confinement Fusion.

Proc. of 14th Int. Conf. on Pl.Phys. and Cont. Nuc. Fus. Res. Wurzburg, 1992. IAEA. Vienna, 1993. V.3, p.481 (IAEA-CN-56/G-3-9).
4. Grigoriev S. F., Zakharov S.V. Magnetohydrodynamics of Strongly Radiative Plasma of Liners. Sov. J. Tech. Phys. Lett. 13, 254 (1987).
5. Vekharev V. D., Zakharov S.V., Smirnov V. P. et. al., Influence of Preionization Effect on the Acceleration Dynamics of Radiative Plasma Flux in High Current Discharges. Sov. J. Pl. Phys. 16, 388(1990).
6. Branitskii A.V., Vikharev V. D., Zakharov S.V. et. al., Investigations of the Liner Compression Initial Stage on ANGARA-5-1 Device. Sov. J. Pl.Phys. 17, 311 (1991).
7. Velikhov E. P., Pis`menny V. D., Rakhimov A.T., Dependent Gas Discharge Exiting Continuous CO_2 Lasers. UFN (Russian), 122, 3, 419.
8. Gasilov V.S., Zakharov S.V., Panin V.M., Influence of Azimuthal Instabilities on the Acceleration Dynamics of Radiating Liners. Preprint IAE 5464/6, 1992.

Radius Scaling of X-Radiation from Gas-Puff Implosions on an Inductive Driver*

D. Mosher, S.J. Stephanakis, J.P. Apruzese, D.C. Black[†], J.R. Boller[††],
R.J. Commisso, M.C. Myers, G.G. Peterson[†††], B.V. Weber, F.C. Young

Plasma Physics Division, Naval Research Laboratory, Wash., DC 20375-5346

Based on simple radiation-scaling models, 150- to 200-ns implosions on Decade Quad (1) will require 4- to 6-cm gas-puff diameters to achieve argon K-shell radiation goals. There is concern that implosions with these parameters may suffer reduced K-shell yield due to 2- and 3-D instabilities and asymmetries. Here, the variation of neon K-shell yield with gas-puff radius over 50- to 300-ns implosion times is systematically studied on the Hawk inductive generator in order to determine the efficacy of the radiation-scaling models and the load-parameter regimes where they fail. Hawk is well suited for such a study because, unlike water-line drivers, the load current is nearly independent of implosion time, allowing load-parameter variations to be performed while preserving key 1-D radiation predictors such as kinetic energy and kinetic energy per unit mass. This capability simplifies interpretation of data and comparisons with scaling models.

On Hawk (2), a vacuum inductor is energized by a 600-kA, 1-µs rise-time current coupled to the imploding load through a fast-opening (< 100-ns) plasma opening switch (3). Neon x-rays can be produced on Hawk with about the same efficiency as is projected for argon on Decade Quad. Annular neon gas-puffs of 2.5-cm length and 2-, 3.5-, and 5-cm dia. are formed by W-Ni-Cu alloy nozzles. For each nozzle, the implosion time is controlled by the plenum pressure and the delay between firing Hawk and gas-valve opening. The injected gas distributions and load masses were determined by laser interferometry (4). Diagnostics for neon-implosion experiments include current B-dot loops in the vacuum coax and load regions, a suite of XRDs covering the XUV and keV regimes, a KAP-crystal spectrometer, and time-integrating XUV and keV pinhole cameras. Primary diagnostics for the data reported on here are the load B-dots, the pinhole cameras, and a 1.8-µm-KIMFOL- plus 2.5-µm-Al-filtered, aluminum-cathode XRD designed to provide near-constant response across the neon K-shell band.

Figure 1 demonstrates the ability to study the variation in radiation output with gas-puff diameter and implosion time while keeping key radiation predictors constant. The load-current and K-shell-XRD signatures for 2- and 5-cm-dia. nozzles are shown on 2 shots with close to the same injected mass, as measured by the product *pD* of plenum pressure and nozzle diameter (assumed proportional to mass for constant nozzle throat width), of 40 and 37.5 psia-cm respectively. For

---

*Work supported by the US Defense Special Weapons Agency.
[†]NRC Postdoctoral Fellow   [††]Current address: JAYCOR, Inc., Vienna, VA 22180
[†††]Current address: Alameda Applied Sciences Corp., San Leandro, CA 94577.

CP409, *Dense Z-Pinches*: Fourth International Conference
edited by N. R. Pereira, J. Davis, and P. E. Pulsifer
© 1997 The American Institute of Physics 1-56396-610-7/97/$10.00

comparable masses and peak currents, the 5-cm implosion time t_{imp} = 226 ns (the interval between linear extrapolations of the current and x-ray rises to the axis) is more than twice the 106 ns for the 2-cm case. The imploded mass m and kinetic energy K are calculated from the load current on each shot by iterating the mass in slug-model implosions until the calculated implosion time matches the data. With an assumed 10:1 compression, m values are 15.8 and 15.6 μg/cm, and K values are 853 and 774 J/cm for the 2- and 5-cm cases respectively.

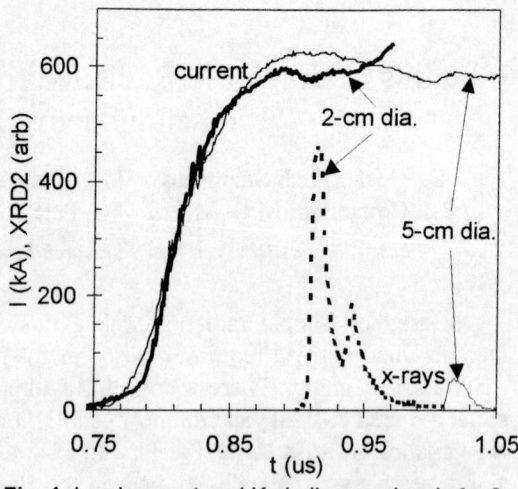

Fig. 1. Load current and K-shell x-ray signals for 2- and 5-cm nozzles with comparable pD.

It will be shown that, unlike the yields determined from Fig. 1, comparable K-shell yields for the two radii are predicted by the simple scaling models for these similar K and m values.

Figure 2a plots m against pD for the three nozzles, and demonstrates the assumed proportionality. Differences between the slug-model masses and the interferometer measurements can be ascribed to initial gas-density distributions (Fig. 6) which are accounted for in the slug and radiation models using radial-density-distribution-weighted initial diameters, 10 to 17% lower than D. Using D to determine mass reduces m by 20 to 30%. As imploded masses are not below the injected masses, it is inferred that the full injected mass is imploded. Figure 2a supports the use of imploded mass calculated from implosion time in the radiation-scaling models. Additional support is provided

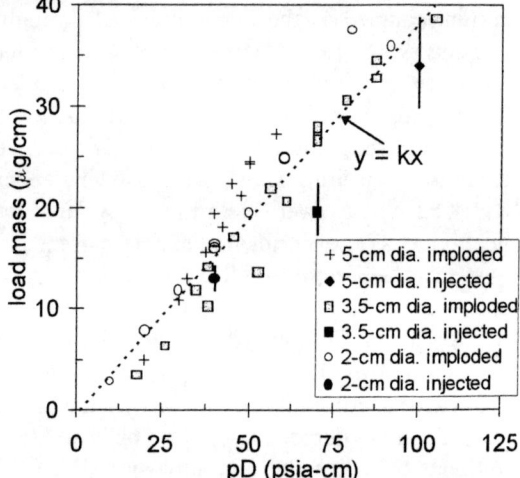

Fig. 2a. Imploded mass vs pD. The solid points represent interferometer measurements.

by Fig. 2b. Implosion dynamics provides the scaling $t_{imp} = Cm^{1/2}D/I$, where I is the peak current and C is a constant that depends on the current variation during implosion. For $m \sim pD$, the scaling $t_{imp} \sim p^{1/2}D^{3/2}/I$ is expected, and is verified by Fig. 2b.

Figure 3a summarizes the K-shell yield vs $1/pD$ for the three nozzles determined from time integration of the K-shell XRD signals. All nozzles have maximum yields in the 30- to 50- psia-cm regime, corresponding to 10 to 20 μg/cm from Fig. 2a. Peak yields for the 2- and 3.5-cm cases were 354 J ± 10% (associated with the spectral uncertainty) while that for 5 cm was only 93 J ± 10%.

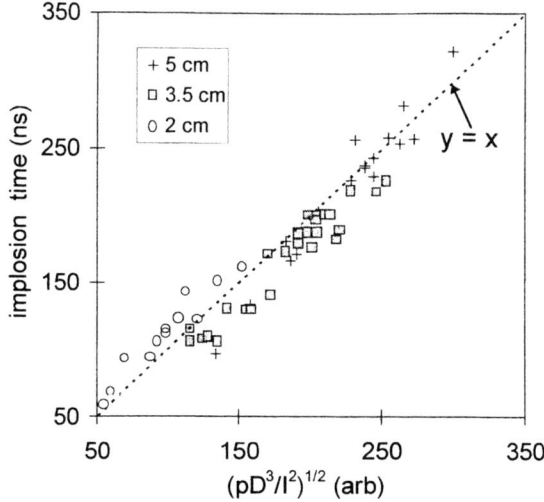

Fig. 2b. Implosion time vs implosion-time predictor for the three nozzle diameters.

Figure 3b plots the ratio of theoretical yield to the kinetic energy K as a function of K/m for a compression ratio of 10:1. For near-constant K and $m \sim pD$ (Fig. 2a), the abscissas of Figs 3a and 3b are proportional. The plotted theoretical yield is an average of two, single-zone radiation-scaling models: a two-level model, and a commonly-used phenomenological (TWG) model employing physics-based extrapolations of selected 1-D rad-hydro code computations (5). The first-principal two-level model (6) determines emission and absorption of principal K-lines using a temperature derived from energy partition of kinetic energy between K-line radiation, internal energy, and losses due to other radiations (typically $K/2$). Radiation escape is computed using a Voigt line-shape profile (7). The time over

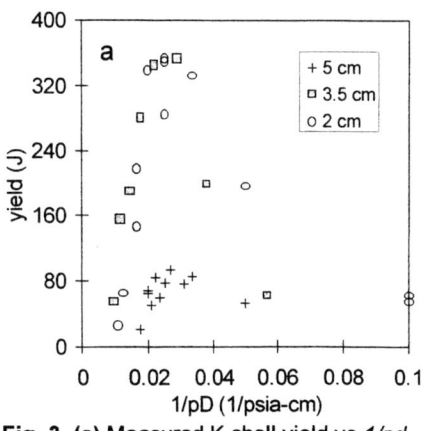
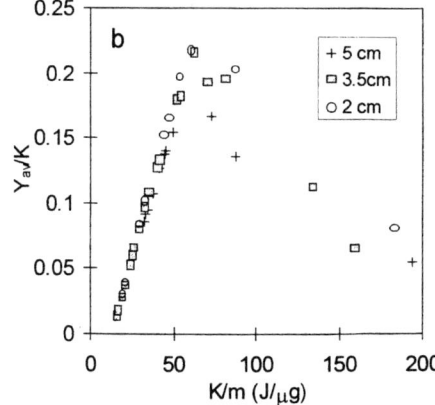

Fig. 3. (a) Measured K-shell yield vs $1/pd$ for the three nozzles. **(b)** Normalized average yield vs K/m from the 2-level and TWG scaling models.

which the plasma radiates is scaled from neon experiments assuming inertial confinement of the stagnated plasma (8). The two radiation-scaling models are usually in rough agreement with regard to optimal K/m and the associated peak radiation. For the present analyses, peak yields from the TWG model are about 30% higher than those of the 2-level model, and fall-off faster with large K/m.

For all radii, Fig. 3b shows maximum yields in the 10- to 20-µg/cm range (using the average K value for each radius), in agreement with Figs. 2a and 3a. For 2- and 3.5-cm, the peak yield from the two scaling models is 440 J, about 25% higher than measured. The predicted yield agrees best with the 2- and 3.5-cm measurements for an 8:1 compression (see below). The predicted 10:1 yield for the 5-cm case is 360 J, about 4-times the measured peak value, suggesting a large-radius transition in pinch quality which cannot be accounted for by the scaling models. Along with the data of Fig. 2, Fig. 3 demonstrates that radius, not implosion time, controls radiation yield, because the ratio of experimental to theoretical yield is a factor-of-three lower for the 5-cm case than for the other two radii in the 130- to 170-ns implosion-time regime where data exist for all radii.

As shown in Fig. 4, the best-fit 8:1 compression inferred from scaling models does not agree with those determined from examination of pinhole images. The data derive from the diameters of XUV pin-hole images (slightly larger than the keV images) measured at the center of the discharge. The highest yields (≥ 340 J) for the 2- and 3.5-cm nozzles are associated with pin-hole compression ratios of 15:1 to 25:1. Higher compressions at greater pD and for the 5-cm nozzle are not supported by the reduced experimental yields. The high 5-cm pin-hole compressions may be due in part to limited film dynamic range, so that larger-volume, weakly-emitting regions are not observed. It is therefore concluded that compression ratios deduced from these pin-hole images are poor predictors of radiation yield when used with the scaling models, as was also observed in Saturn aluminum-wire-array implosions (9).

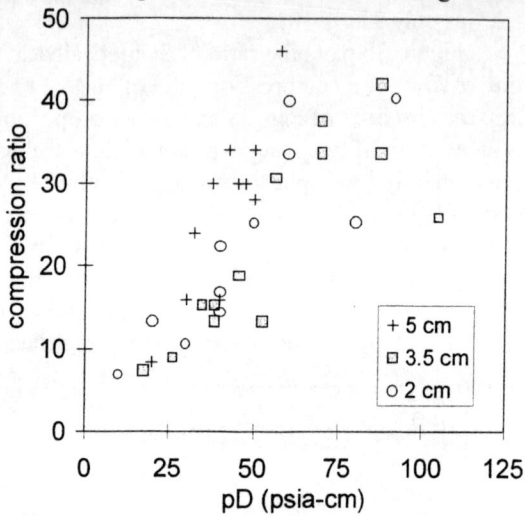

Fig. 4. Compression ratio determined from XUV-image diameters at the discharge center.

The sensitivity of radiation-scaling-model results on compression ratio R_0/r_f is indicated in Fig. 5 for the peak-radiation 3.5-cm shot. The m and K values, computed from slug-model implosions terminated at the minimum radius r_f, is used with that radius to determine the K-shell yield. The computed m, K, and Y_{av}

values are normalized to their values at 10:1 compression and plotted. The sensitivity of yield on compression ratio derives primarily from the variation of $K \sim ln(R_0/r_f)$. The value of m varies weakly with compression. Predicted yields for a 20:1 pinhole compression are about double those for the 8:1 best-fit.

Axial variations in the measured gas distributions and associated variations in pinch diameter suggest a role for the Rayleigh-Taylor instability in the dependence of yield on nozzle diameter. Figure 6 shows the interferometer-measured radial distribution of gas-puff density at the nozzle and cathode ends of the discharge for the 3.5-cm-dia. nozzle at peak-radiation conditions. As emitted from the nozzle, the gas is distributed in a thin annulus that spreads to a filled-in distribution near the cathode. For either distribution, the peak density is about 1.6×10^{17} cm^{-3}. These shapes are similar to those of other nozzles, though the densities vary.

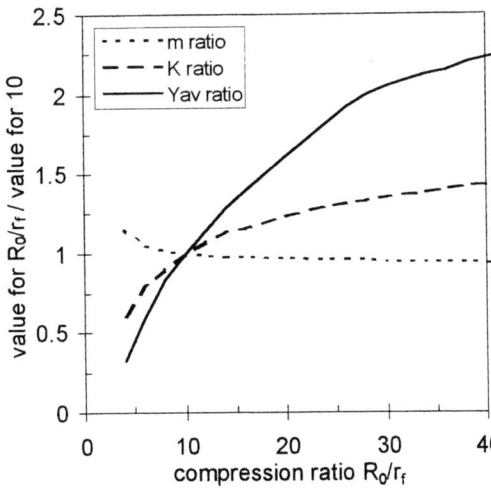

Fig. 5. Dependence of computed m, K, and K-shell yield on assumed compression ratio.

The insert in Fig. 7 shows the keV image produced by the initial gas distribution of Fig. 6 over the 2.5-cm pinch length. The axial asymmetry shown in the image, a larger diameter and less-regular pinch at the nozzle end, is common. Axial asymmetries for the 3.5- and 5-cm data are quantified in Fig. 7, where the ratio of image diameter at the nozzle end to that at the

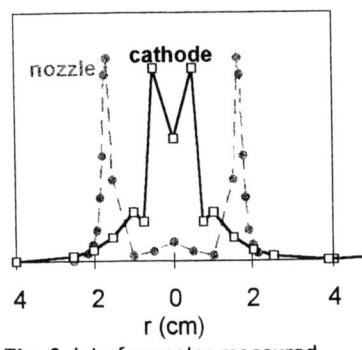

Fig. 6. Interferometer-measured density profiles for the 3.5-cm-dia. nozzle.

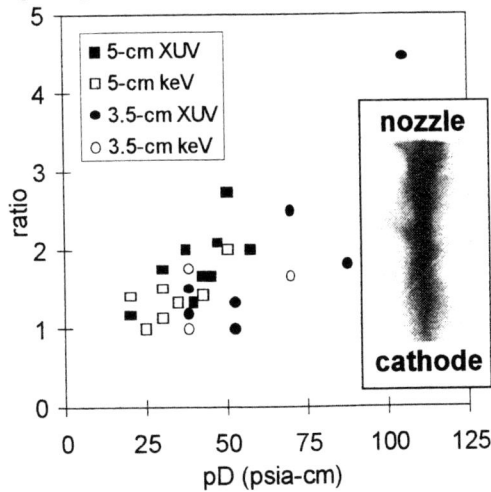

Fig. 7. Ratio of nozzle-end image diameter to that at the cathode end vs pD.

cathode is plotted as a function of pD. If not axially uniform, the pinches appear to be tighter at the end associated with filled-in gas distributions, as evidenced by all of the ratio values in the figure lying above unity. The pD values for peak yield with 2- and 3.5-cm nozzles show ratios near unity. For 5-cm, the ratios are higher, on average, than those for the smaller radii. These observations suggest that the Rayleigh-Taylor instability may reduce pinch quality for 5-cm implosions and that, in support of analysis (10), filled-in distributions may mediate the instability.

In summary, the variation of neon K-shell yield with gas-puff radius over 50- to 300-ns implosion times was systematically studied on the Hawk inductive generator in order to determine the ability of simple scaling models to predict the performance of 200-ns, 4- to 6-cm argon implosions on Decade Quad. Comparison of measurements with analysis was simplified by the ability of Hawk to provide constant kinetic energy over the studied parameter range. For all nozzles, the scaling models correctly predicted the mass range for optimal yield. For 2- and 3.5-cm-dia. gas-puff nozzles, measured K-shell yields agreed with scaling models for about 8:1 compression ratio. For the 5-cm-dia. nozzle, comparable in radius to that planned for Decade Quad, measured yields were 1/3 of that predicted for the 8:1 compression ratio. Compression ratios inferred from pin-hole images were 2- to 3-times larger than required to correctly predict the 2- and 3.5-cm yields. As one-zone radiation-scaling models neglect temperature gradients that lead to K-shell emission from only a fraction of the imploded mass (5), they require smaller-than-observed compression ratios to correctly predict yield using the full mass. Discrepancies are greatest for high pD, where low K/m leads to weak emission from much of the mass. Results demonstrate that gas-puff radius, not implosion time, controls radiation yield. Axial asymmetries in pin-hole images suggest a role for the Rayleigh-Taylor instability that may contribute to reduced 5-cm-nozzle radiation efficiency. These images also indicate mitigation of the insta-bility for filled-in gas distributions. Overall, the results suggest that projected performance based on simple radiation-scaling models may overestimate K-shell yields from large-radius annular loads.

---

(1) Sincerny, P.S., et al., *Proc. 11th International Conf. On High Power Particle Beams*, Prague, 1996, pp.1003-1007.

(2) Peterson, G.G., et al., ibid., pp.749-752.

(3) Commisso, R.J., et al., *Phys. Fluids* **B4**, 2368(1992).

(4) Weber, B.V. and Fulghum, S.F., *Rev. Sci. Instrum.* **68**, 1227(1997).

(5) Thornhill, J.W., et al., *J. Appl. Phys.* **80**, 710(1996).

(6) Seaton, M.J. in *Atomic and Molecular Processes*, New York, Academic Press, 1962, p.374.

(7) Apruzese, J.P., *J. Quant. Spectrosc. Radiat. Transfer* **34**, 447(1985).

(8) Mosher, D., et al., *J. Radiation Effects Research and Engng.*, **12**, 51-61(1994).

(9) Sanford, T.W.L., et al., "Variation of High-Power Aluminum-Wire Array Z-Pinch Dynamics with Wire Number, Array Radius, and Load Mass," these proceedings.

(10) Cochran, F.L., Davis, J., and Velikovich, A.L., *Phys. Plasmas* **2**, 2765(1995).

WIRE ARRAY IMPLOSION EXPERIMENTS ON THE INDUCTIVE STORAGE GENERATORS GIT-4 AND GIT-8

R.B.Baksht, I.M.Datsko, A.A.Kim, V.A.Kokshenev, B.M.Kovalchuk,
S.V.Loginov, A.Yu.Labetsky, A.V.Fedunin, A.G.Russkikh, A.V.Shishlov

High Current Electronics Institute, 4 Academichesky Ave, Tomsk, 634055, Russia

The paper gives the review of the results obtained in the wire array implosion experiments carried out on the inductive storage generators GIT-4 (0.8 MJ, 1.5 MA) and GIT-8 (1.2 MJ, 1.9 MA).

INTRODUCTION

For the last several years the wire array implosion experiments have been carried out in the High Current Electronics Institute on the inductive storage generators with a plasma opening switch of a microsecond conduction time. These experiments were aimed to find out whether an inductive storage generator can be an effective driver for a wire array load, to determine the optimal parameters of a wire array load, to investigate the radiative characteristics of z-pinch plasma, and the possibilities to improve them. The experiments were conducted on the GIT-4[1] and GIT-8[2] facilities that have the following characteristics:

	GIT-4	GIT-8
Stored energy	0.8 MJ	1.7 MJ
Charge voltage	50 kV	50 kV
Upstream inductance	220 nH	160 nH
Downstream inductance	70 nH	28 nH
Upstream current	2 MA	2.4 MA
Load current	1.5 MA	1.9 MA
Load current rise time	120 ns	70 ns

This paper gives the review of the obtained experimental results.

WIRE ARRAY IMPLOSION EXPERIMENTS ON THE GIT-4 FACILITY

The experiments on a wire array implosion were carried out on the GIT-4 generator in the course of the study of the Al-Mg photoresonant X-ray laser scheme in 1991. Those experiments were aimed at obtaining subkilovolt radiation. For a 0.5-cm-diameter, 4-cm-length, and 340-μg/cm-mass tungsten wire array the radiation yield of 30 kJ in the energy range of 0.12÷1.5 keV was measured[3] (an XRD with a gold cathode filtered by 0.1 μm $C_6H_8N_2O_9$ filter). The mass scaling of the radiation power in the energy region of 0.19÷0.29 keV was fulfilled for a 0.6-cm-diameter, 2-cm-length aluminum wire array. The maximum radiation power of 65 GW/cm was observed at the wire array mass of 160÷200 μg/cm[4].

A.V.Shishlov, ash@hded2.hcei.tomsk.su, phone: (382-2) 259-133, fax: (382-2) 258-677

Our further experiments were devoted to the study of the K-shell x-ray emission from aluminum-wire-array implosions. The measurements of K-shell radiation power and yield as a function of wire array initial radius were performed for 68-μg/cm and 102-μg/cm wire arrays. For both wire arrays, the maximum radiation power is in the range from 35 GW/cm to 60 GW/cm and corresponds to the wire array radius of 0.5 cm (Fig.1).

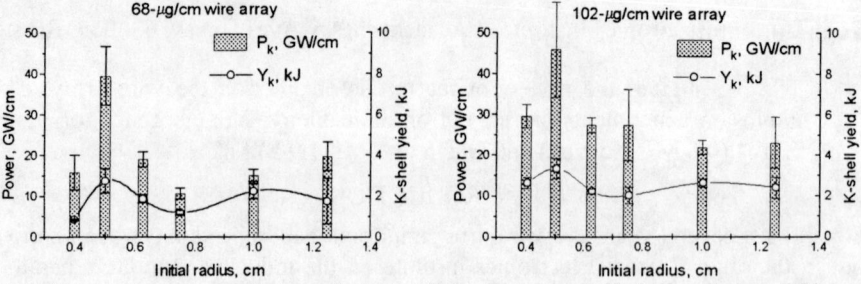

Fig.1 Averaged over several shots K-shell radiation power and K-shell radiation yield as a function of wire array initial radius for 68-μg/cm and 102-μg/cm aluminum wire arrays. The errors bars represent the scatter of measured values in the series of experiments.

In these experiments 68-μg/cm and 102-μg/cm wire arrays consisted of 8 and 12 wires respectively; it means that the distance between the wires was more than 2 mm. The streak camera pictures demonstrate that due to the big distance between the wires they do not form a solid shell after the explosion and move to the axis separately. The implosion of each wire has a streamlike character. Fig.2 shows the implosion of 68-μg/cm, 1.25-cm-radius wire array. The inlet slit of the camera was positioned parallel to and 5 mm off the axis. It is seen that the low-density corona formed in the process of wire explosion is accelerated first and goes 30÷40 ns ahead of the main bulk of the plasma, which is in addition unstable. This can explain a long duration of the radiation pulse (20÷40 ns and even up to 60 ns for a wire array with a big initial radius) observed in the experiments.

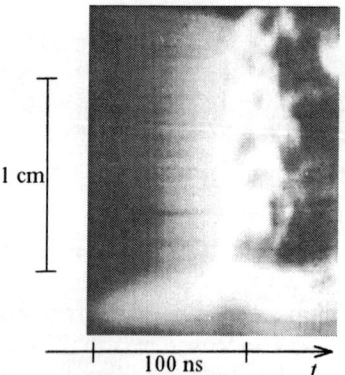

FIg. 2 The streak camera picture of implosion of 68-μg/cm, 1.25-cm-radius wire array.

We have also conducted the experiments in terms of scaling in the radius and mass for the mr_0^2 product being kept nearly constant. Since the wires we had at our disposal had a discrete spectrum of diameters, we were unable to meet precisely the requirement $mr_0^2 = const$.

The value of mr_0^2 ranged from 18.4 µg·cm to 25.5 µg·cm, and the implosion time was within the interval 90÷120 ns. The η value (the ratio of the maximum implosion kinetic energy per ion to the sum of the ionization energies and the ion and electron thermal energies required to reach the He-like stage of aluminum) varied from 0.5 to 2.2. Though the theory[5] predicts no sufficient K-shell yield at $\eta < 1$, in our experiments the maximum K-shell yield was observed at $\eta = 0.7 \div 0.8$ (Fig.3).

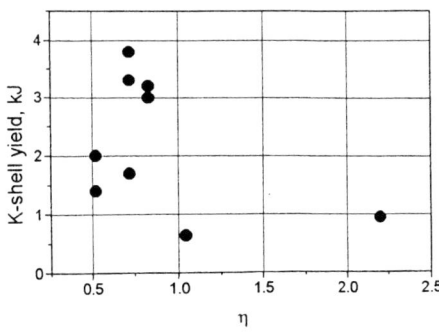

Fig.3 The K-shell radiation yield versus η.

WIRE ARRAY IMPLOSION EXPERIMENTS ON THE GIT-8 FACILITY

An appropriate choice of the delay time between the operation of the plasma guns and the switch-on of the Marx generator determines essentially the performance of an inductive energy store. The experiments on determination of K-shell yield as a function of the delay time were carried out on the GIT-8 generator. In these experiments the mass of Al wire array was 68 µg/cm and its initial radius was 5 mm chosen with regard to the results of the GIT-4 wire array implosion experiments.

As follows from the experiments (see Fig.4), the optimum delay time is 3.3 µs with the switch conduction time being 1.1 µs (the discharge half-time for the Marx generator being 1.9 µs). The load current corresponding to this delay time is 1.9 MA.

For the best POS operation regime, the measurements of the K-shell power and yield as a function of wire array initial radius and mass were performed. The maximum registered K-shell power and yield were 150 GW/cm and 7.9 kJ/cm, respectively (Fig.5). As it was observed in the experiments on the GIT-4 generator, the radiation power and yield (though not so drastically) decrease in passing to larger radii of the load. We associate this phenomenon with the appearance of plasma flows in the run-in phase of the implosion that restrain the compression of the aluminum wire array. The intensity of the plasma flows rises

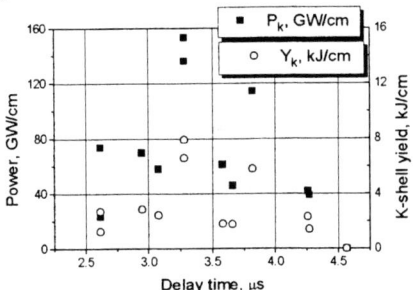

Fig.4 K-shell radiation power and yield as a function of the delay time.

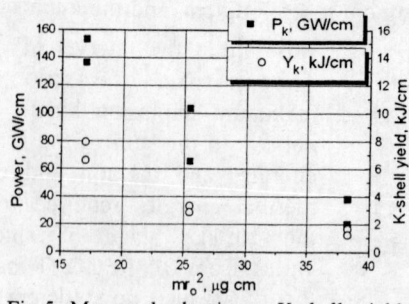

Fig.5 Measured aluminum K-shell yield and power vs the mr_0^2 parameter.

with decreasing of the derivative of the load current. In the experiments the radiation power starts falling when implosion time is close to the current rise time. In our opinion, the increase in the radiation power in passing from the GIT-4 generator to GIT-8 generator is related not only to the increase in the load current but also to the increase in the load current derivative.

GAS-PUFF ON WIRE ARRAY EXPERIMENTS

In order to decrease the load current rise time, a gas-puff-on-wire-array scheme was used in the experiments on the GIT-4 generator[6]. A wire array was installed on an insulating bushing with an outer diameter of 30 mm. The outer argon gas-puff was formed with the help of an electro-magnetic gas valve on the diameter of 80 mm. The gas-puff mass was 9 µg/cm. It was assumed that the current starts to flow through the outer gas shell and then switches on the inner wire array providing a short current rise time in the wire array. The experiments show that a two-fold increase in K-shell radiation power and yield can be achieved using the gas-puff-on-wire-array load. The stability and homogeneity of the radiating plasma are considerably improved. However, the reproducibility of experimental results is low due to the sharing of the total current between the outer gas shell and the inner wire array, which occurs in some shots.

ACKNOWLEDGMENTS.

The research of Al wire array PRS was supported by the Defense Nuclear Agency under contract number DNA 001-94C-0047. The authors would like to thank Dr.D.McDaniel for supporting this work. The excellent performance of the GIT-4 and GIT-8 crews is gratefully acknowledged.

REFERENCES

1. S.P.Bugaev, *et al.*, IEEE Trans. Plasma Science, V.18, No.1, p.115, (1990).
2. S.P.Bugaev, *et al.*, Proc. of the 9th Int. Conf. on High-Power Particle Beams, Washington, DC, V1, p.394, (1992).
3. R.B.Baksht, *et al.*, Laser and Particle Beams, V.11, No.3, p.587, (1993).
4. R.B.Baksht, *et al.*, Laser and Particle Beams, V.12, No.4, p.615, (1994).
5. K.G.Whitney, *et al.*, J. Appl. Phys. 67, p1725, (1990).
6. R.B.Baksht, *et al.*, Plasma Physics Reports, V.23, No.2, p119, (1997).

Comparison of Ribbon and Wire Array Load Characteristics on Double Eagle

E. J. Yadlowsky, R. C. Hazelton and J. J. Moschella
HY-Tech Research Corporation, 104 Centre Court, Radford, Virginia 24141

and B. H. Failor, P. D. LePell, and C. A. Coverdale
Physics International, 2700 Merced Street, San Leandro, CA 94577

and J. P. Apruzese, K. G. Whitney, J. W. Thornhill and J. Davis
Naval Research Laboratories, 4555 Overlook Ave.SW, Washington, D.C. 20375-5000

Abstract

The non-uniform acceleration of partially vaporized and ionized wires in wire array loads results in precursor plasmas that assemble on-axis and soften the subsequent implosion of the remaining mass. One micron thick Al ribbons with an anticipated vaporization/ionization time small compared to the 30 μm diameter wire were fabricated to minimize this load straggling effect. The implosion and radiation characteristics of ribbon and wire arrays were compared on the Double Eagle accelerator. A high resolution x-ray spectrometer in the Johann geometry recorded the energy resolved He_α and intercombination lines of the 1% Mg dopant added for diagnostic purposes. Comparable x-ray yields were obtained with both configurations. The ratio of K-spectrum lines of Al were used to infer comparable electron temperatures (1 - 1.3 keV) and ion densities (7 x $10^{18}/cm^3$) in both cases. The ion temperatures, inferred from the Doppler width of optically thin Mg dopant lines, were a factor of 3 smaller for ribbons. The intercombination/He_α line ratio for Mg was much smaller in the ribbon case, implying a higher density. The inability to reconcile this ratio with the Al spectrum or the radiated power in the ribbon case is not presently understood.

I. Introduction

Ribbon arrays have been investigated in an attempt to reduce the straggling that has been observed in the on-axis assembly of wire array loads[1] and to broaden the x-ray spectrum of these loads through elemental mixing. One possibility for the observed straggling has been attributed to the finite vaporization/ionization time of individual wires of radius, r. The magnetic force on the tenuous current carrying coronal plasma surrounding the core results in a precursor plasma which implodes ahead of the main load mass and softens its implosion. It has been invoked to explain, in part, the longer radiation pulse length, reduced ion densities, and reduced x-ray powers that have been observed relative to the initial theoretical estimates of "hard implosion" which have ignored this precursor plasma.[2] The vaporization/ionization times of ribbons with a thickness t << r is expected to be much less than that of a sparse wire array, possibly resulting in more uniform load assembly. A 1% Mg dopant in the Al wire and ribbon loads provided localized spectroscopic information about the dynamics of these load configurations without the complication of opacity.

II. Experimental Conditions and Results

Two evaporation sources were used to deposit Al alloy ribbons containing 1% Mg. The 0.7 mm wide and 1 μm thick ribbons have the same mass/unit length as 30μm diameter wires containing 1% Mg. The ribbon and wire array loads, 25 mm in diameter, were compared on the Double Eagle accelerator delivering a 4 MA current pulse rising in 100 ns. Filtered photoconducting diodes and calorimeters were used to monitor the K-shell yield. A bent crystal focusing spectrometer in the Johann geometry was used to eliminate source broadening effects.[3] This instrument was used to measure the He_α and intercombination (IC) lines of the Mg dopant in both wire and ribbon loads from which the ion temperature can be determined with an uncertainty of ± 0.1 keV. No entrance slits were used to obtain spatially resolved information about the pinch.

The K-shell yield with 12 wire loads was typically 20 kJ. Although the average yield with ribbons was less than with wires, and the scatter larger (14.5 ± 5 kJ with 12 ribbons and 15.6 ± 3.7 with 14 ribbons), the peak yields were equal to that obtained with wire loads. Mg spectra obtained with the Johann spectrometer are shown in Fig. 1 for the two configurations. Figure 2 presents the Al ribbon spectrum obtained with mixed element ribbons. Figure 3 presents a representative Al survey spectrum obtained with wire loads at another time under similar conditions. The yield in this earlier shot (36 kJ) was larger than those observed here.

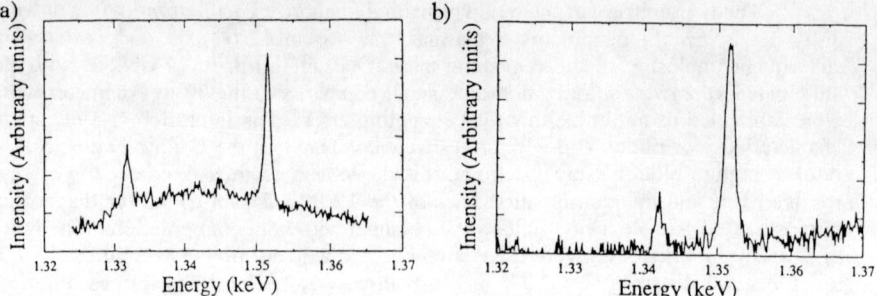

Figure 1. Energy resolved He_α and IC lines of Mg obtained with a) ribbon and b) wire loads. The IC line is absent in 1.a but a Li satellite is apparent.

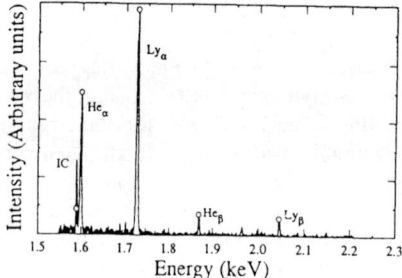 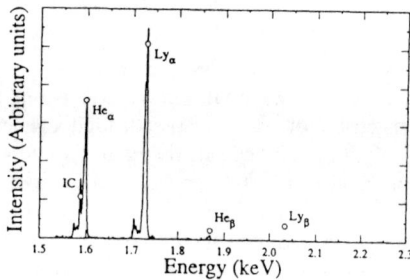

Figure 2. Al spectrum obtained with survey spectrometer and ribbon load. Open circles denote calculated intensities of Al lines using CRE analysis.

Figure 3. Representative survey spectrum of Al wire load obtained at another time. ○ denote calculated intensities of Al lines using CRE analysis.

The Doppler width of the optically thin $MgHe_\alpha$ line in Fig. 1a yields a value of 1.7 ± 0.1 keV for the ion temperature of ribbon loads. Similarly, ion temperatures of 5.6 ± 0.1 and 3.2 ± 0.1 keV are inferred, respectively, from the widths of the $MgHe_\alpha$ and IC lines in the wire case. These temperatures are used in the collisional radiative equilibrium analysis of the Al spectrum where opacity influences the intensities and widths of the lines.

The analysis procedure uses the collisional radiative equilibrium (CRE) model to calculate the x-ray spectrum and radiated K-shell power for assumed density and temperature profiles. Values of N_i, T_i, and T_e which provide a set of x-ray line intensities consistent with the observed spectrum, plasma size, and observed radiated power are sought.[4] The open circles in Figs. 2 and 3 and the values in Table I summarize the spectral features calculated by the CRE analysis under the following assumptions. The ion density and temperature were assumed to be uniform with T_i equal to 1.7 keV for ribbons and 3.2 keV for wires. The ribbon spectra could be fitted with a uniform electron temperature whereas a non-uniform profile was required to fit the He_β/He_α line ratio in the wire case. Best fit for wires was obtained with a hot core (1.3 keV) occupying 70% of the plasma radius that decreased non-linearly to a cooler corona (0.3 keV).

Table I.   Observed and calculated line ratios for assumed condition:
Ribbon Loads: $N_i = 7 \times 10^{18}/cm^3$, $T_e = 1.2$ keV, $T_i = 1.7$ keV, r = 0.16 cm
Wire Loads: $N_i = 7 \times 10^{18}/cm^3$, $T_e = 1.3$ keV (core); 0.32 keV (corona),
               $T_i = 3.5$ keV, r = 0.35 cm

Load	Mg $\frac{IC}{He_\alpha}$	Al $\frac{IC}{He_\alpha}$	Al $\frac{Ly_\alpha}{He_\alpha + IC}$	Al $\frac{He_\beta}{He_\alpha + IC}$	Al $\frac{Ly_\beta}{He_\beta}$	% Initial Mass	Yield kJ
Ribbon (Obs)	0.04 - 0.07	0.32	1.27	0.10	0.80	15	12
Ribbon (Calc)	0.07	0.18	1.37	0.12	0.82		
Wire (Obs)	0.17	0.32	1.23	0.03		50	36
Wire (Calc)	0.12	0.29	1.09	0.04			

The primary difficulty in the wire case lies with obtaining a density which is consistent with both the IC/He_α ratio and the measured K-shell power. This ratio, which is sensitive to the electron density and relatively insensitive to the electron temperature for $T_e > 300$ eV predicts an ion density inadequate to account for the peak radiated power. An adequate fit is obtained for this ratio, and the other spectral features, if the predicted power is fitted to the average observed power ($P_{ave} \approx 0.5\ P_{peak}$). This is a reasonable assumption since the open shutter spectra are a measure of the total energy deposited on the film. The plasma conditions in Table I account for 50% of the initial load mass in the wire case.

A good fit to the Al spectrum and the average radiated power is obtained in the ribbon case even with a uniform electron temperature. Unfortunately, the IC/He_α line ratio for the Mg spectrum in Fig. 1a and Al spectrum in Fig. 2 cannot be simultaneously fit under any assumed conditions. Since these spectra were obtained on different shots, this lack of consistency may be acceptable.

III. Discussions and Conclusions

The optical depth of the Mg He_α line can be calculated for the best fit parameters in Table I. Values of 0.09 and 0.90 were obtained for the ribbon and wire loads respectively. This line can be considered optically thin for the ribbon case. Ion temperatures inferred from its half-width should be relatively free of opacity broadening effects. The higher temperature inferred from the He_α line (5.6 keV) than the IC line (3.2 keV), in Fig.1b, could indicate that some opacity broadening is occurring in the wire shot.

The IC/He$_\alpha$ ratio of these optically thin Mg lines is clearly larger for wire loads than ribbon loads. This would indicate a larger electron density is achievable in the latter case although the ribbons show a large degree of variability. The significance of these spectra vis-a-vis the CRE analysis of the Al spectra is not clear since the data were not obtained simultaneously. Not only does the CRE analysis predict equal ion densities for ribbons and wires, the excellent fit of the observed and predicted spectra in Fig. 2 is irreconcilable with a low IC/He$_\alpha$ Mg ratio for ribbons. Rosmej and Rosmej[5] have shown theoretically that this ratio can be depressed if the electron distribution is non-Maxwellian as a result of plasma instabilities, or if the plasma is rapidly heated or compressed.[6] Some complication such as this, or non-linear saturation of the film response may be affecting the interpretation of the data.

The CRE analysis used here invoked temperature gradients to model all of the spectral features, including the IC/He$_\alpha$ ratio, and predict the average power radiated. This is a slight departure from some previous studies of Al and Mg composite wires, where the model tried to fit only the Ly$_\alpha$ to He$_\alpha$ line ratio and the peak power.[1] Not only is the peak ion density much larger in the earlier studies (1.8 x 10^{20}/cm^3 for r ≤ 0.05 cm, 1.4 x 10^{21} for 0.05 ≤ r < 0.06, and 2 x 10^{18} for 0.06 ≤ r < 0.15 cm), but also the pinch radius is less than half of the 0.35 cm radius inferred here. The good fit that is obtained between the observed lines and their predicted intensities (both Al and Mg) together with the average powers radiated, provides credence to the ion densities and electron temperatures that have been inferred here.

The present results suggest that ribbon array loads result in pinched plasmas having ion densities, electron temperatures, and x-ray yields which could be comparable to those obtained with wire loads. However, these loads, whose ion temperature is inferred to be 50% higher than their electron temperature appear to more closely approach ion-electron equilibrium than wire loads whose ion temperature is between 2.5 and 4.3 times the electron temperature. The low IC/He$_\alpha$ line ratio of optically thin Mg also suggests a high plasma density and may indicate a significant gulf in the present understanding of ribbon pinches. Further experiments are planned with simultaneous acquisition of Al and Mg spectral data.

Acknowledgement

The efforts of E. Carlson and the technical crew at the Double Eagle facility are greatly appreciated. This work is sponsored by the Defense Special Weapons Agency.

References

1 C.Deeney, P.D. LePell, B.H. Failor, S.L. Wong, J.P. Apruzese, K.G. Whitney, J.W. Thornhill, J. Davis, E. Yadlowsky, R.C. Hazelton, J.J. Moschella, T. Nash, and N. Loter, Physical Rev E, 51, Part B, 4823, (1995).

2 K.G. Whitney, J.W. Thornhill, J.P. Apruzese, J. Davis, C. Deeney, P.D. LePell and B.H. Failor, Phys. Plasmas 2, 2590 (1995).

3 B.A. Hammel, D.W. Phillion, and L.E. Ruggles, Rev. Sci. Instrum., 61, 1920, (1990).

4 J.P. Apruzese, K.G. Whitney, J. Davis, and P.C. Kepple, J. Quant. Spectrosc. Radiat. Transfer 57, 41 (1997).

5 F.B. Rosmej and O.N. Rosmej, 21st Conference on Controlled Fusion and Plasma Physics, Montpellier, France, Vol. III, 1292, (1994).

6 F.B. Rosmej and O. N.Rosmej, J. Phys. B: At. Mol. Opt. Phys 29, L359, (1996).

The Experimental Investigations of Imploding Plasma as a Source of Hard X-Ray

Yu.L. Bakshaev, A.V. Bartov, P.I. Blinov, A.S. Chernenko,
S.A.Dan'ko, Yu.M. Gorbulin, Yu.G. Kalinin, V.D. Korolev,
V.I. Mizhiritskii, L.I. Rudakov, A.Yu. Shashkov and S.A. Shibaev

Russian Research Center "Kurchatov Institute", 123182 Moscow, Russia

Introduction. The most critical obstacle to the creation of a bright X-ray source on the base of imploding plasma is the instability of its magnetic compression. It was demonstrated in our previous investigations (1) and by research teams in several laboratories in other countries. The most strong instability appears in imploding shell consisted of the materials including elements of high atomic number. As a result, the shell is disrupted on the rings (strati) and the current penetrates inside the shell.

The conclusions that we can draw from the numerical simulation of these processes and 2D-theory taking into account the contribution of Hall effect into the Ohm's law are: i) due to the high radiative losses the compressed plasma is concentrated in the thin layer with the thickness less than 1 mm; ii) the current flow in such a thin layer of magnetized plasma becomes unstable even up to the its disruption (2); this happens at the initial stages of magnetic acceleration process during the formation of thin plasma shells.

Experimental results. The report presented is a part of systematic study aimed to solving of this problem on S-300 installation (I ~4 MA, t ~100 ns, Fig. 1). As a liner, we used cylindrical gas jet with outer diameter ~4 cm produced by using a pulse-driven valve and an annular supersonic nozzle. In combination with apparatus registering parameters of power pulse we used the following diagnostics:

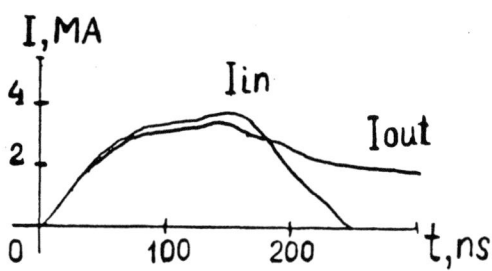

FIGURE 1. Typical current values I_{in} (concentrator) and I_{out} (load).

i) photographing in visible light region, both with streak and with frame image converter cameras; ii) photographing in the region of soft X-ray radiation with frame image converter cameras; iii) registering the radiation pulses with a set of vacuum and semiconductor X-ray diodes shielded with various filters, and iv) convex-mica crystal X-ray spectrograph having two-dimensional spatial resolution.

The main feature of our experiments is that at first stage an outer boundary of the shell is looking in the visible light to move with a constant velocity. This stage may be identified as the compact shell formation in the process of the plasma heating by the shock wave coming and its cooling behind the shock front resulting in the accumulation of the cool plasma in the current layer. Under these conditions the development of Rayleigh-Taylor instability seems to be unlikely. Its typical time of growth is determined by the shell displacement to ~1/2 of the initial radius. In some experiments, we observed an instability of the outer surface of the shell without any acceleration. Such an instability should be identified as one of those of the Hall origin. They are rather typical for the regimes of the fast implosion, i.e., of the small shell masses or of the higher Z materials (Xe in our case). N_2 or SF_6 shells are (statistically) more stable at this stage. Optical frames demonstrating the instability developing of a N_2 jet at the first stage, are shown in Fig.2a. The same process has been illustrated in Fig. 2b for the Xe jet. The formation of the compact shell shown in Fig. 2c occurs without instabilities. This is the case of the SF_6 jet of the great mass. It turns out to be in agreement with the predictions of EMHD theory.

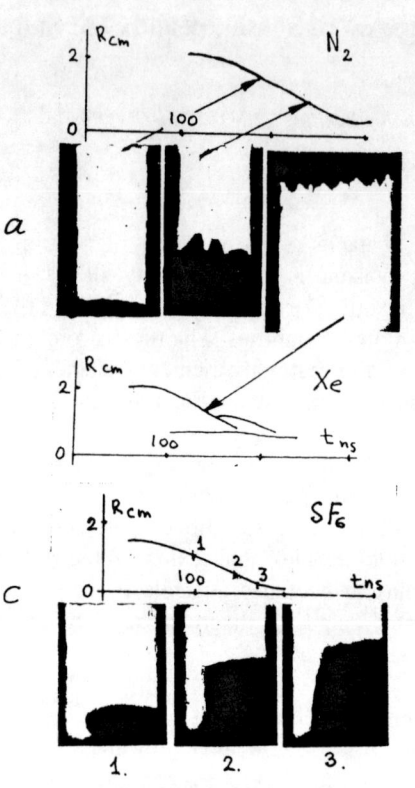

FIGURE 2. R-t diagrams and optical frames demonstrating both unstable and stable regimes of the implosion of the gaseous liners.

Recently Rudakov has proposed an improved scheme of "liner-converter" for the creation of a bright pulse source of X-ray radiation on the base of imploding plasma (which is named as "heterogeneous pinch" (3)). The key point of this scheme is the compression of the plasma cylinder containing the light gas outside it (D,

He) and more heavier one inside it (Ne, Ar). Theoretical considerations and numerical simulation have shown that such a version of Z-pinch allows not only to improve the efficiency of magnetic energy conversion into the X-ray radiation from converter but to increase the stability during the compression as well. The origin of this effect is mainly the absence of Hall effect in the light outer shell; Rayleigh-Taylor instability can develop at less extent, as a result of its large thickness.

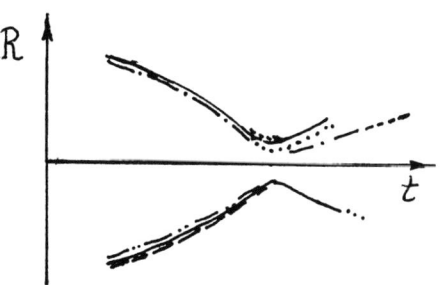

FIGURE 4. R-t diagrams of 5 repeated He liner experiments.

Taking these considerations into account, we have performed another series of experiments using the light thick He jet, accelerated up to the velocities about $5\cdot10^7$ cm/sec. Some examples of streak and frame camera pictures of its compression are presented in Fig. 3a and 3b, respectively. It is important to pay attention to both symmetric form and good reproducibility of the picture presenting the dynamic of the He liner implosion

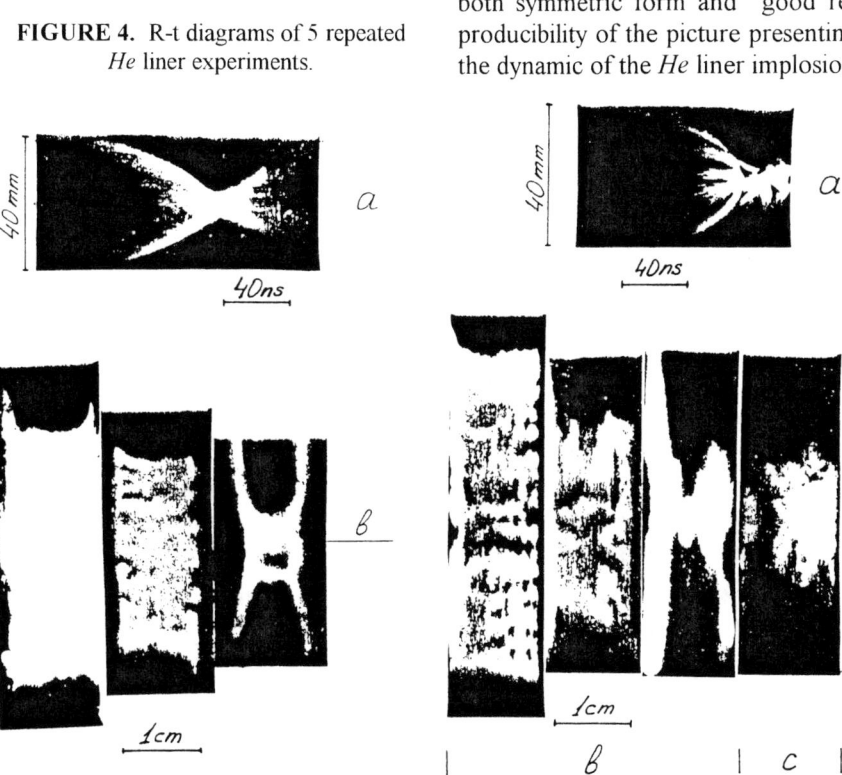

FIGURE 3. Optical streak (a) and frame (b) camera pictures of He liner compression.

FIGURE 5. Optical streak (a) and frame (b - visible, c - soft X-ray) camera pictures of Ar liner compression.

which can serve as an evidence of its stability. The same fact is illustrated by R-t diagrams of 5 repeated experiments performed in the identical conditions (Fig. 4).

In contrast to the previous experiments with He liner, the Ar liner under the conditions of the fast compression demonstrates an evident instability (Fig. 5), in accordance with the theoretical predictions about the significant role played by Hall instability and Rayleigh-Taylor instability for thin liner, created due to strong radiation in snow-plough regime.

However, frame camera pictures of He liner contain some strati (rings) on the surface of the plasma compressed (Fig. 3b). These perturbations do not increase during the compression but decrease in amplitude both absolute and relative - $\Delta r/r(t)$. We believe that it is a manifestation of two-dimensional Hall dynamic of current in rare plasma corona. It develops only in the plasma shell of c/ω_{pi} thickness (4) which decreases as far as the plasma is being compressed. The phenomenon of magnetic field penetration into the plasma and subsequent surface density modulation is a subject of theoretical, numerical and experimental study related to the POS problem (5-7). The common origin of these phenomena in POS'es and in Z-pinch corona was discussed somewhere (8). It is the most important that these phenomena occur on the surface and do not destroy the thick shell of imploding plasma.

Conclusion. We have succeeded to identify the regime of rather stable magnetic compression of the plasma column. It offers the new opportunity of the realization of bright X-ray sources on the base of imploding plasmas by using the scheme "heterogeneous Z-pinch – converter".

Acknowledgments. The authors are grateful to RFBR (grant 96-02-18891) and the Russian Ministry of Science and Technology for financial support.

References

1. Chernenko A.S. et al. *Beams'96 Conf. Proceedings*, Prague, 1996, p. 154.
2. Rudakov L.I., Sevastyanov A.A., *ibid.* p. 779.
3. Rudakov L.I. *This Conf. Proceedings*, report P-4.
4. Kingsep A.S., Rudakov L.I. *Beams'96 Conf. Proceedings*, Prague, 1996, p.729
5. Kingsep A.S. et al. *Sov. J. Plasma Phys.* 1984, **10**, 854.
6. Maron Y. et al. *Beams'96 Conf. Proceedings*, Prague, 1996, p. 41.
7. Zabaidullin O. et al. *ibid*, p. 1237.
8. Chuvatin A. et al. *Phys. Rev. Lett.* 1996, **76**, 2282.

The Dense Z-Pinch Program at the University of Nevada, Reno

B.S. Bauer, V.L. Kantsyrev, F. Winterberg,
A.S. Shlyaptseva, R.C. Mancini, H. Li, and A. Oxner

Department of Physics, University of Nevada, Reno, NV 89557-0058

Abstract. A new program of research into the physics of dense z-pinches is being initiated around a high-repetition-rate two-terawatt generator (formerly Zebra/HDZP-II: 2MV, 1.2 MA, 100 ns, 200 kJ, 1.9 Ω final line impedance) transferred to the University of Nevada, Reno Physics Department from Los Alamos National Laboratory. Areas for study include the early-time evolution of a current-driven wire, the plasma turbulence around and between wires, the suppression or reduction of instabilities, the nature of x-ray bright spots, and the tailoring of the x-ray emission spectrum. Novel loads that introduce a stabilizing velocity shear or density profile will be examined, along with configurations that promise to increase the quantity, hardness, stability, and reproducibility of x-ray emission.

A wide variety of diagnostics are being developed, so as to diagnose the plasma thoroughly and make detailed comparisons between experiment, computer simulation, and theory. These include x-ray, soft x-ray, and extreme ultraviolet space- and time-resolved spectroscopy, with polarization measurements; laser interferometry, collective Thomson scattering, Faraday rotation, and laser-induced fluorescence; ion charge spectrocopy with an electrostatic analyzer; and small-angle x-ray diffraction of hard x-rays from the crystal lattice of the solid state load, using a standard x-ray tube and a double crystal monochromator. Quantities to be measured include the ion charge states present in the plasma; the electron density (n_e) and temperature (T_e); the magnetic field (B); the plasma flow speed (u); extreme ultraviolet (2.5 nm $< \lambda <$ 30.0 nm), soft x-ray (0.3 nm $< \lambda <$ 2.5 nm), and x-ray ($\lambda <$ 0.3 nm) emissions; the energy distribution of electrons with particular attention to the direction and relative concentration of suprathermal electrons; the energy and charge-state distribution of ejected multicharged ions; the ablation of neutral gas and the change of state of solid loads; and the amplitude and spectrum of ion wave turbulence.

The plasma z-pinch is one of the most rapidly developing areas of physics and engineering. Driving a giant current through a tiny load is the most efficient and inexpensive method of producing a hot, dense plasma. Much higher confining magnetic fields are generated by a z-pinch plasma than can be externally imposed. Hot, dense plasmas have a wide array of applications, from x-ray sources for biomedical microscopy, defense, and lithography, to the development of thermonuclear fusion; and from the stripping and focusing of ions in accelerators, to the production of energetic charged particle beams for the modification of materials. However, the z-pinch would come much closer to achieving its enormous scientific and technological potential if control over the dense, self-compressing plasma were improved. This presents a number of interesting, fundamental scientific challenges. First, the dense z-pinch plasma has features that distinguish it strongly from the low-density plasmas that have so far been the primary focus of plasma physics. Particle collisions and radiation transfer are important in a dense z-pinch plasma, resulting in collective behavior considerably different than those predicted by models developed for collisionless, optically-thin plasma. In the densest regions, multiply-charged ions are strongly coupled, and collective modes likely exist, such as ion shear waves, that have yet to be confirmed experimentally. Energy gradually stored in metastable excited ionic states can be suddenly released when the ions converge and are collectively collisionally deexcited. Huge magnetic fields influence the atomic processes. Second, the plasma contains strong gradients and evolves rapidly. Different physical mechanisms dominate in different regions as a function of time, and these regimes interact with each other. Complex magnetic field and flow topologies arise when plasmas merge or are twisted by instabilities. Initial conditions and early development can be important. Third, new diagnostics and more detailed experimental observations are needed to accurately measure the complex three-dimensional plasma state, the turbulence and transport driven by the large z-pinch current, and the broad range of z-pinch emissions. Finally, new z-pinch load configurations must be found that increase plasma stability, density, temperature, and reproducibility. Velocity shear, initial density profiles, finite Larmor radius effects, and viscosity could stabilize the z-pinch[1].

To improve our understanding of the z-pinch and devise new schemes that improve its performance, a new program of experimental, theoretical, and computational work is being initiated at the University of Nevada, Reno (UNR). The plasma, atomic, and radiation physics of the z-pinch will be explored with experiments on a high-repetition-rate, 2-terawatt z-pinch [2] (formerly Zebra/HDZP-II) transferred to UNR from Los Alamos National Laboratory. This pulsed power generator, currently being reassembled, will form the core of the Nevada Terawatt Facility. A Center for High Energy Density Science and Technology is being created to facilitate collaborations and coordinate the multi-user facility.

The 2-terawatt pulsed-power generator consists of four major parts (Figure 1). The Marx bank (far right) consists of 32 $1.3\mu F$, 100kV capacitors, which can store a total of 200kJ of energy. The bank is coupled to a water-filled, 28nF, 2MV intermediate storage capacitor (the horizontal cylinder in the middle of the device). An SF_6-insulated "rimfire" switch connects the intermediate store to a 50ns, 1.9Ω water-filled vertical transmission line (the vertical cylinder on the left). A self-breaking water switch couples the final feed to the load, located in the vacuum chamber on top of the vertical transmission line. An electrical pulse of up to 1.2MA at 2MV is delivered to the load, with a 100ns rise time.

The z-pinch will be thoroughly diagnosed so as to make detailed comparisons between experiment, theory, and computer simulation. All important plasma parameters will be measured, for the most part with spatial and temporal resolution. An overview of the diagnostics planned is shown in Figure 2. Many x-ray measurements will be performed, both because this is the most feasible way of gaining access to the dense z-pinch plasma, and because x-ray emission is closely correlated with plasma dynamics and is important in determining plasma evolution. A novel two-dimensional x-ray imaging spectrometer will use glass capillaries [3] to multiplex a two-dimensional plasma image (from 0.02 to 100 nm in wavelength) into a spatially-separated array of pixels, to be spectrally dispersed, recorded by a temporally-gated imager, and reconstituted as an image by a computer. The plasma

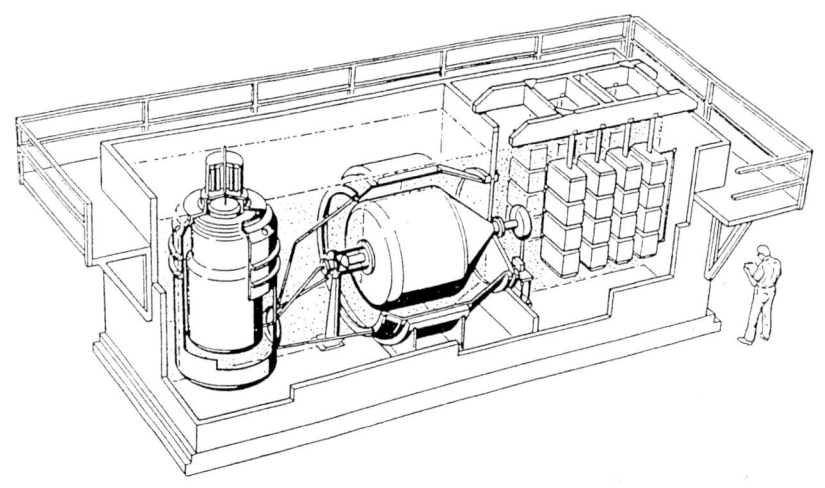

FIGURE 1. Pulsed Power Generator of the Nevada Terawatt Facility (NTF).

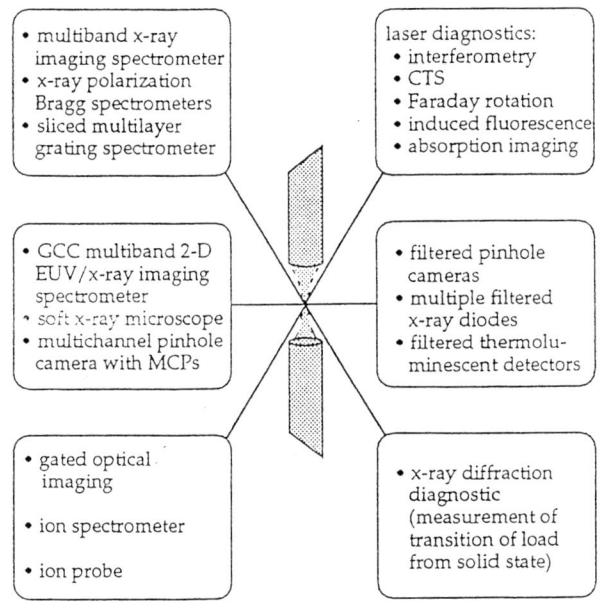

FIGURE 2. Overview of Diagnostics of the Dense Z-pinch Program at the University of Nevada, Reno.

magnetic field and the anisotropy of the energetic electron distribution will be measured using polarization-sensitive x-ray imaging spectrometers, interpreted by detailed spectral calculations [4]. Quantities to be measured include the ion charge states present in the plasma; the electron density (n_e) and temperature (T_e); the magnetic field (B); the plasma flow speed (u); extreme ultraviolet (2.5 nm < λ < 30.0 nm), soft x-ray (0.3 nm < λ < 2.5 nm), and x-ray (λ < 0.3 nm) emissions; the energy distribution of electrons with particular attention to the direction and relative concentration of suprathermal electrons; the energy and charge-state distribution of ejected multicharged ions; the ablation of neutral gas and the change of state of solid loads; and the amplitude and spectrum of ion wave turbulence.

The plasma will be carefully modeled with two-dimensional radiation-magnetohydrodynamic simulation codes (in collaboration with other researchers) [5,6]. In addition, particle-in-cell, Fokker-Planck, and non-local-thermodynamic-equilibrium radiation transport calculations will yield further insight into the interplay between the plasma, atomic, and radiation physics. The effect of collisions and radiation transport on instability thresholds and growth rates will be examined numerically and analytically. Detailed synthetic spectra [4,7] will be computed for the interpretation and analysis of experimental data.

Novel z-pinch load configurations with the potential for greatly enhanced plasma stability or performance will be explored. With improved stability, the z-pinch could yield a solution to the challenge of controlled fusion. Similarly, the quantity, hardness, stability, and reproducibility of x-ray emissions could be increased. The effects on plasma instabilities of velocity shear, initial density profile, finite Larmor radius, and other conditions will be investigated. Velocity shear will be produced by injecting high-speed plasma [8] along and around a wire prior to the discharge.

We hope to improve the understanding of hot, dense z-pinches; develop new z-pinch diagnostics and loads; and help develop z-pinch applications.

REFERENCES

1. M.G. Haines and M. Coppins, *Phys. Rev. Lett.* **66,** 1462 (1991).
2. D.W. Scudder, J.S. Shlachter, J.E. Hammel, F. Venneri, R. Chrien, R. Lovberg, and R. Riley, in "Physics of Alternative Magnetic Confinement Schemes," S. Ortolani, E. Srodoni (Eds), Bologna: SIF, 1991, p. 519.
3. V.L. Kantsyrev and R. Bruch, *Rev. Sci. Istr.* **68**, 770-773 (1997).
4. A.S. Shlyaptseva, R.C. Mancini, P. Neill, and P. Beiersdorfer, *Rev. Sci. Instrum.* **68**, 1095 (1997).
5. J.H. Hammer, J.L. Eddleman, P.T. Springer, M. Tabak, A.T. Toor, K.L. Wong, G.B. Zimmerman, C. Deeney, R. Humphreys, T.J. Nash, T.W.L. Sanford, R.B. Spielman, and J.S. Degroot, *Phys. Plasmas* **3**, 2063 (1996).
6. J. Davis, J.L. Giuliani, Jr., and M. Mulbrandon, *Phys. Plasmas* **2**, 1766 (1995).
7. R.C. Mancini, D.P. Kilcrease, L.A. Woltz, and C.F. Hooper, Jr., *Comp. Phys. Comm.* **63**, 314 (1991).
8. J.H. Hammer, J.H. Eddleman, C.W. Hartman, H.S. McLean, and A.W. Molvik, *Phys. Fluids B* **3**, 2236 (1991).

High Velocity Implosions on PBFA Z

J. S. De Groot[1,2,3]
C. Deeney[2], T. W. L. Sanford[2], R. B. Spielman[2]
K. G. Estabrook[3], J. H. Hammer[3], D. Ryutov[3], and A. Toor[3]

[1]*Plasma Research Group and Department of Applied Science, UC Davis, 95616*
[2]*Sandia National Laboratories, P.O. Box 5800, Albuquerque, NM 87185-1193*
[3]*Lawrence Livermore National Laboratory, P. O. Box 808, Livermore, CA 94550*

ABSTRACT

We consider techniques to produce a significant yield of high energy x-rays (energy > 26 keV) on PBFA Z. Large initial radius loads are required to produce the high implosion velocities and the required high electron temperatures. The classical magneto-Rayleigh-Taylor instability would grow to large amplitude for these loads. We are considering techniques to reduce the total growth of the instability. Two load designs are being considered. The first is a uniform-fill, gas puff load. Snowplow, ion viscosity, and finite ion gyro radius effects reduce the growth rate of the magneto-Rayleigh-Taylor instability. 1-D rad-hydro simulations show that a high energy x-ray yield of 24 kJ is produced by a 5 cm initial radius, 20% xenon/80% argon, uniform fill, gas puff load. The second design uses an aluminum wire array imploding onto a barium loaded CH fiber. 1-D rad-hydro calculations indicate that a high energy x-ray yield of 26 kJ will be produced by a 2 cm wire array imploding onto a 0.3 cm radius CH fiber loaded with 0.003 atomic fraction of barium.

INTRODUCTION

Large radius (~ 5 cm) loads can be used to produce high velocity (implosion velocity, ~ 2×10^8 cm/s) implosions on PBFA Z. These loads are subject to the magneto-Rayleigh-Taylor instability. The perturbation amplitude, A, of the classical M-R-T instability would grow to large amplitude during these implosions, i.e., the exponential growth factor is, $\Gamma = \int \gamma(t) dt$, and for linear growth (1) $\Gamma = \sqrt{3kR_0}$ so that $\Gamma \sim 30$ for a 1 mm wavelength perturbation. Thus we expect that the symmetry of such large diameter implosions would be strongly degraded by the M-R-T instability resulting in a low quality stagnation and reduced radiation yield.

EXPERIMENTS

Large initial radius (3 cm), uniform-fill, krypton gas-puff experiments (2) on the Saturn accelerator at Sandia National Laboratories resulted in very high quality implosions (radius of keV emitting region ~ 0.1 cm) and high x-ray yields (maximum keV x-ray yield ~ 30 kJ). The keV x-ray pinhole framing photographs indicate ~ 2 mm wavelength perturbations, with amplitude, kA ~ 1. The M-R-T instability should be strongly excited in these experiments even though snowplow effects (4) reduce the linear growth factor to $\Gamma \sim 16$. We are therefore led to consider effects beyond the ideal MHD approximation to understand these results.

Viscosity and Finite Gyro Radius Effects on the Instability

Finite ion viscosity (5) could account for the apparent slow growth of the M-R-T instability in the experiments (2). The point is that the ion density is low and the shock heated ion temperature is high resulting in long ion-ion mean-free-paths for large diameter gas puff loads. Calculations (6) indicate that the ion viscosity is large enough that the short wavelength (10-100μm) M-R-T perturbations are stabilized and the growth factor of the ~ mm wavelength perturbations is reduced. This idea agrees with the experimental observation (2, 3) that uniform fill gas puff loads are more stable than annular gas puff loads. The point is that the plasma is shock free after a few shock transits of the annular plasma. The ion gyro radius is also large enough that finite gyro radius effects also reduce the growth rate.

A simple theory demonstrates the key physical points. The maximum growth rate γ_{Max} in a viscous plasma (5) is $\gamma_{Max} = \dfrac{a^{2/3}}{2\nu^{1/3}}$, where a is the acceleration, and ν is the ion kinematic viscosity. Defining a viscosity required to reduce the classical growth rate, $\gamma_{class} = (ka)^{1/2}$ by a factor of two results in $\nu_{1/2} = a^{1/2}/k^{3/2}$. Approximating the acceleration by $a \sim R_0/t_i^2$, where R_0 is the initial radius of the gas puff, $R_0 = 5$ cm, and t_i is the rise time of the current, $t_i \sim 100$ ns for PBFA-Z and using the parallel ion viscosity, we find

$$n_{1/2} = 1.4 \times 10^{21} \frac{T_i^{5/2}}{A^{1/2} Z^4}, \text{cm}^{-3}.$$

where A is the atomic number of the ions. The ions are heated to an approximate temperature of $T_i \sim \dfrac{MV^2}{3}$ by the strong shock so that

$$n_{1/2} \sim 1 \times 10^{17} \left[\frac{A}{135}\right]^2 \left[\frac{V}{10^7}\right]^5, \text{cm}^{-3}.$$

We see that for xenon (A = 135) that velocities as low as $V \sim 10^7$ cm/sec and ion densities as low as 10^{17} cm^{-3} results in a significant reduction to the growth rate.

The Switch Effect

We see that hot ions reduce the growth of the M-R-T instability. The hot ions could also degrade the quality of the stagnation. However, the switch effect can result in a high quality stagnation. The point is that the electron density increases rapidly near the end of the implosion. The switch is thrown when the density is high enough resulting in the rapid cooling of the hot ions and in a high quality stagnation and high x-ray yield. This is seen in the rad-hydro simulation calculations. This may also have been seen in the experiments (2). An abrupt degradation in the quality of the stagnation was observed as the load mass was decreased. The radius of the stagnation plasma increased from ~ mm to ~ 2.5 mm and the keV x-ray yield dropped by a factor of ~ 5 when the load mass was decreased by a factor of 0.6. An experiment was performed to determine if this abrupt transition was due to the lowered electron density. The load was a 50/50 mixture (by pressure) of neon and krypton with a load mass approximately the same as the high quality krypton shot. The krypton mass was approximately the same as the low quality krypton shot, but the electron density was approximately the same as in the high quality krypton shot. The results were close to the high quality krypton shot.

RAD-HYDRO SIMULATIONS

A series of 1-D rad-hydro simulations (6) were performed to compare to the large diameter krypton experiments (2). The idea was that if viscosity and finite gyro radius effects significantly reduce the growth of the M-R-T instability, then the 1-D calculations should roughly agree with the experiments. Non-LTE modeling of the population of the excited ionic states was crucial for these calculations. The point is that the electron densities are so low that the atomic states are not in equilibrium with the electrons. The simulations agree fairly well with the measurements for load masses near the maximum x-ray yield. Calculations of x-ray yields, radii of the plasma at stagnation, and the abrupt transition at low load mass agree with the measured values. Calculations with significantly larger (X2) than optimum masses show cold ions in the acceleration region and radiative collapse in strong disagreement with the measured radii of the stagnation plasma. A series of 1-D rad-hydro calculations were performed to evaluate the high energy x-ray (x-ray energies 26 keV) yield with the PBFA-Z accelerator at Sandia National Laboratories. Two designs were considered. In the first, a 5 cm initial radius uniform fill gas puff load with a mixture of argon and xenon was considered. The optimum yield of high energy x-rays (energies > 26 keV) was 24 kJ for a mixture of 20% xenon and 80% argon. The second design was a aluminum wire array imploded onto a barium loaded CH fiber. The optimum yield of high energy x-rays (energies > 26 keV) was 26 kJ.

SUMMARY

We have considered techniques to produce a significant yield of high energy x-rays (energy > 26 keV) on PBFA Z. Two load designs were considered. The first is a uniform-fill, gas puff load. Snowplow, ion viscosity, and finite ion gyro radius effects reduce the growth rate of the magneto-Rayleigh-Taylor instability. 1-D rad-hydro simulations indicate that a high energy x-ray (energy > 26 keV) yield of 24 kJ is produced by a 5 cm initial radius, 20% xenon/80% argon, uniform-fill, gas puff load. The second design uses an aluminum wire array imploding onto a barium loaded CH fiber. 1-D rad-hydro calculations indicate that a high energy x-ray (energy > 26 keV) yield of 26 kJ will be produced by a 2 cm wire array imploding onto a 0.3 cm radius CH fiber loaded with 0.003 atomic fraction of barium.

ACKNOWLEDGMENTS

Work performed under the auspices of the U. S. Department of Energy by the Lawrence Livermore National Laboratory under Contract W-7405-ENG-48 and Sandia National Laboratories under Contract DE-AC04-94AL85000.

References

(1) J. H. Hammer, J. L. Eddleman, P. T. Springer, M. Tabak, A. Toor, K. L. Wong, G. B. Zimmerman, C. Deeney, R. Humphres, T. J. Nash, T. W. L. Sanford T, R. B. Spielman, and J. S. De Groot, Phys. Plasmas **3**, 2063 (1996)
(2) R. B. Spielman, S. Breeze, J. S. McGurn , T. J. Nash, L. E. Ruggles, J. F. Seamen, K. W. Struve, M. Vargas, and D. Jobe, and J. S. DeGroot, Abstract in B.A.P.S. and presented at the APS/DPP meeting, 1995.
(3) T. W. L Sanford, B. M. Marder, R. B. Spielman, T. J. Nash, M. Douglas, C. Deeney, K. Struve, R. C. Mock, J. F. Seamen, J. S. McGurn, D. Jobe, T. L. Gilliland, M. Vargas, J. P Apruzese, J. W. Thornhill, P. E. Pulsifer, K. G. Whitney, J. Davis, S. L. Cochran, A. L. Velikovich, B. V. Weber, G. Peterson, D. Mosher, J. S. De Groot, and J. H. Hammer, Proceedings of the 23rd IEEE International Conference on Plasma Science, June 3-5, 1996.
(4) S. M. Gol'berg and A. L. Velikovich, Phys. Plasmas B **5**, 1164 (1993); J. S. De Groot, A. Toor, S. M Gol'berg, and M. A. Liberman Phys. Plasmas 4, 737 (1997).
(5) A. G. Gonzalez, J. Gratton, and F. T. Gratton (1989) Dense Z-Pinches, ed. by N.R.Pereira, J.Davis and N.Rostoker (AIP Conference Proceedings, vol. 195; AIP, New York), p. 280.
(6) J. S. De Groot, C. Deeney, K. G. Estabrook, J. H. Hammer, D. Ryutov, T. W. L. Sanford, R. B. Spielman, and A. Toor, UCD PRG -M-276 and to be submitted to Phys. Plasmas, 1997.

CHARACTERISTICS OF X-RAY EMISSION IN A COMPOSITE PINCH

C. Dumitrescu-Zoita, P. Choi, A. Chuvatin, B. Etlicher

Laboratoire de Physique des Milieux Ionisés Ecole Polytechnique, 91128 Palaiseau, France

ABSTRACT

Experiments on the GIT-4 pulsed power generator have been performed in order to find out more about the composite pinch formation, stability and performances [1,2]. Different fibre materials and different external shell masses were tried out, in order to study their influence on the pinch behavior [3]. We present here results obtained from space resolved multi-pinhole, multi-filter X-ray camera under a variety of load conditions. The analysis was performed using a filter analysis [4] code with a full simulated spectra from the high Z materials. The temperature distribution along the composite pinch in those different experimental situations has been estimated, in different spectral regions. The influence of shell mass loading on the pinch uniformity will be discussed.

1. Introduction

In the past few years, a number of experiments have been prepared, in international cooperation, to explore a new concept for generating dense stable plasma. The concept is based on a composite load of a plasma shell surrounding a micron size near solid density plasma. In this paper, we present an observation on the plasma properties in the X-ray region using a multi-pinhole multi-filter camera on a composite load experiment on a MA IES generator GIT-4. The experiments were designed to explore the parameter space where effective current transfer can take place and to study the characteristics of the high energy density plasma produced.

2. Composite Loads

In the composite pinch scheme with a plasma cylinder surrounding the fibre, the initial current flows in the plasma jacket, which is then compressed down towards the fibre as current increases. The large diameter of the jacket provides a substantially lower inductance and therefore better coupling to the driver during most of the current rise. The compression of the surrounding plasma continues until it stagnates onto the fibre, at which point, the current switches over to be shared by the fibre. By optimizing the shell mass and material, two modes of current transfer can be considered. In the first mode the shell plasma remains relatively cold at stagnation, and a large proportion of the total current flowing could be transferred to the fibre in a much shorter time compared with the current delivery time, simply by a consideration of the resistivities of the two plasmas. In the second mode, the shell plasma is expected to disrupt at the final compression and becomes high impedance. This plasma opening switch type operation will transfer a large portion of the total current to the load in a time short compared with the current delivery time [2]. Through finite resistivity and radiation coupling, the central fibre is likely to develop a coronal plasma prior to stagnation of the outer shell. The presence of the jacket, however, serves to constraint the otherwise free expansion of the corona. The end result is that a very large current delivery rate could be established on a small diameter preformed plasma of high density. In the present experiment, only the second mode of operation was attempted.

3. Results on GIT-4

The parameters of GIT-4 and the details of the experimental conditions were reported previously [4]. The load was arranged as a double shell configuration with a wire mounted in the center, Fig. 1. Very low mass was placed on the outermost shell, which serves only to further sharpen the current delivery time. Argon gas was used for the outer shell and wires of different materials and sizes were used as the central load. The amount of shell mass was controlled by the delay time between energizing the injection nozzle and the discharge moment. For delay times as those used during our measurements, between 75 and 450 μsec, the mass of gas varies between 10 - 90 μg/cm. For our purpose it is however enough to mention the delay time employed for every shot, thus showing clearly whether we have a shot with more or less shell mass.

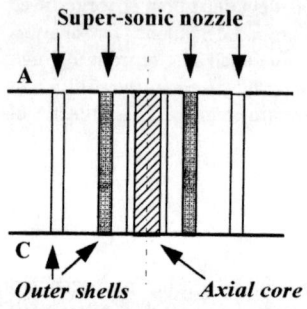

Fig. 1 Schematic of load on GIT-4

As already mentioned before, the experiments on GIT were performed in order to find out more about the composite pinch formation, stability and performances. A range of different wire materials with different shell masses were tried, in order to study their influence on the pinch behavior. There are many aspects related to this subject, but we will present here only some results obtained with the multi-pinhole multi-filter camera we have developed, [4], in order to obtain spatial information about the plasma parameters in multi-component discharges. The analysis was done using the filter code, XRAYFIL, which calculates the expected flux on a detector from a radiation source, often passing through selected filters. The code uses a full K-shell spectra with model, high temperature, high Z plasma [5].

The camera was equipped with 20 pinholes, covered by a variety of filters, selected to give information in different spectral ranges of the elements of interest. As an example Fig. 2 shows the transmission characteristics of three pairs of filters, each of them looking at a certain spectral window: the pair Ti (25 μm) and C (200 μm) is designed to give information about emission from the Titanium K-shell and Argon K-shell recombination, in the 3 - 9 keV region; the pair C (100 μm) + Al (90 μm) and Cu (30 μm), is mainly looking at Titanium recombination emission, in a region that covers the 6 - 12 keV region, while the third pair, Mo (75 μm) - Sn

(70 µm), presented information on hard X-rays up to 30 keV. Other softer filters were used to isolate the Argon K-shell emission or the Aluminium K-shell emission. Stacked films of HP5 and DEF were used to record the signal, providing additional spectral discrimination.

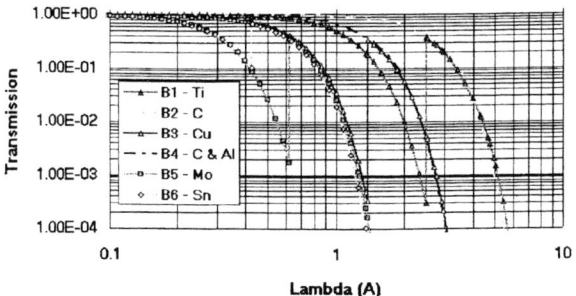

Fig.2 Transmission curve of filter set used for channel B1-Ti(25 µm), B2 - C(200 µm), B3 - Cu(30 µm), B4 - C(100 µm)+Al(90 µm), B5 - Mo(75 µm) & B6 - Sn(70 µm)

Several shots were done without any wire using only argon gas in order to obtain a liner type pinch. These shots present an extremely non uniform, non-symmetric pinch between a delay time as short as 175 µsec, or as long as 450 µsec. Analysis using an argon spectra shows a rather uniform temperature along the pinch of 500 eV. For shots with lower argon mass, this temperature was, as expected, found to be slightly higher, around 650 eV. Operating with pure Ar gas shells gave us the reference point for the composite pinch experiments.

Fig. 3 shows a recorded image and the unfolded temperature distribution along the pinch. In general, higher temperatures are registered in regions of lower emission. It should be emphasized that the average plasma temperature derived from softer filtering is below 1.5 keV.

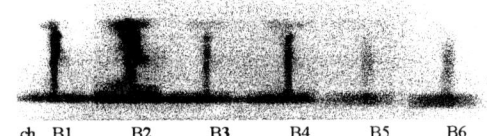

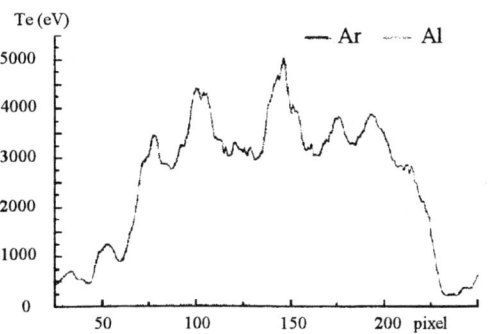

Analyzing results on different shots, we can conclude that, in general, the aluminium wire shots with argon gas shell present a higher average temperature than similar shots done with wires of higher Z.

The influence of the argon shell can be clearly observable throughout the experiments. In the case of a low argon mass, the image looks very much like

Fig.3 Recorded multi-filter multi-pinhole images with filter set in Fig.2 and temperature calculated from filter B3/B4 showing variation along a 2 cm pinch for shot 32, with Al wire and Ar shell, delay time 75 µs

the one obtained for exploding wire, presenting a whole line of small, dense points along the axis, where the initial wire was mounted. When the argon mass is increased, the hot spots disappear and extended region of uniform plasma column begins to form. Under the optimum shell mass / wire material-size combination, we have observed very high quality pinch formation.

An example is shown in Fig. 4, for an Ar on Ni Composite Pinch. Comparing the images recorded on HP5 and DEF film for the Ti filtered channel, which records primarily Ar K-shell radiation, it shows that the softer radiation from the shell exhibits a relatively diffuse structure, Particularly on the anode side. On the other hand, the Mo filtered channel, which rejects the Ar K-shell emission, shows clearly the presence of a high density core embedded inside a high temperature Ni plasma column. The presence of both a hot Ar shell and a hot Ni core emission with good axial uniformity is clearly demonstrated.

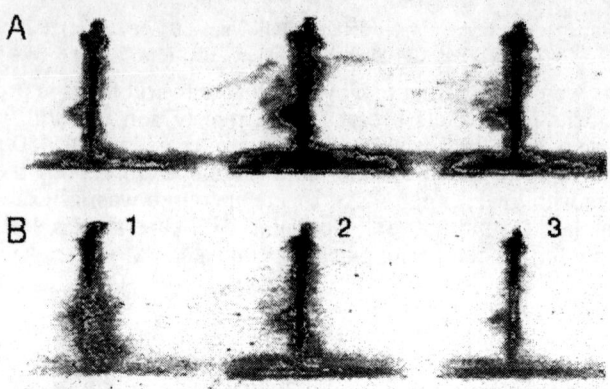

Fig. 4 Recorded multi-filter multi-pinhole camera images for an argon shell on nickel wire Composite Pinch; image A recorded on DEF film and image B recorded on HP5 film, with filter 1 Ti (10.5 µm), filter 2 Al (20 µm) & filter 3 Mo (5 µm).

References

[1] Chuvatin, A., Choi, P., Etlicher, B., Semushin, S., Dumitrescu-Zoita, et al., *Proc. of 10th Int. Pulsed Power Conf., Albuquerque*, **1**, 106-110 (1995)
[2] Chuvatin, A., Choi, P., Etlicher, B., *Phys.Rev.Lett.* **76**, 2282-2285 (1996)
[3] Dumitrescu-Zoita, C., PhD-Thesis, University of Orsay, (1996)
[4] Chuvatin, A., Choi, P., Dumitrescu, C., Etlicher, B. et al., *IEEE Trans.on Pl.Sci.* **23**,2, (1997)
[5] Dumitrescu-Zoita, C., Choi, P., *this proceeding*

IMPROVEMENT OF X RAYS POWER OF GAS PUFF Z'PINCH : DEPENDENCE ON THE INITIAL DENSITY PROFILE.

P.Grua, J.M.Sajer
CEA/CESTA
B.P.2 33114 Le Barp FRANCE

A.Sevastianov
Universite de Bordeaux I
33405 Talence

1 Introduction

Recent Z-pinches experiments[1] conducted on SATURN accelerator have shown that using uniform-fill gas puff loads greatly increases the X-rays radiated power. This enhancement in comparison to the "thin" cylindrical shell case is mostly due to the minimisation of Magneto-Rayleigh-Taylor instabilities .

This paper is devoted to investigate the influence of the initial plasma density profile on the X-rays yield for neon gas. Different profiles ranging from thin shell to filled cylinders are studied with a 1D Radiative-MHD code. In order to compare computed results to experiments, all characteristic scales of simulations are limited to the AMBIORIX generator experimental parameters : maximum electrical current of 2 MA for the implosion of $120\mu g/cm$ in 70ns.

2 Description of the code.

In order to guide Z-pinch experiments realized at CESTA, we have developed the 1D Lagrangian RMHD code PZP-SPECTRO (Radiative MagnetoHydroDynamics). This code operates in cylindrical geometry for Z-pinch applications and takes into account basic processes which intervenes in the MHD description of a two-temperatures plasma (ion and electron temperatures). That two-temperatures description is required to allow the coupling between MHD and atomic physics models, and to provide an accurate reproduction of transport mechanisms for each particle species. The processes integrated in the code are the magnetic diffusion, the Joule heating which is associated, the ion and electron thermal conductions and viscosities. The whole set of Braginskii-Balescu transport coefficients [2, 3] are implemented. These coefficients have the particularity of describing precisely the magnetic field effects on transport. In some configurations the usual Braginskii's approach for transport may fail, in these cases flux limitations based on non local transport theory are adopted. The Richtmyer & Morton's numerical integration scheme is used [4], it contains two numerical viscosities (quadratic and linear).

Ion equation of state is of the ideal gas type, the electrons being described by a partially ionised gas EOS[5] which depends on the ionization energy and the average ion charge. These quantities are computed with a Detailed Configurations Collisional Radiative Model. This model devoted to the K-shell spectroscopy provides also the power radiated in lines and in the continuum. The approximation of optically thin plasma is used. It must be noticed that this approximation is valid only until the end of the first intense radiative phase.

3 Simulation results

Three types of initial density profiles of neon plasmas are considered. The first one corresponds to a thin cylindrical shell, the second to the experimental profile and the third to an uniform-fill liner. For each case, we present the following calculation results : density profiles, operation diagrams, space averaged electron and ion temperatures and radiated energies. The figures 1-4, 5-8, 9-12 are associated to the three types of profiles, respectively. It is observed that the amount of radiated energy increases as the initial density profile widens. The figures 3, 7, 11 show that the radiated energy and electron temperature are related. This is consistent with the fact that for a given density, the radiative power is maximum if the electron temperature lies between 400-500 eV. Since the plasma dynamics is dominated by strong shocks, the ions are predominantly heated. The electron temperature enhancement is related with the density level in the inner part of plasma, which determines the collision rates in this region. Since the collision times are proportionnal to the density, the difference between electron and ion temperatures will decrease as density increases. That is favourable for the plasma to radiate and lose internal energy, and therefore kinetic pressure. For the case of a thin shell, the pressure in the inner part remains very high and does not allow high compression ratios, the radiated energy reaches 6 kJ. The radiated energy is 15 kJ for the experimental profile, which is in good agreement with experiments. The highest radiation yield is obtained with the full cylinder liner and reaches 20 kJ.

4 Conclusion

We have analyzed the influence of the initial density profile on the radiative yield of gas puff Z-pinches using a 1D radiation magnetohydrodynamic code . Results of simulation predict that for a generator like AMBIORIX the highest radiative yield is obtained with a uniform-fill load. On the contrary the classical profile (cylindrical shell) generates the minimum of radiated energy.

References

[1] M.K. Matzen, Bulletin of the A.P.S, Vol 41,No 7 (1996)

[2] S. Braginskii, in *Review of Plasma Physics*, M. Leontovitch (Consultants Bureau, New-York, 1969), Vol. 1.

[3] R. Balescu, in *Transport Processes in Plasmas, Vol. 1*, Elsevier Science Publishers, (1988).

[4] Richtmyer & Morton, *Difference methods for initial value problems*, Interscience Publishers.

[5] R.M. More, *Plasma Processes in non-Ideal Plasma*, UCRL -94360 (1986)

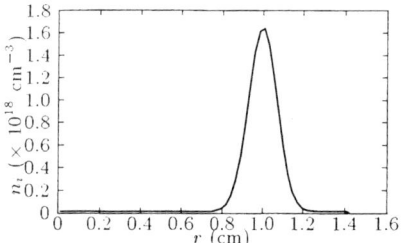

Fig.1 Initial Density Profile.

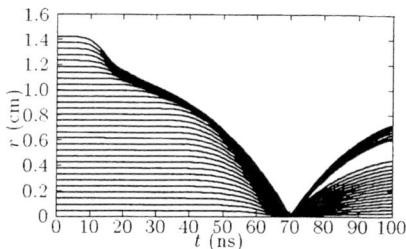

Fig.2 Radius versus time for each lagrangian cell

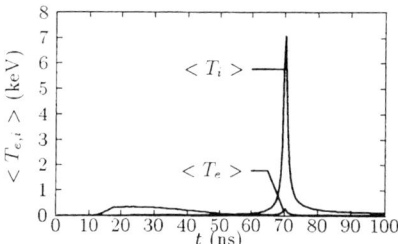

Fig.3 Space averaged electron and ion temperatures.

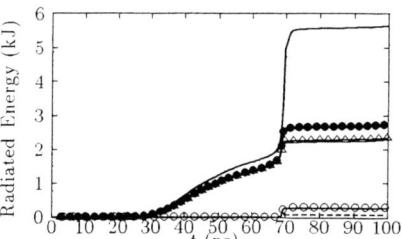

Fig.4 Radiated energy in line H_α (- -), He_α (-○-), in all other lines(mainly LI like lines) (-△-), total energy radiated in lines(-●-) total radiated energy (—)

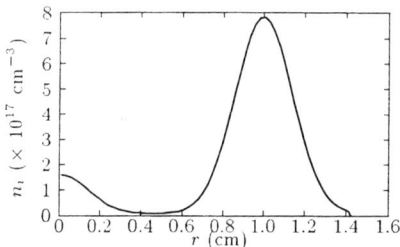

Fig.5 Initial Density Profile.

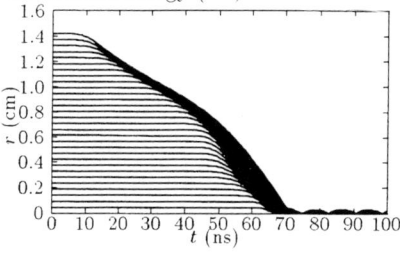

Fig.6 Radius versus time for each lagrangian cell

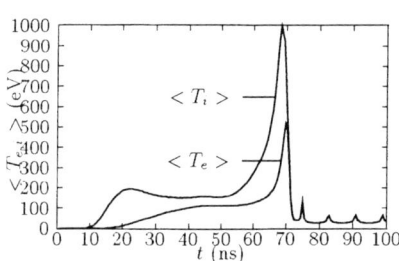

Fig.3 Space averaged electron and ion temperatures.

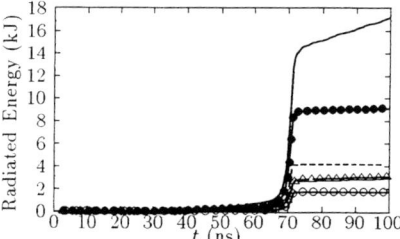

Fig.8 Radiated energy in line H_α (- -), He_α (-○-), in all other lines(mainly LI like lines) (-△-), total energy radiated in lines(-●-) total radiated energy (—)

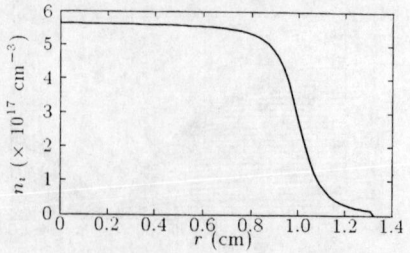

Fig.9 Initial Density Profile.

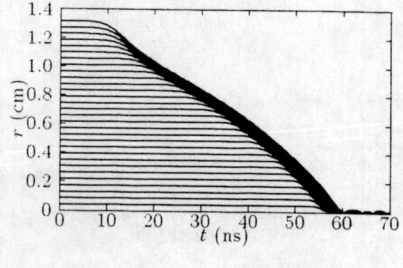

Fig.10 Radius versus time for each lagrangian cell

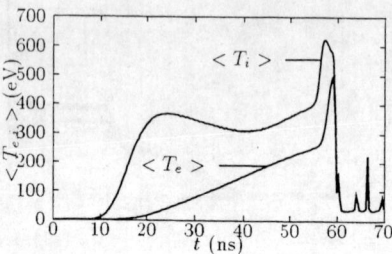

Fig.11 Space averaged electron and ion temperatures.

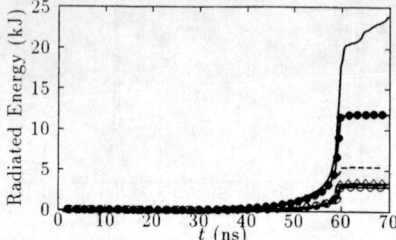

Fig.12 Radiated energy in line H_α (-△-), He_α (-○-), in all other lines(mainly LI like lines) (- -), total energy radiated in lines(-●-) total radiated energy (—)

Plasma Focus Current Shell Implosion onto Foam Liner

L.Karpiński, M.Scholz, W.Stepniewski,

Institute of Plasma Physics and Laser Microfusion, IPPLM, Warsaw, Hery,23, Poland

A.Szydlowski,

Soltan Institute for Nuclear Studies,Swerk, n. Warsaw,Poland

A.V.Branitski, M.V.Fedulov, S.F.Medovschikov, S.L.Nedoseev, V.P Smirnov, M.V.Zurin

Troitsk Institute for Innovation and Fusion Research, TRINITI, 142092, Troitsk, Mosc.reg., Russia.

Abstract. Results of first experiments on powerful plasma focus PF-1000 (IFPiLM) [1], loaded by low mass foam liners, produced for „Angara" (TRINITI) liner implosion program [2], are presented in this report. An interaction of PF current plasma shell and the liner was the problem of main interest in our joint experiment. The main goal is to elaborate new method of plasma liner production for its compact implosion using multiterawatt pulsed power driver.

INTRODUCTION

As the matter of fact, all multiterawatt drivers for liners and Z-pinch implosion use "cold start" of the process. It means that plasma generation and posterior plasma liner implosion are realized by the same current generator. The initial current through cold liner substance is filamented in high extent due-to ionization instabilities. This current contraction is provided by plasma electric conductivity and effects of magnetic pressure are small initially. But these magnetic forces become significant with the current rise. So, the initial current filament could evolve under effect of two alternative factors: expansion due-to Joule current heating and compression due-to the current magnetic pressure. It is possible to propose that the current rise rate, specific for the powerful drivers, is too high to allow the filaments to expand, to flow together and produce azimuthally homogeneous liner plasma; magnetic field around the separate current filament could confine it. Being responsible for liner azimuthal inhomogeneities, the filaments could enhance axial inhomogeneities of initial liner plasma, which are especially dangerous as R-T and MHD- instabilities basis. Moreover, it is shown in experiments on "Angara-5-1" that cold start effects can destroy liner earlier liner implosion beginning [2].

Consequently, an effective preionization of liner substance is the necessary condition for the compact liner implosion. We treat the PF current shell as the one of possible preionizers for multiterawatt driver. The high current plasma shell, being accelerated during some microseconds in PF accelerator, could deliver its kinetic, thermal and magnetic energy in rather short period, ~ 100 ns, to liner, positioned in the shell focus region, providing appropriate initial conditions for fast implosion. It is necessary to mention here that PF current shell fall onto bubble or gas liners was investigated elsewhere [4-6]. We used the microgeterogeneous solid foam as the plasma producing substance, because our foam liner technology allowed to produce liners with different sizes, shaping and radiative dopants. Homogeneity of plasma liner produced was the main problem of interest.

EXPERIMENTAL SCHEME AND DIAGNOSTICS

The experimental scheme is shown on Fig.1. PF current shell imploded onto agar foam liner, positioned on the top of inner electrode of PF-1000 coaxial accelerator. Foam liners were produced by vacuum drying of thin wall cylinders made from frozen agar gel (1.5 - 3 mg agar in 1 g of water). Liners had diameter 20 mm, 15 mm length, 200 microgram/cm or diameter 5.4 mm, 15-20 mm length, 250-280 mg/cm. An axial insulating rod of 5 mm diameter was used to support in horizontal position the ϕ20 mm foam liner with top electrode. The 5.4 mm foam liner had no axial support and no top electrode. The metallic foil current shunt - ϕ 20 mm liner imitator was installed in some shots.

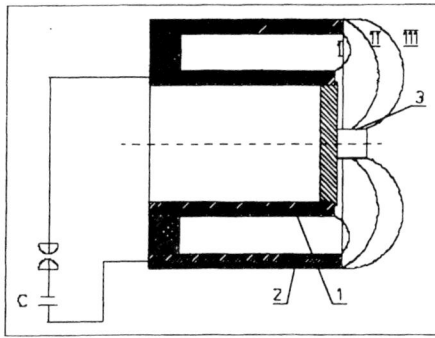

Figure.1. Experimental scheme.
1. - PF inner electrode, anode; 2-PF external electrode, cathode; 3-foam liner; I, II, III- different positions of the PF current shell.

Operating parameters of PF-1000 were: C~ 0.001 F, V_{max}=25 kV, hydrogen pressure ~5 Torr, current I_{max}~1MA, current rise time 4-5 microsecond.

Measurements of **dI/dt** and **V**(t) were done to evaluate the total active power in discharge circuit W_{act} = **VI - LI(dI/dt),** L=L(t)-total inductivity of the circuit with moving current shell. Visible light streak camera with radial slit at the liner middle and optic frame camera (10 frames, 1.6 ns interval between frames) were used for current shell and liner dynamics investigation. Time integrated soft X-ray pinhole camera, filtered by 10-25 microns Be, allowed to analyze soft X-ray radiation during shell - liner contact and the composition implosion. Time resolved VUV and soft X-ray radiation were measured by a set of filtered vacuum photodiodes (XRDs). The VUV and soft X-ray radiation power was computed by linear combinations method, described in [6]. Signals were registered by Tektronix 2430A and THS-700 oscilloscopes and synchronized with of **dI/dt** and **V**(t) traces.

EXPERIMENTAL RESULTS

Typical signals of **dI/dt** (Idot, MA/(mks.div), voltage **V** (10 kV/div), current **I** (MA/div), active power W_{act} (10 GW/div)are shown on Fig.2.

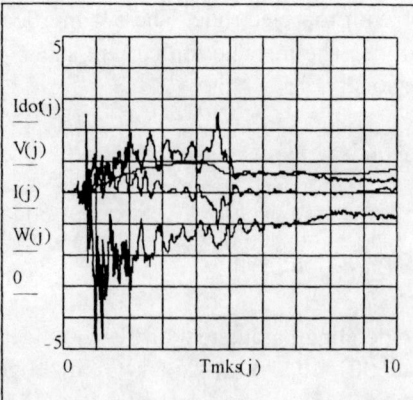

Figure2. Typical signals of dI/dt (Idot, MA/(mks.div), voltage V (10 kV/div), current I (MA/div), active power W (10 GW/div).

Measured maximum of W_{act} during shell acceleration (~4 ms) was ~20 GW, it achieved 20-30 GW in period of the shell - liner collision and implosion, its duration being 150-300 ns. After hydrogen current shell coming to the ϕ20 mm liner the external margin of foam plasma becomes to expand during ~100 ns till ϕ22 mm, after that it implodes to axial rod with specific velocity ~2 cm/ms. The liner plasma reflects from the rod with almost the same velocity after implosion (see Fig.3a). The visible external margin of 5.4 mm liner had no significant compression or expansion during ~0.6 ms after the shell - liner contact (see Fig.3b). Nevertheless, the pretty shaped plasma column of ~2 mm diameter and ~10 mm length with the core of ~1 mm diameter was registered by pinhole camera in quanta energy >600 eV(see Fig4a). No „hot spots" present on this picture.

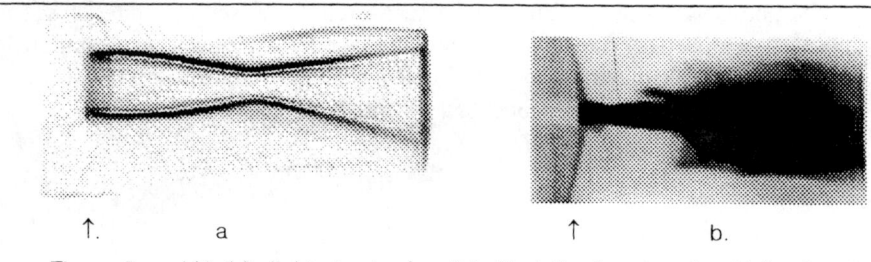

Figure.3. a, bVisible light streak of radial slit at the liner length middle. Streak duration 2 μs. Fig.3a-liner ϕ 20 mm with supporting axial rod of ϕ 5 mm; Fig.3b-liner ϕ 4.5 mm , no supporting rod. ↑- PF current shell image coming to the foam liner.

X-ray measurements.

VUV and soft X-ray radiation arises during 300-500 ns to 10-30% of maximum value. Then intensive narrow (~100-200 ns) peak of emission occurs with front rise time ~50 ns. This is close to frequency band of oscilloscope used. The peak appears in the moment of pinch collapse and corresponds to maximum W_{act}. After the peak emission the radiation intensity continues about 1mks at the level of ~0.3-0.5 of maximal value. This time is some more than the period of "stable" pinch on axis, viewed by optic streak. Then VUV drops during some microseconds

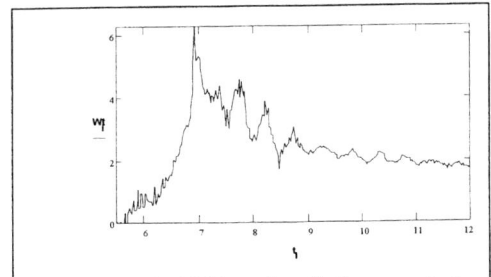

Figure. 4. VUV and soft X-ray radiation power [GW] versus time [μs], for 5.4 mm diameter foam liner

Maximal measured power of VUV and soft X-ray radiation for 5.4 mm diameter foam liner achieved ~2 GW, about 1 GW being measured in 10-120 eV band and almost the same power in 0.12-1 keV region. Taking into account absorption by hydrogen leads to higher power at 10-120eV band. Total power should be estimated as 3.5-8 GW with ~1 GW part harder than 120 eV.

Emission in 0.1-1 keV region should be bound probably with more heated internal region of plasma column, viewed in pinhole photos with 10mm Be filter (hv>0.6keV) (Fig.5a) . This picture shows the pinch of φ 4.5 mm foam liner. The hot core was about 1mm diameter and of almost all initial foam liner length. Plasma corona diameter was about 3-5 mm as measured by optic streak camera, (see Fig.3b). The intensity of hydrogen PF shell radiation , collapsed behind the foam liner of ↔=15 mm length, was relatively small. On the contrary, Fig. 5b shows an alternative case, when the foam liner pinch was absent, but PF current shell pinch was rather strong. We can't explain the reasons of this duality now.

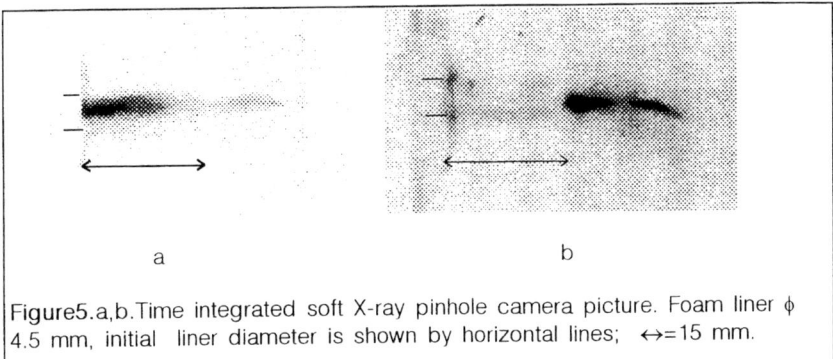

Figure5.a,b.Time integrated soft X-ray pinhole camera picture. Foam liner φ 4.5 mm, initial liner diameter is shown by horizontal lines; ↔=15 mm.

. Preliminary analysis of the radiation spectra gives the plasma temperature evaluation as T=30-50 eV. This value is close to temperature evaluation by Bennett equation at the current of ~ 1MA. Radiation is significantly softer before the maximum peak during the phase of implosion. Power ratio in 10-100 eV and 100-1000 eV regions shows that temperature is lower, probably T<10eV.

DISCUSSION OF EXPERIMENTAL RESULTS

We consider results of this first experiment as optimistic .

1. The low mass PF current shell interaction with higher mass foam liner has produced rather homogeneous foam plasma liner. We couldn't see filaments in liner plasma. Foam liner didn't expand during the process dramatically.
2. The foam plasma produced has temperature T $\geq \sim 20$ eV and , consequently, liner electric conductivity is sufficient for effective implosion beginning by 3-5 MA current with $dI/dt \geq 5.10^{13}$ A/s
3. PF junction with multiterawatt driver and their synchronized operation are possible in principle. We can't couple PF-1000 and "Angara-5-1 " to test preionization effects of ≥ 1 MA current, available on PF-1000. We believe this current level will be good for the drivers, being more powerful than "Angara-5-1". Consequently, testing of this preionization method at ~100 kA current level is planned on "Angara-5-1".

REFERENCES

1. M.Scholz, M.Borowiecki, L.Karpinski, R.Miklaszewski, M.Paduch, W.Stepniewski, K.Tomaszewski, M.Sadowski, A.Szydlowski, V.Gribkov, S.Dubrowski, I.Volobuev. "Investigation of the PF-1000 facility". 2nd National Symposium PLASMA'95,Research and applications of plasmas,Warsaw, Poland, June 26-28, 1995, vol.2, pp.15-21.
2. Branitski A.V., Grabovski E.V. et.al. "Angara-5-1" Program Development on Superfast Liner Implosion for ICF Physics Study and Basic Research". *Proceedings of 11 Intern. Conf. "BEAMS'96"*, Prague, Czech Rep., June 10-14, 1996, vol.1, pp.140-145.
3. R.Appartaim, A.Dangor, H.Kilic, J.G.Linhart, H.Schmidt. "Z-pinch driven foil implosions using liquid film bubbles". *Proc. of workshop on Physics of Alternative Magnetic Confinement Schemes*, Villa Monastero - Varenna, Italy, October 15-24, 1990, pp.985-992.
4. A.Bortolotti, J.G.Linhart, J.Kravarik, P.Kubes. "Plasma driven implosions of a bubble liner". *AIP Conference Proceedings 299*, Dense Z-pinches, 3rd Intern. Conf., London, UK, 1993, pp.372-380.

Dynamic Hohlraum Experiments on SATURN*

T.J. Nash, M.S. Derzon, G. Allshouse, C. Deeney, J.F. Seaman, J. McGurn, D. Jobe, T. Gilliland
Sandia National Laboratory

J.J. MacFarlane and P. Wang
U. Wisconsin

D.L. Petersen
Los Alamos National Laboratory

Abstract

We have imploded a 17.5 mm diameter 120-tungsten-wire array weighing 450 μg/cm onto a 4 mm diameter silicon aerogel foam weighing 650 μg/cm, using the pulsed power driver SATURN. A peak current of 7.0 MA drives a 48 ns implosion to strike time followed by 8 ns of foam compression until stagnation. The tungsten strikes the foam with a 50 cm/μs implosion velocity. Radiation temperatures were measured from the side and along the axis with filtered x-ray diode arrays. There is evidence of radiation trapping by the optically thick tungsten from crystal spectroscopy. The pinch is open to less than a 1 mm diameter as measured by time-resolved x-ray framing cameras. The radiation brightness temperature in the foam reaches 150 eV before the main radiation burst or stagnation.

Introduction

In the concept of the dynamic hohlraum an imploding z-pinch has sufficient mass that it is optically thick to the radiation inside the pinch. The concept is also known by the names flying radiation case(1), imploding liner hohlraum (1a) and double liners (2). The purpose of the dynamic hohlraum is to reach radiation temperatures sufficient to drive ICF capsules and to perform experiments relevant to ICF and high energy density physics.

The dynamic hohlraum temperature scales very favorably with machine current because the pinch mass scales as the current squared and increased pinch mass provides opacity. Also the radiated power from a pinch scales as the square of the current. The scaling of mass and power with current indicate that dynamic hohlraum temperatures will scale more rapidly than the square root of the current. It is important to provide measurements of the radiation temperature on different machines, such as ANGARA (3.5 MA), SATURN (7MA), and PBFAZ (18 MA), to establish the scaling with current. In this paper we present the measurements of dynamic hohlraum radiation temperature on SATURN and discuss the scaling issues.

SATURN is a 20 TW electrical pulsed power machine. (3) It drives up to 10 MA of current into a z-pinch load with a 50 ns rise time. For efficiently radiating z-pinches the load current rises to only 7.0 MA because of pinch inductance.

Recent advances in the stabilization of z-pinches have made possible the demonstration of dynamic hohlraum radiation temperatures well in excess of 100 eV presented in this paper. These advances include the improved radiated power and stability from using large-wire-number arrays (4), and the demonstration of 80 TW of radiation from the accelerator SATURN using a 120-tungsten-wire array (5). The performance of a wire array as a radiator without an inner liner is an indication of its performance with an inner liner. For this reason we used the large-wire-number tungsten array as the outer liner driver for the dynamic hohlraum experiments on SATURN.

Improved stability increases the radiation temperature of a dynamic hohlraum by increasing radiated power and decreasing hohlraum area. Since the radiation temperature goes as the radiated power divided by hohlraum area, increased stability can greatly increase dynamic hohlraum radiation temperature.

Experimental Design

The outer liner is a 120-tungsten-wire array on a 17.5 mm diameter weighing 450 µg/cm. The inner liner is solid cylindrical aerogel foam 4 mm in diameter weighing 650 µg/cm. Both are 2 cm long.

The load current is measured by B-dot monitors placed in the anode plate 4 cm from the pinch axis. The implosion time is measured by the time difference between the current rise and rise of the x-ray pulse as measured by a four channel x-ray diode array.

The x-ray diodes use carbon photocathodes. Four energy cuts are provided by four different filters: 4 µm of kimfol, 1.9 µm of titanium, 2 µm of chromium, and 8 µm of beryllium. The energy cuts of the above filters are at 250, 400, 500, and 800 eV. Two sets of the four-channel x-ray diode arrays are fielded, one views the pinch from the side to record temperatures outside of the outer liner, the other views the pinch from the axis to measure temperatures inside the foam.

For the x-ray diode arrays to measure radiation brightness temperatures the surface area of the imploding pinch must also be measured. The diameter of the pinch as a function of time is measured by two time-resolved pinhole cameras. One is filtered with 5.5 µm of lexan to record emissions in the carbon window at 250 eV. The other is filtered with 8 µm of beryllium to record images of emissions above 800 eV. Both instruments were fielded from the side and both used gated microchannel plates to provide sequential time-resolved images.

A bolometer is fielded from the side of the pinch to measure to total radiated yield. This signal is usually too noisy to be differentiated to give the time history of the total radiated power. A representative x-ray diode channel, typically the kimfol channel at 250 eV, can be integrated in time and normalized to the bolometer yield, to give an estimate of the time-history of the total radiated power.

A time-integrated radially-resolved crystal spectrometer measures the continuum of tungsten emission above 900 eV. It is capable of recording silicon K-shell radiation from the silicon in the aerogel, and can therefore provide information on the optical trapping of the silicon K lines by the tungsten outer liner.

Experimental Results

Figure 1 shows traces of the load current and radiated power versus time for SATURN shot 2261. Peak current is 7 MA and radiated power at stagnation is 54 TW. The power trace has a linear foot before stagnation due to the compression of the foam. Peak radiation temperatures for a dynamic hohlraum are taken at the end of the linear foot just before the inflection point leading to peak power at stagnation. The time from current rise to the striking of the foam is about 48 ns and the time from strike to stagnation is 8 ns.

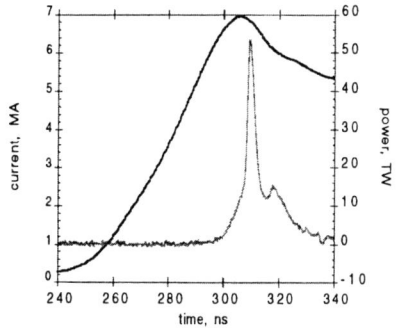

Figure 1. Load current and Radiated Power for SATURN shot 2261

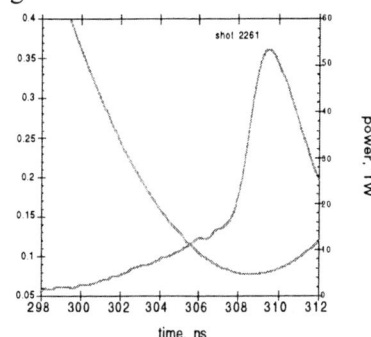

Figure 2. Increasing Power and Decreasing diameter show the concept of the dynamic hohlraum.

In figure 2 we present the same radiated power of figure 1 blown up to show the linear foot and main stagnation burst in more detail. Also plotted in a quadratic fit to the pinch diameter using discrete points from the time-resolved pinhole cameras. With the fit we can produce a continuous time history of the radiation temperatures from the x-ray diodes.

It is worth mentioning that the FWHM of the radiation pulse in shot 2261 is a very sharp 2.5 ns. The fwhm of radiation with the outer tungsten driver alone is 4 ns. (5)

In figure 3 we present the brightness temperatures from each of the x-ray diode channels for both the side-on and axial views. The two soft channels at 250 and 400 eV show temperatures inside the foam significantly greater than the surface temperature outside the foam. This is likely due to the opacity of the tungsten. The data indicate internal temperatures in these channels exceeding 150 eV before compression of the foam is terminated at stagnation. This upper time limit is 307.5 ns on the scale of figure 3. Less trapping is shown in the 500 eV channel and none at all in the 1 keV channel.

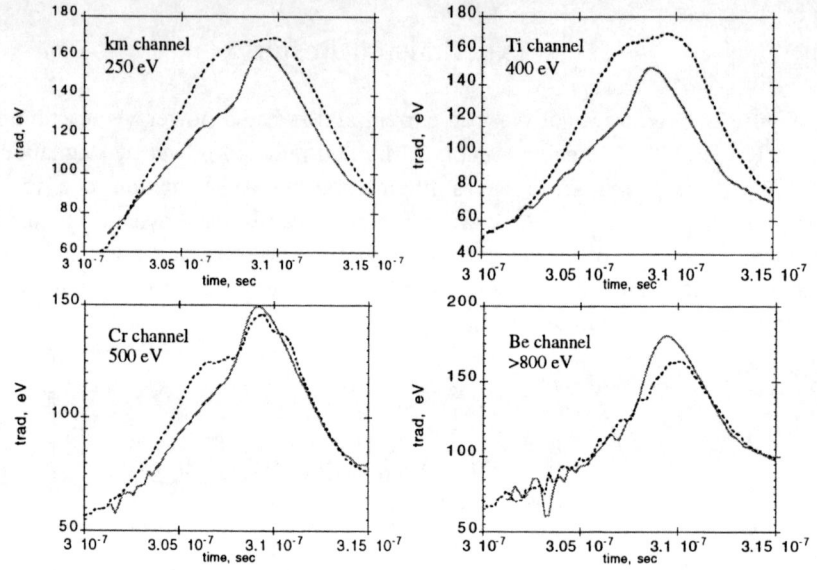

Figure 3. The lower energy XRD channels show radiation temperatures inside the foam exceeding 150 eV before stagnation. Dashed curves are temperatures inside the foam from an axial view. Solid curves are temperatures on the outside of the tungsten from a side view.

The internal radiation temperature, exceeding 150 eV before compression, should scale to temperatures above 200 eV on PBFAZ. Such temperatures are adequate to consider driving ICF capsules. Besides adequate temperature, other issues for driving ICF capsules include stability and symmetry. These issues will be studied on PBFAZ. Rayleigh-Taylor (RT) is the most deleterious instability(6). The large format time-resolved pinhole camera has recorded images with the wavelength of the RT instability during the compression of the foam.

In figure 4 we present the timing gates of frames 4 and 5 of the large format camera with respect to the radiated power trace. Frame 5 is the last frame recorded before stagnation.

In figure 5 we present radial line-outs of the pinch averaged over all z for time frames 4 and 5. One observes the increasing radiation with the increasing foam compression. It is worth noting that the diameter measured from frame 5, 700 µm, actually falls below the value from the quadratic fit, 1 mm, used in the radiation temperature measurements. The large format camera also gives a foam-compression velocity of 25 cm/µs.

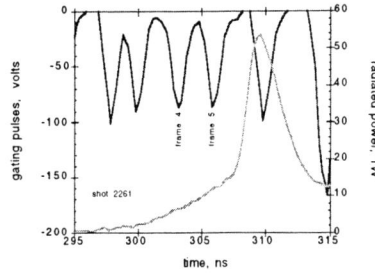 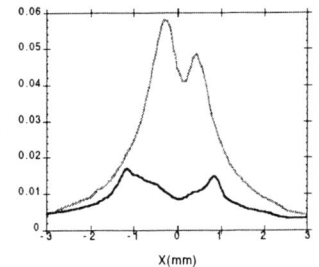

Figure 4. Timing gates of large format pinhole camera with respect to power pulse. Frames 4 and 5 are gated during the compression of the foam.

Figure 5. Horizontal line-outs of large format camera frames 4 and 5. The line-outs show increasing radiation with increasing foam compression.

In figure 6 we present axial line-outs of the pinch averaged over 300 µm in radius at the current sheath for SATURN shot 2261. During frame 5 the RT instability is quite evident with a wavelength of about 400 µm. This instability is the limiting factor in creating a uniform radiation cavity. The RT instability in figure 6 is likely too severe to be useful in driving an ICF capsule. For the dynamic hohlraum to be useful for ICF near the end of the time of foam compression the RT instability must be mitigated. The RT instability may be more severe on PBFAZ than SATURN because the implosion times of 100 ns on PBFAZ are twice as long as the implosion times on SATURN, and this necessitates larger diameter loads.

There is reason to believe that the instability can be overcome from the data of figure 7 which compares the RT of SATURN shot 2261 to that of ANGARA shot 30 taken during US/Russia collaborative experiments in 1993. Both shots had similar strike velocities and foam compression times. For the shot the gating pulse is 1 ns before stagnation. The ANGARA implosion used a xenon gas puff as the outer liner and an annular agar foam doped with molybdenum as the inner liner. The reduced RT of the ANGARA shot may be due to the use an annular foam or due to snow-plow stabilization of the outer liner implosion (7).

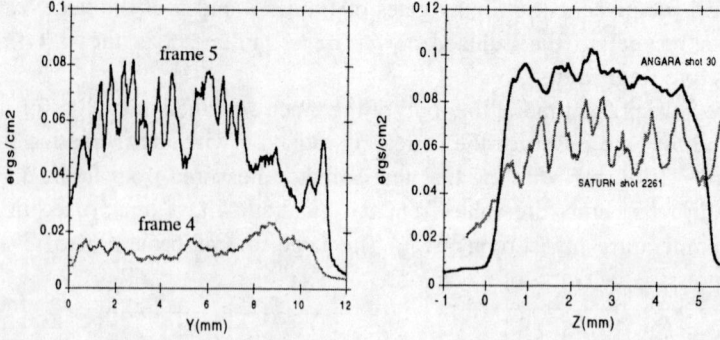

Figure 6. Vertical line-outs along z over 300 μm in radius reveal the growth of Rayleigh-Taylor instability as the foam is compressed.

Figure 7. The annular target of ANGARA shot 30 shows less Rayleigh-Taylor instability than the solid foam of SATURN shot 2261.

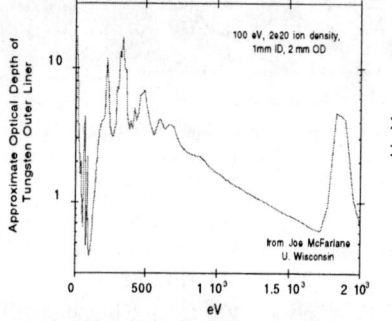
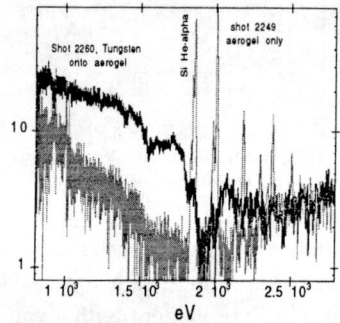

Figure 8. The optical depth of the tungsten should provide some trapping of radiation in the foam.

Figure 9. A time-integrated crystal spectrometer shows that the tungsten is trapping silicon K radiation. The two shots depicted are with and without the tungsten wire array.

An advantageous feature of the dynamic hohlraum z-pinch is that the outer liner will trap radiation inside the foam provided the RT instability does not make it leaky. An estimate of the optical depth of the tungsten for SATURN shot 2261 is shown in figure 8. (8) The trace indicates that significant trapping will occur for energies below 1 keV. This computational prediction is born out by the x-ray diode traces of figure 3 which show radiation trapping in all but the 1 keV channel.

In figure 9 we show data from the time-integrated crystal spectrometer which indicates trapping of silicon K line radiation. This would be due to the tungsten M shell absorption feature at 1.9 keV in figure 8. Figure 9 compares the crystal spectra above 900 eV for SATURN shot 2260 (identical to shot 2261), and

SATURN shot 2249, a shot with only aerogel and no tungsten. Shot 2249 emits silicon K line radiation, while the dynamic hohlraum shot 2260 does not, even at stagnation. Either the tungsten is trapping the silicon K line radiation or the electron temperatures in the dynamic hohlraum are never hot enough (200 eV) to excite the silicon K line radiation.

Radiation temperatures inside dynamic hohlraums have the potential for scaling greater than the square root of the machine current due to the benefit of optical trapping of the internal radiation. This benefit of trapping is predicted to be evident in dynamic hohlraum experiments to be performed on PBFAZ, an 18 MA driver. In figure 10 we compare the predictions of the optical depth of the tungsten for SATURN and PBFAZ.(8) The optical depth will increase as the square of the current. This is in addition the current-squared scaling of radiated power due to the machine power itself.

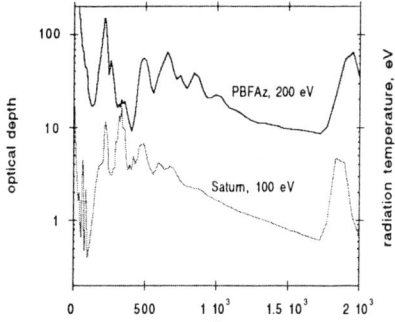

Figure 10. The increased mass loading of the machine PBFAz should provide better trapping of radiation in the foam with respect to the smaller machine SATURN.

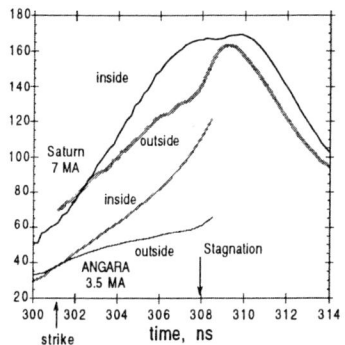

Figure 11. Comparison of radiation temperature on ANGARA and SATURN shows favorable scaling with machine current.

With regard to scaling of dynamic hohlraum temperature with machine current we show inside and outside radiation temperature time histories for SATURN shot 2261 and ANGARA shot 30 in figure 11. Both machines provide similar strike velocities and foam compression times. The peak radiation temperatures before stagnation show scaling slightly greater than the square root of the current. This is to be expected as the mass driven by SATURN is transitional between an optically thin and optically thick radiation case.

Conclusion

We have imploded a 17.5 mm diameter 120-tungsten-wire array weighing 450 µg/cm onto a 4 mm diameter silicon aerogel foam weighing 650 µg/cm, using the pulsed power driver SATURN. Measurements using x-ray diodes and time-resolved pinhole cameras indicate that the radiation temperature inside the foam exceeds 150 eV before stagnation. The narrow 2.5 ns fwhm of the stagnation radiation pulse indicates excellent pinch stability. Stability increases dynamic hohlraum temperature in three ways: increased radiated power, decreased hohlraum area, and increased radiation trapping. Radiation temperature may scale more rapidly than the square root of the current due to optical trapping. Presently Rayleigh-Taylor instability is the limiting factor in achieving high temperature in dynamic hohlraums.

*Sandia is a multiprogram laboratory operated by Sandia Corporation, a Lockheed Martin Company, for the United States Department of Energy under Contract DE-AC04-94AL85000.

References

1. R. Bowers and J. Brownell, private communication
1a. T. Hussey and K. Matzen, private communication
2. V. P. Smirnov, Plasma Physics and Controlled Fusion, **33**, #13, 1697, (1991)
3. R. B. Spielman, M.K. Matzen, M.A. Palmer, P.B. Rand, T.W. Hussey, and D.H. McDaniel, Appl. Phys. Lett., **47**, 229, (1985).
4. T.W.L. Sanford, T.J. Nash, B.M. Marder, R.C. Mock, M.R. Douglas, R.B. Spielman, J.F. Seaman, J.S. McGurn, D.O. Jobe, T.L. Gilliland, M.F. Vargas, R. Humphreys, K.W. Struve, W.A. Stygar, J.H. Hammer, J.H. DeGroot, J.S. Eddleman, K.G. Whitney, J.W. Thornhill, P.E. Pulsifer, J.P. Apruzese, D. Mosher, Y. Maron, Proc. oth the 11th Confg. on High Power Particle Beams, ed. P. Sunka, K. Jungwrith and J. Ullschmied, Prague, June, (1996), paper O-4-2.
5. C. Deeney, T.J. Nash, R.B. Spielman, J.F. Seaman, G. Chandler, K.W. Struve, J.L. Porter, W.A. Stygar, J.S. McGurn, D.O. Jobe, T.L. Gilliland, J.A. Torres, M.F. Vargas, L.E. Ruggles, S. Breeze, R.C. Mock, M.R. Douglas, D. Fehl, D.H. McDaniel, and M.K. Matzen, submitted to Phys. Rev. E., (1996).
6. D.L. Petersen, R.L. Bowers, J.H. Brownell, A. E. Greene, K.D. McLenithan, T.A. Oliphant, N.F. Roderick, and A.J. Scannapieco, Phus. Plasmas **3** (1), P. 368, January 1996
7. Alexander Velikovich, F.L. Cochran, and J. Davis, Phys. Rev. Lett., **77** (5), p. 853, July 29, 1996
8. Joe MacFarlane, U. Wisconsin, private communication

INCREASING A Z-PINCH'S HARD X-RAY YIELD

L. I. RUDAKOV

Russian Research Center 'Kurchatov Institute,' 123182 Moscow, Russia

Introduction

The last decade has seen enormous progress in pulse power techniques to create high currents for making x-rays with hot, dense Z-pinches. In 1996 Sandia Laboratories' PBFA-Z accelerator achieved an impressive 15 MA in 100 ns,[1] creating a macroscopically stable 2 cm long tungsten plasma radiating 1.5 MJ soft black body radiation during 7 ns. So bright a soft x-ray source has many applications, e.g., for ICF as contemplated by Sandia. Other applications need harder x-rays, but even from PBFA-Z the yield of hard x-rays with energy above 10 keV is very small. The problem now is to increase the hard x-ray yield from a compressed plasma.

Radiation from an Imploding Plasma

An experimentalist can not easily upgrade its pulse generator, which is limited to a given pulse time t and peak current I. With I in MA the current's magnetic field can transfer to the plasma at most an energy $\mathcal{E}$ (in MJ/cm) of about $\mathcal{E} = \Delta L I^2 / 2 \simeq 10^{-2} I^2 \ln R_0/R_{min}$. Here R_0 and R_{min} are the initial and the final radii of the plasma cylinder. We want to obtain K_α radiation, for example from krypton, 12.5 keV, so that this type of radiation presents a significant part of imploding plasma energy. This implies that the implosion velocity to the axis must be about 10^8 cm/s so that the energy for each electron of the accelerated atoms can be about 10 keV. Given the machine energy the number of atoms which could be accelerated is determined. The initial radius $R_0 = Vt$ of the plasma cylinder is also determined, because for highest radiation output the plasma must reach the axis at peak current. J. Davis from NRL has shown [2] in a two-dimensional simulation that significant K_α yield is obtained when the energy exceeds 10 MJ and the current 50 MA.

The K_α output per unit length of a Z-pinch with atoms with atomic number $Z \gg 1$ in the helium-like ionization state is

$$Q_r = \int dt P_r = 10^{-32} Z^2 T^{1/2} n_i N A(Z,T) t_0 \propto \mathcal{E}^2/V R_{min} \, (\text{W/cm}), \quad (1)$$

where N is the number of atoms per unit length, n_i is the ion density, and A is a number between 1 to 10. To improve the radiation yield one might try to increase n_i by better compressing the plasma, for example by employing a hollow plasma cylinder. However, compression of a thin plasma shell is unstable to the Rayleigh-Taylor instability.

Liner-Converter Scheme [3]

Increased radiation output is obtained here by separating energy storage and radiation. It is an astonishing coincidence, but at the desired temperature

$T_e = 10$ keV the electron heat conduction is sufficient to conduct the necessary energy from the imploding plasma axially to its edge, where we place the converter: the converter is a layer with a well-defined mass, thickness and atomic composition. The electron heat flux q_e in a plasma is:

$$q_e(\text{W/cm}^2) = \frac{10}{Z}\nabla T_e^{7/2}(\text{eV}), \qquad (2)$$

For $Z = 1$ and $T_e = 10$ keV the heat flux is $q_e \simeq 10^{15}$ W/cm^2 along a 1 cm long plasma, or through each square centimeter in 10 ns about 10 MJ. The approach is efficient for generation of both hard and soft X-rays. We can change the radiation spectrum by picking the converter's atomic composition: thick uranium converter produces soft radiation. For efficient production of hard X-rays the converter must have a specific thickness and density. The thickness d must be chosen close to R_{min}, but the converter will expand because of heating. Acording to Eq. (1) the converter's density should be as large as possible. Energy balance gives

$$\begin{aligned}EL &= \frac{3}{2}(n_e T_e + n_i T_i)\pi R_{min}^2 L \gtrsim \int dt P_r + [\frac{3}{2}n_e T_e \pi R_{min}^2 d]_{conv} \\ &= A_c(Z,T)N_c^2/(\pi V R_{min}^2) + \frac{3}{2}T_e N_e + E_0 d,\end{aligned} \qquad (3)$$

where the number of electrons in the converter N_c must be less than the number of electrons in the Z-pinch. Here E_0 is the energy needed to create a He-like plasma, including ionization and radiation loss. In our scheme, fast heating of the converter by the electron heat flux, E_0 should be much less than that of imploding plasmas. Thus, the optimal radiative capacity of the converter with $d < R_{min}$ becomes

$$Q_r = 10^{-32} A_c(Z,T) E^2 L/(T_e^{3/2} V R_{min} d) \text{W/cm}. \qquad (4)$$

The radiative yield of the converter, Eq. (4), exceeds the radiative yield of the Z-pinch, Eq. (1), by a factor $L/d \gg 1$.

The Heterogeneous Pinch

The more efficient Liner-Converter scheme has two difficulties. One is that the energy of imploding plasma is initially kinetic. When the plasma implodes on axis the energy transforms into ion temperature. Heating of the electrons, which transfers the heat to the converter, is a slow process. Electrons achieve a temperature $T_e = M_p V^2$ only after thousands of collisions with ions. Their heating time, and the ionization time of the He-like plasma, could be longer than the life time of the compressed pinch,

$$\frac{dT_e}{dt} = 10^{-8} n_i Z^2 \frac{M_p V^2}{T_e^{3/2}(\text{eV})}. \qquad (5)$$

Heating could be accelerated by increasing Z, but then the heat flux to the converter decreases according to Eq. (2). This inconsistency can be avoided if the Z-pinch consists of two coaxial cylinders, an internal on made from an intermediate-Z material such as argon, and an external one from deuterium. The electrons will be rapidly heated in the argon plasma and then quickly transfer the energy through the deuterium plasma onto the converter.

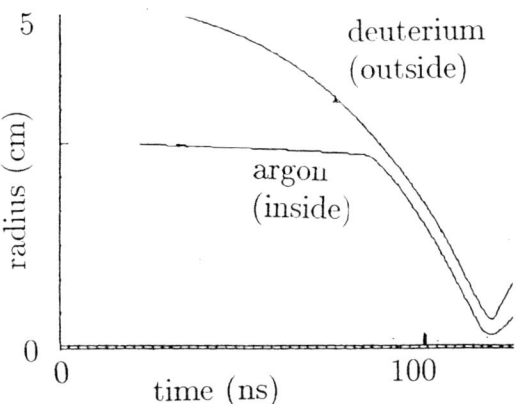

Figure 1: (R-t) diagram of the heterogeneous pinch (parameters in text).

The value T_e estimated by (5) is a lower bound, because after the shock wave hits the axis the electrons are heated not only by ions, but generally as a result of the compression energy provided by the magnetic piston imploding from radius R_1 to R_{min} with velocity V: $T_e(R) = T_e(R_1)(R_1/R)^{4/3}$ With this corection we obtain:

$$q_e = \frac{10}{L}T_e^{7/2} \propto \frac{1}{L}\left(\frac{ZE}{VR_1}\right)^{7/5}\left(\frac{R_1}{R_{min}}\right)^{14/3}. \qquad (6)$$

In this estimate the number of electrons in the internal plasma cylinder is the same as the deuterium atoms.

Results of computer simulation

Figure 1 shows the results of a computer simulation by Yu. Gorbulin [4] for a 1-D radial compression of a plasma with an outer shell of 2 cm deuterium and an inner core of argon with 3 cm radius.. The heat flux in (2) was approximated by replacing the axial gradient by $1/L$. A current rising to 20 MA in 100 ns creates a shock wave, and then compresses and heats first deuterium and later argon. As a result, the plasma cylinder is compressed to a radius 0.5 cm and the argon itself to 0.3 cm. The temperature of the electrons increases to about 10 keV and surpasses the temperature of the deuterium. In this computation 530 kJ of energy is transferred to the converter in 5 ns. Obviously, a 2-D computer simulation is needed to make the results more realistic.

If in this model we exchange places of argon and deuterium the argon shell could be much thinner due to its large losses by ionization and radiation during the acceleration (snow plow approximation). As the computer simulation has shown, its thickness would be close to a skin layer, i.e. less than 1 mm. Such a thin shell is unstable to both the Rayleigh-Taylor and the Hall instability. The theory of Hall instability initiated by the author has been confirmed by the drastic consequences of this instability seen in experiments in many laboratories.

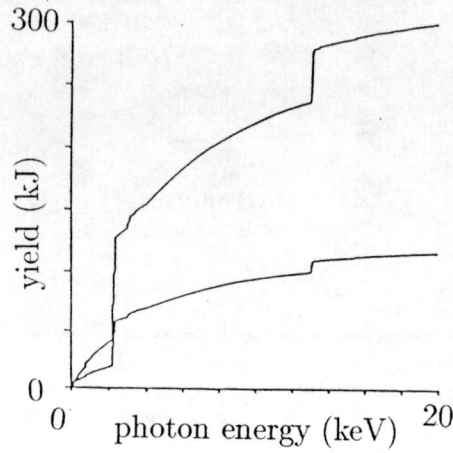

Figure 2: Energy-integrated radiation yield out of a Kr converter for the conventional scheme (bottom curve) and the heterogeneous Z-pinch (upper curve).

The hard X-Ray radiation yield from the converter has been computed by I. Yakunin and A. Starostin (TRINITI).[5] The initial energy and temperature of the deuterium plasma was the peak temperature in Yu. Gorbulin's simulation. The total energy content in the 2.5 cm long and 1 square cm area pinch is 1.25 MJ. The 1-D radiation equation with the time-dependent ionization state and electron heat conductivity was solved for the given geometry with a code by Yu. Kochubey (VNIIEF, Arzamas-16). Figure 2 shows the results. The yield from the heterogeneous scheme proposed here (top curve) is obviously much larger than the conventional yield (bottom curve).

References
1. Quintenz J. P., et al "BEAMS-96, Prague, 1996, V1, p.1.
2. Davis J. Ibid, V.2, p. 709.
3. Rudakov L. I. Patent of Russian Federation N 1223766 M. cl.3 G21B1/100, 33 N 3730677/25, publ.21.04.84;
4. Rudakov L. I. et al., Phys Fluids B, V.3, p.2414, 1991.
5. Rudakov L. I., Starostin A. N., Yakunin I. I., et al., this conference.

Conversion Efficiency of Heat Flux into Hard X-ray Radiation

L. I. Rudakov, A. N. Starostin[†], I. I. Yakunin[†],
A. B. Kukushkin, and V. S. Lisitsa

RRC "Kurchatov Institute", Moscow, 123182, Russia
[†] *TRINITI, Troitsk, Moscow region, 142092, Russia*

Abstract. The conversion efficiency of a heat flux into *hard* X-ray radiation is analyzed for a heat-to-radiation flux converter linked to the edge of low atomic number hot Z-pinch using a time-dependent two-temperature one-dimensional non-LTE-radiation-hydrodynamic numerical model. The model gives a parameter regime with about the same hard x-ray yield as a radially imploding plasma, but with an order of magnitude lower input energy.

FORMULATION OF THE PROBLEM

Pulsed Power Generator-based Z-pinch plasmas are an effective and prolific source of soft X-rays (SXR) [1]. The multiwire array approach to creating high atomic number plasmas offers a highly efficient conversion of magnetic energy into SXR black body radiation. Another ongoing investigation [2] is the radiation yield in the spectral region of L-shell and K-shell transitions in heavy atoms (Ar, Kr, Xe) from a Z-pinch' thermal energy obtained from the kinetic energy of a hollow heavy-atom plasma cylinder (the liner) compressed by a magnetic field. These energy conversion schemes allow limited control over the radiation's spectral distribution, and it is especially difficult to make more hard X-rays (HXR). The reason is the rather slow energy transfer from hot ions to cold electrons that results in the long lifetime of ionization states where multiple transitions in L-shell and higher-lying atomic shells dominate. Various scientific and technological applications under active discussion need a harder X-ray spectrum. Therefore it is worth to look for ways to (i) speed up the ionization process in the HXR radiator and (ii) have some freedom in increasing the electron density (and HXR intensity, respectively) in the HXR radiator.

Pursuing these goals suggests the following two-step progression of energy conversion:

(1) producing and heating an electron plasma in the conventional Z-pinch with *minimal* radiation losses, and

(2) fast "transfer" of the resulting high electron temperature to a converter. This takes the form of a conventional target that is linked to the edge of the above Z-pinch plasma (i.e. the target is situated at the point where the edge of the cylindrical Z-pinch stops at its stagnation stage).

The time ordering of the above two processes is possible because of strong temperature dependence of thermal flux ($\propto T_e^{7/2}$) caused by electron thermoconduction. This scheme appears advantageous due to the following aspects of energy progression. First, the energy cost of atom ionization E_{min}, i.e. a sum of the ionization energies and the ion and electron thermal energies required to reach the He-like stage of the relevant element of atomic number Z, appears, for rather large Z of radiating plasma, to be smaller, due to smaller losses on excitation (and subsequent radiation emission) during ionization. And, second, the profit in decreasing the E_{min} value is closely related to the faster reaching of the relevant ionization degree (i.e. lower "time cost" of a He-like ion) due to higher electron temperatures at initial stage of producing the radiating plasma (cf. thoroughful investigation of time-dependent kinetics effects in [2]).

In this scheme the "load function" is separated from conversion of thermal energy into radiation flux. Such a separation broadens the possibilities for the "generator-load" matching. Indeed, the use of a light-atom gas liner instead of heavy-atom gas liner (typically, krypton) allows (i) to use a thicker liner and thus achieve a more stable regime of compression, and (ii) to "match" the generator and load system with more freedom because of relaxed restrictions on the implosion velocity. This freedom makes it possible to optimize radiation yield by varying the density of the radiating plasma.

This so-called "Liner-Converter" scheme was originally proposed in the Kurchatov Institute [3] for conversion of heat flux into SXR radiation. However it appears that this scheme is most valuable and efficient just for the case of heat conversion into the HXR radiation.

NUMERICAL SIMULATIONS

Evaluating the efficiency of the Liner-Converter scheme and comparing it with that of conventional Z-pinch scheme is done with numerical simulations with the code SS-9 [4], which treats self-consistently level population kinetics, radiative transfer, and gas dynamics (see also [5]). The computations are done for various values of input energy Q, or equivalently, the initial thermal energy of Z-pinch plasma, $W_{pinch}^{(0)}$,

The HXR yield is calculated in a certain spectral region, integrated over a certain time interval Δt_{rad}. The radiation kinetics allows for the ionization states from Ne-like to H-like ions. For H-, He- and Li-like ions the code takes into account the excited atomic levels up to n=5, with allowance for the fine structure of the 2P levels in Li-like ions. Each line is covered by at least 15 points spectral grid. We present here the results of calculations which do not allow for gas dynamics. The simulation are carried out for the following conditions:

(a) conventional Z-pinch scheme: plasma column (cylinder, length L, square S) of a heavy-atom gas (krypton, total mass M_{rad});

(b) Liner-Converter scheme: plasma slab (thickness L) of a heavy-atom gas (krypton, total mass M_{rad}) linked to plasma slab of a light-atom gas (nitrogen, total mass M_{pinch}, initial electron and ion temperature $T_{pinch}^{(0)}$ in the converter at the HXR emission stage (or, equivalently, final temperature in the Z-pinch and converter at thermalization stage)). Slab geometry of the Z-pinch plasma is chosen for simplifying the simulations, even though the final results for the radiation yield are presented for an equivalent cylindrical geometry with the same cross-section S of the Z-pinch.

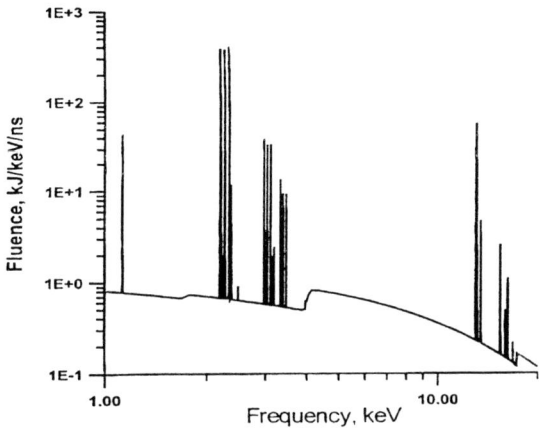

FIGURE 1. Spectral distribution of radiation flux from free surface of Kr converter at $t = 6$ ns for $W_{pinch}^{(0)}$=1.25 MJ, M_{rad}=0.8 mg, l=0.2 cm, M_{pinch}=1 mg, L=2.5 cm, S=1 cm^2, and $T_{pinch}^{(0)}$=15 keV.

Figure 1 gives spectral distribution of radiation flux from the free surface of Kr converter at $t = 6$ ns (in double logarithmic scale) for $W_{pinch}^{(0)} = 1.25$ MJ, $M_{rad} = 0.8$mg, $l = 0.2$ cm, $M_{pinch} = 1$ mg, $L = 2.5$ cm, $S = 1$ cm^2, $T_{pinch}^{(0)} = 15$ keV. Figure 2 shows the spectral distribution of radiation flux Q_{rad} integrated over time (from zero to a current time t) and frequency (from zero to a current value ω), at $t = 10$ ns ($W_{pinch}^{(0)} = 1.25$ MJ) for the Liner-Converter scheme ($M_{rad} = 0.8$ mg, $l = 0.2$ cm, $M_{pinch} = 1$ mg, $S = 1$ cm^2, $L = 2.5$ cm, $T_{pinch}^{(0)} = 15$ keV) (solid curve) and the conventional Z-pinch scheme ($M_{rad} = 1.9$ mg, $L = 2.5$ cm, $S = 1$ cm^2) (dashed curve).

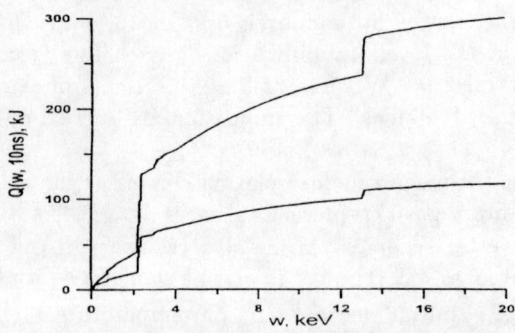

FIGURE 2. Spectral distribution of radiation flux Q_{rad}, integrated in time from time zero to time $t=10$ ns, and in frequency from zero to ω. The solid curve is for the Liner-Converter scheme, with $W_{pinch}^{(0)}$=1.25 MJ, M_{rad}=0.8 mg, l=0.2 cm, M_{pinch}=1 mg, S=1 cm^2, L=2.5 cm, and $T_{pinch}^{(0)}$=15 eV. The dashed curve is for the conventional Z-pinch scheme, with M_{rad}=1.9 mg, L=2.5 cm, and S=1 cm^2.

For a multi-megajoule driver both numerical simulations and semi-analytical estimates show that both the Liner-Converter scheme and the conventional Z-pinch scheme attain the optimal regime of HXR emission within their own frames. Here, for time integrated HXR yield, Q_{HXR}, which allows for the photon energies $\hbar\omega > 12.5$ keV, for the following input parameters of 60 MA driver in conventional scheme we have: $W_{pinch}^{(0)} = 10$ MJ, $M_{rad} = 15$ mg, $L = 5$ cm, $\Delta t_{rad} = 10$ ns, $S = 4$cm^2, $Q_{HXR} = 84$ kJ.

It should be noted that omitting the low-Z ionization states beyond Ne-like ions substantially overestimates the radiation yield in the conventional scheme (cf. [2]), whereas the values of Q_{HXR} in the Liner-Converter scheme are much less sensitive to this approximation. Also, the radiation yield in the conventional scheme appears to be more sensitive to the degree of plasma compression. Thus, in the conventional scheme the 70% population of He-like ions is achieved at 8 ns, whereas for the Liner-Converter scheme this happens at 1 ns. With decreasing input energy the yield Q_{HXR} in the HXR spectral region ($\hbar\omega > 12.5$ keV) during the relevant time period (~ 10 ns) in the conventional Z-pinch scheme goes down faster (and scales approximately as Q^2) as compared with the Liner-Converter scheme (where it scales as Q). Thus, in the domain of lower input energies the Liner-Converter scheme attains its threshold for the optimal HXR radiator: here, $W_{pinch}^{(0)} = 1.25$ MJ, $M_{rad} = 0.8$ mg, $M_{pinch} = 1$ mg, $L = 2.5$ cm, $S = 1$ cm^2, $T_{pinch}^{(0)} = 15$ keV, $\Delta t_{rad} = 10$ ns, $l = 0.5$ cm, $Q_{HXR} = 31$ kJ. In contrast, the conventional Z-pinch scheme gives a much smaller yield.

It follows that in the Liner-Converter scheme about the same HXR yield can be achieved at those values of input energy which are an order of magnitude lower than that in conventional scheme. This makes the Liner-Converter approach a sound canditate for designing a Pulsed Power Generator-based HXR source in the range of moderate input energies.

ACKNOWLEDGMENTS

The authors are grateful to Drs. Yu. K. Kochubey and P. D. Gasparyan for making it possible to use their numerical code SS-9.

REFERENCES

1. Quintenz J. P., et.al., In: Proc. 11-th Int. Conf. High Power Particle Beams ("Beams-96"), Prague: Czech Rep. Acad. Sci., 1996, vol.1, p.1.
2. Davis J., et.al. Ibid., vol.2, p.709.
3. Rudakov L. I., et.al. Phys. Fluids **B3**, 2414 (1991).
4. B. A. Voinov, P. D. Gasparyan, Yu. K. Kochubey, V. I. Roslov, Voprosy Atomnoi Nauki i Tehniki, Ser. Methods and Codes for Mathematical Physics Problems (in Russian), **2**, 39 (1993).
5. G. S. Volkov, V. P. Smirnov, A. N. Starostin et.al., JETP, **74**, 253 (1992).

Decade Quad Load Performance

J.W. Thornhill, F.L. Cochran[†], J. Davis, J.P. Apruzese, and K.G. Whitney
Plasma Physics Division, Naval Research Laboratory
4555 Overlook Ave, SW, Washington, D.C. 20375-5000
[†]Berkeley Research Associates, P.O. Box 852, Springfield Va, 22150

Abstract. The performance of plasma radiation source (PRS) loads imploded by a long current pulse machine is investigated theoretically. The emission tolerant (with oil) Decade Quad design of Pulse Sciences Incorporated is an example of such a long current rise time machine. It has a peak short circuit current of 9 MA that is reached in 300 nanoseconds. Large radius loads must be employed for this type of machine in order to insure that high implosion velocities are reached and that the available energy is utilized. However, large radius implosions are known to be susceptible to the Rayleigh Taylor instability. The extent to which this instability affects the ability of an argon load to radiate K-shell photons is not known and is the subject of this model investigation.

INTRODUCTION

One way to establish a theoretical relationship between the Rayleigh Taylor (RT) stability of an argon load and its ability to efficiently radiate K-shell photons would be to perform two-dimensional (2D) or three-dimensional (3D), time dependent, non-LTE, full radiation transport, MHD calculations. However, this capability does not fully exist in the Z-pinch community primarily because the computer requirements for performing the zone to zone coupling needed to transport radiation in a 2D or 3D plasma are prohibitive. Our 2D MHD model,[1,2] PRISM (which stands for plasma radiating imploding source model), assumes that either the plasma is transparent to radiation or that a local probability of escape model can be used to model some of the self-absorption physics. The LTE radiative diffusion model, used by the 2D Z-pinch community, is also inadequate for modeling the radiation physics of PRS loads designed for K-shell emission because neither the diffusion approximation nor the LTE assumption are valid for this type of plasma. Given these model limitations, we have used the following approach to analyze the K-shell emission capabilities of large radius loads[3]: (1) the pinch is initialized with a density perturbation to seed the RT instability, (2) the PRISM code calculates the evolution of the pinch to the point where radiation effects begin to dominate, and then (3) K-shell scaling relations are employed to assess K-shell emission from each axial slice of the 2D load profile. These scaling relations are based on past experimental results and on 1D MHD calculations.[4,5] The probability of escape radiation transport and collisional radiative ionization dynamics models[6] used in the 1D calculations are more appropriate for approximating the radiation physics of K-shell PRS loads than what is used in a typical 2D calculation. This approach is applied to examining stability and its effect on K-shell yield performance on DECADE QUAD loads that are initially 3.6 cm in radius, and of uniform density.

MODEL

PRISM is a 2D MHD code that models the plasma under the assumptions that its ionization dynamics is in collisional radiative equilibrium. Opacity is not taken into account so the plasma is not photo excited. This treatment simplifies the calculation of the evolution of the RT instability during the runin phase of the implosion. It is probably most valid in situations where the radiative powers are substantially less than the time rate of change of kinetic energy into internal energy during the stagnation phase of the implosion. The 2D portion of the calculation is terminated at the end of the runin phase of the implosion when the change of inductance due to the imploding load is the same as that achieved by a 10:1 compression in a one-dimensional calculation. The choice of cutoff condition is not that critical for the purposes of this investigation since inductance depends logarithmically on radius and many Z-pinch experiments have exhibited similar compressions. The initial perturbation wavelength λ is chosen to be 0.5 cm, which is a typical disruptive wavelength. It is a non-compressible perturbation characterized by: (1) $\rho' = \rho_o \times (0.25) \times cos(2\pi z/\lambda)$; where the mass density ρ equals ρ_o (initial mass density) $+\rho'$ and (2) $u' = -u_o \times (0.25) \times cos(2\pi z/\lambda)$; where the radial velocity u equals u_o (initial radial velocity) $+ u'$. The perturbation is turned on 40 ns after the start of the current pulse and it is limited to zones near the outer radius because it is only applied when $|u| > 0.1 \times |u_{max}|$. The $J \times B$ force that implodes the PRS load is calculated self-consistently in PRISM by solving a lumped circuit equation for the DECADE QUAD current.

Once the 2D portion of the calculation is terminated, K-shell yield scaling relations (see Appendix) are applied to the plasma to determine the K-shell emission from each axial slice of the 2D pinch. These scaling relations express the K-shell yield per unit axial length as a function of the mass per unit length and energy (kinetic + internal) per unit length. These quantities are obtained for each axial slice of the pinch by adding all the radial contributions (orthogonal grid) and dividing by the length of each axial slice. In essence this approach segments the 2D load profile into a series of 1D profiles.

RESULTS

As the RT instability grows it moves mass and energy out of the "bubble" regions and transfers it to the "spike" regions of the plasma. This process is illustrated in Fig. 1 where the mass per unit length of the spike and bubble regions are shown as a function of time. In this example, the 10:1 effective compression is achieved at 240 ns at which time the mass per unit length in the bubble regions is approximately 1/3 that in the spike regions. This feature of the RT instability produces deviations in calculated K-shell yield from 1D behavior due to local deviations in either the mass per unit length and/or the energy per unit length of the plasma. The effect these local deviations have on K-shell emission is shown in Fig. 2 where 0D and 2D scaled argon K-shell yields as a function of initial mass are compared.

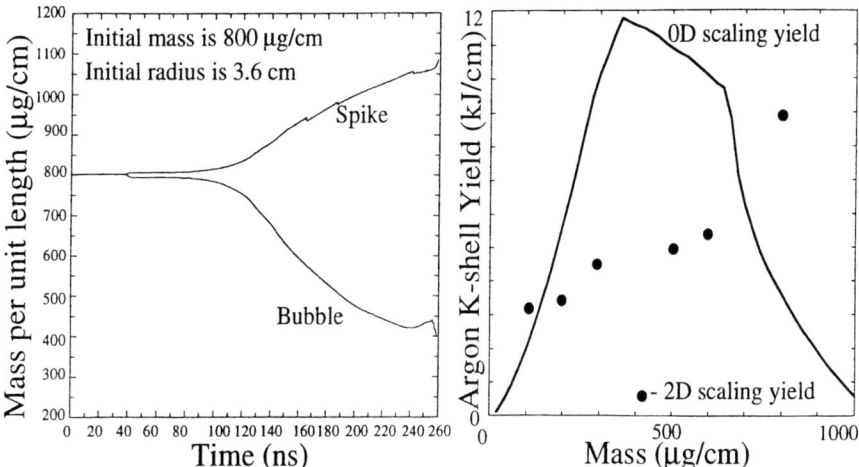

Figure 1. Mass per unit length as a function of time for the spike and bubble regions of the Rayleigh Taylor instability.

Figure 2. 0D and 2D argon K-shell yields as a function of mass.

The reasons for the emission behavior shown in Fig. 2 can be found in the following observations obtained for this load configuration at stagnation: 1) the energy contained in the spike regions is nearly the same as the energy contained in the bubble regions, 2) the spike regions subtends about the same length of the axis as the bubble regions, 3) the spike to bubble mass ratio is approximately 3:1, and 4) the implosion time and peak current are 10 percent lower in the 2D calculation than in the 0D calculation. This latter effect occurs because the effective 10:1 change in inductance is achieved earlier in time in the 2D implosion as a consequence of the earlier arrival on axis of the bubble region of the plasma. The first three of these observations imply that an initially uniform 500 $\mu g/cm$ load evolves into 250 $\mu g/cm$ bubble regions and 750 $\mu g/cm$ spike regions at stagnation. Because of their additional mass the spike regions are not predicted to emit much K-shell emission because their specific energy as characterized by η^* (see Appendix) is too low. The K-shell emission comes almost exclusively from the bubble regions for initial mass loads greater than 500 $\mu g/cm$. For most mass loads the 2D yields are reduced below the 0D yields for the reason just mentioned and also because the reduced current (10 percent) leads to a 20 percent reduction ($I^2 scaling$) in energy coupling to the 2D load. However, the 10 kJ/cm K-shell yield shown in Fig. 2 for the 800 $\mu g/cm$ initial mass load is much larger than the corresponding 0D yield. This yield is due to the larger energy that is coupled to the larger initial mass loads (Fig. 3) and to the high specific energy in the bubble regions of the plasma. The 0D peak currents, η^*s and energies are shown in Fig. 3. Note, there is a large uncertainty in predicting K-shell emission from either scaling calculations or from past experiments whenever $\eta^* < 2$.

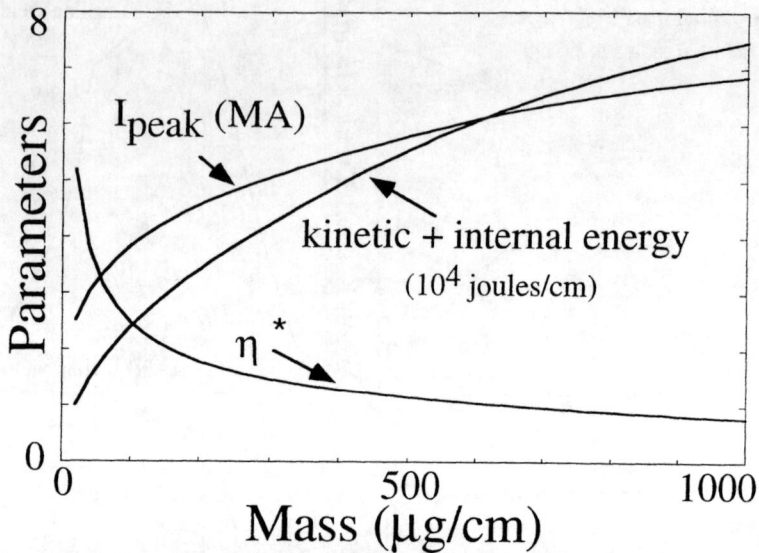

Figure 3. The η^*, energy, and peak currents as a function of mass load. These parameters are calculated from a 0D snowplow model of the 3.6 cm radius, uniform density, Decade Quad PRS load.

CONCLUSIONS

This model study suggests that the Rayleigh Taylor instability has the potential to substantially alter the amount of PRS K-shell emission from that which is expected from purely 1D behavior. As modeled this alteration is due mainly to the flow of mass from the bubble to the spike regions of the instability. The spike regions of the instability were 3 times more massive than the bubble regions for the 3.6 cm radius, uniform density profile, DECADE QUAD argon loads examined here. For optimal 2D loads, this limited the source of K-shell emission to the bubble region of the instability. Although, it was not described in this paper, we also modeled 2.25 cm radius, uniform density, argon Saturn loads. In this case, the spike to bubble mass ratio was approximately 2 and there was little difference in calculated K-shell yield between the 2D and 0D models. On the other hand, we also analyzed a 4.5

cm radius uniform density DECADE QUAD load that never imploded all the way to the axis because virtually all of the mass was pushed out of the bubble region.

APPENDIX: Z Scaling of K-shell Emission

$$\text{Energy (J/cm)} = (\text{kinetic energy} + \text{internal energy}) \text{ (J/cm)}$$

$$\eta^* = \text{Energy} \times 1.1 \times 10^{-5} \times Z^{-2.41}/\text{Mass (g/cm)}$$

$$\alpha = \frac{2.58 \times 10^{-12} \times Z^{5.96}}{e^{-20.6/Z^{0.9}}} \times \max\left(1.0, \frac{\eta^{*2}}{\eta^* + 12}\right)$$

$$\beta = \min\left(0.3 \text{ Energy}, \frac{0.3 \times \text{Energy} \times \text{Mass}}{\alpha}\right)$$

$$\text{K-shell yield (J/cm)} = \begin{cases} \beta, & \eta^* > 1.5; \\ \beta \times (\eta^* - 0.5), & 1.0 < \eta^* < 1.5; \\ \frac{1}{2}\beta \times (1 - [(1-\eta^*)/0.25]^{0.5}), & 0.75 < \eta^* < 1.0; \\ 0, & \eta^* < 0.75 \end{cases}$$

Note, there is no accounting for innershell absorption effects in this scaling. Also, these scaling relations are based on CRE calculations, which do not account inherently for the time dependence of the ionization process. The uncertainty in this scaling is large for $\eta^* < 2$.

ACKNOWLEDGEMENTS

This work was supported by DSWA.

REFERENCES

1. F. L. Cochran, J. Davis, and A. L. Velikovich, Phys. Plasmas 2, 2765 (1995).

2. A. L. Velikovich, F. L. Cochran, and J. Davis, Phys. Rev. Lett. 77, 853 (1996).

3. This approach was initiated by F. L. Cochran

4. K. G. Whitney, J. W. Thornhill, J. P. Apruzese, and J. Davis, J. Appl. Phys. 67, 1725 (1990).

5. J. W. Thornhill, K. G. Whitney, J. Davis, and J. P. Apruzese, J. Appl. Phys. 80, July 15, (1996).

6. J. P. Apruzese, J. Davis, D. Duston, and R. W. Clark, Phys. Rev. A29, 246 (1984).

RADIATION HYDRODYNAMICS

Application of 2-D Simulations to Hollow Z-Pinch Implosions[1]

D. L. Peterson, R. L. Bowers, J. H. Brownell, C. Lund,
W. Matuska, K. McLenithan and H. Oona
Los Alamos National Laboratory, Los Alamos, NM, 87545

C. Deeney, M. Derzon, R. B. Spielman, T. J. Nash, G. Chandler,
R. C. Mock, T. W. L. Sanford and M. K. Matzen
Sandia National Laboratory, Albuquerque, NM, 87185

N. F. Roderick
University of New Mexico, Albuquerque, NM, 87131

Abstract. The application of simulations of z-pinch implosions should have at least two goals: first, to properly model the most important physical processes occurring in the pinch allowing for a better understanding of the experiments and second, provide a design capability for future experiments. Beginning with experiments fielded at Los Alamos on the Pegasus I and Pegasus II capacitor banks, we have developed a methodology for simulating hollow z-pinches in two dimensions which has reproduced important features of the measured experimental current drive, spectrum, radiation pulse shape, peak power and total radiated energy (1,2,3). This methodology employs essentially one free parameter, the initial level of the random density perturbations imposed at the beginning of the 2-D simulation, but in general no adjustments to other parameters (such as the resistivity) are required (1). Limitations in the use of this approach include the use of the 3-T, gray diffusion treatment of radiation and the fact that the initial perturbation conditions are not known *a priori*. Nonetheless, the approach has been successful in reproducing important experimental features of such implosions over a wide variety of timescales (tens of nanoseconds to microseconds), current drives (3 to 16 MA), masses (submilligram to tens of milligrams), initial radii (<1 cm to 5 cm), materials (Al and W) and initial configurations (thin foils and wire arrays with 40 to 240 wires). Currently we are applying this capability to the analysis of recent Saturn and PBFA-Z experiments (4,5). The code results provide insight into the nature of the pinch plasma prior to arrival on-axis, during thermalization and development after peak pinch time. Among other things, the simulation results provide an explanation for the production of larger amounts of radiated energy than would be expected from a simple slug-model kinetic energy analysis and the appearance of multiple peaks in the radiation power. The 2-D modeling has also been applied to the analysis of Saturn "dynamic hohlraum" experiments and is being used in the design of this and other Z-Pinch applications on PBFA-Z.

---

[1] Work Supported by DOE.

INTRODUCTION

Two-dimensional Radiation-Magnetohydrodynamic (RMHD) simulations of hollow z-pinches should provide both physical insight into the important processes leading to radiation production as well as allow for the reliable design of new experiments. To fulfill this role, the RMHD codes must accurately reproduce the important experimental features seen in z-pinch implosions. In this paper we will describe simulations with a 2-D Eulerian RMHD code (1) which has been benchmarked with other codes and experiments on the Los Alamos Pegasus I and Pegasus II capacitor banks, the Procyon explosive generator system (1,2,3) and the Sandia Saturn and PBFA-Z accelerators (4,5).

Imploding z-pinch plasmas are susceptible to the magnetically driven analog of classical Rayleigh-Taylor fluid instabilities. Significant differences from the hydro-only instabilities arise in that instability growth in the azimuthal direction is much reduced from that in the axial direction, allowing for greater applicability of 2-D codes using cylindrical symmetry, and that there is only one fluid (the plasma), with the driving magnetic field (which plays the part of the "light" fluid) penetrating the plasma. Thus there are no interfaces as such and the instability is not easily characterized in spike-to-bubble amplitude. The magnetic field is also not tied to particular fluid elements but to resistivity and the current may flow at various times through different regions of the plasma spikes and bubbles.

The presence of instabilities degrades the performance of the radiation production from that which could be expected from 1-D simulations. Initially, the plasma will be spread over a larger radial extent by the growth of spike and

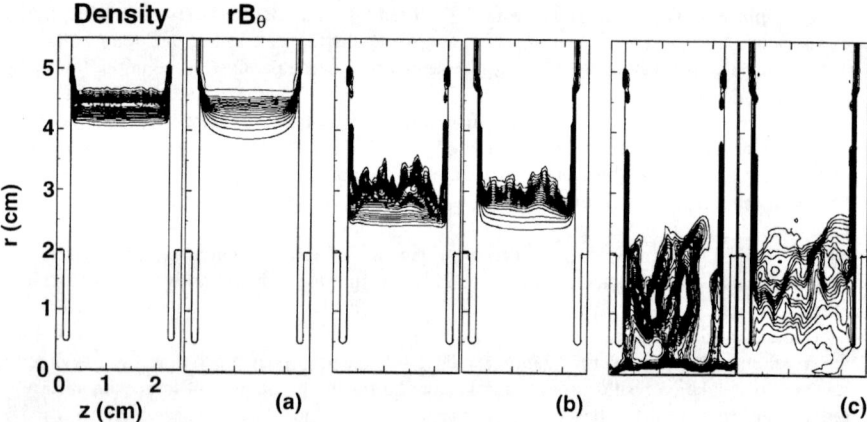

Figure 1. Contours of density and rB_θ for three times in a simulation of the Pegasus II-25 experiment. By the last time shown (c), the bubble region located near the right electrode has burst through the plasma shell. This occurrence is evidenced in the measured load current and is coincident with the onset of the radiation pulse.

bubble regions as well as suffering a loss of axial uniformity. This thickening of the imploding shell is followed by the bubble regions thinning to such an extent that the bubbles may burst through the shell accelerating material and magnetic field to the axis if the disruption is severe enough. This is illustrated in Fig. 1, which shows instability development in density and magnetic field for a simulation of the Pegasus II-25 experiment. This experiment used a 2-cm long, 10-cm diameter, 14.3 mg cylindrical Al foil. The peak drive current of 5.1 MA resulted in a 2 μs implosion time and a radiation pulse with a FWHM of about 200 ns. The simulation reproduced features seen in optical framing camera photographs and the experimentally measured current as well as generating a radiation pulse of the correct total energy and very similar pulse duration and shape as that measured in the experiment (3). The methodology used in such simulations is to begin the 2-D simulations using the current and velocity determined from a 0-D (slug) calculation at a point when the plasma has imploded 2% of the distance to the axis from the initial radius. The 2-D plasma begins as a slab with random perturbations imposed upon a uniform density profile. The perturbation level for the simulation in Fig. 1 is 15% (i.e. the densities in the plasma zones vary between $0.85 \times \rho_0$ and $1.15 \times \rho_0$).

INSTABILITY CHARACTERIZATION

Lacking an interface, it is difficult to define a characteristic amplitude for the instabilities, and in addition, the linear theory employed to examine the instability development is inapplicable after a very short time. A typical implosion time will be many (tens to hundreds) of instability e-folding times which eliminates this description as useful in characterizing the instability growth. With only a single material involved in instability development, measures such as mixing layer thickness are also difficult to apply and of limited utility. In the simulations discussed here, the most important event in the instability development is the breakthrough of the plasma shell. Consequently, we introduce a measure of the instability growth termed "fractional involved mass," ΔM, which allows a characterization of the instability growth, defined as follows:

$$\Delta m = 2\pi \iint |\rho(r,z) - \overline{\rho}(r)| r\, dr\, dz, \tag{1}$$

$$\overline{\rho}(r) = \frac{\int \rho(r,z)dz}{\int dz}, \tag{2}$$

$$\Delta M = \frac{\Delta m}{M}, \tag{3}$$

where M is the total mass of imploding shell. In addition, it is useful to define a "fractional distance traveled," ΔX:

$$\Delta X = \frac{r_0 - \bar{r}(t)}{r_0}, \qquad (4)$$

with $\bar{r}(t)$ being the mass-averaged shell radius. The quantity ΔM can be visualized as a measure of the fraction of the imploding shell which is either contributing to the above-average density in spike regions or missing from the below-average density of bubble regions. In practice, the maximum values of ΔM range between about 0.8 and 1.2, and a peak in this quantity represents a maximum in the instability development at that time. In Fig. 2, the fractional involved mass measure is shown as a function of fractional distance traveled for the calculation pictured in Fig. 1. The breakthrough of the plasma shell by the most developed bubble feature (near the right electrode) is seen as the peak in the fractional involved mass and occurs slightly prior to the time shown in Fig. 1c.

SATURN SIMULATIONS

Using the methodology developed in simulating the Los Alamos Pegasus I, Pegasus II and Procyon experiments, simulations were made of tungsten and aluminum implosions on the Sandia Saturn accelerator. In an experiment using 90

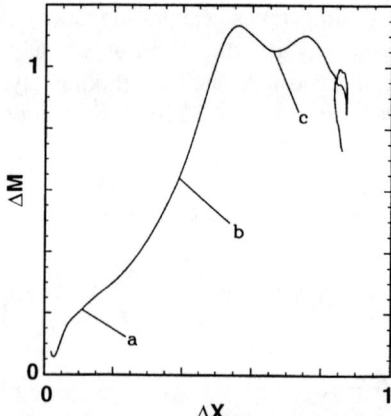

Figure 2. Fractional involved mass vs fractional distance traveled for the Pegasus II-25 experiment simulation. The points marked a, b and c correspond to the times shown in Fig. 1a, 1b and 1c.

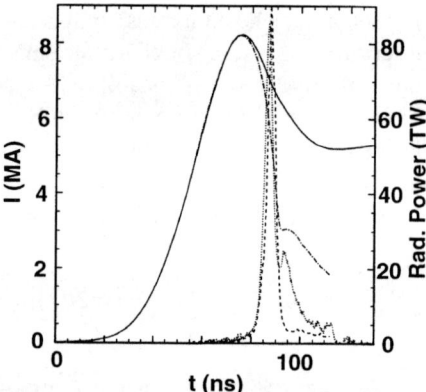

Figure 3. Comparison of experimental (solid, dotted) and simulation (dot-dash, dashed) currents and radiation powers for a 120 wire tungsten Saturn implosion (#2224). Experimental current after the peak is unreliable.

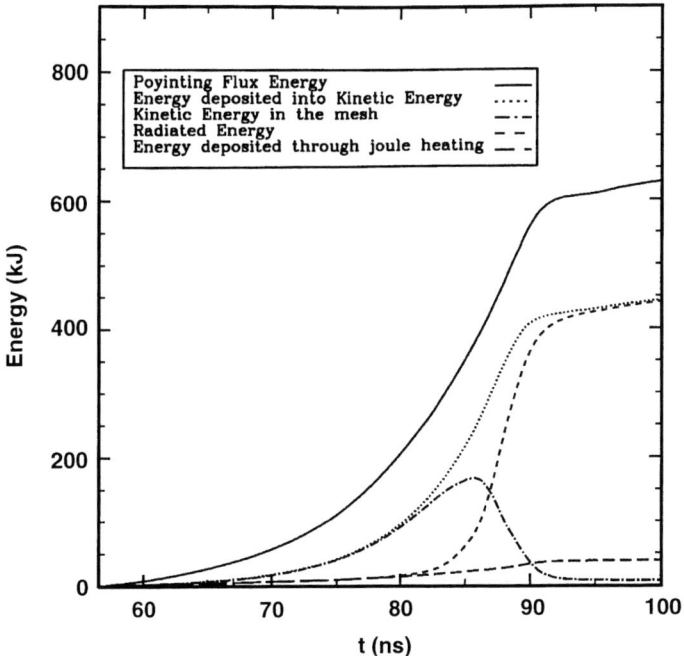

Figure 4. Energy balance and flow in the 2-D simulation of the Saturn 2224 experiment. Total energy (solid) deposited by the circuit model into the 2-D mesh is about 600 kJ. Energy deposited in the plasma through the Lorentz force (dotted) is about equal to the radiated energy (short dash). The kinetic energy (dot-dash) follows the Lorentz force energy until material begins to arrive on-axis which loses kinetic energy while plasma at larger radii continues to gain energy. Joule heating (long dash) is comparatively small.

aluminum wires (previous typical wire numbers had been about 40 or fewer), it was found that the peak power increased and the pulsewidth decreased dramatically. This breakthrough provided opportunities for employing z-pinches in applications which were previously unworkable due to the effects of the instabilities. Simulations with reduced perturbation levels reproduced many of the features of these Al implosions (4,5) and were then done for 120 wire tungsten implosions. A comparison of one such experiment and simulation is shown in Fig. 3. Except for a late time second peak in the radiation, the agreement is good both in current to the peak (the experimental diagnostic fails at or shortly after peak) and in the radiation pulse shape, width, maximum power and total energy. The energy flow in this simulation is illustrated in Fig. 4. It can be seen that resistive (Joule) heating plays only a small role in contributing energy available to be radiated and that the total radiated energy is nearly equal to that provided by the Lorentz force of the current on the plasma, resulting in either acceleration or pdV work. The energy so deposited is about 1.4 greater than that deposited as kinetic energy in a 0-D (slug) calculation when stopped at a reasonable compression ratio of 20:1 (2-D, more than 400 kJ; 0-D, 185 kJ). The explanation for this discrepancy lies in the radially extended nature of the 2-D plasma rather than in a greater compression ratio (a 0-D compression ratio of more than 1000:1 would be required to produce a kinetic energy equal to that radiated). In essence, the front edge of the extended plasma (which reaches the axis early) loses kinetic energy to internal and then radiation energy, while the rest of the plasma and especially the back edge continues to absorb magnetic field energy through accelerated velocity and pdV work. Similar results apply to the calculation shown in Fig. 1 and have been analyzed in more detail for Pegasus simulations in Ref. 3.

PBFA-Z SIMULATIONS

Predictions for PBFA-Z based upon the latest Saturn results were close to the actual performance seen in the device in early shots, except that current delivery was somewhat lower than expected and the observed radiation pulses had reduced pulse widths. The resultant radiation pulses were matched by a 2-D simulation perturbation level about a factor of five lower than for Saturn simulations (Saturn experiments were matched typically at 15%, the latest PBFA-Z experiments at about 3%). The result for PBFA-Z experiment #26 is shown in Figs. 5 and 6 (Fig. 6 compares pulse shape by scaling and time-shifting the experimental pulse within experimental uncertainty). This experiment used 240, 7.5 μm tungsten wires in a cylindrical array 2-cm long and 4-cm diameter (4.1 mg). The implosion was

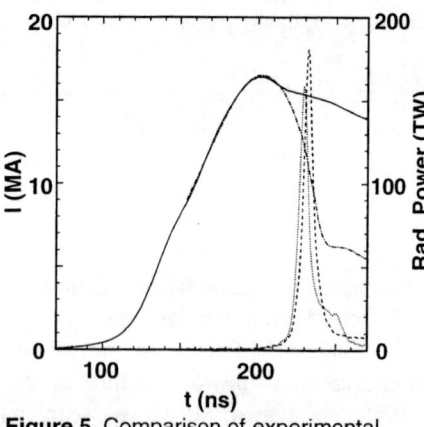

Figure 5. Comparison of experimental (solid, dotted) and simulation (dot-dash, dashed) currents and radiation powers for a 240 wire tungsten PBFA-Z implosion (#26). Experimental current after the peak is unreliable.

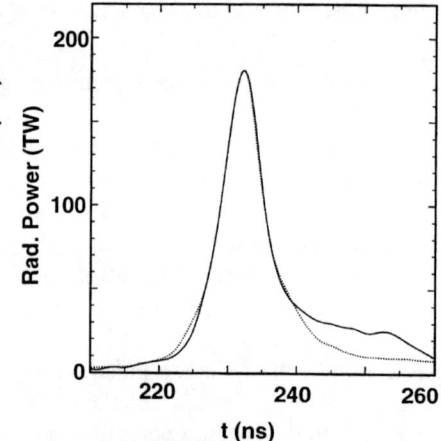

Figure 6. Comparison of experimental (solid) PBFA-Z #26 radiation pulse (scaled to peak 180 TW, shifted +2.7 ns) and 2-D simulation radiation power (dotted).

driven by a peak current of 16.5 MA with an implosion time of about 120 ns resulting in a 7 ns FWHM radiation pulse. The simulation instability growth is illustrated in Fig. 7, which shows the development of a short wavelength phase

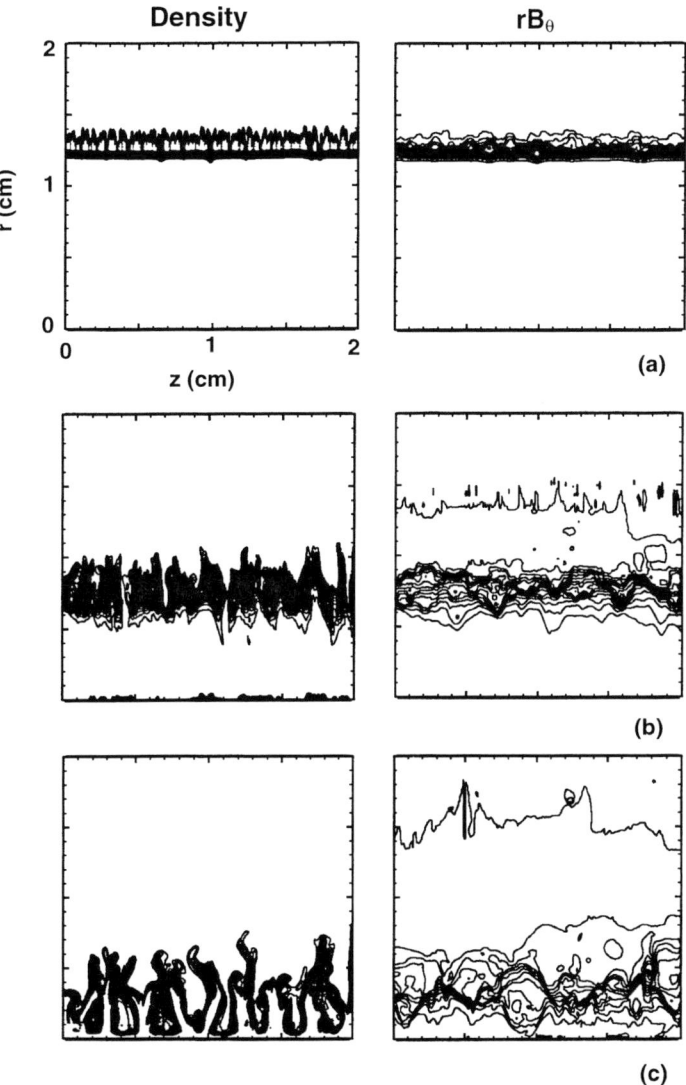

Figure 7. Contours of density and rB_θ for three times in a simulation of the PBFA-Z #26 experiment. The three times shown depict: (a) short wavelength instability development; (b) bubble breakthrough and transition to long wavelength development; (c) maximum instability growth as measured by the fractional involved mass measure.

which saturated without causing the shell to disrupt completely (in contrast to the process described earlier regarding Fig. 1). The initial breakup of the imploding shell is of such short wavelength that the magnetic field does not follow the bubble material being pushed to the axis and instead continues to flow through the spike regions remaining in the shell and the low density regions (remaining from bubble development) between the spike regions. The instability development at this wavelength is now effectively saturated and only a small amount of field and bubble material penetrates inside the plasma shell. Finally, a longer wavelength, comparable in size to the thickened shell, develops the instability growth further. Similar development has been seen in some simulations of Saturn implosions. The fractional involved mass instability measure for this simulation is illustrated in Fig. 8.

Z-PINCH APPLICATIONS

Aside from direct illumination by radiation from the pinch, the electrical return conductor may be used to contain the radiation as a hohlraum, and this may be used to drive experiments. Using both Saturn and PBFA-Z, experiments have been fielded with identical loads in two configurations: "open" (the return conductor had many large slots or return conductor posts were used) and "closed"

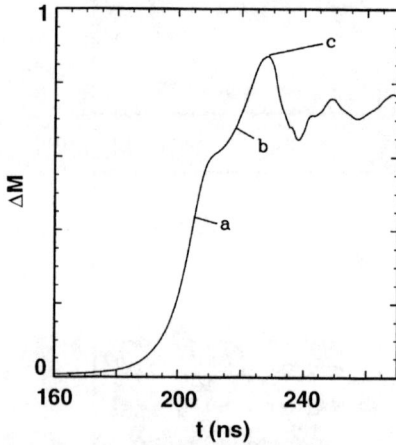

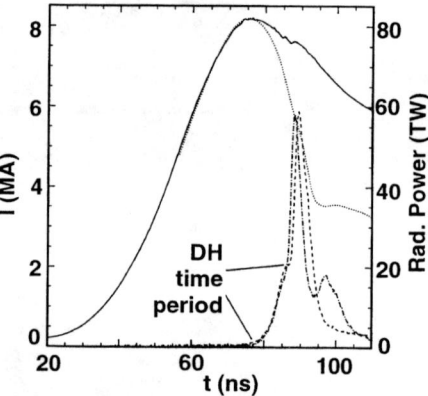

Figure 8. Fractional involved mass as a function of time for the PBFA-Z #26 experiment simulation. The points marked a, b and c correspond to the times of Fig. 7a, 7b and 7c and show short wavelength, long wavelength and the maximum in instability development.

Figure 9. Comparison of experimental (solid, dotted) and simulation (dot-dash, dashed) currents and radiation powers for a Saturn dynamic hohlraum (DH) experiment. The time period when the dynamic hohlraum exists (the "foot" of the radiation pulse) is marked. The tungsten implosion portion of the calculation is the same as that for Fig. 3.

(or "hohlraum," the return conductor was a much as possible an encasing can). The 2-D simulations which included the hohlraum walls (based upon simulations which matched the open configuration experiments) reproduced to a reasonable degree the measured hohlraum temperature. In the latest example, a PBFA-Z 30-mm diameter tungsten implosion in the open configuration was matched by a 2-D simulation (experimental peak power 170 TW, simulation 185 TW). The corresponding closed configuration experiment registered a peak hohlraum temperature of 100 eV and the simulation including the hohlraum walls gave a peak of 94 eV.

Another application of interest is to use the imploding plasma as a radiation case for a hohlraum consisting of a cylindrical (possibly coated) foam on-axis. As the pinch plasma strikes the foam, kinetic energy is converted to thermal energy and radiation fills the foam hohlraum which may then be used for applications experiments. This concept is known as the "dynamic hohlraum" (DH) or "flying radiation case" (FRC)(5,6). The hohlraum will exist for a period of time as the pinch continues to compress the foam. The compression also results in additional heating (pdV work) of the DH. The instabilities play a major role since they can cause bubble material to penetrate the foam early and destroy the integrity of the hohlraum, or cause inhomogeneities in the hohlraum temperature, or cause the radiation case to be leaky because of thinning of the shell in bubble regions. Preliminary experiments of this type were performed on Saturn based upon the tungsten experiments which had produced high radiation powers (85 TW, such at that shown in Fig. 3). Using the simulation which had matched the Saturn tungsten implosions, an aerogel foam was added on-axis in the simulation and the result in shown in Fig. 9. The current again matched the measured current so far as it was reliable, and the radial radiation pulse matched closely in the "foot" region when the dynamic hohlraum exists and also matched the general shape and timing of the pulse as well as the peak power (which occurs after the hohlraum foam has been crushed).

SUMMARY

In conclusion, the methodology of utilizing a 2-D Eulerian RMHD code to simulate an imploding pinch developed for Los Alamos experiments has been successfully applied to Saturn and PBFA-Z implosions using different load materials, masses, radii, implosion times and drive currents. We have obtained results which appear to be in good agreement with measured data, and which explain important features (such as the overall radiated energy). In addition, using the 2-D simulations which have been benchmarked to experiments and then applying them to new circumstances (such as dynamic hohlraum configuration or new drivers such as PBFA-Z) resulted in good agreement with the new experiments, partially fulfilling our desire for design and predictive capability.

There remain limitations in using the 2-D simulations. The code cannot account for 3-D effects when they may become important. As of the present it uses for the radiation treatment a 3-T, gray diffusion model (1). Although such a treatment is not inappropriate for circumstances where the primary effects will come about through MHD dynamics, it is limited in the information which it can self-consistently provide concerning the radiation environment. This limitation is mitigated somewhat in that code "snapshots" in time of density, temperature, etc, can be processed with more sophisticated radiation transport methods (2) to verify consistency in radiation flow with the 3-T model and provide more detailed spectral information. In such a case, we assume that the code has brought us to an appropriate physical configuration and then ask for detailed radiation information which would result from such a configuration. Another limitation is that initial conditions are not known *a priori*, but must instead be determined by finding a perturbation level which will generate the observed data. Although this is not desirable, especially in light of our goal of predictive capability, the circumstance is still one in which physically reasonable conditions early in the simulation develop throughout the implosion in a physically consistent way to produce complex dynamics which generate results that match those of the data.

Despite the limitations, we believe that the 2-D simulations described here have and continue to play an important role in understanding, designing and predicting experimental results in this important field. As our understanding of the complex nature of z-pinches and our ability to simulate the pinches improves, we may expect to make further advances in z-pinch performance and applications by means of such simulations.

REFERENCES

1. Peterson, *et al.*, *Phys. Plasmas* **3**, 368-381 (1996).
2. Matuska, *et al.*, *Phys. Plasmas* **3**, 1415-1429 (1996).
3. Peterson, *et al.*, "Comparison and Analysis of 2-D Simulation Results with Two Implosion Radiation Experiments on the Los Alamos Pegasus I and Pegasus II Capacitor Banks," in *Digest of Technical Papers, Tenth IEEE International Pulsed Power Conference*, 1995, pp. 118-123.
4. Sanford, *et al.*, *Phys. Rev. Lett.* **77**, 5063-5066 (1996).
5. Matzen, M. K., *Phys. Plasmas* **4**, 1519-1527 (1997).
6. Brownell, J. H. and Bowers, R. L., "The Flying Radiation Case," Los Alamos National Laboratory Report LA-UR-97-558, 1997.

IMPLOSION DYNAMICS OF A RADIATIVE COMPOSITE Z-PINCH

R. Benattar, P. Ney, A. Nikitin,
LPMI, Ecole Polytechnique, France
S.V. Zakharov, A.A. Otochin, A.N. Starostin,
A.E. Stepanov, V.K. Roerich
Troitsk Institute for Innovation & Thermonuclear Investigation (TRINITI), Russia
A.F. Nikiforov, V.G. Novikov, A.D. Solomyannaya
Keldysh Institute for Applied Mathematics, Russia
V.A. Gasilov, A.Yu. Krukovskii
Institute for Mathematical Modeling, Russia

2D simulation of a composite Z-pinch was performed by the complete radiative magnetohydrodynamic code ZETA including detailed calculation of EOS, spectral properties of materials and radiation transport in non-LTE multicharged ions plasma.

The load of the generator is made of a central fiber of 30% KCl in agar-agar foam, 1mm diameter and 60µg total mass. Around this fiber is in annular argon gas puff with inner and outer diameter of a nozzle cross section - 3 cm and 3.4 cm respectively and specific mass of 80 µg/cm. Then total length of the cylindrical load is 1 cm. During the plasma implosion the total current reached the value of 3 MA at a time of 85 ns from the voltage start. Such current amplitude is much less than through the consistent load (up to 4 MA as a rule). The plasma implosion is accompanied by the development of different types of short and long wave instabilities (thermal, radiative, non-isothermal and MHD Rayleigh-Taylor modes).

The initial geometry, the substance components and the electric current through the Z-pinch were similar to the joint experiment set up JEX-94 at Angara-5 facility [1].

The R-Z zoning of the code is such as cells 1-10 in the R direction for all Z's are made of KCl+agar-agar and cells 11-40 of argon.

The isocontours of figures 1 show the argon part of the plasma after 122 ns from the beginning of the generator pulse, when the plasma reach the maximum of compression.

The isocontours of figures 2 show the agar-agar+KCl part of the plasma.

This figures exhibit the instabilities of the plasmas components such as the electron and ion temperatures, the mass density and the magnetic field.

We cannot explain the experimental emission spectrum of the plasma exhibiting lines of K, Cl and Ar if we do not reach sufficient temperature for the central fiber. The isocontours in temperatures (keV) show a strong turbulence at the border of the two components. In order to take this effect into account, we built for the cells located at the border of the two components tables of mixtures of agar-agar+KCl+argon (50%-50%). Consequently, the highest electron temperature of 1.2 keV is inside this mixing region and it is possible to get lines of K and Cl in the calculation as in the experiment.

The calculated spectra of figure 3 show these spectra.

[1] Branitskii A.V. et al. Plasma Physics Reports 22, 277 (1996).

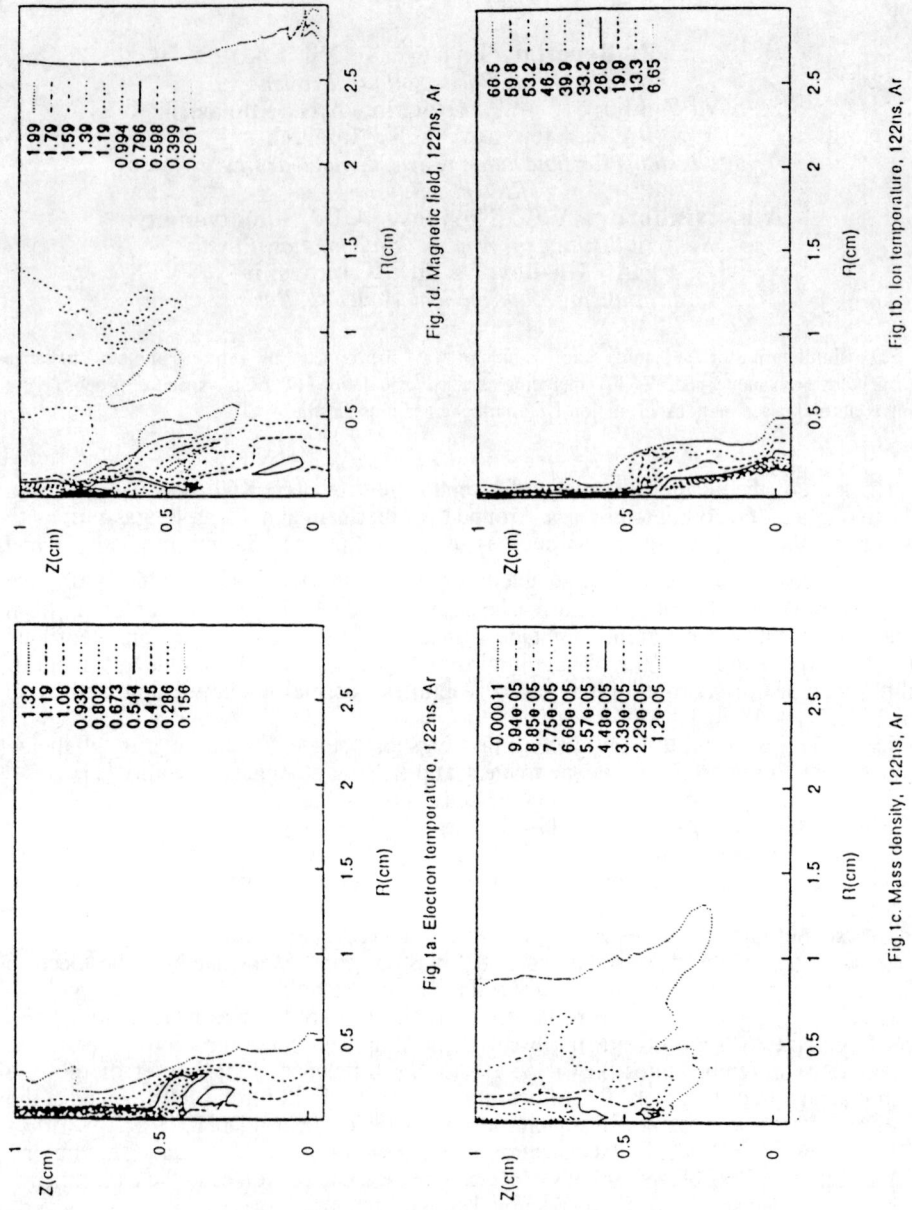

Fig.1a. Electron temperature, 122ns, Ar
Fig.1b. Ion temperature, 122ns, Ar
Fig.1c. Mass density, 122ns, Ar
Fig.1d. Magnetic field, 122ns, Ar

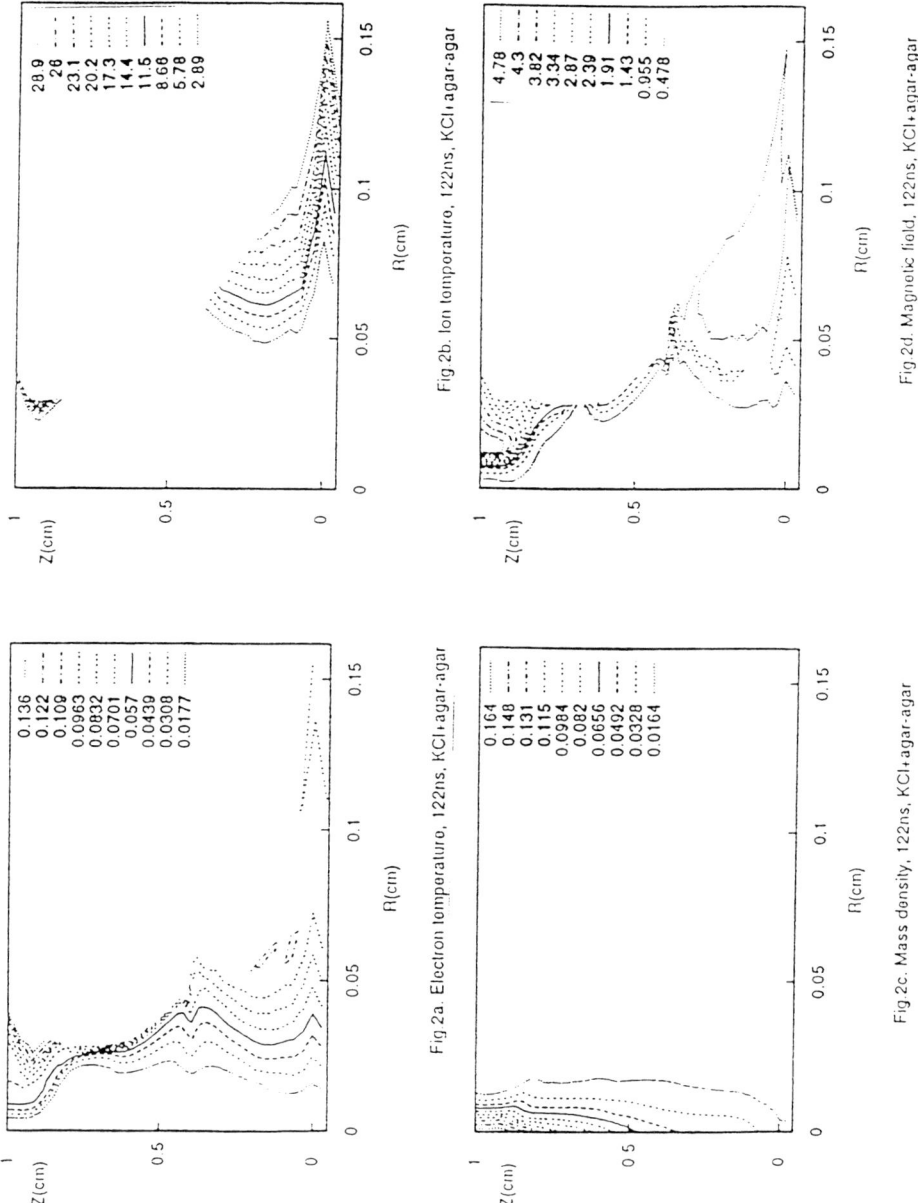

Fig 2a. Electron temperature, 122ns, KCl+agar-agar

Fig 2b. Ion temperature, 122ns, KCl+agar-agar

Fig 2c. Mass density, 122ns, KCl+agar-agar

Fig 2d. Magnetic field, 122ns, KCl+agar-agar

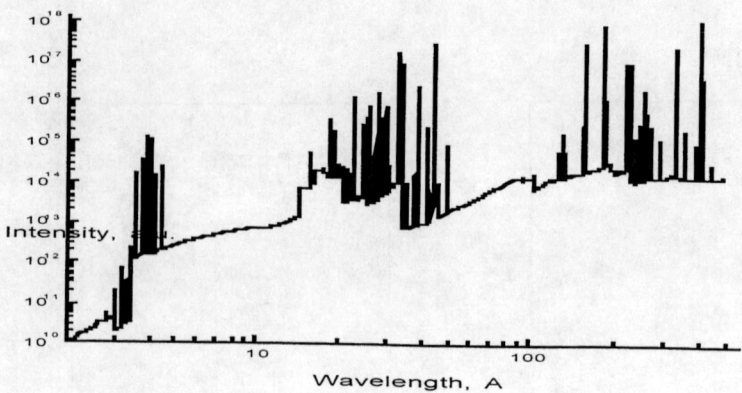

Fig. 3 a. The general view of the radiation spectrum coming from plasma along the ray

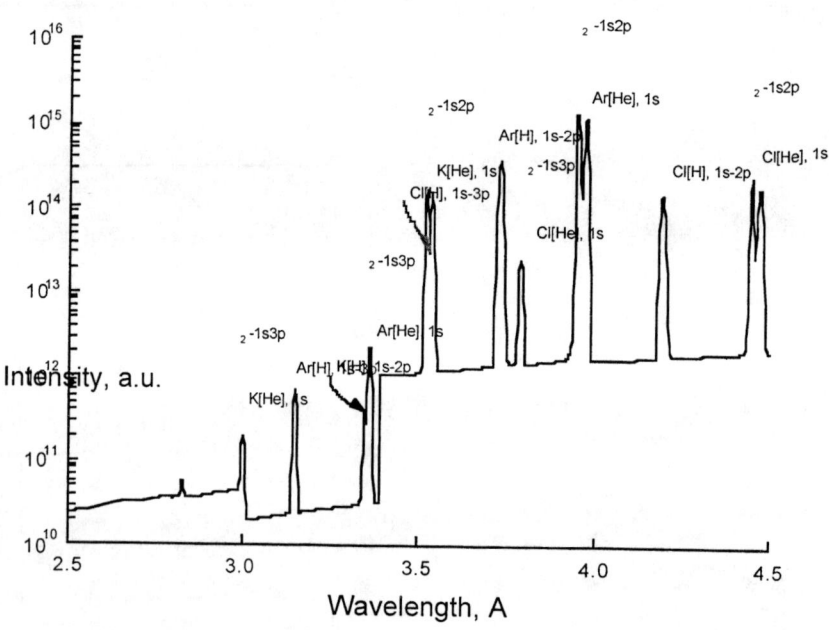

Fig. 3 b. Detailed view of the spectrum with lines of H- and He- like Ar, Cl, K.

Diffusion in Multicomponent Liners

V.I. Oreshkin

High Current Electronics Institute, 4 Akademichesky ave., 634055, Tomsk, Russia

Abstract. The implosion of liners consisting of a mixture of several components has been simulated. It has been shown that during the process of implosion, the components are separated due to the diffusion caused by thermal pressure and magnetic field gradients.

Introduction

In experiments with imploding plasma liners, a situation often arises where a liner consist of a mixture of various materials. This may be a simple liner consisting of a mixture of gases or a multilayer liner. In cases like these, the forces induced by the thermal pressure and magnetic field gradients present in the liner plasma act in different ways on ions of different materials. This results in that both the ions and the electrons acquire diffusion velocities. The diffusion of various materials in the plasma of a liner can be taken into account in the hydrodynamic approximation. To do this, it is necessary to write the Boltzmann equation for each ionic species and for the electrons and to integrate the resulting equations with various weights. Assuming that variously charged ions of one and the same material have the same diffusion velocity and neglecting the viscosity and the thermal force, it is possible to obtain a set of MHD equations presented below.

The MHD equations

The MHD equations taking into account the diffusion of various materials has the form:

$$\frac{dn}{dt} + n\vec{\nabla}\cdot\vec{v} = 0 \quad ; \tag{1}$$

$$\frac{dn_a}{dt} + n_a\vec{\nabla}\cdot\vec{v} + \vec{\nabla}\cdot\left(\vec{V}_a n_a\right) = 0 \quad ; \tag{2}$$

$$\rho \frac{d\vec{v}}{dt} = -\vec{\nabla}(p_i + p_e) + \frac{1}{c}\vec{J} \times \vec{B} \quad ; \tag{3}$$

$$\frac{d\rho_a \vec{V}_a}{dt} = -\rho_a \vec{V}_a (\vec{\nabla} \cdot \vec{v}) - \rho_a (\vec{V}_a \cdot \vec{\nabla})\vec{v} - \vec{\nabla} p_a - \frac{Z_a n_a}{n_e} \vec{\nabla} p_e + \frac{\rho_a}{\rho} \vec{\nabla}(p_i + p_e) +$$

$$+ \left(\vec{R}_a^i + \frac{Z_a n_a}{n_e} \vec{R}^e \right) + \frac{e Z_a n_a}{c}(\vec{V}_a - \vec{V}_e) \times \vec{B} + \frac{1}{c}\vec{J} \times \vec{B}\left(\frac{Z_a n_a}{n_e} - \frac{\rho_a}{\rho} \right) \quad ; \tag{4}$$

$$\rho \frac{d\varepsilon_i}{dt} = -p_i \cdot \vec{\nabla}\vec{v} - \vec{\nabla}(-\kappa_i \vec{\nabla} T_i) + \sum_a \vec{R}_a^{ie} \cdot \vec{V}_a + e n_e \vec{E}^* \cdot \vec{V}_e + Q_\Delta \quad ; \tag{5}$$

$$\rho \frac{d\varepsilon_e}{dt} = -p_e \vec{\nabla} \vec{U}_e - \vec{\nabla}(-\kappa_e \vec{\nabla} T_e + \rho \varepsilon_e (\vec{u} + \vec{V}_e) + \vec{W}_R) + Q_F - Q_\Delta, \tag{6}$$

where the index "a" denotes the a-th material; n_a is the number density of the a-th material; $\vec{V}_a$ is the diffusion velocity of the a-th material being the difference between the macroscopic velocity of the a-th material and the medium-mass velocity of the material; $\vec{V}_e$ is the diffusion velocity of the electrons; Z_a is the average charge of the a-th material; $\vec{U}_e = \vec{v} + \vec{u} + \vec{V}_e$ is the macroscopic velocity of the electrons; $\vec{u}$ is the current velocity of the electrons; $\vec{J}$ is the current density; $\vec{B}$ is the magnetic field; $\vec{E}^*$ is the electric field in the coordinate system moving with the medium-mass velocity; p_e, p_i are the pressures of the electrons and of the ions, respectively; κ_e, κ_i are the heat conductivities of the electrons and of the ions, respectively; ε_i is the internal energy of the ions; ε_e is the internal energy of the electrons; Q_Δ is the heat received by ions in their collisions with electrons; Q_F is the heat released in friction of electrons with ions; $\vec{W}_R$ is the radiation flux; $\vec{R}_a^i = \vec{R}_a^{ii} + \vec{R}_a^{ie}$; $\vec{R}_a^{ii}, \vec{R}_a^{ie}$ are the friction forces acting on the ions of the a-th material from the ions of the rest materials and from the electrons, respectively; $\vec{R}^e$ is the friction force acting on the electrons from the ions.

Equations (1)-(6) are complemented with Maxwell equations and with a generalized Ohm law, which for this case has the form:

$$\vec{E}^* = \frac{\vec{J}}{\sigma} - \frac{\vec{\nabla} p_e}{e n_e} + \frac{1}{e c n_e}\vec{J} \times \vec{B} - \frac{1}{c}\vec{V}_e \times \vec{B} + \frac{m_e}{e}\sum_a \frac{\vec{V}_a - \vec{V}_e}{\tau_a^{ie}} \quad , \tag{7}$$

were σ is the conductivity of the plasma; τ_a^{ie} are the time between the collisions of the electrons with the ions of the a-th material. Moreover, the MHD equations should be complemented with equations of radiation transfer and with an collision-radiation model.

Numerical simulation results

Equations (1)-(7) were solved in the one-dimensional approximation for a cylindrical geometry. To determine the radiation flux involved in Eq. (6), the radiation transfer equations were solved in the diffusion approximation and coefficients in these equations were found based on the multilevel nonstationary collision - radiation model.

The problem on the diffusion in a liner consisting of a helium-argon mixture was solved numerically. Initially, the helium-to-argon content ratio was invariable in radius and made up $\frac{\rho_{He}}{\rho_{Ar}} = 1$. The initial density distribution for the liner materials was specified by a Gaussian function centered at the radius $R_0 = 1$ cm with a standard deviation of 0.4 cm for the liner mass $M = 17$ μg/cm and the liner material temperature 1.5 eV. The liner was accelerated by a sinusoidal current. The amplitude I_m and the current rise time τ_m were varied so that the quantity $\frac{I_m^2 \tau_m^2}{c^2 M R_0^2}$ be invariable and approximately equal to 10, The fulfillment of this condition provides liner implosion at the current maximum.

Two variants were considered: with a comparatively low current and a long implosion time ($I_m = 0.3$ MA and $\tau_m = 500$ ns) and with a comparatively high current and a shot implosion time ($I_m = 1$ MA and $\tau_m = 150$ ns). The distribution of the density and percentage of the materials at the instant the compression was a maximum are given in Figs. 1 and 2. As can be inferred from this figures, during the implosion of the liner, separation of the materials took place: the outer layers of the liner consisted in the main from argon, while helium prevailed in inner layers. The separation was more pronounced for the first variant.

Let us discuss the results obtained. As follows from Eq. (4), diffusion of the materials is caused by two factors: the thermal pressure gradients and the magnetic field gradients. With that, the light materials diffuse in the direction of decreasing thermal pressure, while the materials with high $\frac{Z_a}{m_a}$ ratios diffuse in the direction of decreasing magnetic field pressure. Characteristic of the Z-pinch geometry is an increase in magnetic pressure with radius. Therefore, if diffusion is caused by magnetic forces, the material with the higher $\frac{Z_a}{m_a}$ ratio (helium in our case) will always diffuse into the inner layers of the liner. The situation is somewhat more complicated if diffusion is conditioned by thermal pressure gradients. Thermal pressure gradients reach their maxima at the fronts of shock waves, and in an imploding liner, shock waves may propagate both toward the axis and in the reverse direction. Therefore, at the initial stage of implosion, the lighter material (helium) diffusion into the inner layers of the liner, and, as a

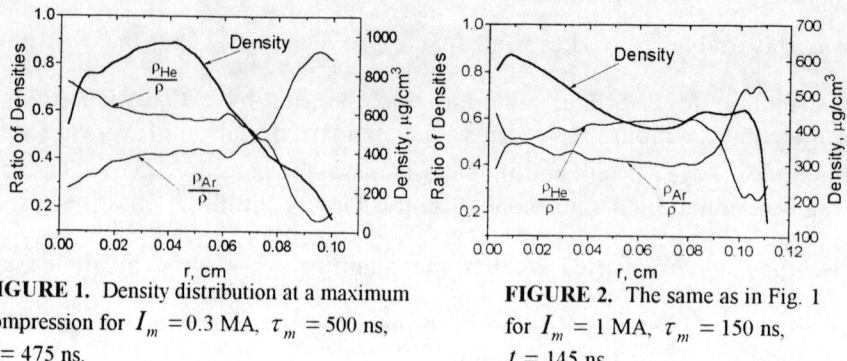

FIGURE 1. Density distribution at a maximum compression for $I_m = 0.3$ MA, $\tau_m = 500$ ns, $t = 475$ ns.

FIGURE 2. The same as in Fig. 1 for $I_m = 1$ MA, $\tau_m = 150$ ns, $t = 145$ ns.

reflected shock wave is formed at the axis, the direction of the diffusion of helium may reverse.

Let us return to Figs. 1 and 2. Comparing the conditions for implosion for two variants under consideration, we see that the implosion time for the first variant is much longer than that for the second variant. This has the result that for the first variant, the magnetic field penetrate in the deeper layers of the liner then for the second variant. This is favored by the fact that for the first variant, the intensity of the shock waves that heat up the liner materials is low, and the electron temperature and, hence, the plasma conductivity, are low as well. This is just the reason why the separation of materials is more pronounced for the first variant.

Conclusion

The numerical simulation performed for an imploding liner consisting of a mixture of dissimilar materials has shown that the implosion is accompanied by diffusion of these materials resulting in their separation. In an imploding liner, the inner layers are enriched with the lighter materials and the outer ones with the heavier materials. The separation of materials is more pronounced for longer current rise times and lower current amplitudes.

Compressional Pinch Simulations and Experiments

N. G. Kassapakis, H. M. Davies and M. G. Haines

Imperial College of Science, Technology and Medicine, London SW7 2BZ, UK

We report on numerical simulations and experiment of compressional Z-pinches. Previous comparisons of 1-D simulation using various models with experimental results, although satisfactory for the initial stages of the pinch, were inconclusive about the fate of the plasma column after first compression.

A recent series of experiments has shown that, far from reaching an unexplained steady radius after the initial compression, the plasma column radius oscillates, as predicted in simulation.

1. Previous Experiments

Previous experiments by Bayley[1] on a 150kA H_2 and Argon pinch confirmed the anomalously long pinch lifetime reported by others[2]. The crowbar model, developed by Culverwell[3], was used to explain this. Bayley reported the formation of an m=0 instability, which grew too slowly compared to ideal MHD in a 500 mtorr H_2 pinch by taking side-on interferograms. The instability saturated at 30% of the pinch radius and the ratio of ion Larmor radius (LR) to pinch radius for this pinch (~0.1) was thought to be evidence of finite LR (FLR) stabilisation effects. This was based upon a time series of interferograms from different shots, claiming reproducibility between shots at the same fill pressure.

2. Simulations

The snowplough[4] and slug[5] models, which are in agreement with experiments up to the time of first compression were used as tests for the 1-D Lagrangian MHD code (written by A.R. Bell). It includes bremsstrahlung losses, an ideal gas equation of state, a single temperature, specified current profile and Spitzer-like resistivity.

All the extractable information from the models about pinch parameters were compared successfully with the code. The time to pinch, the effect of the peak current and variation of pinch time vs filling pressure were examined amongst others. The time to collapse was measured by taking the final radius as equal to the slug model radius (i.e. 3.09mm). As expected, it increases with pressure, because the magnetic force due to the skin current has a larger mass to accelerate to the axis. Some test results are shown in figure 1 and 3.

In figure 4 a typical plot of plasma radius against time is given. There is qualitative resemblance between this and a typical radial streak (fig.2). Unlike the slug model, there is a bouncing of the radius following the collapse which was not experimentally observed before. The plasma collapses to significantly smaller radius than in the slug model, and bounces out to a larger radius than the one predicted by the model (dashed line). Although this was initially deemed to be an artificial code effect physical arguments can account for it. When the plasma is compressed by the magnetic field generated by the current in the skin it is also heated and the number density becomes peaked on axis (fig.5a-c). At this point the compression is higher than expected but the plasma kinetic energy exceeds the magnetic energy containing the plasma, temporarily. As a result it expands, and in doing so cools and becomes less dense (fig.5d). A second compression phase follows due to the constant current and the process repeats itself with falling amplitude since energy is removed from the pinch due to various processes such as resistivity, viscosity or radiation losses.

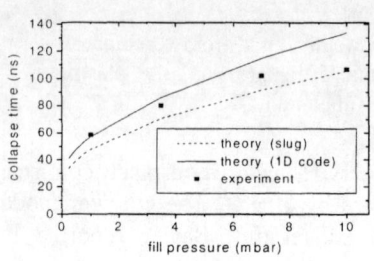

Figure 1: Collapse times for 1-D code, slug model and experiment

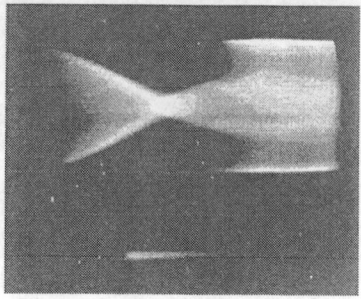

Figure 2 : Streak Photograph, 10mbar H_2 (S0611#02)

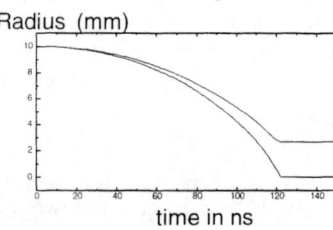

Figure 3 : Piston and Shock radius against time (slug)

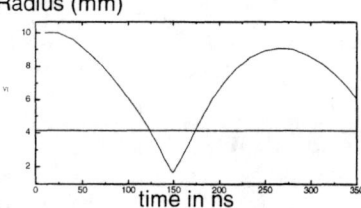

Figure 4 : Plasma radius bouncing in time (MH1D)

As mentioned before the long pinch lifetime was explained by the existence of an internal current crowbar, the current flowing in a loop through the pinch and a wall plasma which had formed following the collapse. Earlier work by Christie[6] suggested that the presence of a highly conducting plasma at the wall of the pinch vessel will have an effect on the vacuum magnetic field between the pinch and the wall, trapping the field, which will not be able to diffuse through the plasma. However, he argued it was a small effect right into the non-linear development of the instability. Indeed, this affects the plasma behaviour. When the pinch expands after crowbar has occurred ($r_p>r_{crow}$) the vacuum B_θ is greater than the value at crowbar, so the plasma is more strongly confined. When $r_p<r_{crow}$, B_θ is less than the crowbar value. The effect is strongly dependent on the time chosen for the crowbar to occur.

When the crowbar is introduced at 85 ns (kinetic energy is high, see pinch profiles, above) the plasma bounces with a very high frequency (6 oscillations in 150 ns) and through a small range of radii. When the crowbar is introduced at later times, the frequency reduces and the range through which the pinch bounces increases. We conclude that the changing frequency is linked to the kinetic energy of the plasma at the time of crowbar, and the range through which it bounces to the value of B_{vac} at this time. The radius to which the pinch is converging, in each case, is the radius which it had reached when the artificial crowbar was introduced. This is, perhaps, predictable. For the case where crowbar is not included the above observations on radial bounce frequency do not apply, and the time taken for the pinch bouncing to decay is far longer.

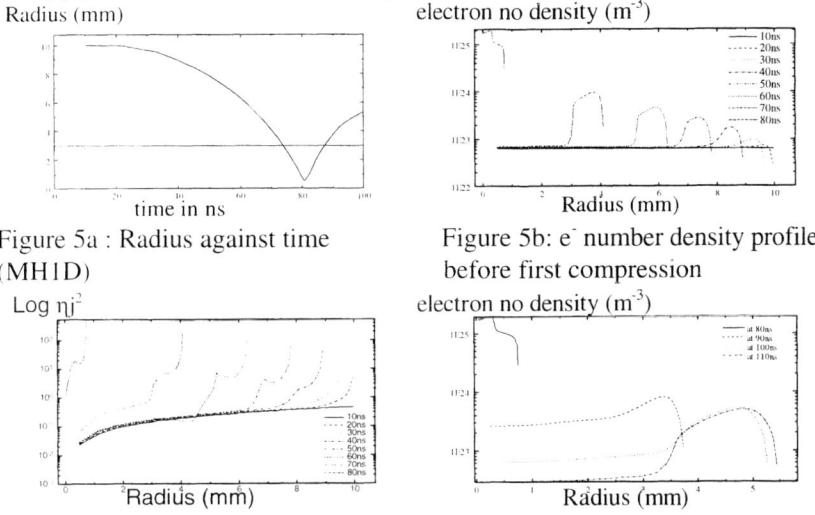

Figure 5a : Radius against time (MH1D)

Figure 5b: e⁻ number density profiles before first compression

Figure 5c : Current density profiles before first compression

Figure 5d : e⁻ number density profiles after first compression

The 1-D MHD code gives radial profiles of electron number density, electron and ion temperatures, and current density, at 10 ns intervals (figs. 5a-d). This collapse sequence shows that many of the assumptions made about the pinch are acceptable: there is a build up of plasma at the wall, which sweeps towards the axis, with a current flowing on the outer surface (cf. the slug model, 'magnetic piston'). Notice the sharp peaking of material on the axis at the time of compression, and the high temperature which is achieved there, which is the basis of the 'high energy' ($=n_e kT_e$) explanation of bouncing. It is worth noting that the pinch does not reach equilibrium with the magnetic field. Even the highest reached temperature does not come close to the quoted Bennett value (175 eV). If this model is to be believed at all, then the calculations of MHD growth times by Bayley, on which the claims of FLR stabilisation were partially based, are questionable.

Moreover Haines and Coppins[7] demonstrated that the conditions of applicability of MHD can be parameterised in a particular form. They show that the parameter space where the experiments by Bayley belong is not that of ideal MHD. Instead they belong to the FLR regime.(The ions in the pinch are magnetised, $\Omega_i \tau_{ii} = 4.4$). It is widely believed that a full modelling of FLR effects cannot be achieved by a fluid code. Work towards this direction is progressing.

3. Recent Experiments

Streak images taken from shots at fill pressures of 10, 7 and 3.5 mbar (figures 2, 8, 9) are presented. The wall plasma is seen to light up, i.e. begin emitting, between 184 and 225 ns after the start of the main current. Furthermore, the plasma radius does not remain constant in time following the collapse. An oscillation occurs in time, which appears to be directly proportional to the fill pressure. This effect could be attributed to the passing of an MHD instability across the radial section being streaked but the evidence is not enough. We believe that this is not an MHD instability, but a genuine oscillation of the plasma radius. The radius reached by the pinch at highest compression was, for all pressures, 4.14 mm. This is *not* the radius predicted by the slug model. Since the slug model radius is dependent on the ratio of the specific heats, γ, it is tempting to infer from this result that the plasma is not pure hydrogen, and contains sufficient impurities to alter γ, such that $\gamma = 1.8$. The slug model curve is for $\gamma = 5/3$, but the MHD curve is plotted for the time taken to reach 4.14 mm. The MHD results agree well with the experiment at lower fill pressures. At higher fill pressure, the slug model is in closer agreement, albeit only for the time to pinch. The lack of agreement between the slug model and the experiment could be attributed to the new value of γ (see above).

Radius (mm)   Radius (mm)

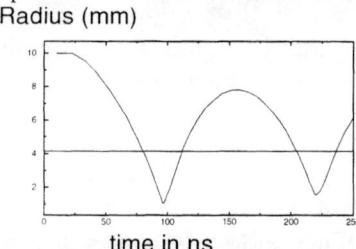 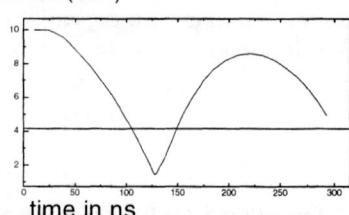

time in ns   time in ns

Figure 6 : 3.5 mbar fill pressure   Figure 7 : 7 mbar fill pressure

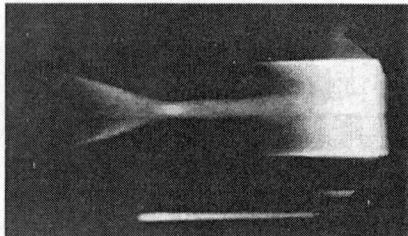 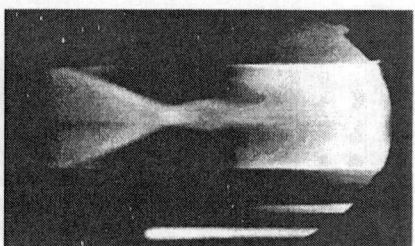

Figure 8 : 3.5 mbar fill pressure   Figure 9 : 7 mbar fill pressure

An alternative account follows from the MHD code simulation. Figures 6 and 7 show 1-D code results for matching fill pressures. The horizontal line on each plot represents the radius observed in experiment (4.14 mm).

All images are 300 ns full sweep. The comparison is summarised in the tables below:

Fill pressure(mbar)	Time from min to max radius (ns)	Max radius(mm)	Time to 2nd min(ns)
3	31.5	6.2	40.5
3.5	40.5	6.2	44
4	45	6.2	47
7*	73.5	8.5	not observed
10*	91.5	8.74	not observed

Table 1a: Experimental values of oscillation parameters (* all values are averages)

Fill pressure(mbar)	Time from min to max radius (ns)	Max radius(mm)	Time to 2nd min(ns)
3	37.2	7.0	46.7
3.5	45.2	7.63	47
4	47.9	7.9	54.5
7	71	8.6	n/a
10	92	9	n/a

Table 1b: Simulated values of oscillation parameters

4. Conclusion

Collapse dynamics of the hydrogen compressional pinch have been observed and compared to simulation. Radial oscillation of the plasma was observed, and has been identified with the plasma heating up to an energy greater than the magnetic field confining it, expanding and cooling, as simulated in the 1-D MHD code. This shows the plasma is not in equilibrium with the confining magnetic field, and does not exactly satisfy the Bennett relation as previously assumed.

References

1. Bayley J.M. (1989) *Phd. Thesis, Univ. London*
2. Haines M.G. (1989) *Proc. 2nd Int. Conf. on High Density Z-pinches, Laguna Beach California*
3. Culverwell I.D., Coppins M., Haines M.G., Bell A.R. and Rickard G.J. (1989) *Plas. Phys. & Cont. Fusion* **31** p387
4. Rosenbluth M., Garwin R., Rosenbluth A., Los Alamos Report - LA 1850 (1954)
5. Potter D.E. (1978) *Nucl. Fusion* **18** p813
6. Christie M.A. *Plasma Phys* **24** (1982) p783
7. Haines M.G. and Coppins M. (1991) *Phys. Rev. Letts.* **66** p1462

N.G. KASSAPAKIS, n.kassapakis@ic.ac.uk, ++0171-5947643 (f ax- 658)

Magnetohydrodynamic Simulation of Capillary Plasmas

N.A. Bobrova[1], S.V. Bulanov[2,3], D. Farina[3], R. Pozzoli[3,4], T.L. Razinkova[1], and P.V. Sasorov[1]

[1] *Institute for Theoretical and Experimental Physics, Moscow, 117259, Russia*
[2] *General Physics Institute of the Russian Academy of Sciences Moscow, 117942, Russia*
[3] *Istituto di Fisica del Plasma, Consiglio Nazionale delle Ricerche Milano, 20133, Italy*
[4] *Department of Physics, University of Milano, Milano, 20133, Italy*

Abstract

The dynamics of capillary plasmas in regimes where soft x-rays amplification has been observed is investigated by means of a magnetohydrodynamic one-dimensional code, which takes into account dissipative processes, ablation, and ionization of the wall material. Redistribution of the electric current between argon and ablated wall plasmas strongly influences the plasma parameters in the core.

It is well known, that short-wavelength lasers demand highly ionized dense plasmas. Direct generation of an amplifying plasma column by a fast discharge through a capillary channel has been successfully used for the production of highly ionized plasmas with parameters suitable for soft-x-ray lasers [1,2]. In a capillary filled with argon the mechanism responsible for the observed amplified spontaneous emission is electron-collisional excitation pumping of argon's 3p-3s transition. Though a variety of spectroscopic observations of capillary plasmas in different discharge conditions have been performed, the dynamics of the discharge plasma in the different regimes is discussed only in a few papers[1,4-7]. In our paper [8] it was shown that the main physical process which had to be taken into account is the ablation of the wall material under the heat flux from the capillary plasma. The heated and ionized ablation products form a plasma with high electrical conductivity that can lead to significant redistribution of electric current between it and the argon plasma in the initial channel, and consequently to differences with the discharge dynamics of a classical pinch.

In the present report, the dynamics of the discharge used to generate the plasma in a capillary filled with initially pre-ionized gas is investigated and compared with experimental results [3]. The results of computer simulations with a one-dimensional magnetohydrodynamic (MHD) code of the plasma behavior inside a capillary channel are presented, taking into account the plasma interaction with the capillary walls, the ionization of the ablated material, and its interaction with the pre-ionized gas filling the channel.

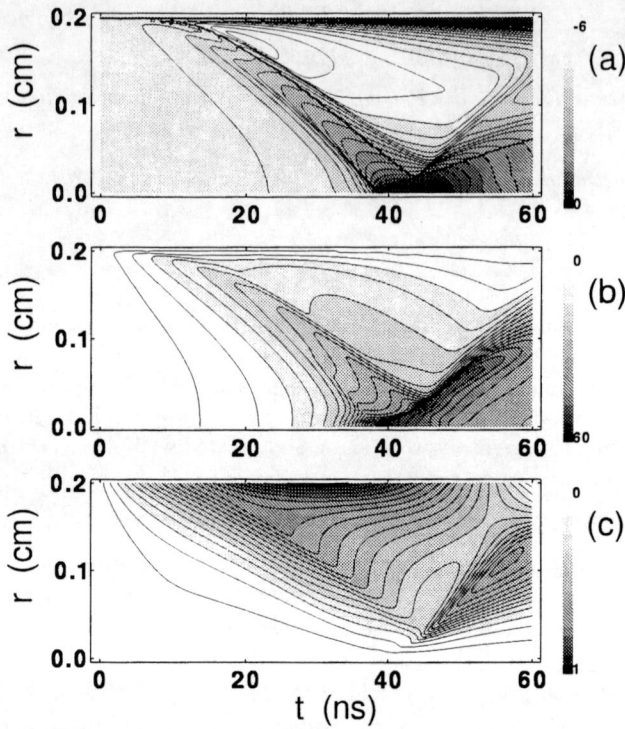

Figure 1: Results of computer simulation of capillary discharge dynamics in a 4 mm diameter channel filled with 0.64 torr of pure Ar at the current $I_0 = 40$ kA, and half-cycle time $t_0 = 60$ ns. (a) lines of constant value of decimal logarithm of plasma density measured in g/cm^3. (b) lines of constant value of the electron temperature measured in eV. (c) lines of constant value of electric current inside the region with radius r, normalized to I_0.

The overall picture of the discharge can be described as follows: the current pulse heats the plasma and creates an azimuthal component of the magnetic field, thus leading to the pinching of the plasma. A region of unionized gas of very high density is formed due to melting and evaporation of the solid matter of the walls. This dense gas is ionized and heated by the heat flow from the plasma region. Then, it implodes radially toward the capillary axis. The basic version corresponds to a capillary with radius $R_0 = 2$ mm, density $\rho_0 = 1.37 \times 10^{-6}$ g/cm^3 (corresponding to a pressure of 0.64 torr at room temperature), and plasma temperature of 0.5 eV. The capillary is of polyacetal $((CH_2O)_n)$.

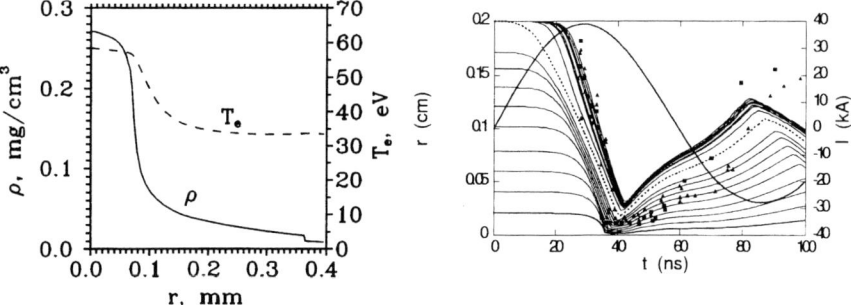

Figure 2: The core's plasma density ρ and electron temperature T_e versus radius r when the shock wave converges.

Figure 3: Comparison of the simulated trajectories of plasma elements (thin solid lines) with the experimentally observed [5] radius of radiative plasma (points of different type). The thick solid line represents the current pulse of Ref. [5], used in the simulation. A good agreement of the observed time of plasma compression, size and lifetime of compressed plasma with the simulation can be seen.

Figure 1 gives the evolution of the capillary discharge. A compression wave propagates from the capillary wall to the channel axis, where it reflects at $t \sim 40$ ns. Around this time, a core of hot and dense plasma forms close to the axis of the channel ($r/R_0 < 0.04$) (see Fig. 2). Plasma density and temperature are uniform inside the core, which lasts for about 10 ns. The maximum plasma parameters reached in the core at $t \approx 40$ ns are $\rho \approx 2.7 \times 10^{-4}$ g/cm^3, $n_e \approx 4 \times 10^{19}$ cm^{-3}, and $T_e \approx 60$ eV. They are in satisfactory agreement with those in an experiment [3] wherein the discharge plasma shows soft x-ray amplification.

The core formation process in the capillary discharge is best understood by looking at the behavior of the current distribution. It is apparent from Fig. 1c that during core formation there is practically no current inside the region close to the axis, and less than $\sim 5\%$ of the full current is inside the core. For this reason, the formation and the evolution of the core, including the converging shock wave, the increase of its amplitude due to the geometry of the capillary, and the resulting increase of the plasma density on the axis, are hydrodynamic processes. It should be noted that the formation of the converging shock wave is due to the action of the ponderomotive force $\vec{j} \times \vec{B}$ in the initial stage of the discharge.

Fig. 3 compares the experimental results presented in [5] with our simulation performed for the same discharge parameters. The experimental and calculated core sizes are in good agreement, and the observed time of plasma compression coincides with the simulation time. We conclude that the approximation used in our code, including the description of radiation and the plasma-wall interaction, is quite adequate. Scaling of the core's maximum density and temperature with discharge parameters was investigated in [9,10].

Let us note that in the absence of the capillary walls argon plasma will contract together with the electric current. As a result the plasma achieves high temperature, density and radiation energy loss. The radiation will melt and evaporate the solid mater of the walls. When we include the ablation of wall material under the heat flux due to electron thermal conductivity, melting and evaporation begin during the initial stage of the discharge. And plasma contraction and hence radiation losses are low. In this case radiation cooling plays a minor role, with the electric current being much less than the Pease-Braginskii current. Taking into account line radiation the Pease-Braginskii current is of the order of several hundred kA, while in our case the peak current is equal to 30-40 kA and the current inside argon plasma at the moment of its maximum compression is several times smaller still.

References

1. Rocca, J. J., et al, *Opt. Lett.* **13**, 565-567, (1988).
2. Steden, C., and Kunze, H.-J., *Phys. Lett. A* **151**, 534-537, (1990).
3. Rocca, J. J., et al, *Phys. Rev. Lett.* **73**, 2192-2195, (1994).
4. Kunze, H.-J., et al, *Phys. Lett. A* **193**, 183-187, (1994).
5. Rocca, J. J., et al, *Phys. Plasmas* **2**, 2547-2554, (1995).
6. Lee, K. T., et al, *Phys. Plasmas* **3**, 1340-1347, (1996).
7. Bulanov, S. V., et al, *Phys. Plasmas* **4**, (1997).
8. Bobrova, N. A., et al, *Fiz. Plasmy* **22**, 387-402, (1996) [Plasma Phys. Reports 22, 349-362, (1996)].
9. Bobrova, N. A., et al, *Fiz. Plasmy* **23**, (1997).
10. Bobrova, N. A., et al, "Magnetohydrodynamic Simulation of Capillary Plasmas," in *X-ray Lasers 1996*, Eds. S. Svanberg & C-G. Wahlström IOP, Bristol & Philadelphia, 1996, pp.197-199.

COMPARISON OF SIMULATION AND EXPERIMENTAL RESULTS FOR A GAS PUFF NOZZLE ON AMBIORIX

J-N. BARNIER, J-M. CHEVALIER, B. DUBROCA,
CEA/CESTA, BP 2, 33114 Le Barp, France

Abstract

One of source term of Z-Pinch experiments is the gas puff density profile. In order to characterize the gas jet, an experiment based on interferometry has been performed. The first study was a point measurement (a section density profile) which led us to develop a global and instantaneous interferometry imaging method. In order to optimise the nozzle, we simulated the experiment with a flow calculation code (ARES). In this paper, the experimental results are compared with simulations. The different gas properties (He, Ne, Ar) and the flow duration lead us to take care, on the one hand, of the gas viscosity, and on the other, of modifying the code for an instationary flow.

INTRODUCTION

The AMBIORIX facility is a pulsed power generator located at CEA/CESTA. It consists of a 300 kJ Marx bank feeding a temporal compression line. The output parameters are : 2.4 MA with a 50 ns rise time on 0.5 Ω, 10 nH vacuum diode. Ambiorix can only implode low weight charge ($\approx$100 µg), hence we use mainly gases like helium, neon, argon.

The gas is initially filled in a small tank. When a valve opens the tank, the gas passes through a tranquillization chamber then through an annular nozzle. The time of opening is about 250 µs to 500 µs. After this time the main current arrives in the gas and a plasma is created. At that time, the total mass of gas is near 100 µg.

The aim of this study is to measure the density profile of the gaseous load before the implosion. For that, various methods based on interferometry were used to achieve this task. In order to improve the quality of the gas shape, a tool is used to simulate our system. Thus, the hydrodynamic ARES code calculates the flow of gas and it is compared with experimental data.

1. DENSITY PROFILE MEASUREMENT

1.1 Experimental set-up

The purpose of this experiment is to obtain the density profile of the gas puff just before the implosion. With this interferometer based on Michelson method [1], we measure the profile only on a shape of the gas puff (one plane).

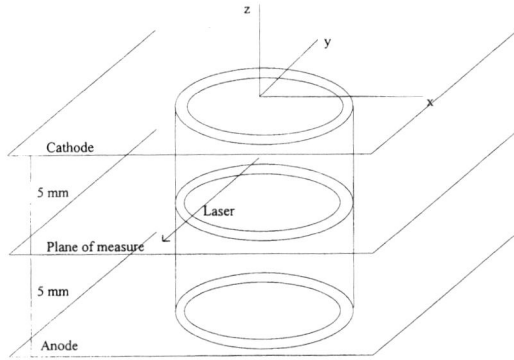

The gas is located between a cathode (nozzle) and an anode (made with wire, and assumed to be presumably transparent to the gas). The distance K-A is 10 mm and the profile is made at 5 mm from the nozzle. The tank is filled with 4 bars of Helium. The radius of the annular puff is about 20 mm. The laser beam is moved manually to obtain a complete profile and more than 100 points are necessary to have the density and for each point. Experimentation thus lasted two weeks.

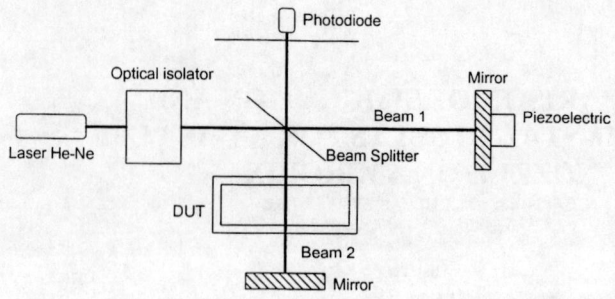

Measurement is made with a Michelson interferometer. The beam of He-Ne laser (632 nm) is splitted in two beams, one passes through the device undergoing test, the other is the reference beam. One mirror is mounted on a piezoelectric to obtain very accurate positioning. The optical isolator prevents the beam from returning into the laser.

The intensity measured by the photo detector is given by $I = \frac{I_0}{2}(1 + V \cos \Delta \phi)$ with V the factor of contrast and $\Delta \phi$ the phase shift between the two optical paths. The phase shift of light ray between two point A,B is defined as the line integral (for a Michelson) : $\Delta \phi = \frac{2\pi}{\lambda} 2 \int_A^B (n(x,y,z) - 1) dy$

Density is connected to the refractive index of the gas by the Gladstone-Dale equation $n - 1 = K\rho$, where K, the Gladstone-Dale constant, is a property of the gas. If we assume that the gas puff is axisymmetric, the equation above becomes : $\Delta \phi = \frac{2\pi}{\lambda} 4 \int_x^{\infty} (n(r) - 1) \frac{r}{\sqrt{r^2 - x^2}} dr$

When applying the Abel inversion [2,3], the index of refraction is :
$n(r) - 1 = \frac{-\lambda}{4\pi^2} \int_r^{\infty} \frac{d\Delta \phi}{dx} \frac{dx}{\sqrt{x^2 - r^2}}$

K (helium) : $0.196 \cdot 10^{-3}$ m^{3}kg^{-1} K (argon) : $0.175 \cdot 10^{-3}$ m^{3}kg^{-1}

1.2 Comparison of calculated and experimental profiles

1.2.1 ARES code

ARES code is a numerical simulation of hydrodynamic [4]. It solves the two dimensional Navier-Stokes equations with the following approximation. For the spatial approximation we have chosen a formulation based on finite volumes. The Euler terms are approximated by a second order upwind scheme constructed using our generalisation of Roe's approximate Riemann solver. In fact, the principal difficulty in extending Roe's method to reactive gases lies in the non-uniqueness of the associated Roe matrix (these choices form a vector space that has a dimension equal to the number of chemical species considered). Indeed, this infinity of choices, as we will note depends implicitly on only one parameter: the velocity of sound. This parameter is then chosen using a gamma-equivalent approximation. This makes our approximation equivalent to the one used for the extension of other upwind solvers like those of Osher or Van-Leer.
The viscous terms are approximated with the help of a standard centered scheme. Convergence towards the stationary state is ensured by an implicit symmetrical relaxation or non-linear method (Gauss-Seidel line relaxation). For the supersonic zones, we performed a non-linear space-marching solver directly derived from the previous solver.
The gas is considered to be perfect. With this code, we compute the flow of different gases used with nozzles mount on the AMBIORIX facility.

1.2.2 Results

The liner of gas is produced with a supersonic annular nozzle. The mean diameter of the nozzle is 40 mm with a shell thickness of 1.5 mm and a length of 10 mm. Each point is taken after a delay of 500 µs after the injection of gas in the nozzle. The annular nozzle feeds on Helium. We compare the flow of a puff of helium with a numerical simulation of ARES code. The code gives different data such as : density, pressure, temperature, etc. With the density, we compute also the index of refraction, i.e. the phase shift. In the figure below, experimental and simulated results are presented. These values of phase profile are in good agreement with numerical simulation obtained when using ARES code.

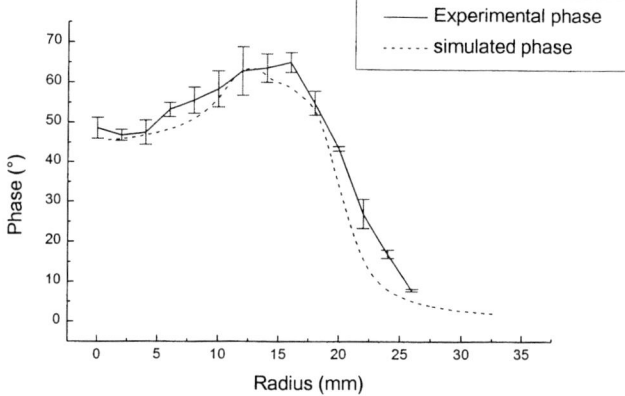

But, taking into account that 100 points are necessary to characterize a complete profile, this method is not useful. Consequently, we developed another diagnosis to get a complete frame of gas puff.

2. GLOBAL RECORD OF GAS PUFF

2.1 Experimental set-up

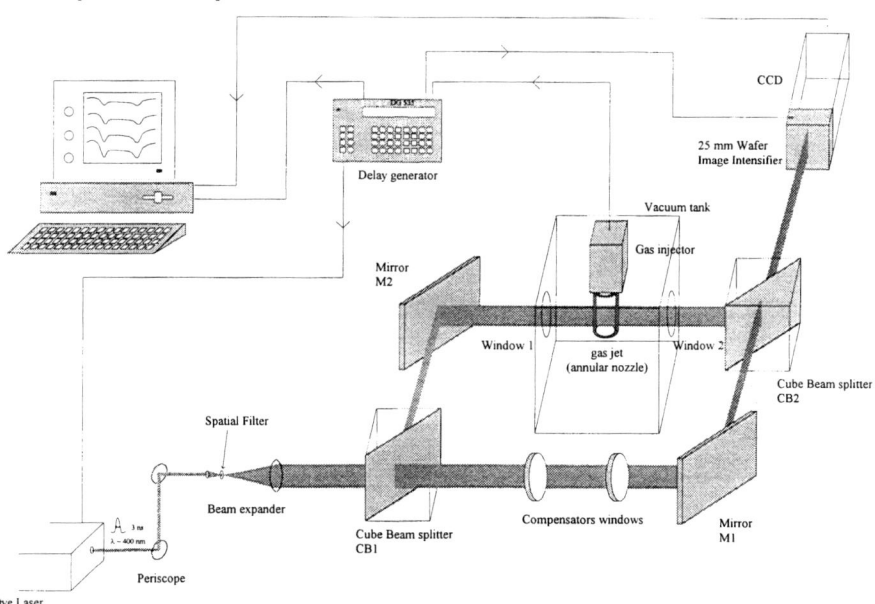

Mach-Zehnder interferometer is adjusted to give wedge fringes [1]. When a disturbance is present within the test section (which is in vacuum tank), the optical path is no longer uniform in this beam. The fringes then are no longer straight, but rather are curved (figure below).

The experimental interferograms are obtained with a 20 mm diameter annular nozzle with argon. Each picture is taken at the same time as the one talken before (Michelson interferometer), i.e. 500 μs after the injection of gas. A laser with a pulse width of 3 ns at a wavelengh of 390 nm is used.

2.2 Comparison of calculated and experimental interferograms

First of all, a static picture (i.e. without gas) is necessary to correct the phase shift. The second picture shows the gas puff. The gas goes from the bottom to the top of the picture. We scan the fringes with a computer and we calculate the phase due to gas flow. The result is given for the first fringe (near the bottom of picture below) and seems to be in good agreement if the gas viscosity is taken into account.

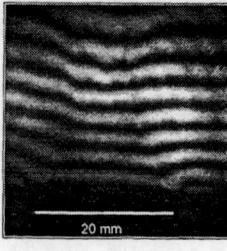
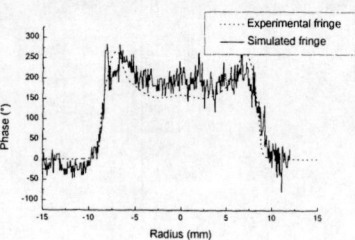

The model is very sensitive to this parameter. So, for the following, all calculations will be made with an instationary viscous flow.

3. Discussion

Two experimental methods have been performed to characterize the gaseous injection for Z-Pinch experiences on AMBIORIX facility [5]. First of them measures the density profile with high accuracy. Another advantage of this method is the temporal resolution of results. But this experiment requires hundred points to characterize a complete profile. The second method provides a complete anode-cathode space measurement with only one shot. Consequently, the gas distribution in AMBIORIX facility may be instantaneously determinated with this method.

Taking into account that different results (for He, Ar and for several nozzle shape's), we have a good prediction of gas flow with the ARES code. The comparison between experimental results and simulations shows off a low discrepancy. Consequently, several hypothesis are made. First of all, the exact nozzle shape's is not drawn in ARES code. ARES is 2D code and the nozzle must be drawn in 3D. Secondly, the opening of the valve takes about 500 μs and it is instantaneous for the code.

Conclusion

The main goal of these studies is to measure the capacity of ARES code to modelize our problem. The comparison of experimental results and simulations leads us to find the most important parameter for the calculation : the gas viscosity. In order to make different nozzle's shape, this code will be efficient to decrease the time of studies and, of course, the cost of them. To improve this result, we want to use another diagnosis based on Rayleigh light scattering. With this methods the density profile will be obtained without the problem of Abel inversion.

References

[1]   R.J. Goldstein, *'Optical systems for flow measurement'*, § 8
[2]   W Merzkirch, *'Current problems of optical interferometry used in experimental gas dynamics'*, Institut für thermo und fluiddynamik, Germany,
[3]   A. Kasperzuk, M. Paduch, J. Tech. Phys, T 19, p 137-150, 1978
[4]   B.Dubroca, G. Gallice, *'Navier-Stokes simulation in chemical non equilibrium'*, Aerothermo chemistry of spacecraft and associated hyper flows, IUTAM symposium, Marseille FRANCE, september 1992
[5]   J.N. Barnier, to submit at IEEE Transactions on Plasma Science Special issue on Z-Pinches, April 98.

PINHOLE IMAGES AND XRD SIGNALS FROM NUMERICAL SIMULATIONS

R. Benattar & A. Nikitin,
LPMI, Ecole Polytechnique, France

2D simulation of a composite Z-pinch was performed by the complete radiative magnetohydrodynamic code ZETA including detailed calculation of EOS. We generate, using the zoning of the code and the plasma parameters and calculating the radiattive transfer across the cells at each step, x-ray pinhole images and x-ray diode signals.

The result of 2d simulations of time evolutions of Z-pinch loads must be represented as real experimental diagnostics. The main diagnostics used in such experiments are x-ray pinhole camera imaging, x-ray diode (XRD) signals recording, in order to determine how the plasma is stable during its time evolution and to know the X-ray yield during the pinch formation.

Calculation of x-ray pinhole images.

We run a simulation with the code ZETA and we use the data files given by the simulations. For each Z section, the zoning in R of the code gives elementary surfaces (Figures 1 & 2). Each surface $\Delta A_{j,k}$ is calculated and memorized. Each volume $\Delta A_{ij,k} * \Delta z$ emits radiation which pass through the other volumes. Each elementary volume has a temperature T, a mass density ρ and opacity and emissivity which correspond to these temperature and density. These values and the zoning are given by the code for each time. For each frequency group, the sum of all the emitted radiations after absorption by the following cells are recorded on the detection plane. The resulting image can be represented for each group of photon energy at each time. Images also can be calculated with the sum of all radiations passing through a beryllium filter of given thickness.

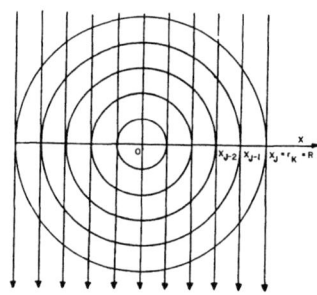 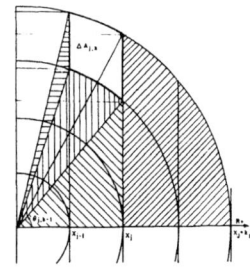

Figure 1   Figure 2

The code gives the opacity CEE in cm^{-1} and the emissivity/ opacity GE in 0.1TW/st/cm^3/frequency group/cm^{-1}
The absorption by an elementary surface occurs on a length (figure 2)

$$L_{j,k=0} = \sqrt{(\frac{1}{2}(x_{j-1} + x_j))^2 + x_j^2}$$

The intensity by cell of surface $\Delta x * \Delta z$ on the detection plane in erg/sec/cm2 is

$$I = \sum coeff \times 1.6 \bullet 10^{-9} \times GE \times CEE \prod^{k} \exp(-CEE \times L_{l_r})$$

The intensity I must be divided by $\Delta x * \Delta z$

coeff $= 2 * \Pi * 1.875 * 10^{22} * 1.10^{18} / \Delta \nu$

Figure 3 shows in its left part the mesh given by the code at 86 ns. The right part of the figure is the pinhole image for the photon energy of 65 eV.

Figure 4 shows at the same time the pinhole image built with all the radiations coming from the plasma and passing through a beryllium filter 1μm thichness

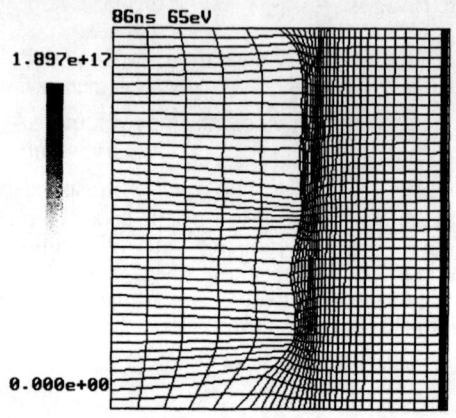

Figure 3

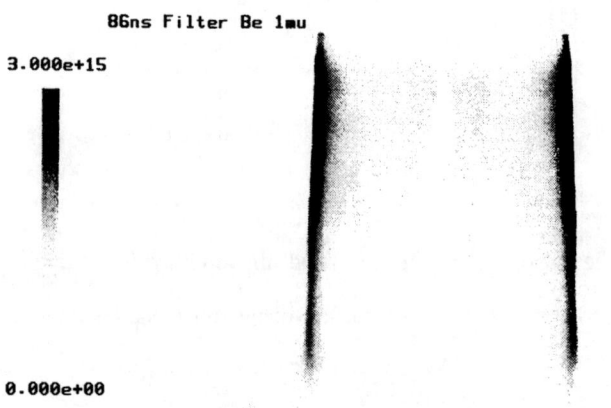

Figure 4.

Calculation of XRD signals.

The intensity I for each frequency group coming onto the diode of surface S at a distance L is

$$I = \sum_{cells} I \cdot \frac{S}{2\pi L^2}$$

The refractive index of filters is $n = (1 - \delta) - i\beta$
The absorption coefficient $\mu(cm^2/g) = \beta\, E(eV) / 987\rho$
where ρ is the filter mass density in g/cm^3

The XRD signal given in ampers at each time is:

$$\sum_{freq.} I \cdot \exp\{-\mu\rho x\} \cdot \varepsilon \cdot 1.6 \cdot 10^{-19}$$

where x is the filter thickness
and ε is the diode quantum efficiency = $10^{-6}\, E(eV)\, \varepsilon_o$ (A/MWt)

An equivalent oscilloscope signal of XRD can be given by a curve plotted from the different values of intensity calculated at each time of the simulation.

In conclusion we have shown how to calculate the x-ray pinhole images filtered as we need and the XRD signals which are very useful diagnostics in Z-pinch experiments.

The electric resistance and electron viscosity of Z-pinch plasma

A. A. Esaulov and P. V. Sasorov

*Institute for Theoretical and Experimental Physics
B. Cheremushkinskaya St. 25, Moscow 117259, Russia*

Abstract. The electron magnetohydrodynamic equations imply that electron gas viscosity can have a significant influence on a plasma's electric resistance. For a long inhomogeneous plasma column with axial symmetry, as in a z-pinch, the viscous part of the resistance can be several orders of magnitude larger than the usual frictional resistance.

Electric current in a high temperature plasma is carried by electrons, hence, the electric resistivity is determined by dissipative properties of the electron gas. Taking as an example the magnetohydrodynamic equations for electron gas obtained by Braginskii [1], it can be seen that there are two types of dissipative processes: "friction" of electrons against ions and viscosity of the electron gas. In electron pressure equilibrium, when $\beta_e = p_e/(B^2/8\pi) \sim 1$, the relative effectiveness of the first process with respect to the second one is of the order of $\sim (a/\lambda_{ei})^2$. Here λ_{ei} is the full mean free path of electrons along their curved trajectories, and a is the characteristic scale of plasma inhomogeneities. Therefore, the relative contribution of the electron viscosity to the resistivity becomes noticeable when $\lambda_{ei} \gtrsim a$. In the latter case, the Braginskii equations [1] become invalid [2,3]. However, the concepts of viscosity and friction retain their meaning even in this case provided that the inhomogeneity scale length a is much larger than the average electron larmor radius ρ_{Be}. In some particular cases - for example when the plasma is axially symmetric in a purely toroidal magnetic field - it is possible to derive equations of the hydro type for the electron gas [3] that are applicable regardless of the value of λ_{ei}/a provided that $\rho_{Be} < \lambda_{lei} < a^2/\rho_{Be}$. Examination of this alternative set of equations sheds a new light on the role of the electron viscosity in giving rise to the resistance of plasma.

According to Ref. [3], the following subset of electron magnetohydrodynamic equations describes the evolution of the electric and magnetic fields **E** and **B** and the electron temperature T in a magnetized plasma with axial symmetry if the electron density n and the bulk velocity of ions $\mathbf{u}_i$ are given functions of $\mathbf{x}$ and t, and the axial magnetic field vanishes:

$$\mathbf{E} = -\frac{\mathbf{u}_i \times \mathbf{B}}{c} - \frac{\nabla p_e}{en} + \frac{\mathbf{j} \times \mathbf{B}}{enc} + \frac{\mathbf{R}_e}{en} + \frac{\nabla(r^3 \Pi)}{nr^3}, \qquad (1)$$

$$\frac{3}{2}n\left(\frac{\partial}{\partial t} + (\mathbf{u} \cdot \nabla)\right) T + p_e \nabla \cdot \mathbf{u} = -\nabla \cdot \mathbf{q_e} + \frac{\mathbf{j} \cdot \mathbf{R}_e}{en} - \Pi r^3 \left(\nabla \cdot \frac{\mathbf{j}_e}{nr^3}\right). \qquad (2)$$

Here $\mathbf{j}$ is the electric current density, $p_e = nT$ is the electron gas pressure,

$$\mathbf{R}_e = en\left(\frac{\mathbf{j}}{\sigma_\perp} + \mathcal{N}\,\mathbf{B} \times \nabla T\right) \qquad (3)$$

is the friction force per unit volume,

$$\Pi = \frac{c}{B^2}\left\{-\alpha_2\,\mathbf{B} \cdot (\nabla \ln Br^2 \times \nabla T) + \alpha_1\,\mathbf{B} \cdot (\nabla \ln n \times \nabla T) - \alpha_1 \frac{B^2 r^3}{c}\nabla \cdot \frac{\mathbf{j}_e}{nr^3}\right\} \qquad (4)$$

is related to viscous components of the pressure tensor, $\mathbf{j}_e = \mathbf{j} - Zen\mathbf{u}_i$ is the electron part of current density j, Ze is the average charge of ions, $\mathbf{q}_e = \mathbf{q}_0 + \mathbf{q}_1 + \mathbf{q}_2$ is electron heat flow:

$$\mathbf{q}_0 = -\frac{5}{2}\frac{c}{e}\frac{p_e}{B^2}(\mathbf{B} \times \nabla T), \qquad \mathbf{q}_2 = -\kappa_\perp \nabla T + \mathcal{N}\,T\,(\mathbf{B} \times \mathbf{j}), \qquad (5)$$

$$\mathbf{q}_1 = \frac{\mathbf{B} \times \nabla(cr^3 T P)}{B^2 r^3} - \frac{cT\Pi}{B^2}\left(\mathbf{B} \times \nabla \ln \frac{n}{T^{3/2}}\right) - \frac{c^2}{B^2}(\mathbf{B} \times \nabla)\left[\frac{T}{B^2}\alpha_4 \mathbf{B} \cdot \left(\nabla \ln \frac{B}{r} \times \nabla T\right)\right], \qquad (6)$$

$$P = \frac{c}{B^2}\left\{-\alpha_3\,\mathbf{B} \cdot (\nabla \ln Br^2 \times \nabla T) + \alpha_2\,\mathbf{B} \cdot (\nabla \ln n \times \nabla T) - \alpha_2 \frac{B^2 r^3}{c}\nabla \cdot \frac{\mathbf{j}_e}{nr^3}\right\}. \qquad (7)$$

A cylindrical coordinate system (r, φ, z) is used, with the z axis coinciding with the axis of symmetry of the system. Eqs. (3) – (7) include a number of kinetic coefficients: $\sigma_\perp$ is the transverse conductivity of plasma, $\kappa_\perp$ is the transverse thermal conductivity, $\mathcal{N}$ is the Nernst coefficient, and

$\alpha_k = p_e \bar{\alpha}_k(Z)/e^2 \nu_{ei}$ (k=1,2,3); $\alpha_4 = Zp_e\bar{\alpha}_4/e^2\nu_{ei}$ are additional kinetic coefficients introduced in Ref. [10], ν_{ei} is the electron-ion collision frequency: the $\bar{\alpha}_k(Z)$ (k=1,...4) are dimensionless coefficients of order unity that depending on Z as given in the Ref. [10].

We assume that plasma configuration adjoins the vacuum. To avoid considering the processes near the electrodes [4,5] we assume that the plasma column has an infinite length and is a periodically corrugated structure with a fixed period l. We will look for a stationary solution, assuming that the total current is constant and ions are motionless.

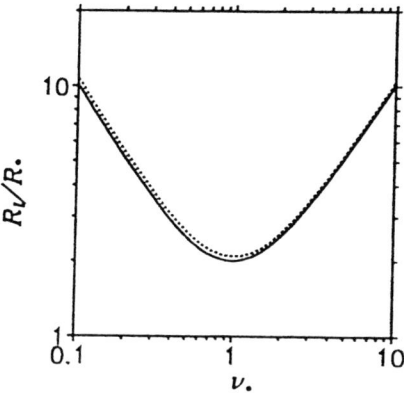

Figure 1. Normalized resistance of the plasma column versus the normalized collision frequency. Plasma density distribution corresponds to Eq. (8). Solid and dotted lines correspond to the cases $\varepsilon = 0.2$ and $\varepsilon \to 0$ respectively

Equation (1) simplifies substantially (but with loss of consistency) if the electron temperature were constant. The form of current flow lines is defined by the topography of contours $nr^2 = const$, and it is possible to find analytically stationary distribution of both electric current and electric field across inhomogeneous plasma column. For the simplest form of the electron density distribution,

$$n(r,z) = n_0(1 + \varepsilon \sin 2\pi z/l), \qquad \varepsilon < 1 \qquad (8)$$

the resistance R_l of plasma column of length l can be expressed in the terms of elementary functions. For example, when $\varepsilon \to 0$, we have

$$R_l = R_*(\nu_* + \frac{1}{\nu_*}), \qquad (9)$$

where $R_* = R_0/\nu_*$ is normalized resistance (that no longer depend upon ν_{ei}), $R_0 = l/\pi r_0^2 \sigma_\perp$ is the usual "friction" resistance of plasma column, and $\nu_* =$

$(l/\varepsilon\lambda_{ei})\sqrt{2/\pi^2\overline{\alpha}_1}$ is the normalized electron collision frequency. Figure 1 shows how the resistance depends on the collision frequency, and sensitivity of $R_l(\nu_*)$ to the parameter ε.

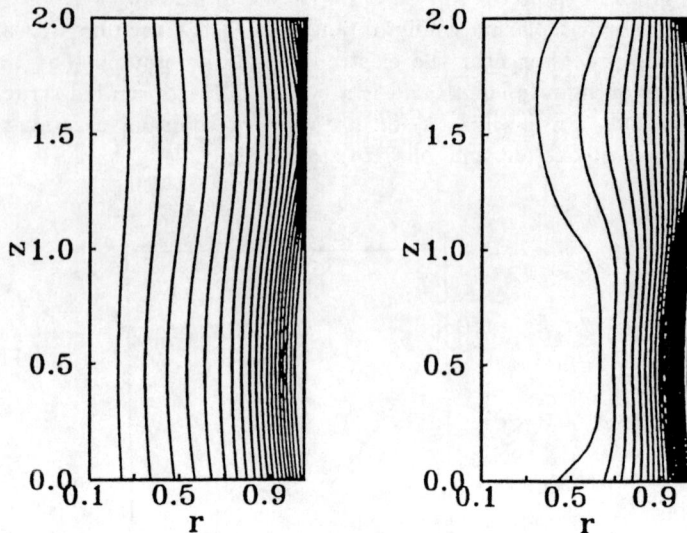

Figure 2. Current flow lines $Br = const$ (left contours) and isotherms $T = const$ (right contours) for the electron density distribution, given by the formula (8), with parameters $\varepsilon = 0.2$, $l = 2$ for $a/\rho_{Be} = 10^3$ and $a/\lambda_{ei} = 0.1$.

We performed a numerical simulation of the full set of Eqs. (1) and (2). Steady-state distributions of magnetic field and electron temperature obtained with the simulation are shown in Fig. 2. The dependence of $R_l(\nu_*)$ in this case is similar to one shown in Fig. 1. The simulation shows that pressure equilibrium is maintained, i.e., $\beta_e \sim 1$.

As to the numbers, it can be seen from Eq. (9) that the viscous resistance can be $10^4 - 10^5$ times higher than the usual resistance in the experiments of Ref. [13], where the electric current (~ 3 MA) passed through a Xe ($Z \approx 10$) plasma column with diameter ~ 3 cm and mass ~ 20 μg/cm.

The similar effect takes place also for the currents along the magnetic field (see Ref. [14]). In the latter case however the resistance is greater than the resistance of uniform plasma by the factor which is only of the order of unit.

REFERENCES

1. Braginskii S. I., in *Reviews of plasma physics*, Leontovich M. A. (editor), New York: Consultant Bureau, 1965, Vol. 1, p. 205.
2. Bobrova N. A., Sasorov P. V., *Sov. J. Plasma Phys.* **19**, 449 (1993).
3. Bobrova N. A., Sasorov P. V., *Sov. J. Plasma Phys.* **16**, 229 (1990).

4. Chukbar K. V., Yan'kov V. V., *Sov. Phys. Tech. Phys.* **33**, 1293 (1988).
5. Isichenko M. B., Kalda J., *J. Moscow Phys. Soc.* **2**, 1 (1992).
6. Batjunin A. N., et al., in Proceedings of the 3rd Inter. Conf. on Dense Z-pinches, Haines M. and Knight A. (editors), New York: AIP, 1993, p. 580.
7. Breizman B. N., Mirnov V. V., Rjutov D. D., *Sov. Phys. JETP* **31**, 948 (1970).
8. Braginskii S. I., Leontovich M. A. (editor), *Reviews of plasma physics*, New York: Consultant Bureau,1965, Vol. 1, p. 205.
9. Bobrova N. A., Sasorov P. V., *Sov. J. Plasma Phys.* **19**, 449 (1993).
10. Bobrova N. A., Sasorov P. V., *Sov. J. Plasma Phys.* **16**, 229 (1990).
11. Chukbar K. V., Yan'kov V. V., *Sov. Phys. Tech. Phys.* **33**, 1293 (1988).
12. Isichenko M. B., Kalda J., *J. Moscow Phys. Soc.* **2**, 1 (1992).
13. Batjunin A. N., Branitsky A. V., et al., in Proceedings of the 3rd Inter. Conf. on Dense Z-pinches, Haines M. and Knight A. (editors), New York: AIP, 1993, p. 580.
14. Breizman B. N., Mirnov V. V., Rjutov D. D., *Sov. Phys. JETP* **31**, 948 (1970).

Numerical Analysis of MHD Instability Suppression in a Double Gas Puff

Igor V. Glazyrin, Oleg V. Diyankov,
Nikolai G. Karlykhanov, Sergei V. Koshelev

*Russian Federal Nuclear Center - All-Russian
Institute of Technical Physics (RFNC - VNIITF)
P.O.Box 245, Snezhinsk, Chelyabinsk Region, Russia, 456770*

The process of double gas-puff implosion is numerically analyzed. Various liner configurations are considered: 1) the outer cascade is ten times lighter then the inner one, 2) the masses of cascades are equal, 3) the outer cascade is ten times heavier then the inner one. The mechanism of liner implosions stabilization is discussed. The comparison of two-cascade and single-cascade schemes is performed. The role of magnetic field penetration at the initial time into liner's body is also discussed.

INTRODUCTION

The purpose of multi-cascade schemes of liner implosion utilizing is to suppress MHD instabilities, which resulting in the X-ray radiation yield increasing. The idea of multi-cascade liners has been discussed in many papers (look references in the paper (1)). While using two-cascade system one should divide into two parts the whole mass of liner in some way. In the paper (1) the ratio of inner cascade mass m_{inn} to the outer cascade mass m_{out} was taken to be equal to $m_{inn}/m_{out} \simeq 10$.

Scheme with this ratio leads to the fast growth of large-scale instability modes in the first stage of liner implosion in the outer cascade, because of its little mass. The small-scale current instability is also growing, which results in the anomalous resistivity. Sometimes due to low mass of the outer cascade the current conducting layer appears in the inner cascade (not in the outer one) (2).

This paper is devoted to the numerical analyses of the mentioned above sources of MHD instability and the methods of its suppression. The MAG code (3) was used for MHD modeling.

THE INITIAL CONDITION FOR MODELING

The geometry of the liner was taken the same as in the experiments, conducted in HCEI, Tomsk (1). When a nozzle is used for gas-puff constructing, the density distribution in the liner is not uniform along the radius. The numeric calculation of gas expansion out of the nozzle has been performed, and the initial profile of density was taken from this calculation.

To analyze the instability developing we took regular perturbations in density along z-axis: $m = 1, 2, 4, 6, 8, 10$; and a random density variation as well. The grid was 60 point along radius and 200 points along the axis z, so we could treat the development of modes up to the 40-th one.

The amplitude of the perturbations has been taken in the range from 10 to 50%. The perturbations were disposed in the inner part of the liner.

THE ANOMALOUS RESISTIVITY AND INITIALLY PENETRATED MAGNETIC FIELD INFLUENCE ON THE STABILITY PROPERTIES OF THE LINER

The light liners are more unstable ones. Especially fast the MHD instability develops when the magnetic field is skinned. If the magnetic field penetrates into a liner, its gradient is much less, and thus the liner becomes more stable. The mechanism, responsible for the fast penetration of magnetic field is the anomalous resistivity. Three calculations have been performed: 1) the magnetic field initially didn't penetrate into the liner, and electric conductivity coefficient was taken in the form of Ermakov-Kalitkin's table (it is equivalent to the initial ionization of the liner); 2) the anomalous electric conductivity coefficients (4) were taken; 3) the magnetic field was supposed to penetrate into the liner.

The results of numerical modeling of these three problems are presented in the Fig.1. One can see, that instability is weaker in the third case, when the magnetic field initially penetrated into the liner. Anomalous resistivity also leads to the stabilizing of liner implosion.

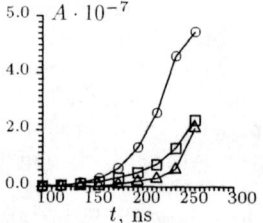

Fig.1: The amplitude of $m = 0$ mode vs. time. Circles - the first case; squares - the second case (with anomalous resistivity); triangles - the third case (initially penetrated magnetic field)

The one more reason for fast penetration of magnetic field may also be considered. When the gas shell is not ionized to the enough level, the discharge may occur, which leads to the appearance of a large number of streamers in the gas (2,5). When the magnetic field around each streamer join into the sole magnetic field of the liner, it would be penetrated completely into the liner.

The density distribution at the final stage of the liner implosion in the calculations, in which the magnetic field penetrated into the inner part of the liner, have been received. One can see a "galo" (low-density shell) around the uniform plasma column in the Fig.2. This picture looks like the "galo" received in the HCEI experiments on uniform compression (taking into account the numerical mesh presition).

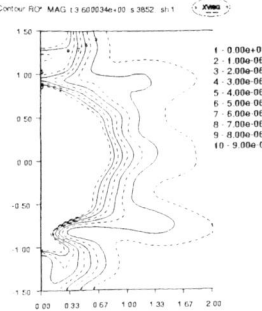

Рис.2: Density for the time of maximum compression

TWO-CASCADE SCHEME WITH DIFFERENT MASS RATIO

Three calculations of two-cascade scheme with different inner cascade to outer cascade mass ratio have been performed: 1) $m_{inn}/m_{out} = 10$ (this data corresponds to the experiments (1); 2) equal densities in inner and outer cascades; 3) $m_{inn}/m_{out} = 0.1$. The results are presented in Fig.1.

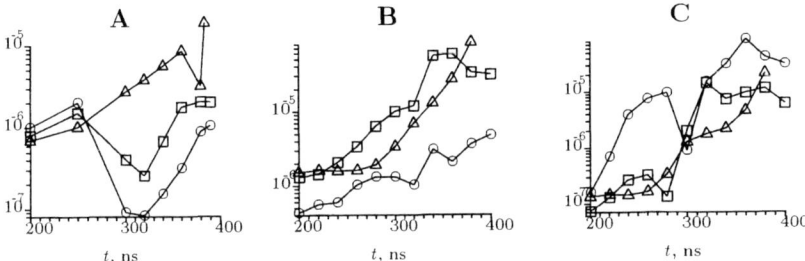

Figure 1: The dynamics of different modes development with respect to time (in nsec). A - $m=0$, B - $m=1$, C the average value of modes from $m = 4$ up to $m = 20$. Circles mark the data for the light outer cascade, squares mark the data for the equal densities two-cascade system, triangles mark the data for the system with the heavy outer cascade

Before the time moment of 300 nsec, at which the outer layer impacts the inner one, the $m = 0$ mode is growing faster for the case of light outer cascade. Then the amplitude of the mode sharply diminishes. The diminishing of the $m = 0$ amplitude is not so significant for the case of equal mass cascades, and can be hardly found for the case of heavy outer cascade (see Fig.1,**A**). We think, that here the mechanism of instability suppression, which was studied for a single mode case in the paper (6), works. When the cavity in the outer cascade, appeared as the result of $m=0$ mode development reaches the inner cascade of the liner, the hot region is produced (the temperature in the region is approximately two times greater then in surrounding plasma). Thus, two shock waves begin to propagate from this region along the boundary, in

z direction, heating plasma at the boundary of the inner cascade. Increasing temperature of the plasma leads to electric conductivity raising and magnetic field gradient at the boundary increases. During delay the backward plasma of the outer cascade reaches the inner cascade, and then the magnetic field continues compressing of the received single more uniform liner in radial direction. The "snowplow" mechanism of instability suppression, described in the paper (7), looks similar to this one. For the massive outer cascade the $m=0$ mode is more dangerous.

The first mode grows slowly for the first case (see Fig.1,**B**). The higher modes (see Fig.1,**C**) are growing rapidly for the first case, this fact may lead to the portion of high temperature plasma in the whole liner increasing and thus the X-ray radiation in K and L-lines yield enlarging.

CONCLUSION

The analyses of MHD instability development for double cascade scheme with respect to different cascades mass ratio is presented. The numerical experiment shows, that the large scale instability developes slightly in a liner with light outer cascade, but small scale modes develop rapidly in this case.

The anomalous magnetic field penetration in the liner may be a reason for "galo" appearing in pinhole images in experiments, hold in HCEI.

ACKNOWLEDGMENT

The authors are grateful to Dr. R.B.Baksht, Dr. A.G.Rousskich, Dr. A.V.Shishlov (HCEI) for the presented experimental information and fruitful discussions of the results obtained, Dr. V.I.Oreshkin (HCEI) for the useful discussions, Dr. S.L.Nedoseev (TRINITI) for the discussions on the anomalous magnetic field penetration phenomena, Ms.A.N.Slesareva and Mrs.N.P.Savina for the help in calcultions conducting.

The work was partially supported by ISTC, project #525.

1. Baksht,R.B., Datsko,I.M., Oreshkin,V.I., et.al., Russ. J. Plasma Phys., Vol.22, No.7, 622-628(1996) (in Russian).
2. Baksht,R.B., Rousskich,A.G., Chagin,A.A., Russ. J. Plasma Phys., Vol.23, No.3, 195-202(1997) (in Russian).
3. Diyankov,O.V., Glazyrin,I.V., Koshelev,S.V., MAG – two-dimensional resistive MHD code using arbitrary moving coordinate system. Submitted to Computer Physics Communications.
4. Glazyrin,I.V., Karlykhanov,N.G., Kondrat'ev,A.A., et.al. AIP Conf. Proc., Dense Z-pinches, Vol.299, pp.139-144 (1994).
5. Branitsky,A.V., Grabovsky,E.V., Fedulov,M.V., et.al., Proc. of the 12th Int. Conf. on High-Power Particle Beams, Prague, Czech Republic, June 10-14, 1996, p.542.
6. Glazyrin,I.V., Diyankov,O.V., Karlykhanov,N.G., Koshelev,S.V., Proc. of the 12th Int. Conf. on High-Power Particle Beams, Prague, Czech Republic, June 10-14, 1996, p.717.
7. Gol'berg,S.M, Velikovich,A.L., Phys. Pluids B **5**(4), 1164 (1993)

Radiation MHD Modeling of a Proposed Dynamic Hohlraum

James H. Hammer[1], John S. De Groot[1, 2], Max Tabak[1], Art Toor[1] and George B. Zimmerman[1]

[1]Lawrence Livermore National Laboratory, Livermore California 94550

[2]Plasma Research Group and Department of Applied Science, University of California, Davis, CA 95616

In this paper we report 2D radiation magnetohydrodynamic simulations of a dynamic hohlraum target designed to be driven by the Z accelerator (1) at Sandia National Laboratory, Albuquerque New Mexico. Z generates currents up 20 MA with a rise time of 100ns and peak electrical power of 40 TW. In this design we attempt to reduce the effects of magneto-Rayleigh Taylor (RT) modes by using a distributed initial density profile. Earlier work (2,3) showed that "tailoring" the initial density profile could reduce the sheath acceleration and the number of e-foldings that the RT instability grows during the implosion . As the sheath moves in radially, fresh material is swept up or "snow plowed", providing a back pressure that counters the J x B force. A special profile can be found in which the unstable outer surface of the sheath implodes at constant velocity, reducing the classical growth rate to zero, although residual Richtmyer-Meshkov type instability (instability of the snow-plow shock front) may be present. In practice, it is hard to create tailored initial density profiles due to the difficulty of machining and otherwise manipulating very low density materials. It becomes easier to manufacture these complex targets as the current, energy and load mass increase with large drivers. Z is the first fast pulse power device with enough energy to consider loads of this type.

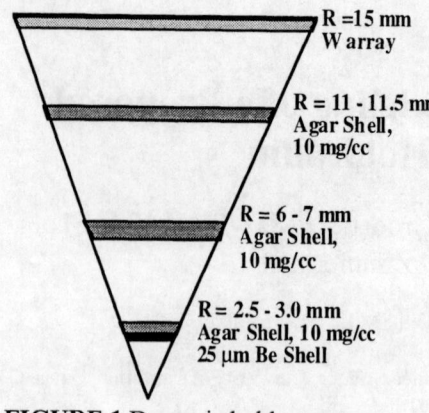

FIGURE 1 Dynamic hohlraum target

The design we have modeled employs a 3 cm diameter, 2 cm long, tungsten wire array of mass 3.8 mg/cm as the high atomic number shell material. The wire array implodes on 3 concentric, annular cylindrical shells of low density (10 mg/cc) agar foam with a 25 μm layer of beryllium covering the interior of the central shell, shown in Fig. 1. The annular shells, with foam between radii of 1.1 and 1.15 cm (outer shell), 0.6 and 0.7 cm (middle shell) and 0.25 and 0.3 cm (inner shell), explode under the influence of the X-rays emitted by the wire array and outer shells. In the process, they create a rough approximation to the zero-acceleration tailored density profile. The use of multiple shells to control RT instability in Z-pinches has been studied previously (4). 1D simulations show the stagnation of the beryllium at a minimum diameter of about 0.16 cm, with a peak temperature of 240 eV and temperatures exceeding 200 eV for 7 ns. The radii of the different layers and the radiation temperature on the pinch axis are shown in Fig. (2).

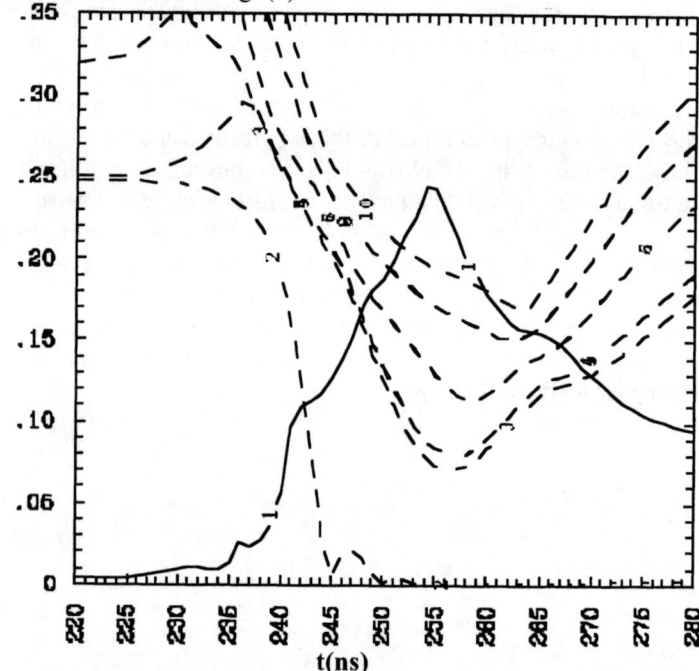

FIGURE 2. T_r (keV) on axis (solid line) and radius (cm) of shells vs. time

The peak temperature occurs 50 ns after the current peaks at 17 MA. In the hot, stagnated beryllium the average distance X-ray photons can travel before being absorbed and re-emitted, the Rosseland mean free path, is several times the pinch diameter, while the agar and tungsten layers have an optical depth of about 20. With these conditions, it is sensible to describe the configuration as a dynamic hohlraum. The beryllium fills the hohlraum at a density of about 0.1 g/cc and pressure of order 50 Mbar. This type of hohlraum is better suited to driving auxiliary experiments than ICF capsules inside the hohlraum that could be affected by the high plasma pressure. The drive temperature seen by an auxiliary load depends on the type of experiment. A highly reflective load on the end sees the full 240 eV, while a highly absorptive load removes significant energy from the hohlraum, dropping the effective drive temperature to 180 eV.

We've also done 2D calculations including the effects of RT modes for this target and similar designs. A calculation for a similar target with a 1% density perturbation and a 2mm wavelength shows significant instability growth in the tungsten, but little perturbation of the inner layers. At stagnation against the inner annulus, the bubble-spike formation in the tungsten is partly reversed. The peak temperature drops from 230 eV to ~ 220 eV, and the hohlraum interior remains optically thin at about the same diameter as the 1D simulation. 2D periodic simulations with 5% initial random zone-to-zone density perturbation show larger amplitude growth in the tungsten with small perturbation of the inner shells. The tungsten is sufficiently disrupted that cracks in the radiation case form and allow large radial losses, causing the radiation temperature on axis to drop to 180 eV. Perturbations in the 1 -5% range have been found to give reasonable pulsewidths for Saturn and Z experiments (2,6), although the perturbation level imposed in this way is zoning dependent.

To reduce the uncertainty in choosing the perturbation level, and come closer to a "first principles" RT calculation, we have done simulations with a very fine mesh. As discussed in earlier work (5), RT modes at short wavelengths can grow to nonlinear amplitudes very quickly from small initial perturbations, then seed more destructive long wavelength modes through a nonlinear cascade. To provide a self-consistent instability seed, we have done a finely zoned (10 micron x 10 micron zones) calculation capable of resolving the short wavelengths that dominate the early phase. This method of initiating RT calculations with a 2D r-z code is only valid when there are enough wires in the initial array to merge into an azimuthally uniform sheath before significant RT growth begins. For this calculation, modeling 3 mm of the pinch length with periodic boundary conditions in the z-direction, random density perturbations of 0.1% were sufficient to stimulate large amplitude RT growth. Once the dominant features were several hundred microns in scale, we linked to a coarser mesh (~ 30 micron zone size) to follow the complete implosion. A series of time snapshots of the plasma density for the multi-shell agar foam + beryllium target are shown in Fig. (3). This calculation showed reasonable integrity of the radiation case so that radiation leakage was not a major effect and the radiation temperature was ~ 220 eV in the hohlraum interior. A plume of optically-thick high Z material reached the axis at one location, however, which could inhibit energy flow to an adjacent load. If the high-Z plume occurs more than a few mm from the output end, the effect on energy flow should be small. The RT calculations do not include the effect of an X-ray-absorbing load, which would cause an additional drop in the radiation temperature.

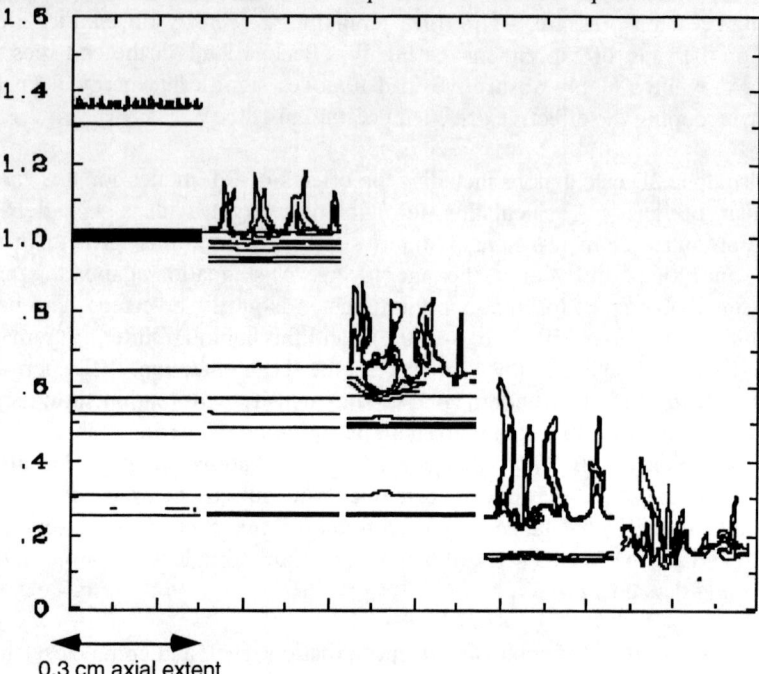

FIGURE 3. Density contours at 20 ns intervals

All of the calculations show significant breakup of the tungsten sheath before it encounters the outermost stabilizing agar shell. This prompts us to consider an initial state that more closely approximates the zero-acceleration density profile at large radius. At present, it is not practical to extend the profile to larger radius with low density foams because of the limited ability to machine and handle these extremely fragile materials. An alternative may be to use concentric, nested wire arrays, with the separation between wires in each shell smaller than the distance the wires expand due to heating by X-rays. With this technique, an arbitrary initial density profile of virtually any material may be possible. The first experiments with simpler loads will tell us if these more complex targets are necessary to control RT instability.

ACKNOWLEDGMENTS

Work Performed under the auspices of the U. S. Department of Energy by the Lawrence Livermore National Laboratory under Contract W-7405-ENG-48 and Sandia National Laboratories under Contract DE-AC04-94AL85000.

REFERENCES

1. R.B. Spielman, et. al., Paper O-4-3 in Proceedings of the 10th International Conference on High Power Particle Beams, Prague, Czech Republic, June 10-14, 1996, ed. K. Jungwirth and J. Ullschmied, pp. 150-153. Copies of the conference proceedings can be ordered from BEAMs96, Institute of Plasma Physics, Czech Academy of Sciences, Za Slovankou 3, 182 00 Prague, Czech Republic.
   R.B. Spielman, G. A. Chandler, C. Deeney, F. Long, T.H. Martin, M.K. Matzen, D.H. McDaniel, T.J. Nash, J.L. Porter, L.E. Ruggles, T.W.L. Sanford, J.F. Seamen, W.A. Stygar, S.P. Breeze, J.S. McGurn, J.A. Torres, D.M. Zagar, T.L. Gilliland, D. Jobe, K.W. Struve, M. Mostrom, P. Corcoran, I. Smith, R.W. Shoup, "PBFA Z: A 20-MA Driver for Z-Pinch Radiation Sources", *Bull. Am. Phys. Soc.* **41**, 1422 (1996).

2. J.H. Hammer, J.L. Eddleman, P.T. Springer, M. Tabak, A. Toor, K. Wong, G.B. Zimmerman, C. Deeney, R. Humphreys, T.J. Nash, T.W.L. Sanford, R.B. Spielman, J.S. De Groot, "Two-dimensional radiation-magnetohydrodynamic simulations of SATURN imploding Z pinches", *Phys. Plasmas,* **3**, 2063 (1996).

3. A.L. Velikovich, F.L. Cochran, J. Davis, "Suppression of Rayleigh-Taylor Instability in Z-pinch Loads with Tailored Density Profiles", *Phys. Rev. Lett.*, **77**, 853 (1996).

4. R.B. Baksht, S.P. Bugaev, I.M. Datsko, A.V. Luchinskii, V.I. Oreshkin, A.V. Fedyunin, Yu.D. Korolev, I.A. Shemyakin, and V.G. Rabotkin "Dense Z-Pinches, Tjore International Conference", ed. by M. Haines and A. Knight (AIP Conference Proceedings, vol. 299, p. 365, AIP, New York), 1994.

5. J.H. Hammer, J.L. Eddleman, M. Tabak, A. Toor and G.B. Zimmerman, "Sheath Broadening in Imploding Z-pinches due to Large-bandwidth Rayleigh-Taylor Instability", in Proceedings of the 10th International Conference on High Power Particle Beams, Prague, Czech Republic, June 10-14, 1996, ed. K. Jungwirth and J. Ullschmied, pp. 150-153. Copies of the conference proceedings can be ordered from BEAMs96, Institute of Plasma Physics, Czech Academy of Sciences, Za Slovankou 3, 182 00 Prague, Czech Republic.

6. D.L. Peterson, R.L. Bowers, J.H. Brownell, A.E. Greene, K.D. McLenithan, T.A. Oliphant, N.F. Roderick and A.J. Scannapieco, "Two-dimensional modeling of magnetically-driven Rayleigh-Taylor instabilities in cylindrical Z pinches", *Phys. Plasmas*, **3**, 368 (1996).

MODELING A DENSE Z-PINCH PLASMA WITH A COLD, DENSE CORE USING A 2D 2-TEMPERATURE MHD CODE.

G.V. Ivanenkov, A.R. Mingaleev, S.A. Pikuz, V.M. Romanova, T.A. Shelkovenko, and W. Stepniewski*

P.N. Lebedev Physical Institute, Moscow, Russia

A 2D (r-z) and 2-temperature ($T_{e,i}$) nonideal MHD numerical model of dense Z-pinch dynamics and its applications to exploded wire plasmas with a cold dense core are presented. The numerical simulation algorithm is based on an Eulerian-Lagrangian methodthat is related to one suggested by D'yachenko for hydrodynamic calculations. As expected, a high density core that appears in the initial stage of the electrical explosion of a metal wire is found to be colder than the surrounding plasma. Model results are in good agreement with experimental data for exploded wires in a high current nanosecond diode.

A 2D (r-z) and 2-temperature ($T_{e,i}$) nonideal MHD numerical model of dense Z-pinch dynamics and its applications to exploded wire plasmas with a cold dense core are presented in this report. In our model, core creation results from the special features of the initial electrical explosion of metallic wires. Estimations[4] have shown that for the short voltage rise time (< 10 ns) a surface evaporation front can reach the wire axis only for wires with diameters < 5 µm. On the other hand, the skin effect can play a major role only for wires with a diameter of 50 µm or more. The portion of the wire core which is not evaporated is superheated, i.e. - the whole volume is heated above the evaporation temperature. After that, the critical parameter is reached and a highly nonuniform density distribution (central core, surrounded by a partially ionized vapor) is formed. The core resistivity is of the order of kΩ's. This resistivity and the skin effect limit core heating, and the center of the plasma column stays relatively cold up to the time of shock wave arrival.

The structure of the plasma shock wave is determined $((m_e/m_i)^{1/2} \ll 1)$ by the slow e-i energy exchange and ionization in the one hand and the large electron mobility on the other hand. Due to these effects, the ions on the front of the shock wave are heated more than the electrons: $\delta T_{e,i} \cong m_{e,i} D^2$, where D is the shock wave velocity. The energy equipartition time is $(m_i/m_e)^{1/2}$ times longer then i-i momentum and energy relaxation in the viscosity of a shock jump layer. These factors and multiple ionization determine the shock wave structure in the dense core. Before shock wave arrival, the hot electrons and radiation form a shock wave precursor. Shock front and precursor length are in the

---

* S. Kaliski Institute of Plasma Physics and Laser Microfusion, Warsaw, Poland

range of tenths of μm. The shock wave stops upon reaching large density gradient at the core boundary, and the precursor slowly (time scale of 10 ns) preheats the massive core. The nonlinearity of electron and radiative conductivity leads to increasing intensity to an preheating. This process becomes faster with the shock wave reflection, and the temperature maximum reaches the axis at the moment of the shock wave arriving at the plasma surface. After this time, most of the current flows in the high density region. As a result, the maximum magnetic field and the bright radiation spots are localized in the core area. Small scale sausage instability growth leads to fine structure formation.

The numerical simulation was carried out using a 2D Eulerian-Lagrangian MHD code based on the modified Free Points Method of D'yachenko. In details, This model and algorithm are described in detail in Ref.5. The physical model is based on the set of dissipative MHD equations including an atomic physics package. An electron viscosity is added to the usual terms of Ohm's law, which includes the classical terms of conductivity and thermoelectricity, as well as the Hall and Nernst effects. Thermal fluxes of ions and electrons are taken into account in the usual conductive, Nernst and Leduce-Rigi effects terms. The kinetics of ionization are described in an average Z approximation and radiation is taken into account at high optical depth with a radiation conductivity approach. Finally, we initially assumed $T_{rad}=T_e \neq T_i$ (2T-model) to simplify calculations, but later we took $T_{rad}=T_e \neq T_i$ (3T-model) . We applied a quasistationary approach for the T_{rad} description and an iterative procedure for its determination at every step of the calculations. The core was simulated by the use of 10-times the average initial density near the axis.

The results of 2T-calculations are shown in Fig. 1 as a series of pictures at a sequence times. The current sheet (high luminosity skin layer and previously formed shock wave) is clear at 17.4 ns on a background of weakly preheated core. At the next time shown, 20.6 ns, the shock wave front reaches the core boundary, and the precursor intensely heats its interior by hot electrons and radiation. The bright linear structures that are parallel to the axis are the result of the shock wave and core interaction (see T_e distribution at the time 24 ns as a typical example). The shock wave is reflected from the core and propagates toward the plasma surface layer, while the electron and radiative conductivity heat the plasma of the core. As a result, the maximum T_e is already on the axis at the time 27.7 ns.

The picture later in time is similar to the usual one for homogeneous plasma pinching, but the magnetic field maximum is inside of the plasma column. The part of the current flowing in the core very rapidly grows with time, and the shape of the necks tends to a cavity form[6]. A finely developed neck is like a minidiode shorted by a thin plasma column. The complexity of the picture is increased because short wavelength m=0 instability of plasma column, which is seen in Fig. 2 as 3-points of fine structure.

The 3T-model give a more detail results. The necks form and its inner structure (Fig.3) are more complicated. The calculated T_e is higher (200 eV or more) then in the 2T-model.

The experiments simulated by the code calculations were carried out on the BIN pulsed power generator (250 kA, 400 kV, 100 ns). The explosion of various metal wires was investigated by the use of several soft x-ray and XUV diagnostics[7]. Time-integrated pinhole images obtained through a diameter a diameter 7 μm are of special of interest for the theme under discussion here. The pinhole camera has a magnification of 4.5 and a filter with cut-off of >1 keV. One can often see double-line parallel structures in these pictures. These structures are not very straight and spread along the axis. The bright spot is clear in Fig.3. The interline distance scale is 10 μm. This is substantially less, than the overall size of bright spota. The luminosity can be connected to early plasma formation phenomena. These structures can be interpreted as regions of shock wave -- core interaction, when the plasma temperature from the line can be about 100 eV for a luminosity duration of $\approx$ 10 ns.

REFERENCES

1. Zakharov, S.M. , Ivanenkov, G.V., Kolomensky, A.A., Pikuz, S.A., Samokhin, A.I. *Sov. J. Plasma Phys.* **9**, No. 3, pp. 271-275 (1983).
2. Aranchuk, L.E., Bogolubsky, S.L., Volkov, G.S., et al. *Fiz. Plasmy* **11**, p. 1324 (in Russian) (1986).
3. Sarkisov, G.S., Etlicher, B., Rouille, C., Attelan, S., Shikanov, A.S., *Pis'ma v ZhETF* **61**, No. 8, pp. 471-471 (in Russian) (1995).
4. Ivanenkov, G.V., Mingaleev, A.R., Novikova, T.A., Pikuz, S.A., Romanova, V.M., Shelkovenko, T.A., *Zh. Tech. Fiz.* **65**, No. 4, pp. 40-45 (in Russian) (1995).
5. Ivanenkov, G.V., Stepniewski, W., *Plasma Phys. Rep.* **22**, No. 6, pp. 479-490 (1996).
6. Vikhrev, V.V., Ivanov, V.V., Rozanova, G.A., *Nuclear Fusion* **33**, No 2, pp. 311-321.
7. Faenov, A.Ya., Pikuz, S.A., Erko, A.I., et al., *Physica Scripta* **50**, pp. 333-338 (1994).

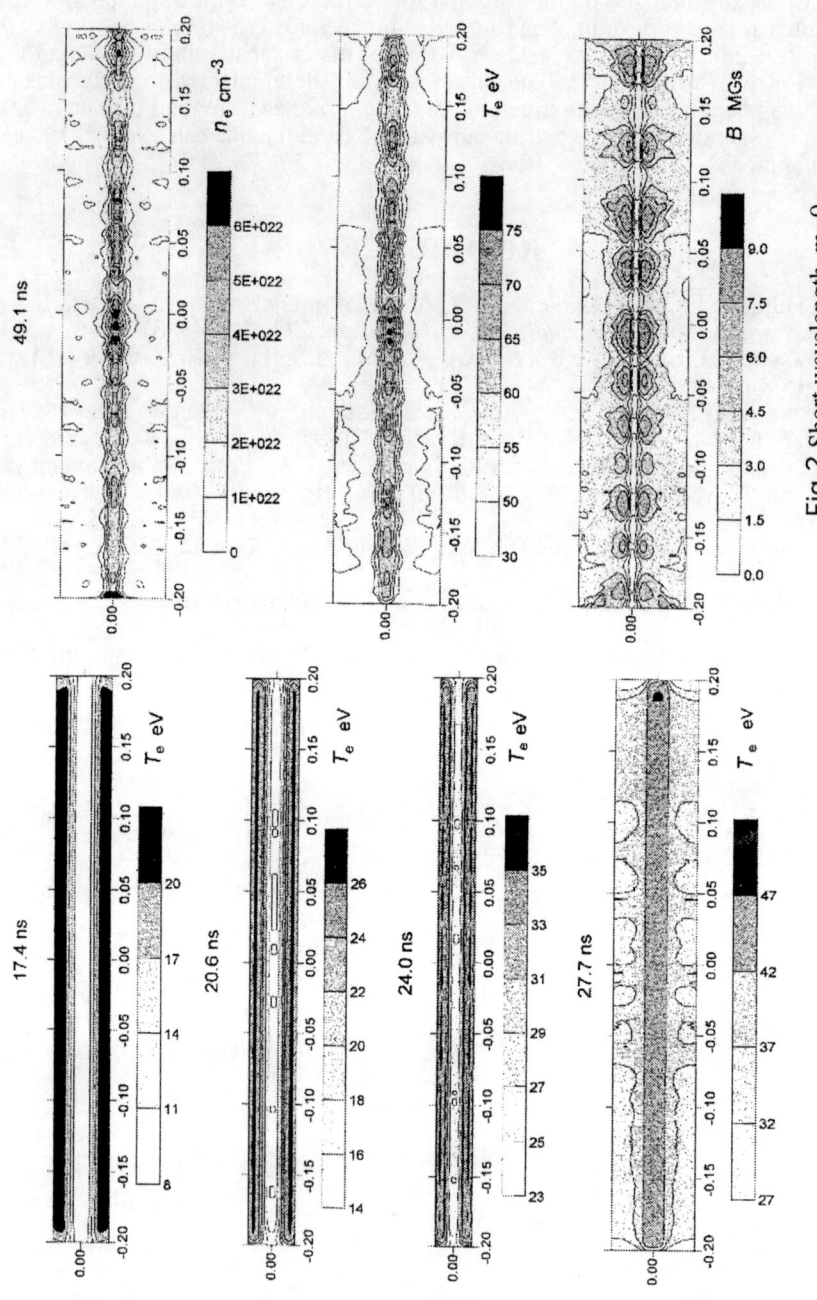

Fig.2 Short wavelength m=0 instability of plasma column

Fig.1 The results of 2T-calculations.

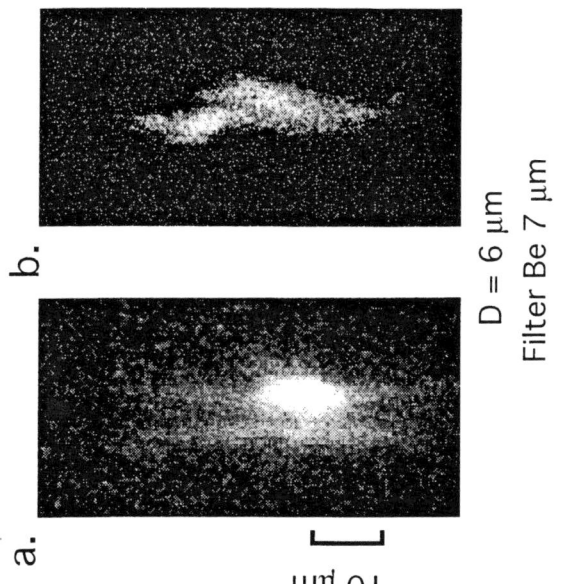

D = 6 μm
Filter Be 7 μm

Fig.4 Time-integrated pinhole image of a Z-pinch bright spot with double structure.

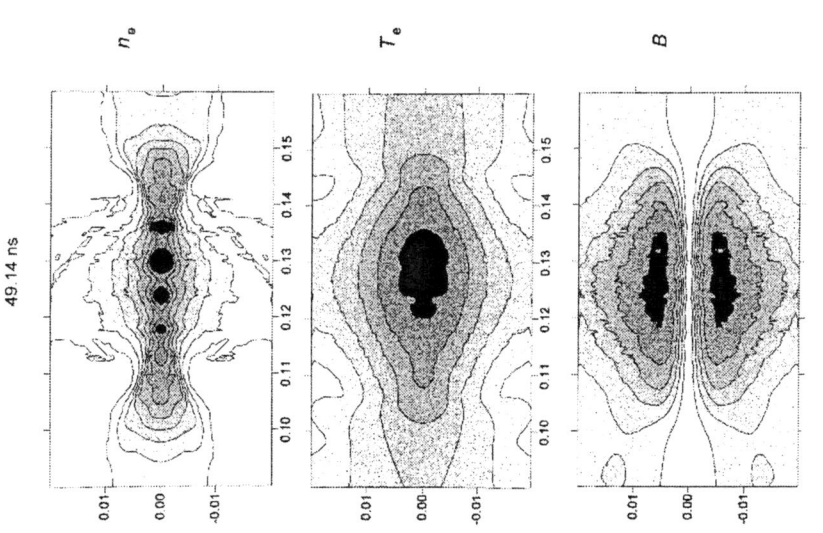

Fig.3 The results of 3T calculations

UCI Staged Z Pinch

P. Ney and H. U. Rahman,

*University of California
Institute of Geophysics and Planetary Physics
Riverside, CA 92521*

and

N. Rostoker and F. J. Wessel

*University of California
Department of Physics and Astronomy
Irvine, CA 92697-4575*

Abstract. The Staged Z-Pinch couples energy to a target, plasma-dynamically in stages. Our analysis considers this multi shell z-pinch configuration driven by the ZOT Facility at the University of California, Irvine with predictions for a 2 MA, 1.8 μs krypton-liner shell and DT target of: $Y \approx 4 \times 10^{15}$ neutrons/pulse, $G \sim 1$ gain, $n\tau \approx 10^{15}$ cm^{-3}-s, $\tau = 2$ ns, $n = 7 \times 10^{23}$cm^{-3}, and $T_{ion} = 10$ keV.

INTRODUCTION

Our prior work considered a plasma liner that implodes onto an embedded, axial-magnetic field and target plasma;[1] that is a Z-θ Pinch. This configuration was predicted to attain high gain thermonuclear conditions with a 10 MA current and 1 MJ energy input. Adding an azimuthal field, inside the liner pinch, allows break-even fusion with a 1 MA current and 10 kJ energy input. This configuration is referred to as a Staged Z-Pinch[2] and is illustrated in Figure 1.

Initially the fiber is ionized by a current pulse, before the z-pinch liner implodes; this injects a B_θ field inside the pinch liner, in addition to an axial-magnetic field that would be provided by external Helmholtz field coils. The fields compress differentially, providing dynamic shear and a means to inductively amplify current in the target plasma, as well as a means to confine fusion-reaction products. For a DT target plasma we predict thermonuclear yields of 4×10^{15} neutrons/shot with a 2 MA, 1.8 μs current pulse.

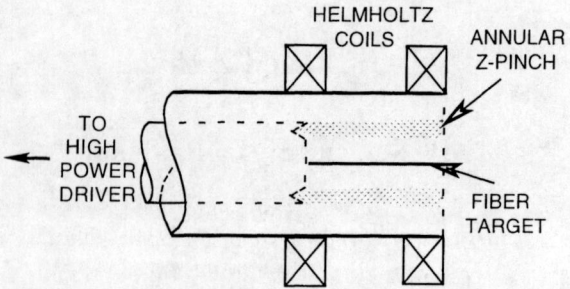

Figure 1: Iillustration of the Staged Z Pinch

THEORETICAL RESULTS

Compressing a B_θ field inside the liner amplifies target current in a few nanoseconds. The current changes according to

$$I(t) = I_0 \frac{ln[r_i(0)/a(0)]}{ln[r_i(t)/a(t)]}$$

where I_0 is the initial current in the fiber, $r_i(t)$ is the inner radius of the liner, and $a(t)$ the outer radius of the fiber. Initially $r_i(0)/a(0) \sim 18$. At maximum compression the ratio, $r_i(t)/a(t)$ would be close to unity, thus $I(t)$ in the fiber plasma could be much greater than I_0.

We have modelled the Staged Z-Pinch with a modified version of the LLNL TRAC2 MHD code. TRAC2 handles different species of interacting plasmas, separated by a vacuum-magnetic field. It is a mixed Eulerian and Lagrangian code, with separate plasma-electron and ion-temperatures, and retains the 3D magnetic field and fluid velocities, including classical and anomalous magnetic-field diffusion; it also models externally-driven capacitor-bank circuits. TRAC2 solves the following equations:

$$\frac{\partial \rho}{\partial t} + \nabla \cdot (\rho \vec{u}) = 0,$$

$$\rho \frac{\partial \vec{u}}{\partial t} + \rho(\vec{u} \cdot \nabla)\vec{u} = -\nabla(P_e + P_i) + (\vec{J} \times \vec{B}) + \nabla \cdot \bar{\bar{Q}},$$

$$\frac{\partial(\rho \epsilon_e)}{\partial t} = -P_e \nabla \cdot \vec{u} + \eta J^2 - P_{rad} + \nabla \cdot (\bar{\bar{K}}_e \cdot \nabla T_e) + \alpha_{ei}(T_e - T_i),$$

$$\frac{\partial(\rho \epsilon_i)}{\partial t} = -P_i \nabla \cdot \vec{u} + \vec{u} \cdot \nabla \cdot \bar{\bar{Q}} + \nabla \cdot (\bar{\bar{K}}_i \cdot \nabla T_i) + \alpha_{ei}(T_i - T_e),$$

$$\frac{\partial \vec{B}}{\partial t} = -\nabla \times \vec{E},$$

$$\vec{E} = \eta\vec{J} - \vec{u} \times \vec{B},$$
$$\vec{J} = \frac{1}{4\pi}\nabla \times \vec{B}.$$

The fusion model includes the deposition of α particle energy and energy loss due to radiation. The power gain from α particles is given by,

$$P_{D-T}(W/cm^3) = 5.6 \times 10^{-13} n_D n_T (\overline{\sigma v})_{D-T},$$

with $(\overline{\sigma v})_{D-T}$ determined from a lookup table and the number density of deuterium and tritium ions (in cm^{-3}) respectively given by n_D and n_T. Bremstrahlung radiation power loss is given by,

$$P_{Br}(W/cm^3) = 1.69 \times 10^{-32} n_e T_e^{1/2} \sum [Z^2 N(Z)],$$

with the electron number density n_e (in cm^{-3}), electron temperature T_e (in eV), charge state Z and number density of the charge state $N(Z)$ (in cm^{-3}). The recombination radiation power loss is,

$$P_r(W/cm^3) = 1.69 \times 10^{-32} n_e T_e^{1/2} \sum [Z^2 N(Z)] \left(\frac{E_\infty^{Z-1}}{T_e}\right)]$$

where the ionization energy is E_∞^{Z-1} in eV. The population densities are calculated using Saha ionization,

$$\frac{dN(Z)}{dt} = n_e\{S(Z-1)N(Z-1) - [s(Z) + \alpha(Z)]N(Z) + \alpha(Z+1)N(Z+1)\},$$

where $S(Z)$ is the ionization rate and $\alpha(Z)$ is the total recombination rate of form $\alpha(Z) = \alpha_r(Z) + n_e \alpha_3(Z)$ where α_r and α_3 are given by,

$$\alpha_r(Z) = 5.2 \times 10^{-14} Z \sqrt{E_\infty^Z/T_e}\{0.43 + 0.5 ln(E_\infty^Z) + 0.469(E_\infty^Z)^{-1/3}\},$$
$$\alpha_3 = 8.75 \times 10^{-27} T_e^{-4.5}.$$

The basic configuration is that of Figure 1. The target plasma is formed from a 100-μm radius DT fiber, pre-ionized by a 105 kA current pulse and expanded to a 2-mm radius (initial density, $n = 2.5 \times 10^{19}$ cm^{-3}, temperature, $T = 11$ eV, and $B_z = 0$). The annular-plasma liner is a singly-ionized krypton shell, 2-cm radius, 1-mm thick, initial density 5×10^{19} cm^{-3}, and a temperature 0.2 eV. Circuit parameters are: $C = 50$ μF, L $= 25$ nH, and $V_0 = 50$ kV, typical of the UCI machine.

Current profiles for the Kr shell and D-T-plasma columns are displayed in Figure 2a. Near peak compression the liner current decreases due to the increasing load inductance. At this time the fiber current is amplified to a value exceeding 3 MA in

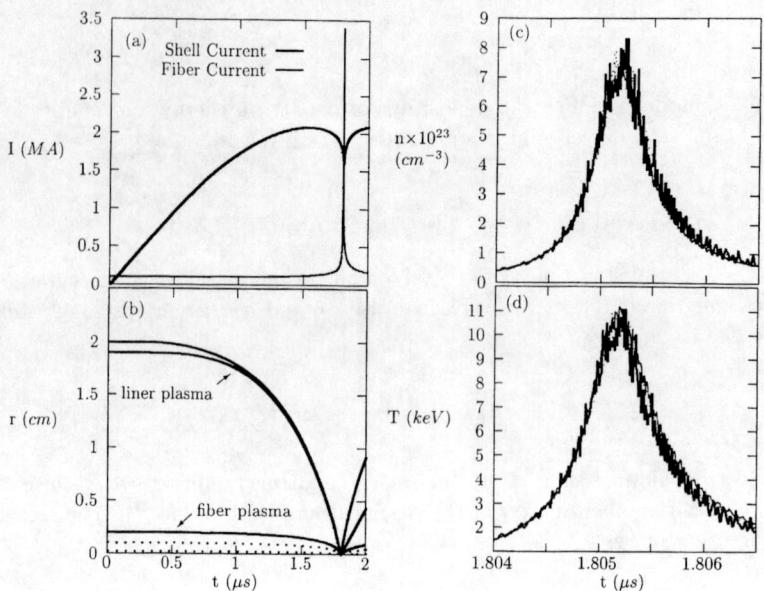

Figure 2: 1D calculations for a Kr liner + DT target plasma: a) liner and fiber currents, b) liner and fiber radii, c) fiber density, and d) fiber temperature.

a few nanoseconds. Radii for the Kr shell and DT plasmas are displayed in Figure 2b; the fiber density and temperature are shown in Figures 2c and 2d, respectively. The peak density of the fiber plasma is 8×10^{23}cm^{-3} and its temperature is 10 keV. Heating is due to adiabatic compression resulting in $n\tau \approx 10^{15}$ cm^{-3}s (i.e., beyond scientific break-even), a yield of, $Y_n = 4 \times 10^{15}$ neutrons/shot, and an energy gain of, $G \equiv E_{thermonuclear}/E_{cap.bank} = 0.2$.

The Staged Z-Pinch requires the plasma shell to implode stably. Rayleigh-Taylor instabilities usually limit the pinch compression. Inhomogeneities grow in the plasma shell with a time dependence, $\xi(t) = \xi_o \exp(\gamma t)$, where ξ_o is an initial perturbation of the shell, $\gamma = \sqrt{gk}$ is the growth rate of the instability with wave number k, and $g = B^2 A/8\pi M$ is the acceleration of the imploding plasma. Typical experimental interventions involve: the application of an axial-magnetic field, to reduce γ;[3-5] initiating the pinch discharge in a pre-ionized plasma,[6,7] to reduce ξ_o; and reducing the liner implosion time to reduce the time during which the instability grows.

Figure 3a shows r-z density profiles for the z-pinch near stagnation ($t = 1.4$ μs), without a target plasma and axial field. In this case a Gaussian-density profile was assumed with a 5 % random, initial-density perturbation in the outer-liner shell. The liner perturbations are due to the Rayleigh Taylor instability.

Axial- and azimuthal-magnetic fields provide stabilizing shear. Figure 3b shows

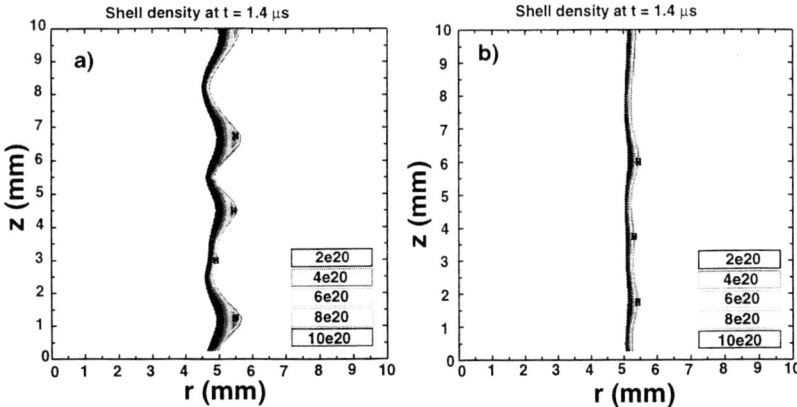

Figure 3: 2D r-z profiles for a) a normal z pinch and b) field-stabilized z-pinch.

the r-z density profiles at the same time as Figure 3a with $B_z = 100$ G and $B_\theta = 10$ kG (measured at the inner-liner surface), demonstrating the significantly improved mode stability.

TRAC2 predictions were compared against the standard LLNL Rad Hydro code (run by J. Hammer, LLNL), including three temperature fluid, full-radiation hydrodynamics, SESAME equation-of-state (EOS), and magnetic diffusion in one dimension. Both codes agreed remarkably well, however the LLNL code predicted a tighter, more compressed final state and an order of magnitude higher yield; probably due to a more realistic treatment of radiation and material EOS for DT.

ACKNOWLEDGMENTS

Special thanks to Drs. J. Eddleman and J. Hammer (LLNL) for their generous assistance with the TRAC2 code. This work was supported by the U. S. DoE, OFES.

References

1. Rahman, H.U., Ney, P., Wessel, F.J., Fisher A., and Rostoker, N. Proc. 2nd Int. Conf. on High Density Pinches, Laguna Beach, April 26-29 (1989), AIP Conf. Proc. #195, N. R. Pereira, J. Davis, and N. Rostoker, Editors, p. 195.

2. Rahman, H.U., Wessel, F.J., Rostoker, N., Phys. Rev. Lett **74**, p. 714(1995);

3. Wessel, F.J., Felber, F.S., Wild, N.C., and Rahman, H.U., Appl. Phys. Lett. 48, 1119 (1986).

4. Felber, F.S., Malley, M.M., Wessel, F.J., Matzen, M.K., Palmer, M.A., Spielman, R.B., Liberman, M.A., and Velikovich, A.L., Phys. Fluids 31, 2053 (1988).

5. Edison, N.S., Etlicher, B., Chuvatin, A.S., Attelan, S., and Aliaga, R., Phys. Rev. E, **48**, 3893, (1993).

6. Goldberg, S.M., Velikovich, A.L., AIP Conf. Proc. 299, Dense Z-Pinches, p. 42(1994).

7. Wessel, F.J., Etlicher, B., and Choi, P., Phy. Rev. Lett. **69**, No.21(1992).

2D MHD Simulations of Rayleigh-Taylor Instability in Z-pinches

D. Zdravkovic, A. R. Bell and M. Coppins

Imperial College of Science, Technology and Medicine, London, SW7 2BZ, UK

Abstract. A two dimensional resistive MHD code (MH2D) is applied to the study of the m=0 instability in z-pinches with diffuse profiles. The effects of sheared axial flow and various initial profiles on the non-linear development of Rayleigh-Taylor and MHD instabilities are explored.

1. Introduction

Since the earliest experiments[1], z-pinches have been known to be violently unstable to the Rayleigh-Taylor instability (RT). There has been a lot of both experimental and theoretical work done on RT in annular liners, annular gas puff pinches and wire arrays (e.g. see Ref. 2-5). The classical RT growth rate:

$$\gamma = \sqrt{gk} \tag{1}$$

describes well the instability growth in the linear regime in such devices. Here, however, we are interested in the z-pinches with diffuse density profiles. These have not received a lot of attention in the way of computer modeling because of their inferiority as x-ray sources as compared to wire arrays say. It is also not completely clear how big a role RT has in such devices since even when they are not imploding they are still unstable to "conventional" MHD instability[6]. For the azimuthally symmetric (i.e. m=0) case it is said that it is easy to distinguish the RT instability in an experiment by the shape of the plasma surface. In the non linear regime of RT thin 'fingers' of plasma and large bubbles should be visible, while m=0 instability makes plasma surface look more like a string of sausages.

In this work we examine a variety of dynamic diffuse z-pinches (here and throughout this paper the word "dynamic" refers to the unperturbed pinch in the non-equilibrium situations). It is restricted to cases with subsonic radial velocities. The main aims are to provide answers to the following two questions: (i) whether it is possible to make a clear distinction between RT and m=0 instabilities or even whether it is useful to do so; (ii) whether sheared axial flow has significant influence on the growth of the MHD instabilities in the dynamic pinch.

2. Two Dimensional MHD Modeling

For our investigation we used a resistive MHD code[7] (MH2D). It solves the continuity, momentum, energy and magnetic field equations in cylindrical geometry in two dimensions. The restriction to the *r-z* plane allows us only to explore the m=0 instability, which is in general most often seen in the experiments. The code itself is Eulerian. The advective parts of the equations are solved explicitly, while the diffusion is done implicitly using a matrix solver based on the biconjugate gradient method. Diffusion coefficients are calculated from Spitzer's magnetic diffusivity and Braginskii's thermal conductivity. The magnetic field, assumed to be purely azimuthal, evolves due to advection and magnetic diffusion. The vacuum is treated as a low density fluid. It is also taken to be highly diffusive. This allows us to limit the time step with respect to the plasma regions rather than the "vacuum".

3. Results

We are primarily concerned with the effects that implosion dynamics have on the growth of the small initial perturbations. We compare the development of the MHD instabilities in the pinch that is in equilibrium to the one that is imploding. The plasma was initialised with the uniform temperature in all cases. The diffuse z-pinch profiles that we examine are given by the current density distribution of the form:

$$J(r) \propto (1 - \alpha r^2) \qquad (2)$$

where r is the radial coordinate normalised to the initial pinch radius, α has values 0, 1, and -1, which determine whether the current is uniform throughout the plasma, peaked at the edge or peaked on the axis respectively. Density profiles are calculated from the MHD equilibrium. For all cases that we examine, density is peaked on the axis.

The plasma is initialised with very small amplitude velocity perturbations in all cases. We consider both random and single mode velocity perturbations. For comparison the density contours for runs initialised with the same amplitude random perturbations are shown in fig.1. The plot on the left represents the evolution of the instability in the equilibrium pinch, while the plot on the right is for the case of the implosion.

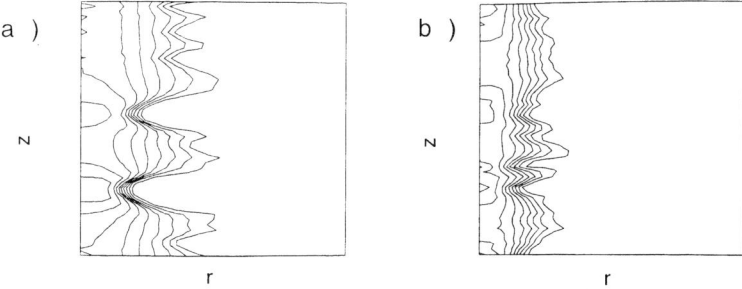

Figure 1: Density contours (equally spaced) for a) equilibrium and b) dynamic pinch.

The growth of the instabilities with time is assessed from Fourier decomposition of the line density of the pinch as a function of z. For comparison plots corresponding to the equilibrium and non-equilibrium runs are given in fig.2a and fig.2b respectively. Both graphs correspond to the non-linear stages of the development. The instability seems to grow nearly exponentially even though the amplitudes have become very large. Dynamics affect the selection of the dominant mode with time. The shift in the dominant wavelength from shorter to longer ones is more prominent in the dynamic case. In both cases however the pinch is destroyed (i.e. the neck of the instability reaches the axis) before any overall saturation of the growth is visible.

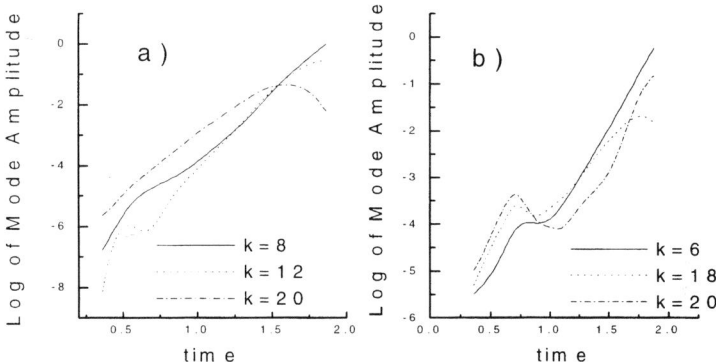

Figure 2: Log of dominant mode amplitude versus time for a) equilibrium and b) dynamic pinch

The runs initialised with a single mode velocity perturbation show no striking difference in the shape of the plasma surface which would allow one to distinguish dynamic pinch susceptible to RT from the equilibrium pinch. The implosion does however significantly alter the internal uniformity of the plasma column.

In order to gain insight into the behaviour of the pinch in the dynamic case, we consider the time variation of the instability growth rate obtained by calculating $d(lnA)/dt$, where A is the instantaneous instability amplitude. This is shown in fig.3. In the equilibrium case the growth rate calculated in the same manner is approximately constant. Thus the well defined variation with time seen in fig.3 must be associated with the pinch dynamics. Two simple physical explanations for this behaviour suggest themselves. Firstly, since in non-dynamic situations the linear MHD growth rate scales as $\gamma = \gamma^* v_{th} / a$, where γ^* is a normalised MHD growth rate and is a function of ka; 'a' is a pinch radius, v_{th} is the thermal velocity, we might expect that variations are basically due to the changes in a and associated changes in v_{th} (due to the adiabatic compression) with γ^* held fixed. The result of a simple minded calculation of this sort is shown in fig.3 (labeled 'Ammended MHD'). The pinch radius 'a' has been calculated from 1D simulation. Alternatively we might suppose that the variation is correlated with the classical RT: at certain times the unperturbed plasma is RT unstable, at others is RT stable. Figure 3 also shows the corresponding classical RT growth rate (eq.1) where the acceleration has been calculated from 1D simulation. All growth rates are normalised to their respective maxima. It is clear that plasma behaviour obtained from the code is strikingly unlike either the classical RT or the simple MHD prediction. It must be emphasized that we are not trying to make detailed quantitative comparisons here. Instead, our aim is to use simple linear theory considerations to attach a qualitative understanding of the numerical results. Unfortunately, as fig.3 makes clear, these interpretations are not tenable.

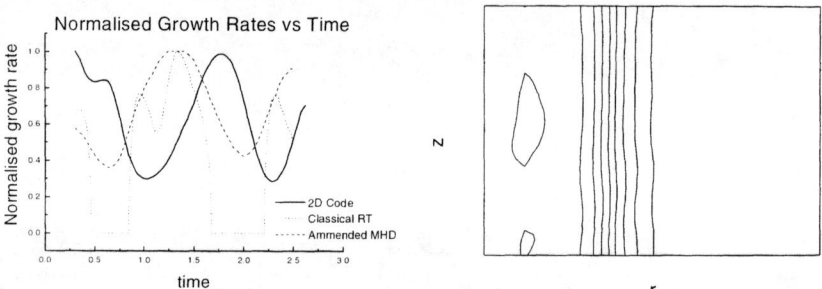

Figure 3: Comparison of normalised growth rates versus time.

Figure 4: Density contours for a run with sheared axial flow.

We have also begun the study of effects of sheared axial flow on the non-linear development of instabilities in the dynamic pinch. The effects of sheared flow are known to be beneficial to the stability of z-pinch. The mechanism behind it in the non-linear regime is in smearing out the initial perturbations[8]. The instability is "broken up" by advecting its structure at different speeds at different radii. Consequently, longer wavelength modes require higher flow speeds in order to be smoothed out. In the case of diffuse z-pinch profiles that we examined preliminary results indicate that a larger amount of shear is required close to the origin since the density is concentrated on the axis. The axial velocity was $v_z=4r^2$ (v_z is normalised to the Alfven speed), with small initial random velocity perturbations added to it. The density contour for the run with the sheared flow presented on the fig.4. The plot is given at the same time as the density plots on fig.1 for comparison. Most of the surface irregularities are smoothed out, however the internal modes are still present. Sheared axial flow delays the onset of the perturbation growth and reduces the growth rates but does not ultimately stabilise the plasma.

4. Conclusions

We have carried out 2D MHD simulations of the m=0 instability for a variety of diffuse profile z-pinches in dynamic situations. From our results we conclude that it is not useful to make a distinction between MHD instabilities and RT. The pinch dynamics affect the growth rate of the instability but not in the way consistent with classical RT behaviour. The shape of the plasma surface is also not consistent with the traditional view. It is therefore more appropriate to refer to this type of instability as Dynamically Enhanced MHD Instability (DEMI) rather than RT. DEMIs affect the internal structure of the plasma more than the m=0 instability in the stationary pinch. The sheared axial flow delays onset of DEMIs and reduces their growth rates, however it does not completely stabilise the z-pinch.

References:

1. F. L. Curzon et al, Proc. Roy. Soc. A, **257**, p386, (1960).
2. J. H. Hammer et al, Phys. Plasmas, **3** (5), p2063, (1996).
3. A. B. Bud'ko et al, Phys. Fluids B, **2** (6), p1159, (1990).
4. D. L. Peterson et al, Phys. Plasmas, **3** (1), p368, (1996).
5. R. B. Baksht et al, Proc. 3rd Int. Conf. on High Den. Z-pinches, London, UK, p365, (1993).
6. M. Coppins, Plasma Physics and Controlled Fusion, **30** (3), p201, (1987).
7. A. R. Bell, Phys. Plasmas, **1** (5), p1643, (1994).
8. T. D. Arber and D. F. Howell, Phys. Plasmas, **3** (2), p554, (1996).

Simulations of Radiatively-Driven Implosions on the PBFA-Z Facility

Joysree B. Aubrey, Richard L. Bowers and Darrell L. Peterson

Applied Theoretical Physics Division, Los Alamos National Laboratory, Los Alamos, NM 87545, USA

Abstract. We have performed two-dimensional calculations of the implosions of thin-walled aluminum cylinders driven by a source of radiation. The source is generated by the stagnation of an imploding plasma liner on to a foam target (dynamic hohlraum or flying radiation case) in the PBFA-Z facility at Sandia National Laboratory in Albuquerque, New Mexico. Both Lagrangian and Eulerian codes are used for the simulations of the compression of the shell by the ablatively-driven main shock.

Introduction

The implosion of a cylindrical wire array (driven by the energy from the discharge of a capacitor bank) has been used to generate a source of x-rays at Sandia National Laboratory[1, 2]. The radiation is produced by the dynamic hohlraum (DH) when the plasma from the wires stagnates. Measurements have shown that peak radiation temperatures within the hohlraum reach 150 ev, with the full-width-at-half-maximum of the radiation pulse being about 5 ns. The x-ray source can be used to study the coupling of radiation to different targets, shock-driven mixing of materials in high-energy-density regimes and fusion experiments.

The dynamic hohlraum and the aluminum target are shown in figure 1. The foam hohlraum is coated with a thin layer of tungsten with a linear density of 2 mg/cm. The foam itself is polyethylene with a density of 20 mg/cc. Coaxial to the hohlraum is the aluminum target with a wall thickness of 200 microns. The inner radius of

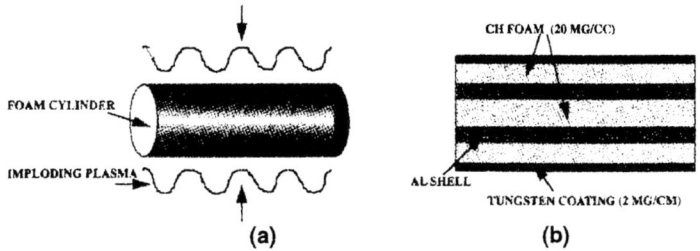

FIGURE 1. (a) The implosion of a cylindrical wire array drives a plasma radially inwards onto a foam target. Stagnation of the plasma on the foam generates the radiation for a variety of experiments. (b) In the simulations, a cylindrical aluminum shell is placed coaxially within the hohlraum.

the shell is 0.10 cm and the cavity is filled with CH foam which also has a density of 20 mg/cc. A Lagrangian radiation-hydrodynamic code was used to simulate the coupling of the radiation to the aluminum and the subsequent generation of an ablatively-driven shock through the material. The plasma is magnetically unstable and perturbations associated with the vaporization of the wires will grow during the implosion, producing a classical spike and bubble structure prior to pinch. These instabilities affect the quality of the radiation front produced during the pinch and of the subsequent shock formed in the target. Therefore, an Eulerian magnethydrodynamic (MHD) code has been used to investigate these effects. The second section describes the results of the Lagrangian calculations. The simulations of an initially perturbed plasma done with the Eulerian MHD code are given in the third section. The last section contains the summary and conclusions drawn from the calculations as well as directions for future research. Both codes use a three-temperature gray diffusion radiation package.

Lagrangian Calculations

The driving source for the Lagrangian calculations is simulated by the impact of a tungsten plasma on the foam hohlraum with an initial velocity of 0.053 cm/ns (figure (2)). The tungsten plasma is initially at a temperature of 5.0 ev and has a density of 4.0 mg/cc. As the tungsten plasma encounters the foam cylinder (which is coated with a thin layer of tungsten with a linear density of 2 mg/cm) it stagnates, creating a source of radiation that fills the hohlraum. The peak radiation temperature inside the hohlraum reaches 155 ev in this model. The electron temperature initially lags behind the radiation temperature. However, temperature equilibration occurs in just under 3 ns. Magneto-hydrodynamic effects have been ignored in the Lagrangian

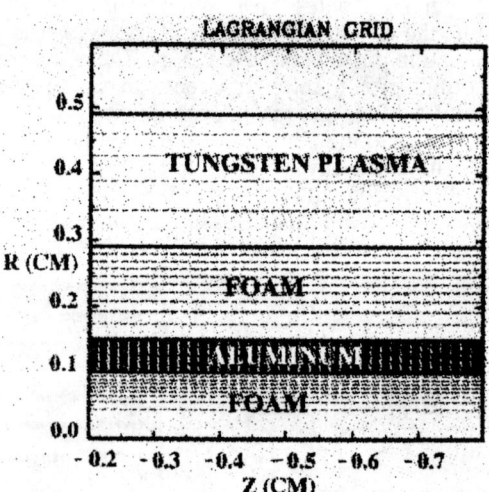

FIGURE 2. In the Lagrangian calculations, the radiation source is simulated by the deceleration of a tungsten plasma (4 mg/cc) which strikes the thin tungsten coating on the foam cylinder with an initial radial velocity of 0.053 cm/ns. As the plasma (which is at a temperature of 5.0 ev) stagnates, the generation of radiation begins to occur.

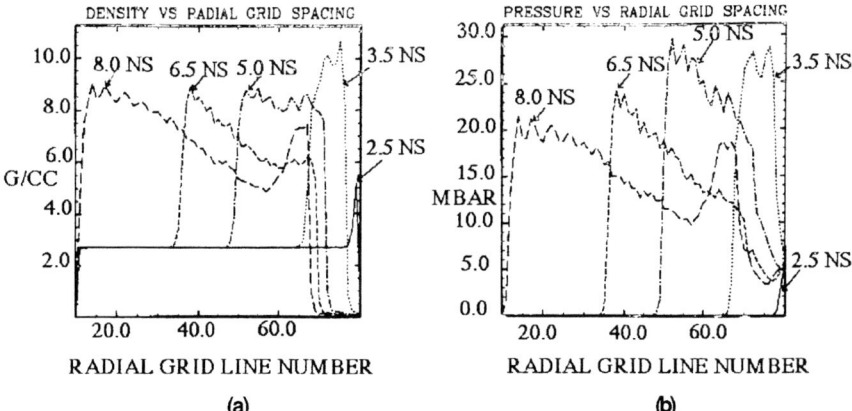

FIGURE 3. (a) Variations of densities with radial grid spacing in the cylinder at different times. (b) Variations of pressures in the aluminum as functions of radial grid spacing at different times.

calculations. The radiation within the hohlraum couples to the outer layers of aluminum, causing them to ablate. This drives a shock in the opposite direction through the aluminum. The ablation velocity reaches a peak of 0.012 cm/ns after which the interface begins to decelerate. The passage of the shock through the aluminum has the effect of compressing the cold material in front of it to densities of ~ 8-10 g/cc or about three times normal density (figure (3a)). The pressures in this region rise to around 25 megabars (figure (3b)). The mean ion charge in the wake of the shock is 11, so the aluminum atoms are stripped of electrons down to the k-shell. Both the radiation and electron temperatures are about 125 ev and the aluminum density in the ablated region drops to around 0.1 g/cc.

Eulerian Calculations

The Eulerian calculations used an initially pertubed model that was calibrated to the implosion of a tungsten wire array which had an initial radius of 2.0 cm and a mass of 4.0 mg. The capacitor bank was represented by a single equivalent circuit loop with a time-dependent voltage source V(t), resistance R=120 mΩ and inductance L=10.24 nH to the load. The tungsten plasma is shown in figure (4) at 212 ns after current delivery to the load. A central cylindrical foam cushion with a density of 20 mg/cc is also represented in the figure. The density contours in the tungsten plasma produced by the wire array are shown and range from 10^{-6} to 10^{-2} g/cc. The leading edge of the plasma has several low-density high-velocity features which may or may not be artifacts of the way in which the calculation was done. The vertical lines at 0.75cm and 1.0 cm indicate the boundaries for a refined radiative implosion calculation.

The refined calculations for the radiatively-driven MHD implosion of the target are begun at 225 ns after current delivery to the load. This time is taken to be the zero time for the results presented here. The stagnation of the perturbed tungsten plasma on the foam cushion generates x-rays which are absorbed by the outer layers of the

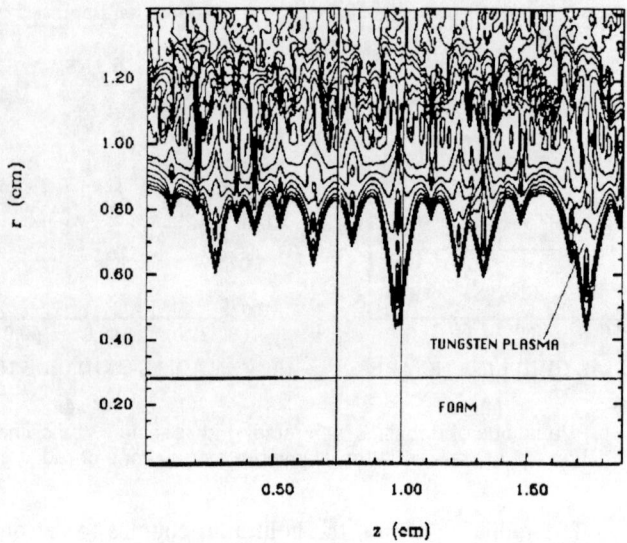

FIGURE 4. The implosion of the tungsten plasma formed by the wire array is shown. The density contours in the plasma range from 10^{-6} to 10^{-2} g/cc. The vertical lines show the "slice" of the initial mesh that is used for the calculation of the radiation coupling to a central target.

aluminum shell. Peak radiation temperatures inside the hohlraum reach 140 ev. The electron temperatures lag behind the radiation temperatures for about 3 ns. The peak densities and pressures are comparable to those in the Lagrangian calculations, but the detailed shapes of the curves are quite different. This may be due to the different driving conditions for the two sets of calculations, the presence of instabilities in the Eulerian model as well as to slightly different equation-of-state and opacity tables used in the two cases. Despite the advanced state of instability growth in the tungsten plasma evident in figure (4), the radiation front produced by impact with the foam cushion was found to be quite smooth. Furthermore, the radiatively-driven shock in the target was found to be nearly uniform in the axial direction (apart from damage associated with the low-density, high-velocity small-scale features). Preliminary results indicate that the latter features, if real, may pose a serious problem for applications experiments. However, if they are not real, or if they can be suppressed, the resulting radiation drive and shock front are insensitive to the large-scale instabilities that arise during the implosion.

Summary

Results from the two-dimensional unperturbed Lagrangian and perturbed Eulerian calculations confirm that radiation generated in the dynamic hohlraum reaches peak temperatures of around 150 ev. The x-ray source has been used in the present calculations to implode a hollow cylindrical aluminum target. The ablation of the outer aluminum layers due to the absorption of radiation drives a convergent shock into the target. The shock strength is calculated to be around 30 megabars and, as a result, the aluminum is compressed to about 2.5 to 3 times normal density.

Future calculations will include a common definition of a radiation source for both the Lagrangian and the Eulerian calculations. A Monte-Carlo photonics package will be used in order to gain an improved understanding of the radiation transport within the dynamic hohlraum. The formation of the high-velocity low-density features in the tungsten plasma that were seen in the Eulerian calculations will be studied in more detail in order to determine their origin. Finally, experimental data are needed to determine whether or not these features actually exist.

References

1. Brownell, J. H. and Bowers, R. L., "The Flying Radiation Case", LA-UR 97-558, 1997
2. Sanford, T. W. L. et. al., *Physical Review Letters* **77**, "Improved Symmetry Greatly Increases X-Ray Power from Wire-Array Z-Pinches", 1996, pp 5063-5066

AXSTRAN : A Non-LTE Radiation Transport and Ionization Dynamics Code in Axisymmetric Two-Dimensional Geometry

Y. K. Chong, T. Kammash and J. Davis*

Dept. of Nuclear Engineering, University of Michigan, Ann Arbor, MI 48109
*Radiation Hydrodynamics Branch, Naval Research Laboratory, Washington, DC 20375

Abstract. A description of AXSTRAN, a fully coupled time-dependent non-LTE radiation ionization dynamics code with the nonlocal opacity effects and the radiation transport for axisymmetric 2-D geometry, is presented. A demonstration of the capability of the code is made through an examination of the radiation field and the ionization state in a finite uniform argon Z-pinch plasma.

I Introduction

Numerous one-dimensional theoretical investigations of hot dense Z-pinch and laser produced plasmas have shown that the radiation plays an important role in affecting the dynamics of the plasmas through an array of emission, nonlocal absorption and energy transfer processes which depend sensitively on the details of the ionization dynamics as well as the temperature, density and size of the plasmas.[1] These plasmas, however, are known to be subject to various sources of multidimensional nonuniformities and inhomogenieties such as the Rayleigh Taylor and MHD instabilities and boundary effects which affect strongly the property of the radiation field as well as the plasma dynamics. In order to accurately predict and understand the nonlinear coupling physics of the plasma and radiation ensemble, therefore, a self-consistent theoretical approach within a full multidimensional framework is required. In this paper, we present a description of AXSTRAN, a fully coupled detailed configuration, time-dependent non-LTE radiation ionization dynamics code with the nonlocal opacity effects and the radiation transport for axisymmetric (r,z) geometry. AXSTRAN can be used as a radiation ionization dynamics postprocessor, and has an appropriate module for efficient coupling to PRISM,[2] the two-dimensional Lagrangian MHD code at the Naval Research Laboratory, for detailed simulations of the dynamic evolution of a hot dense plasma. An application of the code is made by examining a n argon Z-pinch plasma.

II Model

In AXSTRAN, the determination of the radiation field and ionization states that are consistent with the temperature and density at each point in the plasma is achieved through the simultaneous solutions of the ionization dynamics (ID) or rate equations for the atomic level populations together with the radiative transfer equation (RTE) for the specific intensity of the radiation field in conjunction with the equations of state (EOS). An iterative solution technique is employed together with a number of acceleration techniques to facilitate the solution process. The electron and ion temperatures, which are related to the electron and ion specific internal energy densities and the ionization energy density, are determined from the ideal gas EOS. The effective charge $\bar{Z}$ and electron density of the plasma are determined from the charge neutrality condition. The coupling of AXSTRAN to PRISM is made through the EOS and through the electron energy equation by the radiative power density. Furthermore, various transport coefficients in the MHD equations are affected by the internal ionization states.

The rate equation for the population density n_i of atomic level i may be written

$$\frac{dn_i}{dt} = \sum_{j \neq i}(W_{ij}n_j - W_{ji}n_i), \qquad (1)$$

where W_{ij} is the total (collisional and radiative) transition rate coefficient from an arbitrary atomic level j to level i, and the sum is taken over all levels j with a direct transition to level i. An equation of this type is written for each atomic level i included in the atomic model. The atomic and radiative processes considered through a multilevel detailed configuration atomic model are: collisional ionization, photoionization, collisional, dielectronic and radiative (including stimulated) recombinations as well as collisional excitation and deexcitation, spontaneous and stimulated emissions and photoexcitation. The atomic data as well as the collisional rate coefficients are obtained according to the techniques outlined in Duston and Davis.[3] Unless the plasma is optically thin for which the radiation does not affect the levels, the evaluation of the radiative rate coefficients requires the specific intensity of the radiation from the solutions of the RTE. The solutions of Eq. (1) are obtained through a single-step time integration of an implicit integral representation of the equation that is derived from successive substitutions of the formation solution with appropriate closure schemes. Preliminary studies indicated that the method is superior in terms of accuracy and computer cost over the NRL method[4] without the limitation on the timestep. Under the collisional radiative equilibrium (CRE), the explicit time dependent term in Eq. (1) can be dropped and the standard decoupling technique is invoked for the equilibrium solution.

Because the radiation in a multidimensional hot dense plasma can be highly anisotropic and nonuniform (particularly near the optically thin outer region and in strong temperature or density gradient regions) with a strong and complex frequency dependence arising from its typical non-Planckian nature, the accurate description of the field requires, in general, a multifrequency multidirectional radiation transport with sufficient spectral and angular resolutions. In AXSTRAN, this is accomplished through the solutions of an appropriate form of the RTE which may be written in terms of the radiation specific intensity, I_v, as[5]

$$\left[\sin\theta\cos\phi\frac{\partial}{\partial r} - \frac{\sin\theta\sin\phi}{r}\frac{\partial}{\partial \phi} + \cos\theta\frac{\partial}{\partial z}\right]I_v(r,z,\theta,\phi) = -\kappa_v(r,z)I_v(r,z,\theta,\phi) + j_v(r,z), \qquad (2)$$

where the local radiative emission and absorption processes are represented by the total volume emission coefficient or emissivity, j_v, and the opacity, respectively, κ_v. The dominant radiative processes include the bound-bound (b-b), bound-free (b-f) and free-free (f-f) processes (and the inner-shell (IS) photoionization for higher energy photons in a cool dense plasma). Thus, j_v and κ_v can be represented by the sum of their respective contributions (including the stimulation contributions), and are determined from the atomic level populations and $\bar{Z}$ as well as the temperature and density. The spectral lines are modeled through Voigt profile under the assumption of complete frequency redistribution and whose broadening parameter includes all collisional and radiative processes affecting the upper and lower levels of each transition. The IS opacity is incorporated using the technique prescribed in Duston, et al.[6]

The multifrequency discrete ordinates transport methodology is applied to the RTE. Separate adaptive frequency grid techniques are implemented to the lines, edges and continuum radiation in order to take into account the differences in their frequency variation, and with a sufficient grid resolution to differentiate the underlying physics and their overlapping. A Riemann sum quadrature formula is used for the integration of the frequency variable due to the highly nonuniform mesh, and for the angle domain, the double-Gauss quadrature formula,[7] with a position dependent quadrature order $N_\Omega(r,z)$ is employed. The angular quadrature set can be tailored to accommodate the nonuniform anisotropy of the radiation field. The resulting multifrequency discrete ordinate form of the RTE is solved using the accurate ray-based transport technique where the intensity of the rays are obtained by integrating the RTE along each ray characteristic. This requires complex and time-consuming three-dimensional geometric tracking of the rays through the plasma, and a suite of ray transport models are developed; each tailored to meet simultaneously the accuracy requirement of a particular problem and the computer CPU time and storage constraint.[8] In addition, a number of schemes such as the line-doubling and dynamic frequency resolution (dynamical variation of the frequency mesh such as to maintain the desired frequency resolution with the minimum number of frequency points at each iteration step) techniques are employed to speed the transport calculation.

III Application

The code was used for a detailed analysis of the radiation and ionization physics in an isodense isothermal argon Z-pinch plasma of 3.0 cm in length and 0.5 cm radius. It has 1 keV temperature and two ion density cases: 10^{19} and 10^{21} ions/cm^3 were considered. At this parameter regime, the plasma has $\bar{Z}$ of about 16, and we expect it to be collisionally dominated at 10^{21} density while the radiation to play a more significant role at the lower density. The plasma was taken to be in CRE and the electron and ion were assumed to be in thermal equilibrium. The argon atomic model included all 19 ground states and a total of 106 excited levels (chosen based on their energetic and diagnostic significance in balance with the computational cost) distributed more or less uniformly over the H-like Ar XVIII to Na-like Ar VIII ion stages. The transport calculations included a total of 77 b-b lines (seven per ionization stage from the Ar XVIII to Ar VIII consisting mainly of those arising from the transitions between ground states and low lying excited states) in addition to all 124 b-f edges and f-f continuum.

The plasma at 10^{19} density is moderately opaque to lines (κ_ν of the majority of the K- and L-shell lines is in the range of 10 -100 cm^{-1}) and quite transparent to continuum, while at 10^{21}, the lines become extremely optically thick (κ_ν of up to 10^4 cm^{-1}) and the continuum moderate to relatively thick. The source function, $S_\nu = j_\nu / \kappa_\nu$, at the plasma center as a function of photon energy is plotted in Fig. (1) for both densities. The local Planck function, B_ν, is shown by the dotted curve and would represent S_ν for an optically thick collisional LTE plasma. We observe for both densities that S_ν is essentially blackbody at lower energies, where the f-f processes dominate, but substantially below B_ν at higher energies, where the b-f processes persist. S_ν at the regions represented by the lines shows a much greater departure from B_ν than the surrounding continuum even though the lines are much more opaque. This arises as the local ionization state is affected by not only the opacity but also by the completing collisional processes. The higher energy K-shell radiation falls further from B_ν than the L-shell radiation as the K-shell transition levels are more likely to deviate from the LTE state due to their greater energy separations and hence smaller collisional rates. Finally, S_ν at 10^{21} density is closer to B_ν (by about two orders of magnitude for the K-shell edges, for example) than at 10^{19} density as the higher collisional rates and opacity help the plasma toward LTE.

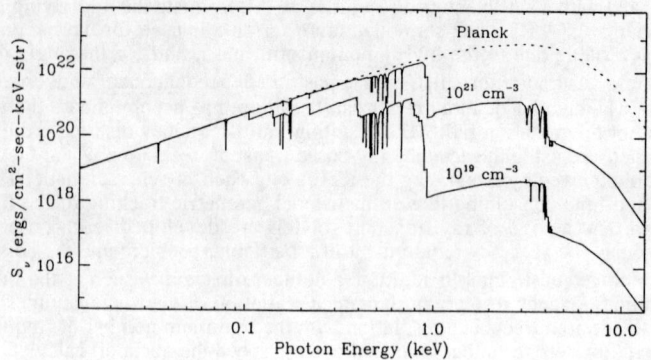

Fig. 1. S_ν at the plasma center at 10^{19} and 10^{21} densities. Planck function is shown by the dotted curve.

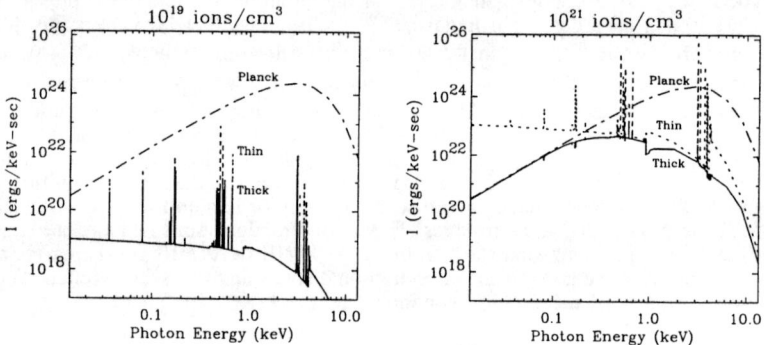

Fig. 2 Total volume integrated emission spectra (thick and thin) for both ion density cases. Blackbody emission curve is also shown for both cases.

The corresponding volume integrated total emission spectra are presented by the solid line curves in Fig. 2 with the blackbody emission limit being represented by the dash-dotted curves. The dotted line curves correspond to the results obtained under the optically thin approximation. The spectrum at 10^{19} density is dominated by lines with many of the K- and L-shell lines exhibiting a noticeable and in few cases substantial intensity decrease from their thin value while continuum is virtually unchanged. The spectrum at 10^{21} density, on the other hand, shows that the lines have virtually diminished down to the continuum level or reversed, leaving the continuum (although being substantially attenuated for the higher energy edges and the f-f radiation at lower energies) as the main contributor. The greater attenuation of the radiation at 10^{21} density is due to the increases in the radiation trapping and collisional quenching (thermalization of photons by collisional rather than by radiative decays of the upper transition state). Relative to the Planck spectrum, we observe that although the spectrum at 10^{21} density is much closer to the blackbody curve in value and shape than at 10^{19} density, it differ markedly from B_v throughout much of the energy range (E > 0.2 keV) due to the reasons stated earlier.

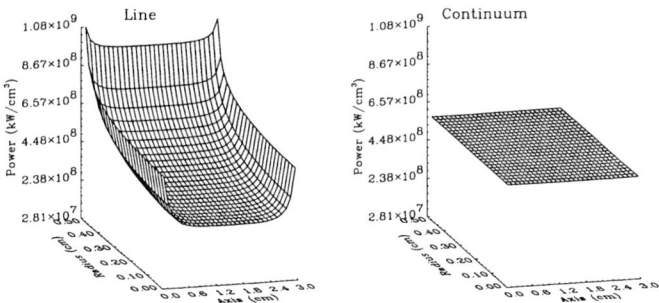

Fig. 3 Line and continuum radiative power densities (kW/cm^3) for 10^{19} density case.

The effects of the nonlocal opacity and transport on the property of the radiation field can be seen in Fig. (3) which presents surface plots of the line and continuum radiative power densities at 10^{19} ions/cm^3. We first observe that the continuum power density profile is practically uniform and flat throughout as the plasma is transparent to the continuum radiation. The line power, on the other hand, is significantly affected by the radiation transport and boundary effects as it is associated with a decrease at the optically thick interior regions and a substantial gradient that increases rapidly toward the optically thin surface regions where the photons are most likely to escape. This is compared to the line specific intensity which is uniform and isotropic in the inner plasma and decreases toward the plasma surface with increasing gradient and anisotropy.

Finally, in Fig. 4 are presented surface plots of the ratio of the optically thick over thin population densities of selected ground states at 10^{19} ions/cm^3 We observe that in addition to affecting the atomic levels via photopumping, the radiation have also introduced varying degrees of spatial nonuniformity to the level populations. The induced gradients in the level populations then act as an additional feedback mechanism to the radiation field gradients. We expect, for a given transition, the upper level population to be enhanced by photopumping while the lower level to become depleted, with a greater enhancement/depletion effect at the plasma center than at the surface. The figure shows that the Ar XVIII ground state (G) is photo-enhanced through out with the peak ratio of 1.134 at the plasma center, while the Ar XVII G is photo-depleted with the minimum ratio of 0.991 at the center and with an inverted ratio profile. The most interesting case occurs for the Ar XIII G Which undergoes the transition between photo-depleted to photo-enhanced stages from the plasma surface to the inner region. The Ar XII G is mildly affected by radiation. Since the b-f radiation is optically thin, we note that the photoionization has virtually no influence on the ground state populations. Thus, the most likely process by which the radiation affects the states is the cascade ionization (photoexcitation of the excited levels followed by collisional ionization). Further, the observed variation in the population ratio for different ground states can then be attributable to the fact that the lower energy K-shell and higher energy L-shell lines are affected most by the opacity.

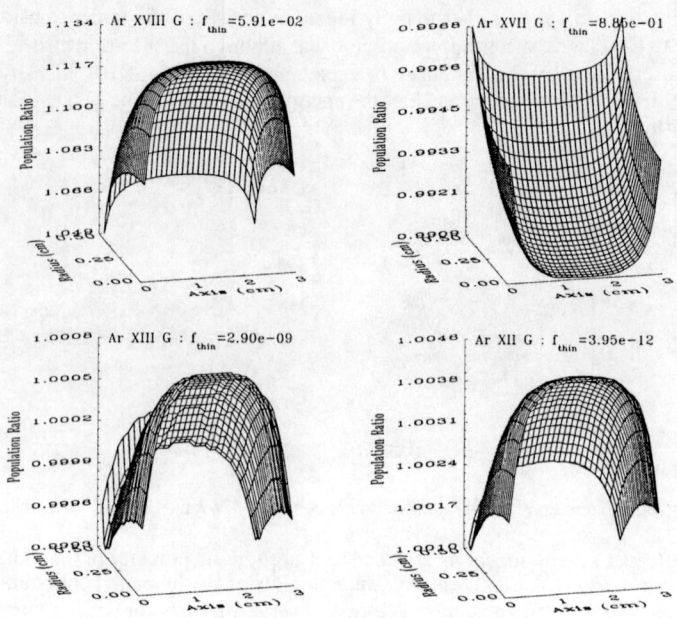

Fig. 4 Atomic population density ratios (thick/thin) for selected ground state ions at 10^{19} ions/cm^3. The corresponding optically thin population density values are noted.

IV Conclusion

We have described a two-dimensional (r,z) time-dependent non-LTE radiation transport and ionization dynamics code, AXSTRAN. The code has the detailed physics and theoretical sophistication to accurately model the non-LTE radiation and ionization dynamics of a multidimensional hot dense plasma throughout its dynamical evolution. An analysis of a uniform argon Z-pinch plasma offered a detailed glimpse at the characteristics of the multidimensional radiation and non-LTE ionization in the plasma. In addition to affecting the local radiation field and ionization distribution, the nonlocal opacity and radiation transport have shown to introduce substantial spatial gradients to the plasma's ionization distribution and to the field itself.

References

1. N. R. Pereira, and J. Davis, J. Appl. Phys. 64, R1 (1988); B. A. Hammel, et al., J. Quant. Spectrosc. Radiat. Transfer 51, 113 (1994).
2. F.L. Cochran and J. Davis, Phys. Fluids B2, 1238 (1990).
3. D. Duston and J. Davis, J. Quant. Spectrosc. Radiat. Transfer 27, 267 (1982).
4. J. W. Thornhill, et al., J. Appl. Phys. 68(1), 33 (1990).
5. D. Mihalas and B. W. Mihalas, *Foundations of Radiation Hydrodynamics*, Oxford Univ. Press, New York, (1984).
6. D. Duston, R. W. Clark, J. Davis, and J. P. Apruzese, Phys. Rev. A27, 1441 (1983).
7. J. B. Sykes, Mon. Not. R. astro. Soc. 111, 377 (1951).
8. Y. K. Chong, et al., to be published in J. Quant. Spectrosc. Radiat. Transfer.

A Detailed Postprocess Analysis of an Argon Gas Puff Z-pinch Plasma Using SPEC2D

Y. K. Chong, T. Kammash and J. Davis*

Dept. of Nuclear Engineering, University of Michigan, Ann Arbor, MI 48109
*Radiation Hydrodynamics Branch, Naval Research Laboratory, Washington, DC 20375

Abstract. A postprocess analysis of a single time frame hydrodynamic profile from the PRISM two-dimensional MHD simulation of an argon gas puff Z-pinch plasma experiment on Double-Eagle generator at Physics Internationals, Co. is presented. In addition, spatially resolved emission spectra and filtered (K- and L-shell radiation) x-ray pinhole images, generated using the SPEC2D code, are examined toward the understanding of the emission characteristics of the hot spots and the formation of the Rayleigh-Taylor instability in the plasma.

I Introduction

The PRISM two-dimensional MHD code has been used to simulate a series of argon gas puff Z-pinch plasma experiments on the 4 MA, 6 TW Double-Eagle facility at the Physics International, Co. in order to predict and understand the effects of nozzle tilting on the stability and uniformity of the plasma as a high energy plasma radiation source [1]. In this paper, we apply the two-dimensional non-LTE radiation ID code, AXSTRAN [2], as a postprocessor toward a detailed understanding of the radiation and ionization physics in the plasma. We focus the analysis on a single time frame hydrodynamic profile from the PRISM simulation of an argon Z-pinch plasma generated from an annular 2.5 cm diameter Mach 4 nozzle with an exit width 0.5 cm that is tilted inward at 10 degree to minimize the "zippering." In addition, synthetically generated spatially resolved emission spectra and filtered (K- and L-shell) pinhole images are examined for the understanding of the radiation characteristics of the hot spots and the formation of the Rayleigh-Taylor instability in the plasma.

II SPEC2D

In order to generate both spatially and spectrally resolved emission spectra and images of an axisymmetric 2-D plasma, a spectrum image synthesis code, SPEC2D, was developed. The code is based on the multifrequency integral transport method wherein the intensity at each spectral value is determined from numerical integration of the formal solution of the radiative transfer equation along the line-of-sight rays propagating through the plasma and then incident on an experimental device. The code allows for any arbitrary observation position and viewing angle of the device, and accommodates the variation in the observables from a narrow band spectrometer to a broad-angle imaging camera. Schematic representation of an experimental setup that served as the model for the SPEC2D is depicted in Fig. 1 which shows a pinhole camera, located at the distance L_d from the plasma axis and Z_d from the plasma midplane and with the viewing angle θ_d relative to the center of the plasma. Once, a family of rays are defined by projecting a line of sight ray from each discretized pixel position (r_c, z_c) on the camera film toward and passing through the plasma region of interest, the measured specific intensity associated with each ray (with an appropriate solid angle factor, ω_c, weighted for the pixel area) may be expressed as

$$I_v^m(r_c,z_c,L_d,Z_d,\theta_d) = \omega_c \int_0^\infty dv' R(v,v') \int_0^{|x_o-x_i|} ds' j_v(\underline{x}_o - s'\underline{\Omega}) e^{-\int_0^{s'} \kappa_v(\underline{x}_o - s''\underline{\Omega})ds''}, \quad (1)$$

where R is the spectral response function of the film, $\underline{\Omega}$ is the direction of the ray and $\underline{x}_i$ ($\underline{x}_o$) is the plasma entrance (exit) point of the ray. The emissivity, j_v, and opacity, κ_v, profiles are deduced from the self-consistently determined radiation field, ionization states and temperature and density of the plasma using the AXSTRAN code. Numerical integration of Eq. (1) through the three-dimensional space occupied by the plasma is obtained exactly using the ray-based radiation transport technique of AXSTRAN.

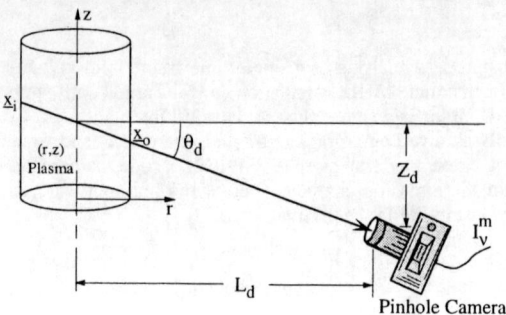

Fig. 1 Schematic representation of an experimental setup serving as the model for the development of SPEC2D.

III Results

The mass and internal energy density profiles of the argon plasma generated by the PRISM code near peak compression time (t = 100 nsec) are analyzed using the AXSTRAN code. The plasma is characterized by the initial dimensions of 3 cm in length and radius 1.25 cm, and a linear mass density of 50 $\mu g/cm$. CRE is assumed for the plasma, and the details of the argon atomic model used as well as the transport calculation can be found in Ref. 2.

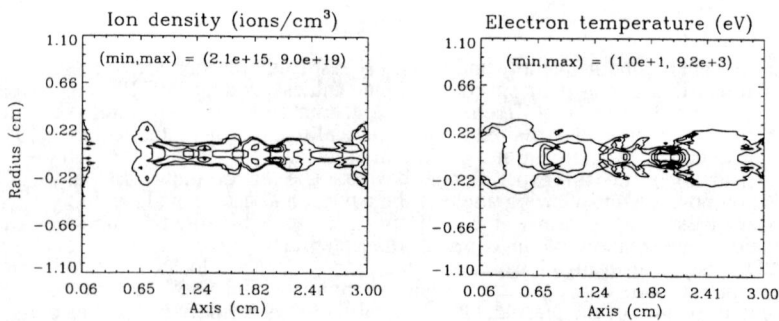

Fig. 2 Temperature and density profiles of the argon plasma at t=100 nsec.

The calculated ion density and electron temperature profiles are presented in Fig. 2 and show that the bulk of the plasma has stagnated near the axis with the typical radius of about 2 mm and the peak ion density of 9×10^{19} ions/cm^3. It is, however, highly nonuniform and may be m=0 unstable. The stagnant plasma has a temperature (over 1 keV) which is sufficient for the emission of high energy K-shell radiation. The plasma at this temperature and density regime should be somewhat optically thick to the bound-bound (b-b) lines and quite thin to the continuum radiation.

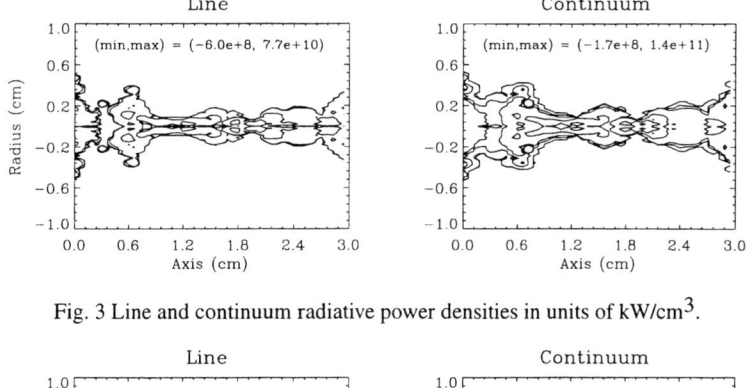

Fig. 3 Line and continuum radiative power densities in units of kW/cm^3.

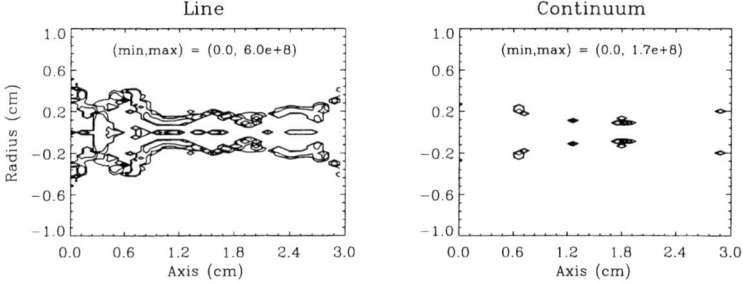

Fig. 4 Line and continuum radiative heating rates in units of kW/cm^3.

The contour plots of the line and continuum radiative power densities in the plasma are presented in Fig. 3 and show that the b-b emission is more significant near the axis and narrower than the continuum emission. This is due to the fact that the core of the plasma is hotter and denser than the outer plasma regions. The total b-b and continuum powers are 0.32 and 0.29 GW, respectively, which are compared to 1.1 and 0.28 GW in the optically thin case. This validates an earlier assertion that the lines are moderately thick while the continuum is thin. The K (L) -shell radiation represents 70 (30)% of the total power, indicative of the relatively high temperature of the bulk plasma. As the high flux of the radiation from the hot and dense core region propagates outward toward the cooler outer region, it provides an efficient means by which the inner plasma cools and transfers the energy to and heats the outer plasma. This is illustrated in Fig. 4 which shows the net radiative absorption or heating rate for the lines and continuum. We notice that the most significant radiative heating occurs at the outer region immediately surrounding the core plasma. The heating is dominated by the b-b processes which represents 97.6% of the total radiation absorbed whereas the continuum constitutes 0.66% and the inner-shell absorption is a non-insignificant fraction with 1.8 % of the total. In this respect, the radiation acts as an energy redistribution agent toward smoothing out of the gradients in the temperature, and hence, in the ionization states as will be seen below.

The effects of the radiation transport on the atomic level distribution are illustrated in Fig. 5 which shows the optically thin and thick fractional population densities for the C-like Ar XVIII ion. When the radiation is ignored, the ion is nonnegligible only at few small spots just beyond the bulk plasma. Once the absorption of the hot core photons is taken into account, however, the photopumping effect on the ion population is immediate and quite drastic. The figure shows that all surrounding outer regions are lit up and highly photo-enhanced by the core photons. The spatial distribution of the resulting ion population is strongly nonuniform due to the gradients and nonuniformities in the plasma temperature and density as well as in the radiation field. The net effect is that the radiation has driven the plasma to a significantly higher ionization state (the effective charge of the outer burnedup region increased on the average by a factor of 2 or more from the local thin value). In general, the radiation effect on the plasma ionization state should lessens as a function of radius because, as the photons from the inner core get absorbed and ionize the plasma, their probability of reaching a larger radius diminishes. This is illustrated in Fig. 6 which presents the radially resolved population density ratio (thick over thin) plots for selected argon ground states at $z = 2.1$ cm. Note the variation in the ratio as a function of radius and for different ion states.

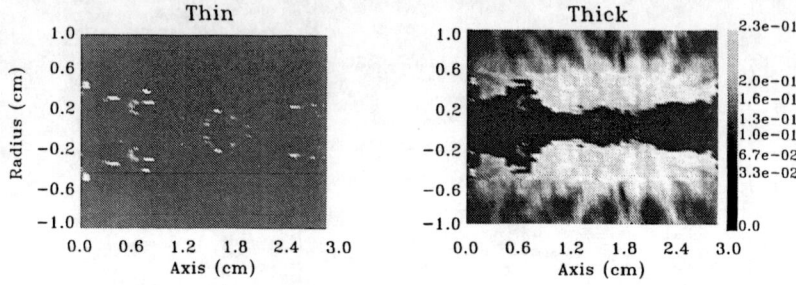

Fig. 5 Optically thin and thick fractional population density plots for the C-like Ar XIII ion.

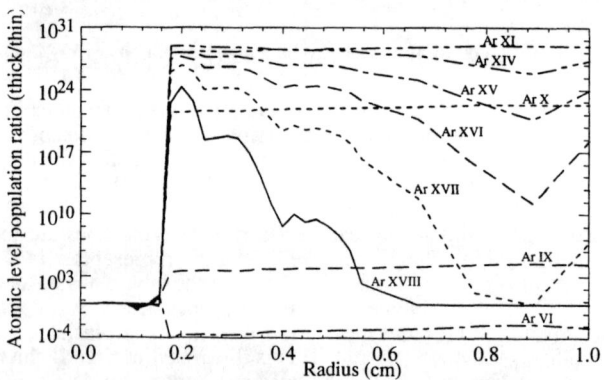

Fig. 6 Radially resolved population density ratio (thick/thin) curves for selected Ar ground states at $z = 2.1$ cm.

Synthetic K- and L-shell x-ray pinhole images of the plasma taken from a camera located at (L_d, Z_d) of (100, 1.5) cm and oriented toward the plasma center are presented

Fig. 7. The images show that both the K- and L-shell emissions are dominated by a small number of hot spots on the order of 1 mm radius and of varying axial lengths from 1 to 3 mm located along the axis. In particular, the hottest and densest hot spot, located at z = 2.1 cm with rough physical dimensions of radius 0.5 mm and 2 mm in length, has the peak radiative power of 217 GW/cm^3 or the emission rate of 0.34 GW. This represents 55.7% of the total radiative power (0.61 GW) in the plasma, and hence a major fraction of the total radiative cooling rate. We also observe that the L-shell emission image is broader than the K-shell image. This is expected since the plasma temperature and density increase sharply near the core while they become broader and lower away from the core. Both the K- and L-shell images exhibit a strong spatial intensity variation, indicative of the underlying structures and inhomogenieties (most likely from the formation of the

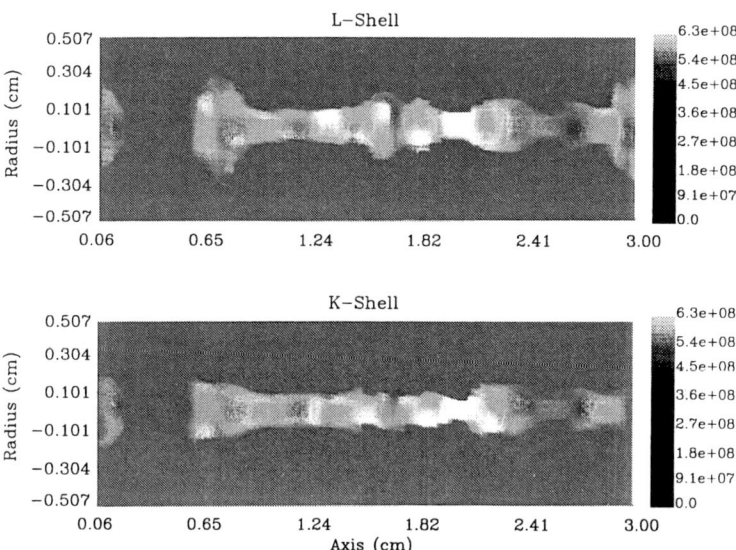

Fig. 7 Synthetic filtered (K- and L-shell radiation) pinhole images of the plasma taken from (L_d, Z_d) of (100, 1.5) cm and viewed along the plasma center.

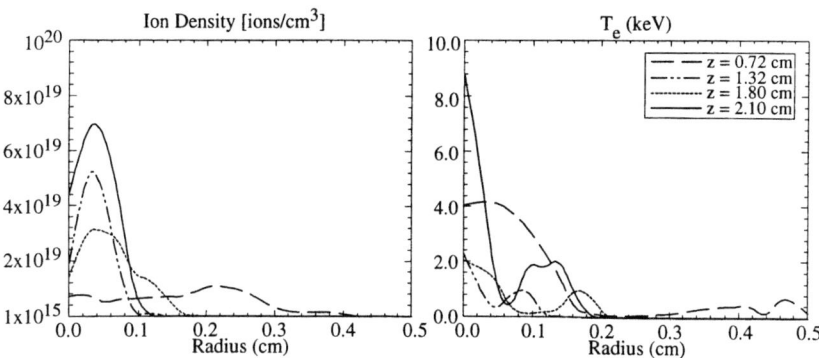

Fig. 8 Radially resolved ion density and temperature for selected hot spots.

Rayleigh-Taylor and/or m=0 instabilities). Unfortunately, the deciphering of the internal structures from the analysis of the images alone would be a difficult task since the images represent spatially averaged quantities along the line of sight rays. In this regard, the radially (and/or axially) resolved temperature and density profiles, as shown in Fig. 8 for selected hot spots, can be useful toward the understanding of their emission characteristics as a function of temperature, density and size. For example, the hot spots at z = 1.32 and 1.80 cm are warm and dense with a hollow core and thus manifest themselves with the corresponding L-shell image being brighter than the K-shell image and with the radial intensity variations which reflect the underlying temperature and density gradients. Finally, the lateral emission spectra across the selected hot spots are presented in Fig. 9. It shows a wide and significant divergence in the number and intensity of spectral features, reflective of different radial variations in the temperature and density profiles and size of the hot spots and the plasma surrounding them.

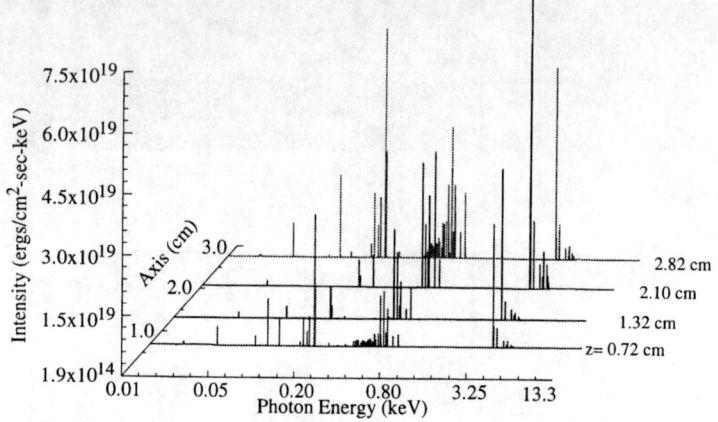

Fig. 9 Calculated lateral emission spectra across selected hot spots.

IV Conclusion

A detailed postprocess analysis of the radiation and ionization physics of an argon gas puff Z-pinch plasma was carried out. It was shown that the radiation and its nonlocal energy transport significantly affects the plasma physics and energetics and whose property is highly influenced by the temperature and density gradients and nonuniform structures in the plasma. An analysis of the hot spots from the synthetic K- and L-shell x-pinhole images generated by SPEC2D confirmed the wide variation in their temperature, density and size characteristics and pointed to the underlying inhomogenieties due to the Rayleigh-Taylor and MHD instabilities.

References

1. C. Deeney, P. D. LePell, F. L. Cochran, M. C. Coulter, K. G. Whitney and J. Davis, Phys. Fluids B 5, 992 (1993).

2. Y. K. Chong, T. Kammash and J. Davis, in this proceeding.

SPECIAL PINCHES (AND LASER SCHEMES)

Compact Plasma Focus Devices: Flexible Laboratory Sources for Applications

R.Lebert[1], A.Engel[1], K.Bergmann[1], O.Treichel[1], C.Gavrilescu[2], W.Neff[3]

[1] Lehrstuhl für Lasertechnik, RWTH Aachen, Steinbachstr. 15, D-52074 Aachen, Germany
[2] University of Iasi, Copou 11, 6600-Iasi, Romania
[3] Fraunhofer Institut für Lasertechnik, Steinbachstr. 15, D-52074 Aachen, Germany

Small pinch plasma devices are intense sources of pulsed XUV-radiation. Because of their low costs and their compact sizes pinch plasmas seem well suited to supplement research activities based on synchrotrons. With correct optimisation, both continuous radiation and narrowband line radiation can be tailored for specific applications. For the special demand of optimising narrowband emission from these plasmas the scaling of K-shell line emission of intermediate atomic number pinch plasmas with respect to device parameters has been studied. Scaling laws, especially taking into account the transient behaviour of the pinch plasma, give design criteria. Investigations of the transition between column and micropinch mode offer predictable access to shorter wavelengths and smaller source sizes. Results on proximity x-ray lithography, imaging and contact x-ray microscopy, x-ray fluorescence (XFA) microscopy and photo-electron spectroscopy (XPS) were achieved.

INTRODUCTION

E**x**treme-**u**ltra**v**iolet (XUV) radiation (0.5 nm-20 nm) offers new opportunities for analysis- and structuring purposes with sub micron resolution [1]. The usability of the XUV spectral range for applications depends critically on the availability of suited laboratory sources. Conventional x-ray tubes are of low brightness when used for XUV production. Therefore, most experiments use synchrotron radiation which is intensely bright. In order to promote XUV-techniques much effort goes into the development of laboratory or table-top sources of higher brightness than x-ray tubes and of lower cost and size than synchrotron sources. Laser-produced plasma sources and pinch plasma sources are promising candidates. Prerequisite for feasibility studies and pre-experimental tailoring is the investigation of the scaling of the emission characteristics with device parameters and the decision whether pinch emission expresses itself as column or micropinch emitters.

DEVELOPMENT OF COMPACT SOURCES

Laboratory XUV sources developed at the Fraunhofer Institut für Lasertechnik (ILT) and the RWTH-Lehrstuhl für Lasertechnik (LLT) in Aachen for single pulse and repetition rated applications are described below. For these applications 2 to 5 kJ plasma focus devices - like the ones shown in Fig.1 were used. These devices are capable of driving currents up to 500 kA through the plasma. Typical volumes of such devices are around 1 m^3. The applications described below demonstrate the flexibility of pinch plasma sources of being optimised to high mean power at repetitive operation and also to high peak spectral brightness at single pulse operation.

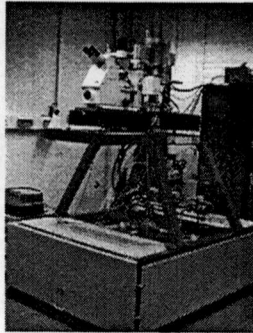

Figure 1: Compact pinch plasma devices are powered by 2-10 kJ capacitor banks and are capable of operation at 1 pulse / 20 s (left) up to 2 Hz repetition rate (right).

In a compact plasma focus devices high reproducibility (< 20 % of X-ray yield in one line) was achieved when using a surface flashover ignition mechanism without any streamers channels occurring in the discharge volume [2]. The run-down phase - which is a speciality of plasma focus discharges - gives the capacitor bank the time to build up its maximum current before pinching starts. This situation, which is different from conventional z-pinches, combines two advantages namely the simple capacitor bank discharge with collapse at maximum current, which is otherwise only achieved with pulse power generators.

With such devices up to 10 % of the plasma energy in the pinch is emitted into the interval between 1 and 10 keV [3,4]. An overall efficiency of about 1 % for the conversion of electrical energy into XUV radiation of more than 1 keV photon energy is achieved. With regard to the yield these devices are comparable to lasers with 100-200 J pulse energy and 10 ns pulses. As shown later, 20 Hz operation of such devices are achieved.

Both micropinch and bulk emission mode is achieved depending on device parameters and elements in the pinch. With respect to applications, XUV production from the bulk is preferred because of higher total yield and reproducibility. The emitting volume is typically about 10 mm in length and 400 µm in diameter. Details are published in these proceedings and elsewhere [5,6,7].

In the case of micropinches - also called "hot spots" - typical dimensions of the emitting volume are some 10 µm in diameter and some 100 µm in length. The density exceeds 10^{21} cm^{-3} and the temperature reaches values of up to 10 keV. Usually a few micropinches are generated in one single pinch event. Micropinches are also subject of intense investigated and many aspects of them are well interpreted by the radiative collapse model. Details are published in these proceedings and elsewhere [8, 9]

APPLICATIONS OF COMPACT PLASMA FOCUS DEVICES

Each application of XUV radiation dictates its individual set of requirements on the radiation. Thus the general development of radiation sources for XUV technology face a wide range of specifications. Helpfully, with pinch plasma devices a variety of technical and physical degrees of freedom can be used to influence the spectrum, yield per pulse as well as the average power to meet such requirements. The spectral distribution of the thermal emission from a plasma is mainly influenced by the choice of the elements. With respect to the emission mechanism the plasma parameters can be matched to optimise the desired radiative transition. Single K-shell line emission offers narrowband radiation with low background. L- or M-shell emission bands of neighboured lines or continuous emission (e.g. radiative recombination) are of use to maximise flux in a given emission band. Pulse duration which is of special interest for x-ray techniques with temporal resolution, can be tuned by both the generator and by internal processes in the plasma. The repetition rate is important for scanning techniques or if the number of photons required can't be provided in a single pulse.

X-Ray Lithography using a Repetitive Plasma Focus Source

Proximity printing XRL is still under discussion as one possible candidate fur future lithography. Large scale production was generally planned around electron storage rings, whose radiation provides nearly ideal properties with respect to spectral distribution, beam divergence and brightness. However, for peripheral tasks low cost point sources being fully compatible with the process at electron storage rings were sought for. In a research program of the Karl Süss Company, Garching, a repetitive pinch plasma XUV source was developed for lithography applications. The major guideline was to achieve full compatibility with manufacturing at electron storage rings for feature sizes of 350 nm. This resulted in the boundary conditions for the development: broadband emission into the spectral range from 0.67 nm-1.2 nm, source size <1mm, source to wafer distance > 40 cm.

A neon plasma focus device (fig. 1) powered by a fast 5 kJ, 2 Hz repetition rate capacitor bank [10,11] and pinch currents up to 400 kA provides about 100 µW/cm^2 of XUV radiation at the resist surface behind a three window beam line of about 15 % transmission. Reproduced 0.2 µm structures have been obtained in 1 µm thick 60 mJ/cm^{-2} sensitivity resist (RAY-PF) Fig. 2 within the full exposure field of one square inch, which reveals that the performance is close to conditions at electron storage rings. Resist exposure is mainly accomplished by recombination radiation of hydrogen-like and helium-like neon ions. About 800 mJ/pulse/sr are emitted into this wavelength range from a average source diameter < 1 mm integrated over several shots.

At Science Research Laboratory in Alameda, USA a similar device emits about 25 J per pulse of K-shell neon radiation at 20 Hz repetition rate [12].

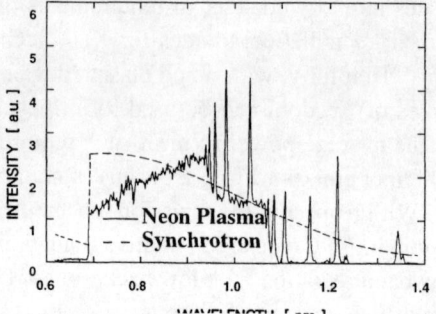 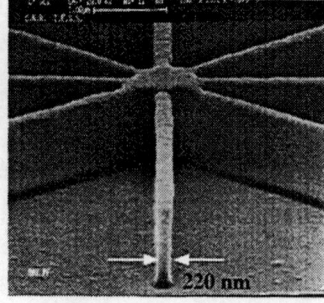

Figure 2: Emission spectrum of the neon source for x-ray lithography (left). About $5*10^{16}$ Photons are emitted per pulse into the spectral range used for x-ray lithography. The average intensity on wafer reaches 0.1 mW/cm^{-2} sufficient to expose an x-ray mask within less than 10 minutes (right: SEM of 0.2 mm structures; Mask: CNR, Rome)

X-ray Microscopy (XRM)

Table-Top laboratory x-ray microscopy is still an unsolved task. In a co-operation with the Forschungseinrichtung Röntgenphysik in Göttingen a pinch plasma source for single pulse full field imaging was developed which uses a mirror condenser for illumination and a zone plate objective for imaging. Radiation within the water window (2.3 nm < λ <4.4 nm) offers high natural contrast on living biological samples. Narrowband radiation is needed due to chromatic aberrations of the optics. High brightness, narrowband ($\lambda/\Delta\lambda = 200$) line radiation without additional background in the spectral neighbourhood makes a monochromator obsolete.

Under these conditions the N VII 1s - 2p (Lyman-α) line was chosen (NVI He-α and C VI Lyman-α are other candidates). A homogeneous illumination of the object is achieved if the spatial emission profile is flat within the 160 µm diameter of the source used for illumination. Pulsed (single pulse) exposure is favourable, in order to capture the image before motion or radiation damage reduces the resolution. In summary sufficient photons have to be emitted in a single pulse from the above defined source as line radiation. As a consequence, the spectral brightness integrated over pulse duration and line profile is the relevant figure of merit.

This number corresponds to the yield per unit source surface, per solid angle and per pulse in the line of interest, and will be referred to as "integrated spectral brightness (ISB)" in the following. For details see [11,13,14,15] and references therein.

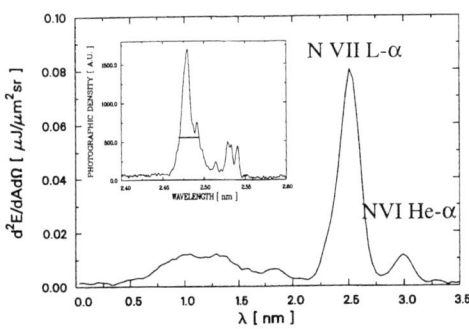
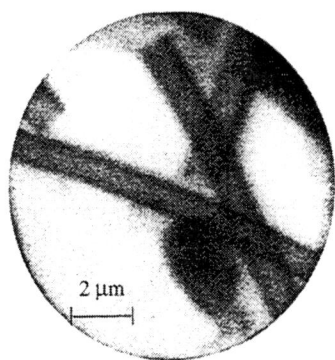

Figure 3: Nitrogen spectrum used for full field imaging microscopy and a single pulse image of dehydrated iron and manganese containing sheaths of bacteria "Leptothrix Oderacea". The image of 40 nm resolution was obtained in an x-ray microscope with a single pulse of the pinch plasma source. (courtesy of G. Schmahl, Forschungseinrichtung Röntgenphysik, Göttingen University; Sample courtesy of K.-H. Jacob, FG Lagerstättenforschung, TU Berlin)

Emission into the line of interest was maximised and continuum content in the vicinity of the line was minimised by both experimental and theoretical investigations. The emission spectrum of this source (Fig.3) shows the nitrogen emission lines in the water window. Recombination continua of hydrogen is at shorter wavelengths that of helium like ions is of low intensity. Reciprocal line widths of the single lines are $\lambda/\Delta\lambda < 210$ as seen from the separation of the N VII 1s - 2p line and the N VI $1s^2$ - 1s3p line. A ratio of line over continuum energy of E_L/E_C of 500 (extrapolated to $\lambda/\Delta\lambda = 200$) was determined. The ISB in the line pair reaches 0.61 mJ/μm^2/sr (standard deviation < 20%) at a pinch current of about 250 kA. Figure 3 shows an image of a wet specimen (resolution about 100 nm) taken with the x-ray microscope at the Forschungseinrichtung Röntgenphysik (Göttingen).

X-ray Contact Microscopy (SXCM)

For Soft X-ray contact microscopy (SXCM) of biological samples broadband radiation in the water window provides natural contrast [16]. The high flux needed to expose the photoresist with a single pulse continuous recombination emission or a band of closely located emission lines are best suited. Argon L-shell emission from the plasma focus device provide many lines into the spectral range of the water window. (E.g., in [17] 75 Argon L-shell lines have "intensities" > 100.) Neon-like Ar IX to Lithium-like Ar XVI contribute to the emission, so that the flux is much larger than for a single emission line and amounts to about 6 µJ/µm^2/sr. The source is about 400 µm in diameter. Lines with wavelengths shorter than the K-edge of oxygen are filtered by oxygen in the beamline.

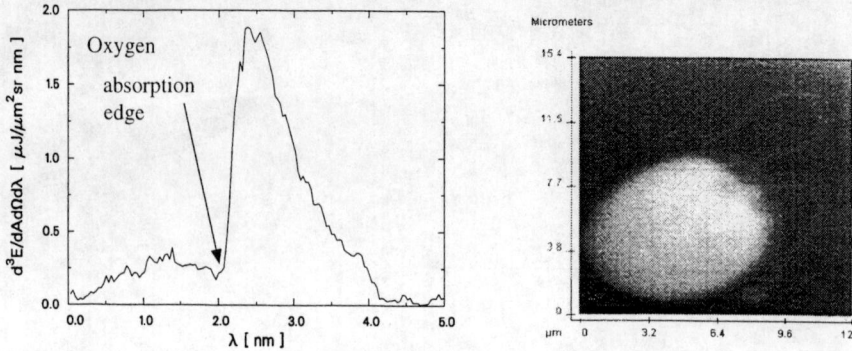

Figure 4: Spectrum of the argon pinch plasma used for x-ray contact microscopy. The spectral distribution is tuned by the use of oxygen gas in the beamline acting as a filter for shorter wavelengths. On the right, a contact microscopic image of Chlamydomonas taken with a single pulse of Argon radiation in the water window.

Using the contact microscope techniques described in [16] in a first experiment images of samples could be obtained, although neither the condenser was optimized for this purpose nor the position of the illumination was controlled to a great extent. The results is shown in fig. 4

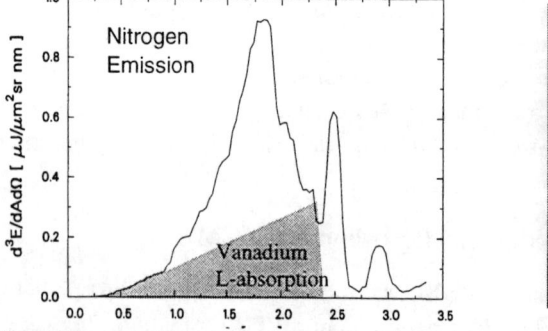 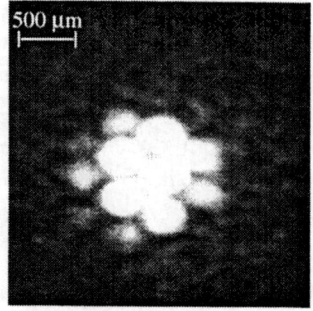

Figure 5: Single pulse recombination radiation of nitrogen VII was used for the excitation of vanadium fluorescence radiation at 2.4 nm. Using a Fresnel zone plate the generated Fluorescence radiation was imaged (1:1) onto a CCD detector (left). Resolution is limited by the pixel size and low photon flux some 100 µm.

X-ray Fluorescence Microscopy (XFM).

Fluorescence X-ray microscopy (XFM) requires intense broadband emission in the high energetic vicinity of the absorption edge of the element to be detected. This technique is demonstrated for Vanadium fluorescence emission at 2.4 nm excited with Nitrogen recombination radiation. The radiation was accomplished using the source for imaging and contact x-ray microscopy. Helium was used as the beamline gas instead of oxygen to avoid the absorption of the continuous radiation. Figure 5 shows the emission spectrum used for the excitation of the x-ray fluorescence of a structured vanadium foil. The fluorescence radiation was then imaged using a condenser zone plate [18] onto a CCD camera showing single pulse resolution of better than 100 µ also shown in fig.5.

X-ray Photoelectron Spectroscopy (XPS)

Photoelectron Spectroscopy is a way of detecting elemental and chemical composition of a sample. Best suited for a pinch plasma source is time-of-flight XPS. By using a part of the condenser mirror of the x-ray microscope a small area of the sample was illuminated with a neon spectrum similar to the one used for x-ray lithography, however tuned for lower content of recombination radiation. The interaction of the radiation with the sample generates electrons with kinetic energies corresponding to the energetic difference between the quantum energy of the exciting photons and the binding energies of the electrons. Using a drift tube of 60 cm in length the electrons are sorted with respect to their kinetic energies and detected using an amplifying MCP and a transient oscilloscope. Figure 6 shows one of the first results obtained on Aluminium covered with soldering tin with a non optimised MCP and the unfiltered neon emission.

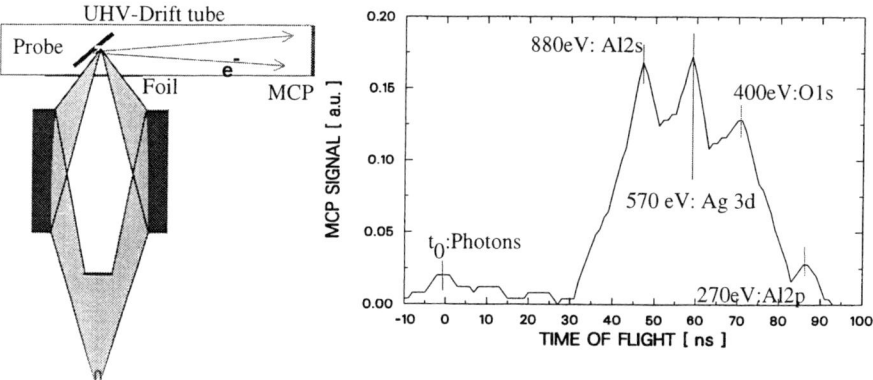

Figure 6: Single pulse XPS-spectrum of Al_2O_3 excited with (broadband) Neon emission. Peaks of the electron lines are due to narrowband excitation with single NeX Lyman-α line.

SUMMARY

Compact pinch plasma devices are promising candidates for technical relevant compact sources which could promote XUV technology to use techniques developed at synchrotron sources in individual laboratories. Their flexibility to tailor their emission characteristics was shown in some examples. Nearly contradicting demands were met with the same source principle.

ACKNOWLEDGEMENTS

This work was supported by the Deutsche Forschungsgemeinschaft (He 979/17-1; Le-904/1-1), the Bundesministerium für Forschung and Technologie (13N53290, 13N5680 and 13N5838), and the Commission of the European Community in the framework of the Association Euratom/IPP. Exchange Atomforum Wien. Discussion with Prof. Kunze and Dr. Koshelev are gratefully acknowledged. The authors wish to express their thanks to Prof. G. Schmahl and Dr. D. Rudolph and Prof. Stead and Prof. Ford for advice, co-operation and the images.

REFERENCES

1. Lebert, R., Rothweiler, D., Engel, A., Bergmann, K., Neff, W. Optical and Quantum Electronics, **278**, 241--259 (1995)
2. Lebert, R., Pinchplasmen als gepulste Röntgenquellen hoher spektraler Strahldichte, PhD Thesis RWTH Aachen (1990), ISBN: 3-86973-004-5
3. Whitney, K. G., Davis, J., J. Appl. Phys. **67** (4), 1735 (1990)
4. Pereira N. R., Davis J., J. Appl. Phys. **64** (3), R1 (1988)
5. Bergmann K., Lebert, R., J. Phys. D: Appl. Phys. **28**, 1579 (1995)
6. Bergmann K., Lebert, R., Neff, W., J. Phys. D: Appl. Phys. **30**, 990 (1997)
7. Bergmann, K., Lebert, R., Neff, W., these proceedings
8. Engel, A, Lebert, R., these proceedings
9. Lebert, R., Engel, A., J. Appl. Phys. **78** (11), 6414 (1995)
10. Richter, F., Eberle, J., Holz, R., Neff, W., Lebert, R., AIP Conf. Proc. **195** (1989) 515
11. Neff, W., Rothweiler, D., Eidmann, K., Lebert, R., Richter, F., Winhart, G., Proc. SPIE **2015**, 32-44 (1993)
12. Krishnan, M., Prasad, R. P., Mangano, J., SRL Technical Memorandum SRL-A-92-05TM
13. Niemann, B., Rudolph, D., Schmahl, G., Diehl, M., Thieme, J., Neff, W., Holz, R. Lebert, R., Richter, F., Herziger, G., Optik, **1** (1989) 35
14. Lebert, R., Rothweiler, D., Neff, W.,J. X-ray Sci. and Technol., **6** (1996) 107-140
15. Rothweiler, D., "Gepulst erzeugte Plasmen als Strahlungsquelle für ein Röntgenmikroskop" PhD Thesis, RWTH Aachen, ISBN 3-89588-067-1, (1994)
16. Ford, T., Stead, T., to be published in **X-ray microscopy** V, Springer Verlag
17. Kelly, R. L., Palumbo, L. J., NRL Report 7599, (1973)
18. Rudolph, D., Niemann, B., Schmahl, G., Christ,O., in G.Schmahl, D.Rudolph (eds.), "X-Ray Microscopy", Springer Ser. Optic. Sci., Vol. 43, Springer, Berlin, 192 (1984)

Control of X-ray Spectrum Emitted from a Gas-puff Z-pinch

Keiichi Takasugi and Tetsu Miyamoto
Atomic Energy Research Institute, Nihon University, Tokyo 101, Japan

Katsuhiro Tatsumi and Takehito Igusa
College of Science and Technology, Nihon University, Tokyo 101, Japan

The axial magnetic field is applied to an annular gas-puff z-pinch for the control of radial dynamics and x-ray emission from the pinched plasma. K-shell and L-shell radiations of Ar ions are detected separately, and only the K-shell radiation is suppressed significantly by the axial field. The radial motion of the plasma is analyzed assuming a simple circuit model. The characteristic radius of the plasma increased with increasing the axial field.

I. INTRODUCTION

The z-pinch is an efficient means of converting electric energy into pulsed radiation. For Ar z-pinch, the K-shell radiation of Ar ions occurs in the range of 0.3 - 0.4 nm, and the L-shell radiation occurs in 3.5 - 4.4 nm.[1] If we eliminate the hard component, we can use it as an intense pulsed radiation source for the Water Window (2.3 - 4.4 nm).

The application of axial magnatic field to z-pinch plasma have been examined to stabilize and sustain the plasma for the controlled nuclear fusion research. The technique was used for high magnetic field generation[2] or creating a new magnetic configuration for plasma confinement.[3] It is also used for control of x-ray emission on a pulsed power z-pinch,[4] and the enhancement of x-ray was reported.

In a compressional z-pinch the formation of hot spots is closely related to the development of Rayleigh-Taylor instability on the contraction.[5] Increasing the uniformity of the z-pinch will not always contribute to x-ray enhancement.[6] So we used this method for suppression of x-ray and control of total emission spectrum of the z-pinch.

II. EXPERIMENTAL SETUP

The experiment was carried out on the SHOTGUN gas-puff z-pinch device. Figure 1 shows the schematic configuration of the device. The energy storage section consists of 24 µF fast capacitor bank, whose storage energy is 7.5 kJ at the charged voltage of 25 kV. The spacing of electrodes is 4.0 cm. Annular gas shell with the diameter of 2.8 cm is produced between the electrodes using a high speed gas valve. Ar gas is used thoughout the experiment.

In order to control the radial motion of the plasma column, a small axial magnetic field was applied. A pair of coils is placed inside the discharge chamber for this purpose. The coil radius is 12.8 cm and the separation is 8.6 cm. Using a slow bank system of 1.6 mF with charged voltage of 100 V, an uniform axial magnetic field of 233 G is produced between the electrodes.

The plasma discharge currents are measured by Rogowski coils placed near the electrodes. The x-ray signal from the pinched plasma ($\lambda < 1$ nm) is detected by a scintillation probe with 10 μm Be foil. The soft x-ray and the extreme ultraviolet (XUV) signal ($0.5 < \lambda < 50$ nm) is detected by a vacuum x-ray diode (XRD) with Ni photocathode.

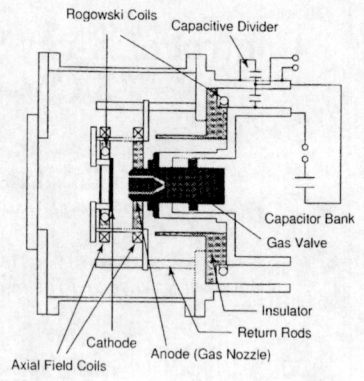

Fig. 1 Schematic view of the SHOTGUN gas-puff z-pinch device.

III. EXPERIMENT

Figure 2(a) shows a typical discharge current, x-ray and XUV signals without the axial magnetic field. Spikes in the x-ray and the XUV signals show the maximum pinch of the plasma. It occurs at the current about 200 kA. A dip is formed in the current signal simultaneous with the pinch, due to rapid change of plasma inductance. When the axial field of 1.17 kG is applied (Fig. 2(b)), the x-ray signal disappears. However, the intensity of the XUV signal is almost the same and its duration becomes longer. The dip in the current signal for this case is not obvious.

The x-ray and the XUV signals are time-integrated and shown in Fig. 3 as a function of the axial magnetic field intensity B_z. The x-ray signal drops sharply

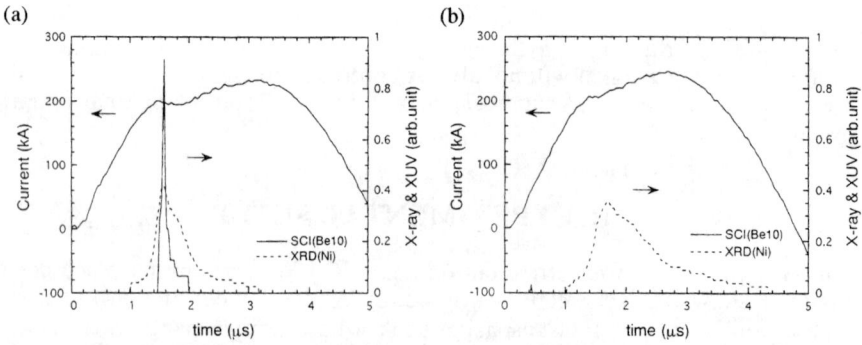

Fig. 2 Typical discharge current, x-ray and XUV signals (a) without and (b) with axial magnetic field.

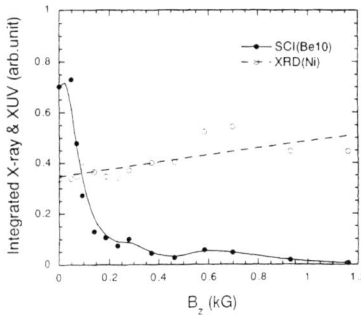
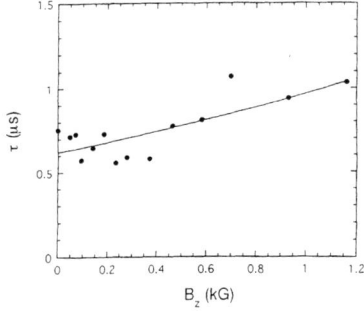

Fig. 3 Time-integrated x-ray and XUV signals vs axial magnetic field.

Fig. 4 Characteristic time of the pinch vs axial magnetic field.

with B_z, and is not significant above 0.2 kG. The XUV signal increases with B_z. In order to see the effect of the axial magnetic field on the sustainment of the pinched plasma, the characteristic lifetime of the pinch τ is evaluated and plotted against B_z. The time τ is defined as

$$\tau \equiv \frac{\int I(t)dt}{I_{peak}}, \qquad (1)$$

where $I(t)$ and I_{peak} are the XUV signal and its peak value respectively. Figure 4 shows the dependence of τ on B_z. The time τ simply increases with B_z, indicating that the plasma is sustained effectively with B_z.

The radial motion of the plasma is estimated from the discharge current signal.[7] The voltage of the capacitor is calculated from the measured current $J(t)$ as

$$V(t) = V_0 - \frac{1}{C}\int J(t)dt, \qquad (2)$$

where V_0 is the charged voltage and C is the capacitance. The circuit is assumed to be composed of inductive L and resistive parts R. Except only the onset of discharge the plasma resistance is low, and it is neglected. So the resistance R is fixed for feeder and electrodes. The inductance L(t) is calculated as

$$L(t) = \frac{1}{J(t)}\int (V(t) - RJ(t))dt. \qquad (3)$$

The change of inductance $\Delta L = L(t) - L_0$ occurs only in the plasma. The parameter L_0 is the initial inductance. From the inductance ΔL we define the characteristic plasma radius as

$$r(t) \equiv r_0 \exp\left(-\frac{2\pi \Delta L(t)}{\mu_0 l}\right), \quad (4)$$

where r_0 and l are the initial radius and the length of the z-pinch respectively. This radius is related to that of an uniform cylindrical z-pinch, if the plasma current flow on the surface of the plasma. The inductance ΔL and the minimum radius r_{min} are shown in Fig. 5 as a function of B_z. The inductance ΔL decreases with B_z and the corresponding radius r_{min} increases with B_z. The radius r_{min} is one order of magnitude larger than that for the magnetic pressure balance $B_z = B_\theta$.

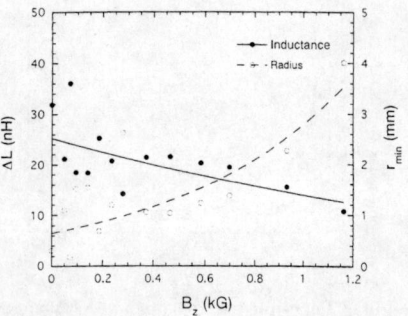

Fig. 5 Change of plasma inductance and characteristic plasma radius vs axial magnetic field.

IV. SUMMARY

The axial magnetic field was applied to the annular gas-puff z-pinch plasma for the control of radial dynamics and x-ray emission from the plasma. The K-shell radiation of Ar ions (0.3 - 0.4 nm) and the L-shell radiation (3.5 - 4.4 nm) were detected separately. The K-shell radiation was suppressed significantly by a small axial magnetic field. The duration of the L-shell radiation was extended by the axial field, and its time-integrated intensity increased. The radial motion of the plasma was estimated from the change of plasma inductance, and the characteristic radius of the pinched plasma increased with the axial field. This radius was one order of magnitude larger than that for the magnetic pressure balance.

REFERENCES

1. Kantsyrev, V.L., Takasugi, K., Tatsumi, K., Miyamoto, T. and Shlyaptseva, A.S. : Proc. 1996 Int. Conf. Plasma Physics Vol. 2, pp.1106 - 1109 (1997).
2. Wessel, F.J., Felber, F.S., Wild, N.C., Rahman, H.U., Ruden, E. and Fisher, A. : Appl. Phys. Lett. **48**, pp. 1119 - 1121 (1986).
3. Rahman, H.U., Ney, P., Wessel, F.J. and Rostoker, N. : AIP Conf. Proc. **299**, pp. 696 - 706 (1994).
4. Edison, N.D., Etlicher, B., Attelan, S. and Rouille, C. : AIP Conf. Proc. **299**, pp. 199 - 209 (1994).
5. Takasugi, K., Miyamoto, T., Moriyama, K. and Suzuki, H. : AIP Conf. Proc. **299**, pp. 251 - 257 (1994).
6. Takasugi, K., Moriyama, K., Shibuya, T. and Miyamoto, T. : Proc. 1996 Int. Conf. Plasma Physics Vol. 2, pp. 1098 - 1101 (1997).
7. Takasugi, K., Suzuki, H., Moriyama, K. and Miyamoto, T. : Jpn. J. Appl. Phys. **35**, pp. 4051 - 4055 (1996).

Stimulated VUV Radiation From Z-Pinch Necks

K. N. Koshelev, P. S. Antsiferov, L. A. Dorokhin and Yu. V. Sidelnikov

Plasma Spectroscopy Laboratory, Institute of Spectroscopy
Troitsk, Moscow Region, 142092 Russia

Abstract. Developing plasma foci and neck-type instabilities in Z-pinches emit beams or jets of high velocity plasma. Their highly charged ions can cause population inversion through selective charge exchange with colder ions in a target. Recent models [1] show population inversion in a disk-shaped region that moves along the discharge axis slightly slower than the plasma jet. The population inversion between n=4 and n=3 of Li-like-like ions of elements with nuclear charge Z=6-10 is high enough to see stimulated emission effects perpendicular to the discharge axis for pinches with currents of about a few hundred kA. Time-resolved and spatially resolved spectra in the vicinity of 4 - 3 transitions (50 - 52 nm) of the Li-like ion O VI were taken on the gas-puff Z-pinch installation "MP-100" [2]. Simultaneous VUV imaging of the plasma column was done using a combination of pinhole and multiframe MCP detector gated with 5 ns. The intensity ratio between 3p - 4d (49.8 nm) and 3d - 4f (52.0 nm) components is close to the equilibrium value 2.7 during compression, but it increases twice or three-fold when the "neck" develops. Spatially resolved measurements show that plasma regions with an anomalous ratio of these two lines are strongly correlated with the position of "neck" type instabilities.

Plasma formed in axial high-current discharges is currently under active investigation as a possible source of stimulated radiation in the soft X-ray and VUV spectral region. Formation of a homogeneous active medium important for several population inversion schemes (collisional excitation, recombination, photopumping) is hindered by instabilities of the discharge plasma column, particularly by neck-type instabilities. On the other hand, neck instabilities in axial discharges (Z-pinches, plasma foci) generate plasma that flows out of the foci in the axial direction. This plasma consists of multiply charged ions, and charge-transfer recombination of these ions with lesser-charged ions in the main plasma column may result in selective population of the excited states of the product ions[1].

In axial non-cylindrical discharges such as the plasma focus, the formation of a focus is the effect of the electrode geometry. Gas puff Z-pinches have neck-type instabilities primarily because of inhomogeneous plasma compression along the discharge axis. Figure 1 shows this schematically for a plasma focus (left) and for

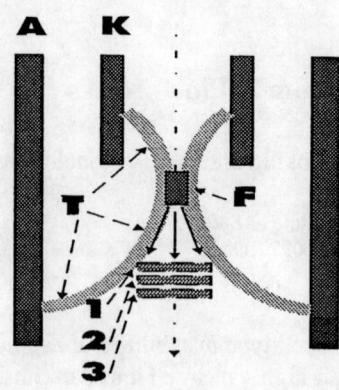 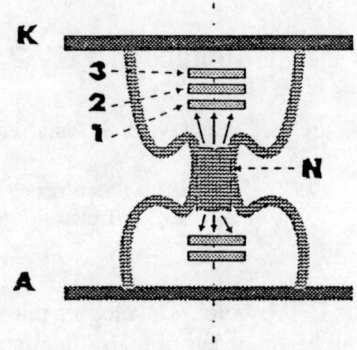

FIGURE 1. Plasma flow and interaction region in a plasma focus (left) and for a neck-type instability in a Z-pinch (right).

a z-pinch (right). The arrows from neck (N) and focus (F) zone show the direction of outflow of a multiply charged ion plasma.

The position of the "active zone," where CS-recombination is taking place, moves along the discharge axis because lesser charged ions are lost from the target plasma. Subsequent positions of the plasma with population inversion are indicated as 1, 2, 3 in Fig 1.

Experiments have been carried out the gas-puff Z-pinch installation "MP-100" described elsewhere (2). Parameters of the discharge circuit are: $C = 96$ μF, $L = 50$ nH, $T/4 \approx 3.4$ μs, $U = 15 - 25$ kV and $I_{MAX} \approx 0.5 - 1.5$ MA. The working gas (here oxygen) is injected by a fast valve into a gap between two plane electrodes. At the beginning of a discharge, the distribution of the working gas density is dome-shaped with a maximum on the discharge axis and an effective radius that depends on the time delay between opening the valve and the start of the discharge current. At maximum compression the current reaches a value of about 300 - 400 kA.

We have reported earlier [2] a strong population inversion between excited levels of H-, He- and Li-like ions of elements from C to Ne in case of quasi "plasma focus" electrode geometry, when a plasma jet was transported to a specially prepared gas target (see paper of A. Engel et al in these Proceedings). FIGURE 2: Time-resolved pinhole simultaneous with the spectra in figure 3.

Experiments presented in this paper are aimed at the observation of non-linear amplification effects due to population inversion conditions in the vicinity of necks in the Z-pinch geometry (right side, Figure 1). Transitions between levels with principal quantum numbers n=3 and n=4 of Li-like ions of O VI have been chosen for observation. Noticeable population inversion between these levels was predicted in [1] as a result of charge transfer recombination of He-like O VII:

$$O\ VII\ (1s^2) + A^{+q} \rightarrow O\ VI\ (1s^2 nl) + A^{+q+1}$$

A disk is azimuthally symmetric, and therefore a disk-shaped active medium has no preferred direction to observe amplification from stimulated emission. On the other hand, amplification may be deduced from changes in the relative intensities of 3p - 4d and 3d - 4f lines, as was done in experiments with very similar geometry[3]. Collisional relaxation between the n=4 levels with different orbital momenta occurs in our plasma much faster than the radiative decay of the levels. Under these conditions of a plasma with sufficiently high gain-length product, the intensity is transferred from the weak to the strong component of the 4 $\rightarrow$ 3 transitions and this gives rise to an anomalous ratio of 3d - 4f and 3p - 4d components.

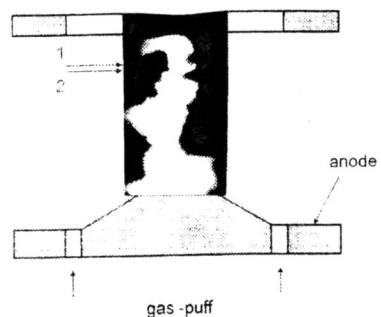

Figure 2. Time resolved pinhole photograph simultaneous with the spectra in figure 3. The point "1" is 1 mm above the "neck," which is at "2". The top electrode, the cathode is 35 mm away from the bottom electrode, the anode.

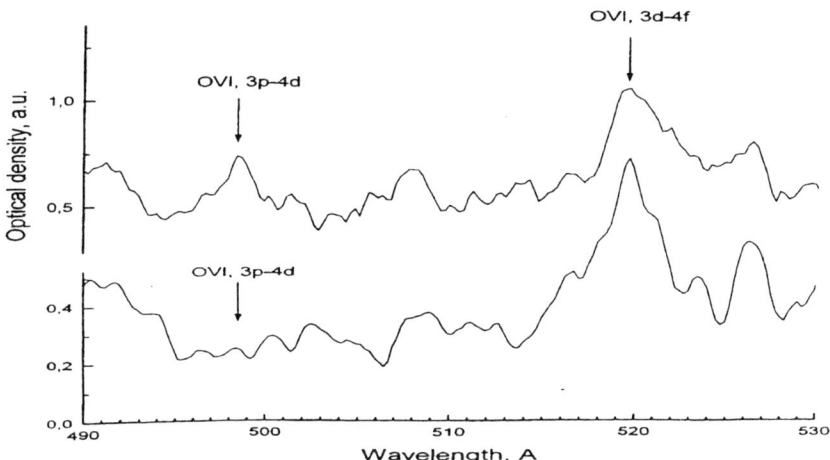

FIGURE 3. (a) Spectrum from position 1, and (b) spectrum from position 2.

A normal incidence 1 m spectrograph DFS-29 and off-Rowland grazing incidence VUV spectrograph GISVUV1 both equipped with multi-strip microchanel plate (MCP) camera with 2 - 5 ns time frame resolution have been used for side-on observation of spectra in the vicinity of the 3p - 4d (49.8 nm) and 3d - 4f (52.0 nm) transitions of O VI. A pin-hole image of the pinch plasma in VUV and soft X-ray radiation has been recorded simultaneously by a gated multi-frame MCP camera. The experimental observations are :
1. Noticeable line radiation in the region 30 - 60 nm appears when the pinch compresses to approximately 1 cm radius. Lines of Li-like O VI appear when the pinch radius is about few mm. They last for about 50 ns.
2. In the column phase, before the necks develop, time-resolved spectra show a ratio between two components $I(52.0nm)/I(49.8nm) \approx 2.5 - 3$, which is approximately equal to theoretical expectations without inversion, from Boltzmann equilibrium between the *nl* components of the n=4 level.
3. An anomalous intensity ratio appears in approximately 50% of the spectra a few ns before the moment of maximum compression: it lasts for up to 10 ns.
4. Axially resolved time frame measurement together with time framed pin-hole pictures show that a measured ratio I(52.0 nm)/I(49.8 nm) has an anomalously high value up to **10** in areas strictly correlated with a position of necks. Integrated over a 5 ns time frame, the plasma region with an anomalous intensity ratio is about 0.5 mm long along the axis.

Figure 2 shows a time-resolved pinhole picture with a neck at the position marked 2. About 1 mm above it is a position marked 1. Figure 3.a is a time-resolved (3 ns frame) spectrum from position 1, at 1 mm outside the neck. This spectrum shows a "normal" intensity ratio. Figure 3.b is a simultaneous spectrum emitted by plasma in close vicinity of the neck (position 2). This spectrum has a weak line 3p - 4d (49.8 nm) below in the noise level. We consider that fact a strong indication of stimulated emission effects for VUV O VI transitions.

ACKNOWLEDGEMENTS
The authors are grateful to Prof. H.-J. Kunze for fruitful discussions. This work was supported by the Russian Foundation for Basic Research.

REFERENCES
1. K. N. Koshelev and H.-J. Kunze, *Quantum Electronics* **27**, 164 (1997).
2. K. N. Koshelev, Yu. V. Sidel'nikov, S. S. Churilov, L. A. Dorokhin, *Phys. Lett A* **193**, 1949 (1994).
3. S. Glenzer, H.-J. Kunze, *Phys Rev E* **49**, 1586 (1994).

INFLUENCE OF PREIONIZATION ON DYNAMICS OF A GAS PUFF IMPLOSION

A.G.Russkikh, R.B.Baksht, A.Yu.Labetsky, A.V.Shishlov

High Current Electronics Institute, 4 Academichesky ave,Tomsk, 634055, RUSSIA mail: russ@hded2.hcei.tomsk.su

The experiments with both single and double Ar gas puff were carried out on the IMRI-4 ($T_{rt}=1.1\mu s$, $I_m=350$ kA) and the GIT-4 ($T_{rt}=0.12$ μs, $I_m=1.7$ MA) generator. Two different system of preionization, a sparking flash of UV radiation and a Planar Magnetron Discharge (PMD) in the crossed E×H fields, were used. The process of current stratification and the dependence of gas puff implosion uniformity on the preionization were investigated.

Previously has been revealed [1,2] that the character of preionization in a gas shell has an essential effect on the formation of the current sheath. For instance, with the help of preionization, one can solve the problem of division of the generator current between an inner and the outer shell of a double gas puff [2]. Varying the characteristics of the preionization system, one varies the probability that the generator current will flow, from the very beginning, only through the outer shell. Without preionization, this probability, depending on the polarity, ranges between 0 and 30%. A similar situation holds when Spark Preionization (SP) is used on the anode side. On the other hand, if we use for preionization a spark on the cathode side or PMD, irrespective of the electrode polarity, the generator current will flow from the very beginning through the outer shell. But what is the dynamics of the formation of a current sheath in the outer shell of the liner ?

First we carried out experiments on a small current generator IMRI-4 [2]. After that, the results obtained were verified on the GIT-4 generator.

In the experiments performed on the IMRI-4 current generator we used both a single- and double- Ar gas puff . The outer shell diameter was 60 mm and the inner shell diameter was 30 mm. The diagnostics was accomplished by ordinary means such as a Rogowski coil, a resistive voltage divider, a pinhole camera, an X-ray vacuum diode with an Al cathode + 3 μm-mylar, an optical streak camera. The streak camera slit was located parallel to the liner axis 5 mm away from the axis. The experiments were performed with two types of preionization. Preionization of the first type is SP initiated on the cathode side [2].

In the case of SP, at the initial stage of liner implosion, the generator current flows through several current channels which are fixed geometrically to the locations of the spark gaps of the preionization system. Optical recording was performed with a streak camera for two cases: the first one in which the liner region where a current channel was formed lies within a streak camera's view and the second one in which this region did not lie within a streak camera's view. For the first case, streak camera pictures show a clearly stratified liner shell, while for the second one no stratification is observed (fig.1). This means that a small number of the current channels have no time to propagate through the entire gas of the outer shell. They run to the center forcing out some material from the outer shell and carrying away some part of the current. They travel through the inner shell, causing a rather intense perturba-tion in the shell. The current chan-nels, while moving to the center and radiating, intensely ionize the remaining gas of the outer shell. Moreover, their inductance grows and, as a result, the generator current is switched to the remained gas of the outer shell. Thus, a second current sheath, and sometimes a third one, is formed. For these shots, pinhole camera pictures show highly nonuniform compression of the liner and a rather large spread in the radiation yield.

An implosion of a 60- mm- single gas puff with SP leads to the gas shell breaks even before the liner is comp-letely compressed. In this case, we observe the flight of only separate parts of the liner.

The second-type preionization is PMD in crossed ExH fields [2]. With PMD the current flows in the upper electrode – grid circuit (fig.2). The current density was approximately equal to 3 A/cm^2 with the ion density being of the order of 10^{14} cm^{-3}. The discharge operation time was 800 μs. The experiments with PMD preionization have given qualitatively different results. At the instant the current generator is switched on, all conditions for the generator current to flow through the liner have already been created. The electron/ion density in the gas shell is ~ 10^{15} cm^{-3}, which is sufficient for the process of electron multiplication to have a Townsend character. The

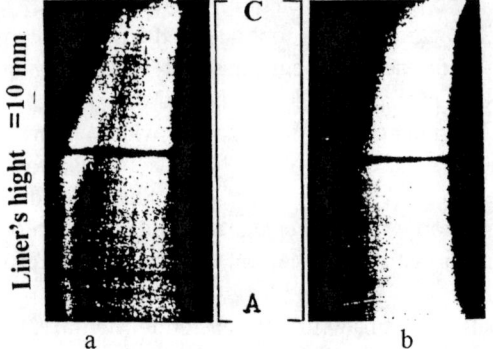

Fig.1. Streak camera pictures a) with and b) without stratification.

implosion process for such a liner can be described as follows: As the current generator is switched on, the high-current glow discharge, retaining its spatial form, starts constricting in radius. For liners of 5 ÷20 μg/cm in mass, at the instant the liner has been compressed from 3 to 2 cm, a current flows through the shell with $j \sim 60 \div 80$ kA/cm^2. These current densities are, of course, too high for this discharge to be considered as a glow discharge. It should be more correctly referred to as an "annular current channel" where the electron emission occurs from a "quasi-single cathode spot". This system is rather stable since it is primarily homogeneous and symmetrical, with a rather high frozen current into it. Further, the generator current, skinning at the external boundary of the plasma cylinder produced, continues to compress this cylinder as a whole. This scenario is indicated by both chronograms and pinhole camera pictures. According to the streak camera pictures obtained, it is possible to uniformly compress a single-gas puff with the initial diameter of 60 mm beginning from the liner mass of 2 µg/cm with the liner shell looking rather undisturbed.

The pinhole camera pictures taken for imploding a single- and double gas puffs with PMD preionization show an improved uniformity of the compression and a decreased compression factor and the radiation power of the liner in the X- ray range for low- mass liners. However, the energy radiated in the same wavelength range remains at the same level for low masses, while for masses over 12 µg/cm it exceeds the output energy of the liner realized with SP. Moreover, the spread in X-ray yield decreases.

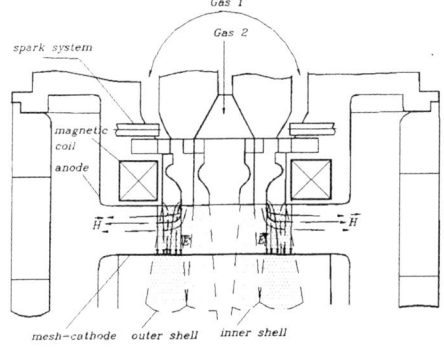

Fig.2 Scheme of a PMD preionization systen on GIT-4 generator.

In the experiments with the PMD preionization we found that the system allows decreasing the current rise time from 1 µs to 120 ns. Moreover, this plasma switch opens at the peak of the generator's current when the current derivative is equal to zero. At present we investigate this phenomenon and the results of the investigation will be published soon.

The experimental results obtained on the GIT-4 system.

The diagnostics was essentially the same as that used in the experiments carried out on the IMRI-4 except for the X-rays measurements. The measurements were performed using a photoconducting detector (PCD) placed behind a 6.35-µm-thick Ti filter. So the energy of the radiation detected ranged between 3 and 5 keV (K-lines of Ar). The SP system was similar to the system described above. The preionization system using PMD was methodically different (Fig.2) because of the specific features of the circuit of the GIT-4 generator. In this case, PMD occurred during ~ 1 µm, while the plasma switches were closed, and thepotential difference across the anode–cathode gap was ~1 kV. To decrease the statistical time delay for this type of discharge, the discharge was initia-ted using spark illumination. This has resulted in a hybrid preionization system combining the spark and the PMD preionization (Fig.2).

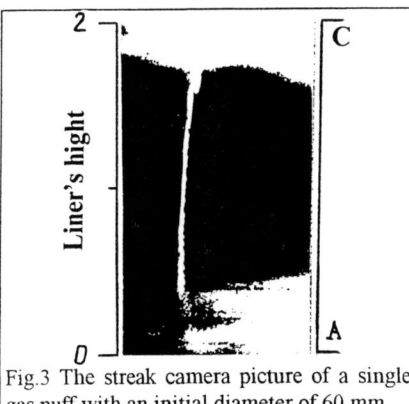

Fig.3 The streak camera picture of a single gas puff with an initial diameter of 60 mm.

The implosion of a single gas puff with an initial diameter of 60 mm.

The streak camera pictures (Fig.3) show that the liner implosion is much more stable when the PMD preionization is used. For low and midle liner masses, a unified, unbroken brightly luminous shell is observed to flow 5 mm away from the axis. The Ar K-lines yield decreased by a factor of ~ 1.5 for PMD, despite an improved homogeneity of the liner. It seems that for compression times of ~ 100 ÷ 200 ns the effect of the penetration of the magnetic field into the bulk of the compressing shell is more pronounced.

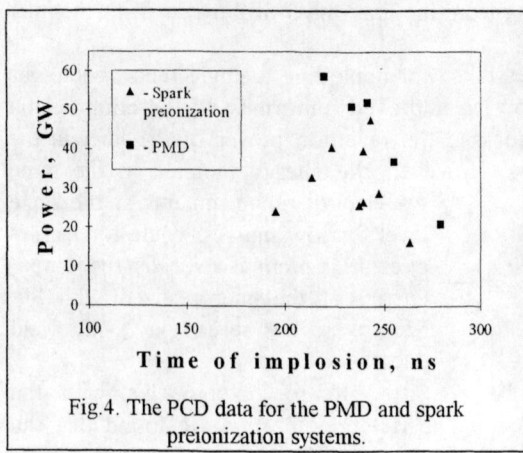

Fig.4. The PCD data for the PMD and spark preionization systems.

The implosion of a double gas puff with an initial diameters of 60 / 30 mm.
The outer shell mass was 40 µg/cm for all shots, while range from 40 to 140 µg/cm. Fig. 4 presents the PCD data for the both preionization systems. When a planar PMD was used, the radiation yield increased on average by 25% and its spread became somewhat lower.

Summary

1. The use of SP on the IMRI-4 generator leads to the formation of a few current channels. These current channels break away from the body of the liner and are com-pressed independently. The current channels remove a probably small portion of the liner mass but this causes strong flow perturbations both in the outer and the inner shells of the liner.

2. Using preionization with PMD in the case of a single gas puff leads to first, sub-stantial stabilization of Rayleigh–Taylor instabilities, second, reduces the liner compres-sion factor and the X-ray yield in the range of the K-lines of Ar (hot spots disappear).

3. Using PMD for preionization in imploding double gas puff, we were able to increase the X-ray power in the range of the K-lines of Ar by 25% and to reduce the spread in compression parameters.

4. In the experiments on the IMRI-4 generator, it was found that under a certain condition the use of PMD causes a decreases in a current rise time from 1 µs to 120 ns.

References

[1] G.G.Peterson, Dissertation, Irvine, 1994.
[2] R.B.Baksht, A.G.Russkikh, A.A.Chagin.// Plasma Physics Reports, vol. 23, No3,1997, pp.175-182.

HIGH DENSITY PLASMOID ACCELERATION BY PHASED IMPLOSION OF CAPILLARY Z-PINCH AND ITS APPLICATION TO HYPER-VELOCITY PROJECTILE ACCELERATION

K.Horioka* , M.Nakajima, T.Aizawa* and M.Tsuchida*

Department of Energy Sciences, Tokyo Institute of Technology,

Nagatsuta 4259, Midori-ku Yokohama 226, Japan

*Faculty of Engineering, University of Tokyo, Hongo, Bunkyo-ku, Tokyo 113, Japan

A Z-pinch plasma was electro-magnetically compressed and axially accelerated by phased implosion in a tapered capillary tube. The implosion was just about timed to the plasmoid drift by the shaped capillary wall. The feasibility of high energy acceleration was experimentally demonstrated using a 100mm long, slightly tapered capillary tube. It was driven by a fast pulse power generator (5 Ω-70nsec). For filling gas of 100Pa of Ar and at current of 80kA, the axial velocity of the plasma was measured to be 7×10^7cm/sec, which corresponds to 70keV argon atoms. The high energy density plasma could drive a small projectile with energy conversion efficiency of about 6%.

INTRODUCTION

We found that capillary z-discharges driven in uniformly pre-ionized gas can compress the plasma upto density of 10^{19}cm^{-3} with very good reproducibility.[1] Based on this results, we have utilized scheduled pinching of high current pulsed Z-discharge in a long, thin and slightly tapered capillary tube, for high density plasma acceleration. The basic operational principle of the plasma acceleration by scheduled pinching is shown schematically in Fig.1. A high current fast Z-discharge of 10^5A current level makes strong azimuthal magnetic field B_θ, which contracts the plasma radially inward down to order of 100 μm in diameter and drives converging shock wave ahead of the current sheet. As it accumulate and compress the plasma up-to pressure of 100GPa level via the snow plow effect, a high density plasmoid is formed at the discharge axis. At a suitable time delay for arrival of the plasmoid, the pinching occurs along the discharge axis. The plasma is radially compressed by the self-B field and axially driven by the B-gradient made by the phased pinching. If the contour of the wall is made so as to strengthen the Mach stem, we can expect that it can drive very intense shock wave in the plasma.

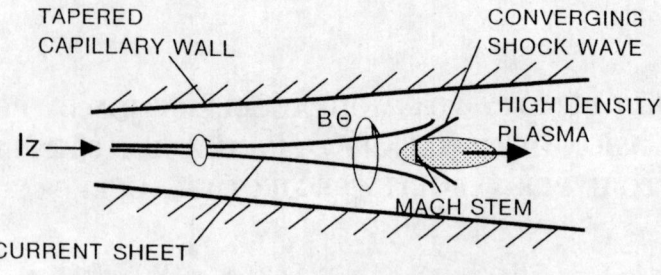

Fig.1 Principle of scheduled pinching and internal structure of plasma

EXPERIMENTAL SETUP AND TYPICAL RESULTS
Time-of-Flight (TOF) Measurement of Plasma Flux

A schematic diagram of the experimental setup is shown in Fig.2. The basic configuration of the experiment is quite similar to the previous experiment[1] except the capillary shape. The capillary of 100mm length has a thin conical wall made of polyacetal, in which Ar is filled as the working gas. The inlet and exit diameter of the tube are 4mm and 8mm respectively. We use a fast pulse power generator to drive the discharge. Typically, it drive load current of 80kA with pulse width of 70nsec. For differential pumping, the discharge section and the diagnostic chamber are separated by a pinhole of 0.2mm diameter. We have measured the plasma flux by a Faraday cup, which consists of a negatively biased

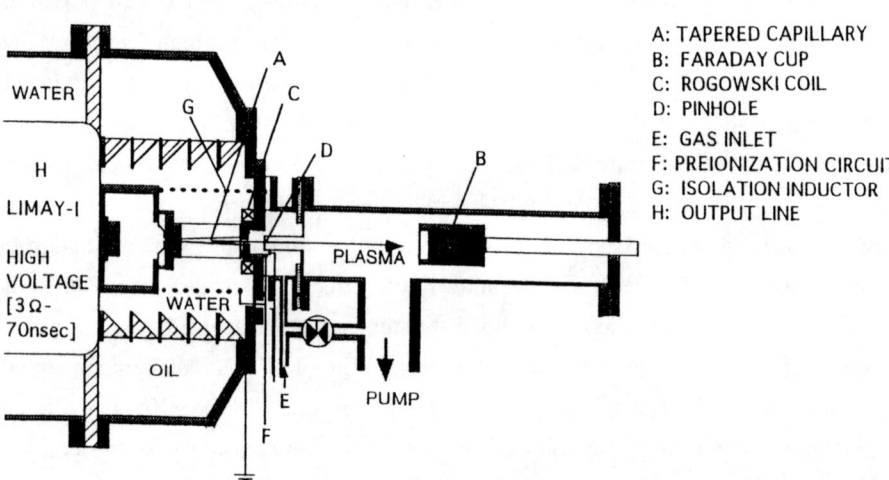

Fig.2 Experimental arrangement for prof-of-principle experiment

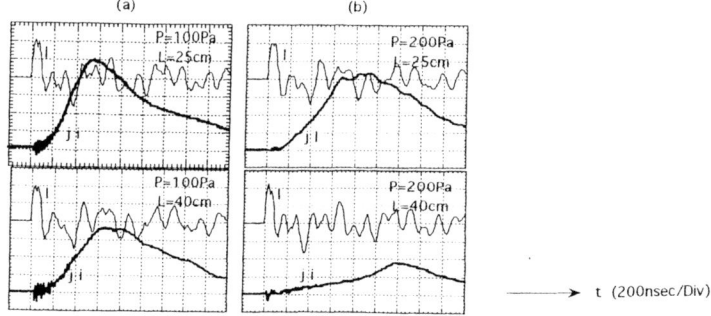

Fig.3 Time-of-flight waveforms of plasma flux ji versus discharge current
I [40kA/Div], at filling pressure of 100Pa (a) and 200Pa (b) of Ar

ion-collector plate and a photo-electron suppresser electrode. Typical TOF waveforms of the plasma ion flux *ji* are shown in Fig.3 with the load current signal *I*. As it indicates, when the initial filling pressure was 100Pa of Ar, the drift velocity of the plasma was $\sim 7 \times 10^7$ cm/sec. Then we found that Ar atoms traveling at well above the sound velocity. To check whether the high energy density plasma might not be produced by the magneto-hydrodynamic instability, the polarity of the pulse power generator was inverted. In spite of the polarity inversion, the plasma signal had almost the same shape. This means that the high energy plasmoid was not accelerated by anomalous electric field.

Flyer Experiments

A preliminary experiment on flyer acceleration was performed. The experimental arrangement is shown in Fig.4(a). Two thin Cu metallic plates of

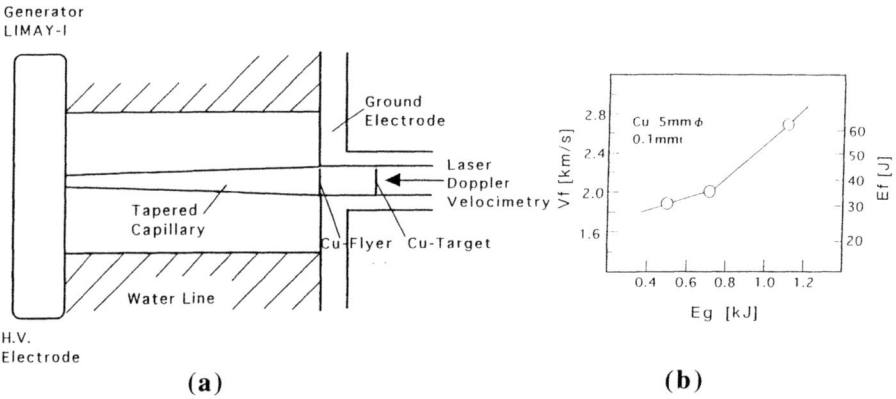

Fig.4 Schematic of flyer experiments (a) and typical results (b)

313

0.1mm thickness and 5mm diameter were placed at the exit of the capillary tube and 30mm from the first plate respectively. The flyer speed was estimated by laser velocimetry (VISAR System). As shown in Fig.4(b), when the charging energy of the pulse power generator ; *Eg* was 1.1kJ, the flyer was accelerated up-to Vf=2.7km/sec, which corresponds to kinetic energy ; *Ek* of more than 60J. This means that almost 6% of the stored electric energy was converted to the kinetic energy of the small projectile.

CONCLUDING REMARKS

In conclusion, a high density plasma was electro-magneticaly accelerated in a tapered capillary. The basic scheme of phased Z-pinch implosion may be related to pinch engine[2] or particle acceleration by zippering[3] in gas puff Z-pinch and plasma focus discharges. However, in contrast to those conventional schemes, the axial drift of the pinching time is a free control parameter in case of the phased implosion. Thus we can independently control the plasma compression and acceleration process by the tapered capillary wall. Parameters influencing implosion stability, initial gas pressure, wall shape are expected to play an important role to develop optimum condition. If those parameters are optimized, we can expect that significant part of the input energy goes into the directional energy of the plasma: this scheme appears capable of compressing the plasma to density of 10^{19}cm^{-3} and/or accelerating it up-to velocities of 10^8cm/s.

The Z-discharge plasma in the tapered capillary tube can accelerate a small metallic plate with 6% of conversion efficiency. This method was found to be useful for shock loading device for compaction[4], at intermediate parameter range between laser and chemically explosive device.

REFERENCES

(1) T.Hosokai, M.Nakajima, T.Aoki, M.Ogawa and K.Horioka; Jap. J. Appl. Phys., **36**, 2327 (1997)

(2) R.Menikoff, K.S.Lackner, N.L.Johnson, S.A.Colgate and J.M.Hyman, Phys.Fluid, **A3**, 201 (1991)

(3) T.W.Hussey, M.K.Matzen and N.F.Roderick, J. Appl. Phys., **59**, 2677 (1986)

(4) T.Aizawa, S.Kamenosono, J.Kihara, T.Kato, K.Tanaka and Y.Nakayama, J. Intermetallics, **3**, 369 (1995)

Carbon Fiber Z-Pinch with Current Prepulse

A. Lorenz, F.N. Beg, J. Ruiz, J.F. Worley and A.E. Dangor

Imperial College, Plasma Physics, Blackett Laboratory, Prince Consort Road, London, SW7 2BZ, U.K.

Abstract. Results of experiments studying the effect of a ringing 10 kA current prepulse on a 160 kA, 7 μm carbon fibre z-pinch are presented. A prepulse duration of 100 - 300 ns was found to radically alter the pinch behaviour. A delayed onset in x-ray and optical emission with respect to the main current was observed. There was a marked improvement of axial uniformity in the pinch. A large soft x-ray pulse, ten to twenty times larger than in discharges without prepulse was recorded. An electron temperature of 250 ± 50 eV was measured which was considerably higher than in discharges without prepulse.

INTRODUCTION

Experiments delaying the onset of rapid expansion and formation of fast growing MHD and Rayleigh-Taylor instabilities found in most z-pinches [1,2,3] have been reported recently [4,5,6]. In this paper we summarise a carbon fibre pinch experiment in which a current was used to produce pre-ionisation. Two previous experiments in which a prepulse current was employed in fibre pinches have been reported in [7,8], but no significant improvement was seen. In a third experiment carried out recently [9], a decrease in the growth of m = 0 instability has been inferred. However, in all three experiments there was no control of the prepulse current. A laser initiated gas embedded pinch [10] in which a prepulse current was used did show radically different behaviour, there being an apparent delay in the growth of the m=0 instability and m=1 being clearly evident. This was ascribed to the instability growing from a lower noise level and a peaked current on axis.

GENERATOR AND CURRENT PREPULSE

The experiment was performed using IMP, a 800 kV pulsed power generator which is capable of driving a fast (60 ns 10-90% rise time) current (up to 200 kA) into a fibre pinch load. A schematic of the generator is shown in Fig. 1. The prepulse current is produced by an independent low voltage (130 kV) Marx circuit, with a large series resistor (R_p=60 Ω) to protect the prepulse generator from the main Marx voltage.

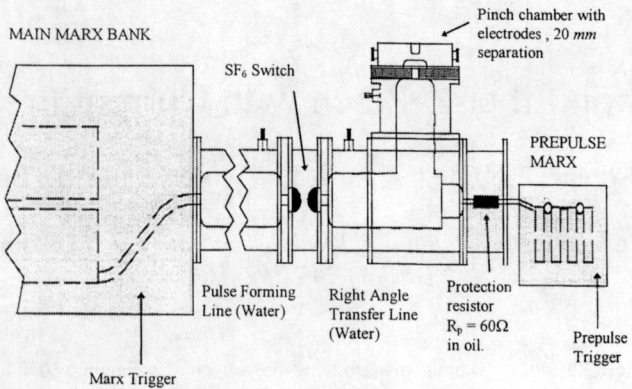

FIGURE 1. The IMP-generator with prepulse Marx and pinch chamber.

A damped oscillatory current (180 ns period) is generated in the 7 μm diameter, 20 mm long carbon fibre, breakdown occurring when the voltage on the transfer line reached ~ 20 kV. A typical prepulse current waveform is shown in Fig. 2. The main discharge current can be switched at any time during the prepulse. Radically different behaviour was observed when the main current was switched on for a prepulse duration, $t_{prepulse}$, between ~ 100 ns and ~ 300 ns, i.e after the first polarity reversal and after 1.5 periods of oscillation. Outside this range, results were comparable to discharges without prepulse reported in detail in [1].

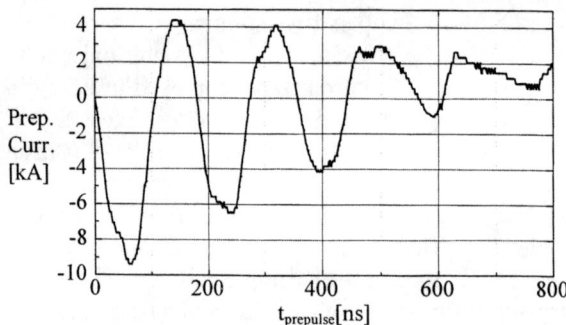

FIGURE 2. Oscilloscope trace of current prepulse into a 7 μm carbon fibre.

PREPULSE-ONLY DISCHARGES

A study of the prepulse only discharge shows that a coronal plasma is produced surrounding a solid core. The fibre was found to survive and remain intact after a prepulse discharge, in some instances for up to four discharges. Optical streak and framing photography shows that the coronal plasma has a radius which varies periodically between 1 mm to 2 mm, the minimum occurring around current peak. There is some non-uniformity of the light emitted along the length of the fibre, which is not reproducible. Estimates based on spectroscopic analysis of the time integrated CII and CIII line emission indicate an electron density of about 10^{17} cm^{-3} and electron temperature of between 5 - 7 eV. The density estimate is in agreement with an estimate of the atomic density of the coronal plasma. This was performed using a measured change in the fibre diameter of ~ 6% and assuming the ablated plasma occupies a radius 2 mm.

MAIN DISCHARGES WITH PREPULSE

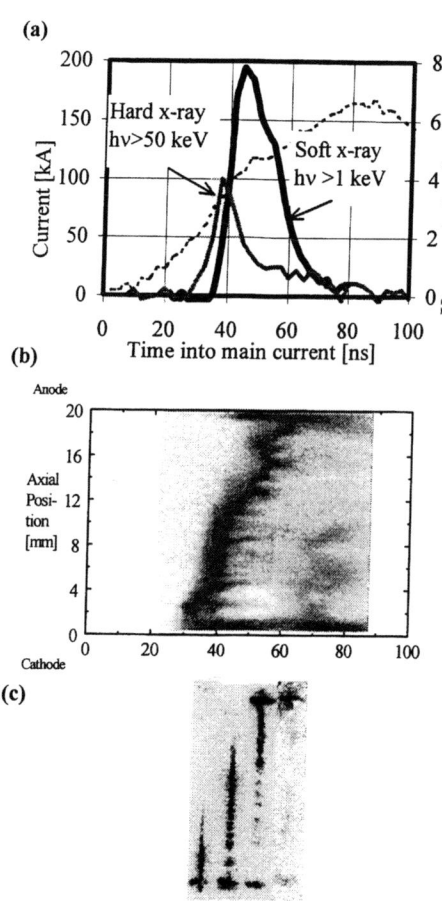

FIGURE 3. (a) Traces of current, soft x-ray and hard x-ray emission, $t_{prepulse}$ = 125 ns, (b) Axial x-ray streak and (c) x-ray four frame photographs of the same discharge, filtered by 1.5 μm Lexan (hv > 200 eV).

The most dramatic changes observed with respect to discharges without prepulse were the delayed onset of x-ray and optical emission and a ten to twenty fold increase in soft x-ray yield. This is illustrated in Fig. 3 (a) which shows the spatially integrated soft filtered (hv > 1 keV) PIN diode and hard x-ray (hv > 50 keV) traces, the latter obtained by a scintillator-photomultiplier combination.

Whereas the soft x-ray emission is delayed by typically 30 - 50 ns, the hard x-ray pulse is always peaking up to 10 ns before or at the same time as the soft x-ray pulse. A typical soft x-ray pulse width of t_{FWHM} = 10 - 25 ns has been recorded.

Fig. 3 (b) shows an axial x-ray streak photograph of the same discharge confirming the duration of the soft x-ray pulse recorded by the diode, although the streak was filtered for softer x-rays. The onset of soft x-ray emission is also delayed until t_{main} = 30 - 50 ns, when a fast radiation front is observed to progress, "zipper" from the cathode towards the anode with a typical velocity of $2·10^6$ ms^{-1}. Closer examination of the x-ray streak photograph shows that the radiation front is

317

immediately followed by a string of bifurcating bright spots. During the zipper the strong x-ray emission persists at any given axial position for 3 - 10 ns. The x-ray streak photograph also reveals that there is some anode emission before the zipper arrives at the anode. The optical streak photograph shows identical structures of both plasma and anode emission.

Fig. 3 (c) displays four 2 ns x-ray frames which are filtered for the same energy range as the streak. The first frame shows uniform plasma emission, but in the second and third frames the x-ray emission becomes progressively structured. The progress of the radiation front can be correlated to the x-ray streak photograph.

Electron temperatures of 250 ± 50 eV were estimated using time integrated pinhole photographs and peaks of PIN-diode signals with the two-filter method [11]. Hard x-ray emission observed from the anode before zippering starts is assumed to originate from energetic electrons incident on the anode. The signals obtained from time integrated pinhole photographs and PIN-diodes are consistent with a 50 keV, 5 kA electron beam.

DISCUSSION AND CONCLUSION

The delay in the x-ray and optical emission can be explained if the main current flows in the coronal plasma and causes it to collapse on to the fibre. The radiation is then emitted only when the implosion reaches the fibre on axis and the current is transferred to it. The snow plough model was used to calculate the collapse time of the coronal plasma using the mass density and radius estimates made in prepulse-only discharges. Good agreement was found between the calculated collapse time and the time of radiation onset obtained from streak photographs.

The zippering is presumably due to the collapse occurring earlier near the cathode. The fibre thus experiences a very rapid rate of current rise which results in a more uniform, better behaved, hotter pinch. The energetic electrons are presumed to be due to runaways in the low density corona.

REFERENCES

1. Beg F.N. *et al*, Plasma Phys. Contr. Fusion, **39**, 1-25, (1997),
2. J.P. Chittenden J.P. *et al*, Phys. Plasmas, **6**(8), 1, (1997),
3. Pereira N.R., Davis J., J. Appl. Phys., **64**, R1, (1988),
4. Edison N.S. *et al*, Physical Review E, **48**(5), 3893, (1993),
5. A. Chuvatin et al, Physical Review Letters, **76**(13), p.2282, (1996),
6. T.W.L. Sanford et al., Physical Review Letters, **77**(25), 5063, (1996),
7. Kies W. *et al*, J. Appl. Phys. **70**(12), 7265, (1991),
8. Riley R. *et al*, Phys. Plasmas, **3**(4), 1314, (1996),
9. Lebedev S.V. et al, ibid p.
10. Choi P. *et al*, Nuclear Fusion, **28** (10), 1771, (1988),
11. Jahoda F.C. *et al*, Phys. Review, **119** (3), 843, (1960).

Enhanced K shell x-ray yield from over-massed Targets[†]

A. Fisher, R.W. Clark, J. Davis, J. Giuliani, Jr.

Plasma Physics Division, Naval Research Laboratory

Washington, DC 20375

Abstract. Pure neon gas-puff solid fill nozzle shots fielded on HAWK at NRL produced surprisingly high K shell x-ray yield. The pinch duration in these shots was about 1 μs. These shots were in the low η regime (where η is the ion kinetic energy divided by the energy needed to reach the K shell) and it is clear that only a small fracton of the imploding mass reached the high temperature needed to produced the K shell radiation. In most of the cases, it is impossible to have a large pulse machine (τ-pinch > 1 μs) work in a large-η regime because of the large radius and low density requirements. The results on HAWK showed that it is possible to get good K shell yield even with over-massed loads. A possible explanation for the HAWK results might be the shock wave evolution in the gas column. The shock wave runs ahead of the magnetic piston and heats the center and then the piston compresses a hot core. In z pinches, the cylindrical convergence enhances the shockwave heating of the center. The results of these experiments and of a numerical parameteric study are presented.

The mass distribution in the z pinch column and the electrical parameters of the driver determine when the collapse of the pinch on the z axis will occur. It is desirable to have the collapse happen near the peak of the current pulse, and so the electrical and hydrodynamic parameters have to be matched. For a shell driven by a current with a ramp shape that peaks at time τ, the following relation holds between the mass per unit length M, the initial radius r_0, and the linear rate of current rise di/dt:

$$M \sim (\mathrm{d}i/\mathrm{d}t)^2 \times \tau^4/r_0^2.$$

Many of the existing z-pinch machines have a short pulse (typically $\tau \sim 100$ ns). Loads imploded on many of these machines can impart enough energy to all ions in the pinch to strip them to the desired ionization stage and still have some extra energy for radiation production. This is what is meant by $\eta > 1$. In long pulse machines ($\tau \sim 1$ μs), the $\eta > 1$ condition is practically impossible to achieve. Both very large radius and low initial density are required to meet this condition, and such pinches have been found both theoretically and experimentally to be hopelessly unstable. For reasonable radii and a machine like ACE4, a matched pinch would have about ten times the mass required for $\eta > 1$ in argon.

Recent neon results obtained on HAWK, employing a 2.5 cm radius solid-fill nozzle, $\tau \sim 1$ μs, and $\eta < 1$, showed K shell radiation yields which were at least as good as the best high-η shots on HAWK, where a plasma opening switch (POS) had been used to reduced the pulsewidth to $\tau \sim 200$ ns. Figure 1 summarizes these results, comparing the yield from the POS shots with the new long-pulse results. Figure 2 shows the current for the solid-fill nozzle shot (solid line), along with output from an XRD filtered to transmit neon K shell radiation. It is clear that in these over-massed shots, only a small fraction of the neon atoms were heated to the temperature needed for K shell radiation. Many groups have seen similar behavior,

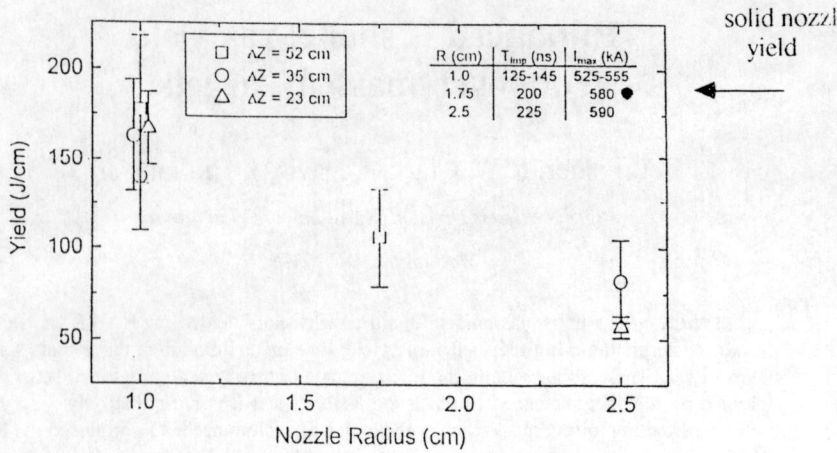

Figure 1. Summary of the POS (200 ns) and solid nozzle (1 μs) shots on HAWK

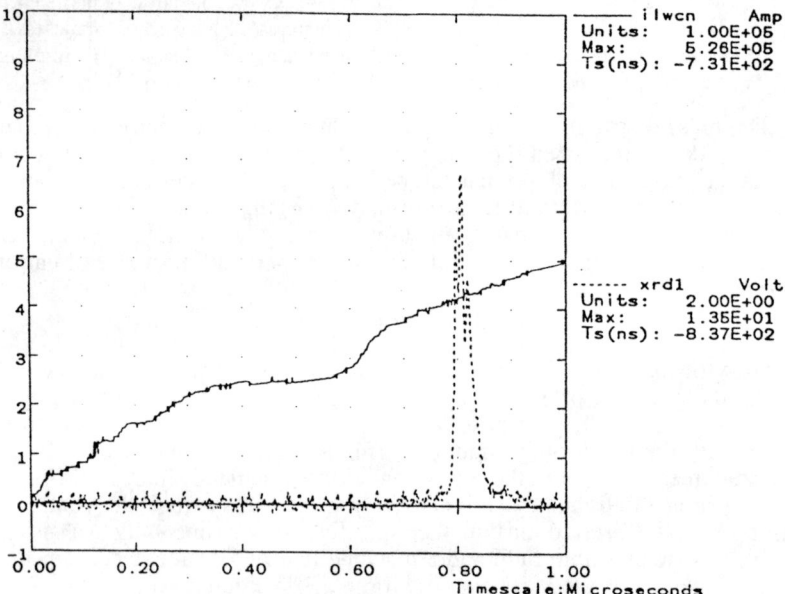

Figure 2. Current (solid line) and shape of x ray pulse (dotted line) for the 5 cm diameter solid-fill nozzle

where only a portion of the total pinching mass participated in the production of K shell radiation.

In an optimized, over-massed z-pinch load, most of the pinch material should stay cold, stagnating against a hot core. Tailoring of the initial density and temperature profiles of the load was considered and studied numerically. Altogether, twelve argon gas load runs were investigated. The simulations were divided into four groups, each consisting of three members. Within each group the initial density profile was held the same and the initial temperature profile was varied.

For the simulations, the pinch was assumed to consist of two uniform parts, each of which could have a different density and temperature: an inner core from $r=0$ to 3 cm and an outer region from $r=3$ to 5 cm. In all cases, the outer region was initially at temperature 0.1 eV and a density ρ_0, chosen so that the pinch would occur at peak current.

The first group was a solid-fill (constant-density) load with 5 cm radius. In the second, third and fourth groups, the initial core density was $\rho_0/3$, $\rho_0/10$ and $\rho_0/30$, respectively. The initial core temperature of the three members of each group was 0.1 eV (i.e., uniform throughout the pinch), 1 eV and 10 eV. Table 1 summarizes the parameters and the results (K shell yield) of each of the runs. The driving current for a typical run is shown in Figure 3.

Table 1. Summary of the simulation runs

ρ	ρ − core	T eV	$(T-$ core$)$ eV	K shell yield (kJ)
ρ_0	ρ_0	0.1	0.1	0
ρ_0	ρ_0	0.1	1.0	0
ρ_0	ρ_0	0.1	10	0
ρ_0	$\rho_0/3$	0.1	0.1	0
ρ_0	$\rho_0/3$	0.1	1.0	0
ρ_0	$\rho_0/3$	0.1	10	0.31
ρ_0	$\rho_0/10$	0.1	0.1	0.18
ρ_0	$\rho_0/10$	0.1	1.0	0.17
ρ_0	$\rho_0/10$	0.1	10	0.79
ρ_0	$\rho_0/30$	0.1	0.1	2.20
ρ_0	$\rho_0/30$	0.1	1.0	2.22
ρ_0	$\rho_0/30$	0.1	10	3.12

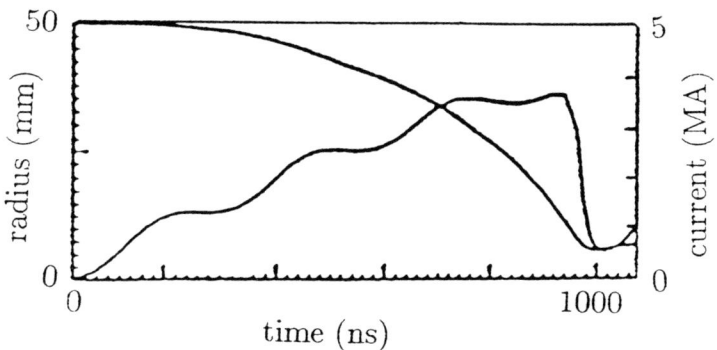

Figure 3. Typical simulation current and radius a function of time

The best K shell yield found by the simulations was for the case where a hot, low-density core occupied the center of the load. A hot core can be realized by more than one method. One can use a laser or a controlled discharge to achieve a few eV temperature. The hot core will expand and its density will drop. Temporal shaping of the driving current (analogous to laser pulse shaping for inertial fusion) can also produce a hot, low-density core. No attempt has been made to optimize the yield in these simulations, and it is clear that additional studies would be beneficial.

REFERENCES

1. Peterson G.G., et al., "Neon PRS Experiments on HAWK," 1996 IEEE International Conference on Plasma Science, Boston, MA.

---

† Work supported by the US Defense Special Weapsons Agency.

Density, temperature and size of a plasma produced in single and double shell liner implosions

S.A.Chaikovsky and S.A.Sorokin

Institute of High Current Electronics
4 Akademichesky Ave., 634055 Tomsk, Russia

Abstract. Single and double shell plasma liner implosion experiments have been performed on the SNOP-3 pulsed generator ($I_{max} \approx 1$ MA, $\tau \approx 80$ ns). The plasma density and temperature were determined with a technique involving collisional radiative equilibrium calculations and measurements of the K-shell radiation power, the K-shell radiating plasma size, and the He-α to Ly-α line ratio. The resonance-to-intercombination and resonance-to-satellite line ratios were also used to determine the plasma density and the temperature, respectively. It is shown in the paper, that all the mass of the inner shell of the double shell liner is assembled into a tight pinch and high radial compressions (≈ 45) of the inner shell occur at the appropriate inner shell initial radius.

INTRODUCTION

The implosion of plasma liners (Z-pinches) driven by a pulse power generator has long been of interest as a means for producing K-shell x-ray pulses ($h\nu \geq 1$ keV). The large scale Rayleigh-Taylor instabilities destroy the plasma liner uniformity, restrict its radial compression ratio, limit the plasma pinch density and, as result, decrease the x-ray production efficiency. To increase the radial compression ratio, some methods of plasma liner stabilization [1] can be used. Tight pinches of radius ≈ 50 µm, as measured with time-integrated x-ray pinhole cameras, were observed in double shell liner implosions. This corresponds to as high as a 100-fold radial compression of the inner shell of a double shell liner [1]. It was not clear whether the whole or only a small fraction of the original inner shell mass is compressed to a tight pinch. If all the inner shell mass is assembled into a tight pinch, the plasma pinch density should be high, resulting in high efficiency of the plasma thermal energy conversion to x-ray radiation.

The experiments performed on the SNOP-3 generator were succeded in producing higher argon K-shell power and yield from double shell implosions as compared with the maximum K-shell power and yield from single shell ones [2]. In this paper, plasma density and temperature measurements in single and double shell implosions are presented and analyzed. The key question needed to be clarified is whether all the inner shell mass of the double shell liner is assembled into a tight pinch and high radial inner shell compressions do occur.

The measurement technique described in detail in Ref. 3 was used to determine self-consistently both spatially and temporally averaged plasma ion density and electron temperature. This technique is based on comparing measurements of the K-shell radiation power P_k and the He-α/Ly-α line intensity ratio with their calculated values using a steady-state collisional radiative equilibrium (CRE) model for an isothermal, iso-dense plasma column with the radius equal to the K-shell emitting pinch radius r_k measured with a pinhole. The CRE code described in Ref. 4 was used. The K-shell radiating mass m_k (i.e., the mass concentrated within the pinch of radius r_k) was obtained using the measured ion density N_i and K-shell emission region radius r_k and compared to the inner shell mass m_2. The mass m_2 was inferred as follows. The outer shell mass m_1 was determined by comparing its measured implosion time (when the inner shell was absent) with that calculated using a 0-dimensional code. Then the inner shell mass m_2 was found from the ratio of the throat cross section areas of the inner and outer shell nozzles (both shells were formed with a gas puff).

The plasma ion density ($N_i^{R/IC}$) was also determined using the resonance-to-intercombination line ratio [5] with the effective ion charge being ≈ 16. The He- and H-like resonance-to-satellite line ratios were used to estimate (using a coronal model) the electron temperature (T_e^{He} and T_e^{H}, respectively). The radiation opacity effects, as the CRE calculations showed, were enough low to allow these techniques to be applied.

EXPERIMENTAL SETUP

The SNOP-3 generator was operated at a 0.9 MA peak current with a rise time of 80 ns. Both single and double shell liners of length 1.5 cm were formed using a

TABLE 1. Initial parameters of liners and measured powers P_k and radii r_k

Implosion No	r_1, cm	r_2, cm	m_1, µg/cm	m_2, µg/cm	P_k, GW	r_k, cm
1	0.6	–	10–15	–	3.4	0.013–0.015
2	0.75	–	8–12	–	12.4	0.04–0.05
3	1.3	0.3	4–5	1.3–1.5	12.4	0.011–0.013
4	1.3	0.4	4–5	1.3–1.5	21.0	0.011–0.013
5	1.3	0.6	4–5	1.3–1.5	5.2	0.04–0.05

r_1 is the initial radius of a single shell liner or of the outer shell of a double shell liner. r_2 is the initial radius of the inner shell of a double shell liner. m_1 is the mass per unit length of a single shell liner or of the outer shell of a double shell liner. m_2 is the mass of the inner shell of a double shell liner.

fast gas valve coupled to supersonic nozzles. The initial radii of the single shell or the outer shell of a double shell liner (r_1), the mean initial radii of the inner shell (r_2), the outer and inner shell masses per unit length (m_1 and m_2, respectively) are listed in Table 1. An x-ray diagnostic set to observe the argon K-shell emission ($hv \approx 3$–4 keV) included (1) x-ray diodes with an aluminum cathode and teflon filters of thickness 12 and 18 µm to measure the K-shell power P_k and yield, (2) time-integrated filtered pinhole cameras to measure the radius of the K-shell radiating pinch r_k, and (3) an x-ray spectrograph with a convex mica crystal to observe the K-shell radiation spectrum.

RESULTS AND DISCUSSION

The measured K-shell radiation powers P_k and the K-shell radiating pinch radii r_k are included in Table 1. Table 2 summarizes the plasma parameters determined in the experiments. Plasma density and temperature measurements was not performed for single shell liners with initial radii of 1.1 and 1.3 cm because they never assembled to tight high-density pinches. In general, the ion densities $N_i^{R/IC}$ derived from the resonance-to-intercombination line ratio are higher than the densities N_i obtained with the CRE analysis. There is a good agreement between the electron temperature $T_e^{He/H}$ obtained by CRE analysis and the electron temperatures T_e^{He} and T_e^{H} for single shell implosions No. 1 and No. 2, and for double shell implosion No. 5 (see Table 2). The electron temperatures T_e^{H} are always slightly higher than T_e^{He} because the maximum emissivity of H-like ions corresponds to higher electron temperatures. The temperatures $T_e^{He/H}$ are well between T_e^{He} and T_e^{H}.

TABLE 2. Measured plasma parameters.

Implosion No	N_i, 10^{19} cm^{-3}	$T_e^{He/H}$, keV	$N_i^{R/IC}$, 10^{19} cm^{-3}	T_e^{H}, keV	T_e^{He}, keV	m_k, µg/cm
1	1.3–1.7	1.2	6–8	1.1–1.5	0.9–1.0	0.5–0.6
2	0.7–0.9	1.25	≤ 6	1.2–1.4	1.0–1.2	3.1–3.7
3	2.2–2.7	1.65	15	0.9–1.0	0.9–1.0	0.7–0.9
4	2.9–3.5	1.5	8–13	1.2–1.4	1.0–1.2	0.9–1.1
5	0.6–0.7	1.05	6–12	–	0.9–1.1	2.4–2.9

N_i and $T_e^{He/H}$ are the ion density and the electron temperature, respectively, measured with CRE analysis. $N_i^{R/IC}$ is the ion density found from the resonance-to-intercombination line ratio. T_e^{H} and T_e^{He} are the electron temperatures found from the satellite-to-resonance line ratio of H-like and He-like ions, respectively. m_k is the mass of the K-shell radiating plasma.

Double shell implosions at the $r_2 = 0.3$ and 0.4 cm produce lower radius and higher density pinches as compared with single shell implosions (see Table 2). The maximum ion density of $(2.9-3.5)\times10^{19}$ cm^{-3} was observed at an approximately 45-fold inner shell radial compression (implosion No. 4). Approximately 70–80% of the inner shell mass was compressed into a tight K-shell emitting pinch of radius 0.011–0.013 cm. In this implosion, higher K-shell powers and yields were observed as compared with other double shells and with the maximum power and yield in single shell implosions (implosion No. 2).

A higher radius, and lower density and temperature plasma pinch was produced at $r_2 = 0.6$ cm (implosion No. 5), that resulted in low K-shell power and yield. The mass m_k exceeds the inner shell mass, indicating that mixing of inner and outer shells occurs, while in implosions No. 3 and No. 4 m_k does not exceed the inner shell mass. A possible explanation is that the azimuthal magnetic field of the generator current partially penetrating into the region between the shells affords shell separation in implosions No. 3 and No. 4 and provides a high inner shell radial compression ratio, as the axial magnetic field does (even at $r_2 = 0.6$ cm) when it is entrained [1].

The electron temperature $T_e^{He/H}$ was measured to be higher than both T_e^{He} and T_e^H in implosions No. 3 and No. 4 (see Table 2). The reason for this difference is not clear. The only remark should be done that the temperature $T_e^{He/H}$ is strongly dependent on the ionization state, while T_e^{He} and T_e^H are not, since $T_e^{He/H}$ is determined from the intensity ratio for lines that belong to ions being in different ionization states. Hence, non-equilibrium ionization (an "overcooled" plasma) state is possibly produced. It is important, that the CRE calculations with the electron temperature $T_e = 0.5(T_e^{He} + T_e^H)$ yield ion densities being a factor 1.6 (implosion No. 3) and 1.3 (implosion No. 4) higher than those obtained at $T_e = T_e^{He/H}$ and, hence, higher K-shell masses m_k. This results in rather good agreement of K-shell radiating mass with the inner shell mass.

CONCLUSION

Plasma densities, temperatures, and sizes have been measured in single and double shell argon liner implosion experiments on the SNOP-3 generator. The ion densities obtained in double shell implosions are a factor of 2–4 higher than in single shell implosions, resulting in higher K-shell powers and yields. It is shown that all the inner shell mass is assembled into a pinch of radius 0.011–0.013 cm, that is an approximately 45-fold radial compression of the inner shell occur at the inner shell initial radius of 0.4 cm. Thus, a high radial compression ratio of the inner shell of the double shell liner has been demonstrated. The maximum ion density of $(2.9-3.5)\times10^{19}$ cm^{-3} providing the maximum K-shell radiation power and yield in double shell implosions is obtained for the inner shell initial radius $r_2 = 0.4$ cm. A higher inner shell initial radius results in mixing of the outer and the inner shells and in a lower radial compression ratio, and thus, the lower density and temperature plasma is produced.

REFERENCES

1. Sorokin, S.A., and Chaikovsky, S.A., *Fizika Plazmy* (in Russian), **19**(7), 856 (1993).
2. Sorokin, S.A., and Chaikovsky, S.A., "K-shell radiation power and yield from double shell plasma liner implosions", in *these Proceedings*.
3. Coulter, M.C., Whitney, K.G., and Thornhill, J.W., *J. Quant. Spectr. Radiat. Transfer*, **44**, 443 (1990).
4. Oreshkin, V.I., and Loskutov, V.V., Preprint No 5 of HCEI (in Russian) (1991).
5. Vinogradov, A.V., Skobelev, I.Yu., and Yukov, E.A., *Usp. Fiz. Nauk* (in Russian), **129**(2), 177 (1979).

About plasma points' generation in Z-pinch

V.I. Afonin, A.V. Potapov

Russian Federal Nuclear Center - Research Institute of Technical Physics
PO Box 245, Snezhinsk, Chelyabinsk region, 456770, Russia

and V.P. Lazarchuk*, V.M. Murugov*, A.V. Senik*

*Russian Federal Nuclear Center - Research Institute of Experimental Physics
Sarov, Hizhnii Novgorod region, 607190, Russia

The streak tube study results (at visible and x-ray ranges) of dynamics of fast Z-pinch formed at explosion of metal wire in diode of high current generator are presented. Amplitude of current in the load reached ~180 kA at increase time ~50 ns. The results' analysis points to capability of controlling hot plasma points generation process in Z-pinch.

Introduction

Despite long-term researches of fast Z-pinches, interest to them is not abated, that is connected with perspective of their use as a compact and inexpensive soft x-rays' source with wide range of application both in scientific researches and in technology. However as against laser plasma, which is originally pointlike, the Z-pinch plasma represents extended linear object, poorly suitable for the purposes of microscopy and microelectronics without the special measures being undertaken. To those, for example, the methods of formation of hot plasma point in given place of pinch can be referred.

In the given work the preliminary results of study of possibility of plasma points formation at pre-established place of electrically exploded composed wire are presented.

Experimental results

The experiments on explosion of wires were conducted on the high current generator SIGNAL with inductive energy accumulator /1/. As a load of the generator there were used the cylindrical targets with length 8 mm and diameter 20..25 micron, whose central part with length 2 mm was made from aluminum and edges - from tungsten.

The time scanning of Z-pinch emission at x-rays with energy of quantum > 150eV was implemented by means of x-ray streak tube /2/. The measurement channel includes streak tube in the mode of slit scanning and slit-hole camera building pinch's image on photocathode of the streak tube. Registration of image from streak tube screen was made by film. Spatial resolution (along pinch's axis) makes 0.15 mm and resolution in time - 0.5 ns.

Registration with resolution in space and time of three Z-pinch's cross sections at optical range was implemented by means of streak tube in the mode of slit scanning also. PS images are arranged relative to the input slit so that middle one finds the slit with its central part and the two other - with shift ±2 mm along PS axis. The streak tube implements scanning in time of three PS cross sections with spatial resolution (across PS axis) 30 μm and resolution in time 0.35 ns.

The working x-ray photochronograms and photochronograms in optical range of emission of explosion of cylindrical composite target with ⌀20 microns are show on fig. 1,2 respectively. The explosion was executed by pulse of current with complex form: first local maximum of current in load reached $J_1 \approx 150$ kA with characteristic increase time $\tau_1 \approx 20$ns, during next $\tau_2 \approx 15$ns the current subsided down to $J_2 \approx 120$ kA and then increased up to its maximum value $J \approx 180$ kA.

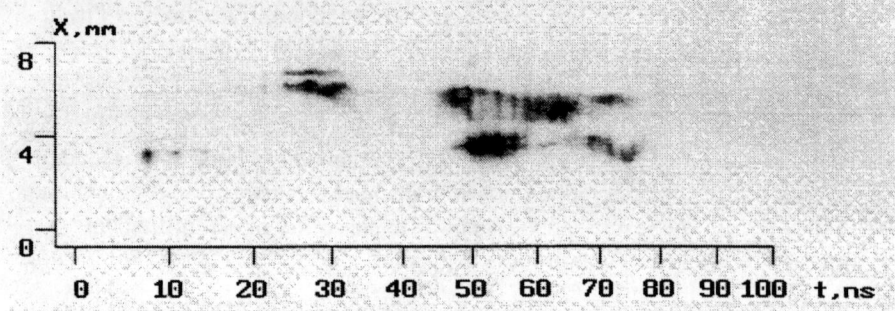

Fig. 1. X-ray photochronograms of composite ⌀20 μm target explosion.

As seen from x-ray photochronogram, in the center of aluminum insert at the time $t_c \approx 50$ ns, the plasma "point" with length ≈ 1 mm radiating in x-ray range of wave length during 10 ns was formed.

In its turn, the photochronograms fig. 2. shows that central cross section (Al) diameter goes through pulsation along sweep line, at the same time the diameters of the other ones, which fall on W, almost monotonously increase with time.

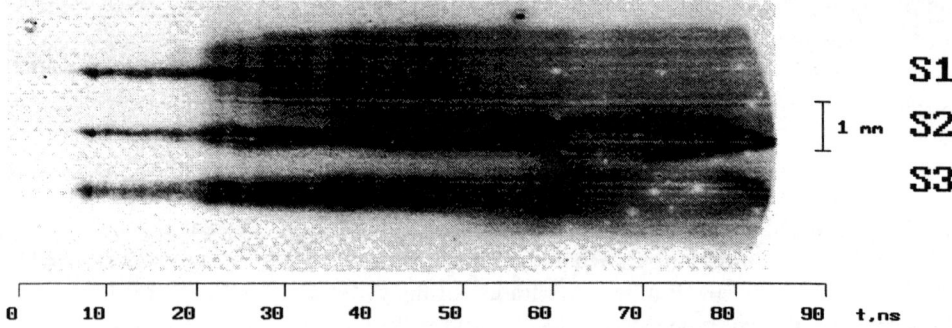

Fig.2. Working photochronograms of composite ⌀20 μm target explosion.

330

As seen from the presented figures, at the moment $t \approx 14$ns aluminum insert radius reaches first local minimum. Afterwards radius of Al-insert makes a number of oscillations, but tungsten radius monotonously increases.

Even though in this experiment we managed to form plasma "hot point" in the pre-set place of the pinch, still it was not single. Together with it others come up. The problem of following experiments was enhancing the single "point" generation effect. With this purpose the current pulse's parameters were changed at the same target's parameters: $\tau_1 \approx 20$ns, $J_1 \approx 150$ kA, $\tau_2 \approx 20$ns, $J_2 \approx 100$ kA, $\tau_3 \approx 30$ns, $J_3 \approx 180$ kA.

As follows from the x-ray photochronograms shown on the fig. 3, in this experiment factual single plasma "hot point" was formed at moment $t \approx 60$ns

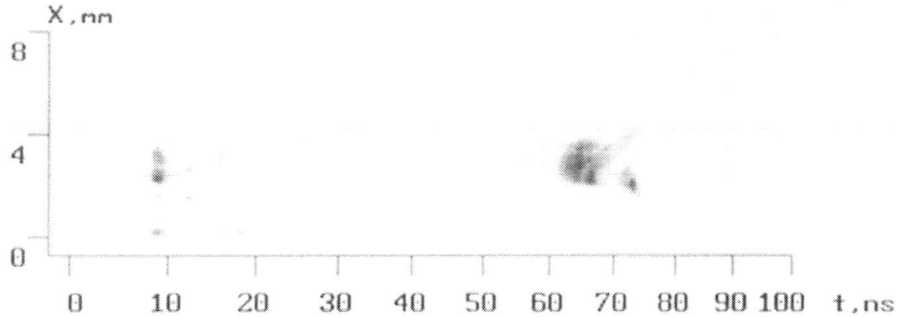

Fig.3. Working photochronograms of composite $\varnothing 20$ μm target explosion.

Discussion on experimental results

We shall consider electroexplosion of conductor, assuming uniformity of its properties and ones of the formed plasma both along length and on cross-section.

Let at the moment $t=0$ through the conductor, described in radius r_0 (cm), density ρ (g/cm^3), length l (cm), conductivity σ (cm$^{-1}\cdot$ohm^{-1}) and nuclear weight A, the current I begins to flow with increase rate β (A/s), and β=constant. We shall characterize the plasma by some average (on time of heating and mass) temperature T_0(eV), average charge of ions Z_0. Then it is possible to find characteristic time of expansion t_0 (s) of plasma string (PS) and the temperature of plasma to the moment t_0 in the form /3/:

$$t_0 \cong 2.5 \cdot 10^7 \cdot (r_0/\beta) \cdot [T_0 \cdot \rho/A \cdot (Z_0+1)]^{1/2}, \quad T_0 \cong 5.7 \cdot 10^{-5} \cdot [A^3/\rho(Z_0+1)]^{1/6} \cdot (\beta/r_0)^{1/6}.$$

For estimation of mean ion charge Z_0 crown equilibrium model can be used /4/: $Z_0 = (T_0/G \, \gamma)^{0.5}$, where $\gamma=1/\ln(10^8 \gamma^3/T_0)$, $G\approx$constant at ionization from shell with main quantum number **n**. At last the expansion radius can be found from formula: $r \approx 5.4 \cdot 10^9 (\rho^{1/3}/A^{1/2})(Z_0+1)^{5/6}(r_0/\beta)^{2/3}$. For obtaining further estimation on plasma collapse by magnet field the "snow plough" model /5/ can be used, according to which the moment of plasma collapse is defined by equation: $t_c \approx 4.6 \cdot 10^5 \, \rho^{5/12}(Z_0+1)^{5/12}(r_0/\beta)^{5/6}/A^{1/4}+t_0$.

At last we take into account the fact that in the similar (with parameters of current and targets) experiments /6/ at electroexplosion of wire the "cold" dense nucleus and the hot plasma corona were observed during long time, and approximately 3% of initial mass of a target passed into corona. It is equivalent to that initial density of a target in our estimations should be replaced: $\rho \to 0.03\rho$. With allowance for these facts, for the conditions of our experiments (β_1=7.5 10^{12} A/s) the estimation of the times of expansion and collapse of Al-target with $\varnothing$20 micron makes $t_0 \approx$6ns, $t_c \approx$15ns. For tungsten target of the same diameter the times of expansion and collapse increase and make 9ns and 30ns, accordingly. This estimation satisfactory agrees with the experimental data. So far as already after 20ns the rapid drop down of current begins so the tungsten plasma actually does not collapse. At the same time despite drop down of the current its magnitude is still enough for confinement of the collapsed aluminum plasma, joule heating of which leads farther on to formation of high temperature plasma point.

Not pretending for completeness of analysis of composed targets' electroexplosion process, the experimental results presented in the work and their interpretation point to principle possibility of controlling plasma points' generation process at fast Z-pinch.

In conclusion the authors render gratitude to maintenance personnel of the installation SIGNAL for technical maintenance of the experiments. The work was fulfilled at financial support by ISTC in frames of the project #009.

References
1. Afonin V.I., Gafarov A.M. et al. Proc. Int. Conf. «BEAMS-96». Prague 1996. P.691.
2. Lazarchuk V.P., Murugov V.M. et al. Plasma Physics, 1994, v.20, p.101.
3. Afonin V.I., Murugov V.M. et al. Proc. Int. Conf. «BEAMS-96». Prague 1996. P.697.
4. Vinogradov A.V., Shlyapcev V.N. Quantum electronics, 1983, v.10, p.509.
5. Luk'yanov S.Yu. *Hot plasma and controlled thermonuclear synthesis*, Moscow: Nauka, 1975.
6. Sarkisov G.S., Etlicher B. et al. JETPh Letters, 1995, v.61, p.547.

Z-pinch Discharges - Bare and Plastic Coated Copper Wires

F. N. Beg, J. Ruiz and A.E. Dangor

*The Blackett Laboratory, Imperial College
London SW7 2BZ, UK*

Abstract. We report on an experiment performed with copper wires on a small generator producing 150 kA in 60 ns with and without prepulse. A comparison is made of 10 μm copper wire with polyurethane coated copper wire of the same diameter. Optical probing shows that the core expands with a velocity of 3×10^3 ms^{-1} for main current discharges. Coronal plasma was not observed, probably the density was below the detection limit $\sim 6 \times 10^{16}$ cm^{-3}. In the prepulse discharges, optical streak photographs show that the corona expands radially at 10^5 ms^{-1}. The optical emission lasts for more than 300 ns for both type of wire. Coronal plasma of density 10^{18}-10^{19} cm^{-3} is observed for prepulse only and for prepulse with the main current discharges. X-ray emission is the same for both types of wire but discharges with prepulse show a larger number of bright spots. A delay of 10 ns in x-ray emission was observed for coated wires.

INTRODUCTION

It has been shown [1] that metallic wires and insulated fibres in Z-pinch discharges have different breakdown properties. It has also been found that the pinch behaviour is radically altered when a prepulse current is applied prior to the main current [2]. The purpose of these experiments is to study the initial breakdown behaviour of copper wire, 10 μm diameter, with and without a polyurethane coating, 5 μm thick, and to compare discharges.

EXPERIMENTAL DETAILS

The generator produced a fast rising current (60 ns, 10-90%) with peak of 150 kA. A separate capacitor circuit, producing a ringing current with a quarter period of 60 ns and peak current of 10 kA was discharged into the wire prior to the main current (at a pre-determined time). A description of the generator is given in Lorenz *et al.* [2].

A variety of diagnostics was used to study the pinch. An optical streak camera was used to study the radial expansion, axial structure and duration of the emission; the x-ray emission from the pinch was observed with a six channel time integrated pinhole camera with different filters and pinhole dimensions to obtain the spectral and spatial resolution; the time evolution of the spatial distribution of the electron density was obtained by Moire deflectometry [3] using a 7 ns ruby laser; the hard x-ray emission was investigated with a scintillator/photomultiplier detector.

RESULTS

Discharges without prepulse

Figure 1 shows Moire deflectograms taken in different discharges at various times during the main current. In the photographs in Fig. 1(a), which are for discharges in bare wires, we see that at 12 ns, the wire has expanded to a diameter of 83 µm which corresponds to a expansion velocity of 3×10^3 ms^{-1}. Perturbations are seen at 22 ns and these develop into well formed m=0 instabilities by 78 ns. Density islands start to appear at 100 ns. The observations with coated copper wires show similar behaviour (Fig. 2(b)), though results seem to indicate that instabilities grow faster. No coronal plasma was observable for either type of wire. Perhaps the density in the coronal plasma was below the detection limit of 6×10^{16} cm^{-3}.

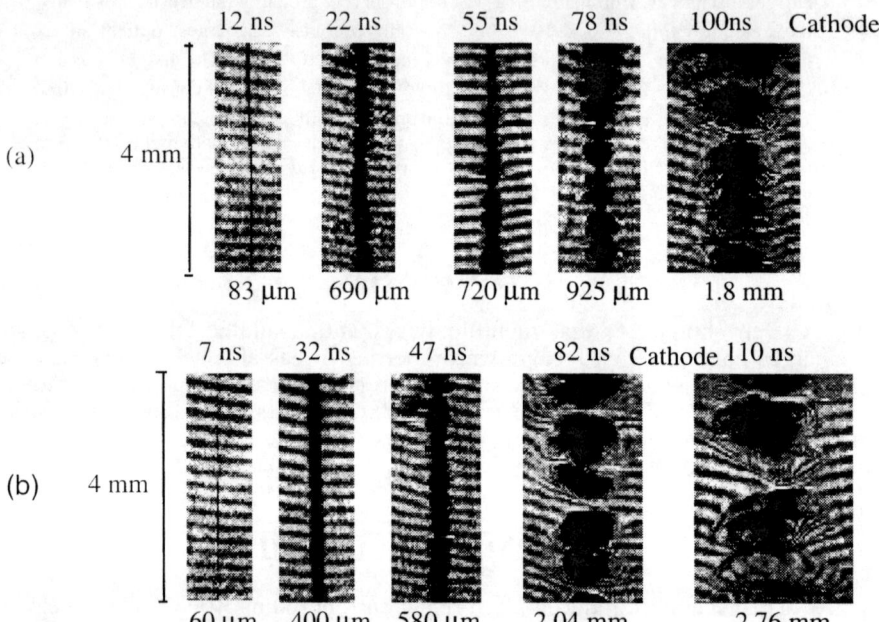

Figure 1. A series of Moire deflectograms for different discharges for a) 10 µm copper and b) 10 µm copper wire coated with 5 µm polyurethane. The time of observations and average diameter of the plasma column are given above and below each photograph.

X-ray emission is observed to be from a series of bright spots distributed along the axis of the pinch as shown in Fig. 2. Discharges with bare and coated wires show similar behaviour. Comparison of images E & F shows that the x-ray emission is mainly in the spectral range of 0.7-1.5 keV (copper L-shell transitions exist in this spectral range). X-ray emission from the anode is also visible in Fig. 2.

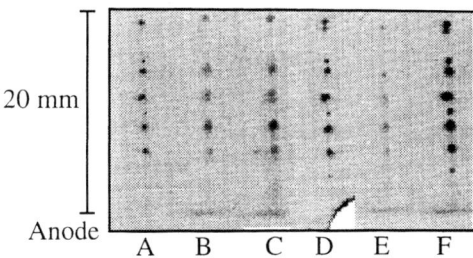

Figure 2. X-ray time integrated pinhole photograph for coated wire with filter and pinhole size, (A) 12 μm Be/50 μm, (B) 5 μm Fe/300 μm, (C) 3 μm Cu/300 μm, (D) 24 μm Be/100 μm, (E) 5 μm Fe/200 μm and (F)10 μm Al/200 μm).

Discharges with prepulse

The pinch behaviour is observed to change when a prepulse current was applied 100-200 ns before the main current. Figure 3 shows radial streak photographs of the prepulse and the main current with prepulse. The streak photograph shown in figure 3(b) was taken with a ND=1 neutral density filter, showing that the emission increases greatly with the main current. The emission appears to start from a much smaller diameter than that observed in the prepulse and expands radially at ~ 10^5 ms^{-1}. Also, Fig. 3(b) shows that the emission is delayed by about 20 ns after start of the main current. Discharges in bare and coated wires behave similarly.

A delay in the emission has also been observed with prepulse discharges in carbon [2] and CD$_2$ [4] fibres. However, in our bare and coated copper wire discharges "zippering" is not observed.

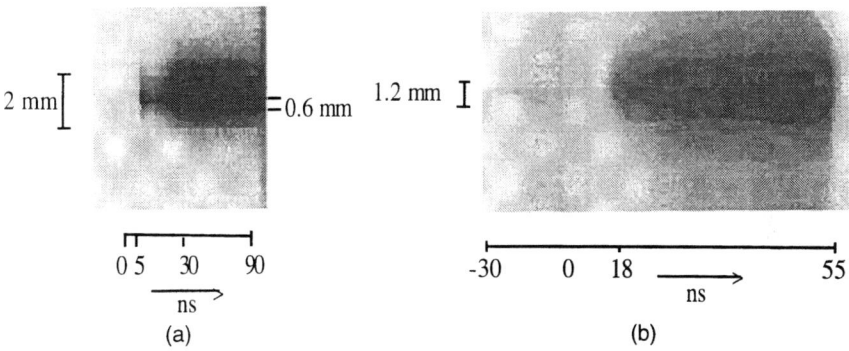

Figure 3. Optical radial streak of 10 μm copper wire a) prepulse only [t=0 initiation of prepulse current] and b) main current with prepulse 107 ns before the main current [t=0, initiation of main current]. The photograph shown in 1(a) is taken without any neutral density filter. The streak photograph in (b) is with a neutral density filter Nd=1.

Figure 4 shows a series of Moire deflectograms. The first taken 177 ns after the start of prepulse current, a), with bare copper wire and a second set at 154 ns, b) with coated wire. Both sets show a core surrounded by a low density corona. The corona is very uniform and has a density ~ 10^{18} cm^{-3}. The core and corona expand radially: at 22 ns in Fig. 4(a), the corona to a diameter of 420 μm and the core to 210 μm. At 100 ns, density islands are observed, but with a diameter of 1.26 mm which is smaller than the density islands without prepulse (c.f. Fig. 1(a)). Coated wires show similar behaviour, although instabilities appear to grow faster. Density islands are seen to emerge at 67 ns and appear to grow to a larger size.

It is of interest to note that the core and surrounding corona expands radially which at early time is at 10^4 ms^{-1}. This velocity is about an order of magnitude smaller than the expansion velocity obtained from the optical streak photographs.

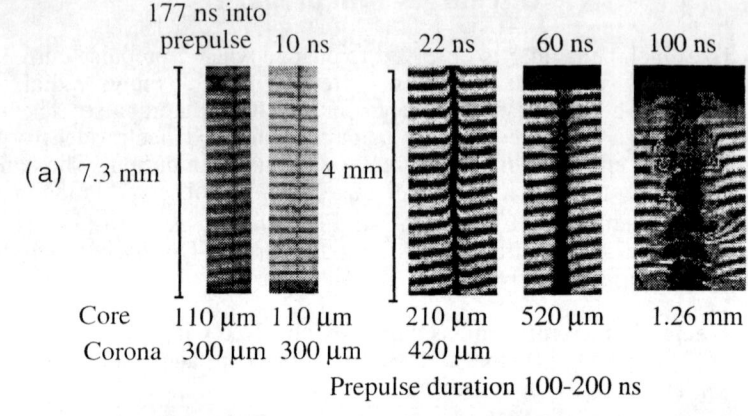

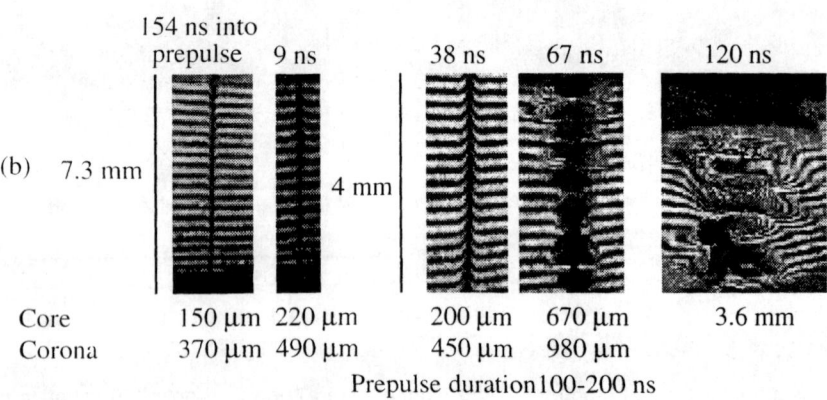

Figure 4. Moire deflectogram with prepulse in (a) 10 μm copper wire and (b) 10 μm copper wire coated with 5 μm polyurethane.

In discharges with prepulse, the x-ray emission is also from a series of bright spots distributed along the axis of the pinch. Copper and coated wires show similar results. However, here, the bright spots are intense and larger in number. As in discharges with main current only, the x-ray emission is mostly in the 0.7-1.5 keV range. PIN diode signals show that the x-ray emission starts at 35 ns for copper and 45 ns for coated wire.

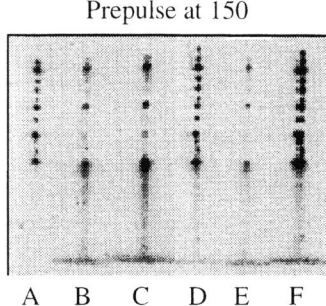

Figure 6. X-ray time integrated pinhole photographs for coated wires for a discharge when the start of current is 150 ns after the prepulse current. The pinhole diameters and filters are as in figure 2.

SUMMARY

Our observations show that discharges in bare and coated copper wires exhibit similar behaviour without prepulse. The column shows expansion (3×10^3 ms^{-1}), m=0 instabilities develop and density islands are formed. The x-ray emission is from bright spots along the axis mainly in the spectral range 0.7-1.5 keV. In discharges with prepulse, discharge behaviour is different if the main current is switched at 100 - 200 ns after the prepulse. A delay in the optical emission is observed, the corona expands faster $\sim 10^5$ ms^{-1} and a low density corona is observed to surround the core. However, the radial expansion observed in the Moire deflectograms appears to be about an order of magnitude smaller than seen in the optical streaks. Coated wire discharges appear to develop instabilities somewhat faster but otherwise discharges are similar to those with bare wires. For discharges without prepulse the x-ray emission is from a series of bright spots, but the spots are brighter and are larger in number. X-ray emission is mainly in the 0.7-1.55 keV spectral range.

It is interesting to note that coating the wire with plastic does not appear to change the behaviour of the discharge, either with or without prepulse. This is surprising, since in carbon [1[and CD_2 [4] fibre discharges, instabilities appear to develop much earlier.

REFERENCES

1. F. N. Beg *et al.*, Plasma Phys. Contr. Fusion **39**, 1 (1997).
2. A. Lorenz *et al.*, these proceedings
3. J. Ruiz *et al.*, these proceedings
4. S. Lebedev *et al.*, these proceedings

X-ray Emission from a Small 2 kJ Plasma Focus

F. N. Beg, I. Ross* and A. E. Dangor

Blackett Laboratory, Imperial College, London SW7 2BZ, UK.
*AWE, Aldermaston, UK.

We report on a study of a 2 kJ, 200 kA plasma focus device as an x-ray source. The x-ray yield from a number of pure gases, deuterium, nitrogen, neon, argon, and xenon, was measured as a function of pressure. X-ray emission is mainly due to line radiation. Maximum x-ray yield of 12.5 J obtained for neon. At lower pressures, electron beams are generated which play an important role.

INTRODUCTION

The plasma focus is a cost effective and compact source of intense soft x-rays, neutrons, ion and electron beams. There is a renewed interest in the soft x-ray emission from a plasma focus as a source for lithography and microscopy. Another potential application is as a backlighter for high density plasmas which is the main motivation of this study.

X-ray emission from a plasma focus depends on current, composition of gases and the anode length. It is well known that when operated in a mixture of deuterium and admixtures of nitrogen, neon, argon, or xenon, the x-ray emission is strongly enhanced, and the pinch behaviour is modified [1-2]; yield depends on the particular gas used and its mass density percentage. It has been reported[3-4] that 10% of the capacitor energy can be converted to x-rays.

In this paper we report a study of x-ray emission from different Z gases viz. deuterium, nitrogen, neon, argon, and xenon, from a small 2 kJ Mather type micro plasma focus designed by Professor Decker of the University of Dusseldorf.

EXPERIMENTAL DETAILS

The micro plasma focus is powered by a 2.6 µF capacitor which can be charged to a maximum voltage of 40 kV. The inner electrode is hollow and has a diameter of 20 mm and length 61 mm. The inner electrode is surrounded coaxially by eight, equally spaced 10 mm diameter, 90 mm long copper rods at a radius of 45 mm connected to the positively charged terminal of the capacitor via spark gap. A schematic of the device is shown in figure 1. For the experiments reported here the capacitor was charged to 38 kV, giving a peak current of 175 kA. The 1/4 period of current is 800 ns.

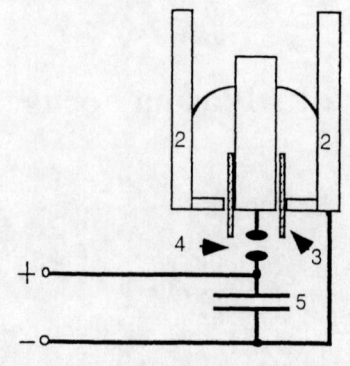
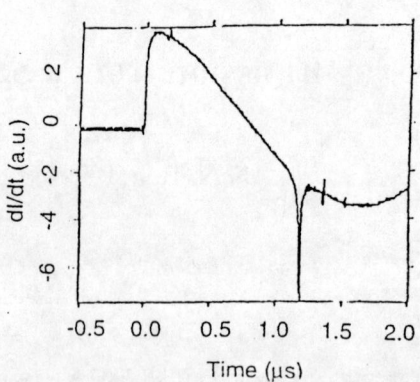

1 Anode, 2 Outer electrodes (8 rods)
3 Insulator sleeve, 4 Spark gap, 5 Capacitor

Figure 1. Schematic of Plasma Focus.

Figure 2. A typical dI/dt waveform, the dip in dI/dt shows the pinching of the plasma.

Figure 2 shows a typical dI/dt trace obtained with a Rogowski coil. The dip in the current shows the focusing and pinching of the plasma at the tip of the anode. For the maximum x-ray signal the dip in dI/dt occurs at 1.15 ± 0.05 μs.

Four 1 cm^2 Quantrad PIN diodes were employed for the x-ray detection, having an intrinsic layer of 250 μm for softer x-rays and 500 μm for harder x-rays. Two PIN diodes were positioned at 57 cm from the pinch to view the plasma a few mm above the anode. Two other PIN diodes were placed at a distance of 37 cm from the pinch at an angle of 45° to view a few mm inside the anode. Another PIN diode with a 50 μm Be filter was placed at a distance of 102 cm from the pinch to observe change in x-ray signals for the different gases. The PIN diodes were cross-calibrated using the same filter. In addition, images of the pinch were recorded on DEF film with a six channel time integrated x-ray pinhole camera. The pinhole sizes were selected and a set of filters chosen to obtain spatial and spectral information. For example, L-edges of high Z materials were used to obtain the information about K-shell emission of neon and argon. The hard x-rays with $h\nu > 100$ keV were detected with a scintillator/photomultiplier setup.

RESULTS

For deuterium, two dips in dI/dt were observed separated by 100 to 200 ns, with corresponding x-ray signals on the PIN diodes. A dip in dI/dt is usually associated with a pinching "focusing" of the plasma column. This is due to the rapid increase in the inductance associated with the pinch. For nitrogen multiple dips over a wide range of filling pressure were observed, While for neon and argon, multiple dips were observed at high pressures only. In contrast, a single focus was observed over a wide range of pressures for xenon.

Figure 3 shows the time integrated pinhole images for different gases. The first image for all the gases shown was recorded with a 2 μm Al filter and a 50 μm diameter pinhole. Filters and pinhole diameters for the other images are shown in table 1. For deuterium, a bright spot is observed near the tip of the anode and faint plasma column extending away from the anode. For nitrogen and neon, a 6-10 mm long plasma column with diameter less than a mm is observed. The x-ray images in argon and xenon show hot spots. The x-ray image for argon with soft filters shows a plasma column but with harder filters which transmit higher energy line radiation always show hot spots. There is a transition of the plasma from a column to a series of hot spots for Z≥18. A similar observation was made by Lebert et al. [5] for different gases.

For neon gas, the 10 μm Al and 5 μm Fe filtered images, 5 and 6, in figure 3(c), show a significant difference in x-ray yield. This clearly demonstrates that most of the emission is in the spectral region of 0.7-1.5 keV. Similarly for argon, images 2 & 3 in figure 3(d) were obtained with 4 μm Ag and 5 μm Mo filters. This filter pair isolates the argon K-shell emission, which is from the hot spots. The images 2 and 6, figure 3(d) are pinhole limited, from which we deduce that the diameter of the hot spots emitting K_α radiation is smaller than 200 μm. Images with softer filters show column like structure but with harder filters hot spots are observed. This is in contrast to what was observed with the soft and hard filters for neon. Similarly, the images for xenon indicate that the emission is due to M-shell transitions and the K-shell emission being absent due to the temperature not being high enough. The size and structure of the hot spots depend on the spectral region being viewed. A larger number of hot spots are observed for xenon than for argon.

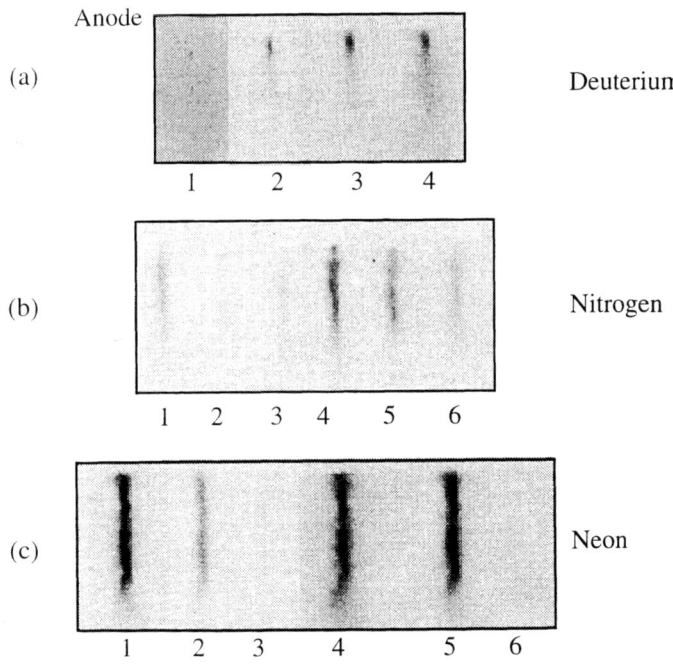

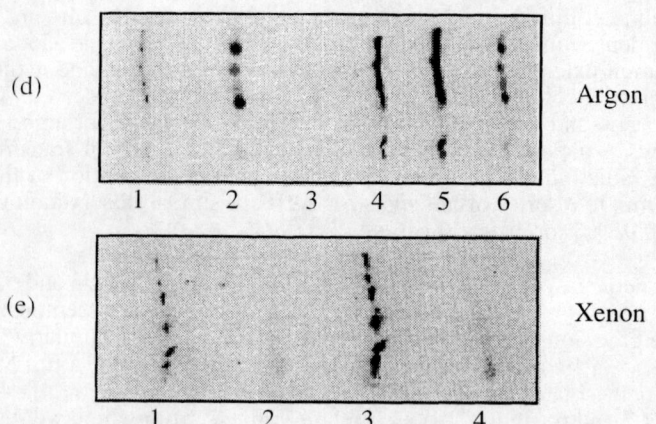

FIGURE 3. Time integrated pinhole photographs through various filters listed in table 1. The first image is with 2 μm aluminium filter and 50 μm diameter pinhole.

TABLE 1. Filters and pinhole sizes used to obtain the images in fig. 3 for different gases (number in parentheses is the pinhole diameter).

Gas	Image 2	Image 3	Image 4	Image 5	Image 6
Deuterium	12 μm Be (100 μm)	24 μm Be (200 μm)	1 μm Ti (200 μm)		
Nitrogen	10 μm Ti (300 μm)	3 μm Cu (300 μm)	12 μm Be (100 μm)	24 μm Be (200 μm)	20 μm Al (200 μm)
Neon	3 μm Cu (300 μm)	5 μm Fe (300 μm)	12 μm Be (100 μm)	10 μm Al (200 μm)	5 μm Fe (200 μm)
Argon	4 μm Ag (300 μm)	5 μm Mo (300 μm)	12 μm Be (100 μm)	10 μm Al (200 μm)	3 μm Cu (200 μm)
Xenon	10 μm Cu 300 μm	12 μm Be 100 μm	10 μm Ti 200 μm		

The x-ray yield calculated from the filtered PIN diode traces for the given spectral range are shown in table 2 (for neon, argon and xenon). The duration of x-ray signals was 10-15 ns. The spectral range which corresponds to K-shell transitions in neon and argon, M-shell transitions in xenon was isolated using the same filters as those used to obtain the time integrated pinhole images in Fig. 3. The total yield was obtained from a PIN diode filtered with 50 μm Be. The total x-ray yield measured for neon and argon was 12.5 and 0.031 J respectively.

TABLE 2. The x-ray yield for neon, argon & xenon and spectral energy range.

Gas	Optimum pressure (mb)	Energy range (keV)	Yield (J)	Pinch type
Neon	1.0	0.7-1.5	4.6	Column
Argon	0.6	2.5-3.8	0.007	hot spots
Xenon	0.1	3.36-4.9	0.0008	hot spots

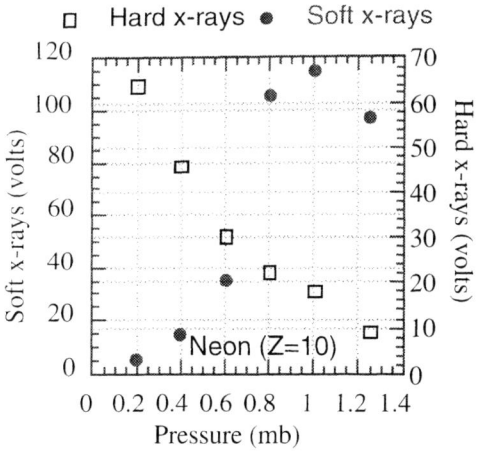

Figure 4. Side-on PIN diode signals (50 μm Be filter) & PIN diode at 45° (10 μm Cu filter) for neon.

It was also observed that for lower Z gases for example deuterium, nitrogen, neon, the x-ray signal was higher on the PIN diode at an angle of 45° even though it was filtered for harder x-rays compared with the side-on PIN diodes. Figure 4 shows that for neon, the x-ray signal on a side-on PIN diode filtered with 50 μm Be (the filled circles) increases with the filling gas pressure but the signal on PIN diode at 45° (the open squares) decreases. This was also observed with the photomultiplier for hv > 100 keV. This indicates that at low pressure electron beam is generated which produces hard x-rays.

SUMMARY

The x-ray emission from various gases in a plasma focus has been investigated. Observations show that for gases Z≤ 18 a plasma column like structure is formed, and for Z≥18, hot spots occur. In argon, the hot spots emit K-shell radiation whereas lower energy x-rays are from a plasma column. This is consistent with the observations of Lebert et al. [5]. The pinch gets tighter with increasing gas Z-number. At low pressure, hard x-rays are dominant indicating the presence of an electron beam. The x-ray yield is greater in neon, the yield being 12.5 J at an optimum pressure of 1.0 mbar. It is interesting to note that this emission has duration of 10-15 ns, giving an x-ray output power of 830 MW.

REFERENCES

[1] K.N. Koshelev et al., J. Phys. D: Appl. Phys. **21**, 1827 (1988).
[2] P. Choi et al., Proceedings of third International Conference on Dense Z-pinches **299**, 288 (1993).
[3] R. Prasad et al., SPIE **2194**, 120 (1994).
[4] Fillipov et al., Phys. Lett. A **211**, 168 (1997).
[5] R. Lebert, A. Engel. and W. Neff, J. Appl. Phys. **78**, 6414 (1995).

Enhanced Propagation Rate of Magnetic Field in Plasmas due to the Hall Effect: Analytic Solutions in Electron MHD

K.V.Cherepanov and A.B.Kukushkin

INF RRC "Kurchatov Institute", 123182 Moscow, Russia

Abstract. The mechanism of the enhanced propagation rate, as compared with ordinary diffusion, of the magnetic field in plasmas, due to the Hall effect in plasmas, is analyzed in the frame of the Electron MHD, under conditions of arbitrary inhomogeneity of plasma density and homogeneous temperature. The analytic results are obtained for the magnetic field front and its effective width.

1. INTRODUCTION

The Hall effect in plasmas [1], which is caused by the frozenness of the magnetic field into electron plasma and by the resulting transfer of magnetic field with electric current velocity, manifests itself in the enhanced rate, as compared with ordinary diffusion, of magnetic field propagation in plasmas in the case of electric current flowing along the gradient of plasma density. This includes both the gradients at the boundaries plasma-conductor and in plasma interior. The major physical mechanism is the "scattering" of the magnetic field, which is transferred by the electric current, at positive gradient of electron density. This phenomenon may strongly influence plasma dynamics both in the case of the so called Electron MHD (EMHD) (immovable ions) and in general case (for observations of 2-fluid effects in plasma focus discharges see, e.g., [2] and this volume).

Here, the enhanced propagation rate is analyzed in the frame of the EMHD, under conditions of arbitrary inhomogeneity of the plasma and homogeneous temperature, and analytic results are obtained for magnetic field front and effective width. The approach used extends a qualitative model [3] which reproduces a number of rigorous analytic results in the EMHD theory.

2. MAGNETIC FIELD FRONT POSITION

The formalism developed is aimed at determination of basic characteristics of the magnetic field propagation for arbitrary profile of plasma density $n=n(z)$ along the (plane) boundary of the magnetic field at time $t=0$ (electric current flows along z-direction within the above-mentioned plane, see Fig.1). The 2D evolution of the magnetic field $\vec{H}(x,z,t) = \{0, H(x,z,t), 0\}$ in plasmas in the frame of the EMHD, with allowing for the Hall effect, is governed — for a homogeneous electron temperature — by the following equation (see, e.g., [4,5]):

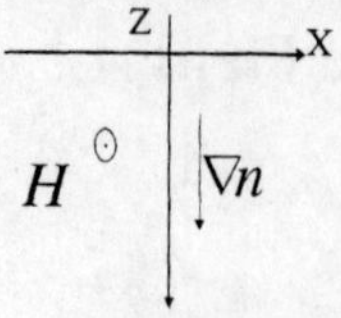

$$\frac{\partial H}{\partial t} = D_\sigma \left(\frac{\partial^2 H}{\partial x^2} + \frac{\partial^2 H}{\partial z^2} \right) + \frac{cH}{4\pi e} \frac{\partial H}{\partial x} \frac{\partial}{\partial z}\left(\frac{1}{n}\right), \qquad (1)$$

where $D_\sigma = \frac{c^2}{4\pi\sigma}$ is coefficient of ordinary diffusion, σ is Spitzer conductivity. At $t=0$ magnetic field occupies space region $x<0$, and the following boundary conditions hold:

Fig.1

$$H(z,x,t)\big|_{x\to-\infty} = H_0; \; H(z,x,t)\big|_{x\to+\infty} = 0; \; \frac{\partial H}{\partial x}\bigg|_{x\to+\infty} = 0. \qquad (2)$$

It appears that it is possible to find a closed equation for the characteristic values of the magnetic field propagation in the plasma, which describe the magnetic field front position and the effective width of this front. This allows to obtain a qualitative description of the exact solution of the original non-linear equation, Eq.(1). Indeed, for the front position $x_0(z,t)$, defined by the following relation:

$$\int_{-\infty}^{+\infty} (H(x,z,t) - H_0 \cdot h(x_0 - x))dx = 0, \qquad (3)$$

where $H(x,z,t)$ is exact solution of Eq.(1) and $h(x)$ is the unit step function, we arrive at the following closed equation for $x_0(z,t)$:

$$\frac{\partial x_0(z,t)}{\partial t} = D_\sigma \frac{\partial^2 x_0(z,t)}{\partial z^2} - \frac{cH_0}{8\pi e} \cdot \frac{\partial}{\partial z}\left(\frac{1}{n(z)}\right). \qquad (4)$$

Equation (4) gives, for initial condition $x_0(z,0)=0$, the following expression for the front position $x_0(z,t)$,

$$x_0 = -\frac{cH_0}{16\pi e} \int_0^t \int_{-\infty}^{\infty} \frac{\partial}{\partial \xi}\left(\frac{1}{n(\xi)}\right) \frac{e^{-\frac{(z-\xi)^2}{4D_\sigma(t-\tau)}}}{\sqrt{\pi D_\sigma(t-\tau)}} d\tau d\xi. \qquad (5)$$

In the case of small density gradient, Eq.(5) gives for the front velocity $u(z,t)$,

$$u = \frac{cH_0}{8\pi e} \frac{\partial}{\partial z}\left(\frac{1}{n(z)}\right), \qquad (6)$$

that coincides with analytic result obtained in Ref. [5] via exact solving the approximate equation for magnetic field propagation (namely, neglecting the term $\partial^2 H / \partial z^2$ in Eq.(1)). In the opposite case of infinite density gradient, i.e. step-like density profile, when $\partial(1/n)/\partial\xi = -\delta(\xi)(n_1 - n_0)/n_1 n_0$, we have:

$$x_0(t) = \frac{cH_0}{16\pi e D_\sigma} \frac{n_1 - n_0}{n_1 n_0} \sqrt{D_\sigma t} = \frac{\sigma H_0}{2ec} \frac{n_1 - n_0}{n_1 n_0} \sqrt{D_\sigma t} \equiv \sqrt{D_{eff} t}, \qquad (7)$$

that corresponds to a diffusion-like propagation with the effective diffusion coefficient D_{eff}. For homogeneous conductivity this result coincides with result in Ref. [4]. In special case of magnetic field propagation along a flat anode (with infinite conductivity: $n_1 \to \infty$), Eq.(7) coincides with analytic formula [6] and the results of numerical modelling [3].

Comparison of Eq.(5) with respective results of the 2D EMHD numerical modelling [4] shows good agreement.

It follows from Eq.(5) that the transition between the above limiting cases, namely Eq.(6) and Eq.(7), can be described approximately by the parameter $\mu = (\partial \ln n / \partial z)\sqrt{D_\sigma t}$ (see also [7]). Application of the qualitative approach [3] to the case of arbitrary inhomogeneity gives:

$$V_x(t) \approx \omega_e \tau_{ei} D_\sigma \frac{\partial \ln n}{\partial z} \frac{1}{\sqrt{1+\mu^2}}. \tag{8}$$

(note that the numerator in Eq.(8) coincides with velocity of Eq.(6)). However, calculations show that Eq.(8) gives substantial error for moderate values of μ. The matter is that the approach [3] do not allow correctly for the diffusion perpendicular to the field propagation front. This shortage can be diminished via 'delocalising' the meaning of parameter μ, via replacing it by the parameter $\mu_L = \sqrt{D_\sigma t}/L$, where L is the characteristic length of plasma density inhomogeneity along plasma-field boundary. This gives a simpler expression, as compared with Eq.(5), for the front motion:

$$x_0(z,t) = \omega_e \tau_{ei} L^2 \frac{\partial \ln n}{\partial z} \left(\sqrt{1+\mu_L^2} - 1\right) \tag{9}$$

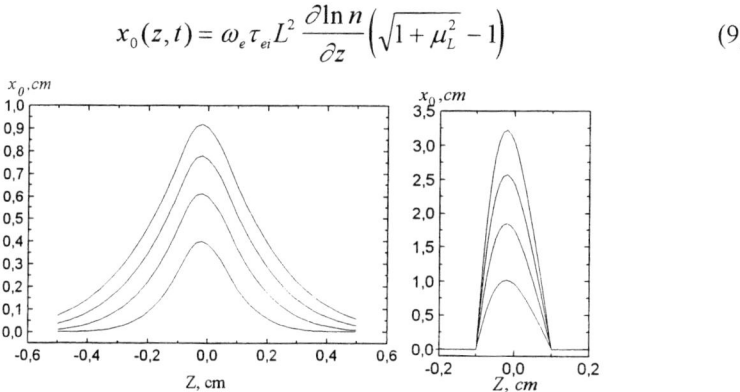

Fig.2. Comparison of the exact front position, Eq. (5), with its approximation, Eq.(9), for $H_0 = 10^4$ G, temperature $T_e = 5$ эB, background density $n_0 = 4 \cdot 10^{16}$ см$^{-3}$ (here, $D_\sigma = 3.7 \cdot 10^5$ cm^2/s and $\omega_e \tau_{ei} = 33,8$). Density fluctuation has the form: $n(z) = n_0(1+0.5\delta(1+\sin(\pi z/L)))$ for $|z|<L/2$, and $n(z) = n_0$ for $z<L/2$, $n(z) = n_0(1+\delta)$ for $z>L/2$; $L = 0.2$ cm, $\delta = 1$. Front positions are given for time moments t=k 25 ns, k=1..4.

3. MAGNETIC FIELD FRONT EFFECTIVE WIDTH

The effective width of the magnetic field front may be defined as follows:

$$H_0\left(\delta x_0(z,t)\right)^2 = \int_{-\infty}^{\infty} (x-x_0)[H(x,z,t) - H_0 h(x_0 - x)]dx. \tag{10}$$

The relevance of such a definition can be illustrated with substituting the exact solution [5] into Eq.(10) and taking $x_0=ut$ with velocity u taken from Eq.(6).

After calculations similar to that of the previous section, we arrive at the following equation for the width δx_0:

$$\frac{\partial (\delta x_0)^2}{\partial t} = D_\sigma \frac{\partial^2 (\delta x_0)^2}{\partial z^2} + D_\sigma \left[1 + \left(\frac{\partial x_0}{\partial z}\right)^2\right] + \frac{c}{4\pi e n H_0} \frac{\partial \ln n}{\partial z} \int_{-\infty}^{\infty} (x-x_0) H \frac{\partial H(x,z,t)}{\partial x} dx, \quad (11)$$

with initial condition taken in the form of Eq.(10) at $t=0$ and for $x_0(0)=0$. Expressing the results in terms of parameter $\omega_{eH}\tau_{ei}$, where the density in τ_{ei} is taken equal to n_0, we have:

$$\frac{\partial x_0}{\partial z} \equiv g(z,t) = -\frac{1}{4}\omega\tau\left(\frac{n_0}{n(z)} - \frac{1}{\sqrt{\pi D_\sigma t}}\int_{-\infty}^{\infty}\frac{n_0}{n(\xi)}e^{-\frac{(z-\xi)^2}{4D_\sigma t}}d\xi\right) \quad (12)$$

It appears that for $\omega_{eH}\tau_{ei} \gg 1$, that is of interest for the problem, the last term in Eq.(11) can be neglected under condition $(L/\delta x_0)\omega_{eH}\tau_{ei} \gg 1$. Interestingly, for the exact solution [5] this term is precisely equal to zero. Thus, we arrive at a closed equation for the front width. Its solution takes the form (here q stands for the initial front width):

$$(\delta x_0(z,t))^2 = D_\sigma t + \frac{1}{\sqrt{4\pi D_\sigma t}}\int_{-\infty}^{+\infty} q(\xi)e^{-\frac{(z-\xi)^2}{4D_\sigma t}}d\xi + D_\sigma \int_0^t\int_{-\infty}^{+\infty} g^2(\xi,\tau)\cdot\frac{e^{-\frac{(z-\xi)^2}{4D_\sigma(t-\tau)}}}{\sqrt{4\pi D_\sigma(t-\tau)}}d\xi d\tau, \quad (13)$$

It should be noted that the above results for the front position and effective width assume that the propagation of the field is faster than washing out of the front's effective position (the latter is certainly true for the initial stage of the magnetic field 'scattering' at an isolated density bump, cf. Sec.1). If this is not the case, one should interpret the quantities of Eq.(5) and Eq.(13) as both contributing to the front position. The latter appears to be the case for the periodic density fluctuations with short enough period, because of the limited quantity of the magnetic field capable of being 'scattered' at an isolated density bump.

REFERENCES

1. Morozov A.I., Shubin A.P., *Zh. Eksp. Teor. Fiz. (Sov. Phys. JETP)* **46,** 710 (1964).
2. Kukushkin A. B., Rantsev-Kartinov V. A., and Terentiev A. R., *Fusion Technology,* **8** (1997); *Transactions of Fusion Technology* **27**, 325 (1995).
3. Kukushkin A.B., *AIP Conf. Proc., #299,* 3rd Int. Conf. Dense Z-pinch (London, 1993), Eds. M. Haines and A. Knight, New York: AIP Press, 1994, p.154.
4. Vikhrev V.V., Zabaidullin O.Z., *Ibid.,* p.165; *Plasma Phys. Reports* **20**, 867 (1994).
5. Kingsep A.S., Mokhov Yu.V., Chukbar K.V., *Sov. J. Plasma Phys.* **10** (1984) 854.
6. Gordeev A.V., Grechikha A.V., Kalda Ya.L., *Sov. J. Plasma Phys.* **16** (1990) 95.
7. Zabaidullin O.Z., *Preprint of the RRC "Kurchatov Institute",* IAE-5828/6, Moscow, 1994.

Capillary X-Ray Laser Research

J. P. Chittenden, M. Michaelis[1], S. N. Bland, M. D. Eaton and J. F. Worley

Imperial College, London SW7 2BZ, UK
[1]*University of Natal, South Africa*

Abstract. A table-top capillary discharge soft X-ray laser system has recently been developed at Imperial College in order to evaluate its potential as a back-lighting source for solid density hydrogen plasmas. 1-D MHD simulations incorporating non-LTE ionization dynamics indicate the generation of the neon-like ionization stage of argon just prior to stagnation of the cylindrically converging shock. The experimental measurements of the collapse dynamics are compared to the simulation. End-on measurements show soft X-ray emission coincident with collapse. At the time of going to press no spectral resolution of this soft X-ray emission had been obtained. Possible methods of increasing output power and operation at different wavelengths, are discussed.

INTRODUCTION

The X-ray laser has a wide range of potential applications in the fields of medicine, biology, semiconductor manufacturing, materials testing, etc. However, of particular interest to the plasma physics community is the use of the X-ray laser as a plasma diagnostic tool.

The use of a capillary discharge to generate the lasing plasma [1] has several advantages over the use of a laser generated plasma. In laser produced plasmas the high ion temperature causes Doppler broadening of the X-ray laser line and the electron number density gradient refracts the laser radiation away from the gain medium. In capillary discharge plasmas, the electrons are heated ahead of a cylindrically converging shock and the lasing occurs in a region of low ion temperature and positive dn_e/dr. There is therefore the possibility of refraction of the laser radiation from this gradient resulting in a self-focusing wave-guide effect. Most importantly the conversion efficiency from wall-plug energy to X-ray output of pulsed power Z-pinch sources results in much smaller and lower cost devices than with laser produced plasmas.

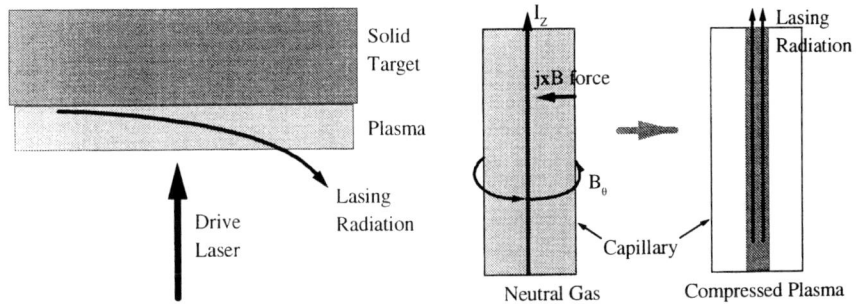

Figure 1a. Laser Driven X-ray Laser   **Figure 1b.** Capillary X-ray Laser

Neon-like X-ray laser schemes have the advantage that the plasma tends to readily adopt the neon-like state because of the plateau in ionization potentials. Collisional excitation populates the 3p level which then decays via a j=0-1 transition to a meta-stable 3s level. Collisional mixing between the 3p and 3s levels sets an upper limit on n_e of $\sim 3 \times 10^{25}$ m^{-3} (ideal 1.5×10^{25} m^{-3}). Collisional excitation from 3p to 3d sets an upper limit on n_{3p} of 8×10^{21} m^{-3} and a maximum gain of ~4 cm^{-1}. Saturation due to stimulated emission depopulating the 3p level is expected at gain length products of ~10-20.

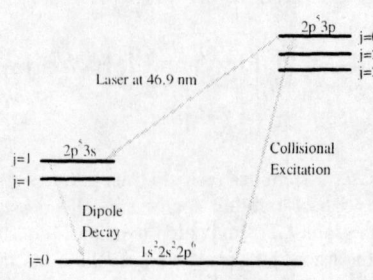

Figure 2. Neon-like argon scheme

BACK-LIGHTING HYDROGEN PLASMA

The diagnosis of radiative collapse and fusion Z-pinch experiments requires the development of a back-lighter to probe keV hydrogenic plasmas at near solid density. This plasma is transparent to K shell radiation from most elements. Figure 3 shows the required wavelength for both 90% inverse bremsstahlung absorption and a 1° refraction for a hydrogen plasma formed from a 100μm diameter fiber with 1MA of current. At wavelengths below 65 Å, absorption is dominant whereas at longer wavelengths, refraction of the probing radiation is dominant and a Schlieren image results. This graph suggests that the 469 Å neon-like argon scheme would produce a back-lighting probe capable of Schlieren photography at up to 15% of solid density. The use of a laser back-lighting probe, being a monochromatic, collimated and coherence source offers further advantages over a thermal emission source.

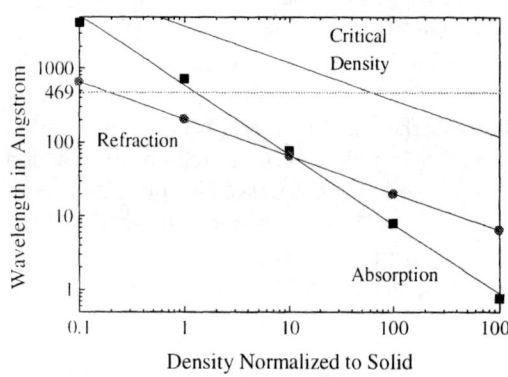

Figure 3. Back-lighter wavelength requirements.

1D, NON-LTE, MHD SIMULATIONS

The capillary discharge conditions described by Rocca [1] were modeled using a 1D Lagrangian, 2 temperature, resistive MHD code. The low collision rates in the low density argon plasma requires a non-LTE ionization model to be used. The NIMP code [2] simplifies the computational problem of calculating all the rates for collisional and radiative ionisation, recombination, excitation and de-excitation by averaging these rates over each atomic shell. Figures 4&5 show radial profiles of

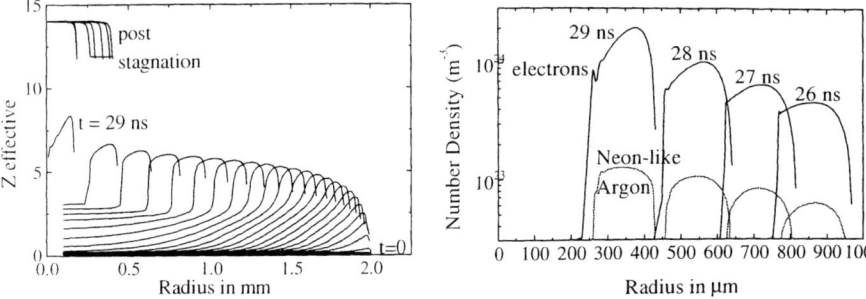

Figure 4. $Z^*(r)$ at 1ns intervals.   **Figure 5.** $n_e(r)$ and $n_{z=8}(r)$ at 1ns intervals.

the effective ionic charge and the number densities of electrons and neon-like ions at 1ns intervals. The plasma becomes neon-like ($Z^*=8$) for a few ns prior to stagnation. Note that the majority of neon-like argon occurs in a region of positive dn_e/dr and hence we might expect refraction of the laser radiation to result in a self-focusing wave-guide effect. However, random ray tracing calculations suggest that in this instance dn_e/dr is insufficient for the effects of refraction to be important.

A TABLE-TOP CAPILLARY DISCHARGE SYSTEM

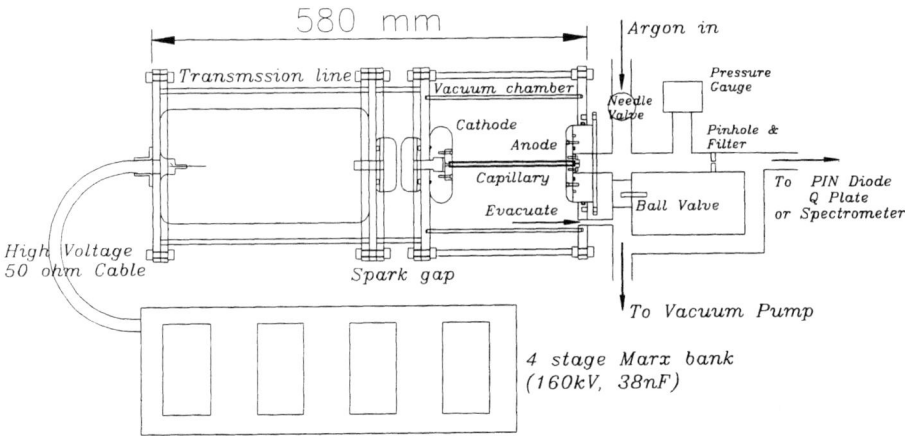

Figure 6. Design of the table-top capillary discharge system.

Figure 6 shows the design for the table-top capillary discharge system recently completed at Imperial College. The four stage Marx bank produces 160KV at erection and charges a 1 Ω water filled transmission line which then discharge via a self-breaking SF_6 filled spark-gap into the 150nH capillary plasma load resulting in a 30kA current with a 10-90% rise-time of 40ns. This rise-time could be halved by moving the return current rods inside the vacuum housing, thereby reducing the inductance at the expense of visibility. The 4mm diameter, 150mm long quartz capillary is filled with 1.5 mbar Argon and differentially pumped through a pinhole in order to maintain a vacuum of 10^{-4} mbar in the end-on X-ray diagnostics.

The collapse dynamics of the plasma were diagnosed using an optical streak camera with a slit aligned perpendicular to the pinch axis. Figure 7 shows the familiar "fishtail" structure of the collapsing shock. The brightness of the post stagnation plasma almost swamps the fainter emission from the collapsing plasma and therefore a schematic representation of the image is also provided for clarity. The collapse time and the apparent piston and shock radii versus time are in rough agreement with slug model calculations and 1D MHD simulations.

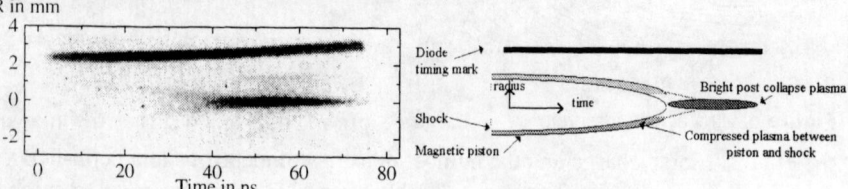

Figure 7a. Optical radial streak.   **Figure 7b**. Schematic of image.

End-on measurements were conducted using soft X-ray sensitive Q-plates. With the pinhole set-up used, it was not possible to distinguish a laser of a few milli-radians divergence from a point source of thermal plasma emission. A silicon PIN diode with a 0.8µm Al filter with a range of sensitivity of 1-30 Å and 150-1000 Å showed a soft x-ray burst coincident with the bright optical emission after collapse. At the time of going to press no spectral resolution of this soft X-ray emission had yet been obtained. The diode was collimated to prevent hard X-ray emission from the electrodes caused by an electron beam at t=0, from causing saturation.

OPTIMISATION AND OTHER WAVELENGTHS

Increasing the pinch length and gain per cm and the use of multiple-pass using mirrors all increase the yield linearly. However once the laser is saturated, at gl ~ 10-20 there can be no further exponential increase. The best way to improve yield is probably to increase the cross-section of the lasing medium or to use a small oscillator stage and a second, larger amplifier stage.

Attaining shorter wavelength neon-like schemes in higher Z elements seems possible with slightly larger drivers. The required temperature is set by the ionization potential of the Sodium-like level (T ~ $E_{na}/3$) which is roughly the Bennett temperature and therefore sets the line density. The requirement that the electron drift velocity be less than the ion sound speed then sets the current (neon-like Xenon requires a minimum current of ~100kA). A fast rising current is necessary to produce a skin current and drive an ionizing shock.

REFERENCES

1. J.J. Rocca, D.P. Clark, J.L.A. Chilla and V.N. Shlyaptsev, Phys. Rev. Lett. **77** 1476 (1996).
2. S.J. Rose (private communication, 1992).

Experimental Studies on a Pulsed Hollow Cathode Capillary Discharge

P. Choi*, M. Favre, C. Dumitrescu-Zoita*, J. Moreno, H. Chuaqui, and E. Wyndham

Facultad de Física, Pontificia Universidad Católica de Chile, Casilla 306, Santiago 22, Chile
*LPMI, Ecole Polytechnique, Palaiseau 91128, France

Abstract. Experimental studies on a fast pulsed capillary discharge are presented. The discharge operates in a 0.8 mm inner diameter alumina capillary, at 10 to 30 kV applied voltage. On axis discharge initiation is achieved by means of the hollow cathode effect. A short, less than 10 ns, XUV pulse is produced. Preliminary time resolved spectroscopic studies indicate that a hot, fast evolving, short duration capillary plasma is produced.

INTRODUCTION

The pulsed capillary discharge is known as a high brightness source in the VUV and soft X-ray region[1], which has also been shown to be a suitable medium for discharge based X-ray laser studies based on recombination[2] or collisional excitation schemes[3]. The efficient operation of a capillary discharge as a high temperature plasma radiation source depends both on the initial formation of a conducting plasma column away from the wall and in the ability to provide a high rate of current rise into a high impedance load. In this paper, we report on time resolved studies on a fast capillary discharge which operates in the nanosecond regime with current in the kA region into a sub-millimeter diameter capillary. To achieve the rapid formation of the discharge column, the initial on axis ionization path is prepared through the transient hollow cathode effect[4]. To produce the required high rate of current rise, the discharge is operated at high voltage, with a very low circuit inductance, which integrates the energy storage medium directly onto the discharge electrodes of the capillary system. Preliminary results indicate that the discharge is initiated by an axial electron beam, which can be associated with the hollow cathode effect. At breakdown, a nanosecond time scale plasma is produced inside the capillary, which is initially detached from the capillary wall and emits in the VUV to soft X-ray region.

EXPERIMENTAL SET-UP

A schematic of the capillary discharge set-up is shown in Fig. 1. An alumina capillary, 0.8 mm diameter, up to 30 mm long, is located on axis between the two electrodes. Current and voltage monitor have been integrated into the ground side of the electrode, which forms part of the energy storage capacitor. A primary DC charged capacitor of 5 nF is used to pulse charge the storage capacitor. At a charging voltage of 25 kV, current above 10 kA is produced through the capillary, with a 10-90% current rise time of below 5 ns. The discharge can be triggered by means of an auxiliary discharge from a small cable plasma gun, located a few mm behind the cathode aperture, with a delay of 80 to120 ns. The cathode region is pumped through the capillary, thus allowing a higher pressure in the hollow cathode region, while maintaining a lower pressure in the capillary to produce a lower line density plasma. The experiments have been performed in Argon at pressures between 80 and 400 mTorr in the hollow cathode region and a few mTorr at the Anode region. Plasma emission is studied in the visible with photomultipliers, in the VUV region with time resolved spectroscopy and in the soft X-ray region with filtered PIN diodes and XRD. For time resolved spectroscopic studies, a VUV grazing incidence spectrometer with a three frame gated intensified microchannel plate detector is used. In these observations the spectrometer was fitted with a 600 groves/mm grating, with a useful spectral range from 70 to 240 Å. The framing time and frame separation can be adjusted by selecting the cable length to the high voltage pulser. A 3 ns gating time was used in these measurements. Electron beams are measured using a Faraday cup and a scintillator-photomultiplier assembly.

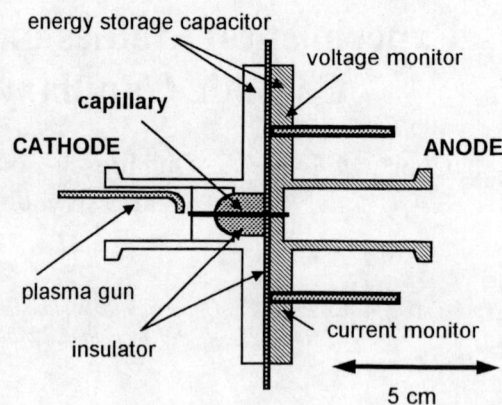

FIGURE 1: experimental set-up

EXPERIMENTAL RESULTS

Figure 2 shows characteristic single shot signals at 280 mTorr. A cable delay of 110 ns is used for the auxiliary triggering discharge. The voltage collapses in less than 10 ns at breakdown. At voltage collapse, a fast current pulse is observed, which is time correlated with an electron beam pulse detected with the Faraday

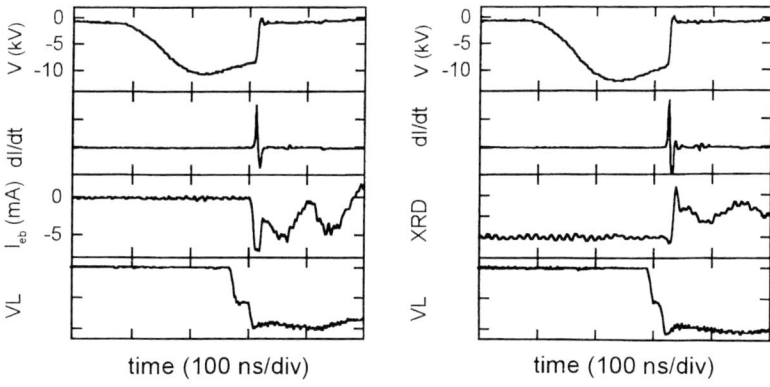

FIGURE 2: characteristic single shot signals

cup. After the initial current pulse, the voltage recovers and grows up to ~1 kV. The visible light emission (VL) is monitored axially, from behind the cathode. The first step observed in the visible corresponds to the auxiliary discharge, and the second one coincides with the onset of the main discharge. The XRD signal shows an initial 10 ns fast pulse. Figure 3 shows two frames of a single shot spectrum, at 320 mTorr. The exposure time is 3 ns with frame separation of 5 ns.

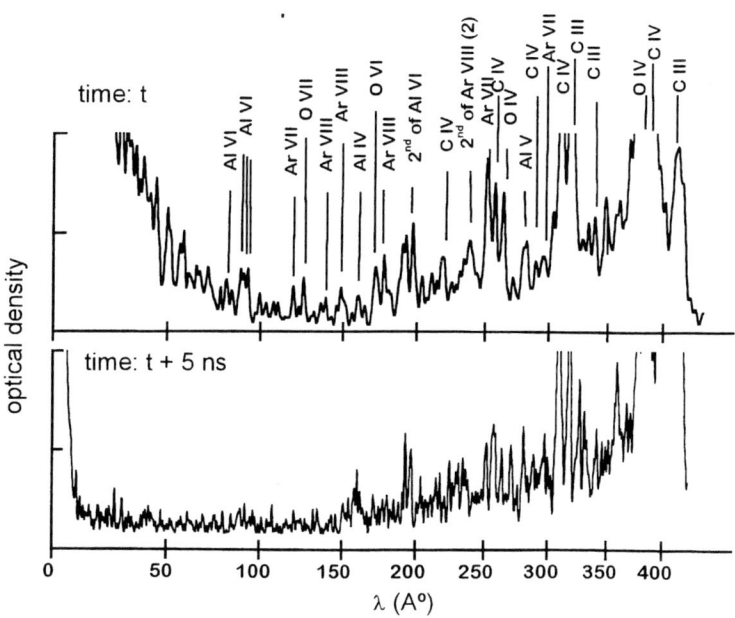

FIGURE 3: single shot, two-frame, 5 ns exposure spectra.

355

DISCUSSION

Transient hollow cathode discharges (THCD) are characterized by a rapid voltage collapse accompanied by the growth of an intense electron beam leading to the formation of an on-axis ionization channel, which develops throughout a moving virtual anode(4). The electron beams originate in a plasma region inside the hollow cathode region, which develops spontaneously following the application of the high voltage across the electrodes. Auxiliary triggering inside the hollow cathode region (HCR) provides an additional source of electrons to enhance the electron beam emission(5). A single electron beam pulse is observed and electric breakdown is achieved in a time much shorter than in self breakdown operation. The spectroscopic data shows clearly the fast evolution of the capillary plasma, as well as the lack of significant continuum radiation over the wave length investigated. The plasma channel is heated up very rapidly, in the sub-10 ns time scale, within a few ns after current rise. The partial voltage recovery after the fast current pulse is an indication of a resistive plasma channel. Through the THCD process, the initial discharge channel is initiated through the action of the self-generated electron beam along the capillary axis, away from the wall. Expansion of the plasma channel results in material being ablated from the capillary wall, thus leading to a fast cooling, with a evolution subsequent increase in plasma resistivity. This is supported by the observation of the of Aluminium, Oxygen and Carbon lines superimposed on the Argon spectrum, indicating the temporal development of wall ablation and axial plasma flow from the HCR plasma gun. As a result of these processes, VUV to XUV light is emitted only in a very short pulse associated with the duration of the current. These preliminary results show that the pulsed capillary discharge can be initiated successfully with the assistance of the hollow cathode effect. The resulting discharge is characterized by a short, less than 5 ns half width current pulse, with an associated XUV light pulse.

ACKNOWLEDGMENTS

The authors gratefully acknowledge K. Koshelev and S. Tchourilov, for help in the identification of spectral lines. This investigation has been funded by FONDECYT project 1950798 and a CNRS-CONICYT collaboration program.

REFERENCES

1. R.A. McCorkle, *Appl. Phys. A*, **A26**, 261 (1981)
2. C. Steden and H.J. Kunze, *Phys. Lett. A* **151**, 534 (1990)
3. J.J. Rocca et al., *Phys. Rev. Lett.* **73**, 2192 (1994)
4. P. Choi et al., *Appl. Phys. Lett.* **63**, 2750 (1993)°
5. H. Chuaqui et al., *Appl. Phys. Lett.* **55**, 1065 (1989)

Observations of Vacuum Spark Dynamics from its X-Ray Emission

H. Chuaqui, M.Favre, R. Saavedra, E. Wyndham, L. Soto*, P. Choi**
and C. Dumitrescu-Zoita**

Pontificia Universidad. Católica de Chile, Casilla 306, Santiago 22, Chile.
*Comisión Chilena de Energía Nuclear, Casilla 188-D, Santiago de Chile.
**LPMI, Ecôle Politechnique, Palaiseau 91128, France

Abstract. The behaviour of a medium energy pulse power driven vacuum spark is shown to depend on the different electrode materials and form of the anode in otherwise similar conditions of operation. The dynamical evolution of the discharge is followed from its soft X-ray emission. The electrode materials compared are titanium and aluminium and with a form of anode that is either conical or tubular. The use of a tubular anode favors a more uniform sheath and a better formation of dense Z-pinch and ensuing hot spots. The Aluminium hotspots are noticeably hotter than the Titanium discharge.

INTRODUCTION

The Vacuum Spark is a rich discharge for experimental observations, especially when operated so as to form a dense Z-pinch(1). For discharge currents in excess of 100 kA, radiative processes from highly stripped ions and plasma instabilities combine to generate short lifetime hotspots or plasmoids whose density may reach 10^{22} cm^{-3} with a characteristic size of a few tens of microns, or even less. We present observations of how the plasma parameters and plasma dynamics depend on both the electrode shape and the Z number, that is for aluminium and for titanium plasmas. By observing the X-ray emission from both the electrodes and from the inter electrode volume over the greater part of the discharge, we are able to characterize both the current sheath during the compression phase and the evolution of the submillimetre dense Z-pinch column in which the micron sized plasmoids form. We find that the form of the current sheath depends on whether the anode is conical or a thin wall tube. The shape of the anode also determines the length of the Z-pinch column and we find that the X-ray burst from the hot spots is considerably increased using a tubular anode. Further differences in the plasma parameters will be presented in the following sections.

EXPERIMENTAL RESULTS

The main experimental considerations have been described elsewhere(2). The variations presented here are the comparison between the use of a conical or a thin walled tube, whose external diameter is 6.3 mm. In both cases a 0.4J laser pulse is focused through anode onto a flat cathode at $4 \cdot 10^{10}$ W·cm^{-2}. A further comparison is made between the use of Ti and Al electrodes. A four frame X-ray framing camera with a 1.5 ns exposure gives spatial and temporal information in different spectral regions as determined by different filters. As the temperatures are of order 100 eV for much of the discharge, the wideband response of a photocathode is optimum. Holographic interferometry provides information on the electron number density.

The plasma temperature is obtained from the relative exposures on the framing camera with the various filter materials employed. The measured ratios are compared to the synthetic spectrum for a given temperature as generated by the RATION(3) or the more recent FLY(4) CRE suite of codes. The spectrum is then convolved with the filter response and the detector response. Some care must be taken in the analysis at lower temperatures, where the spectrum is mainly due to lower ionization states, which are not modeled by the codes. Absolute XUV Si photodiodes, used with wideband filters, allow the X-ray energy to be calculated for a measured temperature, using a synthetic spectrum as referred to above.

In Fig. 1 we present a comparison of the sheath structure over a period of approximately 25 ns for the three electrode configurations used. Images in three different filter thicknesses of Al are shown vertically for the exposures. The images are shown in three groups of three exposures. These exposures were taken as the current is increasing from approximately 50 kA to 90 kA. The sheath is seen to have a well defined spatial structure. Of particular note is the rhombic form in the case of the conical anode geometry in Ti, as seen in the softest filter and at early times. In the two cases of a tubular anode the sheath is conical, opening out towards the anode. The sheath boundary arrives, in all cases, on axis first at the cathode, where the first hot spot is observed. The temperature of the sheath is 50 to 100 eV for Ti and 100 to 150 eV for Al.

The rhombic formation has been observed in earlier holographic interferometry results(2,5). On reworking those results to obtain the development of the electron number density, we find a sharply peaked density profile at the edge. The density increases by nearly two orders of magnitude to reach $1 \cdot 10^{19}$ cm^{-3} in the submillimetre diameter Z-pinch column. The electron line density is observed to increase by a factor four, but the mean ionization level of the Ti plasma, determined principally by the temperature, increases by no more than 50% in the same period. We infer that there is appreciable axial mass transport during the 25 ns compression phase in order to explain the increase in the ion line density.

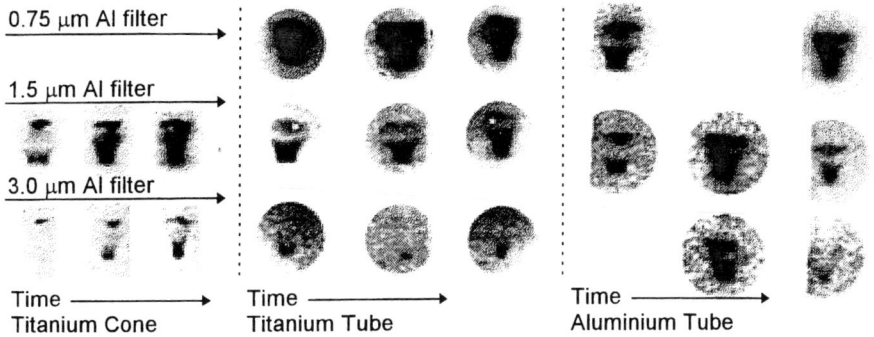

FIGURE 1: The contraction of the sheath for the two geometries used for Al and for Ti.

In Fig. 2 we present the corresponding comparison of the three situations during the phase of the fully formed Z-pinch. Several important differences may

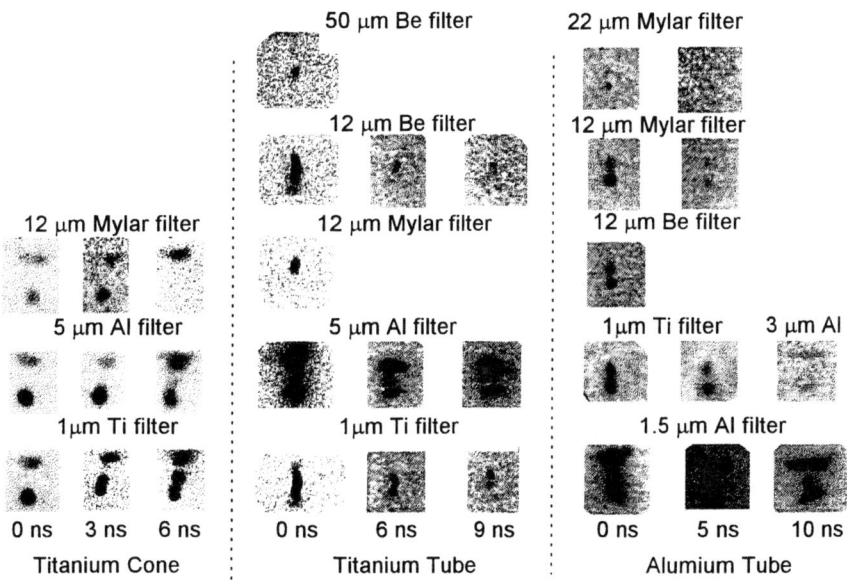

FIGURE 2. The evolution of the dense column Z-pinch from the time of the X-ray burst onwards for the three cases of the previous figure.

be observed. First, the length of the Z-pinch column extends the whole 7.5 mm separation between the electrodes when a tubular anode is used, whereas the length of the column extends slightly more than half way to the anode for the conical case. The main X-ray burst has a duration <3 ns in all cases, however for both Al and Ti tubular anode, a cooler and larger diameter (~ 0.3mm diameter) hotspot persists for up to 10 ns after the burst at 2/3 of the distance to the anode from the cathode. The current reaches its maximum value at this time. A well formed dense Z-pinch is not observed at later times. Analysis of the temperatures reveals that the hot spot plasmas for both Ti cases reach 250 to 350 eV during the X-ray burst and 400 to 500 eV for the Al tube geometry. The hotspot closest to the cathode in the Ti cone and in the Al tube geometry has both the highest temperature and is the brightest emitter, while with Ti tube geometry, the hottest and brightest source is mid-way between the electrodes. For the Ti cone geometry, a hot but diffuse plasma, with a temperature up to 400 eV, is observed just in front of the anode, at a time corresponding to the break up of the Z-pinch column.

The energy of the X-ray burst is twice for the Ti tube geometry as compared with the Ti cone case. As the emitting column length is twice that of the former case, the Ti tube geometry emits a total of four times the Ti cone case. The X-ray output in the Ti tube case is calculated at 2% of the initial Marx stored energy, i.e. 14 J.

CONCLUDING REMARKS

Interferometry results show an optically dense plasma blowing off from the anode and occupying a significant proportion of the inter electrode length as the Z-pinch column becomes fully compressed. This is particularly obvious for the Ti cone case, where the column ends at the edge of this plasma. The rhombic sheath may be explained in terms of the larger surface area of the cone, as compared to a thin walled tube, which offers a much greater cross section of interaction to the electron beams generated by the laser preionization on the cathode. In the cone geometry both electrodes are a significant source of plasma, whereas in the tubular case the cathode dominates, at least until the Z-pinch column has formed.

The substantial increase in the ion line density during the sheath run-in phase underlines axial mass transport as being of importance in the vacuum spark. This discharge remains a challenge for theoretical computer modeling, as the dynamics of the discharge shows a wide range of dense Z-pinch phenomena, even at the rather modest currents of the present experiments.

REFERENCES

1. K.N. Koshelev and N.R. Pereira, J. Appl. Phys. **69**, R21 (1991)
2. E. Wyndham et al., J. Appl. Phys. **71**, 4164 (1992)
3. R.W. Lee et al., J. Quant. Spectrosc. Radiat. Transfer **32**, 91 (1984)
4. R. W. Lee and J. T. Larsen, J. Quant. Spectrosc. Radiat. Transfer **56**, 535 (1996)
5. H. Chuaqui et al., Phys. Plasmas **2,** 3910 (1995)

Interaction of Plasma Jets produced from Pinch Plasma with Neutral Atoms in order to achieve an effective Charge Exchange Table Top X-laser

A. Engel[1], R. Lebert[1], K.N. Koshelev[4], Yu.V. Sidelnikov[4],
S.S. Churilov[4], C. Gavrilescu[3] and W. Neff[2]

[1]RWTH Lehrstuhl für Lasertechnik and [2]FHG-Institut für Lasertechnik Steinbachstraße 15, 52074 Aachen, Germany, [3]University of Iasi, Copou 11, 6600-Iasi, Romania, [4]Institute for spectroscopy, Russian Academy of science, 142092 Troitzk, Moscow region, Russia

Charge exchange recombination is known as an effective scheme to get population inversion in the EUV-range. Highly ionised plasma jets and a neutral target could be very efficient ($\sigma_{CE} > 10^{-15}$ cm^2) to realisation of this atomic scheme. Theoretical estimates and preliminary experiments show that for plasma focus with a stored energy of several kJ one can reach a substantial population inversion for ions of light elements with charge number Z<10. Experimental studies of production of these plasma jets in a 2 kJ plasma focus device is presented. Moreover the optimal properties of possible targets are investigated. Experimental results for the interaction of the plasma jets and targets are presented.

INTRODUCTION

X-lasers are known from laser produced plasma[3] and z-pinches[7]. Plasma jets, ejected from z-pinches, promise to act as a pump source for a recombination scheme[2]. An efficient charge exchange recombination process (impact process of ion and atom while transferring one electron to the higher ionized partner) is given by a highly ionised plasma jet and a neutral partner with adopted density. For given ionization energies the excited level u is populated selectively, so inversion can be achieved. A small Mather type Plasma Focus[1] device PF3 (pinch current 200 kA) -developed as source for x-ray microscopy and x-ray lithography[6]- is used to create this conditions when operate with pure nitrogen. The neutral partner for the process is expected to be the rest gas in the surrounding.

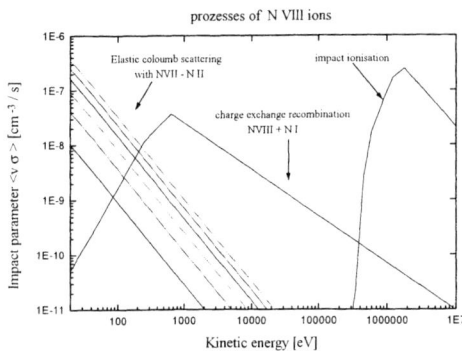

figure 1 Cross sections of processes for NVIII ions. In the range of E_{kin} =500-5000 eV charge exchange recombination is the most probable process.

In figure 1 the cross sections of all possible inelastic processes for totally ionized nitrogen coulomb scattering with other ions, charge exchange recombination and impact ionization are plotted. It is shown that for the expected range of ion kinetic energy between 500 and 5 keV the charge exchange recombination[5] is most important, having more than 2 or 3 magnitudes higher cross sections than other processes. Resulting free path lengths contain values lower than 1 mm for a density of several 10^{16} cm^{-3}. Therefor interaction close to the produced pinch.

OBSERVATION OF PLASMA JETS IN PLASMA FOCUS

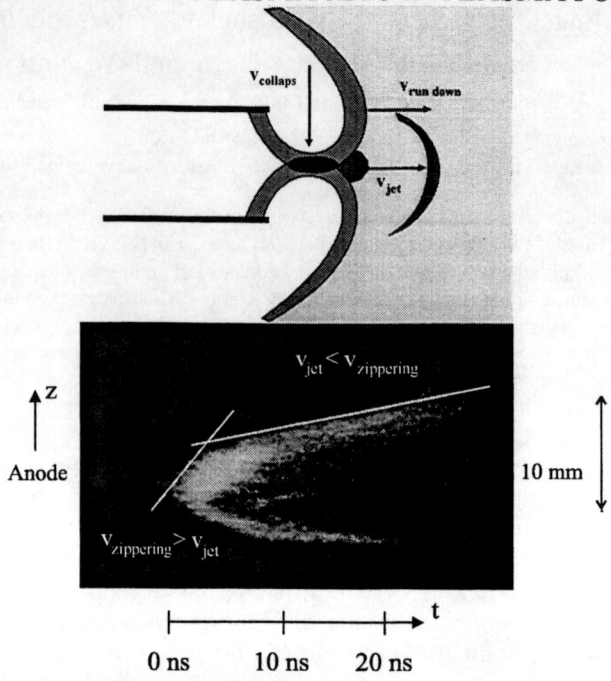

figure 2 **Plasma jet from plasma focus interacts with rest gas only for compression at position with z > 8 mm (take off point), where $v_{jet} > v_{zippering}$. In axial space and time resolved measurement for nitrogen at p = 250 Pa one can see, that the emitting region of soft x-rays is moving to negative z-direction (inside the anode) and to positive z-direction. For high time of t > 12 ns take off point ($v_{jet} > v_{zippering}$) is observed.**

From temporal resolved measurements one can decide whether ions are produced and how they interact with the rest gas or be absorbed in the plasma layer of run down plasma. The temporal resolution was made with two streak cameras (visible and soft x-ray). The observation of the temporal development of hot pinch in the soft x-ray range (corresponding to bremsstrahlung and recombination radiation) shows that during the first stage of pinch development the velocity of the ionization front reaches values up to $3 \cdot 10^8$ cm/s due to superposition by collapsing plasma. After moving 7 mm from the anode the velocity is comparable to the ionic thermal velocity (2-3 keV), so that the recombination process can be initiated. To create an ion beam with nearly totally ionized ions is it necessary to have more than 4 keV kinetic energy per ion. It was shown that this is possible for nitrogen plasmas with the PF3 device with $I_p > 200$ kA[6]. To be sure that the ions can interact with neutral gas or one times ionized gaseous atoms the thermal velocity of the ions must be similar or greater equal to the velocity of the zippering plasma during pinch phase.

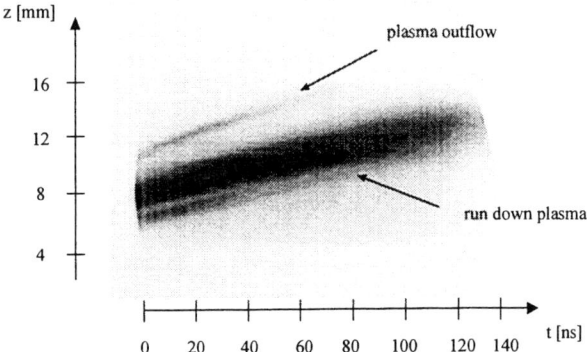

figure 3 Time and in axial direction resolved measurement with nitrogen for p = 400 Pa obtained with a streak camera in the visible range. Two different emission parts can be observed. The first one occurs due to run down plasma whereas the second one appears because of plasma outflow or interaction of plasma jet.

Figure 2 shows the temporal development of the plasma on the z-axis. The maximum of observed plasma extension is 1-2 mm for one time-step. The position where the zippering velocity is high enough to observe outflows plasma is about z ≈ 8 mm. So for z ≥ 8 mm the ions can escape. This point is now defined as „take off" point.

INTERACTION OF PLASMA JET WITH REST GAS

To get information about the interaction of the plasma jet with the rest gas spectra with in axial direction spatial resolution are made. The spectrum in figure 4 obtained with a flat-field spectrograph shows lines from Li-, He, and H-like nitrogen ions in the range from 7 to 17 nm.

The spectrum looks quite different from thermal expected spectra. The spatial resolution indicates that in the later pinch, which is equivalent to a position with higher z, the line of the 2-4 transition is emitted with higher intensity than the 2-3 line. From the intensity-line ratio of 2-3 and 2-4 transition one can deduce a gain G·L of about 1. It is also shown that there is no significant change of spatial resolved line intensities. This is due to the fact that with increasing the neutral gas pressure the outflow velocity is decreased.

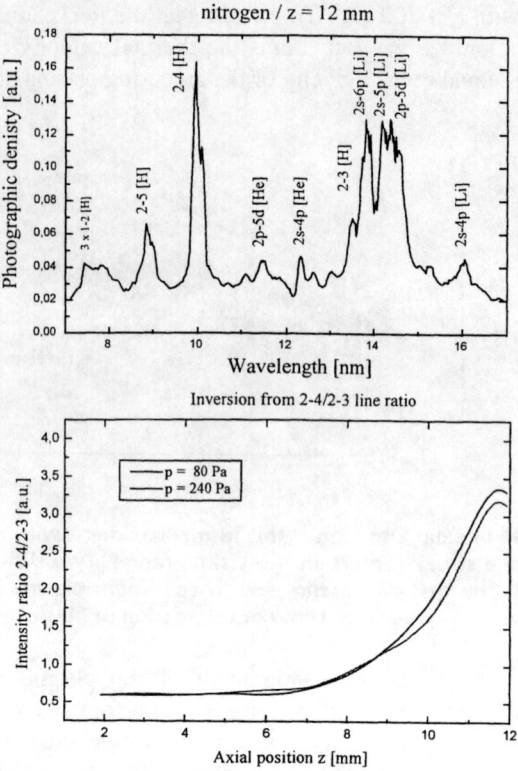

figure 4 Spectrum at z=12 mm. One can see a strong enhancement of 4-2 transition of Hydrogen-like NVII compared with 3-2 transition of the same ion. Enhancement of 2-5 and 2-6 transition in Lithium-like NV is also observed. Relative intensities (right image) of 2-3 and 2-4 transitions of Hydrogen-like NVII do not show any population inversion.

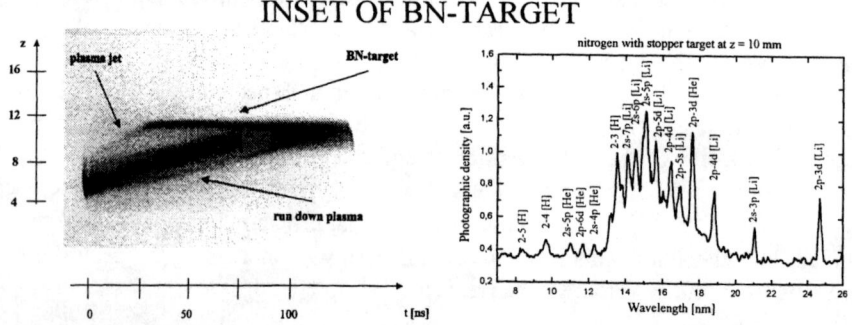

figure 5 Streak image and corresponded spectrum z = 10 mm distance from anode with inset of a BN targets as stopper. The previous seen inversion disappears due to the high density on the target.

To avoid reduction of jet velocity when decreasing the target density a solid BN target is placed at position z=10 mm. It is expected that the plasma jet can interact with neutral nitrogen atoms on the BN target. Spectrum and streak images of this interaction are shown in figure 5. It can be seen that the interaction with the target occurs first with the plasma jet and then with the run down plasma.

The observed spectrum features that transition is stronger that 2-4 one. In this case processes, which act in the opposite way like deexcitation and 3-body-recombination, occur. This is due to 3-body recombination with $n_0 > 10^{19}$ cm^{-3}. By use of an open target (BN target) with 2 mm hole, which is expected to cause lower target inside the hole due to expansion of target material, mainly Lithium like emission is observed.

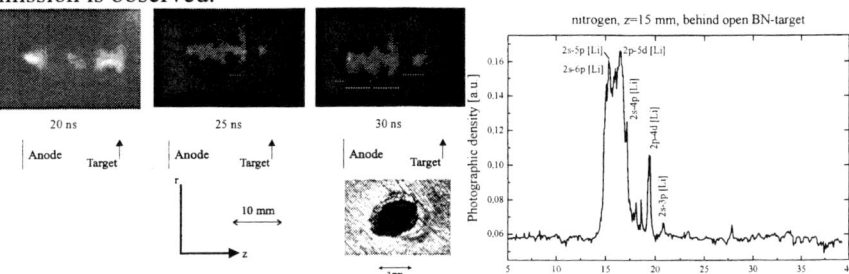

figure 6 Frame obtained with multi frame camera in the EUV range by inset of a BN targets (with whole, diameter: 2 mm, Position: z=10 mm). Emission occurs also behind the target. Multi frame images in the EUV-range. Behind the target intensive emission is observed.

CONCLUSION

It is shown that population inversion by charge exchange recombination for a pinch produced by a small plasma focus is possible. The population of the n=4 level is most effective when thermal energy of the ions is similar to the velocity of the axial plasma movement. The gain, which reaches values $G \cdot L \approx 1$, is expected in radial direction in opposition to other laser schemes which proposed gain in axial direction.

ACKNOWLEDGEMENT

Part of this work has been supported by INTAS under contract number 94-4096.

REFERENCES

[1] Mather J.W. in "Methods of experimental Physics, Plasma Physics", Vol. 9B, H. R. Griem and R. H. Loveberg, (Eds.), p. 187, Accademic Press, New York, 1971

[2] Vinogradov A.V. and Sobel´man I.I., Soviet Physics JEPT, Vol. 36, No. 6, p. 1115 (1973)

[3] Elton R.C., X-ray Lasers, Academic Press Boston (1990)

[4] Koshelev K.N. and Kunze H.-J., Ion beams from axial discharges and the x-ray laser problem, AIP conf. Proc. 299, , 231-235, 1994

[5] Ryufuku H., Sasaki K. and Watanabe T., Phys.Rev. A 21, p. 745-750, 1980

[6] Lebert R., Neff W., Rothweiler D., Journal of x-ray science and technology 6, p. 107-140, 1996

[7] Rocca J.J., Shlyaptsev V., Tomasel, F.G. Cortazar O.D., Hartshorn D., and Chilla J.L.A., Phys.Rev. Lett. **73**, p. 2192 ff, 1994

Transition from Column to Micropinch Regime in Z-Pinches

A. Engel[1], R. Lebert[1], K.N. Koshelev[4], Yu.V. Sidelnikov[4], C. Gavrilescu[3] and W. Neff[2]

[1]RWTH Lehrstuhl für Lasertechnik and [2]FHG-Institut für Lasertechnik Steinbachstraße 15, D-52074 Aachen, Germany, [3]University of Iasi, Copou 11, 6600-Iasi, Romania, [4]Institute for spectroscopy, Russian Academy of science, 142092 Troitzk, Moscow region

Plasma focus and Z-pinches are known to be intensive sources of K-ion radiation. This radiation is observed in two different regimes of compression: column and micropinch. Appearance of these regimes depends on combination of discharge circuit parameter and element composition of plasma. Column regime is typical for low current discharges operating in low Z gases. Micropinch regime, which represents a development of „neck" type instabilities in a presence of strong radiation losses, is typical for heavy ion plasma, i.e. vacuum spark or plasma focus with admixture of heavy gases. Transition from column to micropinch mode has been investigated experimentally. It was found that appearance of either regime can be quantitatively described by a distinction parameter depending on pinch current, particle density and used element.

INTRODUCTION

Pinch plasmas are strong emitters in the EUV-region and find diverse applications (lithography, microscopy and material studies[1]). The emission of these plasmas is observed in column and micropinch mode. Column-like emission and their dynamics are well interpreted by radiation-hydrodynamical[1] models whereas micropinches -further development from neck type instabilities in column-like plasma- are explained by the model of Koshelev and Vikhrev[2]. Micropinches and column of K-ion emission usually occur exclusively. In this paper experiments on the transition regime between the two phenomena with particular regard to argon (Z=18) are described. In order to control the device parameter the Mather type Plasma Focus device[3] PF3 (pinch current about 200 kA) operating with stationary pure gas filling of argon is investigated.

COLUMN AND MICROPINCH

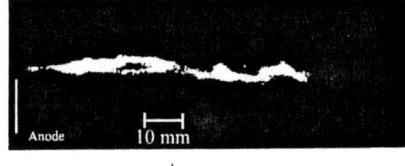
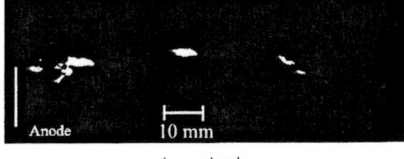

column   micropinch

figure 1 Time integrated in the soft x-ray range taken with pinhole camera. The observed axial and radial extensions are characteristic for either regime of K-radiation.

For light elements (Z ≤ 18) the K-shell emitting region is a column with several 100 μm in diameter and several mm in length as shown in figure 1 for argon. Operating with heavier elements (Z ≥ 18) micropinches are observed. Typically micropinches are about 10 μm in diameter and about 100 μm in length. For high

Z-elements the micropinch regime is most probable. Therefore an intermediate element number Z exists, where a transition between regimes occurs. Table 1 shows that this element number depends on the value of the known parameter I_p^2/N_i, which is build by the pinch current I_p and the line density N_i, if Bennett equilibrium[7] is assumed. This value is highest for plasma focus PF3 and lowest for vacuum spark. Therefore transition regime is found at Z=18 for PF3 and no one for the vacuum spark.

Device	N_i [cm^{-1}]	I_p [kA]	Element [Z]	I_p^2/N_i [normalised to PF3]
PF3[6]	10^{16}	200	18	1
SPEED 2[5]	10^{18}	500-700	14	0.2
MP 100[8]	10^{17}	200-400	10	0.16
Vac. Spark[4]	10^{17}	100	not observed	0.1

table 1 element number where transition from column to micropinch occur.

TRANSITION REGIME FOR ARGON

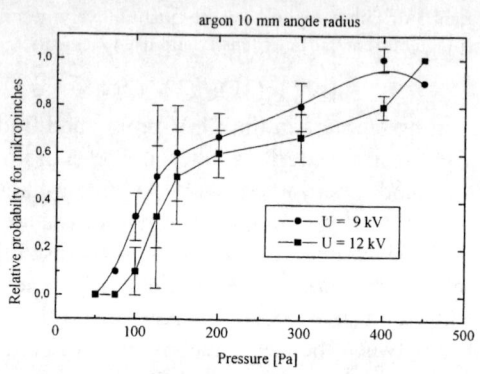

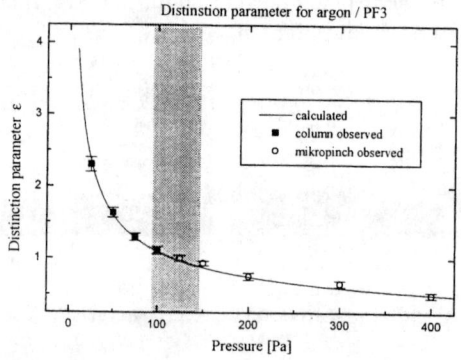

figure 2 transition from column to micropinch working with argon. The transition regime can be moved by varying the device parameters voltage, anode radius and pressure. In principle for given anode radius and voltage the transition regime is selected by the gas pressure. The transition is prognosticated correctly for the dependence of the regime on the pressure like observed in experiment.

For argon both modes can be selected by the controllable device parameters gas pressure, voltage and anode radius. The transition regime has been studied in more detail. In figure 2 the relative probability of the occurrence of micropinches in argon discharges is plotted as a function of the gas pressure. If the anode radius a (a=10 mm) and the pinch current I_p (I_p=200 kA) are kept constant, for p < 100 Pa hardly any micropinch occur. For 100 Pa < p < 150 Pa there is the transition regime and for p > 150 Pa only micropinches are found. The transition regime moves to lower pressure p, when the pinch current is reduced. This happens according to the scaling parameter I_p^2/N_i.

The decision whether micropinch or column mode is expected can be estimated by combination of characteristic currents, which are associated with the processes playing a significant role in the physics of the pinch. The pinch current I_p can be determined by the experiment. The Critical current I_{cr}, which is a generalised Pease-Braginskii current for not totally ionised plasma determines whether radiative collapse can be achieved[3]. For $I_p > I_{cr}$ radiative collapse is possible. The "Shell-current" I_s,

$$I_S = \sqrt{\frac{4\pi}{\mu_0} n_0 \cdot a^2 \cdot T_e \cdot (Z_{eff} + 1)} \quad [1]$$

[n_0: density; T_e: electron temperature]

that is the current estimated from Bennett-equilibrium at a temperature needed for ionising a closed shell ($kT_e \approx 1/5 \times E_{ion}$) determines whether a shell is ionised in the initial plasma. For $I_{pinch} > I_{shell}$ the shell is ionised. There are different shell-currents for each shell. Combination of these currents with special emphasise to the condition for micropinches ($I_p > I_{cr}$), one gets a distinction parameter ε with the property that for ε < 1 micropinch-mode and for ε > 1 column regime occurs:

$$\varepsilon = \left(\frac{I_{cr}}{I_p}\right)^2 \cdot \frac{I_p}{I_S} = \frac{I_{cr}^2}{I_p \cdot I_S} = \frac{I_{cr}^2}{I_p \cdot \sqrt{\frac{4\pi}{\mu_0} n_0 \cdot a^2 \cdot \frac{1}{5} E_{ion} \cdot (Z_{eff} + 1)}} \propto N_i^{1/2} \cdot Z^{-11/4} \cdot I_p^{-1} \quad [2]$$

This parameter predicts for the same device, micropinches are more common with heavier gases. As the line density N_i is proportional to the gas density the parameter ε for each gas depends also moderately on pressure p. Figure 2 shows ε as a function of gas pressure for various working gases with pinch currents used in the experiments described in table 1. Note the observed column- and micropinch-regimes are predicted correctly. For the K-shell emission of argon is predicted that the transition only depends on the gas pressure. It is also predicted, that for shells with larger quantum number n column mode is more probable.

INVESTIGATION OF INITIAL PINCH PHASE

Furthermore, the kinetic energy per ion is high enough to achieve complete ionisation after thermalization. In addition the relaxation-rates between electrons and ions and ionisation rates must be high enough to allow ionisation during the lifetime of the pinch. To determine the properties of either regime initial pinch phase for argon is investigated.

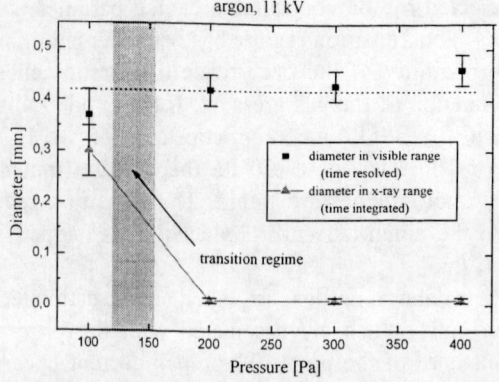

figure 3 Source diameter in visible and x-ray range for argon. In visible range the diameter is more or less constant for both regimes whereas in the x-ray range the difference in size is evident.

Figure 3 shows that in any case a column like plasma with prepinch radius $r_0=0.3$ mm (visible range) is created. Differences are observed the final size (x-ray range) and the heating mechanism (figure 4). In column regime shock heating is most important, whereas for micropinch regime ohmic heating become more relevant. The main point for micropinch regime is that the kinetic energy per ion is not high enough to achieve complete ionisation after thermalization (figure 5).

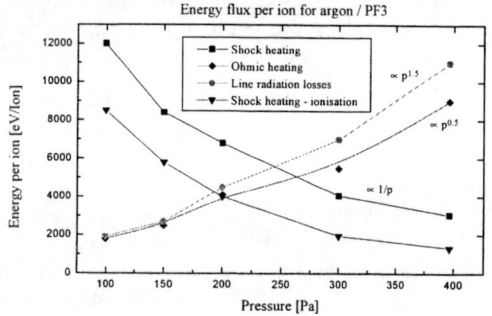

figure 4 Collapse velocity plotted against working gas pressure in order to measure collapse velocity and energy input due to shock heating obtained from measurements of the collapse velocity. For higher pressure p the kinetic energy per ion decreases according to known scaling $\sim p^{-0.5}$. In column regime the temperature is high enough to ionise K-shell, in micropinch regime only L-shell.

Further ionisation is only reached after further local magnetic compression where at least two shells can be ionised during pinch duration. One by means of transfer of kinetic energy and the other by magnetic compression in the micropinch.

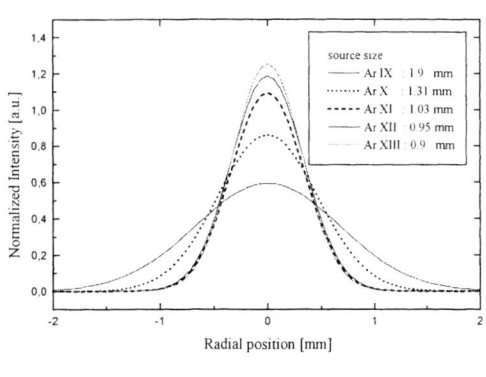

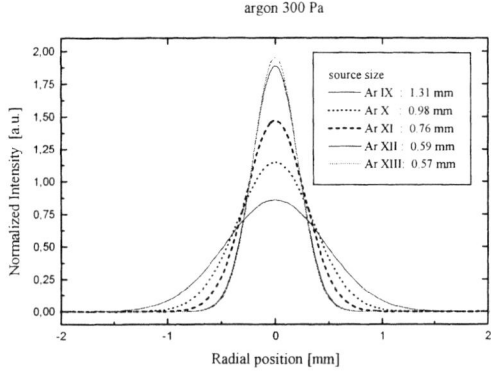

figure 5 Normalised source size for argon in column and micropinch regime. The radius of the source when emitting L-radiation is larger in micropinch regime. This means that the plasma temperature in micropinch mode is lower when reaching the axis and that the emission of L-shell radiation occurs mainly during pinch phase.

CONCLUSIONS

In a pinch plasma the x-ray emission is observed from cylindrical Bulk plasma (column) or from micropinches. A parameter ε is proposed which determines which mode occurs. For Plasma Focus devices with a current of about 200 kA argon features both modes in dependence of the neutral gas line density.

ACKNOWLEDGEMENT

Part of this work has been supported by Deutsche Forschungsgemeinschaft (DFG) under contract number Le 904/1-1 and by Deutsche Gesellschaft für Luft- und Raumfahrt Forschungsgemeinschaft (DLR) under contract number RUS 122-96.

REFERENCES

[1] Pereira N.R. and Davis J, J. Appl. Phys. 64 (3), R1-R27, 1989

[2] Koshelev K.N. and Pereira N.R, J.Appl.Phys. 69, R21 , 1991

[3] Mather J.W., Methods of experimental Physics, Plasma Physics **9B**, Accademic Press, N.Y., 1971

[4] Schulz A., Burhenn R., Rosmij F.B. and Kunze H.-J., , Phys.D.: Appl. Phys. 21 1827-1829, 1988

[5] Decker G., et al., Science & Technology, 5 (1), p. 112-118, 1996

[6] Lebert R., Engel A, Neff W., J.Appl.Phys. 78(11), p. 6413-6420, 1995

[7] Bennett W. H. , Phys. Rev. 45, 1934

Hotspot Features in a Small Plasma Focus Operating in H_2-Ar and H_2-N_2 Mixtures

M. Favre, P. Choi*, C. Dumitrescu-Zoita*, P. Silva, H. Chuaqui, and
E. Wyndham

Facultad de Física, Pontificia Universidad Católica de Chile, Casilla 306, Santiago 22, Chile
*LPMI, Ecole Polytechnique, Palaiseau 91128, France

Abstract. Preliminary experimental results on the investigation of hotspot formation in a small 3.8 kJ Plasma Focus device operating at constant mass density in Hydrogen-Argon and Hydrogen-Nitrogen mixtures are presented, at pressures from below 0.2 Torr upward, with relative percentages of Hydrogen ranging from 80% to 100%. The diagnostics includes voltage and current measurements, multi pinhole and slit-wire X-ray photography, filtered PIN diode array, and an array of small magnetic probes located along the cathode rods to study the current sheath structure during the run down phase of the discharge. A measurement of the hot-spot characteristic size in different spectral regions is obtained from the slit-wire images, whereas the time resolved information from the PIN diodes is used to infer the temperature evolution of the plasma points. Typical results show more reproducible hotspot formation in H_2-Ar then in H_2-N_2 mixtures, with characteristic temperature in the 300 to 570 eV range, and hotspot size between 130 and 260 μm, depending on mixing ratio.

INTRODUCTION

Hotspot formation is a common feature in a variety of transient discharges of the Z-pinch type, such as vacuum spark, gas puff, exploding wire, composite pinch and plasma focus(1-3). Hot-spot phenomena have attracted renewed interest in view of potential applications as soft X-ray radiation sources. Although hotspots have been investigated in different experiments, the physical mechanisms involved in their generation and time evolution, are still not clear. It has been long recognized that overall performance in Plasma Focus (PF) operation depends critically on the quality of the current sheath in the axial phase. The structure of the current sheath depends strongly on the operating pressure and the presence of impurities. We present preliminary experimental results on the investigation of hotspot formation in PFP-I, a small 3.8 kJ PF device operating at constant mass density in H_2-Ar and H_2-N_2 mixtures, at pressures from below 0.2 Torr upward. Constant mass load operation allows effects due to gas admission on radial collapse and subsequent hotspot formation to be decoupled from pure dynamic effects. The main diagnostic for hotspot investigations is a combination of multi pinhole and slit-wire(4) X-ray photography, with filtered PIN diodes. An array of non invasive magnetic field probes is used to investigate correlations between current sheath structure and hotspot features.. The simultaneous recording of pinhole and slitwire images allows both the temperature and characteristic size of the hotspots to be determined. Typical size for the hottest emitting region, at temperatures between 300 and 600 eV, is found to be around 120 µm, with a typical duration of the high temperature phase of less than 20 ns.

EXPERIMENTAL RESULTS AND DISCUSSION

PFP-I, is powered by TORTUGA, a 9 µf, 30 kV capacitor bank. Typical peak current is 140 kA at 20 kV charging. Further details on the experimental arrangement can be found elsewhere(5). An array of small magnetic probes located along the cathode rods is used to study the current sheath structure during the run down phase of the discharge. An array of filtered PIN diodes located side and end-on, is used in conjunction with a multi-pinhole and slitwire X-ray camera. Six pin-holes of 100 and 200 µm diameter are used with six 100 µm slits. X-ray images are recorded in HP5. The same filter combination is used in PIN diodes and pin-hole slit-wire camera. Matched pairs of filters are used in order to discriminate the Ar line emission at around 4 and 20 Å, and the anode K_α emission at 1.38 Å. Thin wires of 25 to 250 µm diameter were used to estimate the measured hotspot size.

Characteristic voltage, dI/dt and Bdot array signals are shown in Fig. 1, for different gas mixtures, at a constant mass density of 0.15 µg·cm^{-3}. Bdot probes are

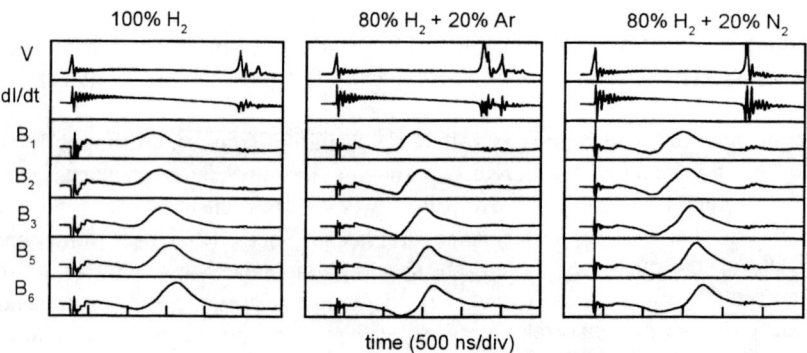

Figure 1: characteristic voltage, dI/dt and Bdot signals

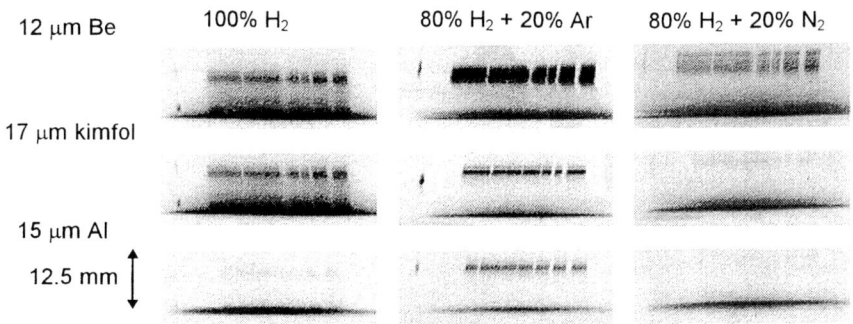

Figure 2: characteristic pinhole and slitwire images

separated 1 cm, and B_1 is located 3.5 cm above the base plate of the PF chamber. Fig. 2 shows pinhole and slitwire images for the same conditions than those in Fig. 1. Pinhole images are on the left hand side of each slitwire image. Fig. 3 shows voltage, dI/dt and X-ray PIN-diode signals, with different filtering, for the same shots than in Fig. 2.

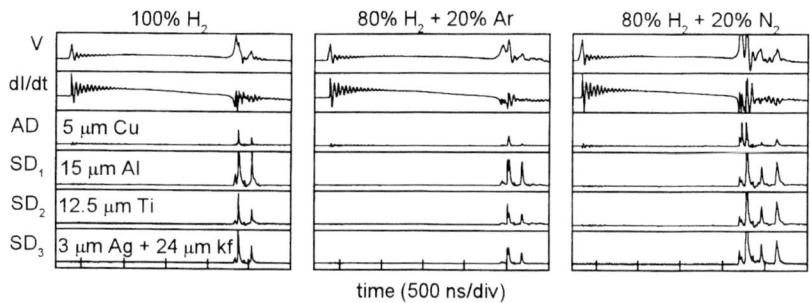

Figure 3: characteristic voltage, dI/dt, and X-ray PIN diode signals. AD is end-on, SD_1 to SD_3 are side-on. The same scale is used for all PIN diode signals.

A simple analytical model is used to interpret the Bdot probe signals(5). The Bdot signals in Fig. 1 show clear differences in the shape and time evolution of the current sheaths, depending on the impurity content. In pure H_2, a nearly constant sheath current density profile is observed to evolve at nearly constant axial velocity, which is not observed with either Ar or N_2 admission, where the characteristic signals evolve continuously, with a changing time dependent profile. This signal behavior can be interpreted as an indication of a transition from a near constant sheath current density profile when operating in pure hydrogen, to a time dependent, increasingly steeper sheath current density profile, as a result of the enhanced ionization due impurity effect, when heavier gases are included. The sudden increase in Bdot signals observed at ~ 250 ns in all cases, but more noticeable in Ar, is probably due to short circuiting to the outer electrodes, through current filaments jumping from the radially expanding sheath detached from the insulator.

The slitwire images in Fig. 2 show clear differences in the final compression phase, depending on the impurity content. In pure H_2 a well defined, narrow ~500 μm diameter pinch is observed, with two brighter spots, one next to the anode front, the other ~8 mm above. With H_2-Ar mixture, a characteristic bright single spot is observed ~8 mm above the anode, which is embedded in a weakly emitting diffuse pinch. For the H_2-N_2 mixture, an elongated, axially structured pinch is seen at the same axial position of the bright spot observed in H_2-Ar and pure H_2. Further features of the X-ray emission can be inferred from the PIN diode signals in Fig. 3. In all cases, a first small X-ray pulse is detected at the time of the first voltage spike, which is followed by one or more X-ray pulses, depending on gas mixture, associated with subsequent high voltage peaks. A change in the relative amplitude of the different X-ray pulses detected with different filters allows emission due to plasma radiation to be discriminated from beam target emission from the anode edge. The H_2-N_2 mixture gives a constant relative amplitude ration regardless of the transmission band of the filters used, in contrast with the situation observed in pure H_2 and H_2-Ar mixtures, were the relative amplitude of the last X-ray pulse decreases as the filtering becomes harder. It can also be seen that the amplitude of the second large X-ray pulse in pure H_2 and all pulses in H_2-N_2 mixtures show little dependence on the filtering used. From these observations we can infer that most of the X-ray emission in pure H_2 and H_2-N_2 mixtures is dominated by characteristic K_α radiation due to high energy electron beams hitting the anode edge, whereas in the case of H_2-Ar mixtures, a significant amount of the X-ray emission is emitted from a hot dense plasma localized in hotspots. This is consistent with the relative intensity of the hotspot and anode emission seen in the slitwire images in Fig. 2. More detailed analysis of slitwire images in H_2-Ar shots give characteristic hotspot sizes ranging from 130 to 250 μm, with temperatures from 430 to 570 eV, as determined using the XRAYFIL(6) code. Further analysis of the existing data is in progress and a detailed characterization of the hotspots and the surrounding plasma, as a function of the impurity content, will be published elsewhere.

REFERENCES

1. C.R. Negus and N.J. Peacock, J. Appl. Phys. D **12**, 91 (1975)
2. J. Shiloh et al., Phys. Rev. Lett. **40**, 515 (1978)
3. S.M. Zakharov et al., Sov. J. Plasma Phys. **10**, 303 (1984)
4. H. Chuaqui et al., "Observations of Dense Plasma Formation in the Vacuum Spark", in *Proceedings ICPP 94*, Foz de Iguazu, Brasil (1994), Vol. III, p. 29.
5. M. Favre et al., "Current Sheath Studies in a Small Plasma Focus Operating in H_2-Ar Mixtures", in *Proceedings VII LAWPP*, Caracas, Venezuela (1997).
6. C. Dumitrescu-Zoita and P. Choi, "XRAYFIL, an Analysis Code for Non-Dispersive X-ray Diagnostics", these proceedings.

Large-Scale Spheromak-Like Magnetic Configuration (SLMC) in High-Current Discharges: Self-Formation and Self-Compression of the SLMC in Plasma Focus Experiments

A.B. Kukushkin, V.A. Rantsev-Kartinov, A.R.Terentiev

Institute for Nuclear Fusion, RRC "Kurchatov Institute",
123182 Moscow, Russia

Abstract. Experimental results are presented which verify the possibility, formerly predicted [1], of the self-formation of a closed, spheromak-like magnetic configuration (SLMC) in a plasma focus discharge.

Experimental results are presented which verify the possibility, formerly predicted [1], of the self-generated transformation of the magnetic field in plasma focus discharge to produce a long-living closed, spheromak-like magnetic configuration (SLMC) of several cm size, confined and compressed by the residual magnetic field of the plasma focus discharge, at a time scale of several hundreds of the respective linear Z-pinch MHD characteristic lifetime. The energy conversion mechanism suggests a possibility of achieving the higher values of plasma power density through utilizing the self-production of a target pre-fusion magnetized plasma in the Z-ϑ pinch at the major axis of the SLMC. Such an approach, if applied to achieving the fusion ignition, gives a 3D magneto-inertial confinement which combines the advantages of a Dense Z-pinch (namely, high peak values of plasma power density [2]) and a quasi-steady-state magnetic confinement (namely, enhanced MHD stability of a closed magnetic configuration, driven and confined by an open magnetic configuration).

The model [1] (see Figs.1,2) and the analysis of experimental results obtained from earlier experiments in various high-current gaseous discharges has allowed the identification of the following characteristics of the SLMC formation [3]:

(1) The self-consistent generation of a poloidal magnetic field (the dynamo effect), solely by the internal dynamics of the magnetic field in the discharge.

(2) Strong filamentation of electric currents, which occurs both in the inner region of the SLMC (i.e. in the combined Z-ϑ-pinch) and in its periphery.

(3) SLMC formation is stimulated by the enhanced propagation rate of the magnetic field along the anode, due to the Hall effect in plasmas.

(4) A magnetic field reconnection process leading to the formation of the SLMC as a closed configuration, appears to occur before the current sheath converges on the axis.

(5) In its final stage, the SLMC takes the form of a squeezed spheromak configuration, confined and driven by the pressure of the residual magnetic field of the plasma focus discharge.

(6) Large space scale (vs. "hot spot") determined by the geometry (and capacitance) of the facility.

(7) The power density in the combined Z-ϑ-pinch at the major axis exceeds the peak power density of a force-free flux-conserver-confined spheromak by several orders of magnitude.

(8) The SLMC exhibits a cyclical, evolutionary tendency to form, be compressed and eventually repelled away from the anode, and reform repeatedly.

(9) Self-organization of the discharge plasma (non-monotonic dependence of input vs. output parameters; "quantization" of the discharge energy).

A great deal of experimental data obtained in earlier neon gas studies carried out at the Filippov-type plasma focus facility [4] supports the SLMC model [1]. Some of the results presented here and in [3] have not been previously published, as they were not fully understood before. Significantly, the identification of the SLMC formation appears to be available essentially from *combining* the results of the following diagnostics with different (and complementary to each other) spatial and spectroscopic scales: namely, (i) motion picture of the evolution of SLMC formation, taken with the help of a ruby-laser-based interferometer operating in the Bates regime (0.01-J laser pulse energy; 2-ns duration) (Fig.3); (ii) visible light photographs taken with the help of an electronic optical converter which is synchronized with the current to an accuracy of <50 ns (Fig.4); (iii) time-integrated SXR spectra from a pinhole camera.

The results are presented for the following discharge conditions: mushroom-shaped anode (11-cm diam.) inside a coaxial metallic chamber 80-cm long and 30-cm high, which acts as the cathode; capacitance, 180 µF; initial inductance, 55 nG; initial voltage, 16 kV; discharge energy, 20 kJ; maximum current, 530 kA; neon gas pressure, 3 Torr. Time zero (t=0) corresponds to the major peak of the time derivative of the current.

REFERENCES

1. Kukushkin A. B. and Rantsev-Kartinov V. A., *Preprint of the RRC "Kurchatov Institute"*, IAE 5646/6, Moscow, June 1993.
2. *AIP Conference Proceedings #299*, Dense Z-pinches 3rd Int. Conf., London, April 1993, Eds. Malcolm Haines and Andrew Knight, New York: AIP Press, 1994.
3. Kukushkin A. B., Rantsev-Kartinov V. A., and Terentiev A. R., *Fusion Technology, 8 (1997); Preprint of the RRC "Kurchatov Institute"*, IAE 5737/7, Moscow, January 1994; *Transactions of Fusion Technology* **27**, 325 (1995).
4. Orlov M. M., Terentiev A. R., and Khrabrov V. A., *Fizika Plazmy (Sov. J. Plasma Phys.)*, **11**, 1268, 1517 (1985).

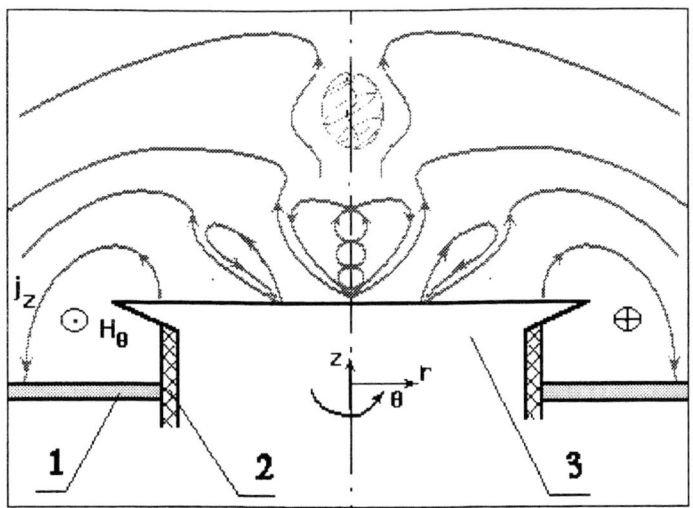

Figure 1. Motion picture of the magnetic field front (and electric current sheath) in a spheromak-producing plasma focus discharge (1, cathode; 2, insulator; 3, anode).

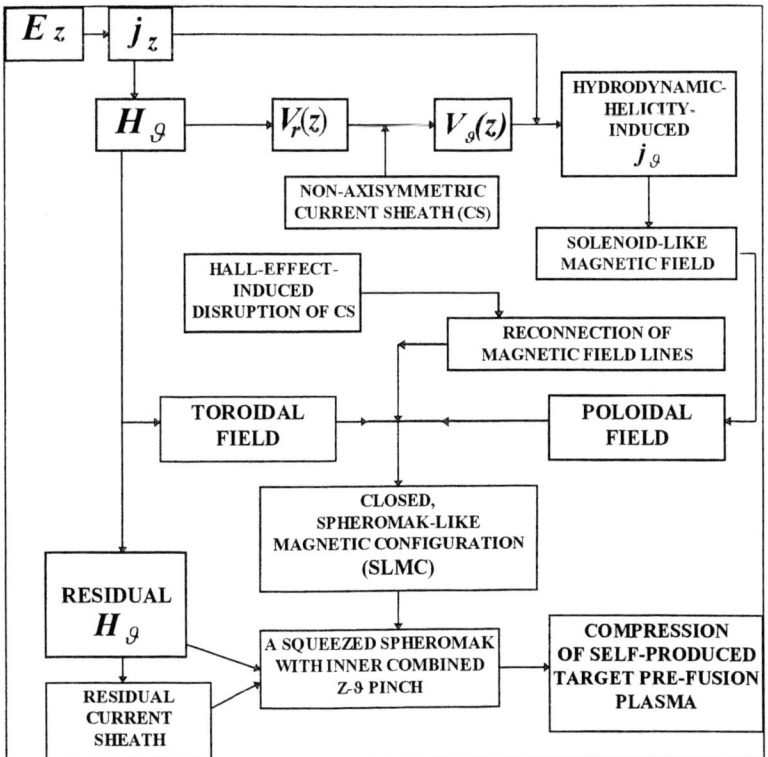

Figure 2. The progression of input energy transformations in a spheromak-producing plasma focus discharge

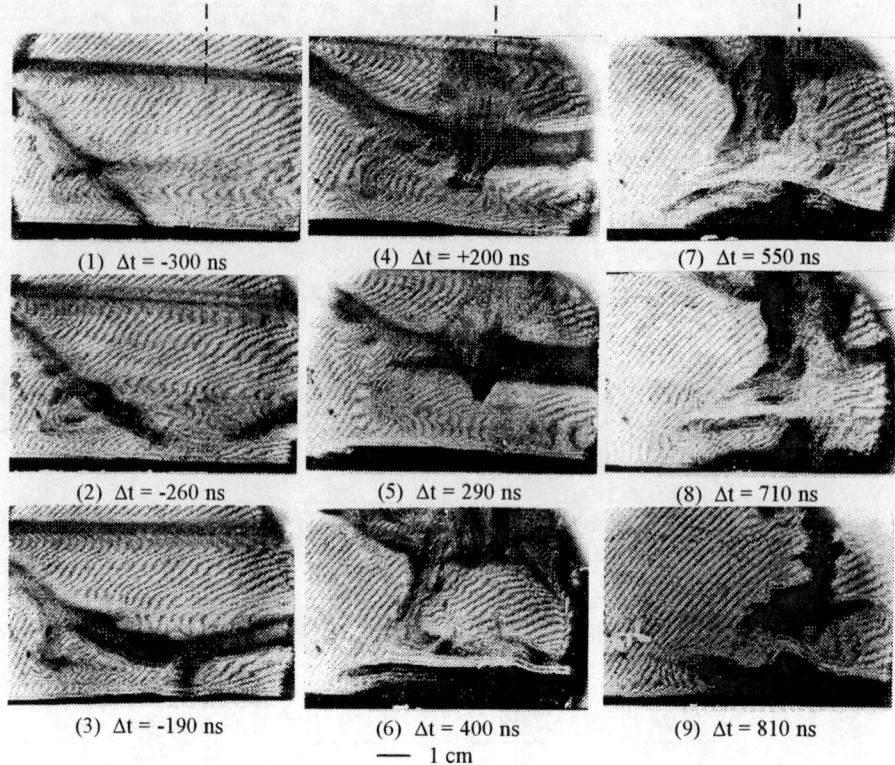

(1) Δt = -300 ns   (4) Δt = +200 ns   (7) Δt = 550 ns

(2) Δt = -260 ns   (5) Δt = 290 ns   (8) Δt = 710 ns

(3) Δt = -190 ns   (6) Δt = 400 ns   (9) Δt = 810 ns
　　　　　　　　　　── 1 cm

Figs.3.1-3.9 are the frames of a motion picture taken with the help of the ruby-laser-based interferometer (2-ns pulse duration; pulse direction perpendicular to the system axis). The system axis location is indicated. The heel-like form of the current sheath (Δt = -300 ns) is caused by the enhanced rate of the magnetic field propagation along the anode, due to the Hall effect. The magnetic field reconnection leading to formation of a closed magnetic configuration take place before the current sheath converges on the axis (Δt = - 260, -190 ns). The latter is suggested by

(i) the *helical* structure of the filaments clearly seen in the middle and especially the lower part of the respective figures; (ii) the detection of a number of *precursors* caused by the reconnection process (viz. short pulses of soft and hard X-rays, visible light etc.) long before (200-300 ns) the major peak of the time derivative of total electric current; (iii) the convergence of the current sheath at the axis *long before* (200 ns) the major peak of the time derivative of total electric current; and (iiii) the *conservation* of the structure, which has detached itself from facility circuit current, while it eventually is forced away from the anode (Δt = +200 and +290 ns). Pictures (6)-(9) show second cycle of forming a closed magnetic configuration. This phase stops with the second peculiarity of the current derivative. The decay of the plasma formation takes plays at Δt = +850 ns.

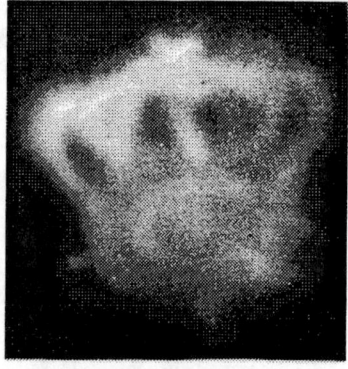

── 1 cm

Fig. 4. Visible light photograph, t=+320 ns (15 ns exposure; 45° below the system axis; anode disc is seen in the background), (cf. thick filaments in Fig.3.5.)

Short-Scale Mixing of the Plasma and Magnetic Field, and Magnetic Flux Ropes in Plasma Focus Experiments

A.B.Kukushkin, V.A.Rantsev-Kartinov, A.R.Terentiev, and K.V.Cherepanov

RRC "Kurchatov Institute", 123182 Moscow, Russia.

Abstract. Experimental results are presented for the formation of complicated 3D structures, both disordered and ordered ones, in plasma focus gaseous discharges. This includes (i) intense short-scale mixing of the plasma and magnetic field, and (ii) cell-like structures built of plasma filaments and magnetic flux ropes, in particular, a heterogeneous magnetic flux rope.

1. INTRODUCTION

The formation of a linear dense Z-pinch as an essentially axisymmetric 2D structure was the major goal of experimental research programs on Dense Z-Pinches, both in cylindrical geometries and non-cylindrical ones, like e.g. plasma foci. Here, all the deviations from axial symmetry were considered as a serious threat to concentrating the energy. However, a lot of data is accumulated that exhibit importance of the essentially 3D behavior of the plasma and magnetic field in the case of successfull enough driving the load in high-current dicharges. In particular, formation of a closed, spheromak-like magnetic configuration, of several cm size, in Filippov-type plasma focus gaseous discharges takes place thanks to (i) strong filamentation of helical electric currents and (ii) enhanced propagation rate of magnetic field (and current sheath) along the anode, due to Hall effect in plasmas (see [1] and this volume). So, the question does exist: to what extent does the 3-D 2-fluid MHD govern the high-current discharges?

Here, we present a number of experimental results which have been accumulated in earlier studies carried out at the Filippov-type plasma focus facility [2]. Such a presentation is focused at illustrating the coexistence and interaction of complicated structures, both disordered and ordered ones, of various space scale, in plasma focus gaseous discharges at conditions typical for the high current discharges.

2. SHORT-SCALE MIXING OF THE PLASMA AND MAGNETIC FIELD

The Hall effect in plasmas [3], which is caused by the frozenness of the magnetic field into electron plasma and by the resulting transfer of magnetic field with electric current velocity, manifests itself in the enhanced rate, as compared with ordinary diffusion, of magnetic field propagation in plasmas in the case of electric current moving along the gradient of plasma density. The major physical mechanism is the "scattering" of magnetic field, which is transferred by the electric current, at positive gradient of electron density. Such a phenomenon is identified in full in the case of steep electron density gradients at plasma-conductor boundaries. For instance, the current sheath slipping along the anode at initial stage of the discharge in plasma focus facility [2] has been reproduced in the 2-D 2-fluid numerical modelling [4] with allowing for the Hall term (see also the database presented in [1]). Contrary to ideal MHD instabilities, this mechanism gives regular, highly reproducible dynamics that agrees quite well with numerous experimental data. Similar phenomena may take place in plasma interior as well. Here, the enhanced propagation rate leads to penetrating the domain of the lower plasma density and subsequent superseding the plasma by the magnetic field. Such a mechanism leads to a stochastic short-scale (as compared with space scales of the current sheath) mixing of the plasma and magnetic field, being thus an alternate for the snow plough regime of current sheath formation and driving.

The experimental setup [2] is a Filippov-type plasma focus with mushroom-shaped anode, 11 cm diameter, located inside a coaxial metallic chamber, which acts as the cathode. The main discharge parameters are: capacitance, 180 µF; initial inductance, 55 nH; initial voltage, 16-24 kV, varying with energy of 20-50 kJ, respectively; maximum current, 600 kA. The pictures shown in Fig.1 (t= -90 ns) and Fig.2 (t= +174 ns), for deuterium gaseous discharge of initial pressure 163 Torr, are the shadowgrams taken with the help of a ruby laser (0.01-J laser pulse energy; 15-ns duration; pulse direction perpendicular to the system axis). The anode is at the bottom of the diagnostic window (4 cm diam.), the plasma focus major axis coincides with the vertical axis of the window; time zero corresponds to the major (first) singularity of electric current derivative.

A great deal of experimental database exhibit cell-like structure of the plasma. Such a phenomenon may be interpreted as the formation of a thin volumetric (three-dimensional) net-like structure of the magnetic field penetrating the plasma. The local values of magnetic field inside this net may substantially

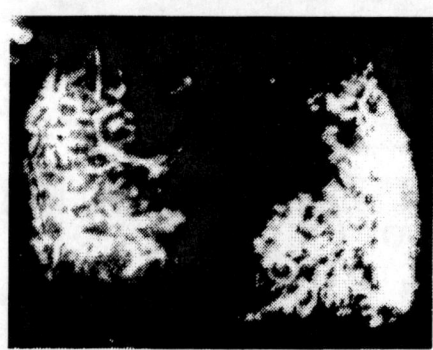
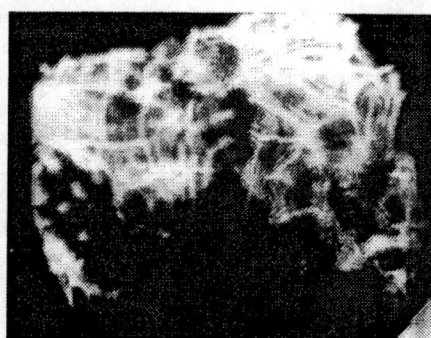

Figure 1	Figure 2

exceed its values averaged over several cells, up to the order of magnitude. Therefore this mechanism of plasma-magnetic field interaction doesn't need large spatially averaged values of the parameter $\omega_e \tau_{ei}$, thus suggesting that the criterion for the onset of the 2-fluid effects may be strongly dependent on the local values of density fluctuation level of the appropriate space scale. Such a 2-fluid instability being developed in a certain small volume, can propagate in space and interact with conventional ideal-MHD instabilities, e.g via triggering the current sheath breakthrough by the Rayleigh-Taylor instability.

3. FILAMENTS AND MAGNETIC FLUX ROPES

The complexity of short-scale structure of the magnetic field penetrating the plasma, leads to existence of a rich background for short-scale self-organization processes. This results in forming the strongly inhomogeneous plasma structures, with the filaments of electric current and the magnetic flux tubes (magnetic flux ropes, according to the space plasma language, see e.g. [5]) being the building blocks of these structures. The filaments are characterised by the enhanced plasma density, due to pinching effect, whereas magnetic flux ropes exhibit substantially lower plasma density, with force free-like configuration of magnetic field. It is combination of these two substructures, under condition of appreciable helicity, that supports long-range, essentially three-dimensional correlations of electric currents and magnetic field.

The filamentation of electric current is well known to characterize plasma behavior at initial stage of gaseous discharges. Being driven by the inflated magnetic field toward system's axis, the filaments may form a quasi-uniform current sheath of a cylindrical/linear Z-pinch. However, this may not be the case when filamented structures form essentially 3D plasma structure as it happens, in particular, in a certain type of plasma focus discharge resulted in the formation of a closed, spheromak-like magnetic configuration of several cm size [1] (see also this

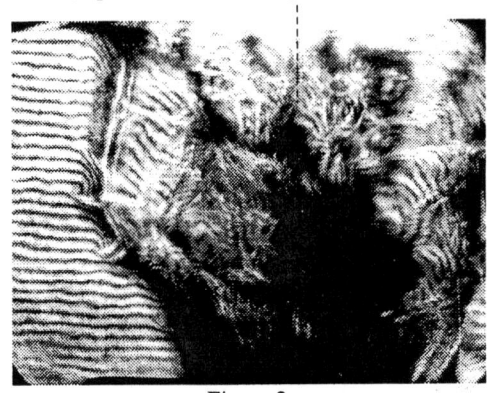

Figure 3.

volume). Here, filamentation of electric currents is needed for the production of the poloidal magnetic field and the respective 3D large-scale ordering of both the plasma incorporated in a closed configuration and the plasma carrying the current of external circuit. Figure 3 (interferogram, Bates scheme, ruby laser, 2ns pulse duration) shows fine structure of the filamented inner, closed mag-

netic configuration and the current sheath formed by the residual magnetic field (D_2, 6 Torr, 24 kV, -150 ns; system's axis is indicated, space scale the same as in Figs.1,2). The interferogram of Fig. 4 (negative; 20 ns; discharge type similar to that of Fig.3) illustrates the braidedness of the twisted filaments.

 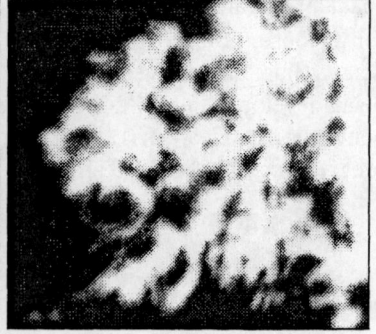

Figure 4   Figure 5

Interaction of filaments and magnetic field lines, via twisting, winding and interweaving, leads to formation of the heterogeneous force free-like magnetic configurations. Such a structure may form a heterogeneous magnetic flux rope. Separate section of the (teared to pieces) rope forms a stick-like, heterogeneous plasma formation embedded into a closed, spheromak-like magnetic configuration. Typical formation of this type is shown in Fig. 5 (see left upper part of the Figure) which is extracted from the right lower part of Fig. 1.

CONCLUSIONS

Experimental results presented illustrate complexity of plasma and magnetic field behavior in high-current discharges. It follows that the 2-fluid effects (the Hall effect in plasmas) may produce more intense (short-scale) mixing of the plasma and magnetic field, as compared with that predicted by the ideal MHD. This, in turn, leads to formation of essentially 3D structures of space scale of the above mixing. Interaction of these processes with electric current filamentation leads to long-range 3D correlations and strong local self-organization. The latter results in formation of closed, strongly inhomogeneous magnetic configurations of various space scale, up to several cm size formations reported in [1].

The results suggest the necessity to allow for the effects of the 3-D 2-fluid MHD in numerical modelling of plasma radiation sources.

REFERENCES

1. Kukushkin A. B., Rantsev-Kartinov V. A., and Terentiev A. R., *Fusion Technology, 8 (1997); Preprint of the RRC "Kurchatov Institute"*, IAE 5737/7, Moscow, January 1994; *Transactions of Fusion Technology* **27**, 325 (1995).
2. Orlov M. M., Terentiev A. R., and Khrabrov V. A., *Fizika Plazmy (Sov. J. Plasma Phys.)*, **11**, 1268, 1517 (1985).
3. Morozov A.I., Shubin A.P., *Zh.Eksp.Teor.Fiz. (Sov.Phys.JETP)* **46**, 710 (1964)
4. Vikhrev V.V., Zabaidullin O.Z., Terentiev A.R., *Plasma Phys. Reports*, **21**, 20 (1995).
5. *Physics of Magnetic Flux Ropes*, Geophysical Monograph 58 (Eds. C.T.Russel, E.R.Priest, L.C.Lee), Am. Geophys. Union, 1990.

STUDIES ON A SMALL MODIFIED PLASMA FOCUS OPENING SWITCH

W.S. Leong, C.S. Wong, *P. Choi and S.P. Moo

*Plasma Research Laboratory, Physics Department, University of Malaya
50603 Kuala Lumpur, Malaysia*

**Laboratoire de Physique des Milieux Ionises, Ecole Polytechnique
Palaiseau 91128, France*

Abstract

The small plasma focus device UNU/ICTP PFF has been modified to test its operation as an opening switch, with a plasma filled diode as the load. Recent experiment results on this modified plasma focus opening switch (MPFOS) showing long conduction opening action have been reported [1,2,3]. In this paper, a series of experiments to characterize the ion beams produced will be reported.

Introduction

In a typical inductive energy storage system, energy stored in a capacitor bank is discharged through the circuit inductance and an opening switch, which is initially in the close state and provides full conduction. When the peak current is reached, the opening switch will be switched, from a low impedance state to a high impedance state. Thus the primary flowing current is transferred to the load, which is connected across the switch. A variety of inductive opening switch technologies have been studied extensively in recent years [4,5].

In a previous paper, the long conduction switching concept of plasma flow opening switch designed based on a modified plasma focus geometry has been described and discussed [1]. This modified plasma focus opening switch (MPFOS) shows an opening action with combination of the plasma flow switch during the long conduction phase lasting for 2.1µs and the dynamic plasma in 60ns [2,3]. In this paper, the ion beam generated by the dynamic plasma be studied using the time of flight techniques.

Experimental Set up And Diagnostics

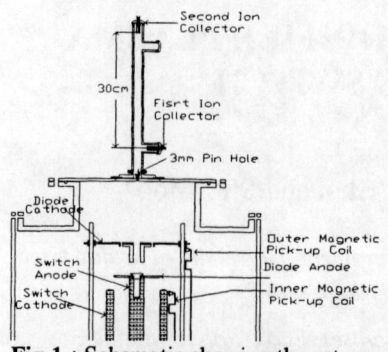

Fig.1 : Schematic showing the set up at the top part of MPFOS.

In the present series of experiments, the MPFOS is operated in argon gas and in nitrogen gas. The system is discharged at 14.5kV at its optimum operating pressure range of 1.0×10^{-2} mbar to 3.5×10^{-2} mbar. The electrode geometry of the UNU/ICTP PFF is modified to redirect the plasma motion in order to avoid the normal plasma focus action [6]. At such a low pressure, a set of twelve additional injection small plasma guns is utilised to initiate breakdown of the discharge. These guns are initiated about 80ns after the voltage is established across the focus electrodes. Plasma filled diode is arranged to sit directly 10mm on top of the switch to act as the load. Two six-turn magnetic pick-up coils are used to monitor the magnetic energy transfer, of which one of these coils is installed at one of the rods of the switch cathode (inner coil) and one is mounted at one of the ground rods of the diode cathode (outer coil).

In order to detect the extracted ion beams, two ion collectors are mounted directly at the top of the chamber. The first collector has a cross section at area of 7.1×10^{-6} m^2 and the second collector is 4.4×10^{-4} m^2. Both of these collectors are biased at -45V. At the end of the axial run down phase, the ion beam will be extracted from the plasma and directed by the induced axial magnetic field to pass through a 3 mm pin hole toward the top of the chamber. This ion beam will be detected by the two collectors.

Experimental Results

Experimental results have been obtained from the MPFOS operated at low pressure of 1.0×10^{-2} mbar to 3.0×10^{-2} mbar. The result from a 14.5kV discharge in 1.8×10^{-2} mbar argon gas is shown in Fig.2. From Fig.2(a), the voltage signal shows that the initial breakdown is under the control of the plasma guns, which is delayed from the start of the voltage by 80ns. After 2.1μs of conduction phase, the voltage is observed to increase sharply within a time <30ns. This voltage is measured outside the back-wall chamber across the electrode. The magnetic probe signals in Fig.2(c) and Fig.2(d) show that the magnetic energy is transferred from the switch cathode to the diode cathode. The inner coil signal shows that the current is rising along the switch cathode before the transfer action takes place. At the same time, the outer magnetic signal illustrates a fast switching effect before the ion beam is formed during the opening action.

From Fig.2(e) and Fig.2(f), the ion beam signals are detected by the two ion collectors. These peak signals show the time different of 196ns between the first collector and the second collector, within a distance of 30cm. By assuming that the extracted ion beam is fully ionized, the maximum energy is 180keV.

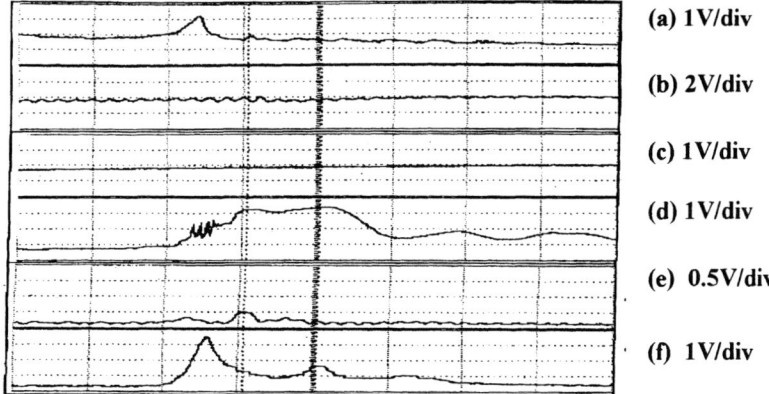

(a) 1V/div
(b) 2V/div
(c) 1V/div
(d) 1V/div
(e) 0.5V/div
(f) 1V/div

Fig.2 : (a)Voltage and (b)current signals measured outside the chamber when opening action is operative. (c)Inner and (d)outer magnetic coil signals measured inside the chamber indicating a magnetic energy transfer. (e)First and (f)second ion collector signals demonstrate the ion time of flight at pressure 1.8×10^{-2} mbar in argon gas; horizontal scale 200ns/div.

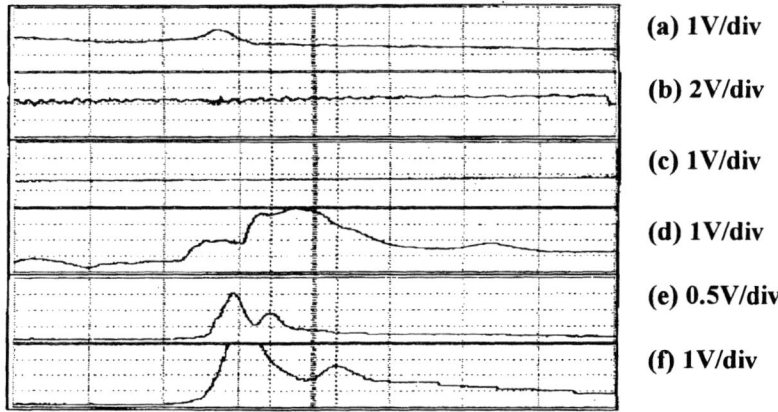

(a) 1V/div
(b) 2V/div
(c) 1V/div
(d) 1V/div
(e) 0.5V/div
(f) 1V/div

Fig.3 : (a)Voltage and (b)current signals measured outside the chamber, showing opening action is operative. (c)Inner and (d)outer magnetic coil signals measured inside the chamber, showing magnetic energy transfer. (e)First and (f)second ion collector signals indicating the ions time of flight at pressure 1.8×10^{-2} mbar in nitrogen gas; horizontal scale 200ns/div

Fig.3 shows a similar set of results obtained from a nitrogen gas discharge of the MPFOS. It can be seen that there was a coincidence and simultaneously peak detected by both ion collectors in Fig.3(e) and Fig.3(f). These first peak signals were caused by the ultra-violet and x-ray radiation generated by the plasma during the switching action in nitrogen gas at pressure of 1.8×10^{-2}mbar. The following two peaks are caused by the ion beams from the plasma. By comparing the time deferent between these two signals, the maximum ion beams energy are 220keV.

Conclusion

The MPFOS is able to couple the axial run down phase of the plasma focus with an inverse pinch phase thus giving raise to the opening switch action. Furthermore, this device has demonstrated that a significant fraction of the current is transferred during the opening action. The current transfer time was measured to be 62ns. In addition, the MPFOS is capable in producing ion beam during the switching action. For a 14.5kV discharge at 1.8×10^{-2}mbar, the experiments of the ion beam produced is 180keV in nitrogen gas and 220keV in argon gas.

Acknowledgements

This project is supported by IRPA Programme under project 02-02-03-0115. P.Choi's participation in this project is partially supported by an ICTP Visiting Scholar Programme. We acknowledge the technical help of Mr. Jasbir Singh.

References

[1] Wong C.S., Choi P., Moo S.P. and J.Singh, "Preliminary Studies On A Plasma Focus Opening Switch", BEAMS 96, (1996)
[2] Leong W.S., Wong C.S., Choi P. and Moo S.P., "Modified Plasma Focus Opening Switch", National Physics Symposium '96 (USM, Dec. 1996).
[3] Wong C.S., Leong W.S., Choi P. and Moo S.P.,"Characterisation Of The Opening Switch Action In A Modified Plasma Focus", IEEE Pulsed Power '97, (London, Mac.1997).
[4] Schoenbach K.H., Schaefer G., Harjes H.C., Leiker G. and Kristiansen M., "Opening Switches", in Proc. 3rd IEEE Int. Pulsed Power Conf. (Albuquerque, NM, 1981) p.74
[5] Special issue on fast opening vacuum switches, IEEE trans. Plasma Sci. Vol6. (1987).
[6] Lee S. et. al. Am. J. Phys. 56, 62 (1988).

Spectral Investigations of Micropinches in a Plasma Focus

M.H. Liu, X. Feng and S. Lee

Division of Physics, School of Science
Nanyang Technological University/NIE
469 Bukit Timah, Singapore 259756

Abstract

Soft x-ray (SXR) spectroscopic structure has been investigated in a 3 kJ plasma focus device -- NIE-SSC-PFF. The focused plasma is generated by an electrical discharge in *neon* gas between electrodes in a Mather-type plasma focus configuration. A crystal spectrograph capable of providing spatially resolved spectrum in the pinching column direction was developed and used to study the output of the pinched plasma region. For neon plasma focus the emission is found to be mainly in the wavelength range from 8 to 14 Å, with the helium alpha line at 13.447 Å and the hydrogenlike alpha line at 12.132 Å being the most intense features. Bright 'micropinches' or 'hot spots' within the plasma column were observed. The electron temperature in the 'spots' was estimated to be ~ 300 eV. For lower pressure these 'hot spots' assumed a longish shape. In the case of higher pressure, point-like spots were more frequently observed.

Introduction

Many pulsed laboratory plasmas are intense sources of x-rays[1]. Recent results with compact plasma focus (CPF) generators show promise for soft x-ray contact lithography[2]. On the other hand, radiation processes are so significant during the evolution of these plasmas that many of their properties can be diagnosed by studying the radiation emitted from them[3].

The emissions of x-ray from a DPF are characterized by high intensity, column-like and with wide spectral emission range. In addition, a strong shock wave from compressions must be taken into account. These properties of the plasma source demand that the crystal spectrograph must have a flexible adjustment system and must be designed to be effective in an environment with shock waves.

In this paper, we describe a crystal spectrograph for the soft x-ray in our plasma focus facility as shown in figure 1.

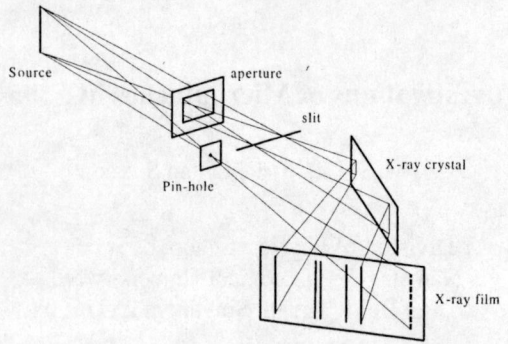

Fig.1 Schematic diagram of the spectrograph and a soft x-ray source.

Spectral Identification

Using Bragg x-ray diffraction equation: $m\lambda=2d\sin\theta$, the relationship between a wavelength λ_x and its position x on the film can be derived as following:

$$MA = \overline{F}\cot(\alpha + \arcsin\frac{\lambda}{2d}) \,, \tag{1}$$

In practice, it is difficult to measure the geometry of the spectrograph accurately. Therefore the observed wavelengths are measured with reference to two or more reference lines for which the wavelengths are known. Assuming that the known wavelength λ_A and λ_B at the A and B positions, according to the geometry of the figure 2, the unknown wavelength λ_x of the spectral line on the film then is given by

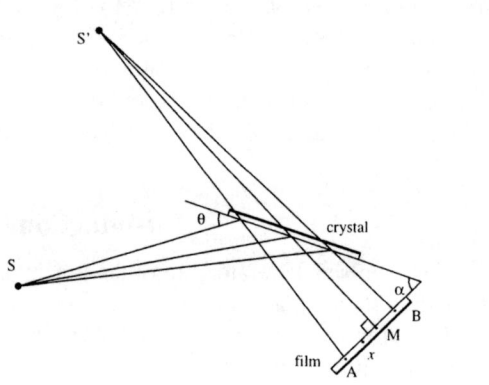

Fig.2 Crystal spectrograph's geometry and X-ray diffraction trace.

$$\lambda_x = 2d\sin[arc\cot(ctg[\alpha + \arcsin\frac{\lambda_A}{2d}] - \frac{L_{Ax}}{F}) - \alpha] \tag{2}$$

$$F = L_{AB}[\cot(\alpha + \arcsin\frac{\lambda_A}{2d}) - \cot(\alpha + \arcsin\frac{\lambda_B}{2d})]^{-1} \tag{3}$$

Experimental Data and Results

Experiments were performed on plasma focus device -- NIE-SSC-PFF[4]. The energy is stored in a capacitor (30μF), which is charged to 11 - 15kV (corresponding energy from 1.8 to 3.3 kJ), with neon gas filling pressure between 2.0 and 6 mbar. The detailed CPF device was described in previous paper[4].

A example using a 4-cm long crystal is shown in Fig.4(a). The image of the CPF x-ray source shows a complex structure as seen in figure 4(a). Spatial distribution of the soft x-ray intensities and the spectrally analyzed image are compared in fig. 4(b) and interpreted as follows.

There is a column of plasma with several localized bright spots. These 100 μm spots are circular and sharp-edged and are larger than the pinhole size of 50 μm. The side-on image of plasma emission is the appearance of a thin column with a 5.0-8.0 mm length. The diameter of the column estimated from the emission image is 400-700 μm.

Table 1 Classification of x-ray emission spectral lines of highly ionized neon plasma.

no	ions	transition
1	Ne IX	$1s^2-1s2p$
2	Ne X	$1s-2p$
3	Ne IX	$1s^2-1s3p$
4	Ne IX	$1s^2-1s4p$
5	Ne IX	$1s^2-1s5p$
6	Ne IX	$1s^2-1s6p$
7	Ne IX	$1s-3p$
8	Ne IX	$1s-4p$
9	Ne IX	$1s-5p$

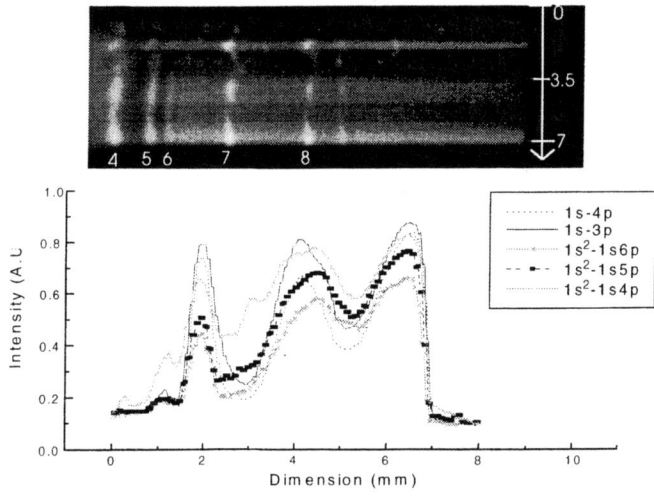

Fig.3. (Upper) The image of the CPF x-ray source with spectral resolution shows a complex structure and (lower) the spatial distribution of the soft x-ray spectral intensities.

Another example using a 8-cm long crystal is shown in Fig.4(a,b). We can see that for neon plasma the emission is found to be mainly in the wavelength range from 8 to 14 Å. The He-α line at 13.447 Å and the (Ly-α) line at 12.132 Å were the most intense features. Bright hot spots were observed. From the spectroscopic information the electron temperature in these 'spots' was estimated to be ~ 250 eV using the method of line intensities ratio and 200 eV by the slope of the logarithm of the continua. In the case of higher pressure, point-like spots were more frequently observed. More discussions are refered to Ref. [5].

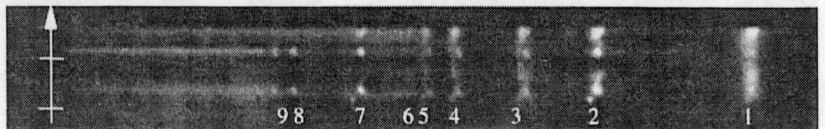

Fig.4(a) Typical spectrum (3.85 mbar, 13 kV, single discharge)1-9: spectrum lines

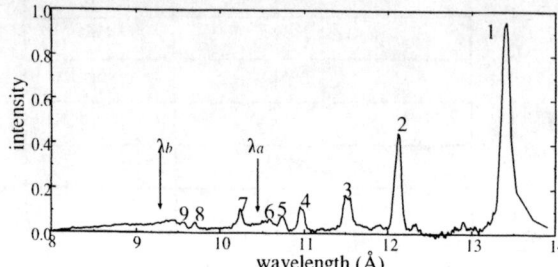

Fig.4(b) X-ray emission spectrum of neon plasma (integrated along the whole column) at charging voltage of 13 kV, pressure of 3.85 mbar)

Conclusions

We have shown that a compact flat crystal spectrograph with spatial resolution along the pinch column can be used to investigate the characteristics of x-ray emission from neon plasma focus. The pinched plasma consists of a column with several localized bright spots. The size, temperature and density of each hot spot can be evaluated.

Reference:

[1] P.G. Burkhalter, G. Mehlman D.A. Newman, M. Krishnan, and R. R. Prasad, *Rev. Sci. Instrum.*, 63, 5053(1992)
[2] Y. Kato, *et al*, *J. Vac Sci. Technol.* B 6, 1951(1985)
[3] Katsumi Hirano, Yoshio Takahama, Min Han, Takeshi Yanagidaira, *J. Physics Society of Japan*, 63, 3657(1994)
[4] S. Lee, T.Y. Tou, S.P. Moo, H.A. Eissa, V.A. Gholap, K.H. Kwek, S. Mulyodrono, A.J. Smith, Suryadi, W. Usada and M. Zakaullah, *Am. J. Phys.*, 56, 62(1988)
[5] M.H. Liu, *PhD Thesis*, "Soft X-rays from Compact Plasma Focus", 1996

EFFECT OF ANODE END STRUCTURES ON PLASMA PINCHING AND NEUTRON YIELD IN A PLASMA FOCUS

Ming-fang Lu

Institute of Physics, Chinese Academy of Sciences, Beijing 100080, China

The anode (center electrode) end structures of the Mather-type plasma focus are found to influence the current sheath (CS) implosion and pinching mainly in two ways: one is that the anode topology determines the motion status of the sliding front of the CS which must always move along and keep contact with the anode surface, by which three interrelated series of anode structures can be grouped with respect to the effect of the anode edge geometry, the hole and the stub structures at the anode end, respectively. The other is by the interaction mechanism of the azimuthal magnetic field introduced in front of the CS by the anode end discharge initiated by the "needle-point"-like anode end erosion structures, which is found to act as a stabilization mechanism to the imploding CS and is essential for plasma pinching.

1. INTRODUCTION

Various anode (center electrode, CE) end structures had been designed and employed in the plasma focus studies from the early years for different purposes, the stub anode for stabilizing and localizing the pinch plasma and for stronger hard X-ray emission[1], the hollow anode for less anode erosion and for stronger soft X-ray emission[2], the hemispherical anode usually for large-scale plasma focus devices[3], and so on. However, a systematic understanding of the influences of the anode end structures on the relevant plasma motion still remains unsettled both experimentally and theoretically. In this paper, this problem has been investigated by experiments on a small Mather-type plasma focus device (DPF-40, 20 kV, 18 kJ). The results fairly reflected the correlation among the anode end structures, the plasma motion and the neutron yield.

2. EXPERIMENTAL SETUP AND ANODE DESIGNS

The Mather-type plasma focus device is shown schematically in Fig. 1. The 250-mm-long coaxial electrode consisted of an anode (oxygen-free copper) and an multi-bar outer electrode (OE) of diameter 64 mm and 118 mm, respectively. The energy source was a condenser bank of C_0=84 μF, inductance L_0=79 nH, current rise time $\tau/4 \approx 4$ μs, stored energy E_0=18 kJ at charging voltage U_0=20 kV and peak current $I_{max} \approx 350$ kA. Deuterium gas was filled to the vacuum chamber to a pressure of 333 Pa for neutron emission. Neutron yield was measured using a calibrated silver activation counter[4]. A double-Wollaston-prism laser differential interferometer with a time resolution of 10 ns and line sensitivity of $\partial n_e / \partial l = (2.58 \pm 0.46) \times 10^{25} m^{-4}$ was used to observe the plasma motion[5].

Three series of the anode end structures were designed and employed in the experiment, as shown in Fig. 2, including the flat-ended 'caved in' series (a), the flat-ended stub series (b) and the hemispherical-ended cavity anode (c). They were all derived from a basic flat-ended solid anode (for $\phi=0$ in Fig. 2 (a)) with a small transition arc of R=10 mm at the anode edge. The 'caved in' anodes were formed by making a cave of 44-mm diameter and 60-mm deep inside the solid anode (the upper wall was 5-mm thick) and a hole at the anode end center. Hole diameters of $\phi=10, 20, 30$ and 44 mm were tested, respectively. In the stub series, a stainless steel stub of diameter $\phi_s=10$ mm was attached to the center of the solid anode end. Stub heights of $l_s=10, 20$ and 30 mm were tested, respectively ($l_s=0$ for the flat solid anode). When the transition arc was enlarged to R=40 mm, the hemispherical-ended cavity anode ($\phi=20$ mm) was formed. The OE and insulator structures were kept unchanged.

3. RESULTS AND DISCUSSIONS

First, the effect of the anode edge geometry is examined, the results can be found by comparing Fig. 3(a) with Fig. 3(c). For the flat-ended anodes with a small transition of R=10 mm, two distinct modes, the unstable mode and stable mode, of CS implosion with different sheath configurations and structures occur randomly in consecutive shots with different probabilities and notably influence the plasma pinching and neutron yield, with the unstable mode always having the lower neutron yield (Table 1)[6]. While for the hemispherical anode with R=40 mm transition arc, the CS is stable due to the smooth axial-radial motion transition and forms better pinch with higher neutron yield. Formation of the instability in the unstable mode (m=0 MHD mode, wavelength $\lambda \sim 3$ mm) can be attributed to the sudden increase in the acceleration of the CS sliding front during transition from the axial acceleration to the radial implosion (a=1.5×10^{16} cm/s^2, which is about one order larger than the stabilization criteria by the anomalous plasma resistivity and the plasma viscosity[7]). Formation of the stable mode under the flat anodes can be attributed to the anode end discharge and related magnetic field in front of the CS initiated by the erosion status at the anode end (see below). For the right-angled anode edge (R ~ 0) in previous studies, the absence of the unstable mode could also be attributed to anode end discharge initiated by the sharp anode edge[8]. Occurrence of the un-stable mode in this paper is due to the fact that the radius of the anode edge is larger enough to have avoided the discharge there, while at the same time, small enough to have caused the inst-ability. The hemispherical anode is a proper one for plasma pinching and neutron emission[7].

Second, the effect of the anode hole is examined (Fig. 3(a)-(b)). The direct mechanism the anode hole influences the plasma motion is the stagnation of the hole edge to the radial movement of the sliding front of the CS, as can be illustrated in Fig. 4. Such stagnation and the hole itself help to form a more axially symmetrical imploding CS just before pinching, and the pinch can develop in both axial directions to produce a perfect Z-pinch plasma which indirectly contacts with the anode through the extended CS at its bottom end and is in a purely magnetically confined state. In this paper, the neutron yield increases with increasing the hole diameter. The other mechanism is the hole-diameter related erosion status at the anode end by the pinch plasma which causes the "needle point"-like discharge there and introduces an azimuthal magnetic field in front of the CS. The magnetic field, after compressed by the rundown CS, acts as a stabilization mechanism to the imploding CS, by which the stable mode is randomly formed due to the fluctuation of the erosion status and thus the anode end discharge from shot to shot. The eroded area (about 2-cm diameter (Fig. 5 (a)) decreases with increasing the hole diameter and the anode end discharge current become smaller. The probability for the unstable mode thereby increases. For the 30-mm hole, the anode end surface completely avoided erosion (Fig. 5 (b)) and it happens almost only the unstable mode (Table 1). For the 44-mm hole diameter hollow anode, the only occurrence of the stable mode is due to the stagnation effect of the hole to the radial motion of the CS sliding front from the very beginning of implosion when the instability has not yet developed.

Thirdly, the effect of the anode stub is examined (Fig. 3 (d)). Under this anode structure, only the stable mode of CS implosion is formed and the pinch is always formed at the stub end, which demonstrated good stabilization and localization effects of the stub to imploding and pinching plasmas[1]. However, since the CS implodes supersonically (magnetic Reynolds number ~ 50) and the instability would occur from the beginning of implosion, so the stub can not directly have stabilized the instability. It needs an interaction mechanism, and this mechanism also can only be the stub end discharge induced magnetic field in front of the CS. In fact, the stub acts as a real needle point at the anode end and caused a stronger anode end discharge which can already be observed in the interference grams (Fig. 3 (d)). However, the existence of the stub at the anode end always has a negative influence on the plasma pinching and the neutron yield (Table 1).

It can be seen from the above experimental results that the various anode end structures employed in the plasma focus can be grouped into three interrelated series according to their influences to the plasma motion, i.e. the anode edge effect, the anode hole effect and anode stub effect. Also, the results reflected the subtle influences of the anode structures on the plasma motion.

4. CONCLUSIONS

The anode end structures in the plasma focus sensitively, both directly and indirectly, influence the plasma pinching. The usually used anode end structures can be grouped into three interrelated series by the physical nature of their effects to the plasma motion. This research made a deepening understanding to the plasma focus and also provided a basis for further theoretical description of the plasma focus.

REFERENCES

[1] Mather J. W., Bottoms P. J., Carpenter J. P., Ware K. D. and Williams H. A., Proc. 4th Int. Conf. Plasma Phys. & Nucl. Fusion Res., (Madison) (Vienna, IAEA, 1971) 561
[2] Kato Y. and Be S. H., Appl. Phys. Lett., 48 (1986) 686
[3] Bernard A., Coudeville A., Jolas A., Launspach J. and Mascureau J. de, Phys. Fluids, 18 (1975) 180
[4] Lu M. F., Han M. and Wang X. X., Nucl. Electron. & Detec. Tech.,14 (1994) 212
[5] Lu M. F., Rev. Sci. Instru., 68 (2) (1997) 1149
[6] Lu M. F., Phys. Rev. 54 E (1996) R1074
[7] Peacock N. J., Hobby M. G. and Morgan P. D., 4th Int. Conf. on Plasma Physics & Controlled Nuclear Fusion Research (Madison) Vol 1 (Vienna: IAEA, 1971) 537
[8] Mather J. W., Methods of Experimental Physics, Lovberg R and Griew H, Eds. (New York:Academic Press, 1971) Vol 9B, 195, 199

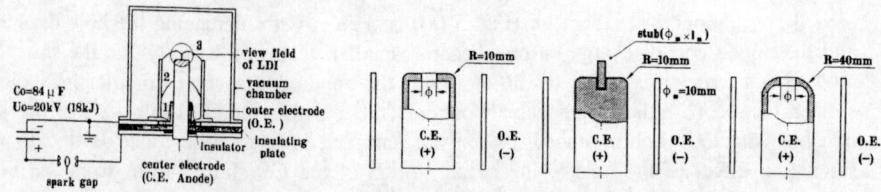

Fig. 1 Schematic diagram of the plasma focus; 1. breakdown; 2. rundown; 3. implosion and pinch phases

Fig. 2 Anode end structures

(a1) t=-85 ns (a2) t=-80 ns (b) t=-90 ns (c) t=-150 ns (d) t=-210 ns

Fig. 3 Interference results of plasma motion. View field 60-mm diameter, sideview, (a) for solid anode; (b) for hollow anode; (c) for hemispherical anode; (d) for stub anode, l_s=30 mm; (a1) stable mode; (a2) unstable mode; (b)-(d) stable mode only.

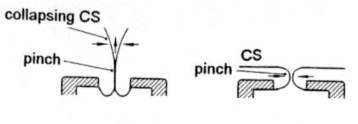

(a) Stable mode (b) Unstable mode

Fig. 4 Pinch formation of the two modes under the cavity anode

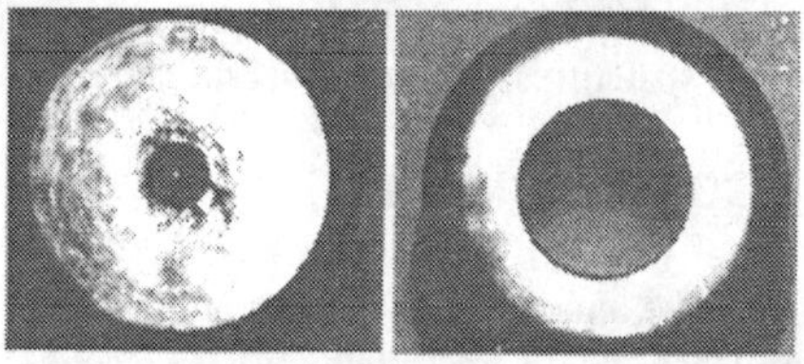

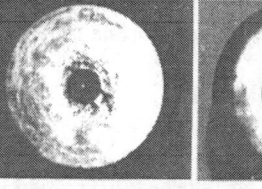

(a) Solid anode (b) 30-mm hole cavity anode

Fig. 5 Erosion status of the anode end

Table 1. Probabilities and neutron yields of the two modes ($Y_n \times 10^8$ per shot)

Mode	Anode	Solid (0 mm)	Hole diameter, φ/mm Cavity			Hollow (44 mm)	Stub l_s=10~30 mm	Hemispherical* R=40 mm
			10	20	30			
Stable	Shots	25	16	7	5	54	214	49
	Probability	52%	48%	23%	0~10%	(100%)	(100%)	(100%)
	Y_n	2.1 ± 0.4	2.0 ± 0.7	2.5 ± 0.8	2.8 ± 0.6	2.2 ± 0.5	1.05~1.53	3.2~4.9
Unstable	Shots	23	17	23	134	—	—	—
	Probability	48%	52%	77%	90-100%			
	Y_n	0.39 ± 0.2	1.3 ± 0.9	2.1 ± 0.7	2.2 ± 0.5	—	—	—
Mixed	Total shots	48	33	30	139	54	214	49
	Y_n	1.3 ± 0.9	1.5 ± 0.9	2.2 ± 0.8	2.3 ± 0.6	2.2 ± 0.5	1.05~1.53	3.2~4.9
	$\Delta Y_n/Y_n$	± 69%	± 60%	± 36%	± 26%	± 23%	±48~67%	± 30%

*The highest neutron yield $Y_{n,max} \sim 10^9$ per shot.

EVOLUTION OF THE FILAMENTARY CURRENT SHEATH IN A PLASMA FOCUS DEVICE

Ming-fang Lu

Institute of Physics, Chinese Academy of Sciences, Beijing 100080, China

The formation and evolution of the filamentary current sheath (CS) in the breakdown and rundown phases in a small Mather-type plasma focus device (20 kV, 18 kJ) have been examined using a high resolution laser differential interferometer. The results show that the thin and quasi-homogeneous CS initially formed along the insulator surface develops into distinct filamentary structures corresponding to the outer electrode (cathode) bars during moving towards the outer electrode driven by the Lorentz force. The sheath then develops into muddled filamentary structures once it crossed the coaxial electrodes. Later, after about 2 μs from the discharge, the filamentary sheath reconstructs to form the distinct and homogeneous CS which accelerates towards the open end of the coaxial electrode and forms the pinch there. Such processes might be explained in terms of the specific ionization energy required to fully ionize the working gas the sheath swept.

1. Introduction

In the plasma focus researches, it has long been realized that a thin and distinct quasi-homogeneous initial current sheath (CS) formed along the insulator surface during the early breakdown phase is essential for forming a good plasma pinch and for producing higher neutron and X-ray yields[1-5]. However, the detailed correlation between the initial CS and follow-up rundown CS and pinch formation in the relatively large spatial-temporal scale has not been clearly fixed. This is mainly limited by the lower resolution of the diagnostics, usually the Image Converter Camera (ICC), once employed[1-5]. The ICC photographic observations did not give detailed information about the fine structures of the CS. In such experiments, observed were mainly the macro filaments either of the direct radial discharges occurring between the coaxial electrodes at the insulator end[1,2], or of the CS during the rundown phase or even lasted to the pinch phase[5,6] determined by the filling pressures. Recently, the macro filamentary CS was explained in terms of the CS plasma energy to be lower than the specific energy required to fully ionize the working gas the CS swept[6]. In this paper, the formation and reunification of the filamentary CS from the early breakdown to the late rundown and pinch phases in a small Mather-type plasma focus device (20 kV, 18 kJ) have been examined using a high resolution double-Wollaston-prism Laser Differential Interferometer (LDI). The results fairly reflected the evolution of the filamentary CS during such processes.

2. EXPERIMENTAL SETUP

The Mather-type plasma focus device is shown schematically in Fig. 1 (a). The 250-mm-long coaxial electrode consisted of an anode (center electrode, CE, oxygen-free copper) and a outer electrode (OE, cathode) of diameter 64 mm and 118 mm. A slotted OE (copper) and a cylindrical multi-bar OE (12 bars, each of 14 mm diameter, stainless steel) were tested. The insulator was made of alumina with diameter of 85 mm and height of 85 mm. A replaceable knife edge groove was attached to the insulator bottom to facilitate the breakdown process. A sharp (0.1 mm thick) and a blunt (2 mm thick) knife edge were tested. The insulator height can be adjusted from 85 mm to 35 mm with axial variation of the knife edge. The energy source was a condenser bank of C_0=84 μF, inductance L_0=79 nH, current rise time $\tau/4$=4 μs, stored energy E_0=18 kJ at charging voltage U_0=20 kV and peak current I_p=350 kA. Deuterium gas was filled to the vacuum chamber to a pressure scope from 1 to 6 Torr with the optimum one for neutron emission of 2.5 Torr. Neutron yield was measured using a calibrated silver activation counter with relative standard error less than ±20%[7]. A double-Wollaston-prism LDI with a time resolution of 10 ns and line sensitivity of $\partial n_e/\partial l = (2.58 \pm 0.46) \times 10^{25} m^{-4}$ was used to observe the plasma motion[8]. The view field of the LDI for the rundown phase is of 14 cm diameter (Fig. 1 (b), for the implosion and pinch phases is 6 cm (Fig. 1 (b) (7)). The laser beam passes through the interelectrode gap and forms the inference images of the breakdown and rundown CSs.

3. RESULTS AND DISCUSSIONS

The results of interference images of the CS evolution in the breakdown and rundown phases at 2.5 Torr Deuterium filling pressure for the 35-mm-long insulator with the sharp knife edge and the bar OE are shown in Fig. 1 (b). It can be seen from the figure that after about 0.1 to 0.2 μs of the discharge the thin and distinct quasi-homogeneous CS can be observed to have formed along the insulator surface and to take off therefrom moving towards the OE driven by the Lorentz force (1). Soon in this process, the CS develops into distinct filamentary structures corresponding to the OE bars ((2), three filaments to three OE bars at the right side). The axial and radial velocities of the CS in this stage are $v_z=1 \times 10^6$ cm/s and $v_r=0.8 \times 10^6$ cm/s. Once the CS crosses the coaxial electrodes, it further develops into muddled filamentary structures, as can be seen in Fig. 1 (b) (3). The CS near the anode has a axial extension of about 1 cm. This filamentary CS lasts to t=2 μs after the discharge (Fig. 1 (b) (4)). In this stage, the CS velocity is $v_z=4.9 \times 10^6$ cm/s. Later when t>2 μs, the filamentary CS reconstructs to form a single and uniform CS (Fig. 1 (b) (5) and (6), the sheath is ~ 1 mm thick) which accelerates towards the electrode end ($v_z=7.4 \times 10^6$ cm/s) and forms the pinch there (Fig. 1 (b) (7)).

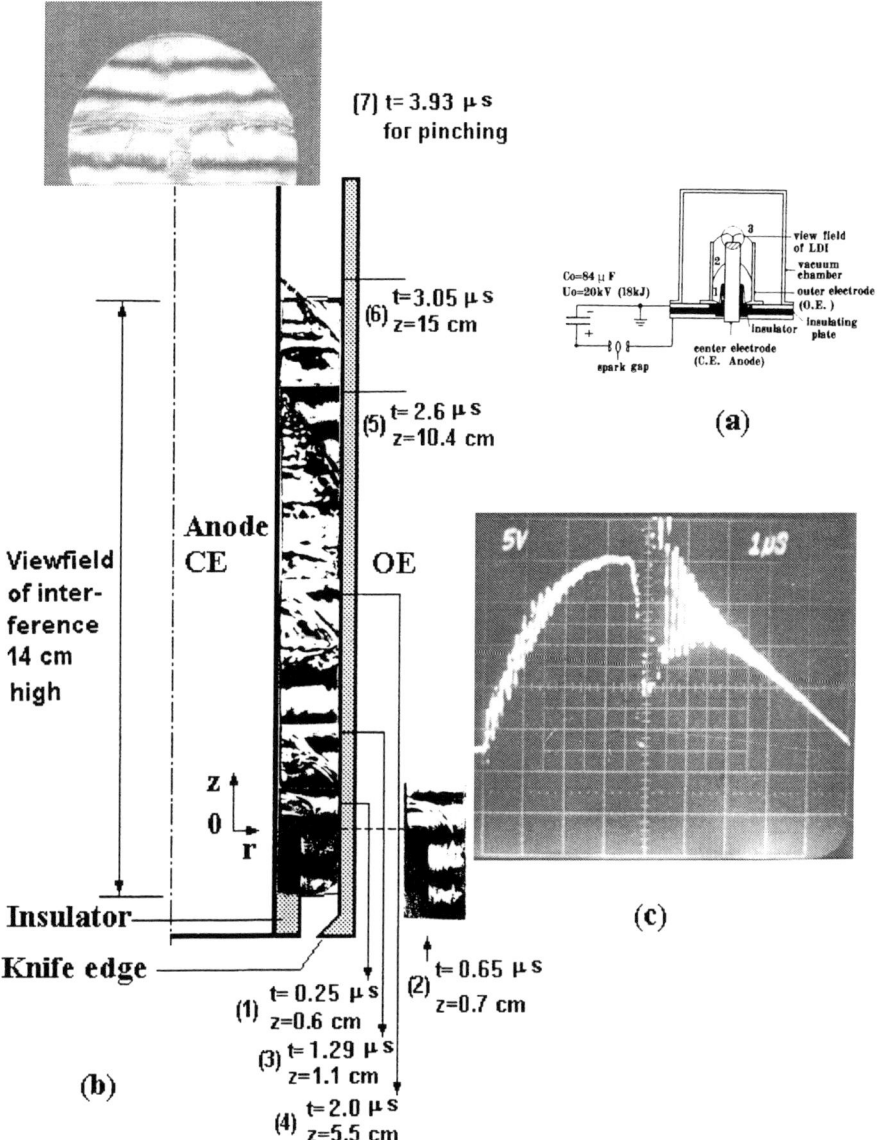

Fig. 1 (a) Schematic diagram of the plasma focus device; (b) interference observations of the CS in the breakdown, rundown and pinch phases; (c) Oscillosgram of the discharge current

When the discharge current I is examined, it can be seen that for t<2 µs, I<310 kA. The noisy signals on the I curve indicate that the filamentary CS is an unstable and fast-changing structure. For t>2 µs, the noisy signals disappeared for the uniform CS. The neutron yield under this condition is in-between $(2-10) \times 10^8$ per shot determined by the anode end structures[9], which is typical for the plasma focus operated at the same discharge current level.

Further cross investigations by employing combinations of different OE structures, insulator lengths and surface status (cleaned or conditioned), and knife edges in a large filling pressure range from 1 to 6 Torr show that the same pattern of CS evolution occurred as mentioned above, except for that at higher filling pressures (4 to 6 Torr) the CS shows more clearly filamentary structures. But all of the CSs reconstruct to the single and uniform CS in the later rundown phase. Also, no direct radial filamentary discharges occurring between the coaxial electrodes were observed on the interference images under all conditions mentioned above. The filamentary CS observed here is thus different from that stated in Ref.[1,2]. However, the much difference in the neutron yields from 1×10^6 per shot to $(2-10) \times 10^8$ per shot for the above different conditions indicates that the initial CS structure does not have a direct influence on the effectiveness of the pinch formation. It needs further investigations on this problem.

According to a similarity evaluation with reference to Refs. [6,10], the filamentary CS in the breakdown and early rundown phases might be explained in terms of the sheath energy to be less than the specific energy required to fully ionize the working gas the CS swept when the discharge current is small at the these stages.

4. CONCLUSIONS

The evolution of the filamentary CS in the plasma focus is a rather complicated process. The relation between the initial CS status and the effective plasma pinch formation still needs further detailed investigations.

REFERENCES
[1] Donges A, Herziger G, Krompholz H, Ruhl F and Schonbach K, Phys. Lett., 76 A (1980) 391
[2] Krompholz H, Neff W, Ruhl F, Schonbach K and Herziger G, Phys. Lett., 77 A (1980) 246
[3] Borowiecki W., Czekaj S., et al, Proc. 9th Europ. Conf. Contro. Fusion & Plasma Phys., Warsaw, 1985, 86
[4] Bruzzone H., Gratton R., Kelly H., Milanese M. and Pouzo J., Proc. 1th Int. Conf. Energy Storage, Compression and Switching. New York, Plenum Press, 1976, 255
[5] Bostick W. H., Grunberger L. and Prior W., Proc. 3rd Europ. Conf. Contro. Fusion & Plasma Phys., Ultrecht, 1969, Vol. I, 120
[6] Milanese M., Moroso R., and Pouzo J., IEEE Trans. Plasma Sci., 21 (1993) 606
[7] Lu M. F., Han M. and Wang X. X., Nucl. Electron. & Detec. Tech.,14 (1994) 212
[8] Lu M. F., Rev. Sci. Instru., 68 (2) (1997) 1149
[9] Lu M. F., Effect of Anode End Structures On Plasma Pinching And Neutron Yield In A Plasma Focus, also presented at this Conference
[10] Vargas J., Gratton F., Gratton J., Bruzzone H. and Kelly H., Proc. 6th Int. Conf. Plasma Phys. and Contr. Nucl. Fusion Res., Berchtesgaden, 1976 (IAEA, Vienna, 1976) Vol. 3, 483

Steady State of Elliptic Z-Pinches

T. Miyamoto

*Atomic Energy Research Institute, Nihon University,
Kanda-Surugadai, Chiyoda-ku, Tokyo, Japan*

Abstract. Steady state of dense Z-pinches with elliptic cross-section is studied. The stability is improved in the elliptic z-pinches, comparing with fiber z-pinches. As a family of the elliptic z-pinch, more stable configurations are shown.

INTRODUCTION

So far most z-pinches have been studied in the cylindrical geometry, in which the plasma is strongly unstable. Recently it was suggested that the lower limit existed for the plasma radius achieved in a quasi-steady state due to the current rearrangement that occurred turbulently[1]. This means that the density is limited. In the cylindrical z-pinches the return current hardly affects the equilibrium. On the other hand, steady state z-pinches with deformed cross-section can exist under influence of the return current conductors. Various types of the deformed z-pinches are possible, corresponding to geometry of the return current. It is sheet z-pinches that are especially interesting in these pinches. Recently, the possibility of fusion based on sheet z-pinches was proposed and investigated[2]. In this paper, we investigate a z-pinch with an elliptic cross-section, which is produced between two parallel plane conductors (see fig.1). When the conductors are far from the plasma, the elliptic z-pinch is reduced to a conventional z-pinch with a circular cross-section. When the conductors are close to the plasma, the plasma is compressed and reduced to an extremely elongated elliptic column, i.e., a sheet z-pinch, as shown in fig 1(c).

EQUILIBRIUM AND STEADY STATE

We suppose that two return current conductors are located at $x = \pm d$ and are infinite in the direction of current (the z-axis) and in width (the y-axis). The pinched plasma produced between them is elliptic in cross section. For simplicity, we

assume that the elliptic cross section is given by
$$x^2/a^2 + y^2/b^2 = 1 \tag{1}$$
and that the thermal conduction is fast enough. Then, the temperature $T_e = T_i \equiv T$ and the current density i are uniform in the plasma. The vector potential A has only z component, and $A = const.$ is an isobaric surface. The pressure and the vector potential are expressed as
$$p = 2nkT = p_0 + iA, \tag{2}$$
$$A = -A_s\left(\frac{x^2}{a^2} + \frac{y^2}{b^2}\right), \quad A_s = \frac{\mu_0 i}{2}\left(\frac{1}{a^2} + \frac{1}{b^2}\right)^{-1} = \frac{\mu_0 i}{2}\frac{a^2 b^2}{a^2 + b^2} \equiv \frac{p_0}{i}, \tag{3}$$
where we choose as $p = p_0$ and $A = 0$ at the origin. As the total current and line density are $I = \pi abi$ and $N = \pi abn_0/2 = \pi abiA_s/4kT$, the Bennett relation is expressed as
$$\mu_0 I^2 f_E = 16\pi kT, \quad f_E = \frac{2ab}{a^2 + b^2}, \tag{4}$$
considering eqs.(2) and (3). The energy conservation law integrated over the whole volume gives the relation $I = (4\overline{P_B}/3\overline{\eta})^{1/2} NT$, where the resistivity is $\eta = \overline{\eta} T^{-3/2}$ and the Bremsstrahlung radiation power is $P_B = \overline{P_B} n^2 T^{1/2}$. Substituting this relation into eq.(4), the current $I_{PB}^{(e)}$ and line energy density $w_{PB}^{(e)}$ in steady state are given as
$$I = I_{PB}^{(e)} \equiv \frac{I_{PB}^{(f)}}{f_E}, \quad I_{PB}^{(f)} = \sqrt{\frac{3\overline{\eta}}{\overline{P_B}}}\left(\frac{8\pi k}{\mu_0}\right) \tag{5}$$
$$w_p = w_{PB}^{(e)} = (3NkT)_{PB}^{(e)} \equiv \frac{w_{PB}^{(f)}}{f_E}, \quad w_p^{(f)} = \frac{9\overline{\eta}}{2\overline{P_B}}\left(\frac{8\pi k^2}{\mu_0}\right), \tag{6}$$
where $I_{PB}^{(f)}$ and $w_{PB}^{(f)}$ are the Pease-Braginskii current and the corresponding line energy density for a fiber z-pinch. Both steady current $I_{PB}^{(e)}$ and line energy density $w_{PB}^{(e)}$ depend on the geometry of the cross-section, differing from the fiber z-pinch. When $f_E \approx 2a/b \ll 1$, the steady current and line energy density are higher than in the fiber z-pinch, but the ratio between them is the same in both cases.

We need the field outside the plasma in order to determine the separation $2d$ between the return currents. We use the fact that the vector potential given by eq.(3) is a sum of A_p due to the plasma current flowing uniformly in the cross section eq.(1) and A_R due to the return current that is assumed to have infinitely width[3]. Vector potentials A_p and A_R are approximated in the vicinity of the axis as
$$A_p = A_{p0} - \frac{\mu_0 i}{2}\frac{1}{a+b}(bx^2 + ay^2) + \cdots, \quad A_R \approx A_{R0} + \frac{\mu_0 iabh}{b^2 - a^2}(y^2 - x^2) + \cdots \tag{7}$$
where A_{p0} and A_{R0} are constant, and h is an infinite series of only d/b for

$a \ll b$. From the relation $A = A_p + A_R$ the separation $2d$ is given by

$$h \equiv \sum_{n=1}^{\infty}(-1)^{n+1}\left[1 - \frac{2nd}{\sqrt{b^2 + (2nd)^2}}\right] = \frac{(b-a)^2}{2(a^2 + b^2)} \approx \frac{1}{2}\left(1 - \frac{2a}{b} + (\frac{a^4}{b^4})\right) \qquad (8)$$

taking into account eqs.(3) and (7). The infinite series h changes almost linearly with d/b, and is approximated as $h \approx 0.5 - (0.25 - 0.27)(d/b)$. The ratio of the plasma thickness to the separation between the return currents hardly depends on the plasma width in the approximation $a, d \ll b$, and is given by

$$a/d \approx 0.25 \sim 0.27 \qquad (9)$$

This means that the return currents must be very close to the plasma to obtain a thin plasma layer. However, an infinite series h is poor in convergency for $d/b \ll 1$. In addition the results are derived using the field expanded near the coordinate origin. Hence, we should notice that eq.(9) gives only a measure of a/d for the z-pinch with the elliptical cross section given by eq.(1).

STABILITY AND STABILIZATION

The elliptic z-pinch is still unstable. The radius of field curvature is an order of b in the central region. When the minimum wavelength of perturbation is limited by the sheet thickness, the growth rate of m=0 mode in this region is given by

$$\gamma_{elliptic} \approx \sqrt{gk} \approx \sqrt{a/b}\gamma_{fiber} \quad , \qquad (10)$$

where $\gamma_{fiber} = 1/t_A$ is the growth rate in the cylindrical z-pinch of radius a, and t_A is the Alfven transit time. We can expect the finite Larmor radius effects for high temperature plasma, and that the growth rate decreases by $(a/a_{Li})^{1/3}(a/b)^{1/6}$. These growth rates corresponding to the elliptic cross-section given by eq.(1) are much lower than the fiber z-pinch, but not still low enough. However, some stabilizing methods exist. When the plasma is wider than the return current conductors, the cross section becomes barbell–like, as shown in fig.2(a). The curvature in the central region vanishes for this barbell-like plasma. This means that the plasma column is stabilized by the finite Larmor radius effect, if both end regions are not taken into account. Another example is to modify the cross section from the elliptic one to a hyperbolic one that makes the field line convex to the plasma in the central region, as shown in fig.2(b). It is easy to obtain the relation corresponding to eq.(4) – (5). The plasmas are unstable at both end regions of these sheet z-pinches as well as in the fiber z-pinch plasma. Even if the instabilities grow at both end regions, however, they will not be important, because of two reasons. First, it takes period $b/v_A \approx (b/a)t_A$ for perturbations to propagate from both ends to the center. This is longer than the growth time given by eq.(10). Second, non-linear effects will be important. Instabilities at the end regions will not grow so as to affect the central region. For example, the current rearrangement in both end regions, which is fatal in the fiber z-pinch, hardly affects the central region.

On the other hand, the tearing mode resistive instability may occur along y-axis. The growth rate γ_R is given as

$$\gamma_R / \gamma_{fiber} \approx (t_A/t_d)^{3/5} \approx 4 \times 10^{-28} (a/\ln \Lambda)^{3/5} (N/b)^{6/5} ,$$

where t_d is the magnetic diffusion time. This growth rate gives $1/\gamma_R \approx 2 \times 10^{-7}$ s for the typical values $N/b = 10^{22}$ m^{-2}, $a = 10^{-4}$ m and $\ln \Lambda = 10$. Hence, the tearing mode is rather important for large values of N/b, but its growth rate is lower by several orders than in the MHD instabilities.

CONCLUSIONS

In conclusion the elliptic z-pinches and their family keep the important features of the fiber or capillary z-pinches. In spite of it, the magneto-hydrodynamic instabilities in the fiber z-pinch are removed in them. Hence, they will give a new method to produce plasmas for the fusion and the soft X-ray laser. They are equivalent to a number about b/a of the fiber z-pinches with the radius a and the same plasma parameters. Hence, they require higher power than the fiber z-pinch, depending on the ratio b/a, which is determined by stability or necessary sustaining time. The return current conductors feel strong repulsive force and explode in high current discharge. So dense sheet z-pinches exist only transiently, but is sustained for long enough period because of inertia of the conductors. The plasma has to be created between the narrow return currents, and to be insulated electrically from them. The methods of plasma production and other problems related with the sheet z-pinches will be given in separated papers.

REFERENCES

1. R.Relay, D.Scudder, J.Schlachter and R.Lovberg, Phys. Plasmas 3,1314-1323 (1996).
2. T.Miyamoto, To be published in NIFS Reports.
3. G.Bateman, "MHD instabilities", The MIT Press (1978).

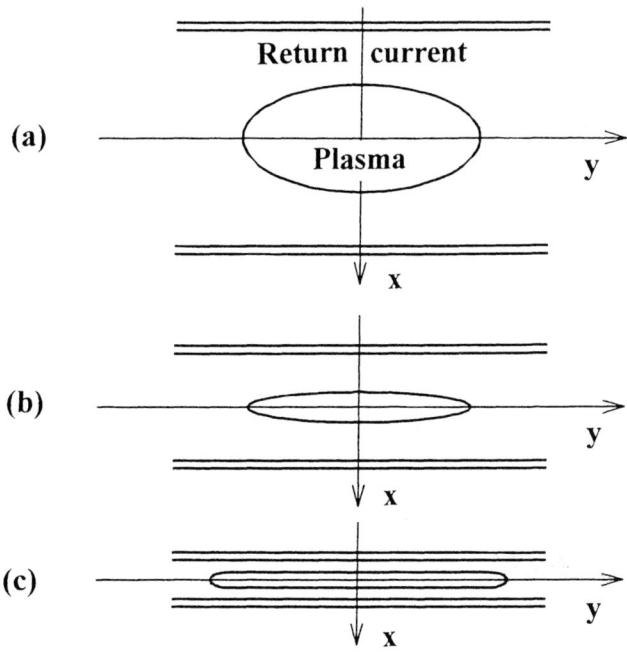

Figure 1 Conceptual figures of the elliptic z-pinches

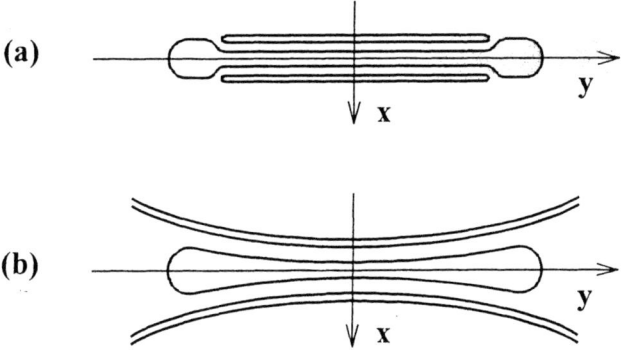

Figure 2 Z-pinches that are stable in the central region: (a) the barbell-like z-pinch and (b) the z-pinch with a hyperbolic cross-section

LASER PROBING OF FIBRE Z-PINCH PLASMAS

J. Ruiz-Camacho, F. N. Beg and A. E. Dangor

The Blackett Laboratory,
Imperial College of Science Technology and Medicine
London SW7 2BZ, UK

ABSTRACT

The Moiré deflectometer and the polarisation interferometer, both based on a pulsed laser, are described and the implementation of these instruments to determine the density distribution in fibre z-pinches discussed.

INTRODUCTION

The linear z-pinch, in which a current carrying plasma column is confined by the azimuthal self-magnetic field, has a number of potential applications in controlled fusion research (1,2), production of ultra high density plasma by radiative collapse (3) and as VUV and XUV radiation sources for microscopy and lithography (4). In all these applications it is necessary to begin with a pinch which is initially at high density as is the case for a z-pinch produced by passing the current through a solid fibre. In this paper we discuss the application of laser probing to determine the electron plasma distribution in a fibre z-pinch.

Several laser probing instruments have been developed and used to diagnose a variety of plasmas, ranging from the low density, slowly evolving, magnetically confined plasmas to the high density, highly transient, laser produced plasmas. The plasma parameters, dimensions and other characteristics encountered in the fibre z-pinch are closer to those in laser produced plasmas. In this paper we discuss the application of two instruments, first used on laser plasmas, to the fibre z-pinch. The two instruments are the Wollaston prism polarisation interferometer (5), which determines the 2-d distribution of the density directly, and the Moiré deflectometer (6) which gives the distribution of the density gradient.

Both these instruments have been used to diagnose the z-pinch plasma produced by a *150kA*, *60ns* rise time, pulsed power generator in a variety of fibres (*7μm* carbon, *25μm* aluminium and *10μm* copper). Details of the generator and current waveforms are in the papers by Lorenz *et al.* (7) and Beg *et al.* (8) both in these proceedings. Examples of the observations made with these instruments can also be found in these papers.

THE MOIRE DEFLECTOMETER

Moiré deflectometry is a technique which maps the deflection of the rays in a collimated beam as it passes through the test object (6). This deflection is proportional to the refractive index gradient transverse to the optical path.

The basic Moiré deflectometer optical system is shown in Figure 1. Collimated light passes through the plasma and then through a pair of transmission gratings, G_1 and G_2, to produce a Moiré pattern. A lens L images the plasma into the camera. An aperture D, located at the focal length from the lens, selects a single diffracted order. A filter F may be necessary to reject plasma light. As the light passes through the plasma, the rays will experience a deflection φ

$$\varphi = \frac{1}{2n_c} \int \nabla n_e \cdot d\ell \qquad (1)$$

where n_e is the electron density, $n_c = 10^{21}/\lambda_{\mu m}^2 \; [cm^{-3}]$ is the critical electron density, ($\lambda_{\mu m}$ is the laser wavelength in microns) and $d\ell$ is an element along the optical path. The plasma is assumed to be underdense, i.e., $n_e \ll n_c$). The deflection distorts the Moiré pattern and from the distortion the density gradient can be obtained.

The Moiré pattern is in the form of a series of straight parallel equispaced fringes separated by $p' = p/\theta$, where p is the grating pitch and θ is the angular rotation between the gratings. In a z-pinch the density gradient is mainly radial, thus the laser beam is passed transversely to the pinch axis. To facilitate interpretation of the fringe distortion the fringes are aligned to be perpendicular to the pinch. If the z-axis is along the pinch and the y-axis along the laser, then a deflection φ in the x-y plane will produce a fringe shift $h = \varphi d/\theta$. From Equation (1), we see that the deflection depends on the density gradient along the x-axis, i.e., dn_e/dx.

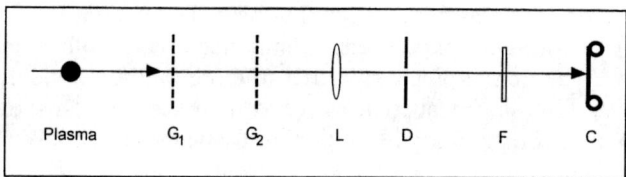

Figure 1. Basic principle of Moiré Deflectometry.

Since the relative fringe shift $\alpha = h/p' = \varphi d/p$, is fixed by measuring accuracy (to about 0.2), we see that the minimum deflection angle that can be detected depends on the ratio of gratings pitch p to the distance d between gratings. However, due to the uncertainty principle, $\delta\varphi \cdot \delta x \geq \lambda/2\pi$, where $\delta\varphi$ is the angular resolution and δx is the spatial resolution (9), increasing the sensitivity decreases the spatial resolution.

The fringe visibility is determined by the spectral width of the laser, the beam divergence, the size of the beam relative to the grating pitch and grating imperfections (10). Also, if the second grating is positioned at the Talbot planes

(9), i.e., $d = kp^2/\lambda$, where k is an integer, the fringes have maximum contrast. Indeed, if the second grating is midway between the Talbot planes, the fringe pattern is totally blurred.

The Moiré deflectometer that we have developed for fibre z-pinches is shown in Figure 2. The beam on emerging from the plasma is telescoped down with the lenses L_1 and L_2. The lenses are separated by the sum of their focal lengths and the plasma position is at the focal plane of L_1. The image, at the focal plane of L_2, is in front of the first grating G_1. The first order diffracted by G_1 is passed at normal incidence through the second grating G_2 to produce the Moiré pattern. The aperture D is at the focal distance from L_3 and is adjusted to select the zero order beam from G_2. The lens L_3 images the plasma onto the film.

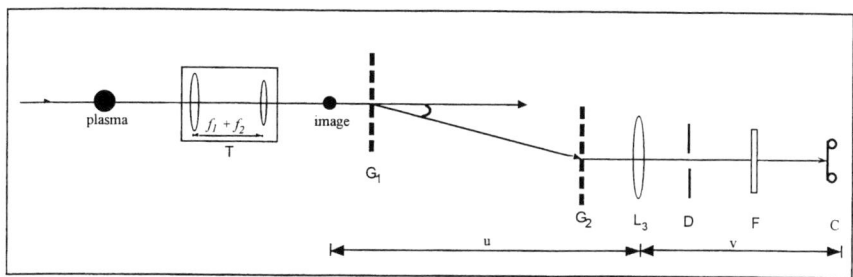

Figure 2. Moiré deflectometer setup used for this experiment.

The advantage of this system is that since the angular deflection is increased by $M = f_1/f_2$ the minimum measurable deflection angle is increased to $\phi_{min} = \alpha p/Md$. The reduction in the spatial resolution is compensated by magnifying the final image produced by lens L_3. Care should be taken to ensure that the image of the plasma produced by the telescope is in front the first grating. By operating only with the first order diffracted beam produced by the first grating, blurring of the Moiré fringes due to the Talbot effect is completely eliminated.

Figure 3.
Moiré deflectogram

Figure 3 shows a Moiré deflectogram of a carbon fibre z-pinch. The parameter values of the optics and layout used are:

Telescope: $f_1 = 400 mm$; $f_2 = 100 mm$; $M = f_2/f_1 = 4$

Gratings: $p = 25 \mu m$; $\theta = 5.4°$; $d = 1.23 m$

Moiré pattern: $h = 53.3 \mu m$

Sensitivity: $\delta\varphi_{min} = 1.0 \mu rad$ (assuming $\alpha = 0.2$)

$n_{e\,min} = 2.3 \times 10^{15} cm^{-3}$ (assuming parabolic distribution)

Note near the core of the pinch there is an inaccessible region of the plasma. This is because of the very large density gradients in this region producing a deflection which is greater than the acceptance angle of the deflectometer.

POLARISATION INTERFEROMETER

This interferometer makes use of the fact that the radial dimension of the fibre z-pinch is small relative to the size of the laser beam. Thus the undisturbed part of the beam, which does not pass through the plasma, can be used to form the reference beam. This is done outside the pinch chamber using a Wollaston polarising prism. As in all conventional interferometers, a set of background fringes are produced which are used as a reference to determine the phase change introduced by the plasma from which the electron density can be calculated using the equation.

$$\phi = \frac{1}{2} \int \frac{n_e}{n_c} \cdot \frac{d\ell}{\lambda_{\mu m}} = \frac{1}{3.21 \times 10^{17} [cm^{-2}]} \int n_e d\ell \qquad (2)$$

for the relative (i.e., the number of) fringe shifts ϕ.

The optical layout of the interferometer is shown in Figure 4. The polariser P_1 ensures that the polarisation of the incident (collimated) beam is at 45°. The lens L focuses the beam so that a divergent beam with spherical wavefront is produced. As the beam passes through the Wollaston prism it is split into two beams with an angular separation ε, one horizontally polarised, the other vertically. These two differently polarised spherically diverging beams appear to come from the point sources S_1 and S_2 in the focal plane of the lens and are separated by $d = l\varepsilon$, where l is the distance between prism and focal plane. A second polariser P_2, orientated parallel (or perpendicular) to the incident polariser P_1, ensures that the transmission is the same for both beams, allowing the beams to interfere and produce fringes at the image plane. The fringe spacing $s = (\lambda/\varepsilon)(z/l)$, where z is the distance from the prism to the image. Thus the fringe spacing can be changed by changing the position of the prism.

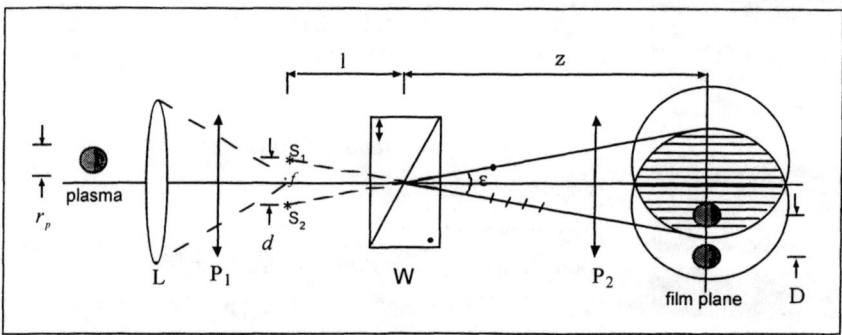

Figure 4. Layout of Wollaston Interferometer.

Each spherical wave produces an image of the plasma. The separation of the plasma images is $D = \varepsilon z$. If, as shown in the Figure 4, the two interfering waves overlap in a region where only one of the plasma images is formed, a single interferogram of the plasma is produced. If this is not the case then two complementary interferograms are produced.

Note that it is important that the two images do not overlap, i.e., $D > 2r_p$ where r_p is the pinch radius. It can be shown that this requires $\varepsilon f > r_p$. Thus care has to be exercised in selecting the angle of the Wollaston prism ε and the focal length f of the image lens.

Figure 5. Interferogram

Figure 5 shows the interferogram of the same carbon fibre z-pinch as in the deflectogram in Figure 4. The particulars of the interferometer were:

focal lens: $f = 566 mm$

Wollaston prism: $\varepsilon = 1.56°$; $l = 6.7 mm$

Image plane: $M = 2$; $z = 70.0 cm$; $D = 8 mm$

Sensitivity: $\int n_e \cdot d\ell = 3.21 \times 10^{17}$ cm^{-2} for one fringe shift.

Assuming that a fringe shift of 0.2 is measurable, we get $n_{e\,min} \cdot r_p \approx 10^{17}$ cm^{-2} for a parabolic density distribution.

Here too, as in the deflectogram, the steep density gradients near the pinch axis, bend the light outside the acceptance angle of the instrument giving a bright field Schlieren image of the edge of the core. The acceptance angle of the two instruments was about the same, $0.13\ radians$.

SUMMARY

Both the Moiré deflectometer and the polarisation interferometer employ a minimum of components and so good quality images with good spatial resolution can be obtained using standard optical components. Also, neither instrument requires a very coherent light source. This means that broadband sources or very short laser pulses can be used. Indeed, in the polarisation interferometer, the path difference between the interfering beams is so small, that it has been utilised successfully in experiments with picosecond laser pulse.

For both instruments there is a region of the plasma near the core which is inaccessible due to steep density gradients. To probe into this region it is necessary to use shorter wavelenghts, which is relatively easily managed for the deflectometer.

REFERENCES

1. Haines, M. G. and Walker, S., *Nucl. Energy* **26**, 361 (1987).
2. Sethian, J., *Phys. Rev. Lett.* **59**, 892 (1987).
3. Haines, M. G., *Plasma Phys. & Controlled Fusion* **31**, 759 (1989).
4. Hagen, P., *et al.*, *AIP Conference Proceedings* **299**, *Dense Z-pinches*, pp. 210 (1994).
5. Benattar, R., *et al.*, *Rev. Sci. Instrum.* **50**, 1583 (1979).
6. Kafri, O., and Glatt, I., *Opt. Eng.* **24**, 944 (1985).
7. Lorenz, A., *et al.*, these proceedings.
8. Beg, F. N., *et al.*, these proceedings.
9. Kafri, O., and Krasinski, J., *Appl. Opt.* **24**, 2746 (1985).
10. E. Keren and O. Kafri, *J. Opt. Soc. Am.* A **2**, 111 (1985).

Capillary and Transient Inversion Table-Top X-ray Lasers

V.N.Shlyaptsev[^,*], J.J.Rocca[$], P.V.Nickles[&], M.P.Kalachnikov[&],
W. Sandner[&], A.L. Osterheld[^] and D.C. Eder[^]

[^]*Lawrence Livermore Nat.Laboratory, Livermore, CA;* [*]*P.N. Lebedev Physical Institute, Moscow;* [$]*Colorado State University, Dept.of Electrical Engineering, Fort Collins, CO;*
[&] *Max-Born Institute, Berlin*

Abstract. Three approaches in the development of efficient small-scale X-ray lasers are considered in this paper. The first is based on electrical discharge in the relatively small capillary during fast current rise-time. The second utilizes modern compact and powerful CPA-lasers for excitation of very large transient inversions. Both of these types of X-ray lasers have recently been implemented in table-top configurations. The third approach is intended to combine advantages of electrical discharge plasma preparation with CPA-laser excitation. It will enable a new kind of X-ray lasers to be built with the ultrashort pulse duration and the high efficiency. The transient population inversion combined with the traveling-wave excitation will also enable to achieve much shorter wavelengths.

During last decade the substantial progress have been made in the development of X-ray lasers and in the understanding of the problems of efficiency. Detailed numerical modeling of MHD and laser produced plasma considerably facilitated the experimental realization of such new lasers as the capillary discharge and transient inversion X-ray lasers[1,2]. Shortly after their demonstrations the achievement of the saturation limits were reported [3,4]. They have a great potential to achieve higher efficiencies and shorter wavelengths than earlier X-ray lasers. Other possibilities for lasing in the short-wavelength region which just appeared, are based on electrical discharges in fibers [5], in hollow cathode configuration [6] etc. And it is expected that many similar X-ray laser schemes will be proposed in future since Z-pinches are a natural medium for such lasers provided its 1D motion will not be strongly distorted. Note, that many Z-pinch designs considered on this conference and oriented toward thermonuclear and other applications, can also be utilized for X-ray laser creation. For example, so called staged pinch [7] with fibers made from frozen gases (Xe, Kr etc) looks perspective for lasing in the sub-100 A range.

In this paper we shortly describe the progress in the physical modeling of two table-top laser schemes [1-4] and present the results of numerical analysis of a novel approach for substantial improvement of their efficiency.

Fast electrical discharge in gas-filled plastic capillaries is an ideal object for X-ray laser investigations. It is a kind of Z-pinches which is simple and has high plug-in efficiency and stability of compression. With just several tens of kA of current it produces hot >100 eV, and dense >10^{19} cm^{-3} high-Z plasma. Analysis of different excitation schemes (recombination, photopumping, charge-exchange etc.) and preliminary experimental investigations in capillary discharge [8] enabled us to concentrate on development of collisional scheme as the most suitable and reliable in conditions of capillary discharge [9]. The numerical model of a fast capillary discharge starting from relatively simple MHD code, represents now one of the most sophisticated physical and numerical programs targeted for modeling both MHD and laser plasma applications [1-4,9-12]. It contains 1D radiation hydrodynamics, with non-LTE all-ion stage multilevel kinetics (for example, 18 stages with total ~600 levels for Argon X-ray laser) and line radiation transport for all included lines (up to $\sim 10^4$ for Ar). All these processes have been taken into account self-consistently so that radiation, excitation, ionization and motion directly influence each other. Table 1 outlines the main physical processes included in the RADEX model. As a reference point, LLNL code LASNEX parameters are also presented.

TABLE 1

Processes Modelled	RADEX	LASNEX
Basic Z-pinch Hydrodynamics	Yes	Yes
Capillary discharge MHD	Yes	No
Thermo-Electric Effects	Yes	Yes
Laser Plasma Hydrodynamics	Yes	Yes
Multicomponent multilayer Target	Yes	Yes
Non-Spitzer Heat Conductivity	Yes	Yes
Line Radiation Transport (RT)	Yes, self-consistent, specifying each line (Biberman-Holstein Escape Method)	Yes, averaging over group (Multigroup Averaged Method) or with line RT in energy balans, not kinetics
Non-LTE Ionization Physics	Yes, Transient Multilevel Kinetics of all ion stages	Yes, Transient Multilevel Kinetics of all ion stages
Multilevel Atomic Kinetics	Yes, self-consistent, in transient approximation	Yes, self-consistent, in transient approximation
QSS 3D Ray Tracing	Yes	Yes
Traveling wave 3D Ray Tracing	Yes	No
Processor oriented	x86, W/Stations	CRAY, W/Stations

We find that the line radiation and evaporation of capillary wall are important effects that principally influence plasma dynamics and which require detailed modeling of line radiation losses specifically in the 300-800 A region of major absorption edges of plastic [9,3,10]. With the simple averaged-atom approximation we have failed to model it correctly. Ionization and excitation, ground and excited states in our model have been represented by the same set of equations. All these equation are treated in the transient approximation. Note, that the transient effects of relaxation of the exited states are not important for the hydrodynamics and radiation of most Z-pinches. But it was found that they become substantial for correct prediction of the X-ray laser population kinetics of some light ions with Z <18 [11]. This is because the transient population inversion here is 1-2 orders larger than the quasi-steady state (QSS) one [2,4,11,12]. Finally, the model includes a sophisticated ray-tracing package as a sensitive tool, without which it would be impossible to compare with experiment and make any definite conclusions about the lasing, and hence the plasma dynamics, ionization, radiation etc. This model enabled us to reliably describe and evaluate the capillary Z-pinch as well as laser produced plasma. Many of the existing hydrodynamics codes, e.g., LASNEX, which were developed for other applications, i.g., inertial confinement fusion and laser driven X-ray lasers, would require significant modification to model these capillary plasmas.

Both RADEX simulations and experiments confirmed that capillary discharge represents an object with unique parameters for X-ray laser design. In this work we describe a new X-ray laser concept which combines the advantages of capillary discharge with large transient inversion obtained with ultrashort pulse excitation by powerful lasers. Note that transient scheme, where plasma excited with two laser pulses appears to be very effective, enabling to shrink pumping requirements and hence facility size and cost by 2-3 orders of magnitude[2,4,12]. The basic idea of combination of capillary preparation method with laser excitation consist of a) pre-forming plasma utilizing capillary discharge for ionization of atoms up to the required stage and, most important, b) preparing the specific electron density profile with the convex index of refraction (or concave density profile) for effective pump laser absorption and X-ray amplification. In last case the pumping laser radiation entering from the end of the capillary will heat plasma and at the same time provide traveling wave synchronization of excitation with X-ray amplification. With RADEX modeling we found several possible ways to realize this concept.

Small initial radius and current configuration. With the capillary diameters in the range 100-300 μm and very small currents of the order of ~1-10kA, and duration ~30-300 ns (active material is externally filled or ablated from the wall), it is possible to achieve saturated laser action, for example, with Ne-like SVII ions utilizing laser pulses with pumping energies of just ~100-300 mJ (1-20 ps duration) in transient and QSS regimes. Gain values in this case may vary from 10 to 100 cm^{-1}.

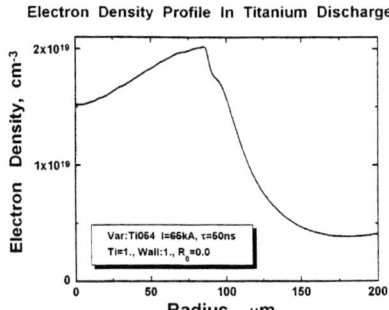
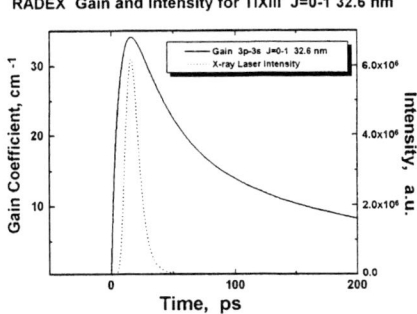

Medium initial radius and current configuration. Here with capillary diameters 1-1.7 mm exist two interesting regimes. One of them delivers required plasma parameters on the axis when the first compression shock reflects from it. Left figure shows RADEX calculated electron density profile of Titanium plasma at this moment (~13 ns, current maximum 65kA, first half cycle pulse duration 50ns). Such density profile exist during ~1ns. Right figure shows the small signal transient gain coefficient and appropriate laser intensity on the axis of capillary vs time. The 3p-3s line intensity plotted is close to the saturation and corresponds to ~0.5cm plasma length. Note well pronounced effect of shortening of laser signal at saturation, ~15ps, compared to the gain duration ~70 ps.

Computations show one more interesting regime which appear later in time, when the plasma pinches and hence forms a ring with concave density profile between the more dense wall and center. In both cases it was obtained that saturated laser action in Ne-like TiXIII at 326 A or Ni-like SnXXIII at 120A can be achieved with as low as ~1 J of laser energy in 0.3-5 ps pulses. This would be a substantial improvement of X-ray laser efficiency, an order of magnitude larger than in the mentioned latest experiments utilizing just lasers for plane target preheating and excitation [2,4,12].

Acknowledgments

One of authors (VNS) acknowledges useful discussions with V.V.Vikhrev on transport phenomena and modeling issues.

References

1. J.J. Rocca, V.N. Shlyaptsev, F.G. Tomasel et al., Phys. Rev. Lett., **73**, 2192 (1994).
2. P.V. Nickles, V.N. Shlyaptsev, M.P. Kalachnikov et al., Phys. Rev. Lett., **78**, 2748 (1997).
3. J.J. Rocca, D.P. Clark, J.L.A. Chilla et al., Phys. Rev. Lett., **77**, 1476 (1996).
4. M.P.Kalachnikov, P.V.Nickles, M.Schnurer et al., to be published.
5. P. Kubes, J. Kravarik, These proceedings.
6. C.S.Wong, P. Choi, S. Tchourilov, ibid.
7. F. J. Wessel, B. Moosman, H. U. Rahman et al, ibid.
8. J.J.Rocca, O.D.Cortazar, B.Szapiro et al., Phys.Rev.E, **47**, 1299 (1993).
8. V.N.Shlyaptsev, A.V.Gerusov, A.V.Vinogradov, J.J.Rocca et al., "Modeling of Fast Capillary Discharge for Collisionally Excited Soft X-Ray Lasers; Comparison with Experiments", SPIE J, **2012**, p.99 (1993).
9. V.N.Shlyaptsev, J.J.Rocca and A.L.Osterheld, "Dynamics of a Capillary Discharge X-ray Laser", SPIE J., **2520**, pp.365-372 (1995).
10. V.N.Shlyaptsev, J.J.Rocca, P.V. Nickles, M.P. Kalachnikov and A.L.Osterheld, "Theoretical Aspects of Efficient Downsized X-ray Lasers", Proc.of 5-th Int.Conf. "X-ray Lasers", Lund, Sweden 1996 (IOP, Bristol, **151**, pp.215-223).
11. F.G.Tomasel, J.J.Rocca et al., Phys.Rev.A, **55**, p.1437 (1996).
12. J.Dunn, A.L.Osterheld et al., To be published.

PRELIMINARY RESULTS ON A PULSED CAPILLARY DISCHARGE

C.S. Wong, P. Choi[*] and Tchourilov Serguei[#]

Plasma Research Laboratory, Physics Department
University of Malaya, 50603 Kuala Lumpur, Malaysia

[*]*Laboratoire de Physique des Milieux Ionisés, Ecole Polytechnique*
Palaiseau 91128, France

[#]*Institute of Spectroscopy, Russian Academy of Science*
Troitz, Moscow Region 142092, Russia

ABSTRACT

A small 17 J pulsed capillary discharge has been set up for studies on its VUV and X-ray emission characteristics. The discharge is triggered by the on-axis electron beam produced by the transient hollow cathode effect which has been integrated into the capillary discharge system. For the present work, a teflon capillary with diameter of 1 mm and length of 10 mm is used. Time resolved spectroscopic diagnostics have been used to study the evolution of the discharge plasma. Two groups of dominant lines in the emission spectrum are observed. One group consists of lines belonging to F-VI, F-VII, F-VIII, O-V, O-VI and O-VII in the 5.5 to 30 nm region, while another group consists of lines belonging to C-III and C-IV in the 30 to 50 nm region. Significant variation of the emission is observed on time scale of several nanoseconds. The operation of the discharge is found to change significantly with the value of the residual pressure in the evacuated capillary. Time resolved observation of the hard X-ray emission reveal the role of the self-forming electron beams in discharge initiation. The relatively slow current rate of rise leads to significant influx of wall materials and the discharge is predominantly in the ablative mode.

INTRODUCTION

The pulsed capillary discharge has been shown to be a copius source of VUV and soft X-ray[1]. Recently, due to the successful demonstration of amplification of the 3s $^1P_1$ - 3p $^1S_0$ line of Ne-like Ar at $\lambda = 46.9$ nm in a capillary discharge[2], this device has attracted much research interests from various researchers[3,4,5].

The results presented in this paper are obtained from a capillary discharge (UMCD3) using eight double ended ceramic capacitors rated at 40 kV, 2700 pF giving a maximum stored electrical energy of 17.3 J. The discharge is triggered by the on-axis electron beam produced by the transient hollow cathode effect[6] which has been integrated into the present capillary discharge system. In this series of preliminary experiments, Teflon (PTFE) has been used as the material of the capillary. This also allows us to derive qualitatively the extend of capillary wall heating by the discharge current through observation of the VUV spectral emission.

EXPERIMENTAL SETUP

The vacuum chamber and the electrode system of the pulsed capillary discharge (UMCD3) is basically identical to that of the UMFX3 flash X-ray tube described earlier[7]. In this case, the inter-electrode gap of the UMFX3 is replaced by a cylindrical block of Teflon with a 1 mm diameter and 10 mm length capillary through its axis. The anode here is a carbon disk while the cathode is made of stainless steel of exactly the same construction as in UMFX3. The capacitor bank is connected to the electrodes through an atmospheric air spark gap. This spark gap acts to hold part of the voltage since the capillary is not able to hold more than 20 kV. On charging up the capacitor to the desired voltage, this voltage will be held across the capillary and the spark gap. The discharge is initiated by triggering the transient hollow cathode effect in a similar manner as in a vacuum spark discharge[8]. An electron beam is produced by the transient hollow cathode effect and it will bombard at the carbon target at the bottom of the capillary. This will vaporize some of the carbon materials which will then expand to fill the capillary. Discharge is expected to occur at this time through the capillary thus heating the carbon vapor to hot and dense condition.

The rate of change of the discharge current is monitored by a 10 turns magnetic pick-up coil inserted into the chamber through the pumping port. The time evolution of emission from the capillary is monitored by a PIN diode (with 22 μm aluminized mylar) to cover the X-ray region of 1 to 10 Å, while a vacuum diode (without filter) is used to cover the UV to VUV region. The emission spectrum in the VUV region of 30 to 800 Å is recorded by using a 1 m Grazing Incidence

VUV Spectrometer (GISVUV1) with Off-Rowland Circle detection. The detector used to register the spectrum is a gated intensified microchannel plate (MCP) whose output image is captured on a MAMIYA 645 camera. The photograph of the setup is shown in Fig. 1. The MCP is divided into 3 frames (stripes) which are exposed to the same spectral image but gated at different times. This enable the emission spectrum of the capillary at three different times to be observed from the same discharge. A timing pulse marking the time when the first frame is recorded is recorded simultaneously with the dI/dt signal.

Fig. 1. Photograph of the UMCD3 capillary discharge setup with VUV spectrograph mounted

However, since the emission from the capillary can only be observed end-on in one direction, the time-resolved PIN diode and vacuum diode measurements cannot be obtained simultaneously with the VUV spectroscopic measurement. These are obtained together with dI/dt measurement in a separate experiment.

EXPERIMENTAL RESULTS AND DISCUSSIONS

1. TIME EVOLUTION OF X-RAY/VUV EMISSION FROM THE CAPILLARY DISCHARGE

The dI/dt, PIN and XRD signals obtained simultaneously for a typical discharge of UMCD3 at 28 kV with a base pressure of 1.8×10^{-2} mbar are shown in Fig. 2. The sharp pulse recorded by the PIN diode is due to hard X-ray produced by intense electron beam bombardment of the carbon target. This electron beam is produced by the transient hollow cathode effect and it occurs just before the main capillary discharge. The XRD signal shows the start of the plasma formation as soon as breakdown of the main gap which corresponds in time with the tip of the sharp e-beam target X-ray signal registered by the PIN diode. This e-beam is sustained by the electric field across the capillary so when the electric field collapses at breakdown, it disappears immediately giving rise to the sharp decay of the X-ray pulse. The subsequent heating of the plasma is relatively slow. This may explain why the plasma is not able to achieve a sufficiently high temperature to produce higher ionization state of C-VI in the present system. It is possible that in this case a high density low temperature Teflon plasma has been formed instead of a hot dense pure carbon plasma column. This is also evident from the observation of strong fluorine and oxygen lines in the VUV spectra obtained.

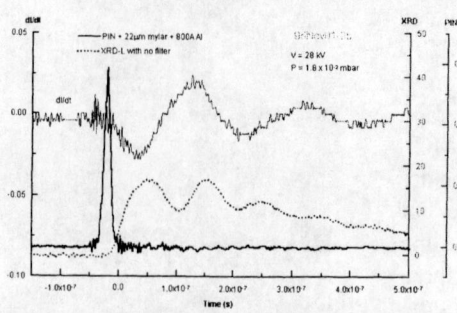

Fig.2. Simultaneous signals of dI/dt, PIN and XRD for a typical UMCD3 discharge

2. VUV SPECTRAL MEASUREMENTS OF THE CAPILLARY DISCHARGE.

The time resolved VUV spectra of a 27.5 kV UMCD3 discharge were obtained and shown in Fig. 3. Since the first frame (t = 0) is similar to the second (t = 7 ns), only the second and third frames are shown. These spectra were recorded using GISVUV1 with a 600 lines/mm grating. The exposure time for each frame is 5 ns. The timing reference t = 0 with reference to the dI/dt signal is shown in Fig. 4.

Comparing the timing of this discharge with that of Fig. 2, it can be seen that the two VUV spectra have been captured at times around the start of the plasma emission (as indicated by the XRD signal). The group of strong lines in the spectrum in (b) are lines in the 55 to 300 Å range belonging to F-VI, F-VII, F-VIII, O-V, O-VI and O-VII. The C-IV and C-III lines which are expected to be in the 300 - 500 Å range, have been found to be relatively week. The present of the fluorine and oxygen lines at t = 20 ns strongly suggests that the discharge may be in contact with the wall of the capillary at this time and vaporize some of its material into the capillary.

CONCLUSION

From the results we have obtained so far, it is evident that upon initiation by the THCD electron beam, the discharge of the UMCD3 starts along the wall of the Teflon capillary. Due to the relatively slow initial current rise, the discharge is expected to have a long wall hang-up time[9]. This results in the heating of the capillary wall and hence a high concentration of impurities such as fluorine and oxygen in the discharge plasma produced. In order to prevent this, it is necessary to increase the rate of current rise to greater than 10^{12} As^{-1}. Further work in this direction is being carried out currently in this laboratory.

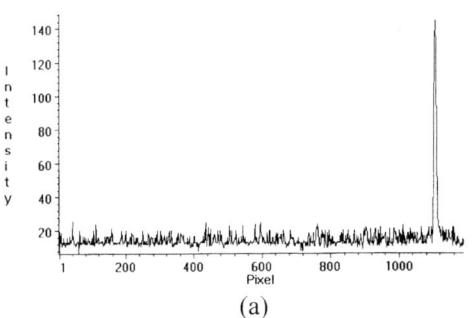

(a)

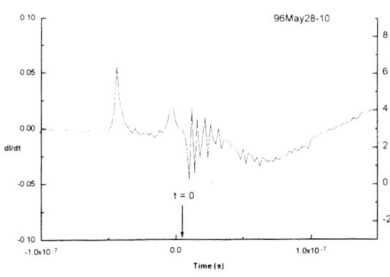

Fig. 4. Simultaneous signals of dI/dt and MCP Frame 1 timing (t = 0).

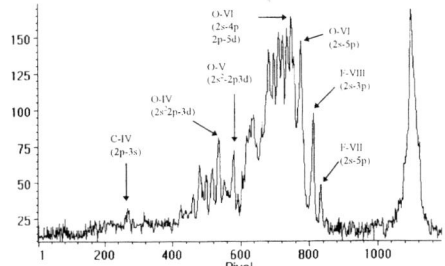

(b)

Fig. 3. Time-resolved VUV spectra at (a) t = 7 ns, and (b) t = 20 ns with reference to t = 0 indicated in Fig. 4.

ACKNOWLEDGEMENT

This project is partially supported by a University of Malaya research grant F621/96. P.Choi 's participation in the project is partially supported by an ICTP Visiting Scholar Programme and IRPA project 02-02-03-0115.

REFERENCES

1. McCorkle, R.A., Appl. Phys. **A 26**, 261(1981).
2. Rocca, J.J. et. al., Phys. Rev. Lett. **73**, 2192(1994).
3. Choi, P., Dumitrescu-Zoita, C., Larour, J., Rous, J., Favre, M., Zambra, M., Moreno, J., Chuaqui, H., Wyndham, E. and Wong, C.S., BEAMS 96.
4. Shin, H.J., Kim, D.E. and Lee, T.N., Phys. Rev. E **50**, 1376(1994).
5. Steden, C. and Kunze, H.-J., Phys. Lett. A **151**, 534(1990).
6. Choi, P. et al., IEEE Trans. Plasma Sci. **15**, 428(1987).
7. Wong, C.S., Ong, C.X., Moo, S.P. and Choi, P., AIP Conf. Proceedings **299**, 637(1994).
8. Wong, C.S., Ong, C.X., Moo, S.P. and Choi, P., IEEE Trans. Plasma Sci. **23**, 265(1995).
9. Wheeler, C.B., J.Phys.D: Appl. Phys. **7**, 363(1974)`

Study of Pulsed Soft X-ray Source Employing a Gas-puff Z-pinch Plasma Device For Lithography Applications

G. X. Zhang, X. M. Guo[+], C. M. Luo[+], S. Lee, X. Feng

School of Science, Nanyang Technological University, 469 Bukit Timah, and Singapore 259756
+ *Department of Electrical Engineering, Tsinghua University, Beijing 100084, China*

ABSTRACT: Employing a gas-puff Z-pinch plasma device, we have developed a bright and reliable X-ray source. The Z-pinch plasma was produced by a capacity discharge, using a fast valve to inject an annulus of Argon gas. The total capacitance of circuit is 24μF, and the peak discharge current is 360kA with a quarter-period of 2μs when the capacitor bank was charged to 25kV. The X-ray spectrum from our gas-puff Z-pinch plasma device is between 2keV and 6keV, and its yield is about 50J per shot. The shape of X-ray source is like a small cylinder with diameter of less than 1 mm and length of less than 2 mm. We have obtained some preliminary results of lithographic patterns by means of this plasma device.

I. INTRODUCTION.

As soft X-ray has wide applications in microfabrication, lithography and microscopy [1], it is important to develop a high quality and strong intensity soft X-ray source. The main sources that can generate such soft X-ray are: synchrotron radiation, laser-produced plasma, plasma focus device, gas-puff Z-pinch plasma and vacuum spark etc. The primary objective of this work is to develop a soft X-ray source suitable for the lithography. The basic requirements for the X-ray source is highly collimated and having strong intensity, so that a clear pattern with considerable depth can be produced. Among the available X-ray source, synchrotron radiation is the best for such purpose because of its high intensity and collimation. However, construction of a synchrotron radiation facility is very expensive, operation and maintenance of it is also difficult. Gas-puff Z-pinch plasma device is more attractive in this regard, because it can emit intense soft X-ray with high efficiency, the cost is low, and its convenient to operate [2]. It has been found that wavelength of X-ray is about 2 ~ 6Å which is commonly used for lithography and microfacration [3]. The intensity of the X-ray source is about 10~50J/shot. So, with this soft intensity X-ray source, it is suitable for the work of lithography.

II. EXPERMENTAL SET-UP

A gas-puff z-pinch is generally compose of the following parts: Charging device, energy storage capacitors, spark gap switches, evacuation system, discharge chamber, fast-acting valve etc. [2]. The configuration of the discharge

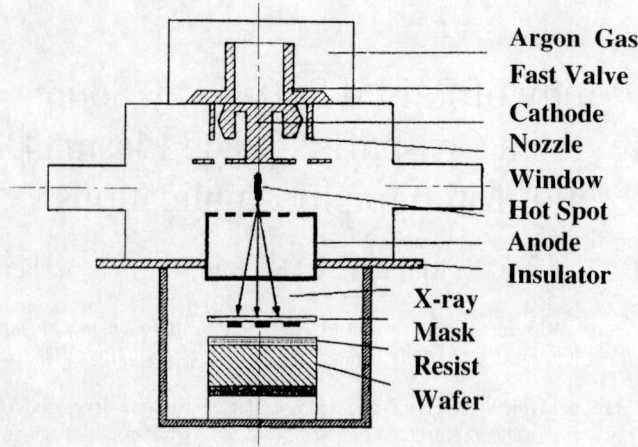

Figure 1. The construction of the discharge chamber and X-ray extraction.

chamber and the associated set-up is illustrated in figure 1. Injecting gas through an annular nozzle from a fast valve produces a hollow gas shell. The mass of the gas shell can be varied by charging the plenum pressure and the delay between the beginning of gas injection and the initiation of the discharge can vary the mass of the gas shell. The structure of the nozzle is shown in figure 2. The characteristic parameters of the device so far obtained are: the total capacitance of circuit is about 24μF, the total inductance of the circuit is about 130nH, the peak discharge current is about 360kA with quarter-period of 2.8μs when capacitor bank was charged to 25kV.

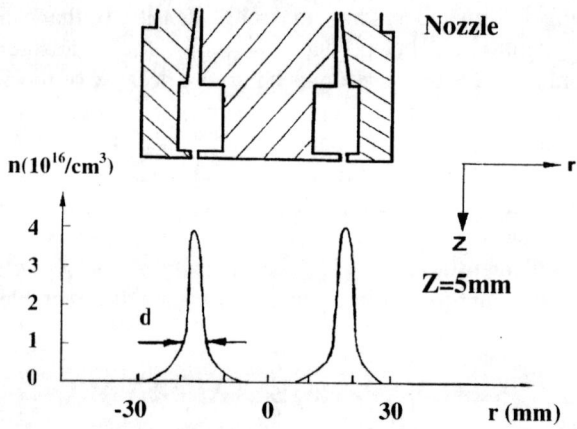

Figure 2. The diagram of the nozzle and argon gas density profile measured at 5mm from the orifice for the nozzle with 5atm argon in the puff valve plenum.

The temporal characteristics of the X-ray emission were measured with a PIN diode filtered with 100μm Be and 1μm Al foil. This PIN detector has a receiving window of 1.2 cm-diameter with a gold layer of 0.3μm thickness, and its response is 3ns. In order to determine the region and distribution of the X-ray emission, two pinhole cameras are employed to take photographs both in the axial and radial directions. The pinhole camera consists of a 500μm aperture filtered with a 5μm thick Al foil [3].

III. EXPERMENTTAL RESULTS

The puffed gas density profiles from the nozzle are measured using fast ionization gauge and shown in figure2. From this figure one can get the maximum

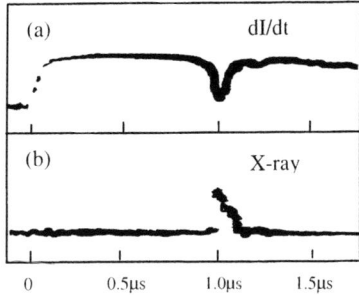

Figure 3. Typical oscillogram (a) dI/dt; (b) x-ray output measured with PIN.

gas density n_{max} which is about $3.6\times10^{16}/cm^3$, the thickness of the hollow gas shell d is about 4 mm. X-ray output waveform from PIN diode is shown in Figure 3. and it can be repeated very well under the same conditions. The X-ray photon energy is estimated to be in the range 2kev to 6keV[2], and the X-ray yield is between 10 to 50 J/shot. Pinhole photographs of the gas-puff Z-pinch X-ray

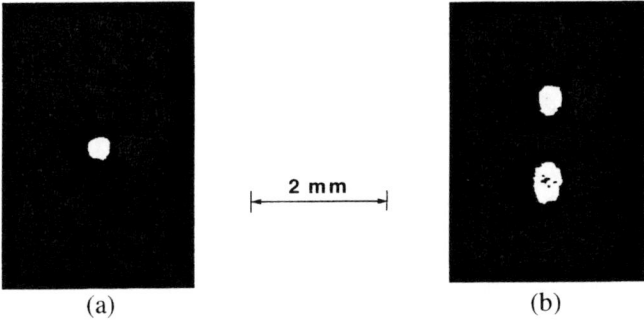

Figure 4 Pinhole photographs of X-ray source in axial (a) and radial (b) directions.

source are shown in figure 4. From this figure, one can see that the X-ray source has a cylindrical shape with a diameter of approximately 1mm and length of 2 mm. So, it is an ideal point X-ray source to reduce the blur caused by penumbral effect in proximity replication. Figure 5 is a preliminary lithographic pattern before final development. The resist material is PMMA, and the mask is a simple mesh with holes of 200 μm–diameter and 100 μm separations.

 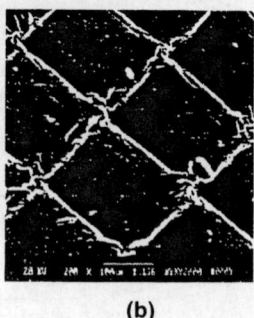
(a) (b)

Figure 5 The preliminary lithography pattern: (a) Mask; (b) lithographic pattern

IV. CONCLUSIONS

Basing on our experimental results, we conclude that gas-puff z-pinch device can work stable and produce X-ray of 10 ~ 50 J/per shot with wavelength 2 ~ 6Å. The size of the X-ray source is very small having a diameter less than 1mm and length less than 2mm. It is therefore potentially suitable for the use in microfabrication and lithography. In the work, a preliminary lithographic pattern has been carried out and more experiments are being under way. The results will be presented in future publications.

REFERENCE:

1. Becher. E.W. Ehrfeld, W. Hagman, A, Maner.A and Munchmeyer D, Microelectronic Engineering 4,35.(1986)
2. Y, Z, Fu, Master Thesis 1991, Tsinghua University, Beijing.
3. C.R. Li, T.C.Yang C.M. Luo, and M. Han, Fusion Technology, 21, 25 (1992)
4. Munchmeyer D. and Langen J. Rev. Sci. Instrum, 63, 713. (1992)

SPECTROSCOPY

Studies of X-pinch Plasma Fine Structure Using High Resolution Optical and Imaging Spectroscopy Methods*

S.A. Pikuz, T.A. Shelkovenko, V.M. Romanova, G.S. Sarkisov

Lebedev Institute, Moscow, Russia

D.A. Hammer and D.F. Acton

Laboratory of Plasma Studies, Cornell University, Ithaca, NY 14853

The predictable position of a constriction and bright x-ray emission at the wire intersection point of an X-pinch enables the use of high resolution visible light and x-ray imaging diagnostics to study the constriction itself and the region around it. Using imaging x-ray spectrographs, an x-ray pinhole camera with a 7 μm pinhole, and visible light diagnostics, we have studied X-pinch structure and dynamics with spatial resolution better than 10 μm and time resolution as good as 1 ns.

An X-pinch is formed by placing two or more fine wires between the output electrodes of a high current pulser so that the wires cross and touch in mid-gap. We are primarily interested in the region near the cross point where a constriction, or "neck", invariably occurs in the discharge. This neck appears in optical photos as a bright spot at the point of wire intersection, and is an intense source of x-ray radiation. The predetermined position of the source of radiation makes the X-pinch preferable for some x-ray source applications to the conventional Z-pinch.

A number of competing physical mechanisms appear to contribute to the formation of complicated source structure of the bright spot. The dependence of some of the features of the source region on the material and linear mass of the wires, and on the configuration of the wires and the quality of their surface is still unclear. Nevertheless, the bright spot has a number of specific distinctive features which are repeatedly observed from shot to shot. These features include the formation of a "minidiode", the formation of two main hot points near the minidiode "electrodes", axial plasma ejections from the wire intersection volume, substantial cathode-anode asymmetry of x-ray images, etc. Some of these specific features have been discussed previously [1]. Further experimental investigation has given us new information about the structure of the bright source region.

The experiments we describe here were performed using the BIN generator at Lebedev Institute (Moscow, Russia) with a current of 250 kA in a 100 ns pulse, and the XP-pulser at Cornell University, operating at about 350 kA peak current in a 100 ns pulse. The experimental setups for the BIN and XP generators are shown schematically in Fig. 1. The main diagnostics used for X-pinch structure investigations are described in [2,3]. The predetermined position of the bright spot makes possible the use of high resolution optical systems [2,4] for investigation of

---

* Research supported by Sandia National Laboratories, Albuquerque, Contract #AJ-6400.

neck-forming plasma dynamics as well as high resolution imaging x-ray spectroscopy [2,5] for studying the internal structure of the neck, as is illustrated in Fig. 1c.

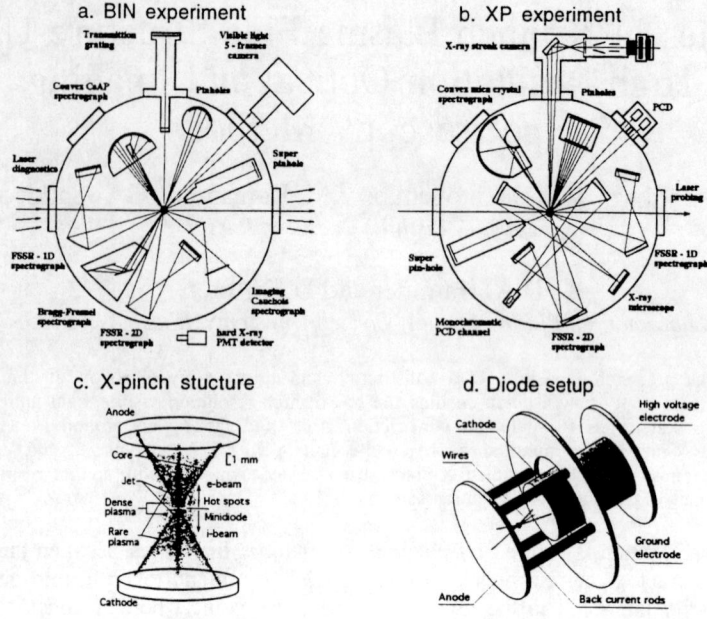

Figure 1. Experimental setups for BIN and XP generators and simplified x-ray spectroscopy set up.

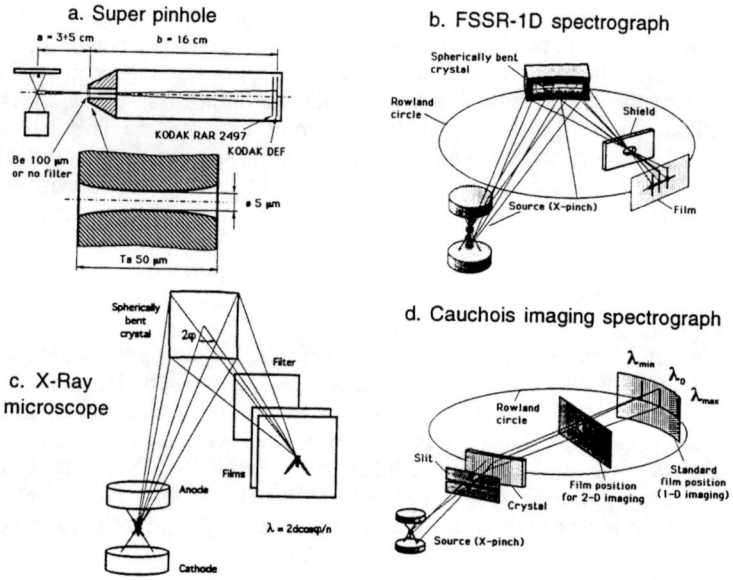

Figure 2. Diagram of the X-ray imaging devices.

Fast photoconducting detectors (PCDs) with a time resolution of 1 ns, and an x-ray streak camera gave the time dependence of some processes.

The position of the neck in relation to the cross-points of the X-pinch wires, the neck structure during pinching, and plasma parameters in the vicinity of the neck and in the plasma around of the wires cores were investigating using pulsed lasers and an optical framing camera. The internal structure of the bright spot near the cross point was studied with a spatial resolution better than 10 microns using a specially-designed time integrated pinhole camera (Fig. 2), as well as in the radiation of individual spectral lines of highly charged ions such as He-like Ni and Al.

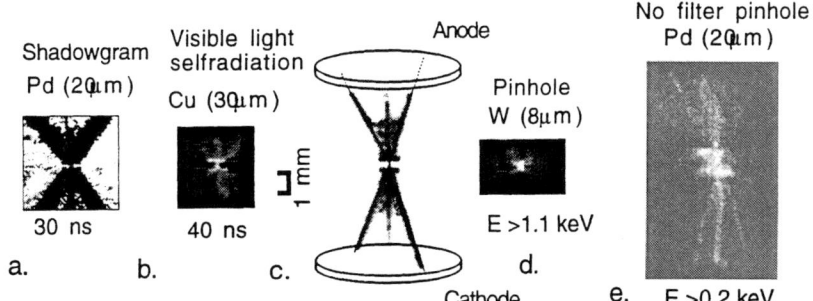

Figure 3. a, b, d, e - minidiode images, c - X-pinch schematic diagram.

Figure 3 shows four images of the neck region of X-pinches (the standard configuration is shown in Fig. 3c) using 3 different wire materials. Although one image is a shadowgram obtained with a pulsed laser (3a), a second is a visible light frame (3b), and two (3d and 3e) are x-ray pinhole images, all show certain common features. In all the pictures one can see indications of a planar gap - a "minidiode" - near the wire intersection point. The minidiode arises approximately 30-40 ns after the start of the current pulse and its lifetime can be up to 50-60 ns, as shown by the data from the visible light multiframe camera shown in Fig. 4. However variation in structure has been seen on a few ns time scale in schlieren images [6]. The minidiode position is a bit shifted to the anode side of the diode as shown in Fig. 5. Often, the minidiode appears to be short-circuited either at its center or at the periphery (see Fig. 3). The size of the gap between the minidiode electrodes depends on both the initial mass and the atomic number of the wires, decreasing with an increase in either of these parameters. The size of the minidiode in frame-camera images and laser shadowgraphs is very similar to that seen in hard X-ray pinhole photographs. The experiments have shown that the minidiode structure and lifetime are not dependent on whether the wires have been cleaned or on the number of wires (Fig. 6).

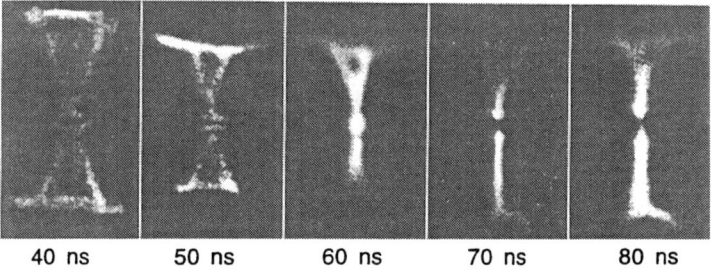

Figure 4. Pd X-pinch images, obtained in the one shot using a visible light multiframe camera.

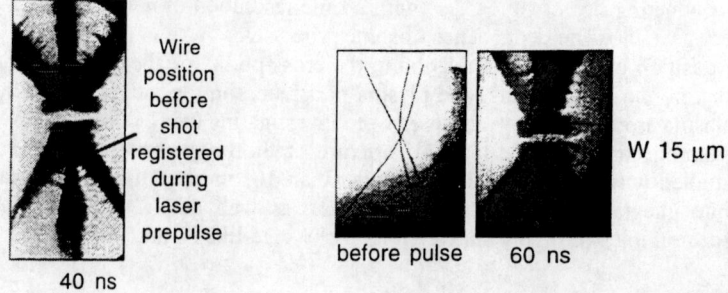

Figure 5. Shadowgraph of an X-pinch showing the minidiode position.

Figure 6. Images of a 3-wire X-pinch.

With the spatial resolution of the diagnostics used some years ago, a single bright spot was observed in the region of the wire intersection point. With the improvement of the spatial resolution of x-ray diagnostics to better than 10 µm, it is very rare to observe a single bright spot. Most often, two bright spots occur on the minidiode axis, one on the anode side and one on the cathode side. The continuum radiation from the anode spot is almost always harder, more intense and emitted from a larger region than that from the cathode spot (Fig. 7). Sometimes, even using diagnostics with modest spatial resolution, many bright spots are seen (Fig. 8). It is almost always possible to identify one of the spots as emitting the hardest radiation. As a rule, this spot is located at the anode of the minidiode. The size of the bright spots depends on the spectral range of the imaging radiation - the harder the radiation the smaller the size (see Figs. 9 and 10). Using diagnostic techniques with very high spatial resolution, almost all bright spots are divided into a few very small bright points (Fig. 9b, 10b and 11a). Sometimes these small bright points also have double structure with sizes of a few µm, as has been seen using the Bragg-Fresnel measurement technique. Recent experiments with fast PCDs have shown that these small bright spots have varying lifetimes ranging down to 1 ns depending upon the wire material and radiation hardness (see Fig. 11).

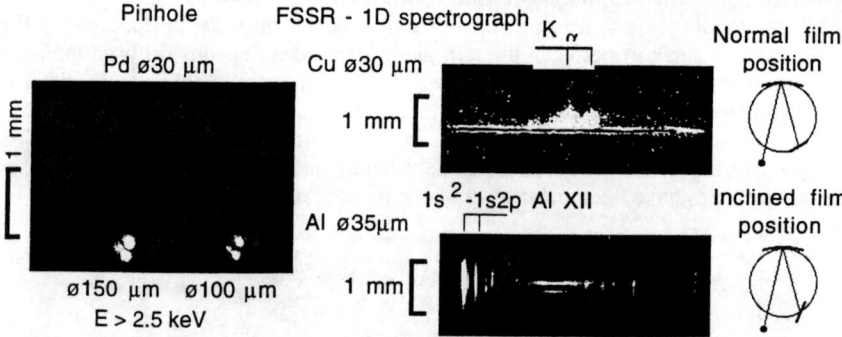

Figure 7. Examples of the double point structure of some X-pinches.

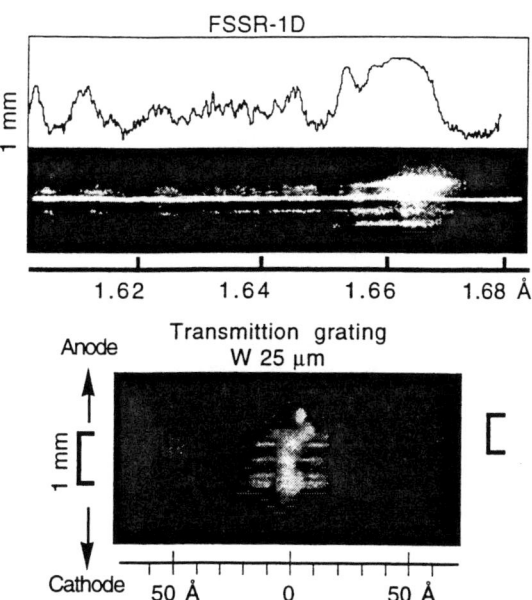

Figure 8. Examples of the multipoint structure of some X-pinches.

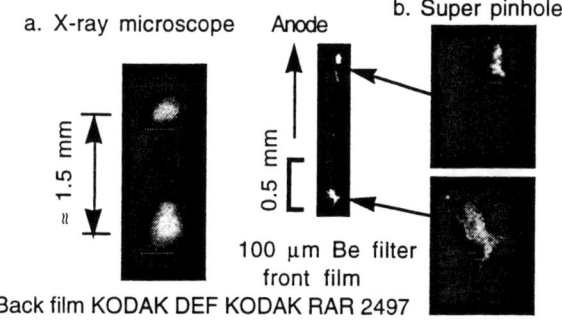

Figure 9. Bright spot structure of an X-pinch from 35 μm Al wires.

In the pinhole, frame and schlieren images of x-pinches (Figs. 3, 4, 5, 6) one can see structures that look like jets of material ejected from the crossing point. Recent experiments have shown that there is a second snake-like structure that is seen in schlieren and frame images and soft x-ray pinhole images and seems to consist of wire surface material blown off the wires early in the discharge which coalesces on the axis between the wires. The "snakes" exist before and after minidiode formation, and are seen even in cases in which the wires are not touching. In the latter case there are no other structures (minidiode, bright spots, etc.) typical of X-pinches. Since the pinhole images do not provide temporal resolution, it is difficult to distinguish snakes from a jet carrying away material ejected from the crossing point. Toward the end of the discharge, these two formation often become intermixed. However, in the harder x-ray pinhole images (Fig. 13a) and X-pinch images obtained using line radiation with very short wavelength (Fig. 14d) one can see only jets. X-ray radiation up to 10 keV emanates from jets, and they move at ~ 10cm/μsec.

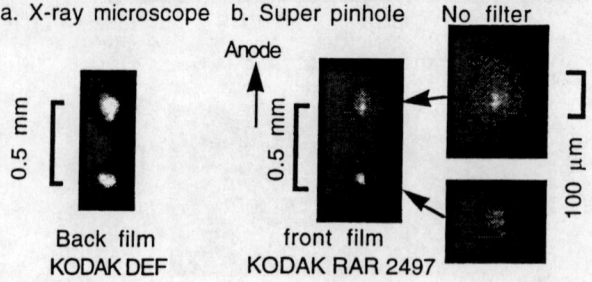

Figure 10. Bright spot structure of an X-pinch from Pd wires.

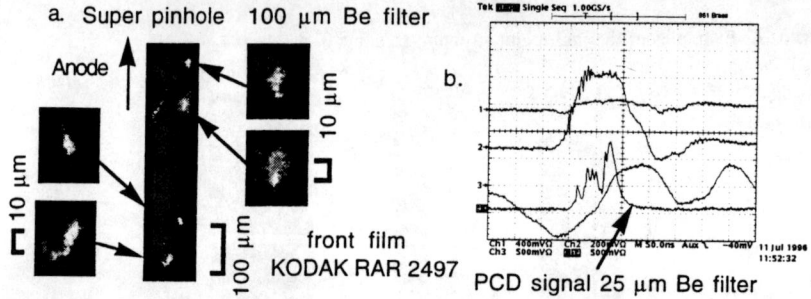

Figure 11. Bright spot structure of an X-pinch from brass wires.

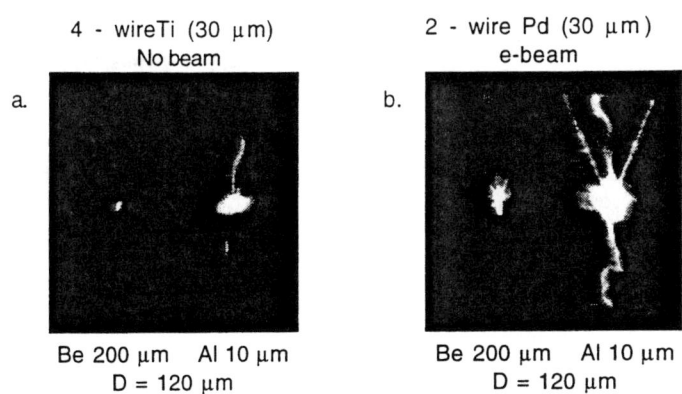

Figure 12. Pinhole images of X-pinches from Ti and Pd wires.

One of the most impressive features of the x-ray images of X-pinches is the presence of a clear anode-cathode asymmetry (Fig. 12, 13, 14). In pinhole images the anode region of the X-pinch is almost without exception more luminous than the cathode region (Fig. 12b). Strong anode-cathode asymmetry also exist in the intensity of satellite lines of He-like light ions (Fig. 13a), as well as in the characteristic spectral lines radiation of heavy ions in the anode region of the X-pinch, (Fig. 14b, c, d). Recent investigation have shown that these phenomena are connected with electron beam generated in the minidiode. The electrons accelerated in the minidiode gap interact with the rather cold plasma surrounding the remains of the wires, and the jets and snakes. This causes continuum radiation and the emission of characteristic spectra. Electron energies may be up to 15-30 keV for light elements and up to 60-100 keV for heavy elements according to pinhole images with very thick filters. We are not yet able to measure the electron beam current directly. Nevertheless, there is some indirect evidence, such as the anode damage due to a shock wave which we believe is excited by an electron beam; the size of the metal evaporation crater in the center of the anode; the intensities of the emitted spectral lines; etc. The influence of the electron beams on the intensity of the satellite lines of He-like ions enables us to calculate the electron beam intensity [7]. Together with a measurement of bremsstrahlung radiation from the electron beams using a scintillator detector we can estimate the relative intensity and hardness of the electron beam (Fig. 14). Using these indirect methods we determined that the electron beam intensity can be up to tens of kA and is dependent upon the wire material and thickness: the heavier the material and the thicker the wire, the more electron beam intensity. The X-pinch geometry is also influenced by the electron beam intensity. For example, the intensity of the electron beam in the case of an X-pinch made from 3 or 4 wires is less than that for the X-pinch from two wires with the same total mass. More intense and harder radiation from the anode bright spot can be explained by the electron beam influence, too. In the case of a weak electron beam, the anode-cathode asymmetry is not observed (Fig. 14a).

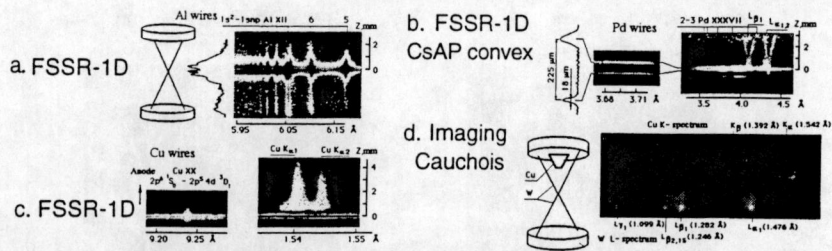

Figure 13. Structure of X-pinch from different materials.

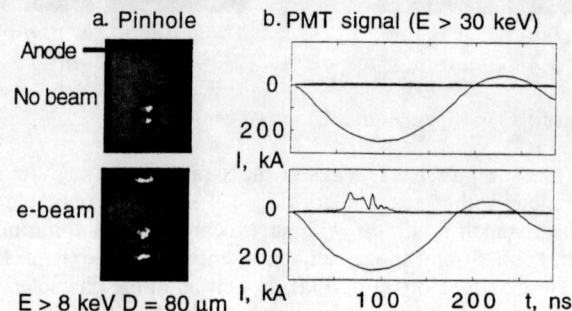

Figure 14. Pinhole images and scintillator detector signals for Al X-pinch without electron beam and with electron beam.

The data presented here make it possible to distinguish a number of regularities of the phenomena accompanying an X-pinch. In general terms, the analysis of the dynamics of various processes in an X-pinch was made in [1] although the complex picture of this plasma in various details are still unclear. We hope that the new results shown here can help to add to this picture.

1. Ivanenkov, G.V., Mingaleev, A.R., Pikuz, S.A., Romanova,V.M., and Shelkovenko, T.A. *Plasma Phys.* **22**, 479 (1996).
2. Skobelev, I.Yu., Pikuz, S.A., Faenov, A.Ya., et al., *JETP*, **81**, 692 (1995).
3. Pikuz, S.A., Brunetkin, B.A., Ivanenkov, G.V., et al., *JQSRT*, **51**, 291 (1994).
4. Kalantar, D.H. and Hammer, D.A., *Phys. Rev. Lett.* **71**, 3806 (1993).
5. Pikuz, T.A., Faenov, A.Ya., Pikuz, S.A., et al., *X.-ray Sci. & Tech.* **5**, 323 (1995).
6. Kalantar, D.H., Hammer, D.A., and DeSilva, A.W., *Rev. Sci. Instrum.* **68**, (7) in press (1997).
7. Abdallah Jr., J., Faenov, A.Ya., Hammer, D.A., et al., *Physica Scripta* **53**, 705 (199).

HIGH RESOLUTION MONOCHROMATIC X-RAY IMAGING SYSTEM BASED ON SPHERICALLY BENT CRYSTALS.

Y. Aglitskiy and T. Lehecka
Science Applications International Corp., McLean VA 22102

S. Obenschain, S. Bodner, C. Pawley, K. Gerber, J. Sethian,
C. M. Brown, J. Seely, U. Feldman
Naval Research Laboratory, Washington DC 20375

G. Holland
SFA Inc. 1401 McCormic Drive, Landover MD 20785

We have developed a new X-ray imaging system based on spherically curved crystals. It is designed and used for diagnostics of targets ablatively accelerated by the Nike KrF laser [1,2]. The imaging system is used for plasma diagnostics of the main target and for characterization of potential backlighters. A spherically curved quartz crystal (2d=6.687 Å, R=200 mm) is used to produce monochromatic backlit images with the He-like Si resonance line (1865 eV) as the source of radiation. The spatial resolution of the X-ray optical system is 3-4 μm. Time resolved backlit monochromatic images of CH planar targets driven by the Nike facility have been obtained with 6-7 μm spatial resolution.

X-ray diagnostics have been proved to be extremely powerful tools in the study of dense laser produced plasmas. They include classical spectroscopic methods as well as more recent self-emission imaging of hot plasmas and backlighting of relatively cold and dense plasma formations.

The focal spot of about 0.75 mm FWHM with a 0.4 mm flat top, requiring a field of view of about 1mm, the acceleration of the targets, the intention to use thick plastic targets, and the necessity of time resolution place special requirements on the X-ray imaging diagnostics designed for Rayleigh-Taylor experiments on the Nike laser. Among well known imaging systems like Kirkpatrick-Baez microscopes [3,4,5], pinhole cameras, Fresnel zone plates, and Bragg-Fresnel structures [6], only spherically bent crystals simultaneously meet our requirements of spatial resolution about 5 μm, large field of view with small backlighter size, high image magnification (at least 20x), high throughput of the system and spectral filtering of imaging radiation from the self-emission of the object under investigation. [7-13].

Described below is a high spectral and spatial resolution instrument based on the spherically curved crystals. Its advantageous features are the following:
1. It can provide 1D imaging with spectral resolution [10].
2. It can also provide 2D self-imaging in the light of an individual spectral line or in the light of an intense continuum with a wavelength band as narrow as 0.01 Å [11].
3. And in this paper we focus ourselves on the 2D monochromatic backlight imaging designed for (but not limited to) the Nike laser plasmas [2,10,11,13].

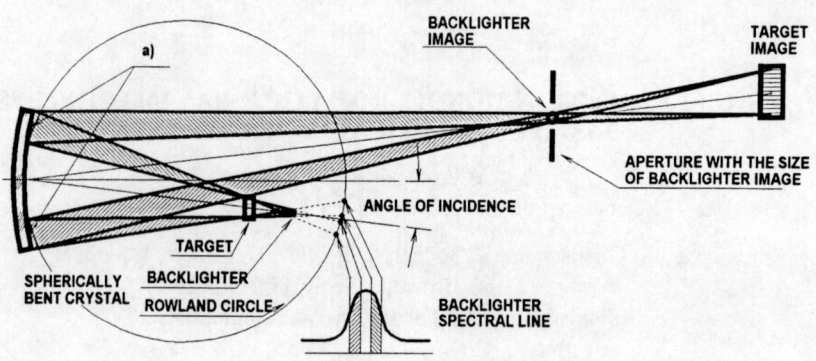

Fig.1. Principal optical scheme of backlighting with the use of spherically bent crystals. a) corresponds to the small area of the crystal which is actually involved into the imaging of a single point of the target (shown for two edge points). A close-fitting light baffle with an aperture stop just larger than the size of the backlighter image has been placed in the reflected beam to block most of the continuum radiation from plastic target. It provides an effective spectral and spatial filtering of the monochromatic imaging radiation from the unwanted broad-band self-emission of the target. Experimental estimation of the contrast achieved by this method has been done in a previous experiment [11]. The results show that the flux of the self-image is at least a factor of 160 smaller than the flux from the He-like Si backlighter.

Scheme of the experiment. Results and discussion.

High resolution X-ray imaging instruments have been used to study foil targets that were irradiated by the NRL Nike KrF laser. The Nike laser produces an energy up to 3 kJ in a 4 nsec pulse and at a wavelength of 248 nm. The beams are smoothed by the Induced Spatial Incoherence (IS1) technique, and up to 44 beams can be overlapped in the focal spot [1,14,15]. The intensity distribution of the laser radiation that was incident on the targets had a FWHM of 750 µm and a flat central region that was 400 µm in diameter. The targets were 40-60 µm polystyrene (CH) with either a smooth surface (surface roughness < 30 Å) or with a pattern on the laser irradiated side. For this thickness a probing radiation energy must be in the range of 1-2 keV in order to provide an absorption picture with the optimal intensity and contrast. A spherical quartz crystal with 2d=6.6870 Å and R=200mm has been used for the X-ray imager, and the resonance line of He-like Si with the wavelength 6.6488 Å (1865 eV) has been chosen as a backlighter source. The backlighter foil of Si was irradiated by 6 overlapping laser beams having a total energy of 300 J in a focal spot about 400 µm in diameter with a flat central region of about 200 µm. The field of view, is determined by the arrangement of the source and the size of the crystal. In the present experimental condition the backlighter target was positioned about 50 mm from the main target, which was 105 mm from the crystal. The crystal dimensions were 5 mm x 10 mm and considering the finite size of the backlighter focal spot, we estimate the maximum field of view as 2 mm x 3.5 mm.

The present instrument has been designed with the idea of having two crystals that deliver two 20x magnified images to the photocathode of a framing camera with four striplines gated at different times. Provided that each of the two images covers two different strip lines, we can record four snapshots of developing instabilities with high spatial and time resolution.

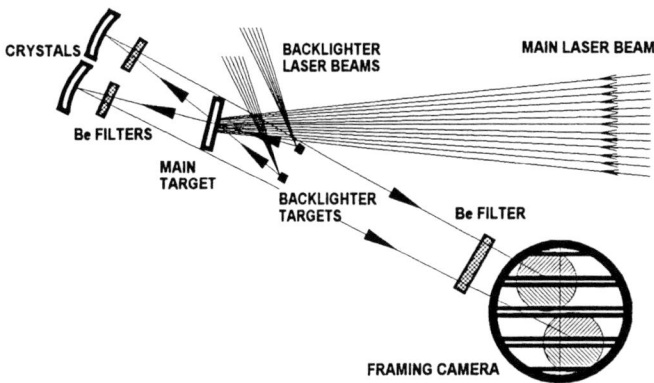

Fig.2. General scheme of the experiment and two crystal design of the planned imaging system.. The main target is turned in order to be in more favorable position for imaging.

The current data were obtained using only one quartz crystal and so only two central photocathode strips were used. The image from the spherically bent crystal was delivered to the assigned vacuum port about 2000 mm away from the target with the accuracy of 2 arcmin. Future development of this diagnostic will enhance the system to use the four strip capability of the framing camera. A second crystal with a separate backlighter and potentially different working wavelength will be added.
Kodak DEF X-ray film has been used to make time integrated test pictures while performing the alignment. A framing camera in the same position has been used for time resolved imaging of plastic targets and for test pictures as well.. The exposure time of each of four striplines is approximately 300 psec and can be timed relative to the laser pulse to ± 200 psec.
A number of test shots were made with meshes as test objects to optimize the optics adjustment. High quality meshes were used. The resolution of the optical system itself was evaluated with DEF X-ray film as a detector.
Very high resolution of 3-4 µm is available within an area 0.8 mm x 0.8 mm, the size being limited mainly by the quality of the crystal. These numbers are close to the limit predicted by theoretical calculations [12]. The rest of the scene has the same intensity and a lower but usable resolution of 5-7 µm. The spatial resolution of the spherically curved crystal imager is determined by the bending quality and by optical aberrations which depend significantly on the Bragg angle. The image quality analysis shows that there are small local variations in the radius of curvature across the crystal surface. The crystal quality criterion therefore is the area of the image where resolution meets the requirements of the experiment. Test shots on the above targets have been taken

with the framing camera as a detector with resultant lower resolution.

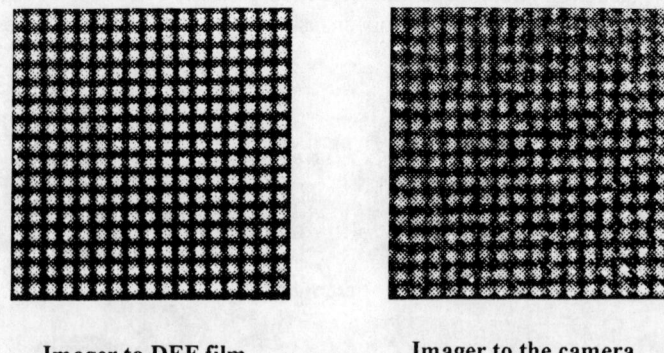

Imager to DEF film   **Imager to the camera**

Fig.3. Enlarged portion of the X-ray film image of a 1500 lpi mesh (17 μm period, 5 μm width of wire) with an ultimate resolution of 3-4 μm on target. An image of the same mesh, with the framing camera as a detector has the resolution of 6-7 μm.

Numerous shots with different targets and different parameters of the laser pulse were taken on the Nike facility [2]. Images of a 40 μm thick CH target with the sinusoidal perturbations (31 μm period, 0.9 μm amplitude) are presented on the Fig. 4.

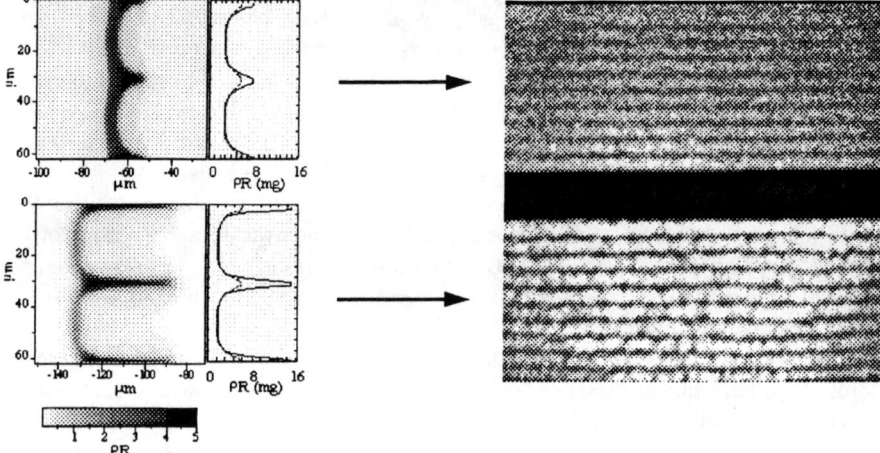

Fig. 4. The qualitative comparison of the 2D simulation code [16] with the monochromatic image of a driven plastic target (initial 0.9 μm amplitude, 31 μm period sinusoidal mass perturbation) taken at 1.7 nsec. and 3.2 nsec. after the beginnings of laser pulse. Significant narrowing and amplification of initially wide ripples is shown The spherically curved crystal imager coupled with the framing camera resolves the width of the high density spike of saturated instabilities. However, the dynamic range of the framing camera needs improvement to fully resolve the minima. Detailed analysis of this and other images has been published [2].

One of the problems of any backlighting scheme is that the field of view with a small backlighter is directly proportional to the size of the input aperture

Imaging systems like Kirkpatrick-Baez (KB) microscopes [3-5], pinhole cameras, Fresnel zone plates and Bragg-Fresnel (BF) [6] structures could be characterized as small aperture devices (several microns for pinholes, about 0.5 mm (the best) for BF elements or zone plates and collecting solid angle of 4×10^{-7} sr. for KB microscope). Compared to that, spherically bent crystal imagers have aperture sizes of several centimeters [12], thus among mentioned above devices only a crystal imager can provide a large field of view with a small backlighter.

There is a difference between large-field-of-view images provided by the crystals and those of KB microscopes where intensity and resolution degrade rapidly when approaching the edges due to the field obliquity aberration [4]. The addition of flat mosaic crystals to KB microscopes makes a KB imaging monochromatic but reduces the field of view and adds smearing to the image [3]. Elliptical Bragg-Fresnel structures are essentially aberration-free for one single wavelength and can provide monochromatic imaging with submicron resolution [6]. But in addition to having a relatively small field of view, they are much dimmer than focusing crystals. Spectrally filtered pinhole cameras have achieved 8 μm spatial resolution [2] but they have very low efficiency and cannot compete with contemporary focusing systems.

ACKNOWLEDGMENTS

The construction and implementation of the Nike laser were carried out by a large team of talented engineers and technicians. High quality spherically curved imaging crystals were provided by Dr. A. Faenov and T. Pikuz of VNIIFTRI, Moscow, Russia. This work was supported by the U. S. Department of Energy.

REFERENCES

[1] S. P. Obenschain et al., Phys. Plasmas **3**, 2098 (1996).
[2] C. J. Pawley et al., Phys. Plasmas **4**, 1969, (1997).
[3] F.J. Marshall, J. A. Oertel, Rev. Sci. Instrum. **68** (1), 735 (1997).
[4] D. Schrimann, Plasma Physics Reports **20**, #2, 113 (1994).
[5] F.J. Marshall, O. Su, Rev. Sci. Instr. **66**(1), 725 (1995).
[6] A. Erko et al., Opt. Comm. **106**, 146 (1994).
[7] A.V. Rode et al., Opt. Comm. **77**, #2,3 , 163 (1990).
[8] S.A. Pikuz et al., JETP Lett. **61**, 638 (1995).
[9] T. A. Pikuz et al. J. X-ray Sci. & Techn, **5**, 323 (1995).
[10] Y. Aglitskiy et al., Phys. Plasmas **3**, 3438 (1996).
[11] C. Brown et al., Phys. Plasmas **4**, 1397 (1997).
[12] M. Sanchez del Rio et al., Physica Scripta , 1997 (in press).
[13] S.A. Pikuz et al., Rev. Sci. Instr., **68**(1), 740 (1997).
[14] R.H. Lehmberg and S.P. Obenschain, Opt. Commun. **46**, 27 (1983).
[15] T. Lehecka et al., Opt. Comm. **46**, 485 (1995).
[16] J. P. Dahlburg et al., J. Quant. Radiat. Transfer, **54**, 113 (1995).

Study of X-ray Polarization and E-beams Generation during Hot-Spots Formation in PF-Discharges

L. Jakubowski, M. Sadowski, E.O. Baronova[*], and V.V. Vikhrev[*]

Soltan Institute for Nuclear Studies (SINS), 05-400 Otwock-Swierk, Poland
[*]*Nuclear Fusion Institute, RRC Kurchatov Institute, 123182 Moscow, Russia*

Strong turbulence phenomena, as observed within dense magnetized plasmas, are of interest for theoretical and experimental studies. This paper concerns the filamentation and formation of "hot-spots" in high current PF-type discharges, as well as the emission of fast electron beams and X-ray pulses. The generation of oriented e-beams can induce the polarization of X-rays, as observed within the MAJA-PF facility at SINS.

1. INTRODUCTION

The emission of intense X-ray pulses, relativistic electron beams (REBs), and energetic ion streams, has been investigated in many plasma experiments of the PF and Z-pinch type. Detailed studies of the main characteristics and correlations of different radiation are important not only for the understanding of the physical phenomena involved, but also for possible applications.

Measurements of a continuous X-ray spectrum, as performed by several authors [1-2], showed its suprathermal character. It suggested that the electron velocity distribution is anisotropic. The non-thermal character of the X-ray spectrum was explained by the appearance of intense electron beams propagating mainly in the axial direction. Such e-beams were measured directly by means of Faraday type collectors [3] and Cerenkov-type detectors [4]. The propagation of the suprathermal electrons can also influence X-ray spectral lines. First of all, it can change relative intensities of such lines due to different excitation mechanisms. The anisotropic velocity distribution of the electrons inducing the X-ray emission, causes polarization effects, and the polarization of various spectral lines depends in a different way on parameters of the electron beams [5]. This paper reports on complex experimental studies comprising space-resolved observations of hot-spots, time-resolved measurements of X-rays and e-beams, as well as measurements of the X-ray spectrum and polarization. The results of these studies are compared with a theoretical model assuming the generation of fast e-beams within hot-spots regions.

2. EXPERIMENTAL SET-UP AND DIAGNOSTICS

In order to perform experimental studies of the phenomena in question the use was made of the MAJA-PF facility equipped with coaxial electrodes of 70 mm and 130 mm in dia., respectively. The experimental chamber was filled up with deuterium and an argon admixture (of 5-35%). That admixture provoked the formation of numerous hot-spots of high-density and high-temperature plasma, which are sources of intense X-ray bursts [6]. The discharges were supplied by a 44 kJ condenser bank charged up to 35 kV. Soft (1-8 keV) X-rays from selected regions of a pinch column have been measured with scintillation detectors. Simultaneously, there were measured fast e-beams emitted through a hollow inner electrode (anode). Those e-beams were registered with Cerenkov-type detectors made of rutil crystals. In order to determine plasma parameters inside the hot-spots observed there was used a spectroscopic technique. The X-ray spectra within the wavelength range from 3.5 Å to 5.0 Å were registered by means of two focusing spectrographs equipped with quartz crystals of 2d = 8.5 Å and 6.67 Å, respectively. Those spectrographs were oriented in such a way that their dispersion planes were perpendicular mutually.

3. EXPERIMENTAL RESULTS

In the experiments described there were registered spectral lines of He-like argon ions, which were identified as intercombination and resonance lines (e.g. ArXVII $^1P, ^3P$). There were also found dielectronic satellite lines (ArXVI) and K_α lines of lower-ionized species (ArXIII-ArXV). On the basis of relative intensities of the intercombination and resonance lines it was estimated that the plasma concentration is $n_e = 10^{20}$ cm^{-3}. Relative intensities of the resonanse and dielectronic satellite lines enabled to estimate an electron temperature value to be $T_e = 1$ keV.

The appearance of the K_α lines gave evidence that there were generated suprathermal electrons, as in many Z-pinch experiments. Such electrons cause that the distribution function becomes anisotropic, and it can induce the polarization of the X-ray emission. Since the reflection of different X-ray lines from a crystal surface depends on the relative orientation of the crystal lattice and polarization vectors, its efficiency might be different for the X-ray lines under consideration. In fact, the X-ray spectral measurements performed within the MAJA-PF device, demonstrated the polarization of the X-ray emission, as shown in Fig.1. An important peculiarity of the X-ray spectra obtained from the PF discharge observed by means of two spectrographs, whose disperson planes were mutually perpendicular, was a considerable difference

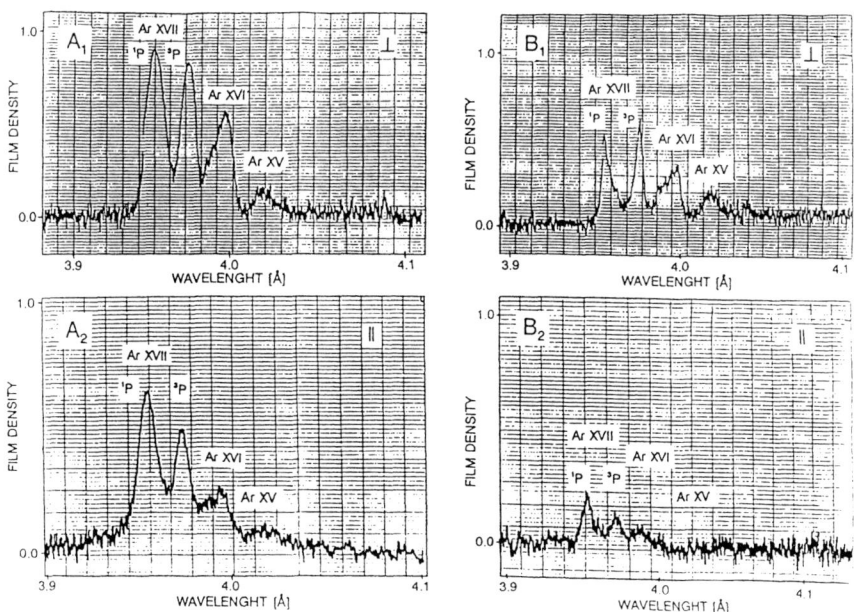

Figure 1. Comparison of the X-ray spectra registered by means of two spectrographs (1&2) with perpendicular dispersion planes, for two diffrent PF-discharges (A&B).

in relative intensities of the intercombination and resonanse lines. This effect can be explained by the polarization of the spectral lines. In that case, the computation of accurate n_e and T_e values would require to determine what portion of electrons has been accelerated to suprathermal velocities and what is their energy and direction.

Studies of correlations between the formation of hot-spot and the emission of fast e-beams are of primary importance for the understanding of the phenomena under investigation. To study such correlations within the MAJA-PF device X-ray signals obtained from two 8-mm-dia. regions, chosen on the z-axis at the distance of 10 mm and 30 mm from the anode, were compared with a signal corresponding to the X-ray emission from the whole pinch column. Simultaneously there were compared signals from the Cerenkov detectors which were used for the registration of fast e-beams. According with the previous experimental studies [6] during recent PF experiments within the MAJA-PF device it was also observed that the hot-spots are formed along the pinch axis, starting from the electrode outlet, as one can see from X_{SN} and X_{SF} signals presented in Fig.2. Signals induced by fast e-beams were registered almost in the same instants when the soft X-ray pulses from the hot-spots were observed. A comparative analysis of numerous oscillograms revealed that the quasi-axial e-beams are emitted from interior or close vicinity of the hot-spots. It should also be noted that also FWHM values of the electron signals and X-ray pulses are comparable (ca. 7 ns).

Considering characteristics of a plasma generated by PF discharges in the MAJA-

PF device, there were performed computations of plasma dynamics within "necking" regions, basing on a model assuming the generation of "run-away" electrons [7]. It was supposed that dense high-temperature plasma regions might be formed inside narrow channels within the pinch column due to a plasma outflow along the z-axis, and the strong compression of the remaining plasma. Breaking of the plasma compression inside the necking could be caused by the plasma heating due to a Joule effect connected with an anomalous resistivity. At the same time, high currents flowing through a plasma might induce high-voltage pulses within the neckings. Strong electrical fields appearing in such regions might accelerate a portion

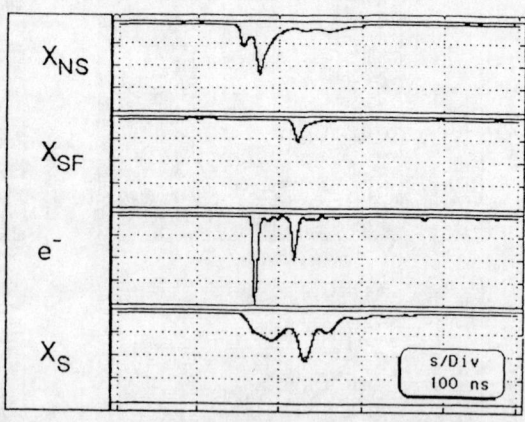

Figure 2. Correlation of the soft X-ray pulses emitted from different hot-spots (near - X_{SN}, and far - X_{SF}) with fast e-beams (e^-) and soft X-rays (X_S) registered from the whole pinch column.

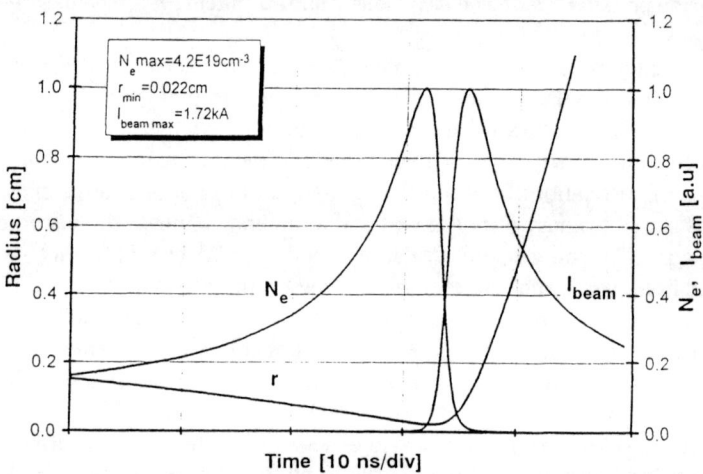

Figure 3. The results of computer modelling of temporal changes in the plasma column radius (r), the electron concentration (N_e) and e-beam intensity (I_{beam}), as obtained for experimental conditions realized within MAJA-PF facility.

of the electron population up to high velocities. Some results of computations, as performed on the the basis of the theoretical model described above, are presented in Fig.3. Those computations were carried out for the experimental conditions realized within the MAJA-PF device, where the maximum discharge current amounted 0.5 MA. It was computed that the maximum plasma concentration reaches 4×10^{19} cm^{-3}, and the minimum radius of the necking (hot-spot) amounts to 0.2 mm. Under such conditions, the generation of e-beams starts during the maximum compression and it reaches the peak value after about 5 ns. The maximum e-beam current intensity is about 1.7 kA, and energy carried by fast electrons is ca. 70 J. It should be noted that such parameters of the generated e-beams correspond to the observed hot-spots characteristics (FWHM ≈ 7 ns).

4. SUMMARY

The main results of the studies described above can be summarized as follows:
1. The studies of X-ray spectra revealed the presence of H- and He-like Ar-lines, and the electron concentration within the hot-spots was found to be about 10^{20} cm^{-3}. The corresponding electron temperature was estimated to be about 1 keV.
2. The most important result of the spectral measurements appears to be an evident difference between relative intensities of the same lines, registered with the two mutually perpendicular spectrographs. Such a difference can be explained by the X-ray polarization, caused possibly by the interaction of fast electron beams.
3. Additional evidences of the appearance of suprathermal electrons were the K_α lines of low-ionized ions, and the time-resolved measurements performed with the Cerenkov-type detectors.
4. Numerical modelling of the electron beam emission gave a relatively good agreement with the experimental results.

References
1. Shilon J., Fiszer J., Rostoker N., Phys.Rev.Letters, 40, 515 (1978).
2. Beier R., Bachmann C., Burhenn P., J.Phys.D, Appl.Phys., 14, 643 (1981).
3. Kania D.R., Jones L.A., Phys.Rev.Letters 53, 166 (1984).
4. Jakubowski L., Sadowski M., Baronova E.O., Experimental Studies of Hot-Spots inside PF Discharges with Argon Admixtures, in Proc. ICPP'96 (Nagoya 1996) Pt.2, p.1326.
5. Krutov V.V, et al., Preprint Lebedev Physical Institute, No. 133 (1984).
6. Jakubowski L., Sadowski M., Studies of hot-spots and their correlation with other phenomena in PF-type discharges, in Proc. EPS Conf. CF&PP (Bournemouth 1995) Pt.2, p.161.
7. Vikhrev V.V, Baronova E.O., Electron Beam Generation in Z-Pinch Discharges, in Proc. ICPP'96 (Nagoya 1996) Pt.1, p.438.

Carbon Fiber X-Ray Lasing

P. Kubeš, J. Kravárik
Czech Technical University, Technická 2, 16627 Prague 6, Czech Republic

Abstract

In this paper the results of visual and X-ray diagnostics of the carbon fiber load located between the copper electrodes of the small 500 J energy pinch devices are presented. A few hundred nanoseconds after breakdown a short XUV burst with the FWHM about 20 ns is emitted in axial direction. The monoenergy wavelength of the radiation, determined both by the filters application and by the method of grazing impact reflection, is in a range of (17-20) nm. It is probable, that the plasma around the fiber occurs the lasing in the Balmer α of CVI at 18.22 nm. The plasma channel of ~50 - 100 μm diameter formed around the fiber explodes during the XUV burst and the electron density in the plasma burst is higher than 10^{25} m^{-3}, in diameter less than ~700 μm. The electron density, electron temperature, magnetic field, magnetic pressure and conductivity of the plasma in channel before during and after burst are discussed.

I. INTRODUCTION

In [1] the first population inversion experiment driven by Z-pinch was demonstrated. A great step in evolution of magnetic pinch lasing was made by fast capillary discharge [2]. In [3] the amplified spontaneous emission of the C VI Balmer α line in capillary discharge for wave length of 18.22 nm was observed.

We presented the results of the visual and X-ray diagnostics of the burst emission of the pinched plasma carbon column around the carbon fiber with electron density higher than ~10^{25} m^{-3} and temperature lower then 10 eV. A hypothesis of magnetic confinement, electron beam acceleration and carbon ions ionization and recombination are discussed.

2. EXPERIMENTAL RESULTS

A small capacitor for pinch devices was described in [6]. The X-ray and visualize diagnostic methods were used with temporal and spatial resolving.

The burst is emitted randomly during the interval (100-400) ns after switching on the current at the intensity of (10-25) kA. The FWHM of this pulse is (10-30) ns. The maximal energy detected by the PIN diode with a sensitive plate of 1 mm^2 in distance of 25 cm reaches value of ~10^{-8} J.

The wavelength of the emitted burst was estimated by using of the grazing impact reflection method. This method is based on the absorption of the XUV radiation of wavelength λ [nm] for the angle of incidence α [grad] (between beam and plate) given by simulated curve [5]. The burst radiation was reflected upon the silicon plate with roughness 1 nm and detected by the PIN diode. The measured and simulated dependence of the reflected intensity on the angle is plotted in Fig. 1. Measured decrease of the burst intensity corresponds to 17-20 nm wavelength.

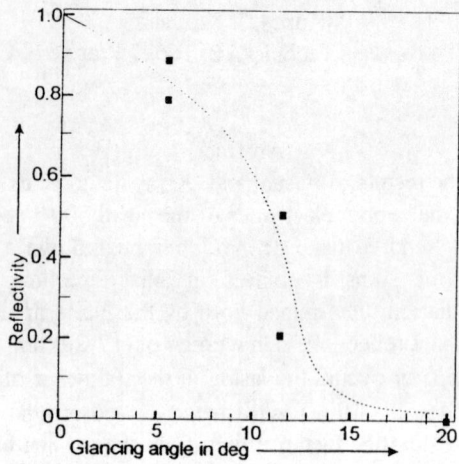

Fig.1. The dependence of the reflected burst intensity on the angle of incidence α. ▪ experimental data. The dashed curve corresponds to the simulated reflection of the 18.22 nm wavelength of the Balmer α of CVI.

The schlieren pictures imagined the evolution of the plasma column and enabled evaluation of the plasma electron density [8].

The diameter of the plasma column of $\sim$ (40 - 60) µm is without great variations during the time (100 - 500) ns after breakdown. The fine shape of the boundary is impossible to measure exactly due to a small diameter and diffraction effects [8].

The relatively stable form of the plasma column boundary is randomly broken due to radial explosion of the part of the column with length of $\sim$ 1 -3 mm. This explosion is in time correlation with both, axial and radial burst radiation. But there are differences between form of explosion in axial and radial direction. The picture of the explosion connected with axial XUV burst has homogeneous and regular cylindrical boundary (Fig.2). The picture of the explosion connected with burst observed in radial direction in visible wavelength is axially irregular and similar to a z-pinch m = 0 and neck instabilities.

The electron density $>10^{25}$ m^{-3} and the diameter ~500 µm for exploding plasma was estimated and the axial length increases and decreases with the intensity of the burst. Before and after the X-ray burst the exploding plasma was not observed. The dense plasma channel with diameter of $\sim$ 50-100 µm in the axis of the cylinder is clear perceptible during the expansion. With Quadro camera diagnostics for window 596 nm the structure like helical ribbon of the discharge channel was observed with 2-3 threads, (400-700) μm diameter and length (2-3) mm [7]. The whole current is conducted only through the narrow plasma column around of the fiber, and the current channel past XUV burst conserves the shape of the origin fiber between electrodes [8]. It means, that the axis part of the fiber remains in a solid state.

Fig.2: Shot 610. Schlieren picture exposed at the maximum of the XUV burst.

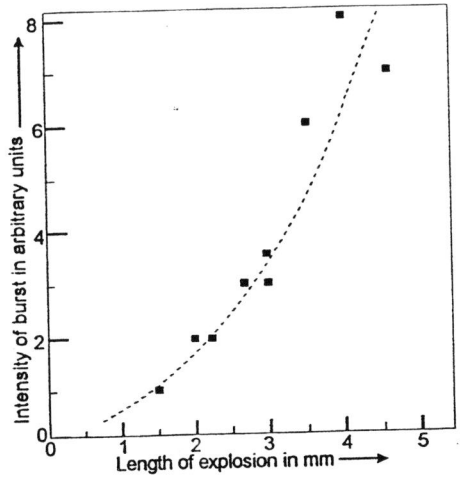

Fig.3: The dependence of the intensity of the axial burst on the length of the exploding plasma column.

The radial velocity of the expansion was estimated at $(1-2) \times 10^5$ m/s, the axial velocity at $(0.5-1) \times 10^5$ m/s.

The intensity of the axial XUV burst is not dependent on the length of the carbon fiber for 5, 10 and 14 mm distances of electrodes. The dependence of the intensity of the axial burst on the length of the exploding plasma column is given in Fig. 3.

3. DISCUSSION

In theory of carbon lasing is possible only for one wavelength, and it is just 18.22 nm. The inverse population of third and second level of CVI could be generated due to narrow neck in probabilities of collision and radiate deexcitation in energy levels for 1keV electrons during recombination. The lasing was observed for example in [3].

Spontaneous 18.22 nm radiation in z-direction could be accompanied with the Lyman α, Pashen series, helium like CV and other UV and visible lines. The observation of only 18.22 nm radiation can be explored by radiation of the Balmer α into a narrow beam along the fiber and radiation of other lines into all directions. The CVI Balmer α line in axial direction in distance 25 cm is than at least 2 order more intensive than in radial direction. On the contrary, the predominant visible radiation in radial direction is composed from more wavelength.

For the lasing and 18.22 nm wavelength observation a mechanism of carbon ionization and excitation of C VI must be realized. Owing the very short lifetime of excited levels this mechanism must operate during the ~ 20 ns XUV burst. Supposing the release of magnetic lines during plasma explosion, so for observed ~ 10^5 m/s velocities and ~ 10 T magnetic field the induction of ~ 10^6 V/m electric field is possible. In that inducted electric field the electrons could be accelerated to 1 keV energies for generation of exited C VI ions.

e) Conductivity of the solid carbon fibber (~ 10^6 Ω^{-1} m^{-1}) and Spitzer's conductivity of the carbon plasma with electron temperature ~ 5 - 10 eV and electron density ~ 10^{26} m^{-3} (~ 10^5 Ω^{-1} m^{-1}) is very low in comparison with conductivity calculated from column magnetic confinement (~ 10^8 Ω^{-1} m^{-1}) and energy estimations (~ 10^8 Ω^{-1} m^{-1}).

4. CONCLUSION

1) In the discharge with carbon fiber of 20 μm diameter the monoenergy burst of radiation ~ 18 nm of 20 ns FWHM was observed in the axial direction.
2) In radial direction the radiation burst with the same FWHM, temporal position and similar intensity is composed mainly from visible wavelength.
3) The burst is emitted from a small energy device with quarter period of ~ 1.2 μs, current intensity of (10-25) kA and the battery of voltage of 20 kV randomly in the time between 100-400 ns after breakdown.

4) The plasma channel of 50 - 100 μm diameter, $\sim 10^{26}$ m^{-3} electron density and the $\sim$ 5 eV electron temperature is formed around the rest of the solid carbon during (100-500) ns after current increasing.

5) At the moment of XUV burst the radial explosion of the plasma from fiber current channel is observed. During the first $\sim$ 10 ns the diameter increases to 100 μm. During the $\sim$ 10 ns around the burst maximum the explosion of the plasma with the electron density higher than $\sim 10^{25}$ m^{-3} to diameter $\sim$ 500 μm is observed. During the last $\sim$ 10 ns of the burst a few narrow dense radial schlieren filaments are perceptible around the plasma column of $\sim$ 50-100 μm diameter.

6) The observed monoenergy wavelength and nonlinear dependence of the burst intensity on the length of the plasma explosion support the idea of lasing of CIV Balmer α line at 18.22 nm.

7) The low a few mm length, long duration of $\sim$ 20 ns, electron density higher than $\sim 10^{25}$ m^{-3} and low incoming energy are different in comparison with other types of observed X-ray lasers, spatially with fast capillary discharge.

8) The long living plasma channel with diameter $\sim$ 50 μm can be explained due to high conductivity of the plasma and by strong magnetic confinement. The important factor for plasma channel stabilizing could be played by solid rest of fiber inside of channel.

9) The plasma explosion during the XUV burst relates to the development of the helical plasma channel and the common transformation of B_z and B_φ [4,6].

10) The magnetic field decrease could induce the electric field for electron beams of $\sim$ 1 keV energy acceleration, CVI ionization and inverse population during relatively long time of $\sim$ 20 ns expansion recombination.

References:
[1] J. R. Porter et al: Phys. Rew. Let. 68, No.6 (1992), 796.
[2] J. J.Rocca et al: 73, No.9 (1994), 2192.
[3] C. Steden, H. J. Kunze: Phys. Letters A, 151 No.9 (1990), 534.
[4] P. Kubeš et all: Proc. Beams 96, Vol.I, 162
[5] L. Pína, personal communication.
[6] P. Kubeš et all: Proc. Conf. Plasma Phys. Nagoya 96, Vol 2, 1102.
[7] P. Kubeš et al: Proc. Beams 96 Vol.II., Prague 1996, 737.
[8] P. Kubeš, J Kravárik: presented on this conference.

This research has been supported by grants GACR No. 202-95-0178 „Stable Structures in Magnetic Pinches" and No. 202-97-0487 „X-ray Source on the Magnetic Pinch Principle".

Spatially resolved Thomson scattering on a gas-liner pinch

Th. Wrubel[1], I. Ahmad[1,2], S. Büscher[1], and H.-J. Kunze[1]

[1] *Institut für Experimentalphysik V, Ruhr-Universität, 44780 Bochum, Germany*
[2] *Department of Physics, Quaid-i-Azam University, Islamabad, Pakistan*

In this report the potentials of spatially resolved Thomson scattering are demonstrated on a gas liner pinch discharge.

MOTIVATION AND EXPERIMENTAL PROCEDURE

The gas-liner pinch is a suitable source for line profile measurements and the determination of the Stark width which is used to diagnose plasmas of a wide range of parameters. In order to test theoretical predictions and scaling of Stark width it is important to measure spectral lines under reliable conditions diagnosing the plasma independently. The Thomson scattering system installed at the gas-liner pinch determines the plasma parameters with high accuracy. The importance of such measurements rises since discrepancies still persist between experimental results and theoretical Stark width calculations even for – from a theoretical point of view – simple resonance transitions (1,2). There is still a lack of a simple and precise Stark width theory which is - for example - required in astronomical line formation calculations (3). Furthermore, 1-D MHD and 2-D MHD models of the implosion of a gas puff pinch performed in order to determine a scaling of K-shell emission predict ion densities on axis too large by up to two orders of magnitude (4).

We measured the plasma parameter electron density (n_e), ion and electron temperature (T_i, T_e) and the radial bulk velocities (v_{mac}) temporally and spatially resolved over the radius of the plasma column with Thomson scattering. This measurements improve the diagnostic of the gas-liner pinch plasma and can also serve to stimulate MHD calculations in order to overcome discrepancies mentioned above.

The gas-liner pinch resembles a large aspect ratio z-pinch (5cm electrode spacing, 18cm in diameter) producing a plasma column of 5cm in length, 1 to 2 cm in

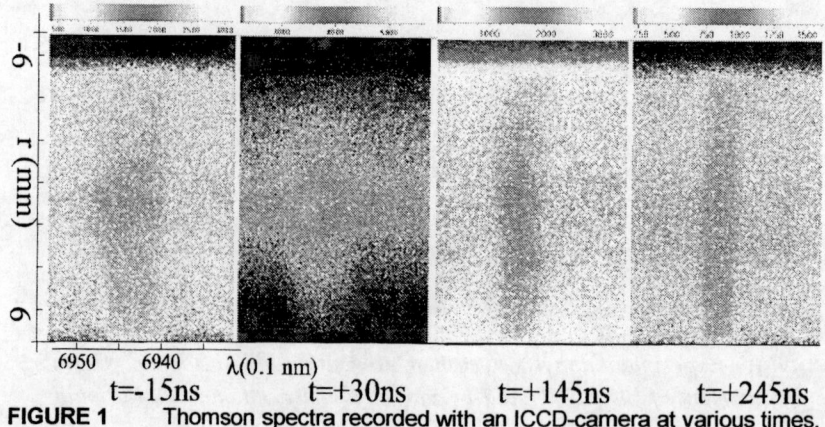

FIGURE 1 Thomson spectra recorded with an ICCD-camera at various times.

diameter reaching typical plasma parameters of $n_e = (0.5...5) \times 10^{18} \text{cm}^{-3}$ and $k_B T_e = (3...50)$ eV depending on both discharge condition and type of test and driver gas. The setup is described in more detail in (5,6). The light of a ruby laser system (Korad K1-Q, 1 J, 25 ns FWHM) is focused to the center of the discharge vessel passing perpendicularly the plasma axis. Using a 1:1 magnification optic the scattering volume is imaged onto the entrance slit of a visible 1m-spectrograph (1200 l/mm grating blazed at 1 µm) in such a manner that every point along the height of the entrance slit corresponds to a radial position in the midplane of the plasma. An ICCD-camera mounted at the exit plane gives a radial resolution of 60 µm (including 3 pixel apparatus profile) and a reciprocal linear dispersion of 6.3pm/pixel in second order. Each recorded Thomson spectrum consists of 578 radial channels and we take the mean of 20 channels in order to improve intensity to noise ratio where the spatial resolution now is 0.46 mm. These spectra are fitted with a theoretical form factor (7). A detailed description of the form factor and the fitting procedure as well as a summary of the theory are given in (5).

RESULTS AND DISCUSSION

The shape of the spectra show that the temperature of the electrons is equal to that of the ions. This can be explained by immediate thermalisation of electrons and ions due to the high collision frequency at such high densities. The Doppler shift of the whole Thomson spectrum gives the projection of the macroscopic velocity v_{shift} onto the scattering vector $\mathbf{k} = \mathbf{k}_i - \mathbf{k}_f$ which is directed radially out of the center of the plasma. Radial compression and decompression velocity can be determined from $v_{radial} = \sqrt{2} \ v_{shift}$. Figure 1 shows four Thomson scattering spectra using a low test gas concentration ($n_{imp}=0.01 n_e$) measured at quoted times. The compression of the plasma can be seen by the growth of the continuum

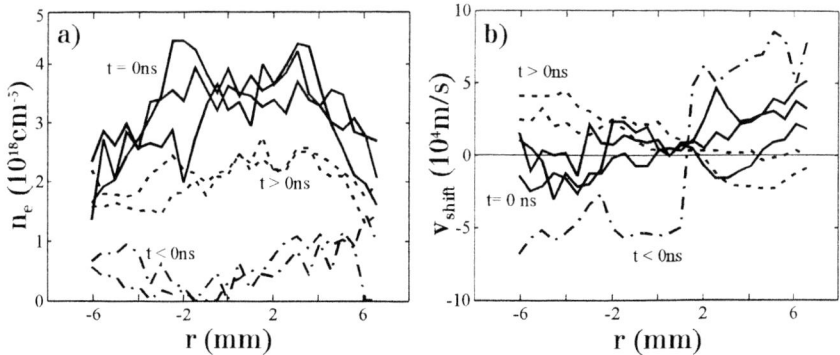

FIGURE 2 The measured time evolution of a) density and b) plasma motion before, at and after maximum pinch compression.

radiation ($\sim n_e^2/\sqrt{T_e}$). The compression and decompression velocity is indicated by the slanting of the spectra. The temperature rise during the implosion phase is mirrored by the increasing width of the spectra where the enhancement of the radiation in the plasma center can be attributed to the emission of the high Z impurity ions on the axis. The spectra indicate that the center of compression is moving very slightly and is – in the mean from shot to shot – about 1mm off the geometrical center of the discharge. These observations confirm the well defined plasma formation when using a small amount of test gas. Such discharge conditions were used in former spectroscopic investigations (6,8).

The time dependence of the radial electron density distribution is shown in Fig. 2a. Before maximum pinch compression it forms a hollow profile in the center whereas at maximum compression (t = 0 ns) it rises up giving a constant density over 1cm in diameter. This density profile remains radially constant till to late times of the discharge, only the absolute value decreases. This confirms former investigations of the homogeneity of the central part of the plasma where the density profile was measured using an OMA system and assuming Salpeter's approximation (9).

The impurity concentration is peaked near the geometrical center of the discharge indicating that the impurity ions are located in a small homogeneous region of the plasma. Therefore, no Abel inversion has to be done and the spectroscopic line shapes do not suffer under cold boundary layers.

The time evolution of the measured radial plasma motion can be seen in Fig. 2b. A fast compression with a velocity of about 10^6 cm/s is observed before maximum compression. This motion stops at maximum compression and the decompression velocity is a factor of 2 lower than that of the compression velocity and decreases with time. Earlier measurements of the magnetic field with magnetic probes giving the velocity of the magnetic piston are verified by the present study (5). Note that at the center of the plasma there is rather no motion over the whole time of the discharge and the corresponding macroscopic Doppler shift is small. Therefore, investigations of even small Stark shifts are reliable (see e.g. (6)).

The situation changes when driving the discharge in a more extreme manner concerning test gas concentration and discharge voltage. In earlier measurements such conditions were used to create population inversion resulting in amplified spontaneous emission in the xuv spectral region ($\lambda = 38$ nm) which occurs together with Rayleigh Taylor instabilities (10,11). Therefore, we increased the test gas concentration by a factor of 10 ($n_{imp} = 0.1\ n_e$) and found a drastical change of the Thomson scattering spectra before maximum pinch compression. First, the position of the center of the plasma varies from shot to shot over some millimeters at t=0ns and there is no uniform plasma formation anymore. Second, before maximum compression the spectra consist of two spatially separated regions with high velocity difference where enhanced, narrow and frequency shifted scattering of laser radiation occurs. This enhanced scattering at $t < 0$ ns can be explained by overthermal scattering at a shock wave running onto the axis. At maximum compression the corresponding ion acoustic waves thermalise. Although unsolved problems still persist and further analysis of these phenomena has to be done, the presented data can serve to test plasma formation calculations.

ACKNOWLEDGEMENTS

This work was supported by the Sonderforschungsbereich 191 of the Deutsche Forschungsgesellschaft. One of us (I.A.) thanks the DAAD for support.

REFERENCES

1. Glenzer, S., Kunze, H.-J., *Phys. Rev. A* **53**, 2225-2229 (1996)
2. Griem, H. R., Ralchenko, Y.V., to be published
3. Werner, K., Heber, U., Hunger, K., *Astron. Astophys.* **307**, 1023-1028 (1996)
4. Thornhill, J.W., Whitney, K.G., *Phys. Plasmas* **1**, 321-330 (1994)
5. Wrubel, Th., Glenzer, S., Büscher, S., Kunze, H.-J., *J. Atmos. Terr. Phys.* **58**, 1077-1087 (1996)
6. Büscher, S., Glenzer, S., Wrubel, Th., Kunze, H.-J., *J. Phys. B: At. Mol. Opt. Phys.* **29**, 4107-4125 (1996)
7. Evans, D.E., *Plasma Phys.* **12**, 573-584 (1970)
8. Wrubel, Th., Büscher, S., Ahmad, I., Kunze, H.-J., "*Line profile measurements in a gas liner pinch*" in Proceedings of the Conference PLASMA 97', Opole, Poland 1997, to be published
9. Glenzer, S., Hey, J.D., Kunze, H.-J., *J. Phys. B: At. Mol. Opt. Phys.* **27**, 413-422 (1994)
10. Glenzer, S., Kunze, H.-J., *Phys. Rev. E* **49**, 1586-1592 (1994)
11. Glenzer, S.H., Wrubel, Th., Kunze, H.-J., Godbert-Mouret, L., *Phys. Rev. E* **55**, 939-946 (1997)

Measurement of Gas Distributions from PRS Nozzles[†]

B. V. Weber, G. G. Peterson,* S. J. Stephanakis,
R. J. Commisso, and A. Fisher

Plasma Physics Division, Naval Research Laboratory, Washington, DC 20375
*Alameda Applied Sciences Corp., San Leandro, CA 94577

Abstract. A high-sensitivity laser interferometer has been used to measure gas distributions from nozzles used in high-power plasma radiation source experiments. These measurements are important for determining experimental parameters and for modeling implosions. The integral of the gas density along the laser beam line of sight is measured as a function of time at one axial distance, z, and one radial displacement, r. The nozzle is moved to scan the (r, z) cross section. The measurements are Abel-inverted to compute the local density $n(r, z, t)$. Several examples are shown to illustrate the technique.

Accurate gas distribution measurements are important for PRS (plasma radiation source) experiments. These measurements indicate how changes in pressure and timing in experiments affect the initial mass distribution. Implosion modeling to evaluate both stability and radiation production requires accurate initial conditions. Gas-dynamics codes used to design nozzles can be tested and improved based on precise measurements. Specific nozzle designs can be tested and modified to obtain desired gas distributions.

The gas distribution is measured using high-sensitivity laser interferometry, depicted in Fig. 1a. The measured phase shift, $\Delta\phi$, is directly proportional to the line-integrated gas density, N:

$$N(y) \equiv \int_{-\infty}^{\infty} n(r)\, dx = \frac{\lambda}{2\pi} \frac{n_o}{v-1} \Delta\phi, \qquad (1)$$

where n is the local density at $r = \sqrt{x^2 + y^2}$, λ is the laser wavelength and v is the index of refraction at density n_o. N is measured as a function of time at one distance, y, from the nozzle axis. The standard deviation of multiple measurements is an

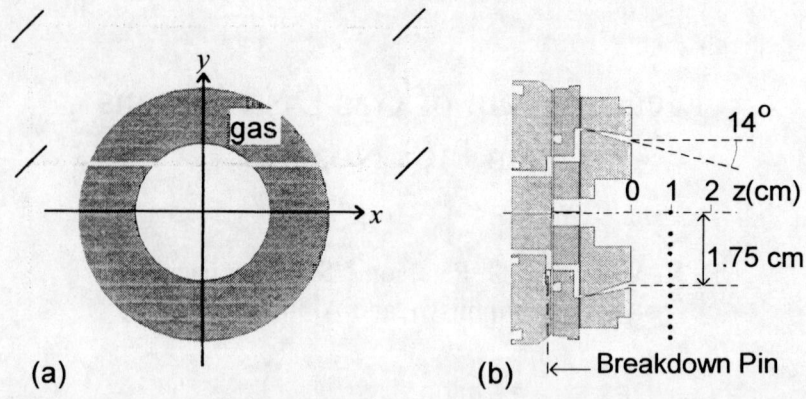

FIGURE 1. Laser interferometry geometry: (a) x-y cross section of a generic gas flow; (b) y-z cross section of the Phoenix nozzle showing measurement locations at z = 1 cm.

estimate of the uncertainty, ΔN. The nozzle is moved to scan the (y, z) space. The local density, $n(r)$, is computed (assuming azimuthal symmetry) by Abel inversion:

$$n(r) = -\frac{1}{\pi} \int_r^\infty \frac{dN/dy}{\sqrt{y^2 - r^2}} dy \pm \Delta n. \qquad (2)$$

The uncertainty in the density, Δn, is determined by propagating the measurement error, ΔN, through Eq. 2.

A high-sensitivity interferometer(1) is used for these measurements. For the laser used here (λ = 532 nm), the typical maximum phase shift from an argon gas puff is 10's of degrees. A sophisticated vibration isolation system and detection scheme allow measurement of phase shifts as small as 0.05 degrees, for a dynamic range as high as 1000. Results from nozzles used on the Phoenix(2) and ACE 4(3) generators will be shown to illustrate the technique.

A cross-section of the Phoenix nozzle is shown in Fig. 1b. The gas distribution was measured for the optimum Ar K-shell yield conditions: 4-cm length, 20.5 psia plenum pressure, and 1.3 ms time delay between actuating the valve and firing Phoenix. One set of measurement locations is shown in Fig. 1b at z = 1 cm from the nozzle exit. To compensate for mechanical jitter in the valve, the breakdown pin signal is used to synchronize the measurements.

Measurements at z = 1 cm and t = 1.3 ms are shown in Fig. 2a. The small error bars, ΔN, demonstrate the high reproducibility of the valve and measurement system. The local density, n, computed using Eq. 2 shows a hollow gas shell, about 1 cm thick, with finite density on axis and a low density tail outside the main gas shell.

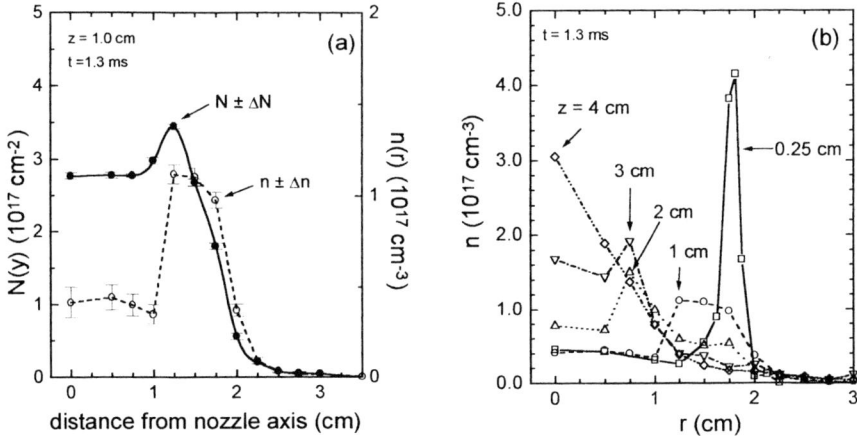

FIGURE 2. Phoenix nozzle measurements: (a) line-integrated density and Abel-inverted density at $z = 1$ cm at $t = 1.3$ ms; (b) radial density distributions at $t = 1.3$ ms.

The error bars for the density, Δn, are larger than for the measurements because of the accumulation of errors in evaluating Eq. 2.

Radial density profiles are shown in Fig. 2b for several z values. Close to the nozzle ($z = 0.25$ cm) the gas forms a high density, hollow shell, while far from the nozzle ($z = 4$ cm), the gas density is peaked on axis. The gas distribution for this nozzle is highly two-dimensional, a fact that should be incorporated in any implosion modeling.

Recent PRS experiments on ACE 4 were performed to compare hollow and uniform gas distributions.(3) To have the same kinetic energy per unit mass at implosion, a 5-cm diam hollow shell and a 7-cm diam uniform fill were required. A nozzle to produce the hollow shell distribution was designed using a Navier-Stokes fluid code(4), and diagnosed with the interferometer. The code results and measurements were synchronized by matching the rising edges of the line-integrated densities along a diameter. The amplitudes were normalized by matching the "plateaus" in the line-integrated densities that follow the initial rise. This normalization was done by artificially increasing the code values by 20%. The resulting radial density profiles at $z = 3.2$ cm are compared in Fig. 3a at three times. The code and measurements agree, within experimental error, during the density rise ($t = 124$ and 224 μs). At later times ($t = 424$ μs), more significant differences are apparent. Detailed comparisons like these will be pursued in the future to refine the parameters used in codes and to refine nozzle designs.

The 7-cm diameter uniform fill ACE 4 nozzle was "designed" empirically, connecting a 12-cm diameter annular ACE 4 nozzle to a 14-cm long, 12-cm inner diameter cylinder and a 7-cm diameter exit aperture. Measured gas distributions 0.2 cm and 3 cm from the aperture are shown in Fig. 3b. Both distributions are fairly

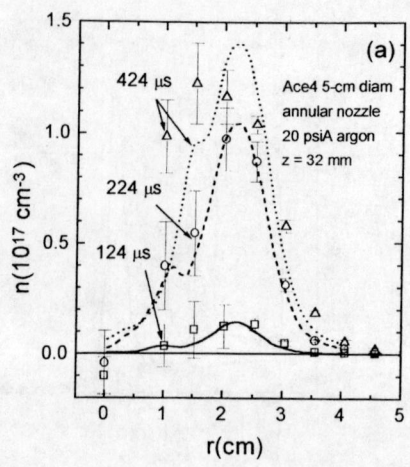
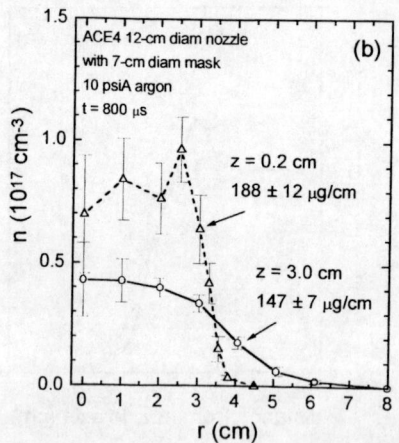

FIGURE 3. ACE 4 nozzle measurements: (a) comparison of measured density distributions (symbols) and code calculations (lines) for 5-cm diam annular nozzle; (b) empirically designed 7-cm diam "uniform fill" nozzle.

uniform. The radius is larger at z = 3 cm than at 0.2 cm, but the mass is smaller, so the implosion times should be about the same at both locations. These examples from ACE 4 demonstrate the capability to test nozzle designs and to engineer particular gas distributions for PRS experiments.

In summary, gas distributions from many PRS nozzles have been measured using a high-sensitivity laser interferometer. The density distributions can be used to choose parameters for experiments, or used as initial conditions for modeling. The measurements have been compared with code predictions of the gas flow, the effects of structures (apertures, walls) have been diagnosed, and nozzle designs have been tested and modified to produce desired distributions. Two enhancements are planned for future applications: two-color interferometry to simultaneously measure gas and electron densities, and a differential phase shift technique to directly measure dN/dy, to improve the precision of the density computed by Abel inversion.

† Work supported by the US Defense Special Weapons Agency, Sandia National Laboratories and the Naval Surface Warfare Center, White Oak.

1. Weber, B. V. and Fulghum, S. F., *Rev. Sci. Instrum.* **68**, 1227–1232 (1997).
2. Nolting, E., *et al.*, "Phoenix Multi-Terawatt Plasma Radiation Source Technology," in *IEEE Conference Record – Abstracts, International Conference on Plasma Science*, 1995, p. 207.
3. Coleman, P., *et al.*, "A Review of Recent ACE 4 Z-Pinch Experiments," these proceedings.
4. Code (2DRZDelta) calculations provided by Hylton Murphy, Maxwell Technologies.

Electron Beam Measurements in Carbon Fibres in MAGPIE

R. Aliaga-Rossel, I. H. Mitchell, J. P. Chittenden, A. E. Dangor and
M. G. Haines

*The Blackett Laboratory
Imperial College of Science, Technology and Medicine
London, U. K.*

and

A. Robledo

*Universidad Autonoma Metropolitana
Mexico City, Mexico*

Abstract: Measurements of electron beam energies calculated from the hard x-ray emission from a carbon and aluminium fibre pinches are presented. This emission occurs at late time (disruption). The x-ray spectrum was measured by an array of detectors (filtered scintillators coupled to photomultipliers) and was used to calculate the energy of the electron beam that produced such x-ray emission. Pulses of x-rays with a duration between 20 ns and 100 ns were observed and electron beam energies around 2 MeV were calculated. Simulation shows that for our experimental parameters, the anode-cathode voltage for a fixed inductive load of 20 nH reaches a maximum value of less than 200 kV. This is one order of magnitude less than the measured beam energy. It was found that the x-ray and the electron beam energies are much greater than the corresponding to the applied anode-cathode voltage. It is likely that energetic electrons are generated neither by a large dL/dt nor by a large dI/dt but by a large and sudden increase in the pinch resistance. The anode-cathode voltage will then be given by d(LI)/dt + IR. As MAGPIE is a high impedance generator, the voltage across the load will increase due to the increased resistance during the disruption phase. This increase in pinch resistance may arise due to the onset of micro-instabilities in the low density plasma regions between the island of dense plasma (called density island). During this experiment MAGPIE was operated at 1 MA, 1.4 MV, 150 ns rise time.

INTRODUCTION

The emission of hard x-rays due to electron beams is a common feature in many Z-pinch plasmas (1,2,3). In most of the cases the energy of the electron beam is much higher than the anode-cathode voltage that is applied. This characteristic of the discharge is often referred as a "disruption". It as been noted in some experiments that the onset of the x-ray emission coincides with a dip in the current signal. This feature, of the discharge, is indicative of the generation of a back emf due to a d(LI)/dt at the load. Several models have been proposed (4,5,6) which explain the generation of MeV beam energies through either a rapid pinching or expansion of the plasma (dL/dt) or a rapid cut-off of the current (dI/dt). We report results obtained in the MAGPIE (Mega Ampere Generator for Plasma Implosion Experiments) generator (7) using carbon and aluminium fibre loads.

EXPERIMENTAL SET-UP

The experimental work was carried out on the MAGPIE generator, charged at 1.4 MV, peak current of 1 MA with a rise time of 150 ns. Figure 1 shows the derivative of the current of the MITL section and the signal from the photomultipliers. Two different loads were used in these experiments, 33µm diameter carbon and 25µm diameter aluminium fibres (both 2 cm long). The electrodes were made of stainless steel (anode plate was 6 mm thick). The principal diagnostics used in these experiments were: a) voltage and current monitors, b) Nd-YAG laser (523 nm, 400 ps) (8) was used to take two schlieren photographs and a self reference interferogram c) a four frame x-ray camera, x-ray streak camera, and hard x-ray detectors. A set of 3 hard x-rays detectors were located outside the vacuum chamber at a radial position 1 m from the pinch. Each of them consisted of NE102A plastic scintillator wrapped with fluorescent fibre and coupled to a fast photo-multiplier tube via a normal plastic optical fibre. Each detector was surrounded by 5 cm of lead with an aperture of 2 cm (9), in which a set of filters composed of aluminium and lead were located, as shown in Table I. With this arrangement it was possible to isolate the photo-multipliers from the adverse effects of electrical noise. The electrical signal produced in the photomultipliers was recorded in a digital oscilloscope with a 400 mega samples per second giving a time resolution of 2.5 ns. The x-ray spectrum measured by the array of detectors was used to calculate the energy of the electron beam that produced the x-ray emission.

RESULTS AND DISCUSSION

In the calculation of the x-ray emission, it was assumed that a monoenergetic electron beam strikes the anode (thick target) and the Kramer's formula (10) was applied. The spectrum obtained is adequated for electrons with energies in the range from tens keV to 5.3 MeV (11). The output voltage v(t) expected from each detector for a given energy of e-beam is calculated using $v(t)=K\int I(E,t)R(E)f(E)dE$ (where R(E) is the filter transmission and f(E) is the detector spectral response. K is a constant (i.e. independent of energy and time) which reflects the conversion efficiency from incident x-ray photons to volts). A relevant filter transmission

TABLE I	
Detector number	Filters
1	3 mm Al
2	3 mm Al + 2.7 mm Pb
3	3 mm Al + 5.4 mm Pb
All the filters have an additional 6 mm of stainless steel due to the wall of the vacuum chamber wall.	

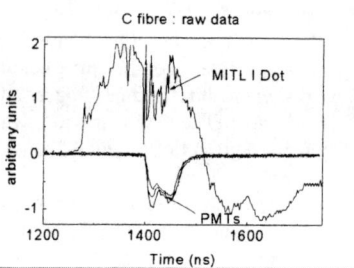

FIGURE 1: dI/dt in the MITL

was used. This was carried out for various energies of electron beams between 300 keV and 4 MeV. Figure 2 shows the ratios of calculated output voltages as a function of electron beam energy for detectors 1 and 2 (E_{12}) and 1 and 3 (E_{13}). The value of K is determined by the overall gain of the system after the scintillator which depends upon the losses due to the coupling of fibers and the PM, as well as the gain of the PM. In practice the value of K is different for

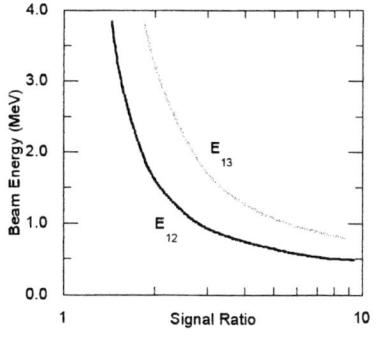
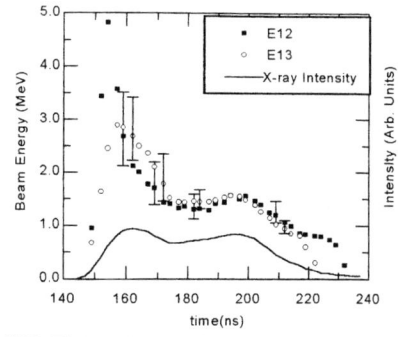

FIGURE 2. Calibration curves for detectors

FIGURE 3: Beam energy for 25 μm Al fibre.

each detector. It should be noted that as signal ratios are used in the evaluation of the beam energies, only the relative gains of the detectors are needed. In order to obtain the relative gains, a series of shots were carried out fitting identical filters to the detectors and comparing the measured signals.

A linear regression was applied to the resulting data giving the relative gains of detectors 1 and 2, K_{12}, and of 1 and 3, K_{13}. Using the ratio of detector signals (compensated for different detector gains) and the curves in figure 2, the energy of the electron beams responsible for the hard x-ray radiation were calculated. Figures 3 and 4 show plots of electron beam energy as a function of time for Al and C shot respectively. (t=0 refers to the start of current rise at the load). Typically energies with maxim around 2 MeV were measured, although in some cases energies as high as 4 MeV were recorded.

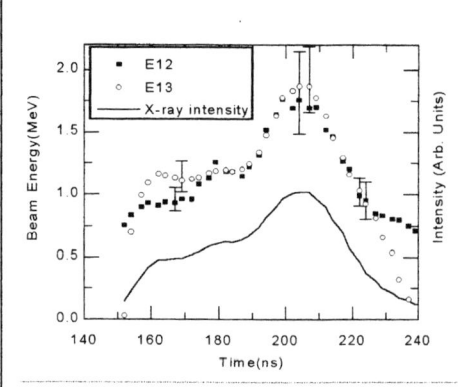
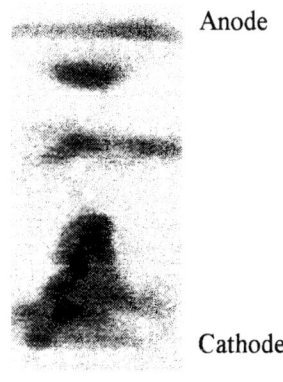

FIGURE 4. Beam energy for 33 μm C fibre.

FIGURE 5. Soft x-ray frame at 185 ns (after disruption)

The hard x-ray pulses are detected between 120 and 200 ns after current starts to flow in the load. The number and duration of pulses vary from shot to shot, however they always coincide with a dip in the dI/dt signal measured at the load with a magnetic pick-up coil. This dip has a duration of only a few nanoseconds. Signals from collimated PIN diodes (spatial resolution < 5 mm, energy response < 100 keV) indicate that the anode is the source of these x-rays. In general the onset of the hard x-ray pulses occur later in the case of the 33µm carbon fibre shots compared to the 25µm aluminium shots. Figure 6 shows an optical streak of a 33 µm carbon shot. The early appearance of optical hot spots and gaps in the plasma column can be observed. At 120 ns the hard x-ray channel shows the first burst of x-ray which is repeated later at about 200 ns. The emission from the anode is detected when the gaps in the plasma start to appear. It is also seen that there is an acceleration of plasma toward the anode at the disruption time (about 200 ns). After disruption the emission of x-ray is mainly from the electrodes. This absence of dense plasma is related to an increase in the plasma resistance which could explain the origin of the high voltage necessary to produce high energy electron beams. This fact is also consistent with the production of high energy neutron emission when CD2 fibres are used (12).

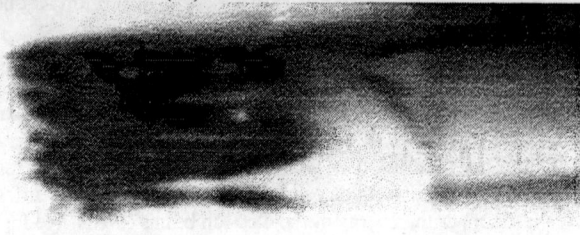

t=0   t=120 ns   t=200 ns

FIGURE 6. Optical streak (anode :bottom ; cathode: top)

References

1. Hirano, K., Kaneko, I., Katsuji, S. and Yamamoto, T., Jpn. J. Apll. Phys. **29**, 1182 (1990).
2. Cohen, L., Feldman, U., Swartz, M. and Underwood, J. H., J. Opt. Soc. Am. **58** 843 (1968).
3. Kies, W., Decker, G., Maizig, M., van Calker, C., Westhiede, J., Ziethen, G., Bachmann, H., Baumung, K., Bluhm, H., Rusch,.D.,Ratajczak, W., Stoltz, O. and Bayley, J., J. Appl. Phys. **70**, 7261 (1991).
4. Uhm, H. S. and Lee, T. N., Phys. Rev. **A40**, 3915 (1989).
5. Fukai, J. and Clothiaux, E. J., Phys. Rev. Lett. **34** 863 (1975).
6. Haines, M. G., Nucl. Instrum. Methods **207** 179 (1983).
7. Mitchell, I. H., Bayley, J., Chittenden, J. P., Worley, J. F., Dangor, A. E., Haines, M. G. and Choi, P., Rev. Sci. Instrum. **57** 1533 (1996).
8. Aliaga-Rossel, R., Bayley, J., mamin, A. and Nizienko, Y., this proceeding.
9. Choi, P. and Aliaga-Rossel, R., Rev. Sci. Instrum. **65** 3034 (1994).
10. Kramers, H. A., Philos. Mag. 46 436 (1923).
11. Robledo, A., Mitchell, I. H., Aliaga-Rossel, R., Chittenden, J. P., Dangor, A. E. and Haines, M. G., Phys. Plasmas **4** 490 (1997).
12. Aliaga-Rossel, R., Mitchell, I. H. and Schmidt, H., this proceeding.

SBS Pulse Compression Applied to a Commercial Q-Switch Nd-YAG Laser

R. Aliaga-Rossel[a], J. Bayley

The Blackett Laboratory, Imperial College, London, U. K.

and

A. Mamin, Y. Nizienko

The Troitsk Institut for Innovation and Fusion Research, Moscow, Russia.

Abstract. In optical diagnosis of dense Z-pinches, sub-nanosecond laser pulses are required in order to freeze the movement of the plasma during the probing. Commercial lasers can provide such type of pulses but they are either very expensive, or they have a very low energy per pulse. A technique that uses Stimulated Brillouin Scattering (SBS) to compress a 8 ns pulse of a commercial Q-switched Nd-YAG laser is reported here. To carry out this passive compression technique, a frequency doubled laser pulse of 10 ns was focused into a single SBS gas cell, 2 m long, filled with a mixture of argon and sulphurhexafluoride (SF_6) at a total pressure of 40 bar. A shorter and high intensity pulse was reflected from the cell (created by SBS) and it travelled back along its original path until it was separated from its original direction by using a dicroic polariser. The pumping volume of the SBS cell, the convergence of the incident beam and the pressure of the gas cell, were optimised to maximise both temporal compression and the output energy. Pulses of 10 ns were compressed to less than 400 ps with a conversion efficiency of 80%. This SBS pulse compression system has been used to make most of the optical measurements of a dense fibre pinch plasma produced in the MAGPIE[1] generator.

INTRODUCTION

The use of SBS[2,3] (Stimulated Brillouin Scattering) phase conjugation as a pulse compression technique provide a low cost way to obtain subnanosecond pulses with output energy over 500 mJ. The use of this technique has been reported[4] in the compression of a ruby laser pulse from a pulse length of 20 ns to 300 ps with an output energy of 1 J. One of the advantages of this passive technique, apart of its simplicity and low cost, is the high quality of the compressed beam and the stability in a pulse to pulse regime. All the distortion and inhomogeneities of the beam, produced due to the imperfection of the optical elements, are removed or compensated by the phase conjugation mechanism.

EXPERIMENTAL ARRANGEMENT

The pulse compression was carried out in a commercial SLM (Single Longitudinal Mode) Q-switched Nd-YAG laser working at a repetition rate of 10 Hz, frequency doubled to 532 nm with a pulse width of 8 ns and output energy of 600 mJ. The output of this laser has a Gaussian profile with diameter of 10 mm and divergence less than 0.5 mrad. No modifications of the laser were required. The compression cell was a stainless steel tube 200 cm long and 4 cm diameter. Both ends were closed by glass windows, 20 mm thick with antireflective coatings at 532 nm. The SBS medium was filled with a mixture of argon and sulphurhexafluoride (SF_6) at a pressure of 40 : 0.3 bar. A lens of long focal length was used to focus the beam inside of the SBS medium. The power density in the focal point has to be low enough to avoid the developing of other competing non-linear process such as stimulated Raman scattering, self focusing or optical breakdown. Details of the set-up are given in figure 1.

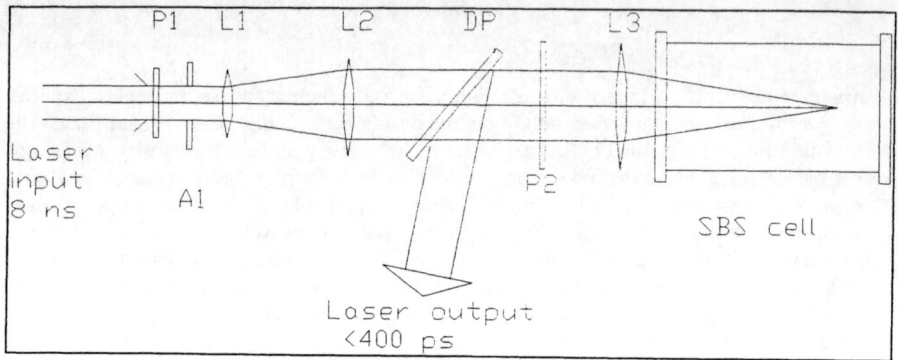

FIGURE 1. SBS pulse compression set-up.

A laser pulse of 8 ns width enters to the system from the left side of the Fig. 1. The plane of polarization of the beam is adjusted by rotating the half-wave plate P1. A set of optical apertures are placed along the path of the beam to avoid any spurious backward reflection and also to select the central part of the beam (only A1 is shown in the figure). L1 and L2 are two lenses that form a telescope. The beam is expanded by a factor of two in order to diminish the power density in the other optical elements and to have the right convergence angle in the SBS cell after crossing the lens L3. DP is a dicroic polarizer which reflects the SBS compressed pulse away from the original optical path. P2 is a quarter-wave plate and L3 is a long focal length lens (1.6 m) that focuses the beam inside the SBS cell. The system works as follows: the polarization plane of the incident beam is adjusted in P1 (vertically polarized), then the beam goes through the telescope formed by L1-L2, expanding its diameter from 10 mm to 20 mm. The beam crosses the dicroic polarizer DP and its polarization is rotated by 45^0 when it crosses the quarter-wave P2. The lens L3 focuses the beam inside the SBS cell. The phase conjugated wave travels back and its plane of polarization is rotated another 45^0 in the same direction of the previous one, which makes its polarization perpendicular (horizontally polarized) to the original direction. When the compressed pulse arrives to the dicroic polarizer DP, is reflected away from its original direction and hence separated from the incoming path. To diminish reflective losses all the components have an antireflection coating at 532 nm.

RESULTS

The laser pulse before compression (FWHM of 8 ns) is shown in the figure 2. The laser pulse was detected with a fast photodiode (1 ns rise time). The SLM laser system produces a temporally homogeneous pulse. Figure 3 shows the laser pulse after compression.

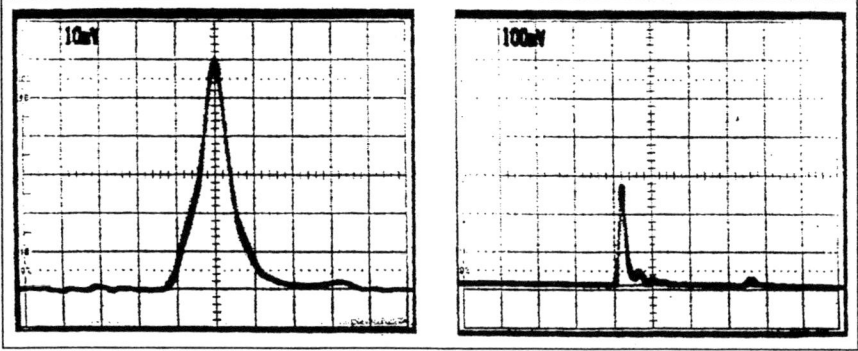

FIGURE 2. Laser pulse before compression (5 ns/division).

FIGURE 3. Laser pulse after compression (5 ns/division).

In this case temporal resolution is limited by the speed limit of the oscilloscope and the photodiode. However it is possible to observe a faster rise time compared with the uncompressed pulse. In order to have a more precise measurement of the compressed pulse an optical streak camera was used. The time base of the streak camera was calibrated using the same compressed pulse. The pulse was split in two pulses, one of the pulses was delayed with respect to the other by 2 ns (an optical path of 60 cm was added). The slit of the streak camera was 200 μm and was placed perpendicular to the time axis. A laser pulse of 10 mm diameter was employed in the measurements. In figure 4, time goes from left to right; the time scale is shown in the picture. Using the calibration mentioned above, the central part of the beam has a width less than 400 ps. The borders of the pulse arrive about 500 ps later than the central part. This could be due to a spatially inhomogeneous pulse. In such a case the pumping intensity in the SBS cell will be also inhomogeneous and time delay or pulse deformation can occur. A calorimeter was employed to measure the beam energy before and after the pulse compression. The energy of the uncompressed pulse was 600 mJ and it falls to 480 mJ after the compression, which gives a compression efficiency of 80%.

SUMMARY

A simple passive SBS compressing system has been tested in a commercial Q-switch Nd-YAG laser. An energy efficiency of 80% was achieved with a compression ratio of about 25 time.
The system is very stable and variations less than 5% were measured in a pulse to pulse regime. Using this scheme, optical diagnostics such as shadowgraphy, schlieren, interferometry and Faraday rotation, are being used simultaneously in the study of the plasma parameters of the dense fibre pinch produced in the MAGPIE generator.

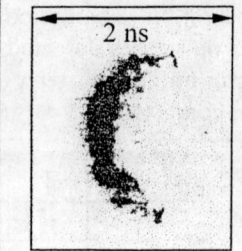

FIGURE 4. Streak photograph of the compressed pulse.

REFERENCES

1. Mitchell, I. H., Bayley, J. M., Chittenden, J. P., Worley, J. F., Dangor, A. E. and Haines, M. G. , Rev. Sci. Instrum. **67** 1533 (1996).
2. Dane, C. B., Neumman, W. A. and Hackel, L. A., IEEE, Journal of Quantum Electronics, **30** 1907 (1994).
3. Fedosejevs, R. and Offenberg, A. A., IEEE Journal of Quantum Electronics, **21** 1558 (1985).
4. Nizienko, Y., Mamin, A., Nielsen, P. and Brown, B., Rev. Sci. Instrum. **65** 2460 (1994).

[a] Electronic mail: r.aliaga-rossel@ic.ac.uk

Optical Multi-Slit And X-Ray Measurements From Carbon And Deuterium Pinches

R. Aliaga-Rossel, S. V. Lebedev, J. P. Chittenden, A. E. Dangor and M. G. Haines

The Blackett Laboratory
Imperial College of Science, Technology and Medicine
London, U. K.

Abstract: Experiments carried out on the MAGPIE generator to study the optical and x-ray emission from carbon and deuterium fibres are presented. The generator was operated at 1.4 MV, peak current of 1 MA, rise time 150 ns. Carbon fibres of diameter of 33 μm and 300 μm were used. The following diagnostics were used: single pass interferometry, two frames schlieren photography, optical and x-ray streak cameras, four frames x-ray camera, single frame optical camera and time integrated pinhole camera. Rogowski coils and capacitive divider were used to monitor the electrical signals of the generator and the plasma. Time resolved measurements of x-ray emission were performed with a four frame x-ray camera. Each frame with a set of three filtered pinholes was used. The novel feature of this measurement is the employment of an optical streak camera with a set of slits arranged along the fibre axis, but displaced in the radial direction. This permitted us to study the axial and radial movement of the plasma regions that are emitting in the visible part of the spectra. Correlation between these regions of the plasma and the location of x-ray hot spots is discussed. Comparison between carbon and deuterium fibres is presented.

INTRODUCTION

The purpose of this experiments is to study the optical emission of a fibre pinch and to correlate them with other regions of the plasma, both in space and in spectral region, particularly x-rays. In the optical streak cameras, it is common to use a single slit, in which is focused the central part of the plasma (axial streak), and the camera records the time evolution of it. This method gives information of how homogeneous is the discharge, or the fibre pinch, along the z axis. Here we used a multi-slit system in which four slits, with their axis parallel to the pinch but displaced along the radius of the pinch. With this arrangement, it is possible to study in a single streak how different parts of the fibre (different radius) evolve in time.

EXPERIMENTAL SET-UP

The experiment was carried out in the MAGPIE generator (1), operated at 1.4 MV, 1 MA peak current and 150 ns rise time. The optical diagnostics was operated with a ND-YAG laser, 400 ps at 532 nm (2). A rotation of 45^0 was introduced to the beam (rotating their polarisation). After passing a dicroic mirror, the beam was separated in two beams with perpendicular polarisation. One beam was delayed respect to the other in 3 ns, after that both beam were following identical path through the plasma. After crossing the plasma, the beams were separated again and sent to CCD cameras. These beam were used to take two schlieren photographs (separated 3 ns) and another was used in self reference interferometer (3). Carbon fibres of 33 μm and 300 μm were used to study the dynamics of the plasma at different radial position as well as the spatial correlation between the x-ray hot spots and the region of necking detected in the schlieren photographs. Deuterium fibres were also used to compare the dynamics in both cases. A pre-pulse scheme was implemented in MAGPIE to study the instabilities of the plasma under different initial conditions (4). The multi-slit was composed of four parallel aligned slits, with their longitudinal axis parallel to the axis of the fibre (z axis). Each slit is 5 mm long and 0.55 mm width. (Slits of different width can be used, according with the resolution that is required.) The first and the last slit are aligned with the central part of the fibre, the second is displaced 2.4 mm from the central axis and the third is displaced 1mm. A two lens system is used to focus the fibre into the input window of the streak camera. The multiple slit was located in the image plane of the first lens.

EXPERIMENTAL OBSERVATIONS

Figure 1 shows two optical streak images in carbon fibres of 300 μm and 33 μm diameter, taken at different times of the discharge. The average expansion velocity for 300 μm carbon fibres is ~1 cm/μs, the fibre shows a more homogeneous expansion in term of the instabilities. Conversely, 33 μm fibres show a higher expansion velocity, about 7 cm/μs, and the appearance of flares indicate an early activity of the m=0 instability.

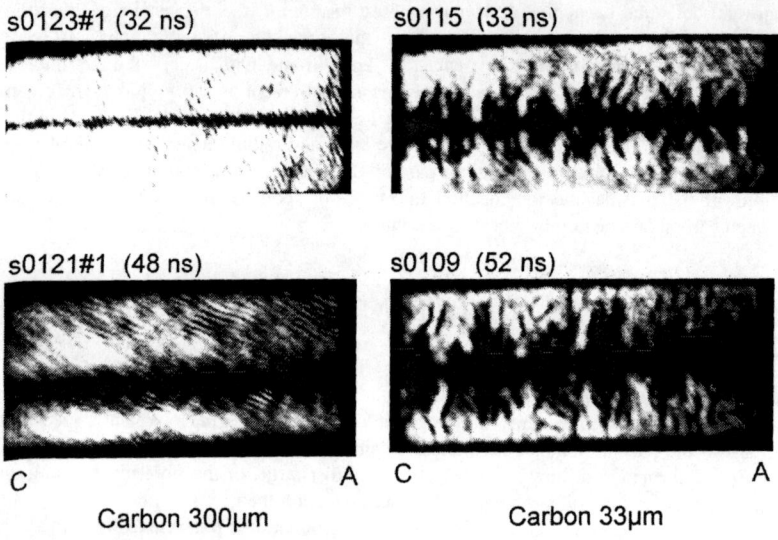

FIGURE 1. Schlieren photographs of 33 μm and 300 μm carbon fibres.

In Figure 2 a multi-slit optical streak is shown for 300 μm and 33 μm carbon fibres. From these images, for the 300 μm fibre, the expansion velocity of the corona is 3.2 cm/μs and 2.6 cm/us for a radius of 0.7 cm and 2.1 cm m. For the 33 μm fibre the velocity is 2.5 cm/μs and 1.7 cm/μs respectively. From this data it can be concluded that most of the visible radiation comes from the central core of the fibre. On the other hand, schlieren photographs show mainly disturbances that occur in the coronal plasma. In Figure 3 a single slit axial streak photograph

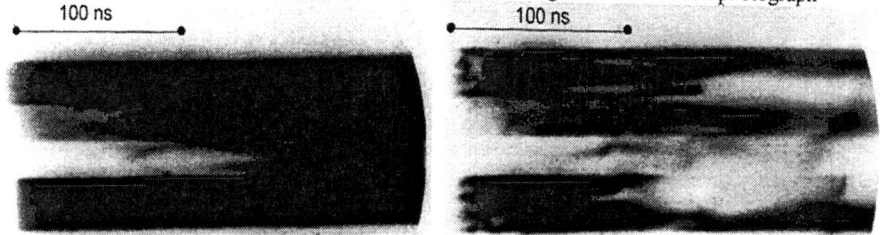

FIGURE 2. Multi-slit optical streak. (Left: 300 μm fibre; Right: 33 μm fibre)

of a 33 μm carbon fibre is shown. The schlieren was taken at 110 ns (L in the streak). There is a correlation between the gaps that appear in the optical streak and those that can be seen in the schlieren. Also the emission of hard x-ray is related to these gaps (disruption) (5).

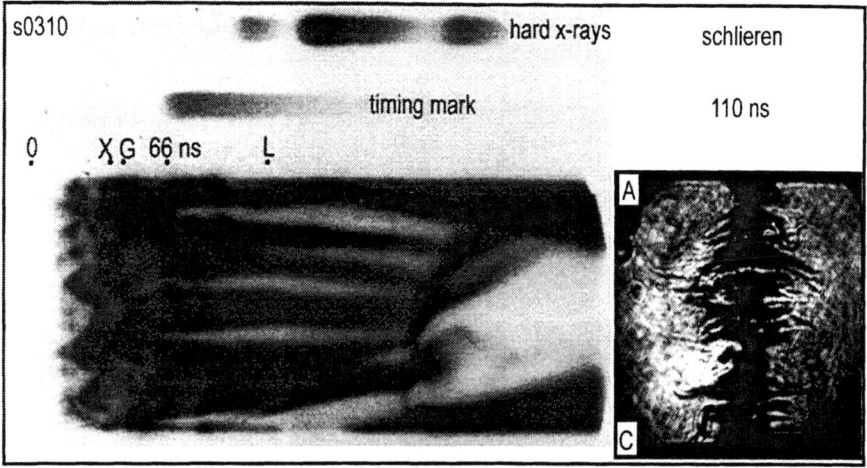

FIGURE 3. Normal optical streak photography and schlieren for 33 μm carbon fibre.

Deuterium fibres of 100 μm diameter, have a similar behaviour in term of the dynamics to the 33 μm carbon fibres. Early instabilities (20-30 ns) leading to a continuous expansion of the fibre and impede the coupling of the current to the plasma. Figure 4 shows the soft x-ray emission from a 100 μm deuterium fibre, filtered with 3.5 μm microfoil. It must be noted that some of these hot spots last for more than 10 ns, which leads to the suspicious of a slow cooling and heating process. A harder channel, filtered with 10 μm of Be does not show hot spots at all. This could be due to either the core of the hot spots is not sufficiently hot to emit in this region or their density is to small, so the radiated flux is not enough to be capture by the camera (or both). The interferogram and the schlieren taken at 70 ns show a black column that could be a dense plasma or is due only to the gradient in the electron density. This shows that conventional optical diagnostics are not useful in the study of dense plasmas.

RESULTS AND DISCUSSION

The use of a multiple slit in the optical streak camera, allows us a better determination of the position of the emitting regions of the plasma and also to have a better evaluation of the axial velocity at different radial positions. To calculate the radial velocity of the plasma by using schlieren photographs, it is necessary to average the different part of the plasma, which it is not always simple. By using a multi-slit optical streak, the velocity that is calculated is already an average value along the length of the slit, on the contrary, in a radial streak the velocity is calculated in a particular point of the fibre. The final resolution of the method, depends of the number of slits.

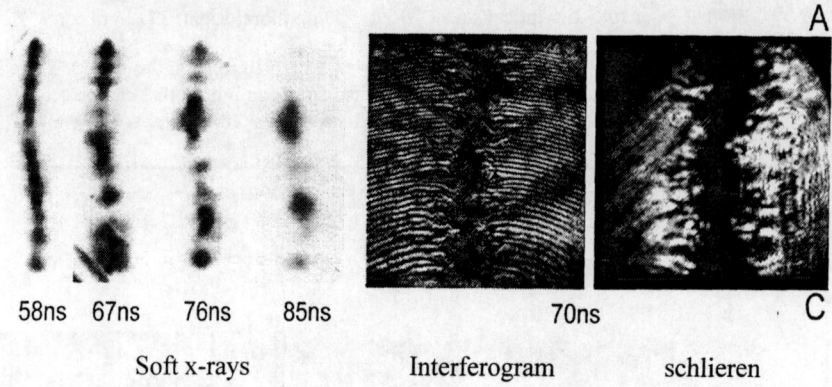

58ns 67ns 76ns 85ns   70ns

Soft x-rays   Interferogram   schlieren

FIGURE 4. Deuterium fibre, 100 µm diameter.

Comparing the multi-slit optical streak with the sclieren photographs, it is was observed that the optical emission comes mainly from the central part of the pinch. The schlieren photographs show only the coronal plasma (due to gradient in density). The expansion velocity of the emitting plasma is about half of the one measured using the schlieren photographs. The optical streak photographs show a series of "optical" hot spots, some of them can evolve to x-ray hot spots. The four frame x-ray camera, shows that the spatial position of the x-ray hot spots, always coincides with the location of the optical hot spots, but the converse is not true. Not all the optical hot spots are sufficiently hot or live long enough to be detected in the x-ray region, as is shown in figure 4.

REFERENCES

1. Mitchell, I. H., Bayley, J. M., Chittenden, J. P., Worley, J. F., Dangor, A. E., Haines, M. H. and Choi, P., Rev. Sci. Instrum. **57**, 1533-1541 (1996).
2. Aliaga-Rossel, R., Bayley, J., Mamin, A. and Nizienko, Y., " SBS pulse compression applied to a commercial Q-switch Nd-YAG laser", this proceeding.
3. Tatarakis, M., Aliaga-Rossel, R., Dangor, A. E. and Haines, M. G. "Optical probing of fibre z-pinch plasmas", Submitted to Physics of Plasmas.
4. Lebedev, S. V., Aliaga-Rossel, R., Chittenden, J. P., Dangor, A. E., Haines, M. G. and Worley, J. F. "Coronal plasma behaviour in C and D2 fibres on the MAGPIE generator", this proceeding.
5. Robledo, A. Mitchell, I. H., Aliaga-Rossel, R., Chittenden, J. P., Dangor, A. E., and Haines, M. G., Plasma Physics, **4** 490-492 (1997).

Polarized x-rays from a Z-pinch Plasma

Elena O. Baronova

*Nuclear Fusion Institute, RRC Kurchatov Institute,
Moscow 123182, Russia*

Abstract. This paper presents the first experimental evidence for polarization of x-rays from He-like argon in a plasma focus, and for x-rays from He-like iron from a vacuum spark. The diagnostic needs two crystal spectrometers with mutually perpendicular dispersive planes parallel and perpendicular to the discharge axis. Resonance lines from He-like ions are always polarized, and this fact has implications for estimating plasma parameters. X-ray polarization also gives information about anisotropy in fast electrons with energy of a few keV.

1. Introduction

Polarization effects must be considered when measuring line emission in the infrared, visible, and ultraviolet [1]. This paper studies the corresponding problem of polarization for x-ray lines emitted from z-pinch discharges.

It is well known that macroscopic electric and magnetic fields split the ionic and atomic energy levels. The resulting spectral lines are also split (Stark and Zeeman splitting), and the different components are polarized. Likewise, microscopic electric fields in electron collisions with ions result in asymmetric excitation of single ions and polarized x-rays in their radiative decay. Radiation from a plasma is polarized if its ions are excited by an anisotropic electron distribution, if the asymmetric excitation is not randomized before radiative decay, and if the resulting polarized x-rays can find their way out of the plasma with their polarization intact.

The first thorough theoretical calculation of x-ray polarization dates from 1927[2]. Reference 3 focuses on polarized lines from He-like ions. These lines are favorites for hot dense plasma diagnostics, because they are abundant over a large temperature range, and they are usually well resolved by modern spectrometers. Lines from He-like ions must be polarized because the lower level of the appropriate radiative transition has angular momentum J=0 [4]. For a single ion the degree and direction of polarization depends on the type of transition considered: the resonance line is usually polarized along the direction of the exciting electron, but the intercombination line may be polarized in the opposite direction due to the long decay time of the corresponding upper level. For a plasma the polarization depends also on anisotropy of the exciting electrons, and on later interactions with isotropic electrons of the appropriate energies.

Figure 1 is a semiclassical picture to illustrate where polarization comes from. It shows an incoming electron with velocity vector **V** going by a He-like ion initially in the ground state $1s^2$, exciting it to the $1s2p$ state by an electric field vector **E**(t) (proportional to [**R**(t) -**r**(t)]/(R-r)^3: **R**(t) describes the incoming electron, **r**(t) the ion's electron). The 1s2p excited state has an elliptical charge distribution that likes to be on the far side of the ion's nucleus, away from the colliding electron. The electron cloud can follow the exciting

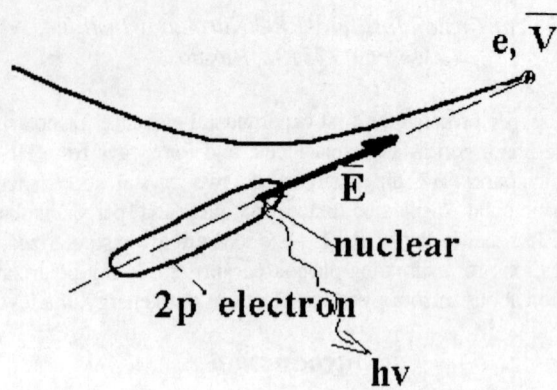

Fig.1. Semiclassical interpretation of polarization of x-ray resonance lines.

electron if it is relatively slow, and in this case the charge cloud remains as shown. The resulting 1s2p-1s^2 resonance line is polarized in the direction of the exciting electron. With increasing electron energy the polarization decreases monotonically: the ion's 2p electron cloud can not follow a fast electron.

In hot dense plasmas the ions are usually excited by electrons with an anisotropic velocity distribution function, and sometimes even by a directional electron beam. Then the x-ray line radiation emitted by the plasma ions is also polarized, but the polarization of the radiation from the plasma may be smaller through interaction with isotropically distributed electrons.

Figure 2. shows different kinds of plasma electrons:
a) ``beam'' electrons with energies comparable to the applied voltage, (e.g., 100 keV). Their runaway mechanism is discussed elsewhere in these proceedings.
b) the low energy part (a few keV) of runaway electrons, who redistribute themselves according to the plasma's electromagnetic fields. Figure 2 illustrates this with the pancake-like distribution found in laser-produced plasmas.
c) the Maxwellian tail of isotropic electrons elsewhere in the plasma.
d) current-carrying thermal electrons. Their velocity distribution function is a shifted Maxwellian or a slightly anisotropic version thereof.

Classical plasma diagnostics usually assumes that relative intensities of the x-ray lines are independent of isotropically distributed suprathermal electrons or on electron beams [5]. However, recent theoretical studies [6] have shown that even one to three percent of hot electrons (belonging, for example, to Maxwellian tail) may influence the line intensities enough to affect plasma density and temperature estimates: fast electrons must be considered explicitly. Moreover, energetic anisotropic electrons (as in electron beams) polarize the x-ray lines, and polarization measurements should give information about these electrons[7].

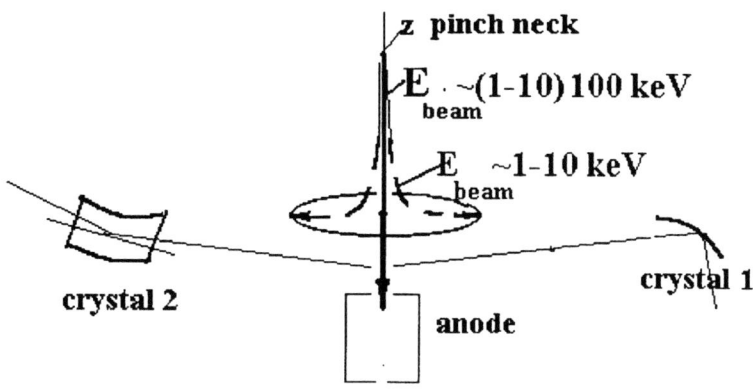

Fig.2. Different kinds of fast electrons and positioning of x-ray crystals to measure polarization of x-ray lines.

Polarized He-like x-rays have been observed in solar flares [8], laser produced plasmas [9] and vacuum sparks [10]. Complete interpretation of these measurements is not yet possible because of theoretical difficulties and because several parameters are insufficiently known. Examples are the radiation's angular distribution, diagnostics calibration as function of polarization and wavelength, and how the data are averaged over space or time.

This study present experimental evidence of polarization of He-like lines from a plasma focus machine, highlights that x-ray polarization leads to a dependence of the spectra on the positioning of the spectrographs, and demonstrates that x-ray line polarization measurements can be used to diagnose suprathermal electrons. Similar work was done for a vacuum spark plasma {10,15], but use of a plasma focus is new. The experiments were part of a fruitful collaboration between Polish and Russian scientists [11,12].

2. Experimental results

Time and space integrated spectra were taken by two focusing Johann spectrographs using thin quartz crystals (dx=0.07 mm) in optical contact with cylindrical substrates (R=500.1 mm). Careful calibration with radiation from an x-ray tube was performed in the crystal reflection's third order, second order (with a Co anode) and first order (with a Si-anode). The spectrometer's resolution is about $d\lambda/\lambda \sim 5*10^{-5}$. Figure 2 shows the spectrometers' orientation, perpendicular and parallel to the discharge axis: theory predicts that maximum polarization is perpendicular. The distance between the source and the crystal is about 700 mm. A 5 micron polyethylene (mylar) filter, covered with 600 A of Al protects the spectrometer from visible light.

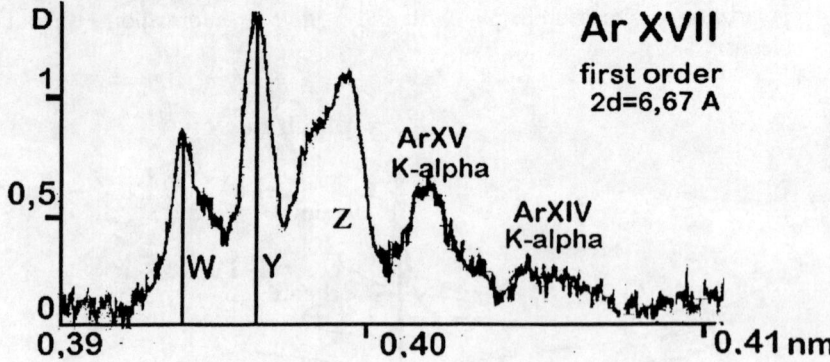

Fig.3. He-like Ar spectra, emitted from the plasma focus device and measured with the crystal 2 (see Fig.2). Dispersion plane of the device is perpendicular to the discharge axis.

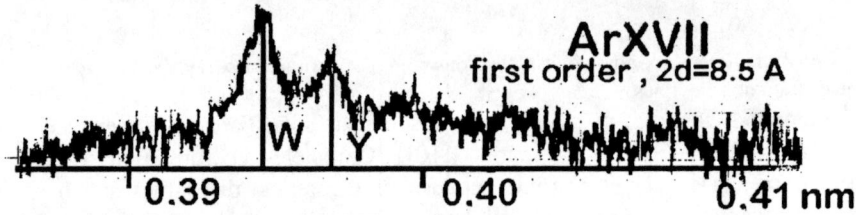

Fig.4. He-like Ar spectra, emitted by the plasma focus device and measured with the crystal 1 (see Fig.2). Dispersion plane of the device coincides with the discharge axis.

Figure 3 is the spectrum of the ArXVII line perpendicular to the discharge axis from a single shot on a Mather-type plasma focus [11]. The same shot gives Figure 4, this time measured parallel to the discharge with a different spectrometer. The perpendicular spectrometer has 2d=0.667 nm and incidence angle 36 degrees, the parallel one has 2d=0.85 nm and incidence angle 25 degrees. The measured intensities differ because (in the first order reflection used here) the reflection coefficient of the first spectrometer is three times larger than that of the second.

Both spectrometers show Ar-XVII's $1s2p(^1P_1)-1s^2(^1S_0)$-resonance line (electric dipole transition), the $1s2p(^3P_2)-1s^2(^1S_0)$-magneto-quadrupole transition, the $1s2p(^3P_1)-1s^2(^1S_0)$-intercombination line (electric dipole transition), and the $1s2s(^3S_1)-1s^2(^1S_0)$-forbidden line (magnetic dipole). In Gabriel's notation these lines are w, x, y, and z as marked in the Figure. They are accompanied by many satellites, which belong to 1s2l3l' transitions in the Li-like ion (ArXVI). These satellites are not resolved in our measurements. Careful theoretical modeling of the spectra makes it possible to estimate the yield of these lines, and to evaluate plasma density from the relative intensities of the y and w lines and plasma temperature from the relative intensities of satellites coinciding with z-line. The line profiles themselves contain additional information that is not at issue here.

The principal feature in Figures 3 and 4 is the relative intensities of the resonance (w) and intercombination (y) lines. Even though they are taken on the same shot, the lines are substantially different only because the spectrometers are placed under different angles to the discharge. In Figure 3 (perpendicular) the w line is less intense than the y-line, but in Figure 4 (parallel) the w-line is more intense. Both spectrometers are aligned accurately: they observe radiation only from a single hot spot. Thus different temperature and density gradients could not be the reason for the intensity differences measured by the two spectrometers. The two different intensity ratios would translate into different electron densities in the plasma if the effects of polarization from the two orthogonal measurements were ignored.

Besides the w/y line ratios there are other features that differ between the two spectrometers. The K-alpha lines of partially ionized argon as marked in Figure 3 are prominent, but they are absent in Figure 4. K-alpha lines come from three processes, inner shell ionization, inner shell excitation and dielectronic recombination. The inner shell ionization rate coefficient for ionization by energetic electrons (e.g., with energy around 30 keV) is two orders of magnitude larger than the rate coefficient for thermal electrons (e.g., those with energy 1 keV): the corresponding excitation rates differ by about an order of magnitude. Thus the presence of K-alpha lines proves the existence of suprathermal electrons, generated in the plasma.

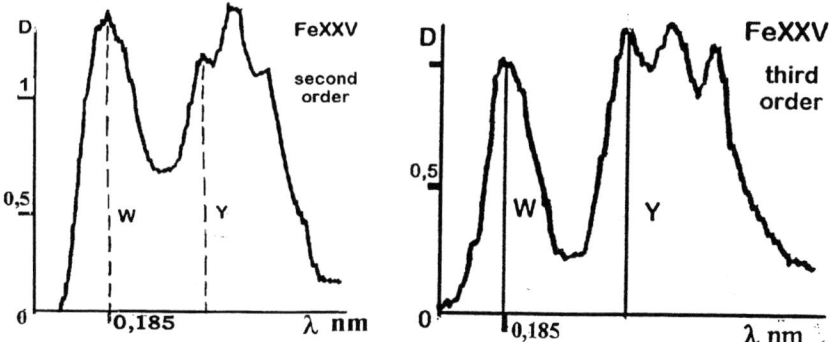

Figure 5. Fe XXV spectrum from a vacuum spark in second order (dispersive plane parallel to discharge axis).

Figure 6. Fe XXV spectrum from a vacuum spark in third order (dispersive plane perpendicular to discharge axis).

Figures 5 and 6 show a FeXXV spectrum averaged over 5000 shots from a vacuum spark device (I=150 kA, U=14 kV, see [15]). Two series of data were taken, one using the second order of reflection with incident angle 25 degrees, Figure 5, and one in third order of reflection with incident angle 36 degrees, Figure 6. Both measurements used the same focusing spectrometers with a quartz crystal (2d=0.85 nm), and in both cases the dispersive plane is perpendicular to the discharge axis. A spectrum taken as in Figure 5, in second order but now with the dispersive plane parallel to the discharge axis (not shown) has differences between the spectra just as for the Ar spectra in Figures 3 and 4, with different relative intensities of the w and y lines in the same experiment depending on the spectrometer's orientation. All this evidence suggests polarization of the appropriate lines. Unfortunately, the Fe lines are wider than the Ar lines, and more difficult to interpret quantitatively.

3. Interpretation

The difference between the relative intensities of the w and y lines in the ArXVII and FeXXV from the two perpendicular spectrometers in the same experiment can not be caused by errors in measuring the x-ray lines emitted by different plasma regions: we are confident that both devices were oriented quite properly to detect the radiation from the same region, in the plasma focus a single hot spot. The plasma focus has additional (2-3) hot spots on axis but maybe $\delta z \sim 20$ mm away from the observation region: radiation from these other spots does not enter the spectrometers. The assumption of spatial isotropy of the line emission, in the radial direction within the cross section chosen for the measurements, is based on results obtained with two identical spectrometers (for the iron spectra), and after analyzing pinhole images (for the argon spectra). Obviously, additional work is needed to provide more careful evidence for this assumption. According to theoretical predictions [7] the degree of polarization of the w, x, and y lines can be equal to 10-60% whilst the forbidden line z is unpolarized. The degree and direction of polarization for these lines depend on the energy of fast electrons in different ways.

The electric dipole line w (resonance line) is known to be the most polarized with its electric field vector parallel to electron beam, or to the principal velocity component in the case of an anisotropic electron distribution. At the same time the degree of polarization of the y line is smaller, and its direction can be parallel or opposite to the electron velocity direction depending on the electron energy. Another effect to be taken into account is hyperfine interaction in highly ionized atoms. These strongly affect the degree of polarization of the y line, but not the degree of polarization of the w-line. The 1s2l3l'satellites belonging to the transitions in Li-like ions should also be polarized, but they are incompletely resolved because of their low intensity, which makes explanation difficult. Thus the most complete interpretation may be done for the w-line.

Taking into account these considerations, the time and space integrated ArXVII and FeXXV spectra clearly show evidence of polarization over the x-ray region caused by an anisotropy in the electron velocity distribution function. This is not so clear with the FeXXV spectra from the vacuum spark because the lines' width is too large. However, considering that the He-like line of Ar in the plasma focus is mainly excited by electrons with energies close to 4 keV, the obvious conclusion is that most of those moderately energetic electrons move perpendicular to the discharge axis. This conclusion is the basis for the disk full of electrons illustrated in Figure 2. Electrons with a few keV energy have quite complicated orbits that are determined by the configuration of electromagnetic fields inside the plasma. At the same time runaway electrons with higher energies move as a directional electron beam along the discharge axis [13]. A pancake-like electron velocity distribution function has already been deduced for laser produced plasma [9]. In [10] it was also shown, that anisotropy of electron velocity distribution function is not determined by the plasma compression anisotropy.

A more quantitative interpretation of the results presented here should be done in the future. It should be noted that the difficulties in doing a more complete analysis are partially connected with construction differences between the two spectrometers, one with 2d=0.667 nm, the other with 0.85 nm and the corresponding differences in geometry and reflection angle. Measurements without this unnecessary ambiguity need two devices with identical crystals (preferably 2d=0.667 nm). Careful quantitative interpretation of polarization phenomena would then be possible when taking into account theoretically the details of radiative decay including the hyperfine interaction of appropriate levels as well as experimental data on reflection coefficients and isotropy of radiation.

The fine spatial structure of any type of z-pinch plasmas (hot spots and micropinches) is well known. In these experiments the hot spots were proven to be the x-ray source, but the spatial extent of the electron beam, its duration, and the probable interaction of spatially anisotropic hot electrons with multicharged ions are still under active experimental and theoretical investigation. However, the experimental evidence of polarization of He-like lines proves that highly ionized ions coexist with fast electrons, at least during some time in the same plasma region. A similar conclusion was reached in [14], where the place of electron beam generation (and also a few kev electron generation) is close to or within the hot spot region, and the starting point of generation process coincides with the breakdown stage of hot spot formation when a sufficient number of ions still exists. It is assumed here that the generated electron beam has an energy distribution from a few keV up to many hundreds of keV, and that the motion of low energy (of a few keV) accelerated electrons from that beam is determined by complicated electromagnetic fields existing inside the plasma.

3. Summary

1. Spectra of He-like Ar lines from a plasma focus, and from Fe lines in a vacuum spark, were taken simultaneously by two carefully calibrated crystal spectrographs oriented with different angles with respect to the discharge axis.
2. An essential feature of the spectra is the difference between the relative intensities of the resonance and intercombination lines observed under different angles on a single experiment. Therefore, the plasma parameters inferred from the relative intensities of these lines differ also.

3. Polarization of the x-ray lines suggests anisotropic electrons. Quantitative interpretation of the observed polarization needs additional modeling.
4 Line ratios depend on the spectrometer's orientation. This affects how to determine plasma parameters from the relative intensities of x-rays from He-like ions.

The author thanks Prof. Kunze and Prof. Lisitsa for fruitful discussion and scientific interest in the work presented here, and N. R. Pereira for editorial advice.

REFERENCES

1. Zaidel A.N., Shreider E. Ya. "Vacuum spectroscopy and its application", Moscow, (1976).
2. Oppenheimer J.R., *Z. Phys.* a, **43**, 27 (1927)
3. Krutov V.V., Korneev V.V., et. al., Preprint 133, P.N. Lebedev Physical Institute, Moscow, (1981).
4. Inal M.K., Dubau J. *J. Phys. B: At. Mol. Phys.* **20**, 4221, (1987).
5. Presniakov L. P. *Uspekhi Fiz. Nauk*, **119**, 49, (1978).
6. Rosmej F.B., Rosmej O.N., AIP. Conf. Proc.#299, 3rd Int. Conf. Dense Z-pinch (London, 1993), New York AIP Press, 560 (1994).
7. Shlyaptseva A.S., Urnov A. M., et. all, Preprint 193, Lebedev Physical Institute., (1981).
8. Korchak A.A. *Sov. Phys. Dokl.,* **12**, 92 (1967).
9. Kieffer J.C. et.al., *Phys. Rev. Letters,* **68**, 480 (1992).
10. Baronova E.O., Vikhrev V.V. et.al. Proc. Eleventh Col. on UV and X-Ray Spectr., Nagoya, Japan, pp 465-467, (1995) and to be published in Physica Plasma, 1997.
11. Jakubowski L., Sadowsky M, Baronova E., Proc.1996 ICPP, Nagoya, v2, . 1326
12. Jakubowski L, et al, present Conf
13. Haines M. AIP Conf.Proc. #299, 3rd Int. Conf. Dense Z-pinch (London, 1993), New York: AIP Press, (1994), p.154.
14. Vikhrev V., Baronova E., Proc. 1996 ICPP, Nagoya, v1, 441.
15. Veretennikov V.,et all, Plasma Phys, 7 , 1199, (1989).

Suprathermal Electron Diagnostics, Based on X-ray Line Radiation Polarization Measurements

Elena O.Baronova

*Nuclear Fusion Institute, RRC Kurchatov Institute,
Moscow 123182, Russia*

Abstract. X-ray polarization measurements of the ArXVll spectra, emitted from PF device, were performed by two crystal spectrographs with mutually perpendicular dispersive planes. An opportunity to make some conclusions, related to fast electrons with energy of a few keV, is shown.

1. Introduction

One of the most remarkable phenomenon typical of Z-pinch discharges is the generation of fast electrons (or directive electron beams) moving towards the anode with energies essentially higher (hundreds of keV) than that, corresponding to the applied voltage. The groups of suprathermal electrons with energies of a few keV, having isotropic (Maxwellian tail) or anisotropic (due to the motion in inner electromagnetic fields) velocity distribution function may also be presented in z-pinch plasmas [1,2]. Such electrons play an important role in heat transport, burning wave and x-ray laser scheme creation as far as the formation of emission spectra in plasma.

At the previous stage of development of plasma diagnostics, based on the analysis of relative intensities of x-ray lines it was stated that the value of relative intensities is not depend on the presence both of isotropically distributed suprathermal electrons and electron beams [3]. Recent theoretical studies [4] has shown that even one-three percents of hot electrons may influence the populations of appropriate levels and lead to the essential mistakes in evaluation of the plasma density and temperature.

Another point should be taken into account is that any spatial anisotropy of electron velocity distribution function (directive electron beams in z-pinch plasmas) causes the polarization of x-ray lines, radiated by multicharged ions. He- like ions, widely used for plasma diagnostics because of it's abundance over the large temperature range, are predicted to be essentially polarized. Thus the polarization measurements of He- like x-ray line emission spectra should contain the information on different kind of suprathermal electrons [5].

The fine spatial structure of any type of z-pinch plasmas (hot spots and micropinches) is well known and whilst the hot spots were proved to be the sources of lines of highly ionized ions, but the place of electron beam generation, it's duration and the probable interaction of spatially anisotropic hot electrons with multicharged ions are under active experimental and theoretical investigation at present. In that sense the experimental evidence of polarization of He-like lines will support the conclusions, which were given in [6,7], where the place of electron beam generation is closed or within the hot spot region and the starting point of generation process coincides with the breakdown stage of hot spot formation when essential number of ions still exists.

The main goal of this study is to demonstrate the principal possibility to make conclusion about the suprathermal electrons, using x-ray line polarization measurements, thus realizing the further step in understanding physical processes in z-pinch plasmas. Such study was presented in [8] for vacuum spark plasma and have never been experimentally demonstrated for plasma focus discharge. Present experiments were carried out due to fruitful collaboration of Poland and Russian scientists and have lighted in other details in [7,9].

2. Experimental results.

The Mather type device (I=500 kA, U=35 kV), described in [6], was used in these experiments. Measurements of ArXV11 lines, performed in the single shot by the device with the dispersion plane perpendicular to discharge axis direction (2d=6,67 A, Q inc=36grad) and by the device, having dispersion plane parallel to discharge axis (2d=8,5A, Qinc=25grad), are shown in fig. 1,2 respectively. Integral reflection coefficient of the first device is for three times grater the same one for the second device.

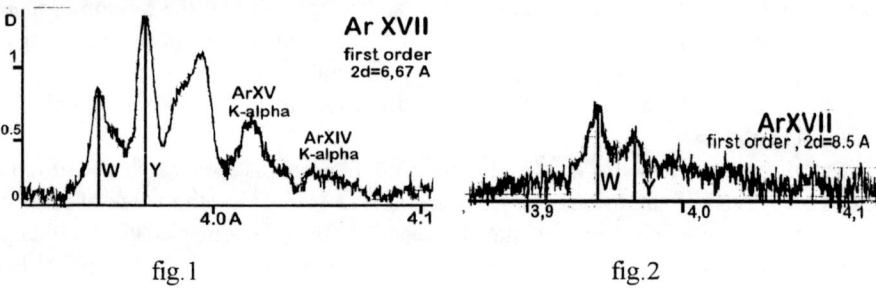

fig.1   fig.2

The following spectral lines of Ar XV11 in the first order of reflection were registered by both the devices: 1s2p (1P1)-1s2 (1S0)-resonance line (electric dipole transition), 1s2p (3P2)-1s2 (1S0)-magneto-quadrupole transition, 1s2p (3P1)-1s2 (1S0) intercombination line (electric dipole transition), 1s2s (3S1)-1s2 (1S0)-forbidden line (magnetic dipole), (w,x,y,z-in Gabriel's notation accordingly). These lines are accompanied by a lot of satellites. K-alpha lines of low ionized argon, observing in the experiments, might be arised as a result of three processes such as inner shell ionization, inner shell excitation and dielectronic recombination. The inner shell ionization rate coefficient due to ionization for enegetic electrons (e.g. for Tbeam=30 keV) is of two orders of magnitude larger than the same one due to thermal electrons (e.g. for Te=1 keV), for exitation rates-about an order of magnitude. Thus the presence of K-alpha lines proves the existence of suprathermal electrons, generated in plasmas investigated.

The main feature of spectra obtained is the difference between relative intensities of w and y spectral lines registered in the same experiment with different devices. Such a feature may be explained by the polarization of adequate x-ray lines, results from the anisotropy of electron velocity distribution function. It should be mentioned that this characteristics is not typical of each discharge, in order to make clear the reason a more detailed study of the phenomenon is necessary.

3. Interpretation

Electric dipole line w is known to be the most polarized, having the direction of electric field vector parallel to electron beam. At the same time the degree of polarization of y line is smaller and it's direction may be both the same and opposite to the electron beam direction in dependence on the electron beam energy. Another consideration that had to be taken into account is hyperfine interaction in highly ionized atoms-the degree of polarization of y line depends strongly but the degree of w line polarization is independent on such interaction. The 1s2l3l' satellites should also be polarized, but have unresolved structure and difficult for explanation also because of the low intensity of these lines. Thus the most complete interpretation may be done for the w-line.

Reflection coefficients of dispersive elements for polarized radiation are depend on the angle between the direction of electric field vector of radiation investigated and the reflection plane of the crystal. Obviously the best reflection takes place for Bruster angle (45 grad for x-rays) and radiation with electric field vector parallel to reflection plane. For the device with Qinc=36 grad the yield of line radiation with electric field vector perpendicular to discharge axis may be estimated as being less than 10 percents. The reflection coefficients for the horizontally arranged device are identical for both electric field vector orientations.

Taking into account these considerations the character of time and space integrated Ar spectra may be treated as a result of polarization over the x-ray region caused by the existence of anisotropy of electron velocity distribution function (the splitting in electromagnetic fields, which might also cause the polarization, has not been observed). When allowance is made for the fact that electrons with energies closed to 4 keV provide the main yield to He-like line excitation process, the obvious conclusion is that the essential part of such electrons moves in the direction, perpendicular to the discharge axis. Actually the comlpicated motion of a few keV electrons is determined by the configuration of electromagnetic fields inside the plasma. At the same time runaway electrons with more high energies move as the directive electron beam along the discharge axis [10]. The existence of pancakelike electron velocity distribution function have been already shown for laser produced plasma [11].

Quantitative interpretation of presented results will be the subject for further consideration, the difficulties are partially connected with peculiarity of Bragg polarimeter, have been used in the experiments. The necessary step to make such investigations more clear for interpretation is to use two devices with identical crystals (2d=6,67A).

The results of numerical modeling of the time dependent process of electron beam generation in comparison with that of density development [7] within investigated experimental conditions is shown in fig.3. An opportunity for the fast electrons with energies of a few keV to be interacted with highly ionized ions may be realized due to reasonable coinsidence of these processes in time.

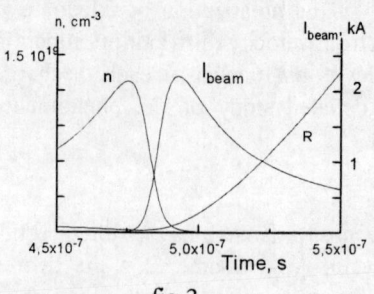

fig.3

Summary

1. The results of time and space integrated spectra analysis of He-like Ar, emitted from Mather type PF-discharge are presented. Measurements were simultaneously performed by two focusing crystal spectrographs with mutually perpendicular dispersive planes.

2. The character of spectra obtained may be completely interpreted with polarization phenomenon in x-ray region, results from spatial anisotropy of electron velocity distribution function.

3. The opportunity of fast electron diagnostics, based on such measurements, carried out in PF, is shown. From the results it is concluded that essential part of hot electrons with energies of a few keV, moves perpendicular to the discharge axis.

References.

1. Bernstein M.J., Meskan D.A., *Phys. of Fluids*, **12**, 2193 (1969).
2. Van Paassen H.L.L., R.H. Vandre, *Phys. Fluids*, **13**, 2606 (1970).
3. Presniakov L.P. *Uspehi fis. Nauk,* **119**, 49, (1978).
4. Rosmej F.B., Rosmej O.N., AIP.Conf. Proc.#299, 3rd Int. Conf. Dense Z-pinch (London, 1993), New York AIP Press, 560 (1994).
5. Shlyaptseva A.S., A.M. Urnov et. all, Preprint n193, Lebedev Physical Institute., (1981).
6. Yakubowski L., Sadowsky M, Baronova E., Proc.1996 ICPP, Nagoya, v2, . 1326.
7. Vikhrev V., Baronova E., Proc. 1996 ICPP, Nagoya, v1, 441.
8. Veretennikov V.,et all, Plasma Phys, 7 , 1199, (1989).
9. Yakubowski L, et all, present Conf.
10. Haines M. AIP Conf.Proc. #299, 3 rd Int. Conf. Dense Z-pinch (London, 1993), New York: AIP Press, (1994), p.154.
11. Kieffer J.C. et all., Phys. Rev. Lett. 68, 480, (1992).

Influence of the transient plasma dynamics on the scaling of the K-shell line emission in pinch plasmas

K. Bergmann, R. Lebert, W. Neff

Lehrstuhl für Lasertechnik, RWTH Aachen and FHG-Institut für Lasertechnik
Steinbachstr. 15, D - 52074 Aachen, Germany

The transient dynamics in pinch plasmas generated in a small plasma focus device has been made visible in the scaling of the brilliance of helium- and hydrogen-like resonance lines. The scaling of these lines was investigated theoretically and experimentally for different low Z - elements (nitrogen, oxygen, neon) considering the whole set of device parameter (current I_o, neutral gas pressure p, anode radius a). Similarity considerations of the transient pinch plasma dynamics suggest that the set of determining parameters for the brilliance of the resonance lines can be reduced to $I_o^2/(pa^2)$ and pa for a fixed element, which is verified experimentally. Furthermore, the theoretical predictions concerning the influence of the transient dynamics on the scaling of the brilliance of the transitions 1s-2p of hydrogen-like ions and $1s^2$-1s2p of helium-like ions are verified experimentally.

INTRODUCTION

Mather type plasma focus devices [1] offer the advantage to study pinch plasmas over a wide range of device parameters using one single device, since the ignition and the compression and pinch phase are separated in space and in time and can be varied and optimised independently. Pinch plasmas generated in small devices (< 2 kJ) having a lifetime of a few nanoseconds are highly transient, since the time scales for plasma processes like ionisation and excitation are of the same order of magnitude. In the present paper it is discussed how the transient ionisation process from helium-like to higher levels leaves its marks in the scaling of the He-α ($1s^2$-1s2p) and the Ly-α line (1s-2p).

SIMILARITY CONSIDERATIONS

Similarity considerations of the magnetohydrodynamics of the compression and the pinch phase and simple model calculations for the radiation physics of the pinch plasma suggest that the determining set of device parameters for the integrated spectral brightness of the K-shell resonance lines and the recombination continuum (ISB), that is the spectral brightness integrated over the time of emission and the line profile, can be reduced to $I_o^2/(pa^2)$ and pa for a fixed element. This is valid in an electron temperature regime which leads to a maximum emission in the hydrogen-like Ly-α line at a given pressure. More details about these considerations are given in Refs. [2,3]. The parameter $I_o^2/(pa^2)$ is a measure for the energy input per ion in the pinch plasma thus determining the achievable electron temperature. The main energy source of the pinch plasma is assumed to be the thermalized kinetic energy of the plasma in the compression phase rather than ohmic heating. The parameter pa is a measure for the confinement parameter $n_e\tau$ (τ pinch lifetime) thus determining the transient character of the plasma dynamics. At the same time, pa is a measure for the optical thickness of the considered lines and so one of the determining parameters for radiation transport.

Furthermore, the theoretical considerations in Ref.[3] predict that the maximum ISB of the Ly-α line at a given pressure, that means choosing the right current or the right electron temperature for maximum emission, scales proportional to pa while the respective ISB of the He-α line shows a nearly constant behaviour. This is valid in the parameter range for pa covered with the device under investigation. The linear increase of the Ly-α line is due to a linear increase of the rate of emission with pa, while the constant behaviour of the He-α line can be referred to a linear increase which is cancelled by a shorter lifetime of the helium-like ionisation level due to faster ionisation. In the temperature regime under investigation most of the helium-like line radiation is emitted during the ionisation process into hydrogen-like and fully stripped ions at an high electron temperature, since the heating of electrons in the pinch plasma is faster than the ionisation process.

The similarity parameter for the ionisation process is given by the product of the ionisation rate and the pinch lifetime $\Gamma_{ion}*\tau \propto n_e T_e/\chi_i^2 *exp(-\chi_i/T_e)$ with $\chi_i \propto Z^2$ beeing the ionisation potential. In an electron temperature regime for maximum emission in the Ly-α line this parameter scales like $\Gamma_{ion}*\tau \propto pa/Z^2$ assuming Te $\propto Z^2$ and $\tau \propto a/Z$. The experimental check of the predictions concerning the scaling of the He-α and the Ly-α line and the ionisation dynamics are already published in Ref. [3]. In this paper some unpublished results will be presented which discuss the role of $I_o^2/(pa^2)$ and pa as similarity parameters.

EXPERIMENTAL SETUP

The pinch plasmas were generated in a Mather-type plasma focus device in an storage energy range of 1.1 kJ - 2.6 kJ (C = 36 µF, L = 30 nH). The intervals for the device parameters were 200 kA - 300 kA for I_o, 50 Pa - 1000 Pa for p, 0.7 cm and 1 cm for the anode radius a. The pinch current was measured directly using a calibrated current probe, which was positioned in the electron system near the end of the anode. The ISB in axial direction was investigated using a pinhole grating spectrograph [4]. In the case of neon a mica spectrograph was used to study the K-shell emission. Both spectrographs were combined with a calibrated CCD-camera [5] as detector allowing to take single pulse spectra.

EXPERIMENTAL RESULTS

Figure 1 shows the integrated spectral brightness in axial direction in the centre of the source profile of oxygen for different sets of pa. The absolute values are approximately the same for the whole spectral range when keeping the parameter pa constant as predicted by the similarity considerations. Furthermore, these considerations would predict a scaling of the optical thin recombination continuum and optical thin K-shell resonance lines according to $\propto (pa)^2$ or $\propto I_o^4$ when matching the current or the parameter $I_o^2/(pa^2) = const$ for maximum Ly-α emission. The results shown in Fig.1, though taken for only two sets of parameters, are consistent with such scaling law for the recombination continuum. A similar scaling law for the total K-shell yield ($Y_K \propto I^4$) is known for other devices [6-8] and can be explained using stationary models [9]. The considerations made above can be understood as a generalisation of this scaling law for plasma focus devices.

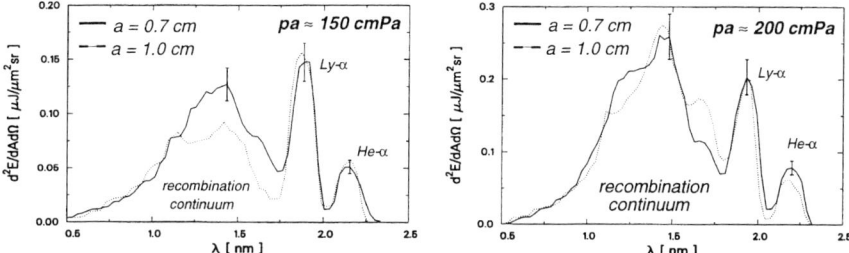

Figure 1: Pinhole grating spectrum of oxygen for different pressures but same pa. The current is matched for maximum emission in the Ly-α line for each parameter set. The data for the recombination continua have to be divided by the spectral resolution of $\Delta\lambda = 0.25$ nm to get the time integrated spectral brightness.

The role of $I_o^2/(pa^2)$ as a similarity parameter is also shown in Fig. 2 where the scaling of the Ly-α and the He-α line for neon pinch plasmas is shown for a fixed parameter pa. There, the parameter E/a^2 is approximately a unique function of $I_o^2/(pa^2)$ as predicted by the theory. Comparison to the theory in Ref. [3], which discusses the brightness rather than the total emission, implies that the parameter E/a^2 can be used as a measure for the integrated spectral brightness of the resonance lines. This would be valid if the radius of the source profiles or the pinch plasma would scale like $r \propto a$ for fixed values of the parameters $I_o^2/(pa^2)$ and pa. Such radius scaling has not been investigated experimentally for neon. However, the results for oxygen shown in Fig.3 support the validity of such scaling law. The source radii of the Ly-α and the He-α line are proportional to the anode radius comparing pinch plasmas at a fixed parameter pa at a current matched for maximum Ly-α emission.

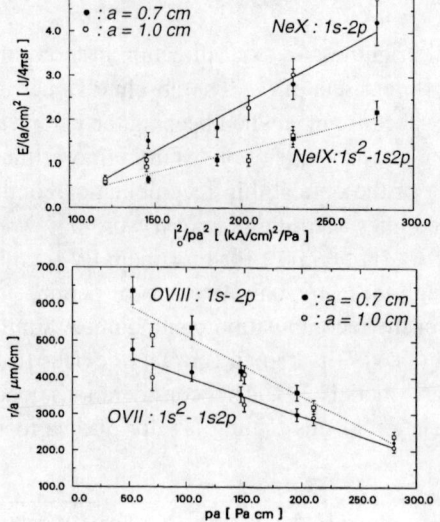

Figure 2: Total emission in axial direction E in the Ly-α and the He-α line divided by the square of the anode radius as a function of the similarity parameter $I_o^2/(pa^2)$ for two anode radii and fixed pa=400 cmPa (neon)

Figure 3: Source radii (FWHM) of the Ly-α and the He-α line as a function of the similarity parameter pa. The source diameter was determined using the pinhole grating spectrograph having a spatial resolution of about 180 μm.

The scaling of the line ratio of the Ly-α and the He-α line shown in Fig.2 for neon support the role of the parameter $I_o^2/(pa^2)$ as a measure for the energy input per ion into the pinch plasma or the achievable electron temperatures. Increasing this parameter leads to a relative increase in the Ly-α line, which can be referred to a higher electron temperature.

ACKNOWLEDGEMENTS

The financial support of the Deutsche Forschungsgemeinschaft (DFG) is gratefully acknowledged.

REFERENCES

1. Mather, J.W., in Methods of Experimental Physics: Plasma Physics vol 9b, H.R. Griem and R.H Loveberg eds., New York: academic) p. 187
2. Bergmann K., Lebert R., J. Phys. D: Appl. Phys. **28**, 1579 (1995)
3. Bergmann K., Lebert R., Neff W., J. Phys. D: Appl. Phys. **30**, 990 (1997)
4. Eidmann K., Kühne M., Müller P., Tsakiris G., J. X-Ray Sci. Technol. **2**, 259 (1990)
5. D. Rothweiler, "Gepulst erzeugte Plasmen als Strahlungsquelle für ein Röntgenmikroskop" PhD Thesis, RWTH Aachen, ISBN 3-89588-067-1, (1994)
6. Deeney C., LePell P.D., Failor B.H., Meachum J.S., Wong S., Thornhill J.W., Whitney K.G., Coulter M.C., J. Appl. Phys. **75**(6), 2781 (1994)
7. Pearlman J.S., Riordan J.C., SPIE-Proc. Conf. Ser., **537**, 1985, p. 102
8. Pereira N.R., Davis J., J. Appl. Phys. **64** (3), R1 (1988)
9. Whitney K.G., Thornhill J.W., Apruzese J.P., Davis J., J. Appl.Phys. **67**(4), 1725 (1990)

XRAYFIL, an Analysis Code for Non-Dispersive X-ray Diagnostics

C. Dumitrescu-Zoita and P. Choi

Laboratoire de Physique des Milieux Ionisés, Ecole Polytechnique, 91128 Palaiseau, France

Abstract. An X-ray filter analysis code, XRAYFIL, has been developed to allow more accurate non-dispersive X-ray plasma diagnosis with absorption filters by using a series of trial spectra of the emitting species. The code calculates a set of emission spectra for a given plasma using a CRE level population model package, RATION[1], and convolves it with the transmission characteristics of the filter set used, as well as the response function of the detector chosen. Comparison of the ratio of the signal through the different filters from these calculated values to that recorded in the experiment allows us to obtain a measurement of the plasma temperature and density. The code incorporates a number of options for various emission scenario and detector choices.

1. Introduction

Analyses of radiation in the X-ray region has long been a well established discipline in the diagnoses of high temperature plasmas. One of the widely used methods of detecting and analyzing the emitted radiation is based on the use of external filters in conjunction with non-dispersive detectors. An unfolding technique is then used to deduce the state of the plasma from which such radiation is emanated. The X-ray Filter Program code, XRAYFIL, was developed for this purpose and is able to handle a variety of different situations.

2. Non-dispersive Spectral Discrimination Method

The non-dispersive spectral discrimination method is based on a combination of multiple broad band edge-filters together with full K-shell spectra calculation. The basic idea originated from the Ross filter measurements, but has been improved and its applicability has been much extended. Most of the disadvantages that Ross filter systems present can be overcome by using the multiple broad band edge-filters diagnostic technique in the study of high temperature plasmas with impurities. This technique consists of using a number of filters to obtain several energy windows with broad spectral resolution. In high temperature plasmas with high Z materials, the dominant radiation originates from the high Z components. The essence of the technique is in selecting K or L edge filters with the absorption edges located about the recombination edge of the high Z radiation. The spectral selection is achieved by additional low Z filter channels having the transmission matched to the long and the short wavelength sides of the K-edge filter.

The filter materials have to be selected carefully according to the radiation we want to look at, and according to the composition of the plasma we are looking at. The transmission through the different filters is then recorded using active detectors, like PIN diodes, or using X-ray film. The deconvolution process is done by fitting a set of simulated spectra, at a number of plasma temperatures and densities, to the filter-detector combination employed and comparing the result with the measured response across the same set of filters. XRAYFIL has been mainly developed to follow all these procedures.

A characteristic of this technique is in the use of a large number of differently filtered channels. Having a large number of filter materials and the resulting artificial spectra generated through them, the plasma parameters (T_e and n_e) can be obtained with good accuracy. This technique is no longer upset by line emission, as this is taken into account when fitting the spectra through the different filters. The method avoids the difficulties in reconstructing spectra using a spectral inversion technique in the presence of strong line radiation and gives good accuracy over a broad temperature range.

3. XRAYFIL

XRAYFIL presents two main advantages over traditional X-ray transmission codes; it considers a realistic plasma source and a realistic detector, instead of the simplified approximations. The code calculates a set of emission spectra for a given plasma and convolves it with the transmission characteristics of the filter set used, as well as the response function of the detector chosen. For the plasma source a full K-shell spectra of the source emission is derived from the population levels of all ions calculated by a separate CRE level population code, RATION [1], together with recombination radiation to the ground state of all ionization stages. In fig. 1 we present an example of such spectra calculated with our code, after convolving with 3 different filters, chosen to analyze an Argon plasma.

For the time being, the code offers three options, as to which type of detectors will be used during the calculations: X-ray film, PIN diodes or diamond photoconducting detector, PCD. For Pin diode and diamond PCD the theoretical response is used. For film detector, however, a complex model is used based on Henke's general film model, [2].

This general film model has been adapted to suit our experimental conditions, imposed mainly due to the characteristics of the pulsed power generators we worked on. Typically DEF film is used to record high energy X-ray emission. As DEF is a double layer film, there are in fact **TWO** over imposed images, one on each side of the plastic base. In a condition where there is a hard X-ray signal on the background, the film is in general uniformly fogged. The light scattering from the first layer could render the desired image undetectable, even if the image is perfectly formed. Removing one of the two layers removes the scattering source and often solves the problem. As a full film model was employed to obtain the response function of the film, the treatment of this new situation, starting from the general equation deduced by Henke, was included.

There were however situations where DEF could simply not be used any more. A commercial visible photographic film, HP5 proved to be suitable as an X-ray recording film, giving very good images, with high spatial resolution, especially in those situations when hard X-ray is strong. Compared with DEF film, this film has smaller developed grain size and an excellent dynamic range. It is practically insensitive to hard X-ray radiation. From the existing literature, there are no information on the characteristics of this film in the region of interest to us. Therefore the characterization of HP5 was attempted by ourselves, following the pattern employed for the DEF analyses. Geometric parameters were obtained from SEM images, absorption coefficients were calculated starting from the chemical formula for each film layer and an adapted form of Henke's film model was used to complete the information needed. In fig. 2 we present the film response curves for HP5 and DEF, calculated using XRAYFIL.

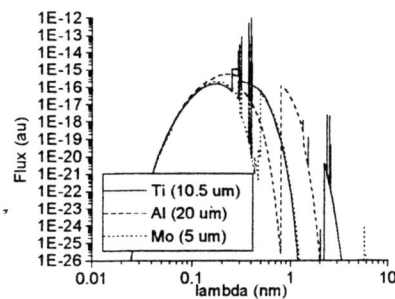

Fig. 1 Realistic argon spectra calculated using XRAYFIL

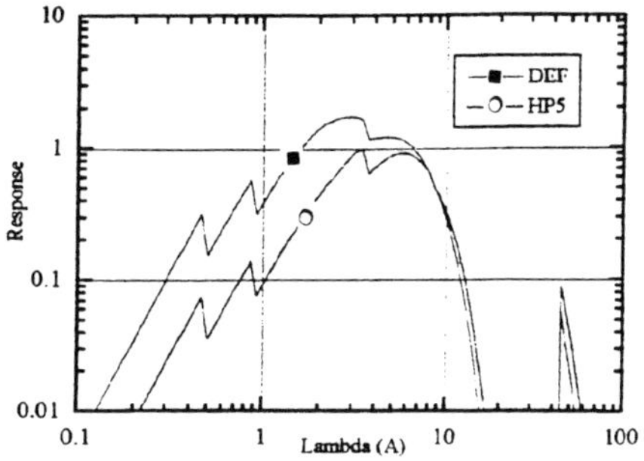

Fig. 2 Film response curves for DEF and HP5 film

We would like to stress the main difference to similar measurements and data analyses done so far:
(i) the spectrum we use in order to fit our transmission curves through the different filters employs a full spectrum, with recombination and lines, and not only a bremsstrahlung spectrum; and
(ii) the filter set is chosen to pick up the radiation in the K-shell spectrum.
These two apparently simple techniques together, can improve results beyond recognition. We will only present two figures to make our point: fig. 3 demonstrates the role of the different emissions: hydrogen Bremsstrahlung, argon Bremsstrahlung, argon recombination radiation and full argon spectra, including Bremsstrahlung, recombination and line emission. One can easily observe the very significant difference it makes for the estimated plasma temperature.

This example has been calculated for an argon plasma. The filters set has been chosen according to the multi-filter technique, to selectively pass or reject the argon K-shell lines and recombination continuum. It is obvious that using the proper emission spectrum is very important: the value for the plasma temperature changes with more complete emission spectra calculation. Errors between the different emission spectra are large. At low temperature, the hydrogen Bremsstrahlung model gives totally erroneous answer. As expected, as the temperature increases, the response curves come closer together and the errors introduced become smaller. While it is true that the results between the Bremsstrahlung spectra and the full spectra gets closer at high temperature, the slope of the curves at this region introduces a very large error for small variation in the ration of detected signal and the filter set in fact no longer become useful.

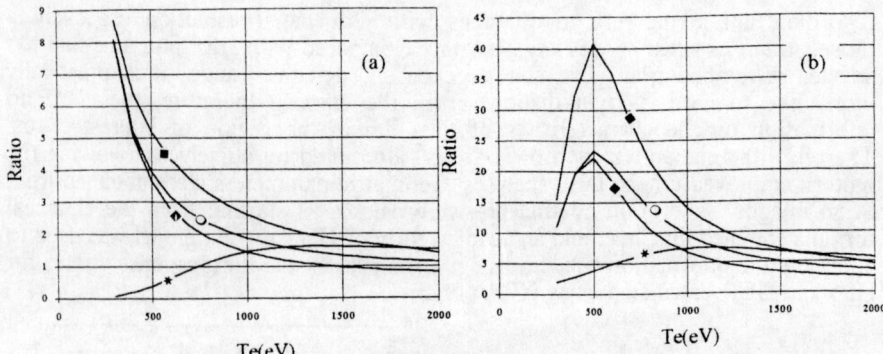

Fig. 3 Ratio of detected signals on film versus plasma electron temperature for filter pair of (a) Ti (10.5 μm)/Al (20 μm) and (b) Ti (10.5 μm)/Mo (5 μm) for 4 types of emission spectra. ✶ - hydrogen bremsstrahlung, ○ - argon bremsstrahlung, ◆ - argon recombination, ■ - full K-shell spectra

4. Practical application

We used in many of our experiments a Multi-Pinhole Multi-Filter Camera, designed to give information from composite plasmas [3]. The code was used both to calculate which filter pairs would give useful results for a given plasma and to estimate the pinhole sizes to be employed, starting from the known sensitivity and dynamic range of the detector, as well as for data analysis after experimental run. This ensures that different degree of filtering could all be recorded with similar signal amplitude, in order to extend the effective dynamic range of the whole system. Using this technique in different spectral bands different temperatures have been observed. The method essentially selects a period in the evolution of the hot plasma when the radiation in the selected region is dominant, even though the overall recording is integrated over the life time of the plasma.

5. Conclusions

The code XRAYFIL used as part of the multi-filter multi-pinhole diagnostic technique for estimating plasma parameters makes it possible to analyze multi-component plasmas with non-dispersive spectral resolution. The use of realistic detector and source models proved to be of great importance, as large differences would be obtained in the temperature range of interest if incorrect emission model is used, particularly when lines and recombination radiation are dominant in high Z plasmas.

References

[1] Lee, R.W., RATION, LLNL 1990.
[2] Henke, B.L. et al., *J.Opt.Soc.Am.B.* **1**, 6, 818-849 (1984).
[3] Dumitrescu-Zoita, C., Choi, P., Chuvatin, A., Etlicher, B., *this proceeding*

Nonlocality of Radiative Transfer in Continuous Spectra and Bremsstrahlung Radiation Transport in Hot Dense Plasmas

V.V. Ivanov* and A.B. Kukushkin**

(*) Institute of Nuclear Physics, Moscow State University,
Moscow 119899 GSP Russia
(**) INF RRC "Kurchatov Institute", Moscow 123182 Russia

Abstract. The importance of nonlocal effects in radiative transfer in continuous spectra is shown in numerical modelling of space profiles of plasma temperature and Bremsstrahlung total power losses in a layer of adiabatically compressed hot dense plasma, via comparing the results of the exact, integral equation formalism and widely used approach of radiation temperature diffusion with Rosseland mean diffusion coefficient.

1. NONLOCALITY OF RADIATIVE TRANSFER IN CONTINUOUS SPECTRA

The nonlocality of radiative transfer, which seems appropriate to define as the irreducibility of original energy transport equation, which is an integral equation over space variables, to a differential, diffusion-like equation, is normally associated with energy transport in discrete energy spectra (see. e.g. the surveys [1,2]). (Here, irreducibility assumes significant difference of the transport calculated in nonlocal and diffusion transport models rather than divergence of diffusion coefficient in formal expansion of original integral equation, see below.) The most pronounced example of such a transport is radiative transfer in resonance atomic lines (RTRAL) of atoms and ions in plasmas and gases. Here, bounded atomic electrons are responsible for holding the atomic excitation energy which is transported by the resonance emission/absorption process. The nonlocality of the RTRAL stems from "multigroup" nature of energy transport that implies coexistence (and strong coupling) of energy carriers -- in given case, photons -- with essentially different scales of the free path in the medium. Normally, the spectral distributions of the source and sink functions for energy carriers exhibit smooth transition between two alternative regimes, namely: (i) frequent emission of energy carriers of a short free path and (ii) rear emission of energy carrier of a long free path. First mechanism represents a diffusion which is, by definition, the statistics of small-step, "non-ballistic" spreading of the energy in a medium. In this case the total energy losses reduce to the nearly-LTE radiation flux integrated over medium surface. In the opposite case, essential contribution to energy transport stems from long, "ballistic" flights of energy carriers, while diffusion mechanism still determines short-range correlations of spatial profiles. The extension of this regime toward smaller optical thicknesses gives smooth transition to the trivial case of volumetric losses.

The nonlocality can be identified in several ways. In particular, one can analyze Holstein function $H(\rho)$, which has been introduced in RTRAL in the case of complete redistribution (CRD) in photon frequencies in elementary act of emission/absorption of the photon by an atom (it is the CRD that assures strong coupling of various space scales in the Biberman-Holstein equation):

$$H(\rho) = \int P(\omega) \, EXP(-\kappa(\omega)\rho) \, d\omega \qquad (1)$$

where $P(\omega)$ is the normalized spectral distribution of source function (for a given radiative transition); $\kappa(\omega)$ is the respective absorption coefficient. Thus, function H describes probability for the photon to pass the distance not less than ρ, without any absorption. In the case of diffusion-like transport function (1) falls down rapidly with growing ρ (exponentially, for pure diffusion). With growing contribution of nonlocal component of energy transport the function (1) falls down slower. In particular, for Lorentz and Doppler line shapes, respectively, one has the following asymptotic behavior at large optical thicknesses ($\tau \equiv \kappa_0\rho \gg 1$, where κ_0 is absorption coefficient value in line center) (see e.g. [1,2]): $H_{Lor} \propto (\tau)^{-1/2}$, $H_{Dop} \propto (\tau)^{-1} (ln(\tau))^{-1/2}$. Correspondingly, the motion of the excitation front from an instant point source in a homogeneous medium substantially differs from the respective diffusion law ($\rho \propto (\gamma t)^{1/2}$) and has the form (see [1,2]): $\rho_{Lor} \propto (\gamma t)^2$, $\rho_{Dop} \propto \gamma t (ln(\gamma t))^{-1/2}$, where γ is the spontaneous emission rate. However, the above properties are still the attributes of the nonlocality rather than definitions. The same appears to be true of the various formal expressions for diffusion coefficient or mean free path. Really, the diffusion coefficient derived from formal expansion of the Biberman-Holstein equation in the RTRAL tends, for Lorentz and Doppler line shapes, to infinity with medium size $L \Rightarrow \infty$. Nevertheless, the divergence of asymptotic expression of this coefficient, which in a homogeneous medium has the form [1,2]

$$D = \gamma \int P(\omega) d\omega / 3\kappa^2(\omega) \propto \gamma \int_0^\infty \rho^2 dH(\rho) \qquad , \qquad (2)$$

has, in general case, no relation to actual degree of nonlocality because such a divergence may come from those far wings of the line shape which do not contribute to H function. Indeed, for each finite value $\tau \gg 1$ there is certain spectral domain contributing to H function (i.e. an approximate one-to-one correspondence $\omega_{ef}(\rho)$ exists). It is the latter property that assures the success of the Escape Probability approaches (see [1-3]). Therefore, the behavior of $P(\omega)$ and $\kappa(\omega)$ in their farthest wings is of no importance for energy transport at given value of optical thickness τ. Correspondingly, the finiteness of D doesn't imply that the transport is a diffusive one. The latter is true of, e.g., Bremsstrahlung radiation: here $dln(1/H)/d\tau$ varies, for $\tau \gg 1$, from 1/4 ($\eta \equiv Ze^2/\hbar v_T \gg 1$, classical spectrum) to 2/9 ($\eta \ll 1$, Born approximation) and the respective (namely, Rosseland mean) diffusion coefficient is finite (cf. [1,2]).

Though for Bremsstrahlung radiation the Holstein function in *homogeneous* media exhibits less strong nonlocality as compared with discrete spectra case, there is an advantage peculiar specifically to continuous spectra: here, the temperature inhomogeneity gives a shift of the spectral line center that may result in substantial enlightenment of the medium. Thus, one may expect an extension of nonlocal correlations in energy transport up to values of τ as large as 10^3-10^4. The latter is confirmed by the results of numerical modelling (see Sec.2).

As to the very definition of the optical thickness τ for continuous spectra, which is not obvious, unlike line spectra case, it is natural to introduce unit optical thickness as the point of the intersection of volumetric and black body power losses: $d\tau = dx \int j(\omega)d\omega / \sigma T^4$, where σ is Stefan-Boltzmann constant, $j(\omega)$ is the source function (power density). For Bremsstrahlung, this gives $\tau = b\ 6.4\ 10^{-48}\ Z^3\ n_e^2(cm^{-3})\ T_e^{-7/2}(keV)\ L(cm)$, with b=1 for $\eta \gg 1$ (this differs by a factor $\pi^4/15 \approx 6,5$ from notations introduced in [4]) and b=1.11 for $\eta \ll 1$.

The irrelevance of the type of energy spectra, either discrete or continuous, to the character of radiation transport can be seen from electron cyclotron (EC) waves transport in hot plasmas. Here, transport of transverse waves, i.e. photons, in homogeneous plasmas makes smooth transition, with increasing temperature, from transport in separate harmonics of gyrotron frequency toward transport in a single broad spectral line composed of many overlapped harmonics. Here, the nonlocality holds true for both discrete and continuous cases (for specific case of ECR transport in a tokamak with highly reflecting walls see [5,6]). Also, the nonlocality holds true for energy transport by the longitudinal waves, i.e. plasmons (for energy transport in a single harmonic of electron Bernstein waves see [7]).

2. NUMERICAL MODELLING

Here we demonstrate the importance of nonlocal transport effects via comparing the results obtained in the frames of the precise, integral equation formalism and the widely used approach of radiation temperature diffusion with Rosseland mean diffusion coefficient D_{BR}.

Figure 1 gives steady-state solution for the electron temperature profile in a layer of hydrogen plasma (3% deuterium) with DD fusion energy release, Bremsstrahlung radiation transport and adiabatic compression at a constant value of pressure (e.g., magnetic pressure of the one-dimensional pinch). Figure 2 gives spatial distribution of the Bremsstrahlung power losses (with allowing for the reabsorption) in a homogeneous plasma layer with a "frozen" electron temperature profile, namely Te = $(T_e^0 - T_b)\{1-(x/x_m)^2\} + T_b$, where $T_e^0 = 10$ keV; x = 0 is the midpoint; $x_m = 8\ 10^5$ cm, T_b=300 eV. Interestingly, for the total power losses we have P=2.1 10^{19} W/cm^2, whereas black body limit and diffusion approximation (D_{BR} (dT/dx) at the boundary) give, respectively, 8.1 10^{14} and 4.4 10^{11} W/cm^2.

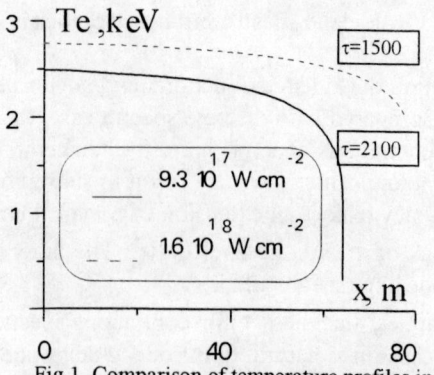
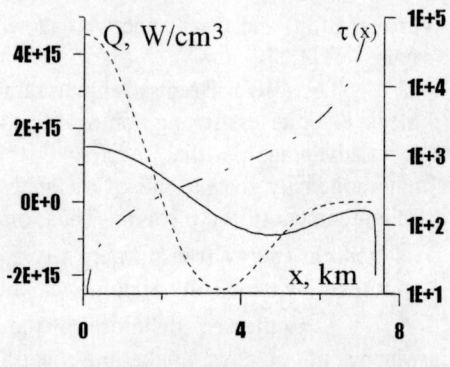

Fig.1. Comparison of temperature profiles in adiabatically compressed plasma layer with DD fusion energy release: exact (solid), diffusion approximation (dashed). The Bremsstrahlung total power loss and total optical thickness are indicated.

Fig.2. Comparison of nonlocal (solid) and diffusion (dashed) profiles of Bremsstrahlung losses for a plasma layer with a "frozen" temperature profile. Optical thickness $\tau(x)$ counted from layer's centre (dash-dotted) is indicated.

The latter suggests also that the above-mentioned enlightenment effects are especially strong in the time-dependent problems of relaxation of a steep temperature gradient (here one should allow for all other short-range energy exchanges, e.g. electron collisional heat conductivity). Numerical modelling shows that the heat propagation from a high temperature source/perturbation is dominated by the long-range correlations, up to $\tau_{cor} \sim 10^3$, rather than strong short-range correlations decsribed by the Rosseland mean diffusion coefficient.

REFERENCES

1. Kogan V.I., In: *A Survey of Phenomena in Ionized Gases (Invited Papers)*, [in Russian], (Proc. ICPIG'67), Vienna: IAEA, 1968, p.583.
2. Abramov V.A., Kogan V.I., and Lisitsa V.S. In: *Reviews of Plasma Physics*, Vol.12 (Leontovich M.A., Kadomtsev B.B., Eds.) New York: Consultants Bureau, p.151.
3. *Methods in Radiative Transfer* (Kalkofen W., Ed.), Cambridge: Cambridge Univ. Press, 1984, ch. 1.
4. Babikov V.V., Kogan V.I., In: *Plasma Physics and the Problems of Thermonuclear Reactions* (Leontovich M.A., Ed.), London: Pergamon Press, 1959, vol.3.
5. Tamor S., *Fusion Technol.* **3,** 293 (1983); *Nucl. Instr. and Meth. Phys. Res.*, **A271**, 37 (1988).
6. Kukushkin A.B., *JETP Lett.* **56**, 487 (1992); *Proc. 14th IAEA Conf. on Plasma Phys. Contr. Fusion* (Wuerzburg, 1992), Vienna: IAEA, 1993, v.2, p.35;
AIP Conference Proceedings #**299,** Dense Z-pinches 3rd Int. Conf. (London, 1993), Eds. M.Haines and A.Knight, New York: AIP Press, 1994, p.519.
7. Kukushkin A.B., Lisitsa V.S., and Saveliev Yu.A., *JETP Lett.* **46**, 448 (1987).

New EUV and x-ray optical instrumentation for hot plasma imaging, polarimetry, and spectroscopy, using glass capillary converters and multilayer mirrors

V.L. Kantsyrev, B.S. Bauer, A.S. Shlyaptseva,
R.F. Bruch, R.A. Phaneuf

*Department of Physics,
University of Nevada, Reno, NV 89557*

Abstract. We have developed a number of methods for plasma diagnostics. Applications of these methods will include: extreme ultraviolet and x-ray imaging of hot plasma with temporal, spatial and spectral resolution; polarization measurements with spatial and temporal resolution; high-resolution spectroscopy of multicharged ions with temporal resolution. Currently-developed instruments include: a pinhole camera with a glass capillary converter using as a hard x-ray filter; an extreme ultraviolet imager and spectrograph unit with multilayer mirrors; extreme ultraviolet polarimeters/spectrometers with glass capillary converters polarizing and focusing element; and a high resolution, a high-sensitivity x-ray or extreme ultraviolet spectrometer with focusing glass capillary converter that also serves as a high- transmission window for differential pumping. New multiband, two dimensional imaging spectrometer will be developed.

INTRODUCTION

A wide variety of advanced extreme ultraviolet (EUV) and x-ray diagnostics are being developed for the Dense Z-pinch Program at the University of Nevada, Reno, USA (see B. Bauer et al. at this Conference). Plasma density, temperature, flow, charge states, and magnetic field must be measured with detailed space- and time-resolved short wavelength spectroscopy.

ADVANCED SHORT WAVELENGTH DIAGNOSTICS

These diagnostics are based on the use of glass capillary converter (GCC) [1,2] (Fig.1) in conjunction with multilayer mirror (MLM) and/or crystal. The GCCs collect, guide, focus, filter and polarize a wide spectrum of short-wavelength radiation. The MLMs and crystal elements are used for dispersing, focusing and polarizing a narrow bandwidth of radiation. The GCCs consist of a bundle of glass or quartz capillaries. EUV, soft x-ray, and x-ray radiation are guided along straight or slightly curved capillaries by multiple grazing-incidence reflections from the inner capillary surfaces[1-3]. The GCCs can filter hard x-ray radiation (at least two or three orders of magnitude [2]). The GCCs were used in plasma diagnostics as hard x-ray filters in a soft x-ray pinhole camera (Fig.1), and proposed as an imaging and filtering x-ray streak camera element [1-3]. We also have experimentally demonstrated for the first time, that a GCC can be used as a new device for enhancing the flux density of x-ray and EUV radiation on the entrance slit of any type of short wavelength spectrometer (coefficient of enhancement of flux density $\eta \approx 10\text{-}100$ in wide spectral region $0.1\text{nm} < \lambda < 100$ nm) (Fig.1) [1-3]. GCC also serves as a high- transmission window for differential pumping [1]. Diagnostics for our z-pinch program will use several x-ray and soft x-ray crystal spectrometers (including spherical crystal imaging spectrometers); EUV ($\lambda \approx 3.5 - 7.0$ nm) concave MLM based imaging system/slitless EUV spectrometer [4]; multichannel pinhole cameras with fast imaging microchannel plate (MCP) detectors; pinhole chambers with hard x-ray GCC filters [1]; x-ray streak camera with GCC elements [1-3]; fast x-ray diodes; a new multiband, imaging, two dimensional time resolution spectrometer and new polarimeters/spectrometers [5].

1. Spectral-Imaging Measurements. Space and time-resolution for spectroscopy are essential, as these plasmas contain strong gradients and evolve rapidly. A broad range of wavelengths must be measured, both to accurately determine the electron energy distribution and to simultaneously observe a variety of atomic processes. For example, to compile two-dimensional (2-D) maps of electron density N_e and electron temperature T_e, an entire 2-D monochromatic plasma image must be captured on each shot, as shot-to-shot reproducibility is limited. These criteria are unfortunately not simultaneously satisfied by current spectroscopic methods. A spatial resolution of better than 10-20 μm can be achieved now, however, for an extended source as a z-pinch, the average distortion of the image generally increases up to 100 μm or more for plasma columns 20 -50 mm in length. One EUV, soft x-ray and x-ray radiation diagnostic stands to redefine the state of the art in the measurement of plasma parameters: a multiband, two-dimensional imaging spectrometer [5]. The scheme that was adopted (Fig.2) includes a multiband flat crystal spectrometer with entrance pinhole imaging camera and new EUV and x-ray wide band optical element, the GCC, to add an extra dimension to the spectrometer throughput. The new spectrometer uses a GCC to multiplex a two-dimensional pinhole-camera image of the z-pinch plasma column into a one-dimensional output arrays. Each linear array is then spectrally dispersed by a crystal (for soft x-ray and x-ray regions) or MLM for EUV and recorded by a temporally-gated imager with MCP and CCD units. The plasma image is then reconstituted by computers. Each element is used in multiple ways, to obtain, for each plasma image point, an array of time-resolved spectra, which covers the full plasma evolution. Spatial resolution of the spectrometer will achieve 30-70 μm. The resolution ($\lambda/\Delta\lambda > 500\text{-}1000$) and spectral range of each channel will be sufficient for simultaneous measurement of several spectral characteristic, for example, of Ne and Mg ions K- and L-spectra. The sensitivity of the spectrometer will be enough for registration of spatial and frame time resolved x-ray and EUV spectra in a single shot.

2. Short Wavelength Polarization Measurements. The method of plasma x-ray polarization measurement has been proposed [6] which employed two identical x-ray spectrometers differing only in orientation with respect to the z-pinch discharge axis. The difference in line intensities and line shapes between two polarization states facilitates discrimination against Doppler and Stark broadening and will yield information on the magnetic field and the energy and relative concentration of suprathermal electrons. We will use this method in our program. For precise measurements of polarization characteristics of separate spectral lines or groups of close placed lines we developed two new polarimeters/spectrometers [7] (Fig.3). Polarization measurements in first device (Fig.3a) for EUV, SXR or x-ray regions are based on the detection of the differences between the coefficient of reflection of radiation with S and P polarization from flat dispersive element (crystal or MLM) surface at an angle of about $45°$ ("Brewster" angle for short wavelength)[8]. Measurements are carried out by using two identical devices oriented at right angle to each other. We improved this scheme by application straight focusing GCCs to enhance sensitivity of registration of the signal by at least an order of magnitude (Fig.3a, for simplicity only one channel has been shown). A more sensitive double focusing EUV or SXR polarimeter (Fig.3b), that using a combination of dispersive focusing (MLM of normal incidence) and polarization sensitive and focusing bent GCC elements was developed. It was predicted theoretically that the degree of polarization of EUV and partially SXR ($\lambda > 2.0 - 2.5$ nm) changes gradually following multiple reflections of beams passing through a curved glass or quartz capillary [9]. Inside the curved GCC (generally with triangular or square capillary cross-section) the relevant P and S components of polarization of the radiation are attenuated differently by multiple reflections, leading to a changed ratio of P and S components. The measurement of polarization of the incident beam is made by comparing the intensity of the signal via two identical orthogonal channels (Fig.3b, for simplicity only one channel has been shown).

1. V.Kantsyrev, R.Bruch, Rev. Sci. Inst. **68**,1(II), 770 (1997).

2. V.L. Kantsyrev, K.I. Kopytok, A.S. Shlyaptseva, AIP Conf. Proc. **299**, "Dense Z-pinches", 612 (1993).

3. V.Kantsyrev, R.Bruch, R. Phaneuf, N. Publicover, J. X-ray Sc. Tech. 7 (1997).

4. V.L.Kantsyrev, K.Takasugi, K.Tatsumi, T.Miyamoto, A.S.Shlyaptseva, Proc. of the Intern. Conf. on Plasma Phys. (ICCP-96, Nagoya, 1996), 1106.

5. B.Bauer, V.Kantsyrev, US Pat. Applic., filed May 29, 1997.

6. A.S. Shlyaptseva, R.C. Mancini, P. Neil, P. Beiersdorfer, Rev. Sci. Inst. **68**,1(II), 1095 (1997).

7. V.Kantsyrev, R.Bruch, A.Shlyaptseva, US Pat. Applic., filed March 30, 1996.

8. P.Dhez, Nucl. Instrum. Methods Phys. Res. **A261**, 66 (1987).

9. M.Watanabe, T.Hidaka, V.Mitsuhashi, J.Lightwave Technol. **LT-5**, 7 (1987).

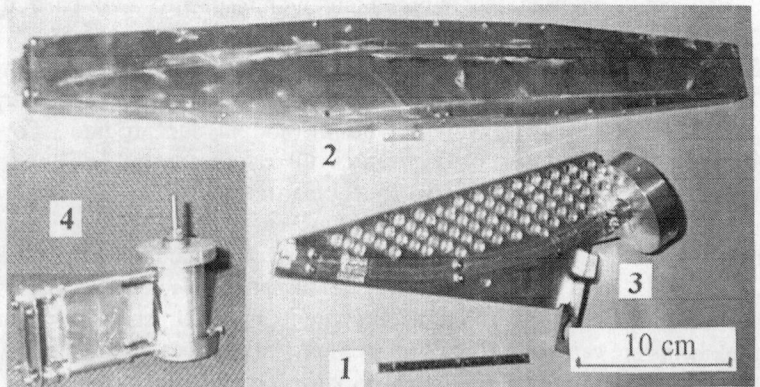

FIGURE 1. Glass Capillary Converters (GCC) from University of Nevada, Reno for EUV and x-ray spectroscopy and imaging plasma diagnostics. **1.** GCC ("beam-line focus" type) collisions spectrometry (η= 4-5, λ=30.0 nm); **2.** GCC ("line source-line focus type" with rotation on 90^0) for EUV ion-atom collisions spectrometry (η= 10, λ<100.0 nm); **3.** GCC for EUV polarization measurements; **4.** Pinhole camera with GCC filter.

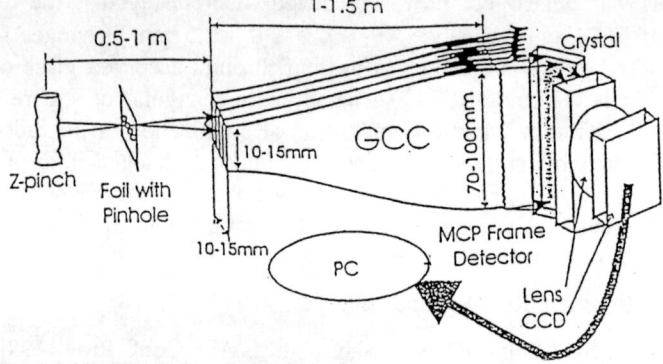

FIGURE 2. Multiband 2-D Imaging Spectrometer.

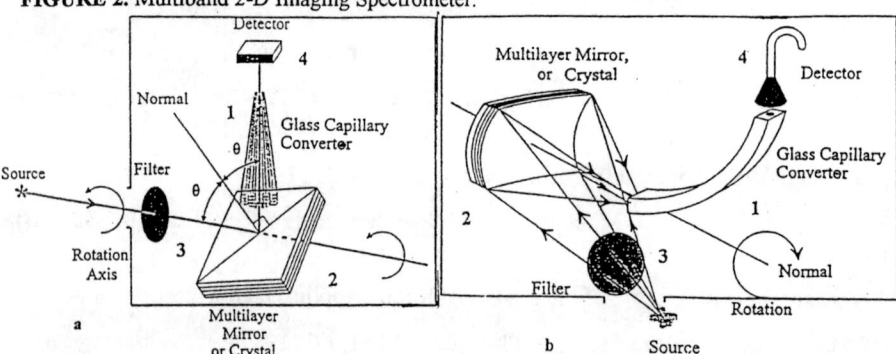

FIGURE 3. **a.** Polarimeter/spectrometer with focusing straight GCC (1); flat, polarization sensitive, dispersive element (MLM or cristal) (2), **b.** Polarimeter/spectrometer with polarization sensitive, focusing GCC (1); concave dispercive element (MLM or crystal) (2). Elements 3 and 4 are filter and detector respectively.

DIAGNOSTICS OF SMALL CARBON FIBER XUV-PULSE

J.Kravárik, P.Kubeš
Czech Technical University
Technická 2, 166 27 Prague 6
Czech Republic

This paper describes last results of XUV and schlieren diagnostics of the carbon fiber discharge. The device of energy capacity of 0.5 kJ and charging voltage 50 kV was able to generate a current pulse of the maximum amplitude of 50 kA in 1.1 µs after the beginning of the current. A carbon fiber of 20 µm diameter was fixed between the copper electrodes with conical and/or flat tops. The XUV emission was registered with a PIN diode in radial and axial directions. It was observed that short XUV bursts of the FWHM ~ 10-30 ns are emitted randomly in Z-direction 100-400 ns after the beginning of the current [1][2][3].

In this contribution the influence of conical and flat electrodes and the orientation of the polarity of electrodes on the intensity of the XUV burst, the evolution and electron densities of the exploding plasma column, the evidence of conduction of the current through the plasma channel and evidence of a rest of the solid fiber in the plasma column during the XUV pulse are presented.

The UV and XUV radiation were detected by PIN diode with temporal resolving ~ 1ns. Two types of diodes were used. The diode with detecting flat coated by 5 nm of Al and the diode without coating but with a possibility of inserting of Al foil of 0.8 µm thickness. The PIN diodes were located in distance 25 cm in radial and axial position of the fiber. For axial observation the electrode with axial bore- 1 µm-diameter was used.

For a visualization by the schlieren method the Nd:Yag laser with length of the pulse 3 ns on the second harmonic wavelength 532 nm was used. The focal screen 2, 4 and 8 - mm-diameter for lens with 1.5 m focal length enables imagination of inhomogeneities with deflection higher than 0.7, 1.4 and 2.8 mrad - it is the imagination of the plasma with the mean electron density higher than ~ 0.5, 1 and 2×10^{25} m^{-3}.

The first important result is illustrated in Fig. 1. In this shot the feasted to cathode fiber did not connect the anode. At the top of the fiber the virtual flat anode is formed with density gradient perpendicular to the fiber. The volume between anode and virtual anode is filled with plasma evaporated from anode. The exposure of this picture was realized 300 ns after breakdown at current 20 kA. It is clear, that all the current of the discharge between electrodes connected by fiber is conducted through the fiber and the plasma column around the fiber. What is not clear is the current conductivity. The conductivity calculated from the time of magnetic confinement of the plasma channel must be ~10^8 Ω^{-1} m^{-1}. The Spitzer's conductivity of the plasma could be ~10^5 Ω^{-1} m^{-1}. The next important result of schlieren diagnostics is presented by unusual form of the fiber in Fig.2. The exposure of this picture was realized 300 ns after breakdown. The curve of the fiber is the same as before the discharge. Fig.3 schowes the schlieren picture of the fiber discharge in time 2 µs after breakdown during the fiber disintegration. The solid fragments of the fiber are visible her. Both Fig. 2, 3 confirm existence of remains of the solid part of fiber in the plasma channel during relatively long time of the discharge (1 µs after current increasing).

In Fig. 4,5,6 are presented the schlieren picture of the explosion of the plasma column during the XUV burst. Fig.4 was exposed at the beginning of XUV burst 20 ns before the burst maximum. On the picture the increased diameter ~ 100 µm near the anode is imagined as a prefase of the explosion. The first 10 ns of the burst is connected with increase of the diameter to ~ 100 μm. Fig.5 is exposured at the maximum of the XUV burst, 10 ns after beginning of the burst. Explosion with nonuniform diameter reaches diameter 250 - 1000μm. The velocity of radial expansion (0.5 - 2) x 10^5 m/s was estimated from dimensions of clouds of explode plasma. The intensive bursts are connected with explosions near the electrodes.

Fig.6 imagines the final phase of explosion The boundary of dense plasma column is similar to m = 0 instabilities. The 50 - 100 μm diameter is similar to that before XUV pulse.

The electron density of the exploding plasma during the burst was evaluated higher than 10^{25} m^{-3} in the cylinder with diameter ~ 500 μm. This value gives an estimation of electron density 10^{27} m^{-3} in the plasma column before and after the explosion. The plasma in the column around the fiber seems denser and relatively colder in comparison with [4][5].

Some plasma explosions have no correlation with axial XUV burst. Typical example is presented in Fig.7. These types of explosions are connected with radiation detected in radial PIN diodes in visible wavelength during ~ 100 ns.

The configuration with flat electrodes produces XUV burst with lower mean intensity. The most intensive bursts correlate with explosion connected with the top of conical electrodes. The influence of the electrode polarity on the rate and intensity of the XUV burst was not observed.

In order to describe correlation between burst and explosion in detail the further experiments are planned for a shorter rise of the current and lower fluctuation of the burst.

References:

[1] P. Kubeš et al: Proc. Beams 96 Vol.II., Prague 1996, 737.
[2] P. Kubeš et all: Proc. Conf. Plasma Phys. Nagoya 96, in print.
[3] L. Jakubowski, M. Sadowski, J. Zembrowski: Proc. 18th SPPT97 in print.
[4] J. J.Rocca et al: 73, No.9 (1994), 2192.
[5] C. Steden, H. J. Kunze: Phys. Letters A, 151 No.9 (1990), 534.

This research has been supported by grants GACR No. 202-95-0178 „Stable Structures in Magnetic Pinches" and No. 202-97-0487 „X-ray Source on the Magnetic Pinch Principle".

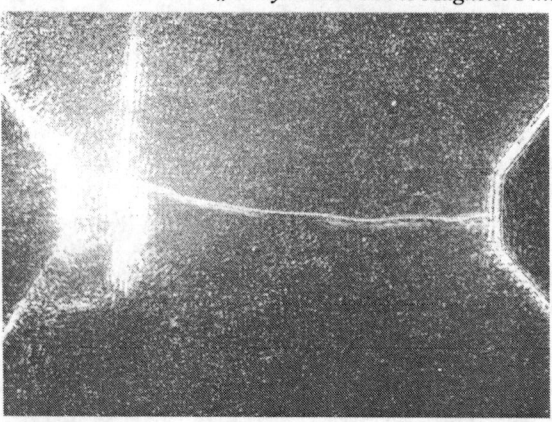

Fig.1: Shot No.817. Schlieren picture of the virtual anode formed by plasma evaporated from electrode at current of 10 kA in time 300 ns after beginning of current.

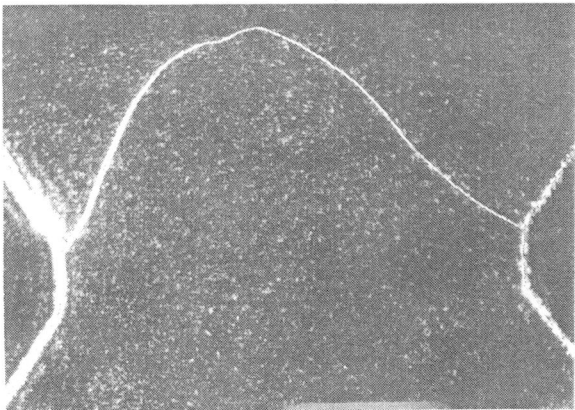

Fig.2: Shot 889. The shape of the plasma channel around the fiber 300 ns after current beginning is the same as before the discharge.

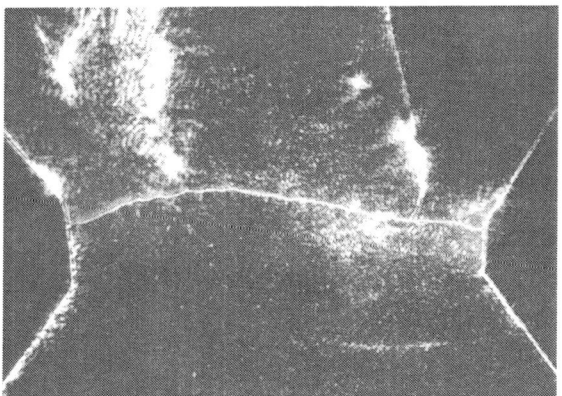

Fig.3: Shot 871. Schlieren picture of the carbon fiber discharge 2 μs after beginning of the current. The schlieren of pieces of evaporated carbon are present.

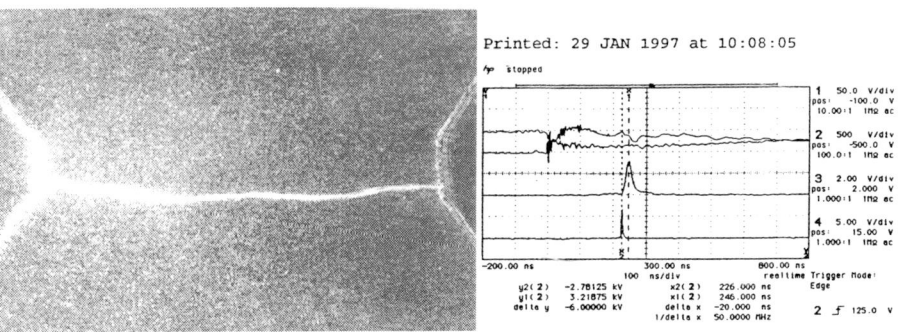

Fig.4: Shot 658. Schlieren picture exposed at the begin of the XUV burst. Trace 1 presents the current derivative, trace 3 PIN diode signal and trace 4 position of the laser diagnostics pulse.

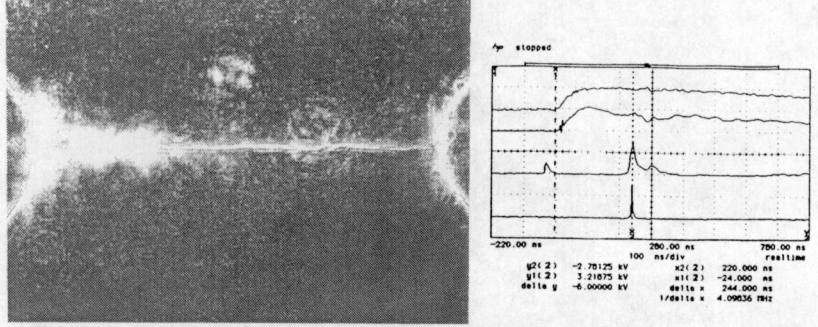

Fig.5: Shot 898. Schlieren picture exposed at the maximum of the XUV burst. Trace 1 presents the current derivative, trace 3 PIN diode signal and trace 4 position of the laser diagnostics pulse.

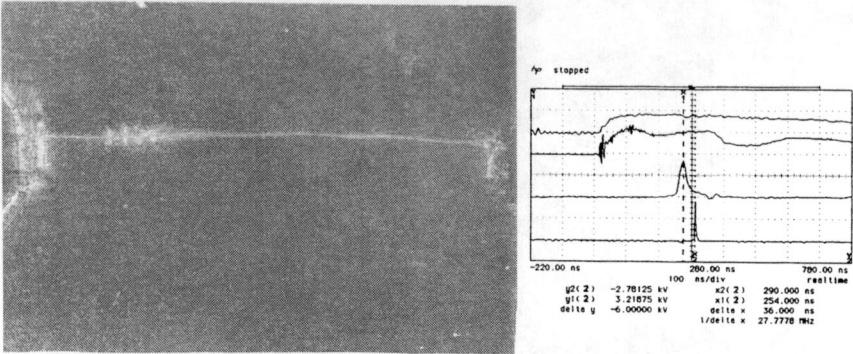

Fig.6: Shot 670. Schlieren picture exposed at the final phase of the XUV burst. Trace 1 presents the current derivative, trace 3 PIN diode signal and trace 4 position of the laser diagnostics pulse.

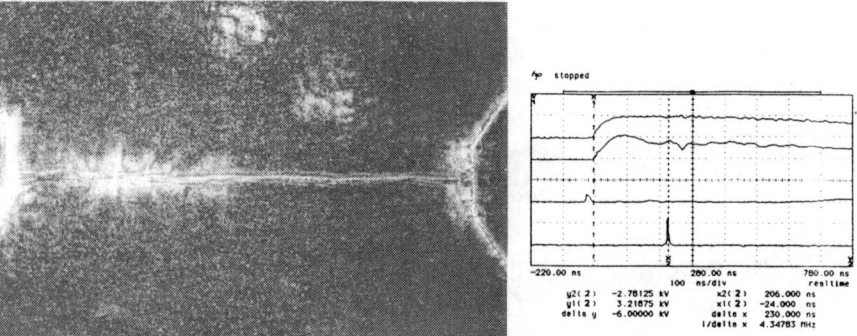

Fig.7: Shot 609. Burst without axial radiation.

Diagnostics of the thick carbon fibre z-pinch

H. Davies[†], A. Lorenz[†], J. Kravarik[*] & P. Kubes[*]

[†] Blackett Laboratory, Imperial College
Prince Consort Rd. London SW7 2BZ

[*] Czech Technical University,
Technicha 166 27 Prague, Czech Rep.

Abstract

Results of a recent experiment on the IMP generator (11 kJ, 600 kV) at Imperial College are presented. 360 μm carbon fibres were used. The plasma was diagnosed with time resolved (from PIN diodes) and time integrated (pinhole camera) x-ray diagnostics, and time resolved optical emission diagnostics. The plasma was found to expand little during the main current pulse. Fragments of fibre recovered from the machine indicated that < 1% of the fibre had been ionised. The x-ray diagnostics revealed a low-temperature, narrow diameter coronal plasma.

Introduction

The experimental series described in this report was motivated by a desire to see if the x-ray lasing action in a carbon fibre z-pinch as described by Kubes et al.[1] could be scaled up to a higher energy machine.

The experiment took place at the IMP facility at Imperial College, London. The IMP generator consists of two Marx capacitor banks. The main unit is a 11 kJ, 600 kV bank, delivering ~160 kA in 80 ns to an inductive load. The smaller, pre-pulse unit is a 180 J, 300 kV unit, delivering ~5 kA with a rise-time of 60 ns to an inductive load. The pre-pulse discharge time, relative to the main voltage pulse, can be controlled. Throughout this experiment the time was 140 - 280 ns. This corresponds to one - two half cycles of the pre-pulse current.

The experiment was carried out on graphite fibres of diameter 360 μm, length 14 mm. This is an order of magnitude thicker than the fibres usually used in z-pinch experiments. The diameter was chosen in an attempt to reproduce the magnetic field conditions at the solid-plasma interface region obtained by Kubes et al [1,2] in experiments on the lower energy machine at the Czech Institute. This value of the magnetic field is thought to be crucial to the lasing action reported.

The choice of diagnostics was influenced by the emphasis placed on the x-ray emission from the fibre. X-ray emission was measured in the axial and radial directions. To measure in the axial direction a novel hollow anode set up was used, allowing the attaching of standard metric vacuum pieces. The x-ray diagnostics used varied between shots; three main instruments were used.

X-ray PIN diodes, active area 1 mm^2, sensitive particularly in the soft X- UV region. These were used in the end-on and side-on directions with two 0.8 μm Aluminium filters. The axial PIN diode was also used in conjunction with a grazing incidence Si mirror, which essentially acts like a filter with a selectable wavelength range, by adjusting the angle at which the PIN diode observes the reflected x-rays. X-ray pinhole cameras were employed in both the end-on direction (replacing the PIN diode), and side-on direction, in this case simultaneously with the diodes. A Photo-multiplier tube (PMT) was employed, above the load chamber, to give some indication of any hard x-ray emission from the pinch.

In an attempt to gain knowledge about the general plasma dynamics during a shot, a Kentech optical streak camera was employed. By means of a single lens imaging system onto a slit, time resolved images of the plasma self- emission were obtained. These were taken radially (to give radius vs time) and axially (to resolve the behaviour of the whole plasma).

The load current was measured by a rogowskii coil, situated in the anode, and a B_θ inductive coil, situated near to the anode. These, and other electrical measurements: of the generator behaviour, and the PIN diode and PMT outputs, were recorded on a Hewlett-Packard digital storage oscilloscope.

Results

X-ray Emission

Measurements with PIN diodes showed a consistent, reproducible, x-ray pulse. This pulse began between 20 and 30 ns after the start of the main current. Its shape was approximately gaussian, with a FWHM of 50 to 60 ns. The pulse was observed from both the end-on and side-on directions, and appeared to be very isotropic, with an intensity of 10^2 Wm^{-1}. The x-rays were estimated (by filter transmission and by spectrometer measurements) to lie in the region 1-10 keV. These soft x-ray pulses were, however, accompanied by two bursts of hard x-rays, recorded by the PMT. These x-rays had energies higher than 50 keV. (see figure 1).

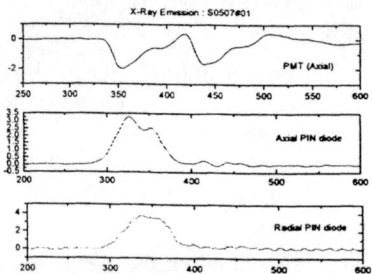
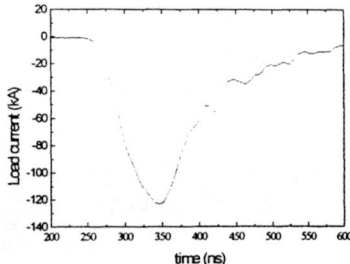

Figure 1.1 PIN diode and Photomultiplier output. PMT output is subject to internal delay of 50 ns (this is corrected for)

Figure 1.2 Current profile from S0507#01, same time as PIN and PMT signals.

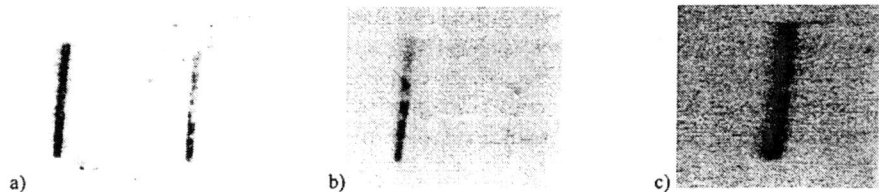

Figure 2 Time-integrated X-ray pinhole images. a) S0509#02: 200μm pinholes filters 0.8μm and 2 μm Al. b) S0506#07: 300 μm pinholes 2 μm & 6 μm Al. c) S0507#02: 300 μm pinhole, 2 μm Al.

The first hard x-ray pulse can be identified with the formation of runaway electrons, formed in the breakdown of the surface of the fibre, which then strike the anode, generating beam-target x-rays. The plasma then emits strongly in the soft x-ray region for 50 ns, as described above. The second hard x-ray pulse, coming at the end of the soft x-ray emission has been tentatively linked with the breaking of the fibre, leading to another electron beam to the anode. It is conceivable that the fibre is broken by purely mechanical means, due to the stresses placed on it by the current flowing in the plasma around the surface. The x-ray pinhole images (figure 2) show that the emission is very low energy. The majority of the column is invisible when filtered by 6 μm Al.

Optical Emission

The plasma was found to not begin self emission in the optical region until late in the pre-pulse current, the last 20 ns (see figure 3). This suggests that the pre-pulse has very little effect on the fibre, unlike results previously obtained with this generator (with 7 and 33 μm carbon fibres). The pre-pulse current is only ~4 kA, which is a fraction of the normal current into a plasma load. The pre-pulse current was detected in the fibre immediately as the generator discharged into the transfer line, with no delay to allow for charging to a breakdown voltage. This behaviour is similar to short circuit conditions, where a steel bolt is used as the load.

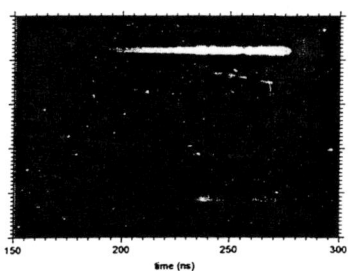

Figure 3.1 Optical radial streak.
100 ns full sweep (nominal) Brighter regions are due to camera malfunction

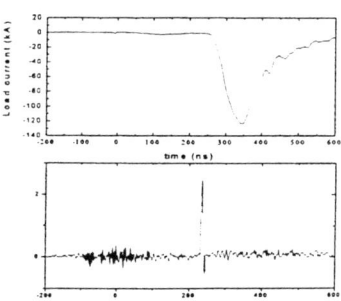

Figure 3.2 Sweep camera timing mark and current Timing mark comes ~20 ns into main current rise N.B. $I_{pre-pulse} << I_{main}$

The fibre is graphite, which has a resistivity, $\rho = 1375$ μΩcm. For 360 μm diameter fibres, this gives: $R \cong 1.9\Omega$. For 33 μm carbon fibre, the resistance is ~240Ω (7 μm C, R=) This difference is the explanation for the lack of effect of the pre-pulse on the fibre. The immediate conduction of the fibre does not allow the pre-pulse to reach the maximum voltage, and consequently the ionisation effect of the pre-pulse is greatly reduced. Joule heating can account for any plasma which forms late in the pre-discharge. The emitting region then expands from 1 to 2 mm in diameter and then remains at a surprisingly constant radius, of ~2 mm for the next 50+ ns (figure 3.1) During this time the soft x-ray emission is taking place. The constant plasma radius at this time, coupled with the relative low energy of the x-ray emission, suggests that the energy of the current is going into ionisation and ablation of the outer surface of the fibre, which is, however, forming a cool plasma, due to its contact with the solid fibre core. During this time, axial streak images show hot spot formation and bifurcation (figure 4) taking place in the coronal plasma. These hot spots move with a velocity $\sim 10^5$ ms^{-1}, typical for such objects.

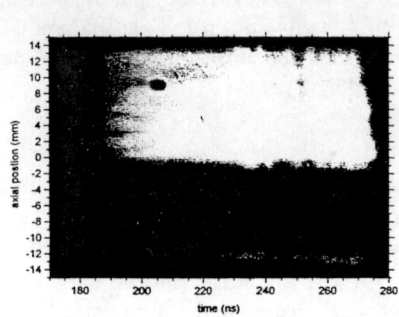 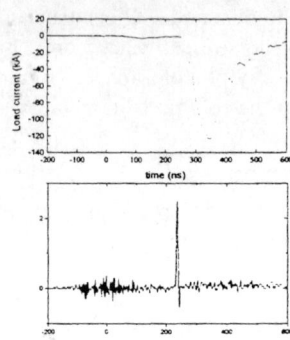

Figure 4.1 Axial streak 100 ns full sweep.(S0509#04)   *Figure 4.2 Current trace and timing mark*

Other Observations

When the chamber was opened, to allow cleaning of the electrodes and perspex insulator, several pieces of carbon fibre were found scattered on the chamber bottom. These fragments were found to have a diameter ~300 μm. This observation, coupled with the existence of the fibre fragments at all, have led us to conclude that the fibres were breaking, and this was the cause of the second x-ray pulse, and that the plasma formed and observed during the experiment was cold due to losses to the solid carbon core, which the current was not penetrating.

Conclusions

The plasma formed is seen to be small diameter, and emitting x-rays during 60 ns of the main current cycle. The flow of current is terminated by the breaking of the solid fibre core by mechanical means. Give that the plasma emits in the XUV region isotropically, there is little evidence for lasing action taking place. It would be unwise to speculate too much as to the plasma temperature, without temperature measurements. However, experiments using the same PIN diodes (filtered with 6 μm Be) looking side on, with a 7 μm carbon fibre load showed considerably higher x-ray yields than in this experiment (filtered 1.6 μm Al). The PIN diode was saturated in the 7μm case[3]. This would point to the 360 μm fibre pinch having a far lower energy.

Acknowledgement

This work was supported by grants GACR No. 2021-95-0178 'Stable Structures in Magnetic Pinches' & No. 202-97-0478 'X-ray Source on the Magnetic Pinch Principle'.

References

1 Kubes P. et al *Proc. Int. Conf. Plasma Physics Nagoya 1996, in print*
2 Kubes P., Kravarik J. *Presented at this conference*
3 Lorenz, A. *Private Communication 1997*

Investigation of the Linear Z-Pinch Plasma by Means of Laser Scattering

V. A. Rantsev-Kartinov* and E. E. Trofimovich°

INF RRC "Kurchatov Institute", 123182, Moscow, Russia
° *International University for Nature, Society and Man of the Dubna, Universitetskaya 19, Dubna, Moscow region, 141980 Russia*

Abstract. Experimental results are presented on the diagnostics of the Dense Z-pinch plasma from laser scattering spectra, which demonstrate the possibility of phase transition to a dense plasma and the importance of dynamical screening effects in interpreting the results. Also, strongly developed Langmiur turbulence is observed in scattering spectra that allows to estimate the fraction of the turbulence-captured electrons.

INTRODUCTION

The scattering of laser radiation in plasmas is known to make transition from incoherent scattering at plasma electrons to the collective scattering (scattering at a dressed particle in plasma). For small values of the Salpeter parameter $\alpha \equiv 1/k\lambda_D \ll 1$, where k is wave number of the laser scattering vector, and l_D is Debye radius [1], the scattering takes place at individual electrons and the spectrum has Gaussian line shape with the half-width determined by the electron temperature. In opposite limiting case, $\alpha \gg 1$, the scattering is a coherent one and the spectrum contains the narrow central peak, with the half-width determined by the ion temperature, and the symmetric satellites shifted to electron plasma frequency. For intermediate values of Salpeter parameter, $\alpha \sim 1$, the spectrum line shape is very sensitive to the value of α and, thus, to the correlation length of the screening of a charged particle in plasmas. For investigating the dynamic screening effects it is worth to study the scattering of laser light in Z-pinch plasma in a scattering geometry when $\alpha \sim 1$.

Here the experimental results are presented on the determination of the linear Z-pinch plasma parameters by means of laser scattering. The results reveal the importance of allowing for the non-static screening effects in interpreting the experimental spectra and determining major parameters of dense plasmas in a high current gas discharge.

EXPERIMENTAL SET UP

The experimental set-up (Fig. 1) includes discharge tube 60 cm long, 20 cm diameter, with flat copper electrodes on the edges, low inductance capacitor, 60 μF, voltage U = 30 kV. Maximum current 350÷400 kA is attained at 4.5 μs after discharge breakdown. The diagnostics uses ruby laser of 15 ns pulse duration and

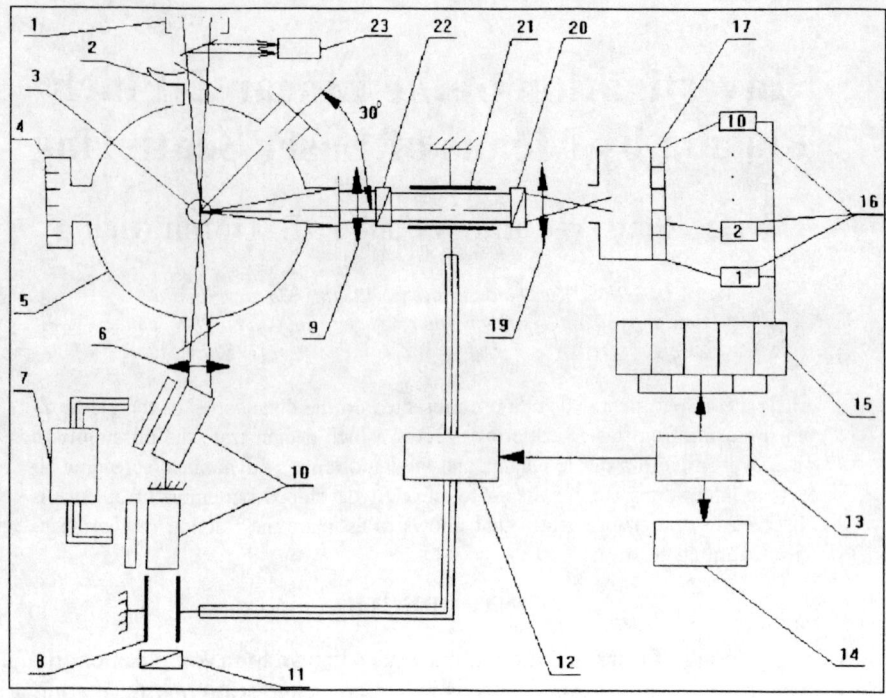

Fig.1 The experimental set-up: 1,2,4 - light's traps; 3 - plasma; 5 - vacuum chamber; 6,9,19 • objective lenses; 7 - laser's energetic block; 8,21 - Kerr cells; 10 - laser; 11,20,22 - polarization filters; 12 - Kerr cells energetic block; 13 - synchronization block; 14 - capacitor; 15 - oscilloscopes; 16 - photoelectron multipliers; 17 - fibrous optics; 18 - polychromator; 23 - coaxial photocell (monitor).

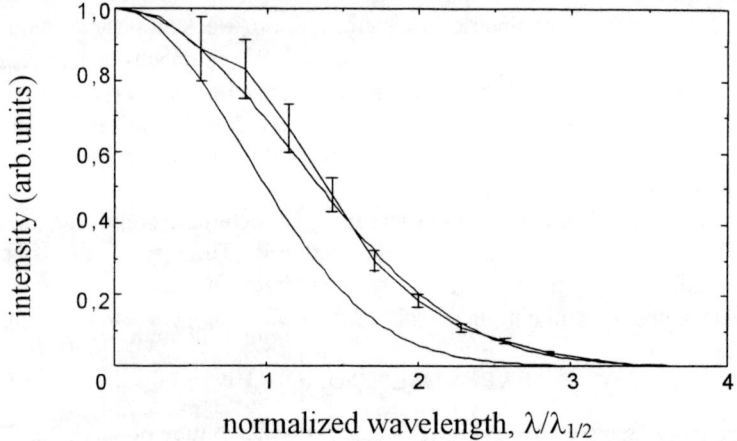

Fig.2. The spectrum of laser light scattered at an angle $\theta = 90°$.

energy $E = 5$ J. The 10-channel differential diagnostic system (DDS) enables us to extract the signal under condition of signal to noise ratio as small as $0.01 \div 0.005$.

RESULTS AND DISCUSSION

Comparison of the experimental results for the scattered light spectrum with theoretical results for scattering angle $\theta = 90°$ is presented in Fig.2. Here, $\lambda_{1/2}$ is the half-width of the spectrum, solid line is experimental curve; dash-dotted is the Gaussian for the temperature value 36 eV which is extracted from interpreting the experimental curve as a Gaussian one; dotted curve stands for the calculation which assumes that 50% of plasma electrons are captured by the turbulence. It follows that the temperature of captured electrons equals to the temperature of the 'background' plasma. The comparison gives the values of α in the range $0.1 \div 0.2$.

Figure 3 shows that the radially averaged plasma electron temperature (T_e) at the moment of the maximum compression ($200 \div 250$ ns before the first singularity of electric current [2]) increases monotonically from 30 eV to 130 eV with the initial working gas (D) pressure (p) varying from 10^{-2} Torr to 0.2·Torr, respectively. Here initial voltage U amounts to 22 kV. The value of T_e attains its limiting value asymptotically at the maximum value of pressure. The asymptotic value is determined by the rate of electron-ion energy exchange, $\tau = 70$ ns, at the stage of maximum compression of the pinch, of duration about 100 ns.

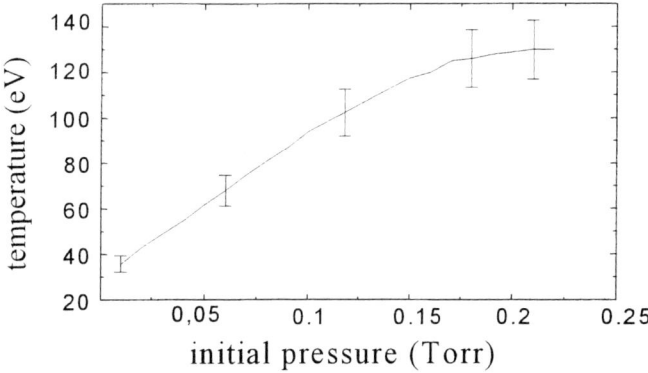

Fig.3 The radially averaged plasma electron temperature at the moment of the maximum plasma compression

The radially averaged plasma electron density (N_e), shown in Fig.4, was determined from the integral intensity of scattered laser light with the corresponding calibration of the DDS, and also from the relative measurements of the scattered laser light intensity at the scattering angles 90°, 60°, and 120°. This gives $N_e = (1 \div 5)\ 10^{17}$ cm^{-3}. Moreover, N_e exhibits stepwise dependence on the pressure, having a jump, by a factor of five, at a certain pressure value (about $7.5\ 10^{-1}$ Torr). This can be interpreted as a sort of the phase transition to a dense plasma.

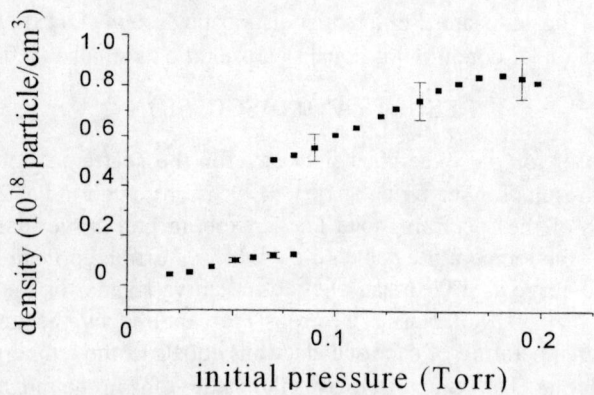

Fig.4 The radially averaged plasma electron density at the moment of the maximum plasma compression.

Thus, the line shape of scattered light gives small values of Salpeter parameter whereas from absolute value of measured we infer values $\alpha > 1$, that would give much different spectral line shapes. This controversy founds its explanation once we allow for the dynamic scattering effect of a charged particle in hot plasmas (substantial descreening as compared with the Debye screening). Really, taking into account the dynamic screening effect of electrons in hot plasmas [3] leads to the size of polarization cloud $l \sim (10 \div 20)\, l_D$. Substitution of this length, instead of l_D, into conventional expression for α resolves this difficulty, and the above discrepancy disappears.

CONCLUSIONS

Experimental results, presented on the diagnostics of the Dense Z-pinch plasma from laser scattering spectra, demonstrate the possibility of the phase transition to a dense plasma. It follows from interpreting the spectra of scattered radiation that the formal use of Salpeter parameter with conventional Debye radius leads to a discrepancy which can be resolved via allowing for substantial dynamic descreening of a charged particle in hot plasmas. Observations of strong Langmiur turbulence allowed to estimate the fraction of the turbulence-captured electrons.

REFERENCES

1. L.N. Pyatnitskiy, "Laser Diagnostics of Plasmas", Moscow, Atomizdat, (1976)
2. V.V. Aleksandrov, A.I. Gorlanov, N.G. Kovalskiy, S.Yu. Lukyanov, V.A. Rantsev-Kartinov, In: Diagnostics of Plasmas (in Russian), v.3, Moscow, Atomizdat, (1973).
3. E.E. Trofimovich, V.P. Krainov, Zh. Eksp. Teor. Fiz.(JETP), **104**, 3971 (1993).

FARADAY ROTATION MEASUREMENTS IN MAGPIE GENERATOR

M. Tatarakis, R. Aliaga-Rossel, A.E. Dangor and M.G. Haines

The Blackett Laboratory, Imperial College of Science, Technology and Medicine, London SW7 2BZ, United Kingdom

Abstract. We report on magnetic field measurements in MAGPIE generator (Mega Ampere Generator for Plasma Implosion Experiments) using the Faraday rotation technique. The generator is operated with a peak current of 1.1 MA rising in 150 ns. The loads are 33 µm diameter carbon fibres.
A measurable Faraday rotation angle is observed only in a time window from 50 ns to 60 ns after the current start due to the fact that this effect depends on the combination of the magnetic field strength and electron number density. A new type of self referencing cyclic radial shear interferometer is used to evaluate the plasma density profiles which are necessary for the reconstruction of the current distribution. It is calculated that 110 kA current is flowing in the plasma at 52 ns after the current start. This value corresponds to 70% of the total current measured with a current monitor.

INTRODUCTION

The angle which a linear polarised wave is rotated when it propagates through a plasma is [1],

$$\theta = 2.63 \times 10^{-13} \lambda^2 \int_l n_e \cdot \mathbf{B} \bullet \mathrm{d}l \qquad (1)$$

Polarimetry has been successfully used for magnetic field measurements in laser produced plasmas [2-6], in tokamaks [7] and in plasma focus devices [8-10]. It has been used rather rarely in fibre z-pinch devices [11]. The difficulty lies in the measurement of very small Faraday rotation angles [11, 12]. To overcome this problem a high dynamic range detection system and a high polarisation extinction ratio optical system is used capable of detecting small intensity variations.

EXPERIMENT

The plasma produced in the MAGPIE generator is probed by a frequency doubled Nd-YAG laser (532 nm), compressed to 400 ps by a SBS (Stimulated Brillouin Scattering) technique [13, 14]. Carbon fibres of 33 μm diameter, 2.3 cm long, are used as a load. A new type of self referencing cyclic radial shear interferometer is used for the electron density measurements [15, 16]

A CCD camera is used as a detector for the polarograms. Rotation angles of the order of one degree are expected. Therefore a resolution of ~0.1 degree is required. The intensity change which has to be detected is of the order of 5% and consequently a high dynamic range camera is required. A 16-bit camera (65000 dynamic range) is then used to record the polarimetry images.

RESULTS

Figure 1(a) shows a polarogram taken with the analyser biased at q=85° clockwise (-85°) from the direction of the electric field which probes the plasma. It can be seen that the intensity is greater on one side of the plasma than on the other. This is due to the reversal of the direction of the magnetic field relative to the direction of the laser propagation.

(a)   (b)
-85°   +85°

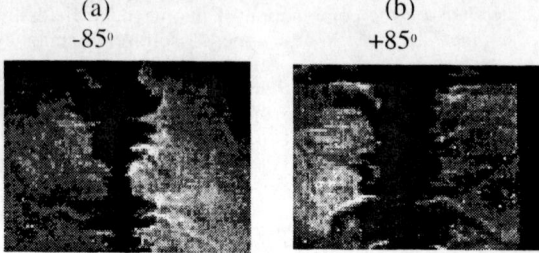

FIGURE 1. Polarograms of central region of 33 μm diameter carbon fibres recorded with the analyser at a) -85°, b) +85°.

Figure 1(b) shows a polarogram with the analyser biased at 85° counter-clockwise (+85°) and the more intense region has now swapped to the other side of the pinch. This is the critical test which confirms that Faraday rotation is being observed. Both images are taken at 57 ns after the current start.

Figure 2 shows a simultaneous polarogram and interferogram taken at 52 ns after the current start. From the polarogram, the Faraday rotation angle is evaluated. From the interferogram the electron density distribution required for the magnetic field reconstruction is found.

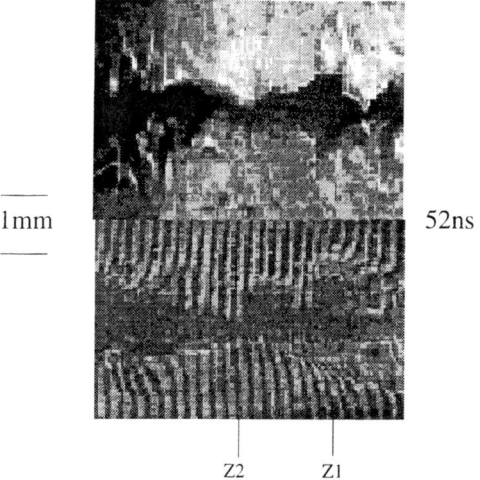

FIGURE 2. Simultaneous polarogram and interferogram 52 ns after the current start.

The black region seen on Faraday rotation and interferometry pictures is due to the absence of laser light arriving on the detectors. This region is called "the plasma core" and is formed due to refraction of laser light out of the acceptance angle of the optical system due to high density gradients existing in the plasma [17].

Abel inversion is used for the reconstruction of the $n_e(r)B(r)$ from the rotation angle distribution and the $n_e(r)$ from the phase shift distribution. Figure 3 shows the rotation angle, density, magnetic field and current distribution for the axial position Z1 shown in figure 3. It can be seen that the rotation angle observed is small (in the window from 0 to 1 degree). The maximum angle occurs at around 1 mm radius. The maximum current found is $\sim(110\pm 20)$ kA. The current measured by the current probe for this particular shot is 160 kA.

The measurement is accurate around radii where the rotation angle is maximum but large errors exist for radii where a small rotation angle is observed. The error in the calculation of the rotation angle is 0.25 degrees coming mainly from the laser beam ununiformity which is 20%.

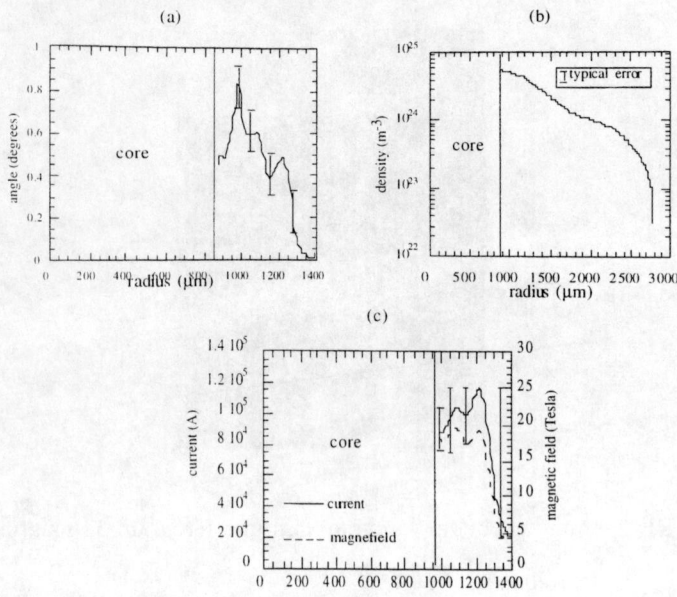

FIGURE 3. (a) Rotation angle, (b) density, (c) current and magnetic field distribution against radius of the pinch at 52 ns from the current start at Z1 axial position.

Direct magnetic field measurement in the neck (Z2) is not possible since the electron density distribution can not be extracted. Straight fringes are observed in this region. The fact that no fringe shift is observed implies that the electron density is below the sensitivity of the interferometer up to the core boundary where refraction occurs.

Alternatively, estimation of the magnetic field in the plasma boundary at the neck region can be performed. The radius of the plasma core in the neck is of the order of 200 µm. The maximum Faraday rotation angle for this position is about half of the one measured in the plasma bulges and is equal to 0.4 degrees. The electron density at 200 µm is estimated from the sensitivity of the interferometer to be 3.5×10^{24} m^{-3}. Therefore a value of ~150 Tesla for the magnetic field is found ($n_e \mathbf{B} \cdot \mathbf{dl} = 2.35 \times 10^{23}$ m^{-2}T for 1 degree rotation). This is about 7.5 times larger than the magnetic field measured in the plasma bulges. The current which corresponds to the above value is of the order 150 kA. Therefore, the current also flows in the neck.

CONCLUSIONS

Polarimetry is used for magnetic field measurements in MAGPIE generator.
It is shown that the current reaches a maximum value of ~110 kA at 52 ns corresponding to ~70% of the value measured with the electrical probe. The Faraday effect could not be observed at times later than 60 ns or earlier than 50 ns. This is due to the fact that this effect depends on the combination of magnetic field strength and electron density. Probably this combination is favourable to give rotation in the above window because the $n_e(r)B(r)$ product is appropriate. The small rotation angles measured in this experiment and the error of 0.25 degrees in the rotation angle make the current measurement accurate only in a small radial window from about 900 to 1250 µm where the rotation angle takes its maximum values.

ACKNOWLEDGEMENTS

The authors gratefully acknowledge the assistance of Dr. P. Lee and Dr. I. Mitchel during the experiment. Also Dr F.N. Beg, Dr. M. Coppins and Dr. J. P. Chittenden for the useful discussions.

REFERENCES

[1] I. H. Hutchinson, *Principles of Plasma diagnostics* (Cambrige University Press), 112 (1987).
[2] J. A. Stamper and B. H. Ripin, *Physical review letters* **34** 138 (1975).
[3] A. Raven, *University of Oxford* PhD thesis (1979).
[4] M. D. J. Burgess, B. Luther-Davis and K. A. Nugent, *Physics of Fluids* **28** (7), 2286 (1985).
[5] V. Y. Bychenkov, Y. S. Kas'yanov, G. S. Sarkisov and V. T. Tikhonchuk, *JETP Letters* **58** (3), 184 (1993).
[6] J. A. Stamper, *Laser and Particle Beams* **9** (4), 841 (1991).
[7] H. Soltwisch, *Controlled Fusion and Plasma Physics* Proc. 11th Europ. conf. Aachen: European Physical Society 123 (1983).
[8] R. E. Kirk, D. G. Muir, M. J. Forrest and N. J. Peacock, *Int. Conf. On Plasma Physics and Controled Fusion Research, Goteborg* 329 (1982).
[9] S. Czekaj, A. Kasperczuk, R. Miklaszewski, M. Paduch, T. Pisarczyk and Z. Wereszczynski, *Plasma Physics and Controled Fusion* **31** (4), 587 (1989).
[10] D. G. Muir, *PhD thesis, Royal Holloway College, University of London, UK* (1983).
[11] G. S. Sarkisov, A. S. Shikanov, B. Etlicher, S. Attelan, C. Rouille and V. V. Yan'kov, *JETP* **81** (4), 743 (1995).

[12] A. V. Branitskii, V. D. Vikharev, A. G. Kasimov, S. L. Nedoseev, A. A. Rupasov, G. S. Sarkisov, V. P. Smirnov, V. Y. Tsarfin and A. A. Shikanov, *Soviet Journal of Plasma Physics* **18** (9), 588 (1992).
[13] M. J. Danson and M. H. R. Hutchinson, *Optics letters* **8** 313 (1983).
[14] D. T. Don, *Optics letters* **5** 516 (1980).
[15] C. B. Edwards, M. Tatarakis and P. Lee, *OE/Laser 95 Laser beam control, diagnostics & Standards* (1995).
[16] F. N. Beg, A. R. Bell, A. E. Dangor, C. N. Danson, A. P. Fews, M. E. Glinsky, B. A. Hammel, P. Lee, P. A. Norreys and M. Tatarakis, *Physics of Plasmas* **4** (2), 447 (1996).
[17] M. Tatarakis, *PhD thesis, Imperial College, University of London, U.K* (1997).

Monochromatic X-ray Backlighting For Application To PBFA-Z

S.A. Pikuz*, T.A. Shelkovenko*, D.A. Hammer, and D.F. Acton

Laboratory of Plasma Studies, Cornell University, Ithaca, NY 14853

Monochromatic x-ray backlighting makes use of a spherically-bent mica crystal as a reflecting surface to enable high resolution imaging of dense plasmas in a single spectral line from a source plasma that is many orders of magnitude less bright than the object plasma. We present a sampling of experimental results obtained with this method using x-ray line emission ranging in energy from 1.8 keV to 7.8 keV. The 7.8 keV line from He-like Ni ions could be appropriate for backlighting z-pinch plasmas generated by PBFA-Z. The field of view is typically a few cm by a few mm.

INTRODUCTION

We describe a new approach to imaging dense plasmas, monochromatic x-ray backlighting. Shadow images of test objects, including exploding wire plasmas, have been obtained with high spatial resolution using a single spectral line from X-pinch plasmas [1] produced using several different wire materials. To accomplish this, we used a mica crystal that has been accurately bent to form a spherical reflecting surface as a lens to form the x-ray shadow image. The crystal provides high spectral selectivity to the focused radiation, enabling a few cm by a few mm field-of-view, while reducing the necessary backlighter source brightness by orders of magnitude relative to the object plasma. This method might be used to image the dense z-pinch plasmas generated by Saturn and PBFA-Z.

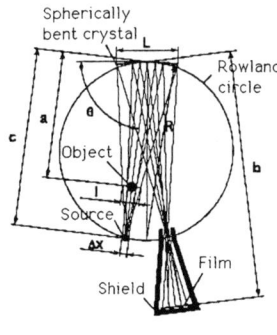

$1/a + 1/b = 1/f = 2/c$

Figure 1. Schematic diagram of the principles of monochromatic backlighting

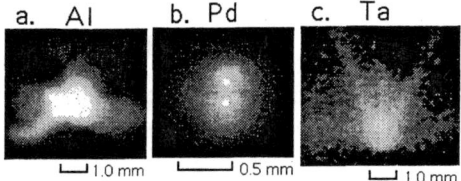

Figure 2. Images of two different x-pinch plasmas obtained with using an x-ray microscope, the first (a) in the 1s(2) - 1s3p line of He-like Al, and the second (b) in 4Å radiation from Ne-like Pd. A 2D image of a Ta-wire X-pinch in Lα line radiation is shown in (c).

The principle of monochromatic X-ray backlighting using an x-pinch as the x-ray source is shown schematically in Fig. 1. (See also Ref. 1, in these proceedings.) An x-pinch is produced by using two or more fine wires mounted so that they cross and touch (in the form of an X) as the load of a high current pulser. The cross point is a predictable location of intense soft x-ray emission, the size of which can be submillimeter, and the

---

* Permanent address: Lebedev Institute, Moscow, Russia

CP409, *Dense Z-Pinches*: Fourth International Conference
edited by N. R. Pereira, J. Davis, and P. E. Pulsifer
© 1997 The American Institute of Physics 1-56396-610-7/97/$10.00

emission spectrum of which is determined by the wire material and diameter, and the current of the pulser. A test object or dense plasma can be imaged in a single line from the x-pinch by using a mica crystal as a reflecting lens and choosing the experimental arrangement so that 2d sin θ = mλ, where d is the crystal lattice spacing, θ is the grazing angle (see Fig. 1), m is the order of reflection and λ is the x-ray wavelength. The spatial resolution of the image is determined by aberrations of the spherical crystal mirror, the diffractive properties of the crystal and the inherent detector resolution. Micron-scale resolution has been achieved. (The source size Δx determines the spectral range of the radiation involved in forming the image.) Spherically bent mica crystals with radii of curvature R of 100, 186, and 250 mm were used in the experiments discussed here. A detailed explanation of the monochromatic backlighting method and its limitations is presented elsewhere [2].

Because the method described here has spectral and spatial filtering of the radiation from the object-plasma [2], and the backlighter may be very small in size (0.1-0.01 mm) compared to bright object plasmas (up to ~ 1cm in high power z-pinch devices), the power required from the backlighter plasma can be orders of magnitude smaller than the power radiated by the object plasma by using a spectral line from the backlighter that is not present in the spectrum of the object-plasma.

EXPERIMENTAL RESULTS

Experiments were performed using the XP-pulser at Cornell University (350 kA peak current in a 100 ns pulse), and the BIN generator at Lebedev Institute, Moscow, (250 kA in a 100 ns pulse). Different metal grids, wires and glass fibers were used as test objects. The radiation from different types of wires, from Al to W, was tested, covering a range of x-ray energies from 1 to 10 keV. An X-ray microscope [3,4] and imaging spectrographs [4,5] were used to monitor the line intensity as well as the size and shape of the backlighter source.

The choice of spectral line for a particular backlighting job depends on the Bragg angle θ, which, for practical reasons, must be between 70 and 88 degrees. Some of the spectral lines we have investigated which satisfy this geometric constraint are presented in the Table 1. In the following paragraphs, we summarize results on each of these lines.

TABLE 1.

Material	Line	λ(Å)	θ	Refl. order (m)
Al	$1s^2$ - $1s3p$ of He-like Al	6.6343	84.9	III
Pd	satellite line of Ne-like Pd	3.990	86.9	V
Ti	$1s$ - $2p$ line of H-like Ti	2.490	85.6	VIII
Ta	$L\alpha$	1.532	85.4	XIII
Ni	$1s^2$ - $1s3p$ line of He-like Ni	1.588	72.5	XII

The $1s^2$ - $1s3p$ line of He-like Al (1.87 keV) is very intense but it is too low energy and too large, more than 1 mm (Fig. 2a), to be used for backlighting Saturn/PBFA-Z pinches. The Pd (3.1 keV) source is small enough (Fig. 2b), but the intensity of this satellite line is barely larger than the continuum radiation, and its energy is marginally adequate for the backlighting task of interest here. The $1s^2$ - $1s2p$ line of H-like Ti (4.98 keV) has a small source size, 30-40μm, but the intensity is not high enough to use as a backlighter. The $L\alpha$ line of Ta (8.11 keV) radiated with high intensity, but the source size is too large, more than 2mm (Fig. 2c). The $1s^2$ - $1s2p$ line of He-like Ni requires a relatively large angle from the normal (17.5°), but backlighting results were very good. The image of the source in this line was obtained using a spherically bent mica crystal with radius 186 mm in a focusing spectrograph with 2 dimensional spatial resolution [3,4]. The principal source point in each of two pulses was under 100μm in radial extent, as shown in Fig. 3.

The best spatial resolution we achieved with any of the backlighter sources, 4μm, was obtained using a mica crystal with R = 100 mm and an Al wire X-pinch [2]. The results of several of the test experiments are shown in Fig. 4.

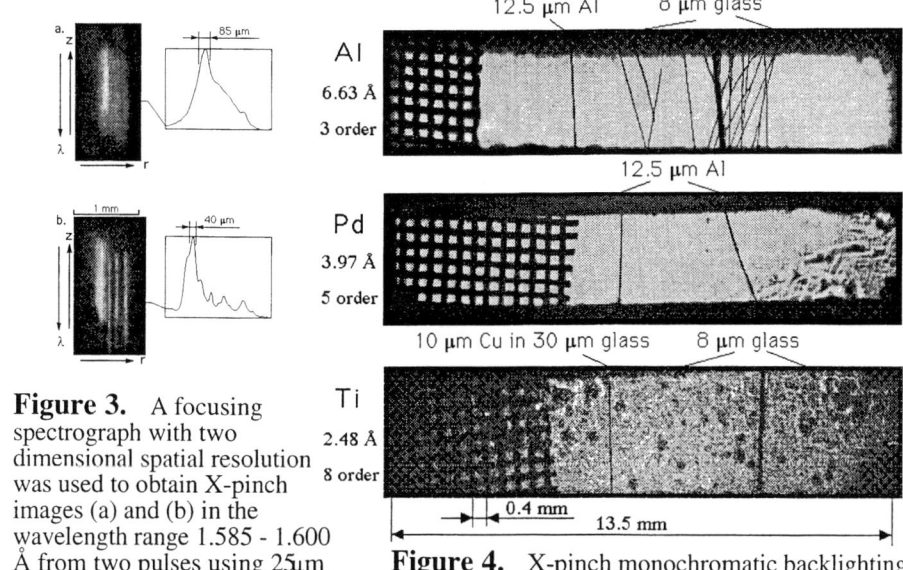

Figure 3. A focusing spectrograph with two dimensional spatial resolution was used to obtain X-pinch images (a) and (b) in the wavelength range 1.585 - 1.600 Å from two pulses using 25μm diameter Ni chrome wires.

Figure 4. X-pinch monochromatic backlighting tests with different wire materials.

Backlighter experiments performed with the $1s^2$ - $1s2p$ line of He-like Ni (1.588 Å; 7.8 keV, m=12) could be done with an 800μm Be filter placed in front of the film. Using a crystal with R = 186 mm, shadow images of test objects, a metal grid and a Cu wire in a glass fiber sheath were obtained (see Fig. 5). In the shadow image we can see the Cu wire inside of the glass envelope, but the glass is practically transparent to 7.8 keV radiation. Spatial resolution in this case is about 10μm.

As a demonstration of using the diagnostic method describe here to image a dense plasma we tested an X-pinch and a z-pinch in parallel in the main pulser output circuit. The X-pinch backlighter source was the $1s^2$ - $1s2p$ line of He-like Al ions from crossed 37.5 μm Al wires. The z-pinch, made from different wires in different pulses, served as the object-plasma. One supplementary wire was not connected from anode to cathode and was used as a test object and for focus control. Shadow images of an exploding 75 μm diam. Mg wire (Fig. 6a) and a 25μm W wire (Fig. 6b) were obtained. In these experiments, the current passing through the Z-pinch at the moment of the image is ~100 kA. The current passing through the X-pinch, was, therefore, about 200 kA. The images of the disconnected wires show that plasma blew off their surfaces also, presumably a result of electron emission from the wire ends. The image of the W wire shows a plasma column of about 1 mm diameter with a smaller diameter dense core inside the plasma column (see Fig. 6b).

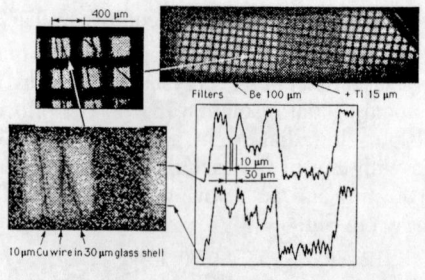
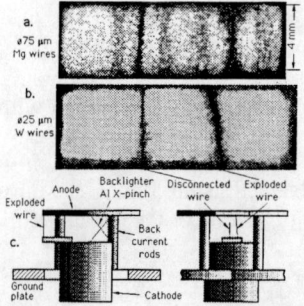

Figure 5. Backlighter images of a grid and 10μm Cu wire in a 30μm diameter glass sleeve obtained with the ~ 1s(2) - 1s2p line of He-like Ni. The blown-up portion of the image is used to demonstrate the spatial resolution.

Figure 6. Backlighter images of Mg (a) and W (b) wires. Two views of the set-up are shown in part (c). Al-wire X-pinches were used as the backlighter.

The shadow image of the exploded 75μm Mg wire shows faint images of 500μm thick shells, presumably material evaporated from the wire surface, with diameters of 3 and 4 mm for the disconnected wire and the z-pinch wire, respectively. In the center part of the z-pinch a neck was forming.

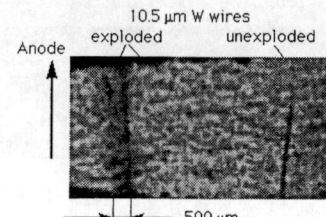

Figure 7. Backlighting image of two W wires obtained with the wires configured in parallel with the Pd X-pinch backlighter source.

Using this diagnostic method and varying the current per wire it is possible to study different stages of the explosion of a wire or wire array. In Fig. 7, a backlighter image of two W wires exploded in parallel with the Al X-pinch backlighter source is shown.

REFERENCES

1. Pikuz, S.A., Shelkovenko, T.A., Romanova, V.M., et al., "Studies of X-pinch plasma fine structure using high resolution optical and imaging spectroscopy methods", to be published in these proceedings.
2. Pikuz, S.A., Shelkovenko, T.A., Romanova, V.M., et al., Rev. Sci. Instr. **68** 740 (1997).
3. Pikuz, S.A., Brunetkin, B.A., Ivanenkov, G.V., et al., JQSRT **51** 291-302 (1994).
4. Boiko, V.A., Vingogradov, A.V., Pikuz, S.A. and Skobelev, I. Yu., J. Sov. Laser Res. **6** 85 (1985).
5. Skobelev, I. Yu., Pikuz, S.A., and Faenov, A. Ya, et al., JETP **81** 692-718 (1995).

Diagnostic of energetic electrons in dense z-pinch plasmas

A.S. Shlyaptseva and R.C. Mancini

*Department of Physics,
University of Nevada, Reno, NV 89557*

Abstract. We discuss the diagnostic of energetic electron beams in z-pinch plasmas using x-ray line polarization spectroscopy. Our previous work in this area has been related to the study of polarization of dielectronic satellite lines of Li-and Be-like Mg and Fe ions in low-density plasmas. Here we extend our work to the case of z-pinch plasmas. We calculate the polarization properties of dielectronic satellite lines of Be-like Ne, Ar and Fe ions. These results can be used to diagnose the energy and directionality of energetic electron beams. This work is motivated by the development of a new Dense Z-pinch Program at University of Nevada, Reno. (see B.S. Bauer et al this conference). However, the results can also be applied to other z-pinch devices.

The presence of electron beams in non-equilibrium plasmas can lead to selective population of M-sublevels and, hence, to the emission of partially polarized line radiation. In this paper we extend our study of polarization properties of X-ray dielectronic satellite spectra driven by electron beams [1] to the case of plasmas characteristic of z-pinch experiments. In z-pinch plasmas densities are high enough and the population of low-lying excited states is not so negligible as compared to that of the ground state. Now electron capture originating from ground and low-lying excited states has to be considered. To our knowledge, the first observation of electron beams in gas-puff z-pinch plasmas was reported in Ref. [2]. Theoretical results for the degree of polarization of He-like Fe lines [3] were used in z-pinch experiments of Ref. [4] to infer the presence of energetic electron beams in the plasma. In our approach we consider the general case of complex spectra in multielectron radiators and calculate polarization-dependent spectra associated with different polarization states. Experimentally, these polarization-dependent spectra can be simultaneously recorded using two X-ray spectrometers [1,3].

We calculate the intensity distribution of dielectronic satellite lines (DS) associated with different polarization states, parallel (PL) and perpendicular (PP) to the electron beam axis, and also total (TOT) intensity. The total intensity of DS was computed and then multiplied by a polarization-dependent factor, which is line and polarization-state dependent. All necessary atomic data were computed using MZ program (see, for example,[5]). This code is based on a perturbation theory expansion over $1/Z$, where Z is the nuclear charge. This is very convenient for isoelectronic sequence calculations. We used LSJ quantum numbers to describe target ion states. Relativistic corrections were taken into account. Each atomic complex with a given J included all LS energy levels with different mixing coefficients C^J_{LS} (if there is no mixing $C^J_{LS}=1$) of all configurations with n=2. For low Z ion, such as Ne, only one case was observed with sufficient mixing (C^J_{LS} <0.6) for $1s2s2p^2$ ^3P and ^3D (J=1). As Z increases, for Fe ion, several cases of considerable mixing are observed for J=1,2.

To obtain the polarization-dependent factor we calculate the degree of polarization using the photon density matrix formalism. We consider electron capture (EC) as the main process for the population of the autoionizing levels of Be-like ions. In low-density plasmas (LDP) EC originates only from Li-like ground level $1s^22s$ which has a zero orbital momentum. Three channels of EC are $1s^22sks$, $1s^22skp$ and $1s^22skd$. The most intense satellite emission is produced via transitions $1s2s2p^2 \rightarrow 1s^22s2p$. Thus only two channels $1s^22sks$ and $1s^22skd$ need to be considered. The emission arising from autoionizing states formed through $1s^22sks$ channel is unpolarized. As a result only one channel $1s^22skd$ contributes to non-diagonal elements of photon density matrix. This simplifies considerably the expression for the degree of polarization. In higher density plasmas (HDP) two Li-like low-lying states $1s^22s$ and $1s^22p$ are populated. All together six channels of EC exist, namely $1s^22sks$, $1s^22skp$, $1s^22skd$, $1s^22pks$, $1s^22pkp$ and $1s^22pkd$. Moreover, in HDP besides the transitions prominent in LDP case, other groups of transitions such as $1s2p^3 \rightarrow 1s^22p^2$ and $1s2s^22p \rightarrow 1s^22s^2$ become intense. Now the elements of photon density matrix are dependent on the value of the amplitude of EC for each of the channels. These amplitudes were calculated using the perturbation theory method [5]. Finally, the polarization properties of these three groups of transitions were calculated for Be-like Ne, Ar and Fe ions including 26, 28 and 29 transitions, respectively.

Theoretical polarization-dependent spectra are displayed in Figs.1 a,b,c (Ne), Figs.1 d,e,f (Ar) and Figs.2 a,b,c,d (Fe). Each spectrum consists of three traces: PL, PP and TOT. The spectra were calculated at different energies of the electron beam. The electron beam energy distribution was assumed to be Gaussian with a FWHM=20 eV (for Ne), 35 eV (for Ar) and 50 eV (for Fe). The values of the electron beam energy, selected from autoionization energies of Be-like transitions, were in the range of 680-720 eV (for Ne), 2220-2290 eV (for Ar) and 4650-4800 eV (for Fe). From these figures it follows that, first, each trace PL, PP and TOT depends on the ion, electron beam energy and polarization state. Second, in general PL depends on the energy of the electron beam in a way similar to that of TOT but different than PP. Third, unpolarized peaks (PL and PP are identical) can be used for cross normalization in the same way as single unpolarized lines.

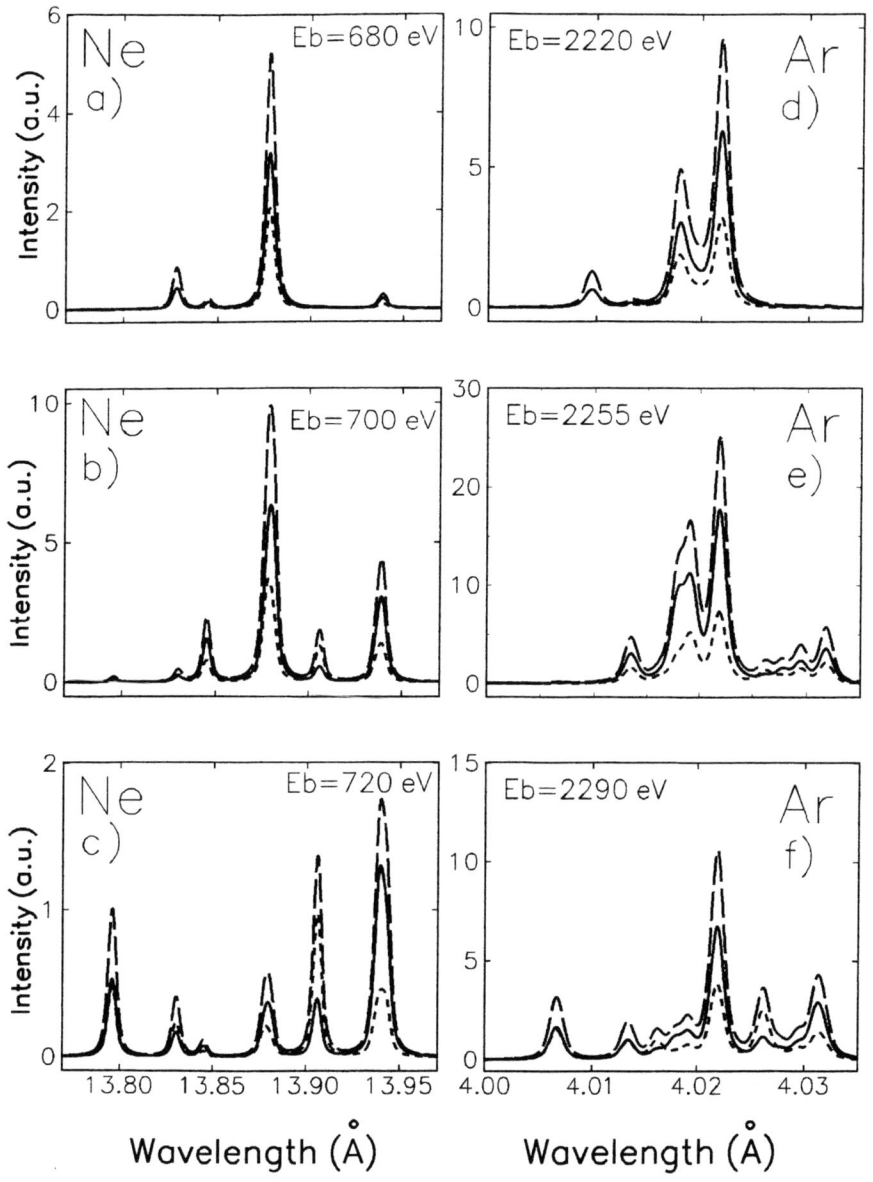

FIGURE 1. Polarization-dependent spectra of dielectronic satellites of Be-like Ne and Ar ions, calculated for different energies of the electron beam and for different polarization states: (———) parallel, (− − − − −) perpendicular, and (— — —) total intensity.

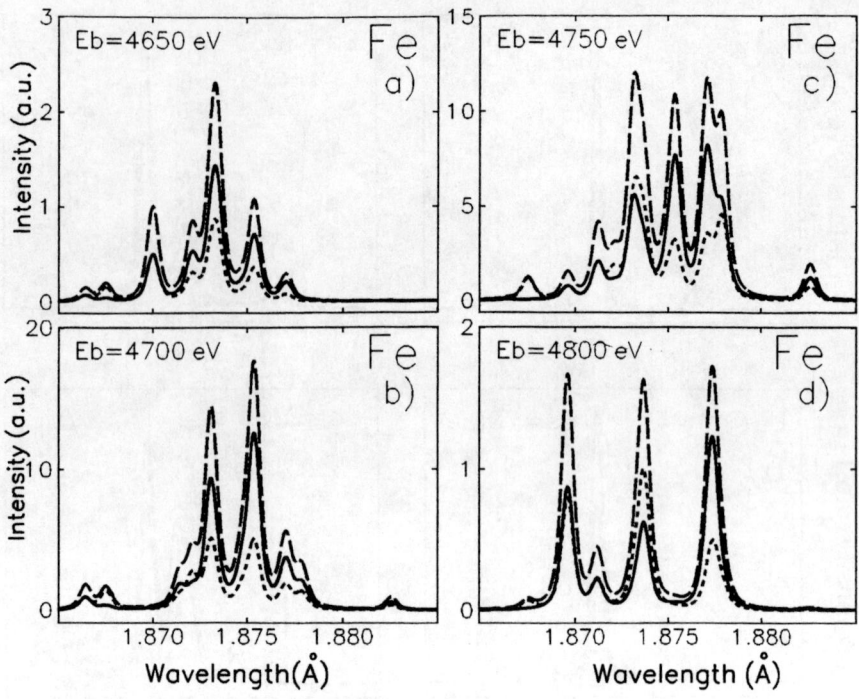

FIGURE 2. Polarization-dependent spectra of dielectronic satellites of Be-like Fe, calculated for different energies of the electron beam and for different polarization states: (———) parallel, (- - - - -) perpendicular, and (— — —) total intensity.

In conclusion, polarization-dependent spectra associated with polarization states parallel and perpendicular to the electron beam have been calculated which can be used to diagnose the energy and directionality of electron beams in z-pinch plasmas. In addition, the use of different target elements is useful to cover a broad range of electron beam energies. This work was supported by NSF grant OSR-9353227 and LLNL contract B336460.

REFERENCES

1. A.S. Shlyaptseva, R.C. Mancini, P. Neill, P. Beiersdorfer,
   Rev. Sci. Instrum. **68**, 1095 (1997).
2. D.R. Kania, L.A. Jones, *Phys. Rev. Lett.* **53**, 166 (1985).
3. A.S. Shlyaptseva, A.M. Urnov, A.V. Vinogradov,
   On diagnostics of suprathermal electrons in high-temperature plasma,
   P.N.Lebedev Physical Institute Report N **193**, Moscow (1981).
4. V.A. Veretennikov, A.E. Gurei, A.N. Dolgov, V.V. Korneev, O.G. Semenov,
   JETP Lett. **47**, 35 (1988).
5. U.I. Safronova, A.S. Shlyaptseva, *Physica Scripta* **54**, 254 (1996).

STABILITY

A REVIEW OF THE STABILITY OF THE Z-PINCH

M. Coppins

Imperial College, Prince Consort Road, London SW7 2BZ, England

Abstract. The poor reputation of the z-pinch with regard to stability is based on ideal MHD linear theory. The conclusions and shortcomings of this historically important theory are reviewed, and an outline of more recent work is presented.

1. HISTORICAL PERSPECTIVE

1.1 Introduction

The magnetic pinching force which underlies all the applications of the z-pinch is also the source of a range of instabilities which render these applications hard to achieve. For this reason stability has been a dominant concern throughout the fifty or so years of detailed research into the device [1]. In this review we try to assess just how much progress has been made in the theoretical understanding of magneto-hydrodynamic (MHD) instabilities in equilibrium z-pinches. These instabilities were the scourge of the pioneering experiments in the fusion programme, and they have continued to plague z-pinches ever since.

The term "equilibrium z-pinch", used above, means one in which the inward magnetic force is balanced by the force due to the plasma thermal pressure. Such a configuration represents the simplest magnetic confinement device. It is the basis not only of all proposed z-pinch fusion schemes but also of various experiments designed to explore aspects of basic plasma physics (e.g., radiative collapse [2]). Experimental examples of equilibrium z-pinches are fibre, gas-embedded, and compressional pinches.

Equilibrium pinches can be contrasted with "dynamic z-pinches". These are usually hollow (at least initially) and the magnetic force drives a rapid implosion which is the basis of the device's technological utility, usually as an intense radiation source. Examples of this type are wire array, gas puff, and composite pinches.

By restricting our attention to MHD instabilities in equilibrium pinches we exclude two important subjects which fall into the category of z-pinch instabilities. The first is microinstabilities. These probably play a significant role in determining the plasma profiles, but, unlike MHD instabilities, they do not lead to a catastrophic loss of confinement. Secondly, we will not discuss the Rayleigh Taylor instability which, of course, is important in dynamic pinches.

In the remainder of Part 1 we consider the following three questions: firstly, what claims does orthodox stability theory make about the z-pinch? secondly, are these claims born out by experiment? and, thirdly, is the orthodox theory actually relevant? As we shall see, the answer to the last of these questions is "hardly ever", and this provides the point of departure for a journey through the wilder regions of z-pinch stability theory in the following parts of the paper. Throughout the journey we will seek an answer to the more basic question: can we make an equilibrium z-pinch stable?

1.2 The Orthodox Theoretical View

Every schoolchild (or at least those who have studied fusion plasma physics) knows that the z-pinch is fatally susceptible to highly destructive instabilities. This was the verdict of the early experiments [3–5], and it was endorsed by the early theory [6,7] using the linearized ideal MHD equations.[1] The z-pinch's poor reputation for stability rests firmly on this early theory, as does all subsequent theoretical work. It is therefore worth considering in some detail just what it actually amounts to.

It is important to recognise at the outset that although it is historically important, conventional ideal MHD linear theory is strictly relevant only in very special situations which almost never occur. Ideal MHD itself has severe shortcomings as a plasma model. These are discussed below (Sec. 1.4). The use of linearized equations restricts us to questions of stability thresholds or instabilities of very small amplitude. Furthermore, the theory assumes that the unperturbed plasma is a perfectly cylindrical, time independent, exact equilibrium, which is stationary everywhere and sufficiently long that end effects can be neglected.

Dubious though these various assumptions are, they simplify the mathematics very considerably by allowing us to consider independent normal modes in which all perturbed variables are proportional to $e^{\gamma t}e^{i(m\theta+kz)}$ (γ is the growth rate, m the azimuthal mode number, and k the axial wavenumber). The stability problem reduces to a straightforward eigenvalue equation which can be solved for both γ and the radial structure of the perturbed variables.

---

[1] Hence the use of the term "MHD instability" to denote the violent, confinement degrading instabilities of interest here, even though ideal MHD is seldom an appropriate plasma model

There is an extensive literature on the ideal MHD linear stability of the z-pinch [6-9], the main conclusions of which are:

1. The most significant instabilities are $m = 0$ (sausage) and $m = 1$ (kink).

2. The growth rate is of the order of v_A/a or v_T/a. (v_A is the average equilibrium Alfven velocity, v_T is the average equilibrium thermal velocity, and a is the equilibrium radius. Because the z-pinch has a plasma β of order unity, $v_A \approx v_T$).

3. The growth rate increases monotonically with k, and the fastest growth always corresponds to $k \to \infty$ (i.e., zero wavelength).

4. The $m = 0$ mode can be stabilized by a suitable choice of equilibrium profiles, but $m = 1$ cannot.

5. The properties of the $m = 0$ mode are strongly model dependent.

Looking in a little more detail at some of these conclusions we first note that with regard to the choice of m (conclusion 1) we cannot be more specific in general. Whether the $m = 0$ or $m = 1$ mode has the highest growth rate depends critically on the exact form of the equilibrium profiles, and on the wavelength of interest. For instance, for an equilibrium in which both the current density and temperature are uniform (the so called parabolic equilibrium) $m = 1$ is dominant at low k and $m = 0$ at high k.

The whole issue is made more confusing by conclusion 3, which implies that the mechanism determining the dominant wavelength is inherently non-linear. Thus linear theory is incapable of making any predictions about the wavelengths of instabilities appearing in experiments.

The fourth and fifth conclusions were discovered by Kadomtsev [8] who obtained $m = 0$ and $m = 1$ stability criteria from the ideal MHD energy principle [10]. He showed that in the limit of $k \to \infty$ absolute stability to the $m = 1$ mode requires a singular equilibrium current density at $r = 0$. Thus any physically realistic z-pinch will be $m = 1$ unstable. However the corresponding result for $m = 0$ is less bleak. It is enshrined in the well known Kadomtsev criterion, which states that absolute stability to the ideal MHD $m = 0$ mode for all k can be achieved if the equilibrium pressure $P(r)$ and magnetic field $B(r)$ satisfy the following inequality everywhere:

$$-\frac{r}{P}\frac{dP}{dr} < \frac{2\Gamma}{(1+\Gamma\beta/2)}$$

where $\beta(r) = 2\mu_0 P(r)/B^2(r)$ is the local plasma beta and Γ is the ratio of specific heats in the adiabatic law.

To satisfy this condition the equilibrium current density and pressure must be strongly peaked on axis and the pressure must be finite at the plasma edge. The latter requirement implies that, strictly speaking, a z-pinch surrounded by a vacuum cannot be made stable to the $m = 0$ mode [11]. However, a z-pinch in which the plasma extends out to a rigid wall can, in principle, be made $m = 0$ stable. There is evidence from 2-D simulations [12] that in such a situation a Kadomtsev stable equilibrium can arise spontaneously. Of course, wall stabilization is not viable for most z-pinches.

Instead of allowing the plasma to extend to a wall we could envisage an $m = 0$ stable pinch surrounded by a neutral gas which could support a small pressure at the plasma edge. We return to this point in Sec. 1.3, below.

The final conclusion in the above list, namely the sensitivity of the $m = 0$ instability properties to the details of the model is readily apparent from the fact that Γ appears in the Kadomtsev criterion (the $m = 1$ condition is independent of Γ). Γ comes into the ideal MHD equations through the adiabatic law. Unlike the ideal MHD equation of motion, which, apart from the assumptions of quasi-neutrality and pressure isotropy, is an exact moment of the basic kinetic equation, the adiabatic law represents a crass approximation, in this case to the exact energy equation. Indeed, the assumptions required to justify the use of this equation together with the ideal MHD Ohm's law are the main reasons for the model's limited validity (Sec. 1.4, below).

Obviously we want to use ideal MHD to study experiments. To be able to do so meaningfully, however, the ideal MHD results must be relatively independent of the details of the inaccurate energy equation it uses. In most situations this requirement is satisfied. The fact that it is not for the $m = 0$ mode in the z-pinch renders the validity of theoretical predictions concerning this particular instability highly questionable.

1.3 The Experimental View

Let us suppose that we were familiar with the orthodox z-pinch stability theory, outlined above, but had no knowledge of any experimental results. What would we expect experiments to show?

Just about the only predictions we could make with confidence would be, firstly, that the z-pinch would be unstable with a growth rate of the order of v_A/a, and, secondly, that the observed instability would probably be either $m = 0$ or $m = 1$. We would be unable to say anything about the wavelength of the instability. Nor would we be able to decide the relative likelihood of the $m = 0$ or $m = 1$ modes, although we might guess that $m = 1$ would occur more frequently because, unlike $m = 0$, it cannot be stabilized. Whatever we might expect on this score, however, we would fervently hope that the $m = 0$ mode would not appear very often, because if it does we are going to need a greatly improved theory.

A quick look at a few experimental papers on z-pinches (e.g., [5,13]) would suffice to dash this hope. The z-pinch nearly always manifests the $m = 0$ instability. The only exceptions are gas embedded pinches [14,15] in which the $m = 1$ mode is regularly seen. This almost universal preference for $m = 0$ on the part of experiments is most unfortunate from a theoretical point of view, not only because it is inexplicable, but also because ideal MHD provides a poor treatment of this instability.

Ideal MHD theory can perhaps provide some qualitative insight into the gas embedded pinch's predilection for the $m = 1$ instability. Unlike pinches in vacuo, devices of this type can satisfy the Kadomtsev criterion for ideal stability to the $m = 0$ mode. Not only do they allow a finite pressure at the plasma edge, but also, and perhaps more significantly, they have an in-built mechanism for setting up the necessary equilibrium profiles. This mechanism is provided by the thermal losses to the surrounding gas, which ensure that the temperature is peaked on axis. The current will tend to flow in the high temperature, low resistivity inner region, resulting in an axially peaked current density profile, as required by the Kadomtsev criterion. Thus gas embedded pinches may spontaneously generate $m = 0$ stable equilibria. If so, it would explain why these devices buck the trend with regard to the choice of instability.

Both orthodox theory and experiment conclude that the z-pinch is highly unstable. At first sight this appears to be a triumphant vindication for the theory. But, in fact, linear ideal MHD stability theory has not been particularly successful when applied to the z-pinch. Its inability to account for many of the most conspicuous features of experiments, notably the predominance of the $m = 0$ instability, prompts one to ask if, in fact, it is an appropriate plasma model. To answer this question we need to consider the ideal MHD validity conditions in detail.

1.4 Applicability of Ideal MHD

Ideal MHD treats the plasma as a perfectly conducting single fluid which behaves adiabatically. For any given plasma to be describable by this model, the plasma must fulfil various conditions, which are most conveniently expressed in terms of certain dimensionless parameters (for a full account of the assumptions as well as the mathematical steps needed to derive the ideal MHD equations see Ref. [16]). The most important of these dimensionless parameters are: (i) S (Lundquist number) $= \mu_0 v_A a / \eta$, (ii) $\tau_i / \tau_m = v_A \tau_i / a$, and (iii) $\epsilon = a_i / a$, where η is the average plasma resistivity, τ_i is the ion-ion collision time, τ_m is the characteristic MHD time, and a_i is the average ion Larmor radius. To be valid ideal MHD requires that the plasma satisfies the following conditions: (a) low resistivity, $S > 1 - 100$, (b) high collisionality, $\tau_i / \tau_m < 1$, and (c) small Larmor radius, $\epsilon \ll 1$.

For the special case of an equilibrium z-pinch it turns out that these three dimensionless parameters (as well as many others) are functions of just two experimental parameters, namely the line density, N (i.e., the number of ions per unit length) and $I^4 a$ (I is the plasma current) [17]. The relevant expressions are:

$$S = \frac{1.54 \times 10^{24}}{A^{1/2} Z (1+Z)^{3/2}} \frac{I^4 a}{N^2}$$

$$\frac{\tau_i}{\tau_m} = \frac{5.85 \times 10^{39}}{Z^4 (1+Z)^{3/2}} \frac{I^4 a}{N^3}$$

$$\epsilon = \frac{8.08 \times 10^8 A^{1/2}}{Z(1+Z)^{1/2}} N^{-1/2}$$

where A is the ion mass number and Z is the degree of ionization. We note that for a given substance, ϵ depends only on N. Thus it is easy to set the value of ϵ in an experiment. For instance, in a pinch formed from a cryogenic hydrogen fibre $\epsilon \approx 2.0/r_f$, where r_f is the fibre radius in microns.

Figure 1 shows contours corresponding to $S = 1$, $S = 100$, $\tau_i/\tau_m = 1$, $\epsilon = 0.1$ and $\epsilon = 1$, in the $(N, I^4 a)$ plane for the case of hydrogen. The region in which ideal MHD is valid is shown shaded. We see that it occupies a fairly small part of the experimentally relevant parameter space.

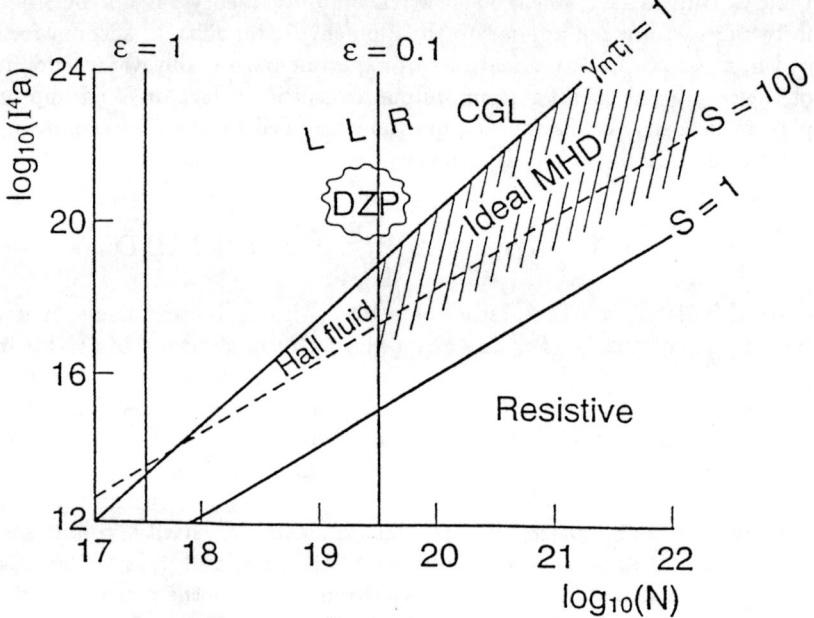

FIGURE 1. The regimes of z-pinch theory

Various other regions are identified on the figure. These are: *Resistive* ($S < 1$), *Hall fluid* (ideal, collisional, but large Larmor radius), *CGL* (collisionless, small Larmor radius, $m = 0$ only), and LLR (collisionless, large Larmor radius). Finally, the intended operating regime of the Imperial College DZP experiment [13] is also shown.

There is no *a priori* reason to suppose that results from ideal MHD theory are at all relevant outside the shaded region. This provides the motivation for the second part of the paper, in which the results of various non-ideal linear stability theories are reviewed. In Part 3 we consider two other aspects of the stability problem which are neglected in the orthodox theory, namely equilibrium flow and the non-linear development of the instability. Finally, the main conclusions are summarized in Part 4.

2. NON-IDEAL LINEAR THEORY

2.1 Resistive

Z-pinch experiments usually employ rising currents. As the current, and temperature, increase the plasma becomes less and less resistive (i.e., S increases also).

Conventional ideal MHD linear theory is valid when $S \gg 1$. In this case not only is resistivity negligible, but also τ_m (the characteristic MHD time, $\sim \gamma^{-1}$) is much less than the equilibrium ohmic heating time. The latter is essentially the characteristic timescale over which the equilibrium is changing. Thus when $S \gg 1$ the usual assumption of a time independent equilibrium is justified.

However, the ideal phase is preceded in the current rise by a resistive phase in which S is small. In this case not only must resistivity be retained in the model, but the assumption of a time independent equilibrium cannot be made. In studying linear instabilities in this phase we can still consider perturbed variables $\propto e^{i(m\theta+kz)}$, but instability growth will not be exponential and may not even be separable in form.

The linear stability of the z-pinch at low S has been studied using an initial value approach [18,19]. Even though the results display convergence to the expected exponential growth at high S, early in the current rise the behaviour is very different. The instability amplitude sometimes rises and sometimes falls, and the radial structure of perturbed variables is not preserved.

The form of the solution can be characterised by a critical Lundquist number S^* which, broadly speaking, measures the duration of the resistive phase. The value of S^* depends on ka: for $ka \sim 1$, $S^* \sim 10$, for $ka \sim 0.1$, $S^* \sim 100$ (this is why the ideal MHD low resistivity condition is quoted as $S > 1 - 100$, in Sec. 1.4, above). The results are in good agreement with 2-D MHD simulations [20].

Broadly speaking, the solution has the appearance of a delayed onset of the instability, and it was originally concluded [18] that the behaviour could be crudely approximated as absolute stability during the resistive phase. However, it is now believed that this phase is distinguished by the predominance of a new instability which essentially grows linearly with time (its radial structure also undergoes evolution). This resistive instability is driven by ohmic heating and is out of phase with the subsequently dominant ideal mode. The interaction between these two instabilities, resistive and ideal, produces the appearance of a delayed onset of the latter.

Further details of this work can be found elsewhere in these proceedings [19].

2.2 Hall Fluid

For a wide range of line densities used in experiments the small Larmor radius condition of ideal MHD, $\epsilon \ll 1$, is not satisfied. In these cases there are several important aspects of the physics missing from ideal MHD. If the plasma is collision dominated the Hall fluid model is appropriate. This is ideal MHD with two additional terms in Ohm's law, namely the Hall term ($\vec{j} \times \vec{B}/ne$) and the electron pressure gradient. Both scale with ϵ, and thus represent large Larmor radius (LLR) corrections.

Most of the work on Hall fluid stability of the z-pinch [21–24] actually uses the cold electron form of the model which includes only the Hall term (although a study of the $m = 0$ mode using the full Hall fluid model is in progress [25]). In this case the magnetic field is frozen in to the electron fluid. Since the electrons carry the equilibrium current this introduces an asymmetry in the axial direction, not present in the ideal MHD case. A further difference from ideal MHD is provided by the fact that the Hall fluid eigenvalue (i.e., γ) is complex. Thus Hall fluid linear instabilities do not merely grow exponentially but also exhibit the characteristics of a propagating wave.

Notwithstanding these differences the Hall fluid linear stability properties are broadly in line with ideal MHD. The growth rate is of the order of v_A/a, and the $m = 0$ behaviour is strongly equilibrium dependent. However, the Hall fluid eigenvalue spectrum undergoes extraordinary changes as ϵ is increased [25]. Roots which were stable in the ideal MHD limit ($\epsilon \to 0$) can be destabilized and take over as the fastest growing mode. For this reason the overall effect of increasing ϵ in the Hall fluid regime (i.e., for a collision dominated plasma) is not beneficial.

2.3 Chew-Goldberger-Low

The Chew-Goldberger-Low (CGL) model [26] is applicable to a collisionless, zero Larmor radius plasma. It employs an energy equation which assumes that the parallel heat flow is negligible. Since the plasma is assumed to be collisionless this requirement usually renders the model untenable, but there are two situations in z-pinch physics in which it is highly useful.

The first concerns equilibria. The CGL equations, unlike ideal MHD, include pressure anisotropy, and the CGL force balance equation is applicable to any collisionless equilibrium regardless of ϵ. Thus we can use the model to construct anisotropic equilibria [27]. Stability calculations based on such equilibria would indicate if significant stabilization could be achieved by a suitable choice of anisotropy. Of course, even if it were found that the z-pinch could be stabilized in this way the practical problem of achieving the required degree of anisotropy might prove insurmountable. We return to this matter below.

The second triumph of CGL in the field of z-pinch theory concerns the $m = 0$ mode. It turns out that the model correctly describes this instability in a collisionless, zero Larmor radius z-pinch [28]. Physically this is because the $m = 0$ mode in the z-pinch has no parallel gradients, and therefore the parallel heat flow is zero, as required by the model. Usually CGL is despised for its absurd energy equation. The fact that it correctly describes this experimentally important class of instability (albeit strictly only in the zero Larmor radius limit) is one of the pleasing curiosities of plasma theory.

The CGL model can be applied to the linear $m = 0$ mode of a z-pinch in which the equilibrium pressure is isotropic. In this case CGL still allows the perturbed pressure to be anisotropic. This would be the case if the plasma was collisionless over the instability timescale, but the equilibrium had been set up earlier in the current rise when the plasma was more collision dominated. Applying CGL to isotropic equilibria allows us to make comparisons with ideal MHD. We find that CGL is more optimistic: for any isotropic equilibrium CGL always predicts a lower growth rate than ideal MHD (although the degree of disparity depends critically on the exact equilibrium considered). It is also more optimistic with regard to stability thresholds. These can be investigated using the energy principle to obtain a CGL version of the Kadomtsev criterion [29]. This states that absolute stability to the CGL $m = 0$ mode for all k can be achieved if the equilibrium pressure and magnetic field satisfy the following inequality everywhere:

$$-\frac{r}{P}\frac{dP}{dr} < \frac{5\beta + 14}{4(1+\beta)}.$$

Unlike the ideal MHD condition (Sec. 1.2, above), this is always satisfied close to $r = 0$ for all physically realistic equilibria. Unfortunately, both conditions require a finite pressure at the plasma edge. Thus the conclusion that a z-pinch in vacuo cannot be made absolutely stable to the $m = 0$ mode, which was obtained from ideal MHD applies equally in the CGL case [11]. However, in situations in which the pinch is not surrounded by a vacuum (e.g., if the plasma extends out to a rigid wall) there is a wider range of CGL $m = 0$ stable equilibria. The fact that CGL and ideal MHD impose such different stability criteria is a further manifestation of the strongly model dependent nature of the $m = 0$ instability.

Turning to anisotropic equilibria, we can use CGL to assess the effect of changes in the degree of pressure anisotropy on $m = 0$ stability in the zero Larmor radius limit [27]. We find that the effect is rather small. It cannot provide absolute stability, and although there is an underlying tendency for growth rates to be reduced when $P_\perp > P_\parallel$ (in qualitative agreement with results for the $m = 1$ mode [30]), the degree of enhancement is severely limited. Paradoxically, this is due to the extremely large effect which anisotropy has on z-pinch equilibria. The correction to the force balance equation introduced by anisotropy is inversely proportional to the radius of curvature of the magnetic field lines. It therefore impinges more strongly on the z-pinch than any other cylindrical configuration. So strong is the effect that it is actually impossible to find z-pinch equilibria which are highly anisotropic, and this is the main reason for the rather small effect on stability.

This result is only strictly valid for zero Larmor radius. However, since it is a reflection of the limited range of accessible equilibria, and the CGL force balance equation is valid for all ϵ, we see that equilibrium anisotropy will have little effect on stability at arbitrary ϵ. Tailoring the equilibrium anisotropy cannot stabilize a collisionless z-pinch. Thus we are spared the necessity of finding some way of doing so.

2.4 Large Larmor Radius

All the models described so far treat the plasma as a fluid. However, there is an experimentally important class of z-pinches for which such a model cannot be used. These are ones which operate in the collisionless LLR regime. For instance, a pinch formed from a 10 μm radius cryogenic hydrogen fibre would be collisionless at currents above about 200 kA and have $\epsilon = 0.2$. For such pinches the ion physics must be treated kinetically. This represents a formidable theoretical challenge.

Most of the work on this topic [31–34] uses the basic Vlasov fluid model [35]. This treats the ions kinetically via the Vlasov equation, while the electrons are treated as a cold background fluid which maintains quasi-neutrality. Thus the ion description is exact (for a collisionless plasma) but the electron physics is rather poorly represented.

The Vlasov fluid model shares features of some of the other theories discussed above. Like the Hall fluid model (which is also an LLR theory, although for a collisional plasma) it includes the Hall term and its eigenvalue, γ, is complex. Like CGL (which is also a collisionless theory, although only valid for $\epsilon = 0$) it allows pressure anisotropy. However, the Vlasov fluid model encompasses a wide range of additional physics, way beyond any fluid theory. It is also a lot harder to solve.

In principle the model can be applied to anisotropic equilibria. However, for the reasons outlined above (Sec. 2.3) such equilibria are unlikely to offer significant advantages, and LLR studies of the z-pinch have concentrated on isotropic equilibria.

With a single important exception the Vlasov fluid stability threshold for arbitrary ϵ is formally given by ideal MHD with Γ set to zero [35]. As we have already seen (Sec. 1.2), the ideal MHD $m = 1$ stability condition in the z-pinch is independent of Γ. Thus the basic ideal MHD conclusion, that all physically realistic z-pinch equilibria are $m = 1$ unstable, also applies in the collisionless case for arbitrary ϵ.

In contrast to $m = 1$, the ideal MHD $m = 0$ stability condition (the Kadomtsev criterion, Sec. 1.2, above) does depend on Γ. Setting $\Gamma = 0$ in this we find that absolute Vlasov fluid $m = 0$ stability at all k requires the equilibrium pressure to satisfy the following inequality everywhere:

$$-\frac{dP}{dr} < 0$$

i.e., the pressure must everywhere *increase* with r. This condition is never satisfied in the z-pinch. Thus we find that neither the $m = 0$ nor the $m = 1$ mode can be Vlasov fluid stable in any physically realistic z-pinch.

The single exception to the usual Vlasov fluid stability threshold is given by the $m = 0$ mode in the z-pinch, in the $\epsilon = 0$ limit. We have already noted that the appropriate model in this case is CGL, which has a totally different $m = 0$ stability condition (Sec. 2.3, above) to the arbitrary ϵ one given here. Once again we see the $m = 0$ instability displaying its customary strong model dependence.

The fact that absolute stability cannot be achieved does not rule out the possibility that Vlasov fluid growth rates could be so low that the plasma is effectively stable. This is precisely what does happen in theta-pinch like devices, with straight or nearly straight field lines. Finite Larmor radius (FLR) calculations, based on fluid-like equations derived by expanding the Vlasov equation, can find that such devices are completely stable even when the Vlasov fluid model predicts instability. In such cases there is a small residual growth rate due to resonant particles [36]. These resonant particle effects are not included in the FLR theory, but do appear in the more accurate, fully kinetic Vlasov fluid model.

Sadly, this hopeful possibility does not survive detailed examination. The Vlasov fluid linear stability of the z-pinch has been extensively studied. No situation has ever been found in which a Vlasov fluid unstable z-pinch displays the very small growth characteristic of a resonant particle driven instability. With hindsight it seems a little over-optimistic to expect that the properties of weakly unstable, low β theta-pinch like devices would be shared by the strongly unstable, high β z-pinch.

LLR effects can reduce linear growth rates by up to a factor of about 5 for $m = 0$ and 8 for $m = 1$. The lowest growth rate occurs for $\epsilon \sim 0.1 - 0.2$. Above $\epsilon \sim 0.2$ the growth rate increases with ϵ [32–34].

The effect of finite electron temperature on the LLR linear $m = 0$ mode has been studied [37,38] using a modified Vlasov fluid model [39]. In general the growth rates are increased beyond those of the cold electron case. In most situations the hot electron Vlasov fluid result shows little variation with ϵ and is actually closer to ideal MHD. The fact that such substantial differences exist between the results of the two alternative forms of the Vlasov fluid model is yet another example of the $m = 0$ instability displaying its tiresome sensitivity to changes in the model.

Further details of this work can be found elsewhere in these proceedings [40].

3. OTHER THEORETICAL ISSUES

3.1 Equilibrium Flow

We have now explored several alternative models and have thus disposed of one of our original criticisms of the orthodox stability theory, namely the lack of applicability of ideal MHD. Another restrictive feature of the orthodox theory is the assumption that the equilibrium plasma is perfectly stationary. Relaxing this assumption has some very interesting consequences.

Various forms of equilibrium flow are possible. Here we consider only sheared axial flow. A uniform axial velocity has no effect on either the equilibrium or its stability. Rotation (azimuthal flow) affects both [25,41], but probably does not provide a significant degree of stabilization. Pinches undergoing radial expansion or contraction fall under the heading of dynamic pinches, and thus lie outside the terms of reference of this review.

Sheared axial flow can be generated either spontaneously during pinch formation as a result of a 1-D LLR mechanism [42], or by means of various proposed technical appliances [13]. The flow does not affect the equilibrium and thus its effect on stability can be determined unambiguously.

The linear stability (to both $m = 0$ and $m = 1$ modes) of z-pinch equilibria with sheared axial flow has been investigated using ideal MHD [25,43]. The results depend sensitively on both the choice of basic equilibrium [i.e., $j(r)$ and $P(r)$] and on the form of the flow profile [$u_z(r)$]. Sheared axial flow nearly always reduces growth rates. The strongest effect occurs when $u_z(r)$ is highly sheared in the outer part of the pinch. In such situations short wavelength modes ($k_a \sim 10$), the eigenfunctions of which tend to be localized near the plasma edge, can be absolutely stabilized for flow speeds above about Mach 4. Longer wavelength modes ($k_a \sim 1$) cannot be made stable. However, they do not have highly localized eigenfunctions, and are therefore probably more strongly affected by non-linear effects (see following section).

3.2 Non-Linear Regime

Hitherto we have been concerned exclusively with linear theory. This is perfectly adequate for questions of stability thresholds and for small amplitude instabilities. In an experiment, however, an instability which has grown enough to be observed has probably passed beyond the linear stage. In order to model experimental instabilities in detail we therefore need to consider the non-linear phase. This involves numerical simulation.

Simulations also allow us to explore the possibility of non-linear saturation of instabilities at finite amplitude. One example of this has already been mentioned, namely the generation of Kadomtsev stable equilibria in wall supported pinches [12].

Non-linear modelling of the z-pinch is a relatively young and expanding field, and there is no coherent overall picture, as there is for linear theory. For this reason a rather selective approach will be adopted here. The three examples discussed below all originate from Imperial College.

Most z-pinch simulation work at Imperial College uses a 2-D resistive MHD code (MH2D) originally written by A.R. Bell. This code can only handle $m = 0$ ($m = 1$ require a 3-D code; see below), but, of course, this is the instability actually seen in experiments. The code, with considerable augmentations, has been used to simulate the so called "hot-spot bifurcation" seen in experiments [13]. The possibility that this is associated with the extreme non-linear development of an $m = 0$ instability was originally proposed by P. Jaitly [44]. Recent work by J. Chittenden [45] has confirmed this to be the case, but has shown that the phenomenon is surprisingly complex, involving an axially propagating ionization wave.

MH2D has also been used, this time in an ideal manifestation, to study the effect of equilibrium flow. The ideal assumption allows comparisons to be made with both the linear theory (Sec. 3.1, above) and results from a 3-D ideal MHD code (MH3D, also written by A.R. Bell) which can be used to model the $m = 1$ instability [46]. The beneficial effect of sheared flow found in the linear calculations is further enhanced in the non-linear regime. For instance, simulations reveal saturation of both $m = 0$ and $m = 1$ instabilities for axial flow speeds above about Mach 2.5 (as opposed to Mach 4 for the linear case). This is probably due to the instability structure becoming smeared out by the flow.

Finally, we turn to the non-linear LLR regime. Recently a 2-D hybrid code (PHY2D), based on the Vlasov fluid model, has been written by T.D. Arber [47]. This can in principle model large amplitude $m = 0$ instabilities in a collisionless z-pinch, in which ion kinetic effects are important. First results show no sign of saturation. Instead, the instability continues to grow exponentially until it reaches the axis. However, there is reason to suppose that an improved treatment of the plasma edge would produce a significantly modified result [40]. This is not surprising when we recall how fastidious the $m = 0$ instability is with regard to details.

4. DISCUSSION

Can we make an equilibrium z-pinch stable?

It is relatively easy to stabilize the experimentally dominant $m = 0$ mode if the plasma is supported by a rigid wall (although this would give rise to other problems, associated with losses and impurities). However, stability is much more elusive in the usual arrangement in which the plasma is surrounded by a vacuum.

The single mechanism studied so far which can stabilize the z-pinch is sheared axial flow, Admittedly absolute linear stability occurs only in a rather restricted range of situations, but simulations suggest that the linear theory is probably unduly pessimistic in this respect. Furthermore, there is a wealth of unanswered questions on this topic, for instance what role does the Kelvin-Helmholtz instability play? and what effect does sheared flow have in the LLR regime? This is a fascinating subject and one which will repay detailed study.

The other area which deserves, and is receiving, further exploration is non-linear modelling. We can expect this to yield a deeper understanding of instability development, and to establish if non-linear saturation is the hoped for panacea.

The future may show that sheared flow and non-linear effects can stabilize the z-pinch. But putting these promising possibilities to one side, and confining ourselves to the present state of knowledge of z-pinch stability theory, as reviewed here, it is interesting to ask how much has changed since the early days. Our survey of the various non-ideal theories has revealed a richness in the details of the stability behaviour quite lacking in ideal MHD. But, in spite of this diversity of detail, there is an extraordinary unanimity with regard to the basic overall properties. Linear ideal MHD's main conclusions concerning z-pinch stability are listed in Sec. 1.2, above. Remarkably, most of them apply equally in all other regions of parameter space.

Conclusion 5 (the strongly model dependent nature of the $m = 0$ instability) has turned up with irritating regularity throughout this review. It probably represents the greatest obstacle to obtaining a full theoretical understanding of the $m = 0$ instability.

It is however conclusion 2 which has the most serious implications. All equilibrium z-pinches are unstable with a growth time of the order of a/v_T. Over nearly the entire range of physically realistic parameters this time is very much shorter than the desired duration of the experiment. For instance, the Imperial College DZP experiment is designed to achieve radiative collapse at a current of about 1.5 MA reached after a current rise lasting for 150 ns. Although this is quite a short time to achieve such a high current it represents approximately 4×10^4 instability growth times for the case of a z-pinch formed from a 10 μm radius hydrogen fibre (this would correspond to $\epsilon = 0.2$, i.e., the optimum value for LLR effects, at which the growth rate can be reduced by up to a factor of 8).

This is worrying enough. But the universal validity of conclusion 2 is the source of a more profound concern. The instability growth time, a/v_T, is also the timescale for setting up the equilibrium itself. Under these circumstances it is very hard to see how equilibria can actually form.

Thus the theory's most significant result is not an answer to our original question but the realisation that we should have asked a different one. Not "can we make an equilibrium z-pinch stable?" but "can we make an equilibrium z-pinch at all?".

ACKNOWLEDGEMENTS

I am indebted to the following for their help (which included doing much of the work described here): T.D. Arber, A.R. Bell, S. Channon, J. Chittenden, M.G. Haines, D.F. Howell, N. Kassapakis, S. Lucek, P.G.F. Russell, J. Scheffel and D. Zdravkovic.

REFERENCES

1. J. Sethian, "The quest for a z-pinch based fusion power source: an historical perspective", these proceedings.
2. M.G. Haines, Plasma Phys. and Contr. Fusion **31**, 759 (1989).
3. R. Carruthers and P.A. Davenport, Proc. Phys. Soc. B **70**, 49 (1957).
4. I.V. Kurchatov, J. Nucl. Energy 4, 193 (1957).
5. O.A. Anderson, W.R. Baker, S.A. Colgate, H.P. Furth, J. Ise, R.V. Pyle, and R.E. Wright, Phys. Rev. **109**, 612 (1958).
6. M.D. Kruskal and M. Schwarzschild, Proc. Roy. Soc. A **223**, 348 (1954).
7. R.J. Tayler, Proc. Phys. Soc. B **70**, 31 (1957).
8. B.B. Kadomtsev, in *Reviews of Plasma Physics*, ed M.A. Leontovich (Consultants Bureau, New York, 1966), Vol 2, p 153.
9. M. Coppins, Plasma Phys. and Contr. Fusion **30**, 201 (1988).
10. I.B. Bernstein, E.A. Frieman, M.D. Kruskal and R.M. Kulsrud, Proc. Roy. Soc. A **244**, 17 (1958).
11. J. Scheffel and M. Coppins, Nuclear Fusion **33**, 101 (1993).
12. P.T. Sheehey, R.A. Gerwin, R.C. Kirkpatrick, I.R. Lindemuth and F.J. Wysocki, "Computational modelling of wall supported dense z-pinches", these proceedings.

13. M.G. Haines, "An overview of the DZP Project at Imperial College", these proceedings.
14. E.A. Smars, Ark. Fys. **29**, 97 (1964).
15. P. Choi, M. Coppins, A.E. Dangor and M.B. Favre, Nuclear Fusion **28**, 1771 (1988).
16. J.P. Freidberg, *Ideal Magnetohydrodynamics* (Plenum Press, New York, 1987).
17. M.G. Haines and M. Coppins, Phys. Rev. Lett. **66**, 1462 (1991).
18. I.D. Culverwell and M. Coppins, Phys. Fluids B **2**, 129 (1990).
19. M. Coppins and I.D. Culverwell, "Resistive stability of the z-pinch revisited", these proceedings.
20. F.L. Cochran and A.E. Robson, Phys. Fluids B **2**, 123 (1990).
21. R.J. Tayler, Nucl. Fusion Suppl. (part **3**), 877 (1962).
22. U. Schaper, J. Plasma Phys. **29**, 1 (1983).
23. U. Schaper, J. Plasma Phys. **30**, 169 (1983).
24. M. Coppins, D.J. Bond and M.G. Haines, Phys. Fluids **27**, 2886 (1984).
25. D. Howell, "The effect of equilibium flow on the stability of the z-pinch", PhD Thesis (Univ. of London, in preparation).
26. G.F. Chew, M.L. Goldberger and F.E. Low, Proc. Roy. Soc. A **236**, 112 (1956).
27. M. Coppins and J. Scheffel, Phys. Fluids B **4**, 3251 (1992).
28. H.O. Akerstedt, J. Plasma Phys. **44**, 137 (1990).
29. M. Coppins, Phys. Fluids B **1**, 591 (1989).
30. M. Faghihi and J. Scheffel, J. Plasma Phys. **38**, 495 (1987).
31. T.D. Arber and M. Coppins, Phys Fluids B **1**, 2289 (1989).
32. T.D. Arber, M. Coppins, and J. Scheffel, Phys. Rev. Lett **72**, 2399 (1994).
33. T.D. Arber, P.G.F. Russell, M. Coppins, and J. Scheffel, Phys. Rev. Lett **74**, 22698 (1995).
34. P.G.F. Russell, T.D. Arber, M. Coppins, and J. Scheffel, "Linear stability of the collisionless, large Larmor radius z-pinch", Phys. Plasmas (in press).
35. J.P. Freidberg, Phys. Fluids **15**, 1102 (1972).
36. C.E. Seyler and J.P. Freidberg, Phys. Fluids **23**, 331 (1980).
37. T.D. Arber, Phys. Fluids B **3**, 152 (1991).
38. J.Scheffel, T.D. Arber, M. Coppins, P.G.F. Russell, Plasma Phys. Contr. Fusion **39**, 559 (1997).
39. R. Gerwin, Informal Report LA-6130-MS (Los Alamos Scientific Laboratory, New Mexico, 1975)
40. M. Coppins, "Recent progress on Large Larmour radius theory", *these proceedings*.
41. A.L. Velikovich and J. Davis, Phys. Plasmas **2**, 4513 (1995).
42. T.D. Arber (private communication).
43. T.D. Arger and D.F. Howell, Phys. Plasmas **3**, 554 (1996).
44. P. Jaitly, "Self-similar z-pinch equilibria and their stability", Ph.D. thesis (Univ. of London, 1993)
45. J. Chittenden (private communication).
46. S. Lucek (private communication).
47. T.D. Arber, Phys. Rev. Lett. **77**, 1766 (1996).

Stabilized Z-pinch loads with tailored density profiles

A. L. Velikovich,[*] F. L. Cochran,[*] and J. Davis[†]

[*]*Berkeley Research Associates, Inc., Springfield, VA 22150*
[†]*Radiation Hydrodynamics Branch, Plasma Physics Division, Naval Research Laboratory, Washington, D. C. 20375*

Abstract. We discuss the design of structured Z-pinch loads capable of mitigating the detrimental effect of Rayleigh-Taylor (RT) instability on the performance of fast Z-pinch devices used as plasma radiation sources. The stabilizing effects of density tailoring in both the radial and the axial directions are considered. Our 2-D numerical simulations demonstrate that using a structured gaseous load with a radial density profile specifically tailored for a given current wave form, it is possible to delay the onset of the RT instability development while a shock wave propagates through the load. Once the acceleration of the magnetic field/plasma interface is inverted, perturbations are shown to oscillate rather than to grow exponentially. Our simulation results indicate the possibility of high-quality implosions producing significant Ar K-shell yield from initial radii in the range between 4 and 8 cm, with current pulse duration of 250 ns and longer. Axial density tailoring can actually mitigate the RT instability, or even suppress it completely, at the expense of decreased hydrodynamic efficiency of acceleration. Seeking the best trade-off between stability and performance, it would be natural to combine the two approaches.

The Rayleigh-Taylor (RT) instability is known to be a major problem for a wide range of applications, from pulsed power technology to inertial confinement fusion. This instability could be detrimental to the Z-pinch implosions in a similar way as it is to the implosions of laser fusion pellets. Mitigation of the RT is particularly important for developing long-implosion plasma radiation sources. To be an efficient K-shell radiator, the Z-pinch column driven by a 0.3 to 0.5 µs long current pulse should be imploded from a large initial radius, 6 to 8 cm. Unless something is done to mitigate the RT, no gaseous load imploded from a ~8 cm radius would reach the axis in one piece and be an efficient radiator.

Here we discuss the concept of the RT mitigation by density tailoring. This means distributing the initial mass of the load in space according to a certain prescribed law. Appropriately tailored initial density profiles of Z-pinch loads could be produced using specifically designed nested nozzles or low-density foams.

The mitigation scheme involving radial density tailoring has been studied the most. Observations of a substantially increased radiative output in numerous experiments with double gas-puff and puff-on-puff loads [1-4] stimulated theoretical studies that explained enhanced stability of the double gas puffs by the snowplow effect, which is basically stabilization by the shock wave propagating inward [5], and by the reversal of acceleration of the outer shell when it impacts the inner shell [6]. Further analysis [7, 8] has shown that essentially the same physics of stabilization applies to relatively large diameter uniform fill gas puffs, whose lower efficiency (compared to annular shells) in transforming magnetic to kinetic energy is more than compensated for by their superior stability. This conclusion was confirmed by the experiments done at Sandia on Saturn [9].

A uniform fill load, being more stable than an annular shell, is still not stable enough to survive an implosion from ~8 cm. Two similar methods for making a solid-fill-type load more stable than a uniform fill were suggested at about the same time in Refs. 10 and 11. Both involve the use of "tailored" [10] or "peaked" [11] radial density profiles, with density increasing toward the pinch axis. With magnetic pressure driving a shock wave into increasing density, the shock slows down, and the acceleration of the plasma/magnetic field interface decreases. In Ref. 10, we suggested tailoring the radial density profile in order to reverse the acceleration of the interface, producing an essentially stable "light-fluid-supported-by-heavy-fluid" configuration, whereas the idea of Ref. 11 is to make this acceleration exactly zero. This might be more difficult to achieve and less effective for stabilization - the initially perturbed plasma/field interface moving at constant speed would certainly exhibit a linear Richtmyer-Meshkov instability growth, in agreement with the numerical results of Ref. 11. On the contrary, a cylindrical load with appropriately tailored radial density profile is demonstrated to be stable as long as the acceleration is inverted [10]. Although magnetic pressure continues to perform work, accelerating an increased plasma mass, the interface feels a deceleration, and this is all that counts.

Figure 1 compares our simulation results for loads with differently tailored density profiles that had radial variation according to $1/r^n$, with $n = 0$ (a thick annular shell, almost a uniform fill), 1, 2, and 3. Each density profile extended from an inner radius of 2 cm to the outer radius of 9 cm. All loads were normalized to have an Ar line mass of 200 µg/cm and imploded by a current which rises linearly to 4 MA over 120 ns. The current then remains constant during the remainder of the implosion. The simulations were run for a 10% initial random density perturbation and a wavelength 3 mm. In Fig. 1, the line density variation along the axis is shown for various loads at two times during the implosion. At the first time, 120 ns, when the current switches from linear rise to a flat current, this ratio is nearly the same, close to unity, for all cases. The density variation was also measured 25 ns before the load stagnated on the axis. Obviously, this time varies between the load types. Near stagnation, the density variation rapidly decreases with increasing exponent n. For $n = 1$ (slightly decreased acceleration), the stabilizing effect is small; for $n = 2$ (this resembles the zero-acceleration case of Ref. 11) perturbation growth is slowed down, but still noticeable. For $n = 3$ (inverted acceleration) there is virtually no line mass variation near stagnation. This is nearly a 5-fold reduction in the perturbation amplitude compared to the case of the uniform fill, $n = 0$.

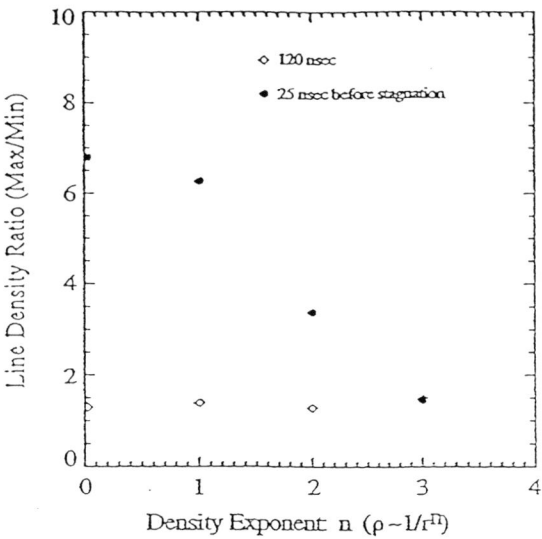

FIGURE 1. Line density variation along the axis (measured as the ratio between the maximum and the minimum line densities for all the computational z-slices) vs. the density tailoring parameter, n, for two instants of time.

Mitigation of the RT instability by axial density tailoring was discovered at Sandia [12]. For the hourglass-shaped uniform fills, a remarkable mitigating effect of the initial load curvature on the RT instability during the implosion was demonstrated. This effect was shown to be related to the sheared axial mass flow along the outer edge of the load.

The "hourglass" stabilizing effect is not yet fully understood. Our analysis indicates that this effect could be a rather unexpected manifestation of an effect which is well-established in hydrodynamics. As known since the 1950s, the development of a single-mode RT instability produces a smooth bubble rising at a constant speed and a sharp spike falling with a constant acceleration. The bubble evolves to a steady state and becomes stable due to convection, to the flow of the particles along the bubble surface (just as observed in Ref. 12). The RT instability develops in the fluid particles flowing past the bubble interface, which are dumped to the spike before any appreciable growth can take place. For a supersonic acceleration realized with uniform fill Z-pinch loads, the rising bubble drives ahead of it a shock wave, and, consequently, a relatively uniform and stable layer of accelerated plasma between the interface and the shock front. The fluid particles that constitute this layer flow out of it along the interface (which keeps it stable all the time), being constantly replaced by new particles entering the flow through the shock front. Mass losses due to plasma flow through the accelerated layer constitute the price for stability of its acceleration.

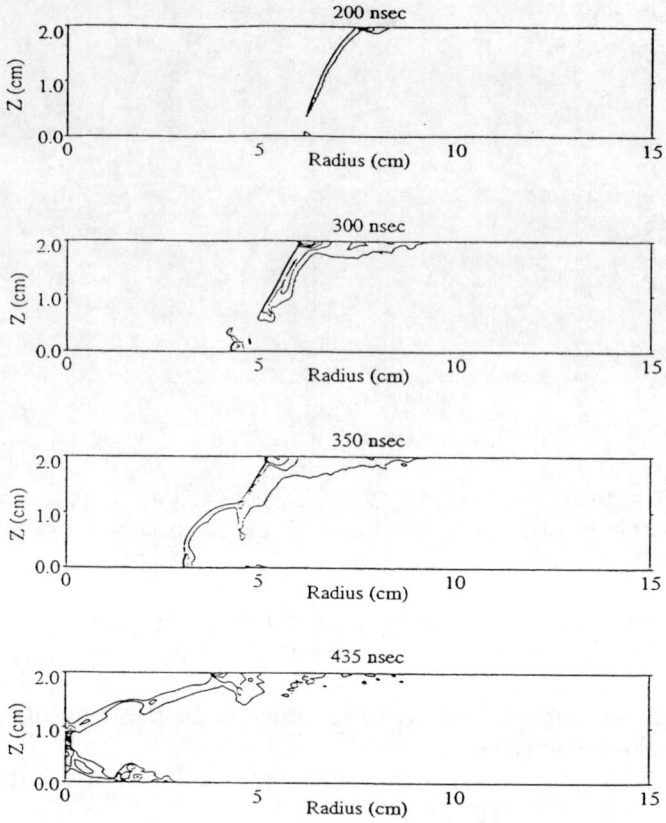

FIGURE 2. Density contours for a load with axially tailored density.

The above considerations might open an attractive opportunity for designing instability-free large radius gas-puff or foam Z-pinch loads. The goal of this design would be to produce a stable large-scale bubble with an appropriately curved shape from the very start, thus avoiding competition between small-scale bubbles that generate undesirable localized non-uniformities which degrade the quality of the implosion. A simple numerical study of such a load configuration is illustrated by Fig. 2. The initial uniform fill Ar load has a parabolic outer boundary extending from $r = 8$ cm to 11 cm, its line mass, 5 mg, corresponding to implosion time about 500 ns for a constant current $I = 5$ MA. Figure 2 shows the density contours at four different times. At about $t = 200$ ns, the accelerated bubble attempts to break up near the tip. Due to the intrinsic stability of the large-radius bubble, the break-up produces essentially the same bubble, with appreciable mass dumped into the spike (see the density contours for $t = 300$ and 350 ns). Finally, at $t = 435$ ns, the mass accelerated in a stable way near the tip of the bubble stagnates at the axis. This mass does not consist of the fluid particles that started their acceleration from 8 to 11 cm (these are long gone into the spike), but the acceleration is nevertheless continuous throughout the implosion. The stagnated mass is a small fraction (less than 10%) of the total initial mass of the load, which is the price paid for the stable acceleration from 8 cm over 435 ns.

In contrast with the radial density tailoring, the axial density tailoring could actually mitigate the RT instability (even suppress it completely, at the expense of accelerating a small fraction of the load mass) rather than delay its onset. Seeking the best trade-off between stability and performance, it would be natural to combine the two approaches. Radial density tailoring could be used to suppress perturbation growth in the lower-density outer layers of the load during the current rise time, whereas the axial density tailoring would help in slowing down the growth during the load acceleration by the peak current.

REFERENCES

1. 1. P.Sincerny *et al.*, Proc. of the 5th IEEE Pulsed Power Conference, June 10-12, 1985, Washington, D.C., p. 781.
2. T.-F. Chang, A. Fisher, and A. Van Drie, J. Appl. Phys. **69**, 3447 (1991).
3. R. B. Baksht, A. V. Luchinskii, and A. V. Fedyunin, Sov. Phys.-Tech. Phys. **37**, 1118 (1992).
4. R. B. Spielman, T. Nash, and M. Krishnan, Bull. Am. Phys. Soc. **37**, 1578 (1992).
5. S. M. Gol'berg and A. L. Velikovich, Phys. Fluids B **5**, 1164 (1993).
6. S. M. Gol'berg and A. L. Velikovich, in *Dense Z-Pinches*, 3rd International Conference, London, UK, 1993, ed. by M. Haines and A. Knight, AIP, New York, 1994, p. 42.
7. F. L. Cochran, J. Davis and A. L. Velikovich, Phys. Plasmas **2**, 2765 (1995).
8. N. F. Roderick *et al.*, Proc. of the IEEE International Conference on Plasma Science, Madison, Wisconsin (IEEE, Piscataway, NJ, 1995), p. 252.
9. R. B. Spielman *et al.*, Bull. Am. Phys. Soc. **40**, 1845 (1995)
10. A. L. Velikovich, F. L. Cochran, and J. Davis, Phys. Rev. Lett. **77**, 853 (1996).
11. J. H. Hammer *et al.*, Phys. Plasmas **3**, 2063 (1996).
12. M. R. Douglas, C. Deeney, and N. F. Roderick, Phys. Rev. Lett. **78**, 4577 (1997).

STABILITY AND K-SHELL RADIATION OF Z-PINCHES

R.B.Baksht, A.V.Fedunin, A.Yu.Labetsky, V.I.Oreshkin,

A.G.Russkikh, A.V.Shishlov

High Current Electronics Institute, 4 Academichesky ave, Tomsk, 634055, RUSSIA

Experiments of GIT-4 facility are shown that Rayleigh-Taylor instabilities destroy the cylindrical gas shell at low mass and the X-ray yield falls sharply. Double gas puff is the efficient load designs for plasma radiation source. A double gas puff can be efficiently only if the magnetic field diffusion is absent at the onset of the current. The problem of the skin layer production in the outer shell was decided using magnetron discharges. As a result a good reproducibility of K-shell generation was achieved for Ar double gas puff. The final plasma column demonstrated the absence of bright spots and produced 0.3 kJ/cm of the K-shell yield at the 1.7 MA current.

In plasma radiation sources, Z-pinch loads are imploded to generate large amounts of soft X-rays with 1-10 keV photon energy (K-shell radiation)[1]. Such photons can be produced only in the plasma with H-like and He-like ions. The performance of plasma K-shell radiation source is known to be very sensitive to stability. Better stability means tighter Z-pinch at the stagnation phase, more uniform radiating plasma and, therefore, higher radiative output and a greater K-shell yield. Rayleigh-Taylor (RT) instabilities are the greatest danger at the Z-pinch implosion. The experimental study of RT-instabilities were started in connection with the Ar gas puff implosion in 93[2]. In this paper some new experimental results are reported. All experiments were carried out on GIT-4 (1.5 MA) facility. Typical current and X-ray traces are shown in Fig.1. The gas puff length is 2 cm. The η value in our experiments is 0.5-1. Thus, only a part of the total amount of ions take part in the K-shell radiation.

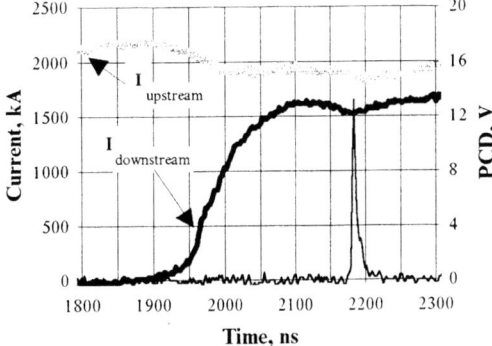

Fig.1. Implosion of Ar double gas puff, R_{out} = 3 cm, R_{in} = 1.4 cm, Sh.No.398.

According to our experimental data, the RT-instability is very dangerous for a gas puff with a small mass and a large diameter. Experimental parameters and calculated kinetic energy are shown in Table 1 for some Ar gas puffs. Gas puff stability was determined by a visible streak camera. According to the table, a K-shell yield and K-shell power depend on the stability of the implosion and does not depend on the kinetic energy.

Table 1

Radius, cm	Gas puff mass, μg/cm	Kinetic energy, keV/ion	K-shell yield, kJ	K-shell power, GW	a_{RT}/Δ^*	Stability
3	33	113	23	1.9	>1	very unstable
3	100	37	170	9.6	≈1	unstable
3	150	25	182	10	0.5	unstable
1.4	60	45	77	6.2	0.7	unstable
1.4	85	32	165	17.7	0.4	unstable
1.4	180	15	348	33.4	0	stable

* a_{RT} is the wave amlitude; Δ is the thickness of the luminous gas puff shell.

We studied the X-ray yield as a function of mass for Ne K-shell radiation and Kr L-shell radiation (Fig.2). The picture is similar for all the studied gases: RT-

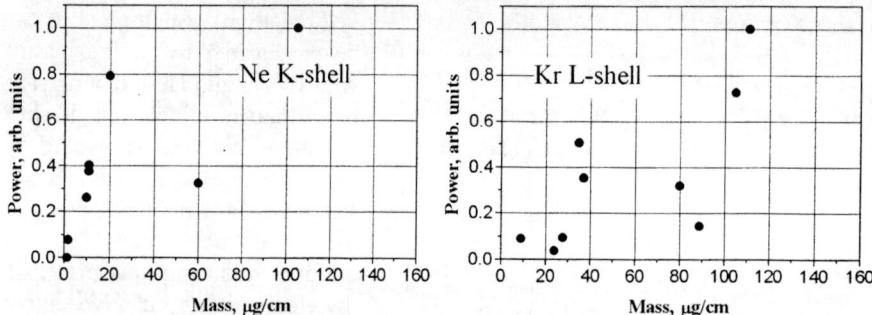

Fig.2. Ne K-shell radiation and Kr L-shell radiation versus gas puff mass

instabilities destroy the cylindrical gas shell at low mass and X-ray yield falls sharply[3]. One of the possible reason for the low mass instabilities is the fast rise of an increment of the volume mode instabilities for low gas puff[5].

A double gas puff Z-pinch load is known to mitigate the RT-instability[4,5]. Application of the improved preionisation assembly[6] allowed suppressing the current division between outer and inner gas shell. As a result we obtained the stable increase in K-shell power and yield (Fig.3). The transfer from a single gas puff to double gas puff changes the final plasma column produced in the stagnation phase: the implosion approaches to 1D-like dynamics. Comparison between the two kinds of gas puff (Table 2 and Fig.4) demonstrates that in spite of the low temperature stagnation phase of a double gas puff is more uniform. A plasma column is like a structure to be observed in Sandia wire array experiments[7] and consists of two parts: a stable core with high efficiency of K-shell radiation and a halo with a low temperature. The core having a 1-mm FWHM (full width at half maximum) surrounds a 3.4-mm FWHM halo.

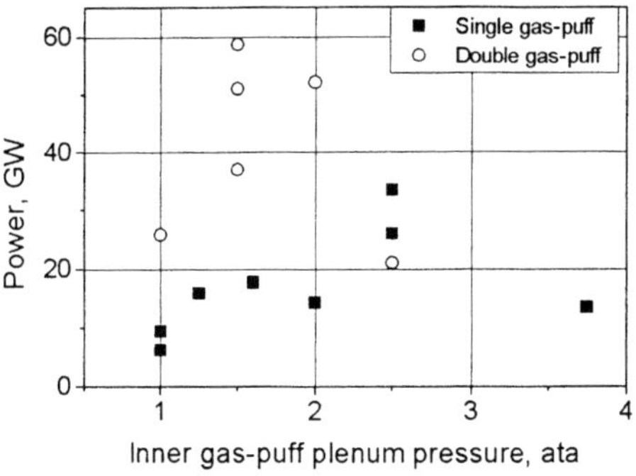

Below we will shortly discuss the experimental results. First we believe that RT-instability suppression in the double gas puff is related to a rapid switching of the current to the inner gas puff. Indeed, magnetic work done on the outer plasma is small up to the switching moment as the outer gas puff mass and radius relation are small ($m_{out} = 33$ μg/cm, $r_{out}/r_{in} = 2$). It results in the modest role of the outer shell energy in the final plasma energy balance. On the other hand, the estimates show that the experimental

Fig.3. Ar gas puff K-shell radiation.

Table 2

	Sh.No.378	Sh.No.398
m_{in}, μg/cm	180	80
r_{in}, cm	1.4	1.4
m_{out}, μg/cm	0	25
r_{out}, cm	0	3
P, GW	33	51
T, keV	1	0.65
n_i, cm^{-3}	$0.8 \cdot 10^{19}$	$5 \cdot 10^{19}$

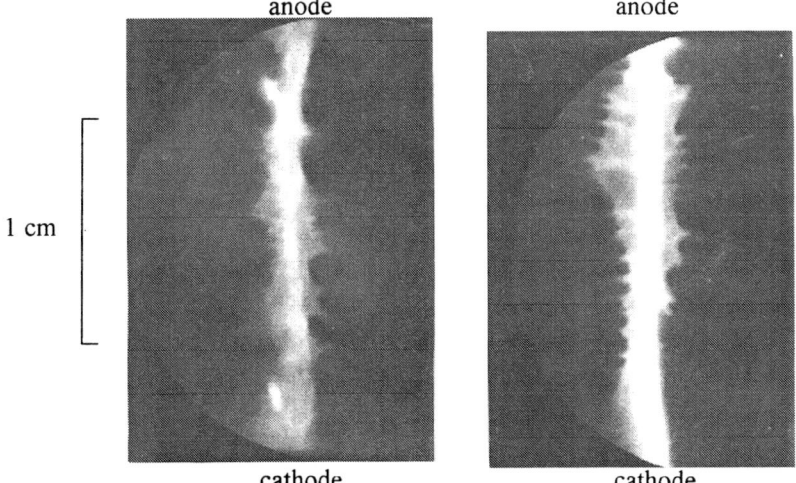

Fig.4. Pinhole pictures for Sh.378 and Sh.398. Mean photon energy is 0.8 keV.

imploding time of the double gas puff coincides with the switching time of 10 ns approximately. Short rise time of the current does not allow developing a volume mode of the instability.

To explain the phenomenon of halo let us assume the existence of an anomalous conductivity connected with current instabilities at low plasma density[8]. We assume that the anomalous conductivity is determined by the low hybrid instability. Then

$$\sigma_{turb} = \frac{n \cdot e^2}{\varepsilon_0 \cdot (v_{coul} + k \cdot v_{an})} \quad (1)$$

where

$$v_{an} = \left(\frac{\pi}{2}\right)^{1/2} \cdot \omega_{LH} \cdot \left(\frac{u_{dr}}{c_s}\right)^2 \quad (2)$$

and k is the arbitrary coefficient.
The low hybrid frequency ω_{LH} is determined as

$$\omega_{LH}^2 = \Omega_p^2 \cdot \left(1 + \frac{\Omega_p^2}{\omega_c \cdot \Omega_c}\right)^{-1} \quad (3)$$

Using (1)-(3) we can write for an imploding plasma with comparatively low density

$$\sigma_{turb} = \sigma_{spitz} \cdot \left\{1 + k \cdot \left(\frac{u_{dr}}{c_s}\right)^2 \cdot \omega_c \tau_{ei} \cdot \left(\frac{m \cdot z}{M}\right)^{1/2}\right\}^{-1} \quad (4)$$

One can show from formulae (4) that for

$$\left\{k \cdot \left(\frac{u_{dr}}{c_s}\right)^2 \cdot \omega_c \tau_{ei} \cdot \left(\frac{m \cdot z}{M}\right)^{1/2}\right\} \gg 1$$

$\sigma_{turb} \sim n_e^3 \cdot j^{-2}$. Thus, there are conditions for a change of the current distribution: the current begins to flow along the internal layers of the imploded plasma as the density increases. Therefore, the internal layers are compressed more than the outer Z-pinch layers.

Calculating the implosion of a gas puff in terms of the formulae (4) we found that the 1D-simulation density distribution is close to the experimental one for the stagnation phase. In Fig.5b the simulation results are shown at the moment of maximal compression, in Fig.5a the simulation results are shown at 5 ns before compression. One can see that before the compression the most part of current flows along the outer plasma layers. During the stagnation phase flows along the inner plasma layers with high density. The simulation shows also that in our case most part of the internal plasma energy is the result of the Lorenz force work during the stagnation phase.

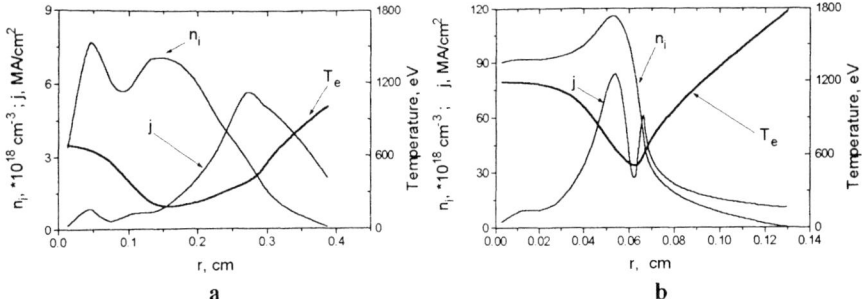

a   b

Fig.5. The 1D-simulation Sh.398; the ion density n_i, the current density j and the electron temperature T_e; a) 5 ns before the moment of maximal compression; b) the moment of maximal compression.

At the nearest future we plan to perform an experiment with the multiple layers load with a microsecond implosion time and a megaampers current. We hope that the availability of the skin layer in the outer gas puff shell[6] will allow producing a 1D-like implosion for the last layer and sharpening the driver power without POS. This effect will be most pronounced at PRS for total radiation.

The work was partly supported by the International Science and Technology Centre, Project #525.

References

1. K.G.Whitney, J.W.Thornhill, J.P.Apruzese and J.Davis, J.Appl.Phys **67**, p.1725, (1990).
2. R.B.Baksht, I.M.Datsko, A.A.Kim, B.M.Kovalchuk *et al.*, Proceeding Beams 94, p. 748-751, (1994).
3. R.B.Baksht, I.M.Datsko, A.A.Kim *et al.*, Proceeding Beams 96, (1996).
4. A.Velikovich, S.Golberg, Phys.Fluids B **5**, 1164, (1993).
5. R.B.Baksht, I.M.Datsko, A.A.Kim *et al.*, Reports on Plasma Physics **21**, p.959-965, (1995).
6. A.G.Russkikh, R.B.Baksht *et al.*, this proceeding.
7. T.W.L.Sanford, T.J.Nash, R.C.Mock *et al.*, 11th APS Conf. on Plasma Diagn., Monterey, CA (1996).
8. I.P.Guzbaud, A.I.Pjatakch, V.L.Sizonenko, JETF, **64**(5), p.2085-2096, (1973).

Variation of High-Power Aluminum-Wire Array Z-Pinch Dynamics with Wire Number, Load Mass, and Array Radius

T. W. L. Sanford, R. C. Mock, B. M. Marder, T. J. Nash, and R. B. Spielman
Sandia National Laboratories, Albuquerque, NM 87185

D. L. Peterson and N. F. Roderick
Los Alamos National Laboratory, Los Alamos, NM 87545

J. H. Hammer and J. S. De Groot
Lawrence Livermore National Laboratory, Livermore, CA 94550

D. Mosher
Naval Research Laboratory, Pulsed Power Physics Branch, Washington, DC 20375

K. G. Whitney and J. P. Apruzese
Naval Research Laboratory, Radiation Hydrodynamics Branch, Washington, DC 20375

Abstract. A systematic study of annular aluminum-wire z-pinches on the Saturn accelerator shows that the quality of the implosion, (as measured by the radial convergence, the radiated energy, pulse width, and power), increases with wire number. Radiation magnetohydrodynamic (RMHC) xy simulations suggest that the implosion transitions from that of individual wire plasmas to that of a continuous plasma shell when the interwire spacing is reduced below ~ 1.4 mm. In this "plasma-shell regime," many of the global radiation and plasma characteristics are in agreement with those simulated by 2D-RMHC rz simulations. In this regime, measured changes in the radiation pulse width with variations in load mass and array radius are consistent with the simulations and are explained by the development of 2D fluid motion in the rz plane. Associated variations in the K-shell yield are qualitatively explained by simple radiation-scaling models.

INTRODUCTION

Increasing the symmetry of cylindrical aluminum-wire arrays by significantly increasing wire number (and decreasing the associated interwire gap spacing and wire size) has resulted in the highest x-ray power (~ 40 TW, Fig. 1) and narrowest x-ray pulse-width (~ 5 ns, Fig. 2) measured for aluminum-wire implosions (1) on the 20-TW Saturn accelerator (2). For gap spacings less than 1.4 mm (Fig. 1), a dramatic increase is seen in the peak power when plotted against the interwire gap. The similar results seen in Fig. 1 for different initial implosion radii (R) may

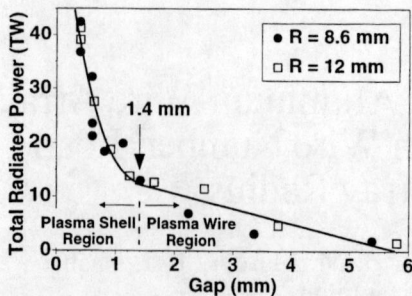
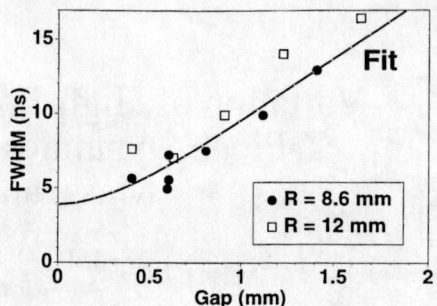

FIGURE 1. Total radiated power measured by bolometer (corrected for pulse shape distortion, spectral response, and viewing angle) versus interwire gap for two initial array radii R.

FIGURE 2. Estimated total-radiated-power pulse width (measured in XRD filtered by 1-μm Kimfol) versus interwire gap (g) corresponding to data of Fig. 1. Fit = $\sqrt{(3.9\pm2)^2 + [(8.8\pm2)g]^2}$.

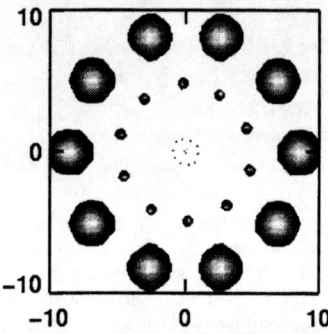
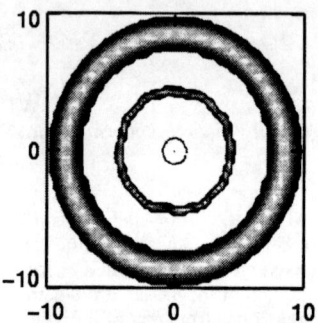

FIGURE 3. xy-RMHC simulations for (A) a wire-plasma regime (N=10 wires) and (B) a plasma-shell regime (N=40 wires) 8.6-mm implosion at 86, 11, 3, and 0 ns before stagnation. Azimuthal orientations at different times are not correlated.

be explained by the fact that the larger radius implosions have both increased total energy and pulsewidth (Fig. 2), resulting in similar powers to those at smaller radius. For loads with large number of wires and small interwire gaps (<1.4 mm), the dramatic improvement in pinch quality may be associated with the effect seen in xy ("end-on") RMHC calculations (1) of the merger of the separate wire plasmas to form a single plasma shell (Fig. 3). Here such loads are designated as being in the plasma-shell regime, and those with larger interwire spacings as being in the plasma-wire regime. In general, our measurements and analyses show that as the wire number increases, the quality of the implosion and the radiated power increase monotonically. Similar effects are found for tungsten wire implosions as discussed in Refs. 3 and 4 and papers at this conference.

Prior to these aluminum experiments, wire-array loads had been optimized for x-ray energy and power output only in the small-wire-number (N ≤ 40) wire-plasma regime (5-7), where the calculations now indicate (1) the array may implode inward as a set of discrete wire plasmas. The substantial improvement in implosion quality and associated radiated power that occurs in the plasma-shell regime suggested that a reexamination of the radius and mass conditions that maximize the radiation output in this regime would be instructive. Measurements of the more-nearly-2D, azimuthally-symmetric plasma-shell-implosion could then be more realistically compared with rz ("side on") RMHC simulations to gain insight into fundamental physical processes, free of the large azimuthal in homogeneities that have limited previous detailed comparisons. Moreover, 1D-RMHC simulations that utilize detailed radiation models not employed in the 2D codes, and often used to infer the measured plasma conditions from the measured x-ray spectral data, become more realistic.

Additionally, a need existed to assess the effect that longer implosion times would have on the pinch quality expected for the recently commissioned, 45-TW PBFA-Z accelerator (8,9). For PBFA-Z, the optimal coupling of accelerator energy to the load was designed to occur for implosion times on the order of 100 ns, roughly twice that found for loads on the Saturn accelerator. Therefore, after the initial wire-number experimental series (1), two additional experimental series were conducted both to re-examine the conditions that optimize the XUV and K-shell x-ray output and to explore the longer-implosion-time regime using large-wire-number, plasma-shell-regime loads. In one series, the radius was held fixed, and the mass was varied (10). In the other series, the mass was held fixed near the mass that optimized the x-ray output, and the radius was varied (11). These two new experimental series enabled the x-ray production to be measured in conjunction with the implosion quality over implosion durations of 40 to 90 ns.

A schematic of the generic arrangement used for all of the experiments and the equivalent-circuit model that was used as driver to the RMHC simulations are shown in Fig. 4. The particular arrangements and circuit model are discussed in detail elsewhere (1,12, and 13). The experiments were simulated with the multi-photon-group 2D-Lagrangian-code (L-RMHC) (14), which modeled 1 mm of the pinch length (12); the three-temperature 2D-Eulerian-code (E-RMHC) (15,16), which modeled the entire pinch length; and the detailed-multi-spectral 1D-Lagrangian-code (1D-RMHC) (17). In the initial wire-number scan, the array mass and radius were fixed at either ~ 0.62 mg and 8.56 mm or 0.84 mg and 12 mm for the 2-cm-long loads. The wire number was changed incrementally from 10 to 192 by varying the diameter of the individual wires from 37 to 10 μm, permitting interwire spacings from 6 to 0.4 mm to be explored. For the mass scan, the radius was fixed at 12 mm and the mass was varied between 0.42 and 3.36 mg by changing the diameter and the number of wires. For masses of 0.84 mg and greater, the number of wires was held at 192 (0.4 mm interwire spacing), and the wire diameter was varied between 10 and 20 μm in 2.54-μm steps. Because the

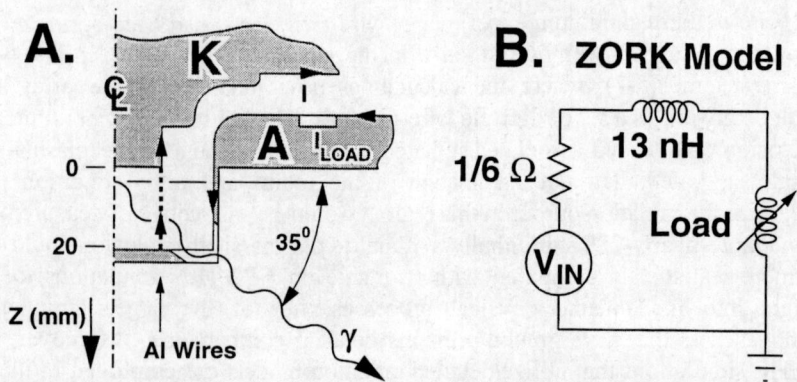

FIGURE 4. (A) Schematic of experimental arrangement showing (B) equivalent circuit. The 13-nH corresponds to the system inductance up to the initial wire location.

smallest diameter aluminum wire available was 10 μm, the mass region below 0.84 mg was necessarily explored by reducing the number of wires, resulting in an increase in the interwire spacing to 0.8 mm. For the radius scan, the array mass was held fixed at 0.6 mg (136, 10.2-μm-diameter wires), where high powers had been produced in the mass scan. These arrays had interwire spacings that varied from 0.4 mm for the smallest, to 0.94 mm for the largest array diameter.

In this paper, the results of the initial wire-number scan are briefly reviewed. The plasma and radiation generated in the plasma-shell regime are then characterized and compared with the L-RMHC, E-RMHC, and 1D-RMHC simulations. Lastly, the variation of these plasma and radiation characteristics with mass and radius in the plasma-shell regime is discussed and contrasted with results produced by the E-RMHC and two K-shell models.

WIRE NUMBER VARIATION

The improvement in pinch quality with wire number in both geometries (R = 8.56 and 12 mm) is evident in three ways: (A) the x-ray pinhole camera images that show increased radial convergence (Fig. 5A) and associated total increased radiated energy (Fig. 5B), (B) the load-current traces that show an increased inductive current notch at peak compression, and (C) the x-ray pulses that show reduced radiation rise time and pulse width (Fig. 2), and the associated increase in radiated power (Fig. 1). The increase in measured radiated energy is consistent, within a scale factor (1.3±0.2), with that expected from the increased kinetic-energy input predicted by the ZORK coupled circuit and slug-model code (18) (Fig. 5B), using the measured radial convergence from x-ray pinhole photographs (Fig. 5A). For high-wire-number loads, the current and main radiation pulse shapes are consistent with those simulated by the rz RMHCs (1,10,11,12,14), as is discussed next.

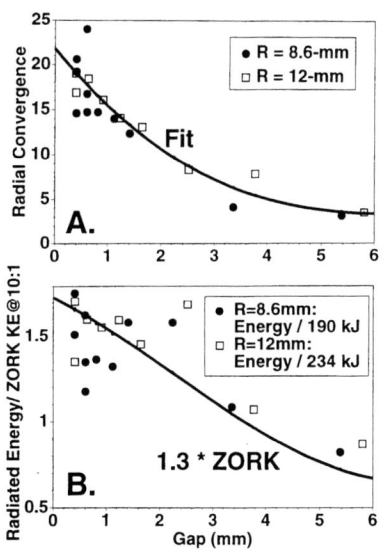

FIGURE 5. (A) Radial convergence determined from time-integrated pinhole images filtered to view x rays greater than 1 keV versus interwire gap (g): Fit = 22 - 7.2 g + 0.9 g^2 - 0.0034 g^3. Comparison of (B) total radiated energy measured in bolometer (uncorrected for spectral response and viewing angle) normalized by calculated ZORK kinetic energy for a 10-to-1 radial convergence (190 and 234 kJ for R = 8.6 and R = 12 mm, respectively) with 1.3 times the kinetic energy predicted by ZORK using measured radial convergence of Fig. 2A.

PLASMA-SHELL REGIME CHARACTERISTICS

Loads in the plasma-shell regime having interwire spacings of less than ~ 0.6 mm exhibit both a strong first and weaker second radiation pulse (Fig. 6A) that correlates in time with strong and weaker measured radial convergences (Fig. 6B). The radial convergence is obtained from radial lineouts of time-resolved x-ray images such as those shown in Fig. 7 for Shot 2235 (m = 1.3 mg, R = 12 mm, N = 192 wires) taken during the mass scan, or in Fig. 8 for Shot 2085 (m = 0.6 mg, R = 8.6 mm, N = 90 wires), taken during the wire-number scan.

In general, by adjusting a random density perturbation (14,15) in a given initial plasma-shell thickness, one can produce rz-RMHC simulations that replicate many of the features measured at the first implosion. This perturbed density seeds the growth of a magnetic Rayleigh-Taylor (RT) instability in the rz plane, which limits the radial convergence. An emission opacity approximation in the simulations, however, allows greater radiation cooling to occur than is physical (12). The cooler plasma has reduced radial expansion following stagnation and the simulated plasma rapidly reaches a quasi-equilibrium without going through the second compression seen experimentally. A reduction in the plasma emissivity, as is done in some of the E-RMHC simulations (to account for the reduced emission from aluminum), nevertheless, permits more realistic peak powers to be simulated and simultaneously generates a second convergence as shown in Figs. 6A and 6B.

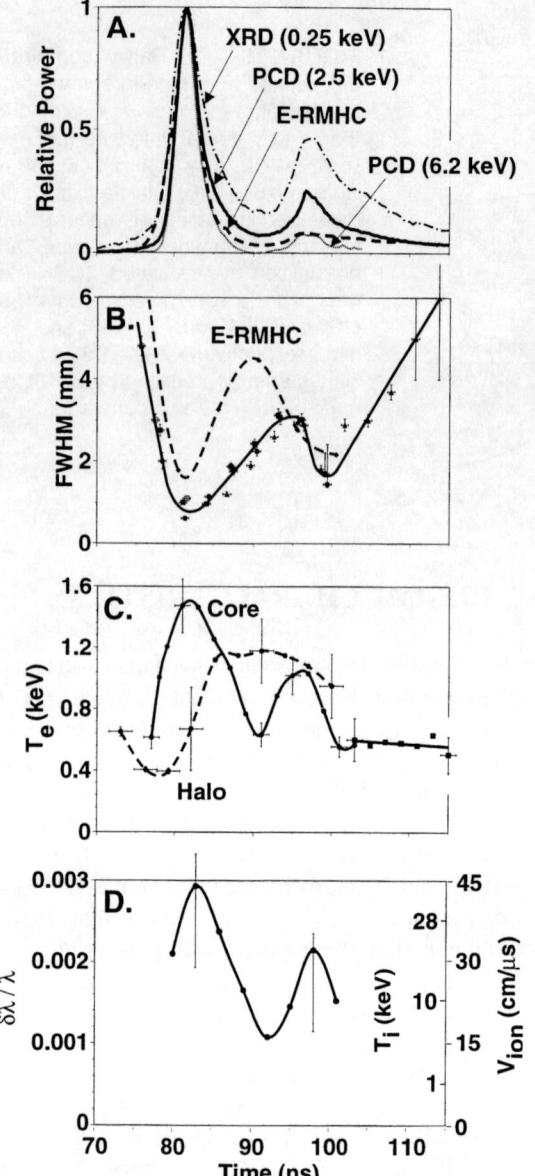

FIGURE 6.
(A) Comparison of radiation pulses measured in XRDs and PCDs sensitive to x rays in bands about 0.25, 2.5, and 6.2 keV with that simulated by the E-RMHC using reduced plasma emissivity for the 90-wire 8.6-mm radius Shot 2094.
(B) Comparison of the pinch diameter measured from fast-framing x-ray pinhole cameras (sensitive to x rays greater than ~ 1 keV) with the E-RMHC FWHM of the z-averaged radial mass profile associated with the Fig. 6A simulation, for three 90-wire, 8.6-mm radius Shots 2085, 2094, and 2095.
(C) Comparison of the core electron temperature derived from free-bound slope for Shot 2094 with the average halo electron temperature determined from the ratio of He-like to H-like transitions for Shots 2085, 2094, and 2095.
(D) Opacity corrected, Doppler broadened, Lyman-beta, relative line-width averaged over Shots 2085, 2094, and 2095. Right-hand scales indicate inferred generalized ion temperatures and associated velocities.

Near peak emission, the radial lineouts of x-ray images like that shown in Fig. 8A (0 ns) are well fit by the sum of an intense Gaussian-like core surrounded by a diffuse Gaussian-like halo, when integrated over the axial length of the pinch. Radially-resolved images of the x-ray spectrum show that K-shell free-bound

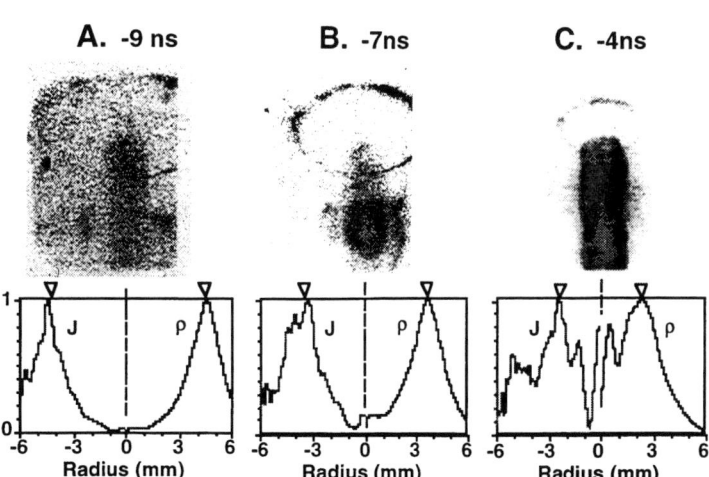

FIGURE 7. Comparison of fast-framing x-ray camera images [top] (sensitive to 0.2 to 0.3 keV and greater than 1 keV x rays) with E-RMHC simulations [bottom] of mass and current density, −9, −7, and −4 ns before peak power for Shot 2235. Arrows correspond to RMS averages.

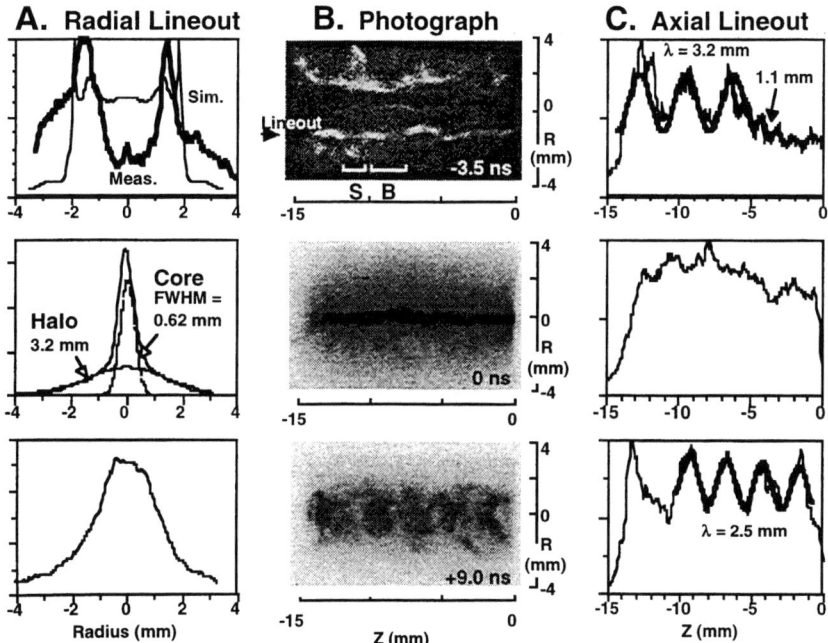

FIGURE 8. Comparison of fast-framing x-ray camera images (center) and their associated lineouts −3.5 (including the L-RMHC simulated radial lineout), 0, and 9 ns with respect to the time of peak power for Shot 2085. Images are sensitive to x-rays greater than 1 keV. The −3.5-ns image was also sensitive to 0.2 to 0.3 keV x rays.

continuum emission originates from the core, whereas K-shell line emission originates from both the core and halo (13), in agreement with 1D-RMHC simulations (19). The time-resolved slope of the optically-thin free-bound emission measured with filtered photoconducting detectors (PCDs) gives a spatially average temperature of the electrons in the core (13), and the ratio of the time-resolved, optically-thick hydrogen-like to helium-like line emissions gives a measure of the electron temperature in the halo (19) (Fig. 6C). The Doppler-broadened line emission provides an estimate of a generalized ion temperature corresponding to a combination of line-emission shifts caused by the coherent radial motion of the plasma, motion from fluid turbulence, micro instabilities, and intrinsic ion thermal motions (Fig. 6D) (12). The measured hot followed by cooler peaks in the core electron and ion temperatures/velocities correlate in time with the two radiation bursts and radial compressions, as expected for multiple implosions (Fig. 6).

At peak convergence, the temperature of the halo (~ 0.4 keV) is estimated to be three to four times cooler than that of the core (~ 1.4 keV) (Fig. 6C), indicating the presence of strong temperature gradients. While both line and continuum emissions are generated in the hot plasma core at the time of peak convergence, only the optically-thin continuum x rays escape the plasma directly. The optically-thick line emission is absorbed and reemitted from the cooler regions until the pinch expands and the plasma opacity declines. The high temperatures inferred from the line analysis at times when the plasma is expanding and the core electrons are cooling (Fig. 6C) suggests a lag in recombination (19). That is, the ionization state is likely no longer in equilibrium with the cooled electrons, but reflects instead the earlier state of the plasma when the electrons were hot and the plasma was in equilibrium. The core electrons cool too rapidly in this case for recombination processes to maintain ionization equilibrium. Alternatively, the rz L-RMHC simulations show rapid stirring of the plasma after stagnation, with development of a boundary between a radiatively dominated, high-density, cooling core and a low-radiating, low-density, hotter halo. The boundary between the hotter and cooler regions has a reversed pressure gradient ($\partial p/\partial r > 0$) and is stable, preventing convective mixing across the boundary. Convection does not cross the boundary, allowing a temperature gradient to exist as suggested by the temperature difference measured between the cooler core and hotter halo midway between the two implosions (Fig. 6C).

The early time characteristics of the plasma-shell implosions are illustrated in Fig. 7 and Fig. 8A (–3.5 ns). The rings evident in Fig. 7 are hypothesized to be from a radiating plasma produced by the electron-current flow into the surface of the imploding plasma shell (Fig. 4A). In support of this hypothesis, the radii of the rings correlate with both the mass-averaged and current-density averaged radii calculated by the E-RMHC simulation (Fig. 7). In the simulation, the calculated radially-converging plasma precursor from instability bubble regions (15) begins to stagnate on axis 9 ns prior to peak emission (Fig. 7A), reaching electron temperatures of ~ 1 keV. As indicated by the simulation and measured, this precursor becomes detectable just at this time. As the stagnation evolves in the simulation, plasma accretes on axis, and together with the inward moving shell the appearance of a hot inward-moving luminescent shell is generated (Fig. 7C). The L-RMHC simulation of such a shell, which takes account of camera sensitivity and 35° viewing angle (Fig. 4A), is shown in Fig. 8A (–3.5 ns, Sim.) for Shot 2085 (12).

These observations are consistent with the 1D-RMHC analyses (19) that required a cool shell of plasma to implode onto a small amount of precursor plasma to explain the total to K-shell power ratio. In this model, the cool outer shell of plasma is needed to explain the greater than 4-to-1 power ratio measured, and it is also consistent with the spectral analysis (13,19) which found that only 20% of the mass contributes to the K-shell radiation at the time of peak power. Moreover, the experimentally observed size of the K-shell emission region being smaller than the calculated mass-averaged plasma size (Figs. 6B and 7C) is consistent with this low-mass fraction contributing to the K-shell emission.

In Fig. 8B (–3.5 ns), the bubble and spike structure associated with the RT instability is seen. Axial lineouts of the image show the presence of ~ 1-mm wavelengths that spatially merge with ~ 3-mm wavelengths (Fig. 8C (–3.5 ns)). These wavelengths are characteristic of those simulated by the L-RMHC (12) and E-RMHC (1), respectively. As the pinch expands after stagnation, an m=0 sausage instability sets in (Fig. 8B (9 ns)). The correspondence between the axial structure measured prior to (Fig. 8C (–3.5 ns)) and after stagnation (Fig. 8C (9 ns)) suggests that the RT instability seeds the sausage instability. These observations are consistent with the rz-RMHC simulations.

MASS AND RADIUS VARIATION

The variation in the widths of the total radiation pulse that was measured in the mass and radius experimental sequence is well described by E-RMHC simulations in which the magnitude of the random density perturbations (10,11) is kept constant (Figs. 9A and 9E). The dashed curve in Figs. 9A, 9B, 9E, and 9F corresponds to the use of the nominal plasma emissivity in E-RMHC (15). The associated solid curve corresponds to calculations using a reduced emissivity, applied to match the measured radiation peak power at ~0.6 mg load mass (Fig. 9B), and which allows the appearance of a second peak in the radiation pulse (Fig. 6A). Differences between the two curves provide an estimate of the uncertainty introduced by the approximation used for the Planckian emission-absorption opacity in the E-RMHC. As shown in Figs. 9B and 9F, the measured peak power is bracketed by either calculation, which, together with the agreement in pulse width, provides credibility to the average hydrodynamics of the E-RMHC simulations.

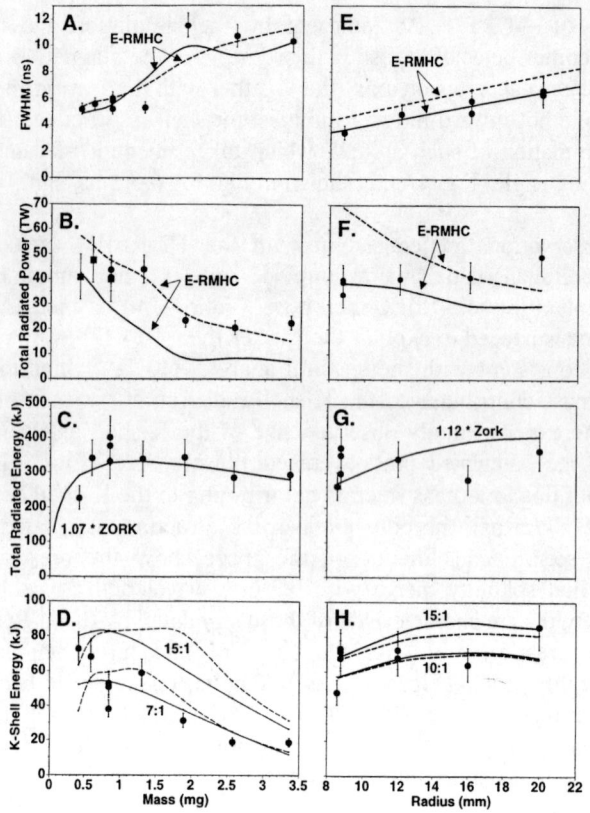

FIGURE 9. (A) Measured (uncorrected) and E-RMHC calculated (dashed curve–nominal emissivity, solid curve–reduced emissivity) total-radiated power pulse width versus load mass. (B) Measured (uncorrected) and E-RMHC calculated (dashed curve–nominal emissivity, solid curve–reduced emissivity) total radiated power versus load mass. ■ corresponds to loads with less than 192 wires corrected for gap effect (Fig. 4). (C) Measured total radiated energy (uncorrected) and (1.07±0.14) times the ZORK calculated kinetic energy using measured radial convergence of 20:1) versus load mass. (D) Measured K-shell yield and that simulated by the Mosher two-level model (solid curve) and the Whitney-Giuliani scaling model (dashed curve) versus array mass for 7:1 and 15:1 radial convergences. (E) Estimated (measured using XRD filtered by 5 μm Kimfol) and E-RMHC calculated (dashed curve–nominal emissivity, solid curve–reduced emissivity) total radiated power pulse width versus array radius. (F) Measured (uncorrected) and E-RMHC calculated (dashed curve–nominal emissivity, solid curve–reduced emissivity) total radiated power versus array radius. (G) Measured total radiated energy (uncorrected) and (1.12±0.28) times the ZORK calculated kinetic energy using measured radial convergence (which increased from 19:1 to 37:1 as radius increased from 8.6 to 20 mm, respectively) versus array radius. (H) Measured K-shell yield and that simulated by the Mosher two-level model (solid curves) and the Whitney-Giuliani scaling model (dashed curves) versus array radius for 10:1 and 15:1 radial convergences.

These simulations show an evolution of the RT instability from short to long wavelengths similar to that seen previously (15). When a wavelength of the order of the shell thickness is reached, the bubble bursts through the shell and the shell thickness increases. As the bubbles burst, current immediately begins to flow in the low-density material between the spikes. This flow results in a continued acceleration of the shell, with only a small amount of bubble material being thrown ahead of the main body of plasma. Since the shell is still accelerating, the instability growth continues. Evolution to longer wavelengths continues again until a wavelength of the order of the new shell thickness is reached. At this time, the shell again bursts, now at longer wavelengths, which results in a significant amount of material being accelerated to the axis (Fig. 7A) and the beginning of the radiation pulse (Fig. 6A).

The total radiation yield in the simulations varies only slightly as was observed experimentally (Figs. 9C and 9G). The change in radiated power (Fig. 9B and 9F) results primarily from changes in the pulse width (Figs. 9A and 9E). In zeroth order, the pulse width is the shell thickness just prior to stagnation divided by the final average velocity. From the simulations, the lower load masses results in higher accelerations, higher final average velocities, and an increased instability growth. This growth results in a thicker shell at stagnation. At increased accelerations, however, the two-stage development of the instability produces a proportionally smaller increase in the shell thickness compared to the increase in the final average velocity. This combination results in the higher peak radiated power observed at the lower masses (Fig. 9B). Simulations of the radius variation series show that the load acceleration, the average final velocity, the instability growth, the implosion time, and the energy coupling to the load all increase as the radius increases. In the simulations, the shell thickness (due to instability growth) increased proportionately more than did the velocity, thereby lengthening the x-ray pulse width as the array radius increased. For the simulations that used the reduced plasma emissivity, the combination of increased pulse width (Fig. 9E) and increased energy coupling to the load resulted in a radiated power that was relatively flat as the radius increases (Fig. 9F).

For the experimental data, including those of the wire number scan, the kinetic energy calculated from the ZORK model (using the measured radial convergence obtained from the FWHM of the radial lineouts of the time-integrated images of the K-shell emission) accounts for ~ 85±16% of the total x-ray emission measured by the bolometer (uncorrected for spectral sensitivity and viewing angle) (Figs. 5B, 9C, and 9G). L-RMHC simulations suggest that the bolometer is sensitive to ~ 85% of the total x-ray emission (12). E-RMHC simulations suggest that the K-shell lineouts overestimate the radial convergence by a factor of two (see Figs. 6B and 7C, for example) and thus the ZORK kinetic energy by ~ 30%. Therefore, about 60% of the actual total radiated energy can be accounted for by the 0D model of kinetic energy delivered to the load. The remaining 40% is likely due to the effect of two-dimensional features including increased late-time pdV work and a small increase in the amount of Joule heating (10-12).

In Figs. 9D and 9H, the two-level (10) (solid curves) and the 1D Whitney-Giuliani (20) scaling-model (dashed curves) of K-shell yields are in qualitative agreement with those measured. These K-shell models depend only on the kinetic energy delivered to the load, the effective stagnation radius, and participating mass, and ignore the multi-dimensional dynamics of the implosion and plasma radiation after stagnation. The calculations showin in Figs. 9D and 9H used ZORK kinetic energies derived from convergence ratios varying between 7:1 and 15:1 with the full imploded mass. The calculated variation of yield with mass is due to the change in both temperature and density for the nearly-fixed kinetic energy (Fig. 9C). The calculated yield with radius is due primarily to the slight variation in the kinetic energy (Fig. 9G). The constancy of yield with radius reflects the optically-thick nature of the emission for aluminum and the nearly constant temperature. The need for compression ratios that are less than those calculated from lineouts may reflect unaccounted for temperature gradients (Fig. 6C) that reduce the experimental mass participation, a common problem with the one-zone scaling models. This conclusion is also reached in analysis of neon gas-puff implosions (21). This argument is supported by the L-RMHC (12) and E-RMHC (Fig. 6B) simulations indicating that radial K-shell line-outs overestimate radial plasma convergence.

SUMMARY

Increasing the wire number has produced significant improvements in pinch quality, reproducibility, and x-ray output power (1). The mass variation experimental series, which was done for high-wire-number plasma-shell loads, shows that a factor of two decrease in pulse width (and an associated doubling of the radiated x-ray power) and a greater-than-two increase in K-shell yield occurs when the load mass transitions from masses greater than 1.9 mg to those less than 1.3 mg, in qualitative agreement with the E-RMHC and simple K-shell models, respectively (10). In the high-power, low-mass, plasma-shell regime, the peak power and K-shell yields remain relatively flat with increased radius, again in agreement with the E-RMHC and the K-shell models, respectively (11). Importantly, the high quality of the implosions generated by the large-radius loads supported using such high-wire-number, plasma-shell loads for PBFA-Z (22,23). The large radius enabled the initial electrical stress on PBFA-Z due to load inductance to be kept to a minimum, as the power flow in the upstream portion of the accelerator was improved. In conclusion, these experiments have conclusively demonstrated the necessity of using a large number of wires within a wire array in order to optimize the x-ray energy and power output from z-pinch implosions, and they represent a breakthrough in z-pinch implosion load configurations for large pulsed power generators (4).

ACKNOWLEDGMENTS

We thank T. L. Gilliland, D. O. Jobe, J. S. McGurn, P. E. Pulsifer (NRL), J. F. Seamen, K. W. Struve, W. A. Stygar, J. W. Thornhill (NRL), M. F. Vargas and the Saturn crew for their dedicated technical support; G. O. Allshouse, Y. Maron (Weizmann Institute), A. Toor (LLNL), M. Tabak (LLNL), and C. Deeney for useful discussions; M. K. Matzen, J. E. Maenchen, and D. H. McDaniel, M. E. Jones (LANL), A. Toor (LLNL), and J. Davis (NRL) for vigorous programmatic support; M. K. Matzen for carefully reviewing and L. O. Peterson for typing this manuscript. Sandia is a multiprogram laboratory operated by Sandia Corporation, a Lockheed Martin Company, for the United States Department of Energy under Contract DE-ACO4-94AL85000.

REFERENCES

1. Sanford, T. W. L., et al., *Phys. Rev. Lett.* **77**, 5063 (1996).
2. Bloomquist, D. D., et al., *Proc. 6th Int. IEEE Pulsed Power Conf.*, Arlington, VA, ed. P. J. Turchi and B. H. Bernstein (IEEE, New York, 1987), p. 310.
3. Deeney, C., et al., "Power Enhancement by Increasing Initial Array Radius and Wire Number of Tungsten Z Pinches," submitted to *Phys. Rev. E* (1996).
4. Matzen, M. Keith, *Phys. Plasmas* **4**, 1519 (1997).
5. Pereira, N. R. and Davis, J., *J. Appl. Phys.* **64**, R1 (1988).
6. Spielman, R. B., et al., *Dense z-Pinches*, 3rd Intern Conf, London, UK, 1993, AIP Conf Proceed 299, editors H. Haines and A. Knight, American Institute of Physics, New York 1994, L.C. No. 93-74569, p. 404-420.
7. Deeney, C., et al., "Improved Large Diameter Wire Array Implosions from Increased Wire Array Symmetry and On-axis Mass Participation", to be submitted to *Phys. Plasmas* (1997).
8. Spielman, R. B., et al, *10th International Conference on High-Power Particle Beams*, Prague, Czech Republic, June 10-14, 1996, ed. K. Jungwirth and J. Ullschmied, p. 150.
9. Spielman, R. B., et al., *Bull. Am. Phys. Soc.* **41**, 1422 (1996).
10. Sanford, T. W. L., Peterson, D. L., Roderick, N. F., Mosher, D., et al., "Increased X-ray Power Generated from Low-Mass Larger-Number Aluminum-Wire-Array Z-Pinch Implosions," to be submitted to *Phys. Plasmas* (1997).
11. Sanford, T. W. L., Peterson, D. L., Roderick, N. F., Mosher, D., et al. "Symmetric Aluminum-Wire Arrays Generate High-Quality z-Pinches at Large Array Diameters", to be submitted to *Phys. Plasmas* (1997).
12. Sanford, T. W. L., et al., *Phys. Plasmas* **4**, 2188 (1997).
13. Sanford, T. W. L., et al., *Rev. Sci. Instrum.*, 68, 852 (1997).
14. Hammer, J. H., et al., *Phys. Plasmas* **3**, 2063 (1996).
15. Peterson, D. L., et al., *Phys. Plasmas* **3**, 368 (1996).
16. Matuska, W., et al., *Phys. Plasmas* **3**, 1415 (1996).
17. Thornhill, J. W., Whitney, K. G., and Davis, J., *J. Quant. Spectrosc. Radiat. Transfer* **44**, 251 (1990).
18. Sanford, T. W. L., et al., Sandia National Laboratories Technical Report SAND94-0694 (June 1994).
19. Whitney, K. G., et al., "Analyzing time-resolved spectroscopic data from an azimuthally symmetric, aluminum," to be published, *Phys. Rev. E* (1997).
20. Whitney, K. G., et al., *J. Appl. Phys.* **67**, 1725 (1990).
21. D. Mosher et al., these proceedings.
22. R. B. Spielman et al., these proceedings.
23. C. Deeney et al., these proceedings.

Instabilities in Z-pinch and Liner Systems

P. V. Sasorov[1], A. A. Esaulov[1] and S. L. Nedoseev[2]

[1] *Institute for Theoretical and Experimental Physics, Moscow, 117259, Russia*
[2] *TRINITI, Troitsk, Moscow region, 142092, Russia*

Abstract

Recent results concerning plasma instabilities and their influence on Z-pinches and liners dynamics are considered in the report. Three-dimensional two-fluid instabilities (TFI) can be responsible for the high value of anomalous electric resistance of plasma column. Two-dimensional two-fluid MHD simulation, taking into account such anomalous resistance, was performed to evaluate a threshold of sausage instability stabilization. This threshold is expressed in the terms of Z-pinch mass per unit length. Competition of the TFI's and usual Rayleigh-Taylor instability (RTI) for thin plasma liner is also discussed. The effect of an extremely fast redeployment of current from liner to a central load which had been discovered previously by the ANGARA-5 team is examined from the viewpoint of liner instabilities.

It is well known that different plasma instabilities are the most striking feature of dense Z-pinches. The most significant instability of Z-pinches leads to formation of sausage necks and to disruption of plasma column. Results of several tens of works devoted to this problem are summarized for example in [1,2]. The main conclusion presented there is that there is actually no region for stability of Z-pinches in the frame of one fluid ideal or dissipative magnetohydrodynamics (MHD), when collision frequencies are high. Recent investigations of Z-pinch instabilities in collisionless regime, see for example Refs. [3-5], show that collisionless effects don't stabilize Z-pinches.

When electron current velocity u is comparable with or exceeds alfven (c_A) and sound velocities in Z-pinch plasma, two-fluid effects become very important and determine growth rates of instabilities [6-8]. In this case shortwavelength disturbances turn out to grow much faster than the global one. This situation can lead to the global instabilities of Z-pinches be strongly affected by fully developed small scale turbulence. It was shown in Ref. [8] that mutual excitation of drift helicons and fluctuations of ion density (DHIF) causes an anomalous resistance of the plasma column, which can be estimated in terms of effective electron-ion collision frequency as $\nu_{\text{eff}} \sim \omega_{Be}$, where ω_{Be} is a characteristic value of the electron gyrofrequency. This mechanism is similar to the mechanism of anomalous resistance on low hybrid drift (LHD) turbulence

[9-11], but much more effective. Moreover it is the most effective one in the relevant parameter region. Taking into account the LHD anomalous resistance, a simplified 1.5-dimensional simulation of sausage neck dynamics [12] showed the possibility of saturation of the instability for sufficiently low line mass of the neck. It was argued in Ref. [8], that replacing LHD resistivity by DHIF resistivity can lead to sausage instability saturation for $u \sim c_A$ or for neck radius comparable with the ion skin depth.

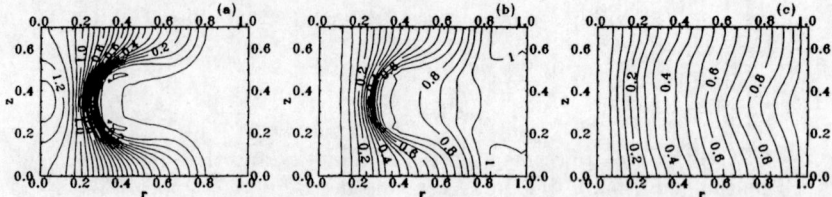

Figure 1: Simulation of the sausage instability. Figs. (a) and (b) correspond to the case $k \equiv \pi\rho(0)(r_{1/2}eZ/Am_Ac)^2 = 3 \cdot 10^4$ at the moment t defined by $c_s t/r_{1/2} = 2.6$. Fig. (c) corresponds to the case $k = 153$ at the moment t defined by $c_s t/r_{1/2} = 5.6$. Fig. (a) shows the lines of constant plasma density normalized over axial density at $t = 0$. Figs. (b) and (c) show electric current lines, that are the lines of constant Br normalized over external boundary value.

We have performed two-dimensional two-fluid simulation of the sausage instability to check the suggestions mentioned above. We used the code recently elaborated and used for simulation of plasma opening switches (POS) [13]. Ion and electron plasmas are simulated simultaneously for the axially symmetrical geometry. To take approximately into account the shortwavelength three-dimensional DHIF turbulence anomalous resistivity and anomalous heat conductivity are introduced, applying that locally $\nu_{\text{eff}} = \omega_{Be}$. Fig. 1 shows the simulated sausage instability development for two plasma columns which have different initial masses, whereas initial column radiuses ($r_0 = 0.5$ in dimensionless units) and relative amplitude of initial disturbances are the same. Note that the lighter mass Z-pinch was simulated for a significantly longer time period measured in ion transit time. Fig. 1 shows that electric current lines do not tend to be compressed in the case of lighter mass plasma column. Thus we conclude that our model demonstrates a saturation of the sausage instability for $\pi\rho(r=0)r_{1/2}^2 \leq k_{cr}(A/Z)^2(m_A c/e)^2$, where $r_{1/2}$ is the radius where the plasma density is half of its axial value $\rho(r=0)$, A and Z are the mass number and mean charge of the plasma ions respectively, and m_A is the atomic mass unit. Our model gives $k_{cr} \approx 150$.

Turn now to the 2nd item of the report. Experiments [14] with double (compound) Z-pinches (2÷3 MA, 100 ns) were performed recently by the ANGARA-5 team. Plasma liners with low mass per unit length ($\sim$10÷40 μg/cm) and with initial diameter of the order of 3.5 cm were used as a first cascade of the double Z-pinch. Axial wires or foam bars were used as an inner load. The most striking feature of these experiments is the fast brightening of the inner load in x-rays which lasts a few nanoseconds. This brightening occurs at early stage of liner compression, when the liner flies over only a few of its initial thickness. The liner is destroyed near this moment. Two possible explanations of this phenomena were discussed [14]. The 1st of them assumes that impact of inner layers of the liner (accelerated by penetrating magnetic field) on the inner load is responsible for its excitation. The 2nd one assumes that significant portion of the total electric current is redeployed suddenly from the liner to the inner load as a result of some quick process which may be similar to the mechanism of POS operation considered in Ref. [15,14]. This mechanism implies a substantial heating of current carrying plasma due to the DHIF anomalous resistance mentioned above and explosive ejecting of the plasma onto cold electrodes.

We examine here the latter explanation of the ANGARA-5 experiments. One dimensional dynamics of liner, background plasma and inner load is simulated by modification of two codes NPINCH [16] and MAGDA (by A. V. Gerusov, [17]). Since u is comparable to c_A for such low mass liners, the anomalous resistivity considered above is taken into account for liner and background plasma. Ionization is treated as non steady-state process. Line emission and reabsorption in resonant lines of most abundant ions are treated using the Biberman-Holstein method.

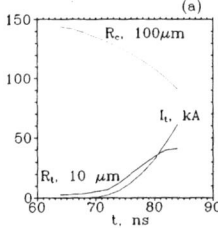
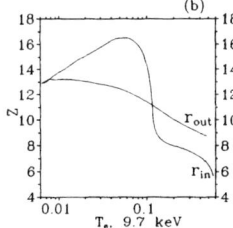

Figure 2: Results of one-dimensional MHD simulation of Xe liner (20 μg/cm, initial outer diameter = 3.5 cm, I_{max} = 3.5 MA at t = 100 ns) with background Xe plasma (5% of the liner mass) and with wire as the inner load. Fig. (a) shows radius of the maximum liner density (R_c), inner load radius (R_t) and electric current through the inner load versus time. Parametric plot of electron temperature (T_e) and mean ion charge with the radius as a parameter at t = 75 ns is presented on Fig. (b). It indicates evidently that ionization is not steady-state.

From this simulation one may conclude the followings. Though the resistance of the liner turns out to be about order of magnitude more than the Spitzer one, the rate of plasma cooling due to emission is so high, that liner plasma has relatively low temperature and high density. The liner thickness on the acceleration stage becomes about order of magnitude less than its initial value, and is of the order of several tenths of mm. Liner plasma pressure is much less than the magnetic one. For these reasons the rates of ionization are high enough that leads to the high degree of ionization and to considerable decreasing of the anomalous resistance. These properties of liner dynamics prevent the liner to operate as POS in accordance with [15].

Nevertheless owing to the anomalous resistance electric current penetrates the inner load and explodes it always earlier than the shock wave strikes it. See Fig. 2. However 1d-models give too low amplitude of central load current.

Strong decreasing of liner thickness during its acceleration opens the possibility of its very fast disruption due to the RT instability. Our estimations show that this disruption should occur when the liner flies over only $1 \div 2$ its initial thickness (in accordance with the experiment) and that combination of the RT instability and the anomalous resistance of liner and background plasmas can give considerable and sufficiently fast current redeployment to the inner load. The results reviewed briefly in this paragraph will be published elsewhere in more details.

References

1. Haines, M. A., et al., *Phys. Rev. Lett.* **66**, 1462 (1991).
2. Bobrova, N. A., et al., "Dynamics and Stability of Dense Z-pinches", in *Dense Z-Pinches*, Eds. M. Haines & A. Knight, AIP Press, NY, 1993, p. 10.
3. Isichenko M. B., et al., *Fiz. Plasmy* **15**, 1064 (1989) [*Sov. J Plasma Phys.* **15**, 723 (1989)].
4. Arber, N. D., et al., *Phys. Fluids B* **3**, 1152 (1991).
5. Arber, N. D., *Phys. Rev. Lett.* **77**, 1766 (1996).
6. Gordeev, A. V., et al., *Fiz. Plasmy* **18**, 3 (1992) [*Sov. J Plasma Phys.* **18**, 3 (1992)].
7. Sheffel, J., et al., "Linear Stability of Large Larmor Radius Z-Pinches" in *Dense Z-Pinches*, Eds. M. Haines & A. Knight, AIP Press, NY, 1993, p. 75.
8. Sasorov, P. V., *Fiz. Plasmy* **18**, 275 (1992) [*Sov. J Plasma Phys.* **18**, 138 (1992)].
9. Davidson, R. C., et al., *Phys. Fluids* **18**, 1327 (1975).
10. Davidson, R. C., et al., *Nuclear Fusion* **17**, 1313 (1977).
11. Huba, J.D., et al., Phys. Fluids **B 2**, 1676 (1990).
12. Korzhavin, V. M., et al. *Fiz. Plasmy* **4**, 735 (1978) [*Sov. J. Plasma Phys.* **4**, 357 (1978)].
13. Esaulov, A. A., et al., *Fiz. Plasmy* **23**, (1997)[*Plasma Phys. Rep.* **23**, 357 (1997)]; accepted for publication.
14. Branitskij, A. I., et al., *Fiz. Plasmy* **22**, 272 (1996)[*Plasma Phys. Rep.* **25**, 272 (1996)].
15. Sasorov, P. V., *Pis'ma Zh. Eksp. Teor. Fiz.* **56**, 614 (1992) [*JETP Lett.* **56**, 599 (1992)].
16. Bobrova, N. A., et al, *Fiz. Plasmy* **18**, 517 (1992) [*Sov. J. Plasma Phys.* **18**, 269 (1992)].
17. Gerusov, A. V., "Numerical Simulation of a Hollow Z-Pinch ...", in *Dense Z-Pinches*, Eds. M. Haines & A. Knight, AIP Press, NY, 1993, p. 129.

Metallic Wire Pinch Instability

P.L. Auer and D.D. Ryutov

Laboratory of Plasma Studies, Cornell University, Ithaca, NY 14853

Abstract. Pinch experiments by D.A. Hammer[1] and associates on metallic wires show clear evidence of an m=0 instability, though of a non-disruptive nature, but no sign of an m=1 instability. The experiments also indicate that the pinch plasmas contain an inner core of dense material that may still be solid. This has prompted us to re-examine the original calculations of pinch instability by R.J. Tayler[2] and to modify his model by including a solid core to the plasma. Details of our calculations for the m=0 and m=1 instability will be presented here. We find that the presence of a solid core does affect the growth rates, as may be expected, but does not appear to suppress the m=1 mode relative to the m=0 mode.

INTRODUCTION

The simplest and most unstable model of a Z-Pinch assumes a uniform plasma, ideal MHD, and a surface current such that B vanishes inside the plasma but pressure equilibrium exists so that $p_0 = B^2/2\mu_0$. Here, B is the vacuum value of the magnetic field at the plasma edge and p_0 is the unperturbed plasma pressure. The plasma is surrounded by a vacuum and one looks for perturbations that vary as $f(r)\exp\{i(m\theta+kz) + \omega t\}$.

PLASMA MODEL

The perturbed field in the vacuum satisfies both

$$\nabla \cdot \mathbf{B}_1 = 0 = \nabla \times \mathbf{B}_1 .$$

So $\mathbf{B}_1 = \nabla \psi$

$$\psi = A\, K_m(kr)$$

where A is an arbitrary constant and both m and k are assumed to be real, positive quantities and K_m is a modified Bessel function. Inside the plasma the perturbed pressure, along with all other plasma variables, obeys

$$\nabla^2 p_1 - (\omega/c_s)^2 p_1 = 0$$

where $c_s^2 = \gamma p_0/\rho_0$ is the speed of sound. Accordingly, one finds

$$p_1 = C\, I_m(qr)$$

where C is an arbitrary constant, $q^2 = k^2 + (\omega/c_s)^2$ and I_m is also a modified Bessel function. The perturbed plasma displacement ξ and outward normal $\hat{n}$ may be obtained from p_1 as follows

$$\xi_r = -\frac{1}{\omega^2 \rho_0} \partial_r p_1$$

$$\hat{n}_\theta = -\frac{1}{r} \partial_\theta \xi_r|_a$$

and $r = a$ is the outer radius of the unperturbed plasma. The dispersion relations result from the boundary conditions at $r = a$ that require continuity in the normal component of **B** and the total pressure.

$$B_0 \hat{n}_\theta + B_{1r} = 0$$

$$p_1 - \frac{1}{\mu_0} B_0 B_{1\theta} - \frac{1}{2\mu_0} \xi_r \partial_r B_0^2 |_a = 0.$$

This leads to the following dispersion relation

$$ka + m^2 \frac{K_m(ka)}{K_m'(ka)} = ka\, \frac{\omega^2 a^2}{qau_A^2} \frac{I_m(qa)}{I_m'(qa)}. \quad (1)$$

where prime denotes differentiation with respect to argument, $u_A = B_0/\sqrt{\mu_0 \rho_0}$ is the Alfven velocity and the above is in agreement with the results of Tayler (1957).

SOLID CORE

One now assumes the plasma is bounded on the inner surface by a solid core at $r = b$, while the outer edge is still at $r = a$. The solution for p_1 has to be modified so that now

$$p_1 = C\, I_m(qr) + B\, K_m(qr)$$

and one eliminates the arbitrary constant B through the boundary condition that $\xi_r = 0$ at $r = b$. Thus, one finds

$$ka + m^2 \frac{K_m(ka)}{K'_m(ka)} = ka \frac{\omega^2 a^2}{q a u_A^2} \frac{I_m(qa) K'_m(qb) - I'_m(qb) K_m(qa)}{I'_m(qa) K'_m(qb) - I'_m(qb) K'_m(qa)} \quad (2)$$

for the case of a plasma with a solid core at $r = b$.

RESULTS

We plot the growth rate, $\omega a/u_A$, against the wave number ka. The $m=0$ and $m=1$ instabilities are treated for both a compressible and incompressible plasma. In general, the compressible plasma has the greater growth rate. Although the growth rate for $m=1$ is greater than for $m=0$ at long wavelengths ($ka<<1$), the reverse is true as the wavelength decreases. The results for the incompressible case with $m=0$ is given in Figure 1, and by comparison Figure 2 is for $m=1$.

FIGURE 1. Incompressible $m=0$; b=ratio of inner to outer radius

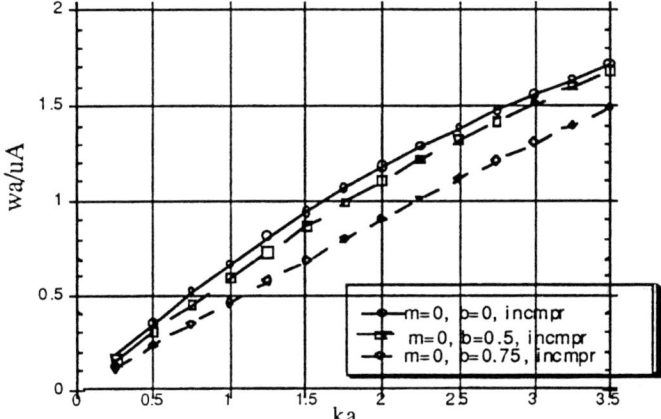

As indicated here, b measures the ratio of inner to outer radius. Accordingly, $b=0$ represents the uniform plasma without an inner core.

FIGURE 2. Incompressible m=1; b=ratio of inner to outer radius

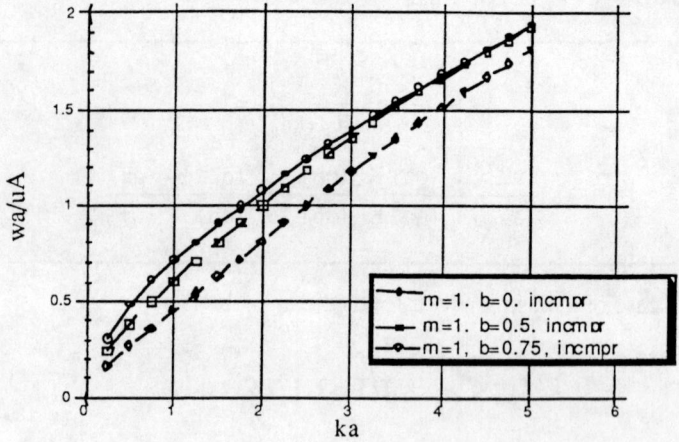

The case of the compressible plasma with m=0 and m=1 is shown in Figures 3 and 4, respectively.

FIGURE 3. Compressible m=0; b=ratio of inner to outer radius

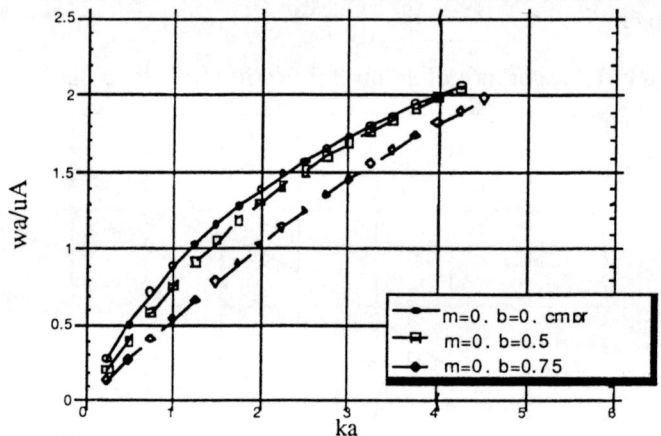

FIGURE 4. Compressible m=1; b=ratio of inner to outer radius

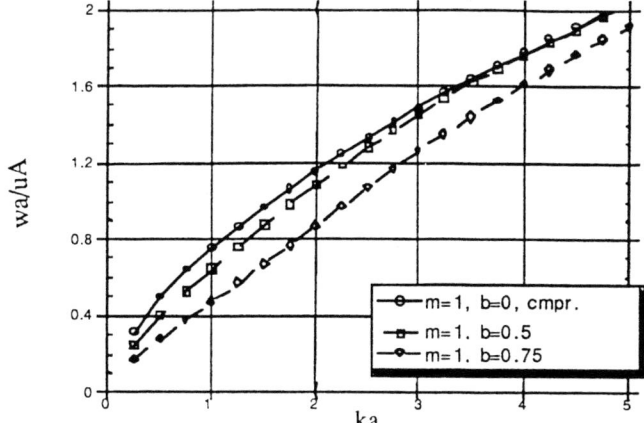

In Figure 5 we plot the ratio of the m=1 growth rate to that of the m=0 growth rate for an incompressible plasma and compare the results for a plasma with and without a solid core. As expected, the ratio of growth rates exceeds unity at long wavelengths, but then decreases below unity as the wavelength decreases.

FIGURE 5. Incompressible Ratio of Growth Rates w(m=1, b)/w(m=0, b)

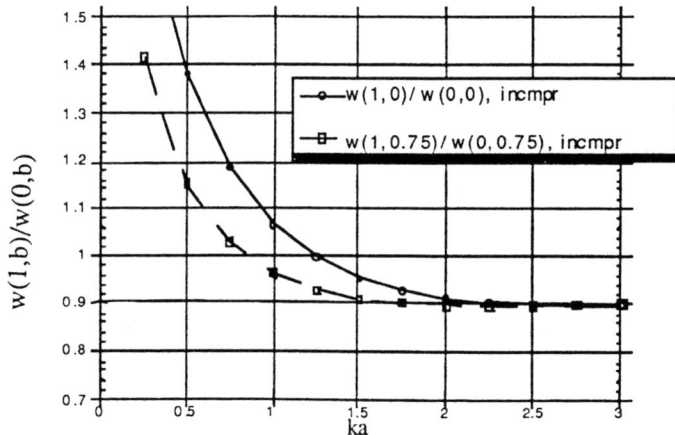

The ratio of the m=1 growth rate to that of the m=0 growth rate for a compressible plasma is given in Figure 6, and the results for a plasma with

and without a solid core are compared as above. Note the difference between the compressible and the incompressible cases, in that the effect of a solid core is less pronounced when the plasma is compressible.

FIGURE 6. Compressible Ratio of Growth Rates w(m=1,b)/w(m=0,b)

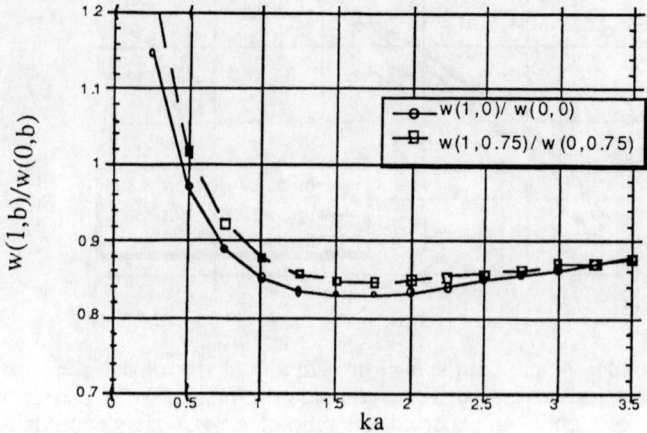

CONCLUDING REMARKS

The results shown here indicate that assuming the plasma to have a solid core is not sufficient to explain the absence of m=1 instabilities in metallic wire Z-pinches as suggested by the experimental observations. The cases examined here treated the plasma as an ideal MHD fluid with surface currents. It is known that departures from this simple model tend to decrease growth rates, by comparison. We have also examined the case of a plasma with a uniform current distribution, but the results are not given here. The effect of a solid core in that instance is rather similar to what is shown here and does not offer any greater support for the notion that a solid core, as such, will suppress the m=1 instability. It should be noted that these conclusions are based on linear analyses of the m=0 and m=1 instabilities. The observed difference in these two modes of instability, however, may be due in part to non-linear effects.

1. D.H. Kalantar and D.A. Hammer, Phys. Rev. Letters **71**, 3806 (1993).
2. R.J. Tayler, Proc. Phys. Soc. **B70**, 31 (1957); Phil, Mag. **2**, 33 (1957).

RECENT PROGRESS ON LARGE LARMOR RADIUS THEORY

M. Coppins, T.D. Arber[1], P.G.F. Russell[2], J. Scheffel[3]
Imperial College, London SW7 2BZ, England
[1] University of St. Andrews, Fife, Scotland
[2] TA Consultancy, Farnham, Surrey, England
[3] Royal Institute of Technology, Stockholm, Sweden

Abstract

An overview of theoretical work on large Larmor radius stability of the z-pinch is presented, highlighting two recent innovations. Firstly, finite electron temperature has been included for the linear $m = 0$ instability. Compared to the usual cold electron case, growth rates are increased and are closer to those of ideal MHD. Secondly, a 2-D hybrid code has been written to study the non-linear development of the $m = 0$ instability. First results provide no evidence of instability saturation.

1. Introduction

Ideal MHD predicts that an equilibrium z-pinch is highly susceptible to both the $m = 0$ (sausage) and $m = 1$ (kink) instabilities, with linear growth rates of the order of v_T/a (v_T is the ion thermal velocity, a is the pinch radius). However, at high currents the z-pinch can operate in the collisionless, large Larmor radius (LLR) regime, in which the plasma is not described by ideal MHD or any other fluid model. Here the ion physics must be treated kinetically, and the behavior depends on the value of ε (the ratio of the average ion Larmor radius to the pinch radius), which, in turn, depends only on the pinch line density [1,2]. Imperial College's DZP experiment [3] is designed to operate in this regime. For instance, a pinch formed from a 10 μm radius hydrogen fiber would be collisionless at currents above about 200 kA and have $\varepsilon \simeq 0.2$. Clearly, the stability of the z-pinch in this experimentally relevant regime is an important issue. This paper provides an overview of some recent theoretical work on this topic [4-7].

2. Theoretical Background

Z-pinch stability in the LLR regime has been studied using the Vlasov fluid model [8]. This involves an exact kinetic description for the ions and a simple fluid treatment for the electrons. Quasi-neutrality is imposed. For the linear theory (Sec 3, below) the ions are treated using the Vlasov equation. For the non linear modeling (Sec 4, below) collisionless ions are moved in self-consistent fields, as in a PIC code. Most of the work is based on conventional Vlasov fluid theory in which the electrons are assumed to be cold and satisfy the equation $\vec{E} = -\vec{u}_e \times \vec{B}$. However, the effect of relaxing the $T_e = 0$ assumption for the linear $m = 0$ instability has been explored.

Two approaches to the Vlasov fluid linear stability problem have been developed. Firstly, a variational method in which the Vlasov fluid linear eigenvalue equation is reformulated as a dispersion functional. Eigenfunctions are expanded in a truncated series of orthogonal functions (e.g., MHD eigenfunctions, Bessel functions, etc) and solutions are obtained by minimizing with respect to the coefficients. The second approach is embodied in FIGARO, a linear initial value code. It starts with a random perturbation, advancing the linearized equations in time until the solution converges to the fastest growing mode. The linear Vlasov equation is advanced by integrating along the equilibrium trajectories of an ensemble of particles (usually 4×10^4). The two methods give good agreement.

Before considering detailed results for specific equilibria, it is worth mentioning two general results from Vlasov fluid theory. Firstly, the $\varepsilon = 0$ limit is given by ideal MHD except for the z-pinch' $m = 0$ that is given by the Chew-Goldberger-Low (CGL) equations [9]. Secondly, for all finite values of ε the Vlasov fluid stability threshold is formally given by ideal MHD with Γ (the ratio of specific heats) set to zero. This implies that absolute Vlasov fluid stability to either the $m = 0$ or the $m = 1$ mode cannot be achieved for physically realistic z-pinch equilibria.

Even when a Vlasov fluid is absolutely unstable, LLR effects effectively stabilize configurations with mostly straight field lines. There remains a very small residual growth rate due to resonant particles. However, this phenomenon does not occur in the z-pinch.

In our equilibria the ion distribution function is Maxwellian with a uniform temperature, implying an isotropic equilibrium pressure. Vlasov fluid equilibria of this type conform to the usual MHD force balance condition $\nabla P = \vec{j} \times \vec{B}$. Results are given here for two equilibria: (a) *Parabolic*, corresponding to a uniform current density and a parabolic (falling) density profile, and, (b) *Bennett*, corresponding to a uniform electron fluid velocity. For the Bennett equilibrium the maximum magnetic field occurs inside the plasma, rather than at the edge.

3. Linear Theory Results

(a) Fixed Boundary Modes

The $m = 0$ results are strongly equilibrium dependent. For the parabolic case the growth rate can be reduced by up to a factor of 3. The maximum reduction occurs for values of ε in the region of 0.1 to 0.2, above which the growth rate increases with ε. The Bennett equilibrium, on the other hand, shows LLR destabilization over the whole range of ε. This fact is related to the occurrence of the CGL solution in the $\varepsilon \to 0$ limit. The $m = 1$ results are less equilibrium dependent. The growth rate can be reduced by up to a factor of 5, the maximum reduction again occurring for $\varepsilon \sim 0.1$-0.2.

(b) Free Boundary Modes

Free boundary modes show a greater reduction in growth rates by LLR effects, by up to a factor of about 5 for $m = 0$ and 8 for $m = 1$. Again the lowest growth rate occurs for $\varepsilon \sim 0.1$-0.2.

(c) *Effect of Finite Electron Temperature*

The conventional Vlasov fluid model assumes cold electrons. The effect of finite T_e on the LLR linear $m = 0$ mode has been investigated with a modified Vlasov fluid model [10], which uses an electron fluid equation of the form $\vec{E} = -\vec{u}_e \times \vec{B} - \nabla P_e$ and an adiabatic electron equation of state. Figure 1 shows results for the parabolic and Bennett equilibria. Growth rates are increased from their $T_e = 0$ values and are closer to those of ideal MHD.

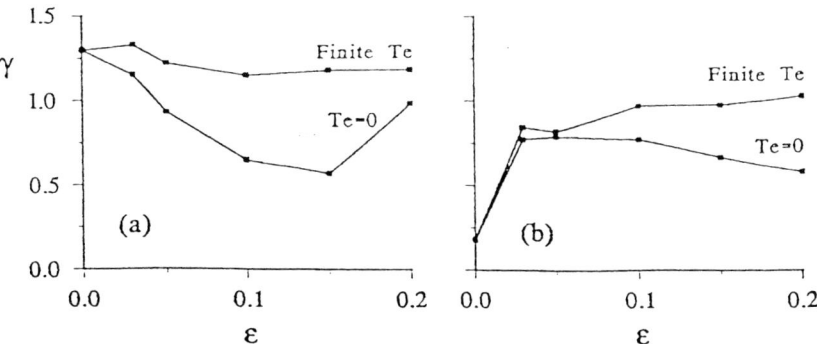

Fig 1. Growth rate (normalized to v_T/a) against ε for $m = 0$, $ka = 10$, fixed boundary mode, with and without finite T_e, (a) Parabolic, (b) Bennett.

4. Non-Linear Modeling

PHY2D, a 2-D (r-z) hybrid (particle ions/fluid electrons) code, has been developed to study the non-linear evolution of the $m = 0$ instability in the LLR regime [11]. The code is periodic in the axial direction and includes a vacuum region outside the plasma. It uses the conventional cold electron Vlasov fluid model. At present it is restricted to $Z = 1$ (i.e. H, D, or T).

Maxwellian ions (initialized consistent with a chosen equilibrium) move in self consistent fields. The magnetic field update is split into an advective part and a diffusive part. In the vacuum the field diffuses with a large artificial vacuum resistivity, for computational reasons, in the plasma the field advects with the electron fluid. Particles can move anywhere in the computational domain. Cells with a particle density below some specified cut-off value are vacuum, and their magnetic field is updated diffusively. The results are insensitive to the choice of cut-off density. The code conserves energy and momentum, and has been tested by comparison with Vlasov fluid linear growth rates. Full runs use 4×10^5 particles and a 32×128 grid.

The code is intended to establish if conditions exist in which the instability saturates at finite amplitude without significantly degrading the plasma confinement. First results from PHY2D do not show this happening. Instead, the instability continues to grow exponentially until the neck reaches the axis.

An earlier version of the code handled the plasma/vacuum coupling erroneously. Those early computations suggested instability saturation. Clearly, the results are sensitive to the details of the treatment of the plasma edge, but precisely here the model itself breaks down. In the existing code arbitrarily large electron flows can occur in the low density outer region, whereas, actually, the electron velocity would be limited by microinstabilities, i.e., some anomalous resistivity. Thus there is good reason to suppose that improving the treatment of the plasma edge produces a significantly modified result. This issue needs to be resolved before final conclusions can be drawn concerning the stability of the z-pinch in the LLR regime.

5. Conclusions

The stability of the z-pinch in the collisionless LLR regime has been studied using the Vlasov fluid model. Absolute stability to either the $m = 0$ or the $m = 1$ mode is not achieved and although LLR effects can reduce linear growth rates by up to a factor of 8 (for $\varepsilon \sim 0.1\text{-}0.2$): the pinch is not "effectively stable." Finite electron temperature effects increase the LLR growth rates to values close to the MHD ones.

Even in the LLR regime the linear instabilities persist, and their non-linear development is important. First results from a 2-D hybrid code developed to study the nonlinear sausage mode in the LLR regime do not show instability saturation. However, an improved treatment of the plasma edge is needed.

References

1. M.G. Haines and M. Coppins, Phys. Rev. Lett **66**, 1462 (1991).
2. M. Coppins, "A review of the stability of the z-pinch," these proceedings.
3. M.G. Haines, "An overview of the DZP Project at Imperial College," these proceedings.
4. T. D. Arber, M. Coppins, and J. Scheffel, Phys. Rev. Lett **72**, 2399 (1994).
5. T. D. Arber, P.G.F. Russell, M. Coppins, and J. Scheffel, Phys. Rev. Lett **74**, 22698 (1995).
6. J. Scheffel, T.D. Arber, M. Coppins, P.G.F. Russell, Plasma Phys. Contr. Fusion **39**, 559 (1997).
7. P. G. F. Russell, T. D. Arber, M. Coppins, and J. Scheffel, "Linear stability of the collisionless, large Larmor radius z-pinch," Phys. Plasmas (in press).
8. J. P. Freidberg, Phys. Fluids **15**, 1102 (1972).
9. H. O. Åkerstedt, J. Plasma Phys. **44**, 137 (1980).
10. R. Gerwin, Informal Report LA-6130-MS (Los Alamos Scientific Laboratory, New Mexico, 1975).
11. T. D. Arber, Phys. Rev. Lett. **77**, 1766 (1996).

RESISTIVE STABILITY OF THE Z-PINCH REVISISTED

M. Coppins, I.D. Culverwell[†]

Imperial College, London SW7 2BZ, England

[†] The Metereological Office, Bracknell, Berkshire, RG12 2SZ, England

Abstract:

Further studies of the linear stability of the z-pinch in the resistive regime have been carried out. These elucidate the previously reported [I.D. Culverwell and M. Coppins, Phys. Fluids **B2**, 129 (1990)] "stability" at low Lundquist number. The effect is due to the development at early times of an instability whose spatial structure is out of phase with the subsequently dominant exponentially growing ideal mode. The new resistive instability has a basically linear time dependence and its spatial structure undergoes continuous evolution.

1. Introduction

Resistive stability theory concerns MHD instabilities in situations in which the infinite conductivity assumption of ideal MHD is inappropriate. In other magnetic confinement devices this usually means that the ideal equations fail in a small region of the plasma close to a "singular (flux) surface". This is not the case here, however, as there are no singular surfaces in the z-pinch (since $B_z = 0$). Instead we are concerned with instabilities early in the current rise.

During the current rise the z-pinch passes through several phases: (1) *pinch formation* (plasma formed and equilibrium set up), (2) *resistive phase* (plasma has low temperature and is highly collisional), (3) *ideal phase* (plasma still collisional but hot enough for resistivity to be negligible), (4) *collisionless phase* [see Ref 1]. Throughout this development the importance of resistivity is parameterized by the Lundquist number $S = \mu_0 v_A a/\eta \propto \tau_R/\tau_m$, where v_A is the average Alfvén velocity, τ_R is the equilibrium resistive diffusion time (= equilibrium ohmic heating time = characteristic timescale over which the equilibrium is changing), and τ_m is the radial ion thermal transit time (= characteristic timescale for MHD instabilities). As the current rises S increases.

Conventional ideal MHD linear theory is valid when $S \gg 1$ (i.e., in the ideal phase). In this case not only is resistivity negligible, but also $\tau_m \gg \tau_R$, implying that MHD instabilities develop faster than changes in the equilibrium. This allows us to assume a time independent equilibrium, which gives rise to the familiar independent normal modes in which the space and time parts of the perturbed variables are separable. Everywhere in the plasma the perturbation is growing exponentially with some growth rate γ. In the case of the z-pinch the space part of all normal modes are harmonic functions of θ and z. Thus all perturbed variables $\propto \exp i(m\theta + kz) \exp \gamma t$.

However, the ideal phase is preceded by the resistive phase in which S is small. In this case not only must resistivity be retained in the model, but the assumption of a time independent equilibrium cannot be made (since

$\tau_m > \tau_R$). In this phase the perturbed variables can still be $\propto \exp i(m\theta + kz)$, but instability growth will not be exponential or even separable.

Previous studies of the z-pinch' linear stability in the resistive phase [2,3] concluded that the behavior could be crudely approximated as absolute stability below some critical Lundquist number. Here we examine the matter more deeply and offer a new interpretation of this interesting phenomenon.

2. Previous work

Without radiation early in the current rise there is a well known [4] class of time dependent self-similar z-pinch equilibria (i.e., states satisfying force balance $\vec{\nabla} P = \vec{j} \times \vec{B}$ at all times) with a rising current $I \propto t^\alpha$. Simulations show that such states arise spontaneously [5].

We base our calculations on such equilibria with the following simplifying assumptions: (1) $\alpha = 1/3$, implying that the equilibrium plasma radius, a, is constant, (2), the temperature is exactly uniform, and, (3) atomic number $Z = 1$ (similar behavior would occur for higher Z). Assumption (1) implies $S \propto t^{4/3}$.

Previous calculations [2] with the linearized resistive MHD equations (including $\eta \vec{j}$ in Ohm's law and ohmic heating in the energy equation) solved for the $m = 0$ instability in such a time dependent equilibrium. To simplify the analysis the perturbed temperature was assumed to be uniform in r and the radially integrated energy equation was used.

Because the usual normal mode approach is untenable at low S, we used an initial value method. This technique is straightforward in the ideal case, with a time independent equilibrium. An initial perturbation is applied, the linearized equations are advanced in time, and after a few growth rates te behavior becomes dominated by the fastest growing mode with a well defined structure, growing exponentially with a well defined growth rate. However, in the low S regime of interest here the equilibrium variables had to be advanced in parallel with the linearized equations. This made interpreting the results difficult.

Eventually the equilibrium reaches the ideal regime ($S \gg 1$) in which it is evolving slowly and the instability follows a WKB-type time dependence $\propto \exp \int dt \gamma$ (where γ is the instantaneous ideal MHD growth rate). Remembering that $\gamma \propto 1/\tau_m$ it is easy to show that $\int dt\gamma = 0.064 \gamma^* S(t)$ (where γ^* is the ideal MHD growth rate normalized to τ_m).

The convergence to the expected high S behavior was clearly seen by monitoring the evolution of the instability amplitude as a function of S. Early in the current rise, however, the behavior was very different, the amplitude rising at certain times and falling at others. Furthermore, during this early non-ideal phase the spatial structure of perturbed variables was not preserved. Instead it evolved continuously. Only at high S did the solution converge to a well defined mode structure.

The form of the solution can be characterized by a critical Lundquist number S^* which, broadly speaking, measures the duration of the resistive phase. The value of S^* depends on ka: for $ka \sim 1$, $S^* \sim 10$, for $ka \sim 0.1$, $S^* \sim 100$. Overall, the behavior can be crudely approximated as absolute stability during the resistive phase. These results agree with 2-D MHD simulations [3].

3. A New Interpretation

To gain further insight into this behavior the equilibrium time dependence was removed and the code used as a conventional initial value code, with a fixed equilibrium corresponding to some specified value of S. Now the solution converges to the fastest growing exponential mode, defining an instantaneous growth rate γ for every S. The results agree with the full calculation at high S, when the equilibrium time dependence is unimportant.

The properties of the instability at low S are clarified if we examine both the fastest and the second fastest growing modes (the initial value code can obtain both if their growth rates are sufficiently close). Figure 1 shows the two normalized growth rates (γ^*) at $ka = 1$ as function of S.

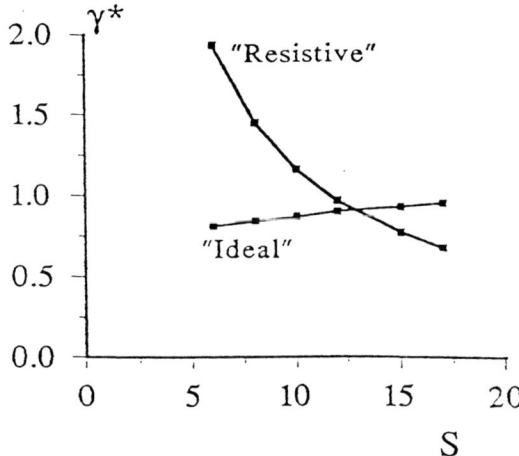

Figure 1. Normalized growth rates over a range of Lundquist numbers.

There are two distinct types of solution, which cross over at $S \simeq 12$: a "resistive instability" whose γ^* falls rapidly with S, and the "ideal instability" which eventually dominates, and whose γ^* shows little variation with S.

At any S the code can determine the form of the mode structure. To compare the different modes we set the phase by defining the perturbed temperature (uniform in r) to be real and negative (i.e., we chose the value of z such that the plasma is cooling). The spatial structure of the resistive instability varies strongly with S, whereas the ideal instability shows little variation. Furthermore, in the resistive case cooling is associated with radial contraction, whereas in the ideal case cooling is associated with radial expansion. Thus the two modes are out of phase.

For all values of ka we find these two distinct types of solution: a resistive instability which dominates at low S and an ideal instability which dominates at high S. Furthermore, the normalized growth rate γ^* for the resistive instability is independent of ka and proportional to S^{-1}. If we assume a WKB-type time dependence we can substitute for γ^* in terms of S and evaluate $\int \gamma dt$, Remarkably, we find $\int \gamma dt = \ln t$. This suggests that if we could isolate the resistive instability and follow its evolution as the equilibrium S rises, it would display not exponential growth ($\propto \exp \gamma t$) but linear growth ($\propto t$).

Of course, the situation is more complicated because as S increases the resistive instability's radial structure undergoes considerable evolution. Space and time parts of the instability are no longer separable. In the ideal instability cooling is associated with radial expansion; this is just adiabatic cooling. In the resistive instability, on the other hand, cooling is associated with radial contraction. This suggests that it is driven by ohmic heating.

A linearly growing resistive instability that is out of phase with the subsequently dominant ideal instability probably causes the low S resistive behavior seen in the original initial value code, which included the time dependence of the equilibrium.

We can still characterize the duration of the resistive phase by a critical Lundquist number, S^*. Instead of the previous definition in terms of a geometrical construction, $S^* = S^*(ka)$ is now defined as the value of S at which the ideal instability grows faster than the resistive one. For $ka = 1$, for instance, $S^* = 12$. The values of S^* defined in this way agree well with those obtained previously.

4. Conclusions

Earlier [2] we concluded that there was a resistive phase early in the current rise, in which the stability behavior of the z-pinch differed qualitatively from the final ideal behavior, and that the duration of this phase could be characterized by a ka dependent critical Lundquist number, S^*. These conclusions still stand. However, we would no longer claim that the z-pinch is stable during the resistive phase. Instead, we now believe that the resistive phase is distinguished by the predominance of a new instability which essentially grows linearly with time, and is associated with ohmic heating.

4. References

1. M. Coppins, "Recent progress on large Larmor radius theory", *these proceedings*.
2. I. D. Culverwell and M. Coppins, Phys. Fluids **B2**, 129 (1990).
3. F. L. Cochran and A. E. Robson, Phys. Fluids **B2**, 123 (1990).
4. S. I. Braginskii and R. D. Shafranov, in *Plasma Physics and Problem of Controlled Thermonuclear Reactions*, ed M.A. Leontovich (Pergamon, London, 1959) p. 39.
5. Coppins, J. P. Chittenden, and I. D. Culverwell, J. Phys. D **25**, 178 (1992).

K-SHELL RADIATION POWER AND YIELD FROM DOUBLE SHELL PLASMA LINER IMPLOSIONS

S.A.Sorokin and S.A.Chaikovsky
*Institute of High Current Electronics,
Akademichesky Ave., 634055 Tomsk, Russia*

Abstract
Experiments have been performed on the SNOP-3 pulse generator to increase the argon K-shell radiation yield and power. A double shell liner structure with and without an initial axial magnetic field was used to increase the plasma radial compressions and pinch plasma density. The K-shell radiation yield and power were measured on varying the inner shell initial radius and the initial axial magnetic field. The measured yields and powers are maximized for the inner-to-outer shell radius ratio $r_2/r_1 = 0.3$. For comparison, the K-shell radiation yield and power versus the initial radius of a single shell liner were measured.

1. Introduction

Kilovolt radiation sources are of great interest for applications in x-ray lithography, x-ray laser pumping, inertial confinement fusion, and material studies. Intense x-ray radiation pulses are formed when a plasma liner accelerated by the magnetic field of the generator current implodes upon itself. Typically more than 50% of the plasma energy can be radiated in a wide wavelength range. However, most of the radiated energy appears as sub-keV radiation. Attempts to increase the x-ray hardness result in a decrease in efficiency of the plasma energy-to-x-rays conversion. X-rays above 1 keV are emitted from the K-shells of moderate-atomic-number materials. To provide efficient conversion of the liner plasma energy into the energy of the K-shell radiation, it is required to fulfill the conditions, as follows: (1) $\eta = E_k/E_m \sim 1$, where E_k is the kinetic energy per ion and E_m is the minimum energy required to reach the helium-like stage of the ion [1, 2]. (2) The pinch plasma radiative cooling time τ_r should be less than the confinement time τ_i. To have the pinch formation time close to the time at which the current peaks, the liner mass per unit length m_1 and the initial liner radius r_1 should satisfy the condition $m_1 r_1^2 = $ const for a generator with given peak current I_m and current pulse duration τ. For a fixed compression ratio $r_1/r_f = 10$ we have for maximum ion density $n_i \propto m_1/r_f^2 \propto m_1^2$. The cooling time is $\tau_r = 3 n_i T/(2K(T) n_i^2) \propto n_i^{-1} \propto m_1^{-2}$ (for an optically thin plasma). The kinetic energy per ion is $E_k \propto v_f^2 \propto (r_1/\tau)^2 \propto I_m^2/m_1$. High current generators are required to fulfill conditions (1) and (2). The minimum I_m required to fulfill these conditions depends on the atomic number of the element and is approximately equal to the "breakpoint" current I_n at which the transition occurs from I_m^4 to I_m^2 scaling for the K-shell radiation yield. The "breakpoint" current I_n is about 2 MA for neon and about 4 MA for argon [3].

If the radial compression ratio is not fixed and compressions $r_1/r_f > 10$ can be achieved, condition (2) can be fulfilled at a lower peak current value. Actually, setting $I_n = I_m$ for $\tau_r = \tau_i$ and $\tau_i \approx 2r_f/v_f$, we have $1 \sim \tau_r/\tau_i \propto r_f/m_1 \propto r_f/I_n^2$, so that $I_n^2 \propto r_f$. So, the higher the plasma radial compression ratio, the lower is the "breakpoint" current for which efficient conversion of the plasma energy into the energy of the K-shell radiation occurs. On the other hand, when the generator peak current I_m is fixed, the higher compressions correspond to the higher atomic number Z for which the efficient regime can be achieved. It has been shown [4] that the liner compression ratio can be increased sufficiently using some methods to stabilize the liner implosions. In particular, 100-fold radial compressions of the inner shell of a double shell argon liner were observed [4]. The objective of this work was to study experimentally the possibility of the argon K-shell radiation power P_k and yield W_k increasing, when high radial compressions of the inner shell of a double shell liner (DSL) are achieved. To compare the optimum powers and yields from DSLs and single shell liners (SHLs), measurements of the power and yield versus SHL initial radius were carried out. The reasonings for choosing the DSL initial parameters were as follows: (1) The initial outer shell radius r_1 should ensure to provide $\eta \approx 1$ for the outer shell plasma. (2) The outer shell mass m_1 should provide the pinch formation time at the peak current. (3) The inner shell mass $m_2 \approx m_1/3$ should provide $\eta > 1$ for the inner shell plasma (an elastic collision of the shells was supposed).

2. Experimental arrangement

The experiments were performed on the SNOP-3 generator (I = 1MA, τ = 100 ns). A puff-on-puff nozzle (Fig. 1 in Ref. 5) was used to form the argon double shell liner. An initial axial magnetic field was produced by Helmholtz coils. Copper coils reduced the initial magnetic field in the region of the inner shell. The liner length was 15 mm. The experiments were conducted with argon. Aluminum cathode XRDs with 12–18 μm teflon filters were used to measure the argon K-shell radiation power and yield. The size of the K-shell emitting regions was measured with filtered pinhole cameras.

3. Results and discussion

A solid nozzle of initial radius 6 mm and annular nozzles of mean initial radii 7.5, 11, and 13 mm were used to induce SSL implosions. According to the 0-dimensional calculations of the SSL implosion dynamics (10-fold radial compression), these nozzles provide the normalized kinetic energies $\eta \approx 0.5$, 0.9, 1.4, and 2.1, respectively. Figure 1 illustrates the measured K-shell yields W_k and peak powers P_k versus the SSL initial radius. Both the yield and the power have a maximum at $r_1 = 7.5$ mm ($\eta \approx 1$). The yield and the power are maximized at 110 J and 12 GW. In DSL implosions, the outer shell had a mean initial radius of 13 mm and the radius of the inner shell was varied from 3 to

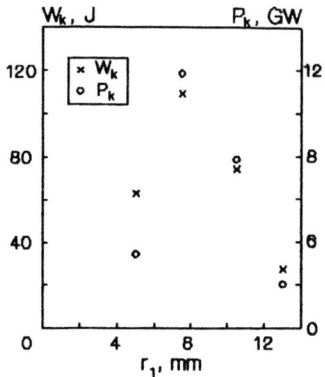
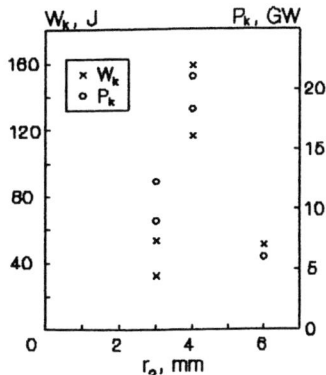

Fig. 1. K-shell yields W_k and peak powers P_k versus the radius r_1 of the SSL.

Fig. 2. K-shell yields W_k and peak powers P_k from the DSL versus the inner shell radius r_2 (for $B_0 = 0$, $m_2 = m_1/3$, $r_1 = 13$ mm).

6 mm. The K-shell peak powers and yields versus the DSL inner shell radius for implosions without an initial axial field are shown in Fig. 2. For the inner shell radius of 4 mm, the peak power and yield are maximized at 21 GW and 160 J. These power and yield exceed the optimum power and yield from SSL by a factor of 1.8 and 1.5, respectively. K-shell X-ray pinhole pictures for an SSL implosion with $r_1 = 7.5$ mm (*a*), and for DSLs with $B_0 = 0$ and $r_2 = 4$ (*b*) and 6 mm (*c*) are presented in Fig. 3. It is particularly remarkable that the x-ray images observed for the DSL implosions with a relatively small r_2 (3–4 mm) differ substantially from the x-ray images observed for the SSL implosions and for the DSL implosions with $B_0 = 0$ and relatively large r_2 (≥ 6 mm). The pinch diameter of 0.2 mm at $r_2 = 4$ mm corresponds to a 45-fold radial plasma compression.

DSLs of inner shell radius 6 mm and more are observed to form x-ray pinhole images that are closely similar to the SSL images. The K-shell peak power and yield in this case are approximately equal to the power and yield for SSL implosions with an initial radius of 13 mm (equal to the DSL outer shell radius). The powers and yields decrease as the initial axial magnetic field is increased. This is because some part of the liner kinetic energy is delivered to the compressed magnetic field, rather than to the pinched plasma. The results of the plasma temperature and density measurements are presented in a companion paper [6].

A composite plot of argon K-shell yields versus currents from 1 MA up to 10 MA is

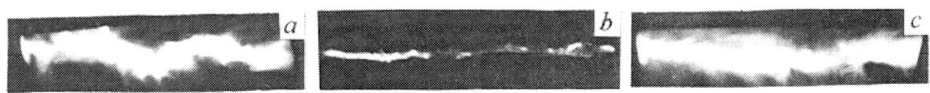

Fig. 3. Time-integrated pinhole photographs from implosions of SSL with $r_1 = 7.5$ mm (*a*), DSL with $B_0 = 0$, $r_2 = 4$ mm (*b*), and DSL with $B_0 = 0$, $r_2 = 6$ mm (*c*).

shown in Fig. 4. The SNOP-3 data on the yields for SSL implosions with initial radii of 7.5, 11, and 13 mm and for DSL implosions with the inner shell radius of 4 mm are given as well. The higher yields for the SSL implosions realized on the SNOP-3 and on the Double-Eagle are due to the optimization of the initial radius and the suppression of the "zippering" effect accomplished for these generators [7]. The transition from I_m^4 to I_m^2 behavior occurs at $I_n \approx 4$ MA. Let us draw the I_m^4 scaling line through the point that corresponds to the yield for DSL implosions on the SNOP-3. This gives a rough estimate of the new transition current $I_n \approx$ 1.3–1.5 MA. Though the new "breakpoint" current value found in this way is rather optimistic, nevertheless, there is reason to

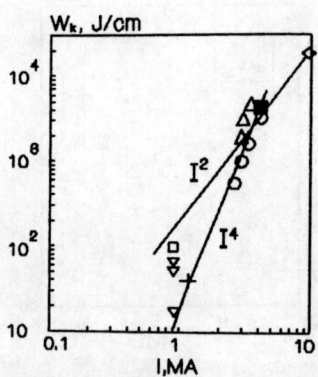

Fig. 4. Argon K-shell yields versus currents in gas-puff Z-pinches: ◊ – Saturn, ■ – Blackjack 5, Δ – Double Eagle, + Gamble II, o – Pithon, ∇ – SNOP-3 (SSL), □ – SNOP-3 (DSL with $r_2 = 4$, $B_0 = 0$).

think that the current at which the transition to an efficient regime occurs can be reduced using a double shell liner.

4. Conclusions

High radial compressions of the DSL plasma enable an increase in K-shell radiation power and yield. Power and yield strongly depend on the inner shell radius r_2 (and peak at $r_2 = 4$ mm). At $r_2 = 4$ mm ($r_1 = 13$ mm, $B_0 = 0$) the power and yield from a DSL exceed the optimum power and yield from an SSL by a factor of 1.8 and 1.5, respectively. Further experiments should be performed to study the power and yield dependencies on the DSL outer radius r_1, on the inner–to–outer shell mass ratio, and on the liner length.

References

[1] K.G. Whitney, J.W. Thornhill, J.P. Apruzese, and J. Davis, J. Appl. Phys. **67**, 1725 (1990).
[2] C. Deeney, T. Nash, R.R. Prasad et al. Phys.Rev. A **44**, 6762 (1991).
[3] M. Krishnan, C. Deeney, T. Nash, P.D. LePell, and K. Childers, Proc. of the 2nd Int. Conf. on Dense Z-Pinches, AIP Conf. Proc. **195**, 17 (1989).
[4] S.A. Sorokin, and S.A. Chaikovsky, Proc.of the 3rd Int. Conf. on Dense Z-Pinches, AIP Conf. Proc. **299**, 83 (1993).
[5] S.A. Sorokin, and S.A. Chaikovsky, these proceedings
[6] S.A. Chaikovsky and S.A. Sorokin, these proceedings.
[7] Deeney, P.D. LePell, F.L. Cochran, et al., Phys. Fluids B **5**, 992 (1993).

DOUBLE SHELL LINER IMPLOSIONS

S.A.Sorokin and S.A.Chaikovsky
Institute of High Current Electronics
4 Akademichesky Ave., 634055 Tomsk, Russia

Abstract
Experiments on the double shell liner (DSL) implosions with and without an initial axial magnetic were performed on the SNOP-3 pulse generator (1.1 MA, 100 ns). In implosions of a DSL without an initial axial magnetic field, high radial compressions of the inner shell were observed, as in previous experiments with an initial axial magnetic field. Possible mechanisms for the formation of the initial azimuthal magnetic field are discussed.

1. Introduction

Numerous experiments on plasma liner (Z-pinch) implosions have shown that the radial compressions of the fast liner (Z-pinch) plasma achievable in the experiment are substantially lower than those predicted by one-dimensional MHD calculations. There are various approaches to the explanation of the low degrees of liner compression [1, 2]. For liners accelerated to velocities $v \geq 5\times 10^7$ cm/s within a time of the order of 100 ns, the most probable mechanism for the limitation of the bulk plasma radial compression is the disruption of the cylindrical structure of the liner in the run-in phase due to the growth of the Rayleigh-Taylor (R-T) instability. The seeding of an initial axial magnetic field improves the stability of the imploding plasma. However, this means fails to increase substantially the plasma radial compression [3, 4]. The initial magnetic field B_0 necessary for stabilization is too high, and the plasma compression ratio is limited by the compressed axial magnetic field. The contribution of the R-T instability to the formation of a plasma pinch depends on the final velocity of the liner and on the dynamics (time) of its acceleration. Thus, for the classical rate of R-T instability growth $\gamma = (kg)^{0.5}$ at the acceleration g ($\lambda = 2\pi/k$ is the perturbation wavelength) invariable for a time τ we have $g\tau = \tau(kg)^{0.5} = (kv/\tau)^{0.5}\tau = (kv\tau)^{0.5}$. That is the number of e-foldings that determines the growth of the perturbation amplitude in the linear approximation is proportional to the square root of the time required for the liner to be accelerated to a given velocity. Thus, shortening the liner acceleration time should favor the stabilization of the liner compression and hence the increase in plasma compression ratio. A double-shell liner (DSL) with an initial axial magnetic field can be used to attain stable high radial compression of the liner plasma [5]. In this case, the current generator and the outer shell stabilized by the axial magnetic field serve as a generator of rapidly rising azimuthal current for the inner shell. 100-fold stable radial compressions of the inner shell were observed. However, in a case like this, some part of the kinetic energy of the inner shell is converted to the energy of the compressed axial magnetic field thereby decreasing the pinch plasma heating.

It is of interest to investigate whether it is possible to implode a double shell liner with an initial azimuthal magnetic field between the shells. This magnetic field can be created by a certain priming current passed through the inner shell with no field present inside this shell. When the liner is produced with the use of two coaxial gas puff nozzles, the creation of an initial azimuthal field by an auxiliary current generator involves some problems (since the region between the shells is filled with a rarefied gas). Meanwhile, an azimuthal magnetic field in the intershell region can be induced by a portion of the main generator current penetrating in the inner shell at the early stage of the acceleration of the outer shell. Although this way to induce the "priming" field is in fact uncontrollable, it is reasonable to suppose that several percents of the main generator current can penetrate in the inner shell and this is sufficient to produce a "magnetic sheath" between the shells. Some possible reasons for the current penetration in the inner shell are as follows:

(1) almost identical conditions for the current passage through both shells at the stage of the gas breakdown and the formation of the conducting current sheath;
(2) diffusion of the magnetic field through the outer plasma shell (including due to a possible anomalous resistivity of the plasma);
(3) the formation of isolated toroidal cells of the magnetic field as a result of the evolution and external reconnection of the spikes which float into the intershell region.

Another difficulty involved in the compression of the inner shell through the sheath of the azimuthal magnetic field is concerned with the absence of the magnetic field shear that stabilizes the implosion of the outer shell. The perturbations developing in the outer shell can affect substantially the uniformity of the implosion of the inner shell.

The experiment whose goal was to establish whether it is possible to achieve high radial compressions of the inner shell in the absence of an initial axial magnetic field was carried out as follows: A series of shots was performed where the initial axial magnetic field B_0 was decreased beginning from 1.5 T at which stable implosions of the DSL were realized [4]. A high degree of compression of the inner shell was detected with a time integrated x-ray pinhole camera and judged by the presence of an intense pulse of argon K-radiation. To determine selfconsistently both spatially and temporally averaged plasma electron temperature and ion density, the measurement technique proposed by Coulter et al. [6] was used. In addition, the electron temperature was estimated from the resonance-to-satellite line intensity ratio for H-like and He-like ions and the ion density was measured from the resonance-to-intercombination line intensity ratio for He-like ions [7].

2. Experimental arrangement

The experiments were performed on the SNOP-3 generator. The generator provides a current rising up to 1 MA in 80 ns in the diode with an inductance of 25 nH. A puff-on-puff nozzle (Fig. 1) was used to form an argon double shell liner. An initial axial magnetic field was produced by Helmholtz coils. Copper coils reduced the initial magnetic field in the region of the inner shell. The liner length was 15 mm. Aluminum cathode XRDs with 12-18 μm teflon filters were used to measure the argon K-shell radiation power and yield. The size of the K-shell emitting regions was measured with filtered pinhole cameras. A mica crystal spectrometer was used to measure the time-integrated K-shell radiation spectra. To determine the plasma parameters we used the analysis technique proposed in [6].

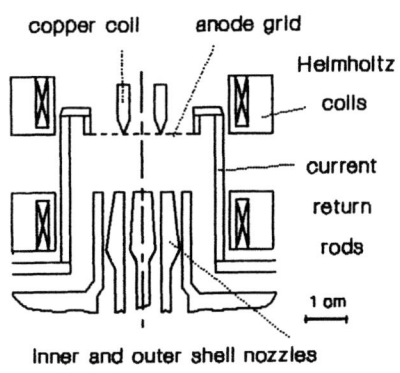

Fig. 1. Schematic of the load region.

Fig. 2. K-shell peak powers P_k from the DSL versus the initial axial magnetic field B_0 (for $m_2 = m_1/3$, $r_1 = 13$ mm, $r_2 = 4$ mm).

Fig. 3. Time-integrated pinhole pictures from implosions of DSL with $r_1 = 13$ mm, $r_2 = 4$ mm, and $B_0 = 0$ (*a*), 0.6 (*b*), 1.2 T (*c*).

3. Experimental results

Figure 2 presents the argon K-shell peak powers from the DSL (for $m_2 = m_1/3$, $r_1 = 13$ mm, $r_2 = 4$ mm) versus the initial axial magnetic field B_0. The x-ray pinhole pictures obtained for DSL implosions with the outer shell radius $r_1 = 13$ mm, the inner shell radius $r_2 = 4$ mm and a varied axial magnetic field B_0 ($B_0 = 1.2$ T, 0.6 T, and 0) are shown in Fig. 3. It can be seen that the high radial compression of the pinch is retained the as the initial axial field is decreased to $B_0 = 0$. With decreasing B_0, sausage perturbations become more and more pronounced whose inherent wavelength is of the order of the pinch diameter. The great number of hot spots (3-4 per 1 mm) favors the increase in K-radiation power and yield. For $r_2 \geq 6$ mm, a dense pinch observed at $B_0 \geq 1$ T is not observed at $B_0 = 0$ (Fig. 3 in Ref. 8). To check whether the mass of the dense pinch corresponds to the mass of the inner shell, the ion density of the pinch plasma was measured [9]. The K-shell emitting mass of the pinched plasma was than estimated from the pinch diameter measured using x-ray pinhole pictures. The ion density ($n_i \approx 10^{20}$ cm^{-3}) inferred from the resonanse-to-intercombination line intensity ratio for He-like ions was two or three times greater than n_i measured by the technique described in Ref. 6. This substantial difference is most probably related to the nonsimultaneous compression of the inner shell along the pinch axis. The compression nonsimultaneity decreases the K-shell radiation power and hence n_i determined by the technique [6]. The width ($\tau_r = 7$ ns) of the K-shell radiation pulse observed to be emitted by the pinch throughout its length is greater than the width ($\tau_r = 2$ ns) of the radiation pulse observed for a pinch section of length $\Delta z = 1$ mm. The pinch mass estimated by the density inferred from the resonance-to-intercombination line intensity ratio is about 2 µg/cm, which is in reasonable agreement with the mass m_2 of the inner shell inferred from the dynamics of the liner implosion.

4. Conclusion

For a certain ratio of the shell radii of a double shell liner ($r_1/r_2 = 3-4$), high radial compressions for the inner shell plasma can be achieved without seeding an initial axial magnetic field. It can be assumed that in this case the inner shell is compressed through the azimuthal magnetic field induced by a portion of the generator current penetrating into the inner shell. The fact that high radial compressions of the inner shell are invariably observed as B_0 is decreased from 1.2 T to zero suggests that these high radial compressions are not related to the current switching to the inner shell resulting from the breaking of the outer shell by R-T instabilities. The implosion of a DSL in the absence of an initial axial magnetic field provides an increase in K-shell radiation power and yield.

References

1. J. Davis, J. Giuliani, M. Mulbranton, and F.L.Cochran, Proc. of the 3rd Int. Conf. on Dense Z-Pinches, AIP Conf. Proc. 299, 112 (1993)
2. C. Deeney, T. Nash, R.R. Prasad, et al. Phys. Rev. A 44, 6762 (1991).
3. F.S. Felber, F.J. Wessel, N.C. Wild, et al. J. Appl. Phys. 64, 3831 (1988).
4. S.A. Sorokin, and S.A. Chaikovsky, Proc. of the 2nd Int. Conf. on Dense Z-Pinches, AIP Conf. Proc. 195, 438 (1989).
5. S.A. Sorokin and S.A. Chaikovsky, Proc. of the 3rd Int. Conf. on Dense Z-Pinches, AIP Conf. Proc. 299, 83 (1993).
6. M.C. Coulter, K.G. Whitney, and J.W. Thornhill, J. Quant. Spectr. Radiat. Transfer, 44, 443 (1990).
7. A.V. Vinogradov, I.Yu. Skobelev, and E.A. Yukov, Usp. Fiz. Nauk (in Russian), 129(2), 177 (1979).
8. S.A. Sorokin and S.A. Chaikovsky, in these proceedings.
9. Chaikovsky and S.A. Sorokin, in these proceedings.

ANALYSIS OF MAGNETIC INTERLAYER STAGED PRS LOADS[†]

R. E. Terry, R. W. Clark

Plasma Physics Division, Naval Research Laboratory

Abstract

Recently, yield enhancements have been reported [1,2] with a double puff that entrains an externally generated axial magnetic field between the annuli.

Implosion of the outer layer with an external current pulse enhances the interlayer field through flux conservation. The axial field strengths obtained at the inner puff, as the outer one stagnates, might reach the threshold needed to show a strong compression of the inner puff.

A survey with 0-D and 2-D MHD models applied to this configuration shows that strong compressions are quite possible, but a reliable knowledge of thin plasma resistivity at the boundary of the interlayer and a reliable model of end effects must be developed.

[†]Work sponsored by DSWA.

1 Self–similar Interlayer Solutions

The self–similar flows available to an adiabatic screw pinch [3] can be characterized by ordinary differential equations with three distinct frequencies: ν_S, measuring a sonic transit time, and two Alfven frequencies ν_{B_θ}, ν_{B_z}, associated with transit times involving the azimuthal and axial magnetic field components.

1.1 High compression flows

Denote the pinch compression as $\alpha = r/r_0$. Because the radial force from an axial field scales as α^{-3}, a stronger power than the $\alpha^{-7/3}$ scaling of (adiabatic) retarding pressure, a self–similar flow driven by exterior axial fields can be driven to extreme density when a sufficiently high initial field value forces the flow to compress below a critical radius. The critical radius depends upon the relative strength of the three frequencies, c.f. Fig. 1 below. In the case shown, [ν_S^2=1.0, $\nu_{B_\theta}^2$=1.5, $\nu_{B_z}^2$=0.271742988]. The pinch is set to just graze and stop at the edge of the strong collapse domain, viz. $\alpha \approx 0.2$. A slightly smaller axial field for this case would cause the pinch trajectory to bounce, whereas a slightly stronger axial field would send the pinch trajectory to the origin with no way to stop. Such strong compression is quite idealized, it can only be sustained so long as the axial field outside the pinch is able to rise and track the increasing field strength required in the plasma.

The existence of a very sharp boundary in frequency ratios arises from a local maximum in the effective potential for the flow, a potential which is otherwise monotonically attractive. The bump's existence persists with the inclusion of further dynamics such as viscous heating or radiative cooling, although the location shifts.

1.2 Homogeneous Compressions — Two Variations

Homogeneous compression in a radial flow (c.f. Fig. 2) can have one of two dependencies on initial position, r_0. The more familiar (top) one is radial ($\alpha(t)r_0$) scaling, but the equally accessible (bottom) one has areal ($\beta(t)r_0^2$) scaling. These flows differ in the time development of the Lagrangian marker $R(t)$, but provide identical scaling for density and field quantities. Each flow leads to a distinct separation of

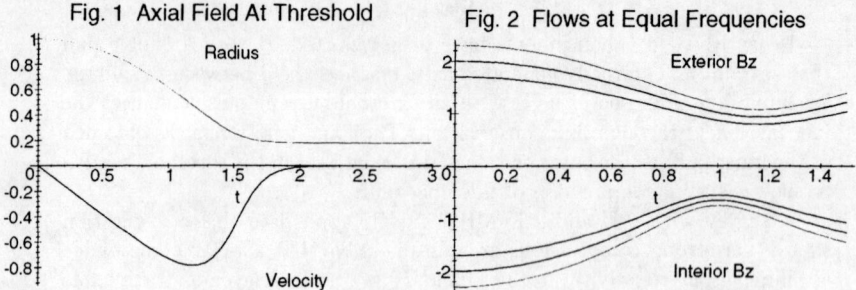

Fig. 1 Axial Field At Threshold

Fig. 2 Flows at Equal Frequencies

the screw pinch momentum transfer equation. With $\mathcal{Z}$ the charge, τ the ion relaxation time, and T the temperature in energy units, one finds upon using an arbitrary scale radius r_s, and defining: $\nu_S^2 \propto c_s^2/r_s^2$, $\nu_{B_z}^2 \propto c_{A,z}^2/r_s^2$, $\nu_{B_\theta}^2 \propto c_{A,\theta}^2/r_s^2$, $\mathcal{H} = [\alpha(t) \text{ or } \sqrt{\beta(t)}]$, and $\mathcal{G} = \frac{1}{3}\nabla \cdot V$, that this momentum transfer relation becomes

$$\nu^2(t)r_o = \frac{2\nu_S^2}{\mathcal{H}}\left[1 - \tau\mathcal{G}(t)\right]\left(-\frac{r_s^2}{2}\partial_{r_0}\ln n_0(r_0)\right)_{F_1}$$

$$+\frac{\nu_{B_z}^2}{\mathcal{H}^3}\left(-\frac{r_s^2}{2n_0(r_0)}\partial_{r_0}B_{0z}^2(r_0)\right)_{F_2}$$

$$+\frac{\nu_{B_\theta}^2}{\mathcal{H}}\left(-\frac{r_s^2}{2r_0^2 n_0(r_0)}\partial_{r_0}[r_0 B_{0\theta}(r_0))^2]\right)_{F_3} .$$

The separation constants, $\{ F_1, F_2, F_3 \}$, determine the profile shape factors and can be chosen freely on piecewise domains in r_0 so long as the acceleration profile remains spatially continuous. The two flow types obey distinct differential equations for the exterior axial field $\alpha(t)$, and interior axial field $\beta(t)$:

$$\ddot{\alpha}(t) = \nu_S^2\left[1 - \tau\frac{2\dot{\alpha}}{\alpha}\right]\alpha^{-1} - \nu_{B_z}^2\,\alpha^{-3} - \nu_{B_\theta}^2\,\alpha^{-1},$$

$$\ddot{\beta}(t) = \frac{1}{2}\frac{\dot{\beta}^2}{\beta} - 2\nu_S^2\left[1 + \tau\frac{\dot{\beta}}{\beta}\right] + 2\nu_{B_z}^2\,\beta^{-1} - 2\nu_{B_\theta}^2 .$$

Here the separation constants have been set to unity. The leading term in the expression for $\ddot{\beta}$ arises as a purely kinematic constraint, analogous to a Coriolis force. The distinct trajectories obtained in each case, exterior or interior, are compared in Fig. 2 for **equal** frequency parameters. The interior axial field flow shows a slightly sharper, earlier, and deeper compression.

1.3 Profiles for the flows

The dynamics shown above is supported by a broad class of piecewise self–similar profiles, with the interior magnetic field flow appropriate to the region $[r_>, r_p]$, and exterior magnetic field flow impressed for $r \leq r_<$ and $r \geq r_p$. Here a new variable r means r/r_s. The example profiles in Fig. 3 set $r_< \to 1$, $r_> \to 2$, and $r_p \to 2.5$.
On the **interior**:

$$n_0(r)/n_p = exp - \mathcal{F}_1 r^2 \ , \ r \leq r_< \ ;$$

while on the **exterior**:

$$n_\pm(r)/n_p = exp \pm [\mathcal{F}_1 \frac{(r^2 - r_p^2)}{(r_p^2 - r_>^2)}] \ ,$$

with a "+" taken for $r_> \leq r \leq r_p$ and a "−" for $r_p \leq r$. The (unlabeled) density profiles are constructed with $\mathcal{F}_1$ to keep a continuous radial acceleration across the boundary r_p.

Fig. 3 Example Load Profiles

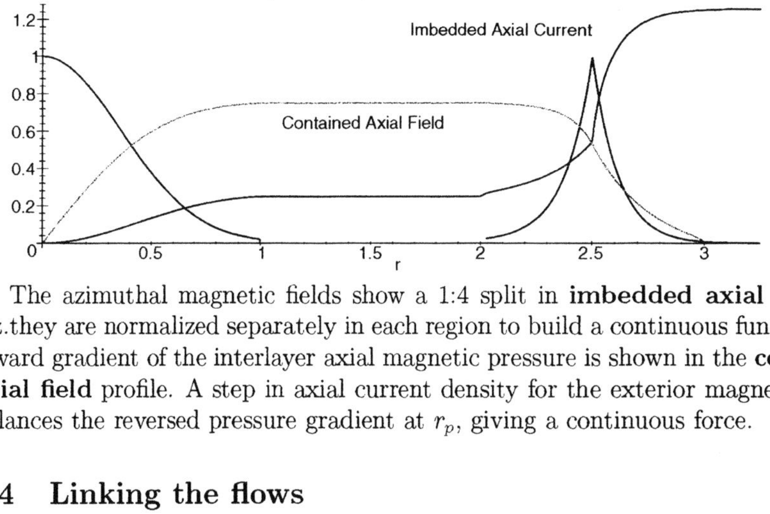

The azimuthal magnetic fields show a 1:4 split in **imbedded axial current**, viz.they are normalized separately in each region to build a continuous function. An inward gradient of the interlayer axial magnetic pressure is shown in the **contained axial field** profile. A step in axial current density for the exterior magnetic fields, balances the reversed pressure gradient at r_p, giving a continuous force.

1.4 Linking the flows

Connecting the axial field frequency for the two flows by a free field compression, the exterior and interior flows can be coupled with an interlayer axial field term scaling $\propto area^{-2}$, viz.

$$\nu_{B_z}^2(t) = \nu_{B_z}^2(t_0) \left[\frac{r_{>,0}^2 - r_{<,0}^2}{r_>^2(t) - r_<^2(t)} \right]^2 \ .$$

When a strong interlayer magnetic field is used, an interior pinch that begins in near equilibrium and shares no axial current is imploded firmly after little interlayer compression, $\delta r/\delta r_0 \approx 0.35$ with $\delta r = r_> - r_<$. When the interlayer field is weakened, the magnetic interlayer must be heavily compressed, $\delta r/\delta r_0 \approx 0.10$ to drive the interior pinch, c. f. Fig. 4. The interior pinch compression is thus delayed in a controlled way, the rate $\nabla \cdot V$ is shown as the curve labeled "Compression".

Fig. 4 A Compressed Interlayer Field

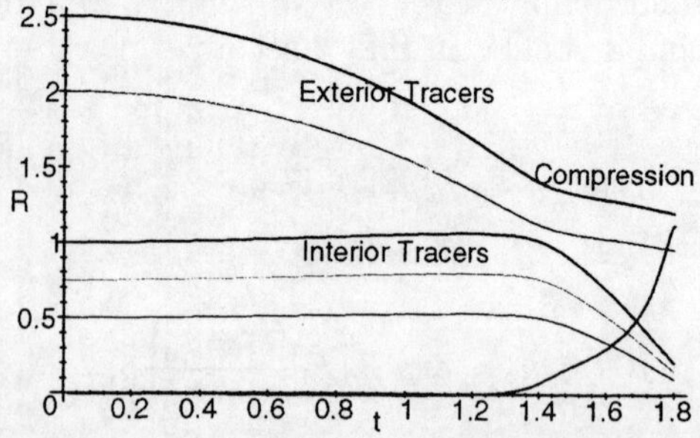

When the interior pinch shares more current and the interlayer field is made weaker still, then the interior pinch may approach an asymptotic equilibrium only to be abruptly compressed by the gathering interlayer field pressure.

2 Mach2D MHD Survey of the Interlayer

Either the self–similar 0D model just discussed or even a more general 1D picture has three rather fundamental limitations: no connection of the axial field at the ends of the pinch to its source currents, no direct access to the tradeoff between compression and stability, and no definitive treatment of field diffusion and decay.

2.1 Initial Conditions and Anomalous Resistivity

For simplicity the Mach2D calculations were performed with a specified initial value for the axial field on annular portions of cathode and anode planes spaced 2 cm apart. In keeping with previous experience, the "$\omega_{p,e}$" anomalous resistivity model was selected because it tracks the plasma frequency in assigning the resistivity[4]. The two annular puff regions were loaded with ion densities of $1.25 \cdot 10^{17}$ cm^{-3} at 5 cm radius and $7.53 \cdot 10^{16}$ cm^{-3} at 8 cm radius in order to match well, in pure z-pinch mode, to a driver risetime of 1.25 μs with a 5.6 MA peak current.

2.2 Variations in Implosion Dynamics with initial B_z

Three initial field strengths were investigated, $B_{z,0}$ = 0.005T, 0.75T, 1.5T. The low field case showed a prompt annihilation of the axial field bubble, and no long term effect on the rundown. The middle field case showed a very energetic final compression emerging from a plasma channel created by end effects, c. f. $n_i(r, z)$ contours in Fig. 5. The low density is about $1.5 \cdot 10^{16}$ cm^{-3}, while the highest is $1.3 \cdot 10^{17}$ cm^{-3}, with intervals of about $1.4 \cdot 10^{16}$ cm^{-3}. Here the central mass is shown beginning a rapid ($\approx$ 125 ns) implosion to the axis, while the two "belt" concentrations will remain behind and migrate to positions next to the electrode planes. Radial distortion of the (initially purely axial) interlayer field creates this focusing effect.

Fig. 5 Interior Load Emergence Off The Compressed Interlayer

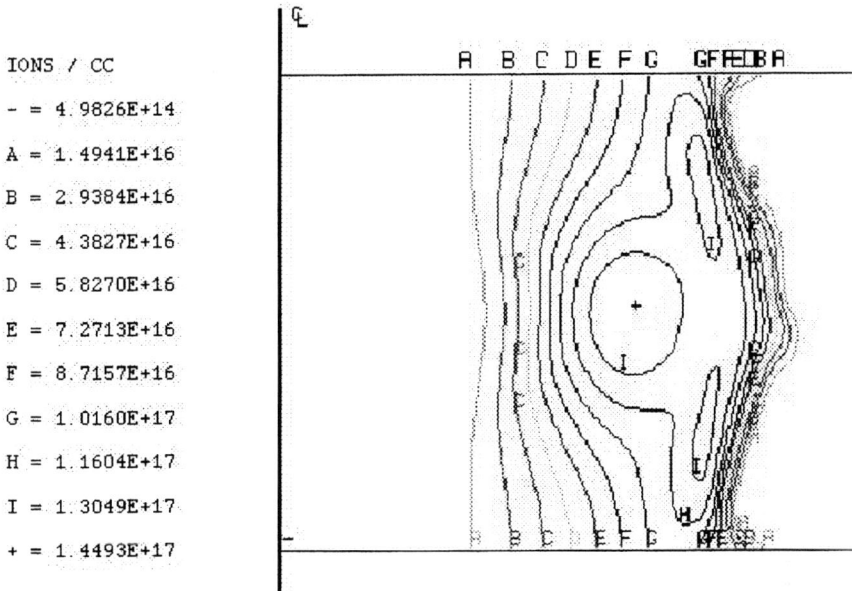

IONS / CC

- = 4.9826E+14
A = 1.4941E+16
B = 2.9384E+16
C = 4.3827E+16
D = 5.8270E+16
E = 7.2713E+16
F = 8.7157E+16
G = 1.0160E+17
H = 1.1604E+17
I = 1.3049E+17
+ = 1.4493E+17

The high field case could not form such an energetic compression because the end effects filled the interlayer with thin plasma, whereupon resistivity absorbed the bubble.

3 Conclusions

When considering the use of a magnetic interlayer, the optimistic implications for strong compressions, and hence improved yield, apparent in a simple 0D model are strongly modified by end effects and the role of plasma resistivity. The calculation of a lifetime for any axial field "bubble" demands a reliable model for plasma resistivity and an accurate picture of the amount of thin plasma injected into the interlayer region. If the self-focusing implosion mode seen above can be consistently realized, then an improved implosion quality and kinetic energy may result.

References

1. S. A. Sorokin and S. A. Chaikovskii, Plasma Physics Reports **22** 11, 1996
2. S. A. Sorokin and S. A. Chaikovskii, Dense Z-Pinches, AIP Conf.Proc. 299, 83 (1993)
3. F. S. Felber, Phys. Fluids **25**, 643 (1982)
4. R. E. Terry, Phys. Plasmas **1**(7), 2189 (1994)

Effect of Enhanced Thermal Dissipation on the Rayleigh-Taylor Instability in Emulsion-Like Media

A. Toor, D. Ryutov,[*]

Lawrence Livermore National Laboratory, Livermore, CA 94551, USA

Abstract. Rayleigh-Taylor instability in a finely structured emulsion-like medium consisting of the two components of different compressibility is considered. Although the term "emulsion" is used to describe the structure of the medium, under typical fast Z-pinch conditions both components behave as gases. The two components are chosen in such a way that their densities in the unperturbed state are approximately equal. Specific emphasis has been made on the analysis of perturbations with the scale λ considerably exceeding the size of the grains a. Averaged equations describing such perturbations are derived. The difference in compressibility of the two components leads to the formation of temperature variations at the scale a, and increases the rate of the thermal dissipation by a factor $(\lambda/a)^2$. The strongest stabilizing effect of the thermal dissipation takes place when the thermal relaxation time is comparable with the instability growth rate.

In this paper we consider a phenomenon that may be helpful in mitigating the Rayleigh-Taylor instability of imploding liners, namely a phenomenon of enhanced thermal dissipation in emulsion-like media. We assume that the fluid (or gas) is two-component, with fine grains of one component randomly, but on the average uniformly, distributed in the other component. With respect to motions with the scale λ considerably exceeding the scale of the fine structure, a, this fluid behaves more or less as a uniform fluid, with some average density and average thermodynamic functions. In particular, it may have a gradient of the average density that would drive the Rayleigh-Taylor instability. We use the word "emulsion" in a somewhat loose sense, just because its visual image is close to the structures we are studying. In fusion-related and pulse-power applications, the initial state will most probably be that of a heterogeneous solid. In fusion-related experiments, this solid usually experiences fast heating and becomes a gas (inhomogeneous in our case) early in the pulse. For this reason, all our analysis will be based on the hydrodynamics equations, without accounting for elastic forces.

Our approach to the mitigation of the Rayleigh-Taylor instability is based on the observation that the presence of a fine internal scale may considerably enhance the dissipation rate (Cf. [1]). The origin of the enhanced dissipation is as follows: the two components that form the emulsion, have, generally speaking, different compressibilities. Therefore, when the pressure perturbation associated with the Rayleigh-Taylor instability develops, the temperature perturbation becomes unequal in the two components. Then the thermal dissipation begins, but at the scale a which is much smaller than the scale λ of the perturbation; accordingly the dissipation rate becomes $(\lambda/a)^2$ times higher than for a homogeneous fluid. Clearly, the just-discussed mechanism works only in compressible media.

In the case of wire arrays, it might be simpler to produce a kind of 2D emulsion. The simplest way would be to use the wire array where the neighboring wires would be made of different materials (Fig. 1a). This would produce a "coarse-grained" emulsion. A more sophisticated way would be to use bunches of wires of different composition, possibly interwoven within every bunch. The whole array

---

[*] On leave from Budker Institute of Nuclear Physics, Novosibirsk 630090, Russia.

would then be assembled of a large number of such bunches (Fig. 1b). After evaporation at early stages of the discharge, separate conductors gradually merge and form a continuous shell of the liner, but the initially embedded non-uniformities should survive this process because of a very low diffusion coefficient. Of issue may be the convective mixing of the two components. However, one should expect that under optimum conditions, where the instabilities are kept under control, there won't be too violent convective motions in the liner.

One can expect that the maximum dissipation will occur if the grain size is chosen in such a way that the characteristic heat exchange time between the grains,

$$\tau \sim a^2/\chi \qquad (1)$$

(where χ is thermal diffusivity) is of the order of a time-scale ω^{-1} of development of the instability. Indeed, if compression (rarefaction) occurs too rapidly, so that heat exchange between the neighboring grains doesn't have time to develop, the process is purely adiabatic and dissipation is absent. On the contrary, if the change of the volume occurs too slowly, the temperature remains uniform and dissipation vanishes again. The compressibilities of the medium, $\delta\rho/\delta p$, in these two limiting cases are different: in the first case the compressibility is adiabatic, in the second case it is isothermal. By δp and $\delta \rho$ we mean perturbations of the pressure and of the density of macroscopic volumes, i.e., the volumes containing many grains. One has:

$$(\delta\rho/\delta p)_{fast} = 1/s^2_{fast}; \quad (\delta\rho/\delta p)_{slow} = 1/s^2_{slow}, \qquad (2)$$

where s is a sound speed for the respective process ($s_{fast} > s_{slow}$).

As was shown in Ref. [2], one can use the following interpolation that covers intermediate frequencies (where thermal dissipation *is* important):

$$\delta p_L = \hat{s}^2 \delta \rho_L; \quad \hat{s}^2 \equiv F(\omega) s^2_{slow} \qquad (3)$$

where

$$F(\omega) = \frac{1 - i\omega\tau(s_{fast}/s_{slow})^2}{1 - i\omega\tau}. \qquad (4)$$

The subscript "L" in Eq. (3) shows that δp and $\delta \rho$ are *Lagrangian* perturbations, related to a particular element of the fluid. The Lagrangian and Eulerian perturbations are related to each other in a standard way:

$$\delta\rho = -\xi \cdot \nabla\rho + \delta\rho_L; \quad \delta p = -\xi \cdot \nabla p + \delta p_L, \qquad (5)$$

with $\delta\rho_L = -\rho\nabla \cdot \xi$, and ξ being a displacement of a certain macroscopic element with respect to its unperturbed position.

In the stability analysis, we assume that the gravity acceleration is directed downward along the z axis, $g_z = -g$, with $g > 0$. Equilibrium pressure distribution obeys the barometric law:

$$p' = -\rho g \equiv -p/h \qquad (6)$$

where the prime designates the z-derivative, and h is a scale-length of the unperturbed pressure and density variation. We consider perturbations of the form $f(z)\exp(-i\omega t + ikx)$ and assume that the scale-length of the perturbations satisfies condition $\lambda \gg a$. In this case, the basic set of equations describing small perturbations reads as:

$$\omega^2 \rho \xi_x = ik\delta p \, , \tag{7}$$

$$\omega^2 \rho \xi_z = \delta p' + g\delta\rho \, , \tag{8}$$

$$\delta\rho = -\rho(ik\xi_x + \xi_z') - \xi_z \rho' \, , \tag{9}$$

$$\delta p = -p'\xi_z - F(\omega)\rho s_s^2 (ik\xi_x + \xi_z') \, . \tag{10}$$

All the quantities entering Eqs. (6)-(10) are volume-averaged over a scale much greater than a but much smaller than λ. There is a subtlety at this point: although the densities of the two components are equal in the unperturbed state, they become unequal in the perturbations. The gravity force causes then mutual displacements of the elements of different density, and displacement ξ that enters Eqs. (7)-(10) should be understood as the average over these elements. However, as shown in Ref. [2], in the typical situation, mutual displacements of the two components are insignificant, and a simple interpretation of ξ as a displacement of the whole macroscopic volume remains essentially correct.

Consider stability of localized modes, with the wave-length much less than the scale-length h of the unperturbed state. For such modes one can use the eikonal approximation in the z direction by taking the z dependence of the unknown functions in the form $\exp(iqz)$, where q is the wave number of the perturbation in the z direction. For the fastest growing modes, one has $k \gg q$ (see, e.g., [3]). For such modes, the dispersion relation reads as:

$$\omega^2 = -g\left(\frac{\rho'}{\rho} - \frac{1}{F(\omega)} \frac{p'}{\rho s_{slow}^2}\right) . \tag{11}$$

At this point, it is convenient to introduce, instead of the complex frequency ω, the complex growth rate Γ, $\Gamma = -i\omega$; $Re\Gamma > 0$ corresponds to an instability.

It is convenient to present the resulting dispersion relation in the dimensionless form:

$$\tilde{\Gamma}^2 = 1 + \frac{\eta}{1 + \tilde{\Gamma}\tilde{\tau}} \tag{12}$$

where

$$\tilde{\Gamma} = \frac{\Gamma}{\Gamma_{fast}}; \quad \tilde{\tau} = \Gamma_{fast}\tau\left(\frac{s_{fast}}{s_{slow}}\right)^2; \quad \eta = \left(\frac{\Gamma_{slow}}{\Gamma_{fast}}\right)^2 - 1 \, , \tag{13}$$

and

$$\Gamma_{fast}^2 = g\frac{\rho'}{\rho} + \frac{g^2}{s_{fast}^2}; \quad \Gamma_{slow}^2 = g\frac{\rho'}{\rho} + \frac{g^2}{s_{slow}^2} \, . \tag{14}$$

When switching from Eq. (11) to (12), we used the equilibrium condition (6). The growth rate Γ_{fast} corresponds to fast (purely adiabatic) perturbations, for which τ can be considered infinite. The growth rate Γ_{slow} corresponds to slow perturbations, for which τ can be considered zero. We imply that both Γ_{fast}^2 and Γ_{slow}^2 are positive (i.e., the system is unstable with respect to both slow and fast perturbations). Note that $\Gamma_{slow} > \Gamma_{fast}$.

The solutions of the dimensionless dispersion relation (12) vs. the dimensionless relaxation time are presented in Fig. 2. The effect of the thermal relaxation processes on the unstable root becomes noticeable at η's exceeding a few tenths. Note that, in a broad range of the dimensionless relaxation times $\tilde{\tau}$, there appears a mode of an oscillatory damping ($Re\omega = -Im\Gamma \neq 0$). The presence of this mode may considerably change the nonlinear behavior of the system leading to a slower nonlinear growth (because additional -- and damped -- degree of freedom appears in the system). Analysis of this part of the problem goes beyond the scope of the present paper. Note also that the appearance of this new mode is a very robust phenomenon: it exists in a broad range of relaxation times, even if the difference of slow and fast compressibilities (characterized by the parameter η) is as low as 5%.

At higher $\tilde{\tau}$'s, there appears a weakly damped mode (the A-B branch of the dispersion curve in Fig. 2). This mode should also have a stabilizing effect at the nonlinear stage.

ACKNOWLEDGMENTS

The authors are grateful to Drs. J. De Groot, J. Hammer, P. Springer and P. Wheeler for valuable comments. This work was performed under the auspices of the U.S. Department of Energy by Lawrence Livermore National Laboratory under Contract W7405-ENG-48.

REFERENCES

1. Landau, L.D., Lifshitz, E.M., *Theory of Elasticity*. Pergamon Press, 1986.
2. Ryutov, D., Toor, A. "Enhanced Thermal Dissipation and the Rayleigh-Taylor Instability in Emulsion-Like Media". LLNL Report UCRL-JC-127007, 1997 (submitted to *Phys. Plasmas*).
3. Kull, H.J., *Physics Reports*, 206, 197 (1991)

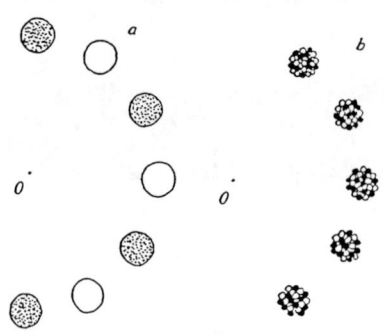
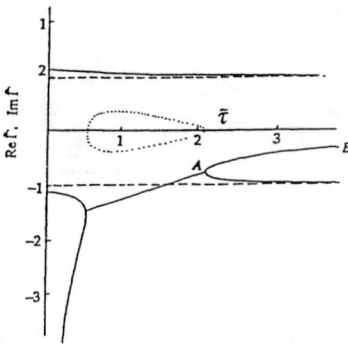

Fig.1 Wire arrays consisting of the wires of two different materials (a) and a wire array consisting of many bunches of interwoven thin wires of two materials (b); O is the axis of the array.

Fig.2 Solutions of the dispersion relation (12): $\eta=0$ (dashed lines); $\eta=0.25$ (solid lines). In the latter case there appears a broad interval of relaxation times ($0.53<\tau<2.1$) where $Im\Gamma \neq 0$ ($Im\Gamma$ is shown in the dotted line). Note the presence of weakly damped mode at large τ's (the segment AB)

Electron beam generation in the turbulent plasma of Z-pinch discharges

Victor V.Vikhrev and Elena O.Baronova

*Nuclear Fusion Institute, RRC Kurchatov Institute,
Moscow 123182, Russia*

Abstract. Numerical modeling of the process of electron beam generation in z-pinch discharges are presented. The proposed model represents the electron beam generation under turbulent plasma conditions. Strong current distribution inhomogeneity in the plasma column has been accounted for the adequate generation process investigation. Electron beam is generated near the maximum of compression due to run away mechanism and it is not related with the current break effect

1. Introduction

The presence of electric fields in a plasma of z-pinches is generally known. The plasma electrons are accelerated in these fields, acquiring the velocity component, directed along the field. Due to collisions of electrons with ions and plasma oscillations (i.e. due to the existence of a friction force proportional to the electron velocity and effective frequency of collisions) this directed velocity becomes chaotic. In case of an equilibrium between the friction forces and the electric ones some constant electron current velocity or a drift velocity, directed along the electric field, takes place. In a purely Coulomb plasma the frequency of collisions between electrons and ions (i.e. a friction force too) is significantly reduced with a rise in the velocity. It is evident that there are rather fast runaway electrons having an opportunity to be accelerated. Such electrons are present in any fields, their amount depends on the electric field magnitude: in strong fields all the electrons run away; the weaker the field, the smaller part of electrons is accelerated.

If the plasma interacts with the electron beam, generated outside that plasma and injected into it, the appearance of some instabilities has been shown [1]. The opposite situation is considered in this paper, when the development of plasma instabilities leads to the electron beam generation in the frame of run away mechanism. Really if the current velocity exceeds the thermal ion velocity the development of plasma instabilities was stated. The presence of instabilities may be represented as the existence of additional frequency of electron collisions, which causes an anomalous rise in plasma resistivity, accompanied by corresponding rise in voltage, thus providing the condition for electron beam generation.

Estimation shows, that the main part of electrons in the z-pinch discharges is not accelerated, they are moving with the drift velocities. This takes place when the electric field strength in the plasma is essentially lower than some critical value E_c [2], and the acceleration of a part of electrons up to high velocities occurs due to the electron runaway mechanism. That mechanism has been developed for the plasma in which the main role is played by Coulomb collisions. At a final stage of the z-pinch discharge development, the frequency of collisions of electrons with plasma waves is much greater than the frequency of Coulomb collisions. Therefore it is necessary to use the mechanism of running away for electrons in a turbulent plasma to represent the electron beam generation in z-pinch discharges.

The analysis of electron beam generation in z-pinch discharges were also presented in [3,4,5].

2. Theoretical model

A simple model of "run away" electron generation under turbulent plasma conditions is given in [6]. It is shown there that the frequency of turbulent collisions depends on the velocity of electrons as v^{-3}, i.e. similar to the dependence for Coulomb collisions. Due to this, the theoretical approach, developed in [7] can be used for the turbulent plasma of z-pinch neck, replacing the frequency of Coulomb collisions by sum of one and an effective frequency of turbulent collisions between electrons and plasma.

$$G = 0.4 \, (v_{ei} + v_{eff})(E/(E_c))^{1/2} \exp[- 1/4(E_c/E) - (2E_c/E)^{1/2}]. \quad (1)$$

In [8] a list of probable instabilities taking place in the z-pinch is given, and the instability, giving the greatest contribution to the anomalous z-pinch resistance, the so called lower hybrid drift instability, is selected. Then

$$v_{eff} = \omega_{LH} (v_d/v_s)^2 \quad (2)$$

In spite of such the dependence of the frequency of turbulent collision, the electron velocity ($\sim v^{-2}$), we use the results of [7] for our numerical consideration.

The detailed simulation of the development of the plasma in z-pinch neck has shown that allowing for the electron collisions with a turbulent ripple turns out to be insufficient for explaining the electron beam production in the z-pinch discharges. For explaining the observed electron beam parameters it is necessary to take into account of the so-called geometry factor of the current passage through the pinch. Introduction of this factor is related with a non-uniform current distribution across the plasma column at the final stage of z-pinch development.

The probable reason for non-uniform distribution of the current in radial direction might be the presence of density inhomogeneity along the plasma column under the condition $\omega_{Be}\tau_{ei}>1$. It was shown in [9,10] that when electrons move from the low density region to the more dense plasma region the essential concentration of the current in near by axis region due to the Hall effect takes place. Just the opposite-the concentration of the current has been found to be near the boundary of plasma column when electrons transfer from the dense plasma region to the low dense region. Thus it is stated that the electrons path exceeds the distance which they overcomes along the pinch and the effective cross section for current passage is less than cross section of plasma column.

So the typical situation may be represented with the help of two geometry factors, allowing the increase of the electrons path along z-axis g_l and the decrease of cross section of current passage g_s. This "geometry factors of the current passage" considerably increases the plasma column resistance, average electric field strength along the pinch and the power of the Joule heat release in it.

It should be mentioned, that the concentration of the current in the vicinity of z-axis leads to the essential rise in the strength of the local electric field and electron beam might be generated in such the region.

Taking this factor into account electric field along the Z-pinch is:
$$E^* = g_s g_h (I - I_{beam})/\sigma S \qquad (3)$$
where I is total pinch current, I_{beam} is electron beam current, $\sigma = ne^2/m_e(v_{ei} + v_{eff})$, $S = \pi r^2$, r - plasma column radius in the neck.

We used simple model of final stage of the Z-pinch for description of plasma dynamics in the neck of the Z-pinch [11,12]. According to this model the plasma temperature T is defined from radial equilibrium:
$$T = I^2/4c^2 N, \qquad (4)$$
where N is number of ions per unit height of neck.

The loss of plasma from the neck can be described by the equation
$$dN/dt = -N/\tau, \qquad (5)$$
where t is the characteristic time for escape of plasma out the neck, $\tau = h/2v_s$, h is length of neck, $h \approx 20r$ [11,12], v_s is the sound velocity. The energy balance in the neck is written as
$$3d(NT)/dt = -5NT/\tau - r B^2/4 \, dr/dt + Q_J, \qquad (6)$$
where $B = 2I/cr$, Q_J is Joule heat release, $Q_J = (I - I_{beam})E$.

The equation, describing the generation of electron beam is as follows
$$dI_{beam}/dt = (G N e h/g_h - I_{beam}/h) v_b, \qquad (7)$$
where e is electron charge, v_b is the electron beam velocity ($m v_b^2/2 = E^* h e$).

3. Results of modeling

The results of numerical simulation of Z-pinch discharge at the current 1 MA and deuterium gas pressure 1 Torr. $g_s = 20$ and $g_h = 1$.

Maximum of density is arise 10^{20} cm^{-3}, minimum radius of plasma column - 0.5 mm, maximum of voltage is 380 kV and maximum beam current - 10 kA.

3. Summary

Our theoretical investigation of electron beam generation and comparison with experimental data leads to following conclusions:
1. Electron beam generation takes place at the neck, the emerging beam of runaway electrons is moving towards the anode.

2. The beam generation starts directly before the maximal compression and it is terminated at the stage of the pinch expansion.
3. The electron energy in the beam is determined by the resistance of the plasma column.
4. Acceleration of electrons in Z-pinch plasma takes place due to the runaway mechanism under turbulent plasma conditions.

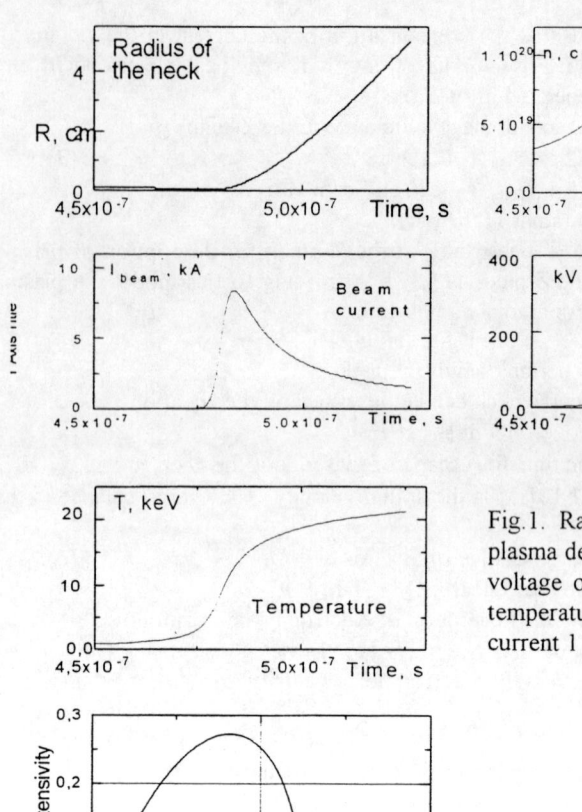

Fig.1. Radius of plasma column r, plasma density n, beam current I_{beam}, voltage on the neck U and plasma temperature T vs the time at the current 1 MA.

Fig.2. Spectra of electron beam for plasma focus discharge at the current 1 MA.

References
1. Citovich V.N., Fizika plasmy, **32**, 157, (1966).
2. Knopfel H. and Spong D.A., *Nuclear Fusion*, **19**, 785 (1979).
3. Cohen R.H., Phys Fluids, **19**, 239, (1976).
4. Haines M., J. Phys.D: Appl. Phys. **11**, 1709,(1978).
5. Chittenden J.P., Haines M.,J. Phys.D :**26**,1048,(1993).
6. Benford G., *Appl.Phys.Lett*, **33**, 983 (1978).
7. Gurevich A.V., JETP, **39**, 1296. (1961).
8. Vikhrev V.V. and Korzhavin V.M., *Sov. J. Plasma Physics*, **4**, 735, (1978).
9. Kingsep A.S. and Rudakov L.I., *Plasma Physics Reports*, **21**, No.7, 576 (1995).
10. Zabajdullin O.Z., Vikhrev V.V., *Phys. Plasmas*, **3**, 2248 (1996).
11. Vikhrev V.V, *Sov. J. Plasma Phys.*, **3**, 539 (1977).
12. Vikhrev V.V. and Braginskii S.I., *Reviews of Plasma Physics*, **10**, Plenum, 425 (1986).

Resistance of Z-Pinch Current Sheath

Victor V.Vikhrev

*Nuclear Fusion Institute, RRC Kurchatov Institute,
Moscow 123182, Russia*

Abstract. The resistance of plasma depends on the frequency of electron Coulomb collisions in pure Coulombian plasmas. The results of numerical studying of the resistance of z-pinch plasma column allowing the presence of kinetic instabilities and the inhomogeneities of plasma electron density along pinch axis are given. The total resistance are found to be reduced due to the electron beam presence and to be independent on the Hall effect. The maximal neck resistance is determined in the case of low energy radiation losses and beam current.

1. Introduction

The resistivity of z-pinch plasma column is one of the most important characteristics of z-pinch discharges, which determines in particular the plasma column voltage, the penetration of magnetic field into plasma, electron beam generation, etc.

The plasma layer resistance in a completely -ionised plasma is determined by Coulomb collisions of electrons with ions. However, with a rise in the current and with that in the magnetic field produced by that current, the presence of kinetic instabilities becomes essential. An analysis [1] shows that low scale - high frequency - instabilities are being developed in the Z-pinch plasma when current electron velocity is considerable. Reduction in the plasma conductivity is a macroscopic sequence of such instability emergence in the plasma. In that case, the additional friction force - provided by scattering of electrons by the oscillations - affects of electron flux, producing the electric current. As a result, the plasma conductivity is determined by the expression:

$$\sigma = \frac{ne^2}{m(\nu_{ei} + \nu_{eff})} \quad (1)$$

where ν_{ei} is the frequency of Coulombian electron collisions, ν_{eff} is the effective frequency of collisions related with kinetic instabilities.

The contribution of various instabilities into the anomalous plasma resistance has been analysed for the Z-pinch in [1]. It has been shown that the greatest contribution into the anomalous plasma resistance is done by a low hybrid drift instability and by electron-sonic one. Both these instabilities emerge under passage of a current across the magnetic field, and at the electron current velocity, $u < v_{Ti}$ they provide the frequency of collisions for electrons equal :

$$\nu_{eff} = (u/v_S)^2 (\omega_{Be}\omega_{Bi})^{1/2}. \qquad (2)$$

Here u is the drift velocity, v_S is the sound velocity, ω_{Be} is the cyclotron frequency of electrons, ω_{Bi} is cyclotron frequency of ions.

One should note that the relationship (2) includes nonlocal value of u/v_S and those a magnetic field but their values averages over some plasma volume. It is related with the fact that the plasma oscillations are propagating in the nearly-located plasma regions, therefore the amplitude of plasma oscillations at the some place is determined by generation of these oscillations in the nearby-located regions.

It has been many times mentioned in the scientific literature that a small fraction of plasma electrons in the existing Z-pinch discharges can make a transfer into the process of acceleration due to a runaway effect. High energy electrons emerging as a result of that phenomenon interact less with ions and with plasma oscillations. Since a friction force affecting these electrons in the beam is reduced, the current layer conductivity is increased. As a result, the plasma column resistance of a pinch drops because of runaway electrons.

Moreover, in the magnetized plasma ($\omega_{Be}\tau_{ei}>1$) there is phenomenon of current concentration near axis. That phenomenon occurs due to the Hall effect, and it takes place in the presence of electron density in homogeneity along the plasma column. This pinching is more brilliantly manifested in the near electrode layers of a discharge since the plasma-electrode layer has the greatest electron density gradient.

2. Theoretical model

The following problem has been considered for analyzing the Hall effect influence on the plasma column resistance. Let us have an immobile plasma column of the Z-pinch confined by intrinsic magnetic field of a current.

Let us find the stationary current distribution in the plasma column of a Z-pinch, when the electric density depends on the direction along the Z-axis only in such a way that:

$$n = n_0 + n_1 \cos(z/\lambda), \qquad (3)$$

where n_1 is the plasma density disturbance along the plasma column ($n_1 \ll n_0$), λ is the wave length of the disturbance.

For this purpose, let us use an equation for the magnetic field within the framework of the magnetic electron hydrodynamics which, in the absence of an electron pressure gradient, has the form:

$$d\mathbf{H}/dt = \text{rot}[\mathbf{u}, \mathbf{H}] - \text{rot}(D\,\text{rot}\,\mathbf{H}), \qquad (4)$$

where the velocity of a current, $\mathbf{u} = \dfrac{c^2}{4\pi e n}\text{rot}\mathbf{H}$ and the coefficient of magnetic field diffusion, $D = \dfrac{c^2}{4\pi\sigma}$ is coefficient of ordinary diffusion. In the cylindrical coordinates, taking account of sinusoidal change in the electron density along the axis, equation (3) for the φ - component of magnetic field is written in the form:

$$\frac{\partial H}{\partial t} = \frac{\partial}{\partial z}\left(D\frac{\partial H}{\partial z}\right) + \frac{\partial}{\partial r}\left(\frac{D}{r}\frac{\partial rH}{\partial r}\right) + \frac{cH}{4\pi er}\frac{\partial rH}{\partial r}\frac{\partial}{\partial z}\left(\frac{1}{n}\right) - \frac{cH}{4\pi e}\frac{\partial H}{\partial z}r^2\frac{\partial}{\partial r}\left(\frac{1}{nr^2}\right),$$
(5)

3. Results of modeling

We analyzed stationary current density distribution being a solution to equation (5) under the condition $\partial H/\partial t = 0$ and long wave disturbance ($\partial H/\partial z \ll \partial H/\partial r$).
This stationary distribution is shown in Fig.1a ($n_1/n_0 = 0.09$) and Fig.1b,2 ($n_1/n_0 = 0.18$) for I = 1 MA, R = 1 cm, T = 100 eV. That figures shows, that the disturbance

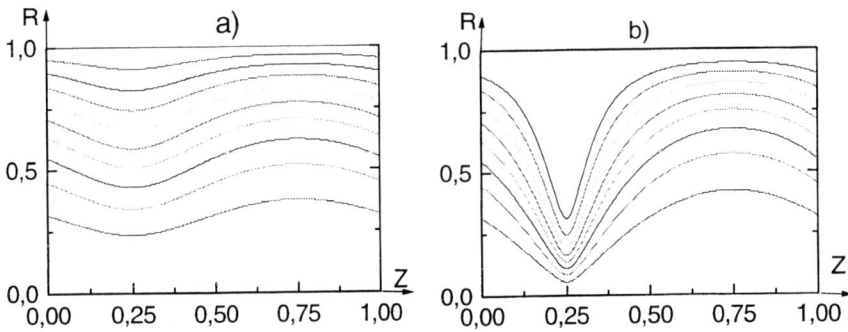

Fig.1. Current lines for I=1 MA, R=1 cm, T = 100 eV, a) $n_1/n_0 = 0.09$, b) $n_1/n_0 =$

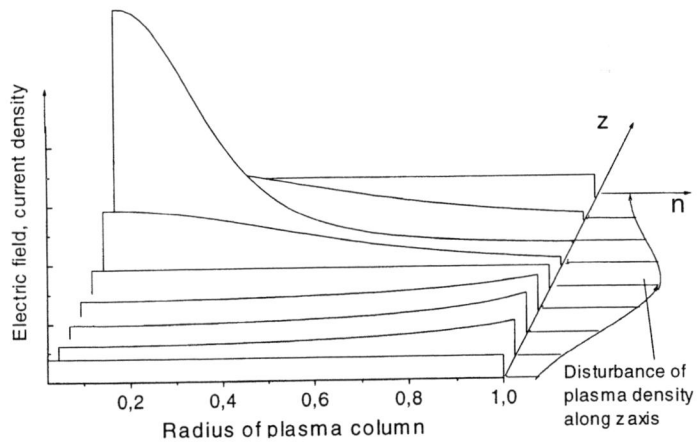

0.18

Fig.2. Current density (or electric field) distribution for I = 1 MA, R = 1 cm, T = 100 eV, n_1/n_0 = 0.18.

of plasma density leads to the curvature of current lines. The more the disturbance-the more concentration of current lines near the axis (in our case) can be achived.

From the solution it follows that the total plasma column resistance is insignificiantly changed in comparison with the case without Hall effect, meanwhile the local electric fields in the pinch abruptly rise. At the places of enhanced electric field., the electron acceleration mode is realized [3].

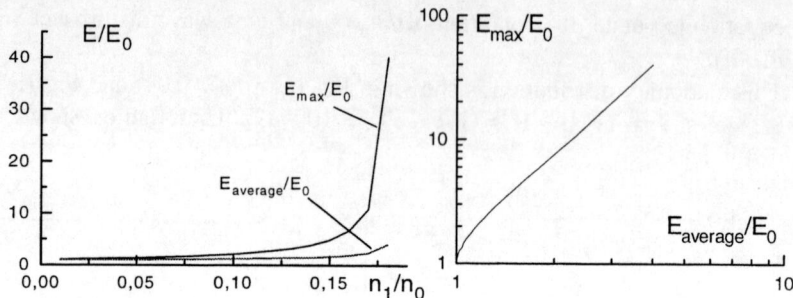

Fig. 3. The increasing of maximal and of
average values of electric field owing to
Hall effect in dependence of n_1/n_0 for
I = 1 MA, R = 1 cm, T = 100 eV

Fig.4. Dependence of increasing
maximal electric field on average
onew may be approximated as
$E_{max}/E_0 = (E_{average}/E_0)^2$

In Fig.3 $E_0 = I/\pi R^2 \sigma$, where R-pinch radius,

σ-pinch conductivity, E_{max}-maximal electric field value in plasma column, $E_{average}$ - average electric field, which is equal to the plasma column voltage devided by the length of the column. Fig.4 shows, that $E_{max}/E_0 = (E_{average}/E_0)^2$.

One should also note that, according to [2], it is possible to estimate the maximal resistance of one neck in the Z-pinch which occurs at an instant of maximal compression. At that instant, the Joule heat release, $Q_{Joule} = U(I-I_{beam})$, is equal to the energy losses from the pinch plasma due to radiation and due to the plasma outflow:

$$U(I - I_{beam}) = 5NTh/\tau + Q_{Rad} \qquad (6)$$

where U is the neck voltage, I is the total current, I_{beam} is the beam current, τ is the characteristic time of plasma outflow from the neck ($\tau = h/2v_s$). At the smallness of radiation energy losses ($Q_{rad} < 5NTh/\tau$) and $I_{beam} \ll I$ this equation determines the resistance of a neck in the rather simple form:

$$R = U/I = (5/2c^2)v_s . \qquad (7)$$

Summary

Thus the resistance of Z-pinch current sheath is determined by three factors:
1. Frequency of collisions between the current electrons, ions and plasma oscillations;
2. Hall mechanism deviating the current passage path, resulting in a reduction in the cross-section, along which the current passes through plasma column;
3. presence of the electron beam because of which the plasma column resistance is reduced.

The maximal neck resistance in the Z-pinch is determined by a balance between the Joule heat release in neck and the energy losses due to the plasma outflow from the neck and due to radiation. In the case of low energy radiation loses and beam current, the maximal neck resistance is determined by the expression (7).

References

1. Vikhrev V.V., Korjavin V.M., «Fizika plasma», **4**, 735, (1978).
2. Vikhrev V.V. and Rozanova G.A., *Phys. Plasma Rep.* **19**, 40 (1993).
3. Vikhrev V.V., Baronova E.O., present conference.

Email List of All First Authors

First Author	Email address
Afonin, V. I.	ovd@dep26.ch70.chel.su
Aglitsky, Y.	aglitskiy@ssd0.nrl.navy.mil
Aliaga-Rossel, R.	r.aliaga-rossel@ic.ac.uk
Atchison, W. L.	wla@lanl.gov
Aubrey, J.	jba@lanl.gov
Auer, P. L.	pla1@cornell.edu
Bakshaev, Yu.L.	potap@potap.msk.su
Baksht, R. B.	baksht@hded2.hcei.tomsk.su
Barnier, J-N	boisard@greco-prog.fr
Baronova, E. O.	baronova@qq.nfi.kiae.su
Bartov, A. V.	kingsep@tapdki.ips.ras.ru
Bateson, W. B.	bateson1@llnl.gov
Bauer, B. S.	bbauer@physics.unr.edu
Beg, F.	f.beg@ic.ac.uk
Benattar, R.	benattar@chausey.polytechnique.fr
Bergmann, K.	k_bergmann@llt.rwth-aachen.de
Bobrova, N. A.	bobrova@vxitep.itep.ru
Branitsky	angara@fly.triniti.troitsk.ru
Chaikovsky, S.	chertov@hded3.hcei.tomsk.su
Cherepanov, K. V.	chekonst@qq.nfi.kiae.su
Chittenden, J. P.	j.chittenden@ic.ac.uk
Choi, P.	pchoi@lpmi.polytechnique.fr
Chong, Y. K.	chong@ppdrh0.nrl.navy.mil
Chuaqui, H.	ewyndham@lascar.puc.cl
Clothiaux, E. J.	ejc@physics.auburn.edu
Coleman, P.	coleman@scubed.com
Commisso, R. J.	commisso@suzie.nrl.navy.mil
Coppins, M.	m.coppins@ic.ac.uk
Davis, J.	davisj@ppdu.nrl.navy.mil
Decker	exphysik@rz.uni-duesseldorf.de
Deeney, C.	cdeene@sandia.gov
DeGroot, J. S.	jdegroot@raphael.engr.ucdavis.edu
Diyankov, O. V.	ovd@dep26.ch70.chel.su
Engel, A.	engel@ilt.fhg.de
Esaulov, A. A.	esaulov@vxitep.itep.ru
Favre, M.	mfavre@lascar.puc.cl
Fisher, A.	fisher@ppdu.nrl.navy.mil
Furuya, S.	ishii@ee.titech.ac.jp
Garanin, S. F.	garanin_072E@spd.vniief.ru
Glazyrin, I. V.	ovd@dep26.ch70.chel.su
Grua, P.	boisard@greco-prog.fr
Haines, M. G.	m.haines@ic.ac.uk

Hammer, J. H.	hammer2@llnl.gov
Horioka, K.	khorioka@es.titech.ac.jp
Howell, D. F.	d.howell@ic.ac.uk
Ivanenkov, G. V.	ivanenkv@sci.lpi.ac.ru
Ivanov, V. V.	ivv@mics.msu.su
Jakubowski, L.	p05msa@cx1.cyf.gov.pl
Kantsyrev, V. L.	victor@physics.unr.edu
Karlykhanov, N. G.	ovd@dep26.ch70.chel.su
Karpinski, L.	angara@fly.triniti.troitsk.ru
Kassapakis, N. G.	n.kassapakis@ic.ac.uk
Kies, W.	kies@bennett.exphy.uni-duesseldorf.de
Koshelev, K. N.	koshelev@lls.isan.troitsk.ru
Koshelev, S. V.	serge@dep26.ch70.chel.su
Kravarik, J.	kravarik@feld.cvut.cz
Kubes, P.	kravarik@feld.cvut.cz
Kukushkin, A. B.	kuka@qq.nfi.kiae.su
Kunze, H. J.	kunze@ep5.ruhr-uni-bochum.de
Larour, J.	larour@lpmi.polytechnique.fr
Lebedev, S. V.	s.lebedev@ic.ac.uk
Lebert, R,	lebert@ilt.fhg.de
Leong, W. S.	wcs@fizik.um.edu.my
Liberman, M. A.	mishal@snobben.teknikum.uu.se
Lindemuth, I.	irl@lanl.gov
Liu, M. H.	NF2135236P@acad2l.ntu.edu.sg
Lorenz, A.	a.lorenz@ic.ac.uk
Lu, M. F.	mflu@aphy202.206.1
Miyamoto, T	miyamoto@phys.cst.nihon-u.ac.jp
Mosher, D.	mosher@ppd.nrl.navy.mil
Muravich, A.	muravich@shotgun.phys.cst.nihon-u.ac.jp
Nash, T. J.	tjnash@sandia.gov
Rahman, H. U.	rahman@ucr.edu
Oreshkin, V. I.	root@hded3.hcei.tomsk.su
Peterson, D.	dlp@lanl.gov
Pikuz, S. A.	pikuz@lps.cornell.edu
Politov, V. Yu.	ovd@dep26.ch70.chel.su
Rantsev-Kartinov, V. A.	rank@qq.nfi.kiae.su
Roewekamp, P.	roewe@bennett.exphy.uni-duesseldorf.de
Rudakov, L.	rudakov@dapinf.msk.ru
Ruiz, J.	j.ruiz@ic.ac.uk
Russkikh, A. G.	baksht@hded2.hcei.tomsk.su
Sanford, T.	twsanfo@sandia.gov
Sasorov, P. V.	sasorov@vxitep3.itep.ru
Sethian, J.	sethian@this.nrl.navy.mil

Sheehey, P. T.	pete@lanl.gov
Shlyaptseva, A. S.	alla@physics.unr.edu
Shlyaptsev, V. N.	slava@lamar.colostate.edu
Smirnov, V.	smirnov@dir.iiapp.msk.su
Sorokin, S. A.	chertov@hded3.hcei.tomsk.su
Soto, L.	hchuaqui@lascar.puc.cl
Spielman, R.	rbspiel@sandia.gov
Takasugi, K.	takasugi@phys.cst.nihon-u.ac.jp
Tatarakis, M.	m.tatarakis@ic.ac.uk
Terry, R. E.	terry@nrlfs1.nrl.navy.mil
Thornhill, J. W.	thornhil@ppdrh0.nrl.navy.mil
Toor, A.	toor1@llnl.gov
Velikovich, A. L.	velikov@bra4a.nrl.navy.mil
Vikhrev, V. V.	vikhrev@qq.nfi.kiae.su
Weber, B. V.	weber@suzie.nrl.navy.mil
Wessel, F. J.	fwessel@uci.edu
Wong, C. S.	wcs@fizik.um.edu.my
Yadlowsky, E. J.	EDYHTR@aol.com
Yap, S. L.	wcs@fizik.um.edu.my
Zabaidullin, O.	oleg@qq.nfi.kiae.su
Zakharov, S.	suzakar@fly.triniti.troitsk.ru
Zdravkovic, D.	j.chittenden@ic.ac.uk
Zhang. G.	nf24281301@acad21.ntu.edu.sg
Zoita, C. Dumitrescu-	czoita@lppmi.polytechnique.fr

AUTHOR INDEX

A

Acton, D. F., 429, 523
Afonin, V. I., 329
Aglitskiy, Y., 437
Ahmad, I., 455
Aizawa, T., 311
Aliaga-Rossel, R., 47, 55, 61, 71, 79, 463, 467, 471, 517
Allshouse, G., 175
Antsiferov, P. S., 303
Apruzese, J. P., 135, 145, 193, 561
Arber, T. D., 585
Aubrey, J. B., 271
Auer, P. L., 579

B

Bakshaev, Yu. L., 149
Baksht, R. B., 141, 307, 555
Barnier, J-N., 229
Baronova, E. O., 67, 443, 475, 483, 611
Bartov, A. V., 149
Bauer, B. S., 153, 499
Bayley, J., 55, 467
Beg, F. N., 71, 315, 333, 339, 407
Bell, A. R., 71, 265
Benattar, R., 211, 233
Bergmann, K., 291, 487
Black, D. C., 135
Bland, S. N., 349
Blinov, P. I., 149
Bobrova, N. A., 225
Bodner, S., 437
Boller, J. R., 135
Bowers, R. L., 201, 271
Branitsky, A. V., 125, 169
Breeze, S. P., 101
Brown, C. M., 437
Brownell, J. H., 201
Bruch, R. F., 499
Bulanov, S. V., 225
Büscher, S., 455

C

Chaikovsky, S. A., 323, 593, 597
Chandler, G. A., 101, 201
Cherepanov, K. V., 345, 381
Chernenko, A. S., 149
Chevalier, J-M., 229
Chittenden, J. P., 71, 79, 349, 463, 471
Choi, P., 51, 75, 161, 353, 357, 373, 385, 417, 491
Chong, Y. K., 277, 283
Chuaqui, H., 47, 353, 357, 373
Churilov, S. S., 361
Chuvatin, A., 161
Clark, R. W., 319, 601
Cochran, F. L., 193, 549
Coleman, P., 119
Commisso, R. J., 135, 459
Coppins, M., 265, 533, 585, 589
Coverdale, C. A., 145
Culverwell, I. D., 589

D

Dangor, A. E., 71, 79, 315, 333, 339, 407, 463, 471, 517
Dan'ko, S. A., 149
Datsko, I. M., 141
Davies, H. M., 219, 507
Davis, J., 145, 193, 277, 283, 319, 549
Decker, G., 21, 71
Deeney, C., 101, 157, 175, 201
DeGroot, J. S., 157, 247, 561
Derzon, M. S., 175, 201
Diyankov, O. V., 243
Dobryakov, A. V., 89
Dorokhin, L. A., 303
Douglas, M. R., 101
Dubroca, B., 229
Dumitrescu-Zoita, C., 51, 75, 161, 353, 357, 373, 491

E

Eaton, M. D., 349
Eder, D. C., 413
Engel, A., 291, 361, 367

Esaulov, A. A., 237, 575
Estabrook, K. G., 157
Etlicher, B., 161

F

Failor, B. H., 145
Farina, D., 225
Favre, M., 47, 353, 357, 373
Fedulov, M. V., 125, 169
Fedunin, A. V., 141, 555
Fehl, D. L., 101
Feldman, U., 437
Feng, X., 389, 423
Fisher, A., 319, 459

G

Garanin, S. F., 93
Gasilov, V. A., 211
Gavrilescu, C., 291, 361, 367
Gerber, K., 437
Gerwin, R. A., 17
Gilliland, T. L., 101, 175
Giuliani, Jr., J., 319
Glazyrin, I. V., 243
Gorbulin, Yu. M., 149
Grabovsky, E. V., 125
Grua, P., 165
Guo, X. M., 423

H

Haines, M. G., 27, 71, 79, 219, 463, 471, 517
Hammer, D. A., 429, 523
Hammer, J. H., 157, 247, 561
Hazelton, R. C., 145
Holland, G., 437
Horioka, K., 311

I

Igusa, T., 299
Ivanenkov, G. V., 253
Ivanov, V. V., 495

J

Jakubowski, L., 443
Jobe, D. O., 101, 175

K

Kalachnikov, M. P., 413
Kalinin, Yu. G., 149
Kammash, T., 277, 283
Kantsyrev, V. L., 153, 499
Karlykhanov, N. G., 243
Karpiński, L., 169
Kassapakis, N. G., 219
Kies, W., 21
Kim, A. A., 141
Kirkpatrick, R., 17
Kokshenev, V. A., 141
Korolev, V. D., 149
Koshelev, K. N., 303, 361, 367
Koshelev, S. V., 243
Kovalchuk, B. M., 141
Kravárik, J., 449, 503, 507
Krukovskii, A. Yu., 211
Kubeš, P., 449, 503, 507
Kukushkin, A. B., 187, 345, 377, 381, 495
Kunze, H.-J., 455

L

Labetsky, A. Yu., 141, 307, 555
Lazarchuk, V. P., 329
Lebedev, S. V., 71, 79, 471
Lebert, R., 291, 361, 367, 487
Lee, S., 389, 423
Lehecka, T., 437
Leong, W. S., 385
LePell, P. D., 145
Li, H., 153
Lindemuth, I. R., 11, 17
Lisitsa, V. S., 187
Liu, M. H., 389
Loginov, S. V., 141
Lorenz, A., 71, 315, 507
Lu, M.-F., 393, 397
Lund, C., 201
Luo, C. M., 423

M

MacFarlane, J. J., 175
Mamin, A., 467
Mancini, R. C., 153, 527
Marder, B. M., 561
Matuska, W., 201
Matzen, M. K., 101, 201
McDaniel, D. H., 101
McGurn, J. S., 101, 175
McKenney, J. L., 101
McLenithan, K., 201
Medovschikov, S. F., 169
Michaelis, M., 349
Mingaleev, A. R., 253
Mitchell, I. H., 47, 61, 71, 463
Miyamoto, T., 299, 401
Mizhiritskii, V. I., 149
Mock, R. C., 101, 201, 561
Moo, S. P., 75, 385
Moosman, B., 39
Moreno, J., 353
Moschella, J. J., 145
Moses, R., 17
Mosher, D., 135, 561
Muravich, A., 83
Murugov, V. M., 329
Myers, M. C., 135

N

Nakajima, M., 311
Nash, T. J., 101, 175, 201, 561
Nedoseev, S. L., 125, 169, 575
Neff, W., 291, 361, 367, 487
Ney, P., 39, 211, 259
Nickles, P. V., 413
Nikiforov, A. F., 211
Nikitin, A., 211, 233
Nizienko, Y., 467
Novikov, V. G., 211

O

Obenschain, S., 437
Olejnik, G. M., 125
Oona, H., 201
Oreshkin, V. I., 215, 555

Osterheld, A. L., 413
Otochin, A. A., 211
Oxner, A., 153

P

Pawley, C., 437
Peterson, D. L., 101, 175, 201, 271, 561
Peterson, G. G., 135, 459
Phaneuf, R. A., 499
Pikuz, S. A., 253, 429, 523
Porter, J. L., 101
Potapov, A. V., 329
Pozzoli, R., 225

R

Rahman, H. U., 39, 259
Rantsev-Kartinov, V. A., 377, 381, 513
Rauch, J., 119
Razinkova, T. L., 225
Rix, W., 119
Robledo, A., 463
Rocca, J. J., 413
Roderick, N. F., 201, 561
Roerich, V. K., 211
Romanova, V. M., 253, 429
Romeas, P., 47
Ross, I., 339
Rostoker, N., 39, 259
Rudakov, L. I., 149, 183, 187
Ruiz-Camacho, J., 315, 333, 407
Russell, P. G. F., 585
Russkikh, A. G., 141, 307, 555
Ryutov, D. D., 157, 579, 607

S

Saavedra, R., 47, 79, 357
Sadowski, M., 443
Sajer, J. M., 165
Sandner, W., 413
Sanford, T. W. L., 101, 157, 201, 561
Sarkisov, G. S., 429
Sasorov, P. V., 225, 237, 575
Scheffel, J., 585
Schmidt, H., 61

Scholz, M., 169
Seamen, J. F., 101, 175
Seely, J., 437
Senik, A. V., 329
Serguei, T., 417
Sethian, J., 3, 437
Sevastianov, A., 165
Shashkov, A. Yu., 149
Sheehey, P., 17
Shelkovenko, T. A., 253, 429, 523
Shibaev, S. A., 149
Shishlov, A. V., 141, 307, 555
Shlyaptsev, V. N., 413
Shlyaptseva, A. S., 153, 499, 527
Sidelnikov, Yu. V., 303, 361, 367
Silva, P., 373
Skowronek, M., 47
Smirnov, V. P., 125, 169
Solomyannaya, A. D., 211
Song, Y., 39
Sorokin, S. A., 323, 593, 597
Soto, L., 47, 357
Spielman, R. B., 101, 157, 201, 561
Starostin, A. N., 187, 211
Stein, S., 21
Stepanov, A. E., 211
Stephanakis, S. J., 135, 459
Stepniewski, W., 169, 253
Struve, K. W., 101
Stygar, W. A., 101
Szydlowski, A., 169

T

Tabak, M., 247
Takasugi, K., 299
Tatarakis, M., 517
Tatsumi, K., 299
Terentiev, A. R., 377, 381
Terry, R. E., 601
Thompson, J., 119
Thornhill, J. W., 145, 193
Toor, A., 157, 247, 607
Torres, J. A., 101
Treichel, O., 291

Trofimovich, E. E., 513
Tsuchida, M., 311

V

Van Drie, A., 39
Vargas, M., 101
Velikovich, A. L., 549
Vikhrev, V. V., 67, 87, 89, 443, 611, 615

W

Wagoner, T., 101
Wang, P., 175
Weber, B. V., 135, 459
Wessel, F. J., 39, 259
Whitney, K. G., 145, 193, 561
Wilson, R., 119
Winterberg, F., 153
Wong, C. S., 75, 385, 417
Worley, J. F., 79, 315, 349
Wrubel, Th., 455
Wyndham, E., 47, 353, 357, 373
Wysocki, F., 17

Y

Yadlowsky, E. J., 145
Yakunin, I. I., 187
Yap, S. L., 75
Young, F. C., 135

Z

Zabaidullin, O. Z., 87, 89
Zagar, D. M., 101
Zakharov, S. V., 125, 211
Zdravkovic, D., 265
Zhang, G. X., 423
Zimmerman, G. B., 247
Zurin, M. V., 169

AIP Conference Proceedings

	Title	L.C. Number	ISBN
No. 249	The Physics of Particle Accelerators (Upton, NY 1989, 1990)	92-52843	0-88318-789-2
No. 250	Towards a Unified Picture of Nuclear Dynamics (Nikko, Japan 1991)	92-70143	0-88318-951-8
No. 251	Superconductivity and its Applications (Buffalo, NY 1991)	92-52726	1-56396-016-8
No. 252	Accelerator Instrumentation (Newport News, VA 1991)	92-70356	0-88318-934-8
No. 253	High-Brightness Beams for Advanced Accelerator Applications (College Park, MD 1991)	92-52705	0-88318-947-X
No. 254	Testing the AGN Paradigm (College Park, MD 1991)	92-52780	1-56396-009-5
No. 255	Advanced Beam Dynamics Workshop on Effects of Errors in Accelerators, Their Diagnosis and Corrections (Corpus Christi, TX 1991)	92-52842	1-56396-006-0
No. 256	Slow Dynamics in Condensed Matter (Fukuoka, Japan 1991)	92-53120	0-88318-938-0
No. 257	Atomic Processes in Plasmas (Portland, ME 1991)	91-08105	0-88318-939-9
No. 258	Synchrotron Radiation and Dynamic Phenomena (Grenoble, France 1991)	92-53790	1-56396-008-7
No. 259	Future Directions in Nuclear Physics with 4π Gamma Detection Systems of the New Generation (Strasbourg, France 1991)	92-53222	0-88318-952-6
No. 260	Computational Quantum Physics (Nashville, TN 1991)	92-71777	0-88318-933-X
No. 261	Rare and Exclusive B&K Decays and Novel Flavor Factories (Santa Monica, CA 1991)	92-71873	1-56396-055-9
No. 262	Molecular Electronics—Science and Technology (St. Thomas, Virgin Islands 1991)	92-72210	1-56396-041-9
No. 263	Stress-Induced Phenomena in Metallization: First International Workshop (Ithaca, NY 1991)	92-72292	1-56396-082-6
No. 264	Particle Acceleration in Cosmic Plasmas (Newark, DE 1991)	92-73316	0-88318-948-8
No. 265	Gamma-Ray Bursts (Huntsville, AL 1991)	92-73456	1-56396-018-4
No. 266	Group Theory in Physics (Cocoyoc, Morelos, Mexico 1991)	92-73457	1-56396-101-6

	Title	L.C. Number	ISBN
No. 267	Electromechanical Coupling of the Solar Atmosphere (Capri, Italy 1991)	92-82717	1-56396-110-5
No. 268	Photovoltaic Advanced Research & Development Project (Denver, CO 1992)	92-74159	1-56396-056-7
No. 269	CEBAF 1992 Summer Workshop (Newport News, VA 1992)	92-75403	1-56396-067-2
No. 270	Time Reversal—The Arthur Rich Memorial Symposium (Ann Arbor, MI 1991)	92-83852	1-56396-105-9
No. 271	Tenth Symposium Space Nuclear Power and Propulsion (Vols. I–III) (Albuquerque, NM 1993)	92-75162	1-56396-137-7 (set)
No. 272	Proceedings of the XXVI International Conference on High Energy Physics (Vols. I and II) (Dallas, TX 1992)	93-70412	1-56396-127-X (set)
No. 273	Superconductivity and Its Applications (Buffalo, NY 1992)	93-70502	1-56396-189-X
No. 274	VIth International Conference on the Physics of Highly Charged Ions (Manhattan, KS 1992)	93-70577	1-56396-102-4
No. 275	Atomic Physics 13 (Munich, Germany 1992)	93-70826	1-56396-057-5
No. 276	Very High Energy Cosmic-Ray Interactions: VIIth International Symposium (Ann Arbor, MI 1992)	93-71342	1-56396-038-9
No. 277	The World at Risk: Natural Hazards and Climate Change (Cambridge, MA 1992)	93-71333	1-56396-066-4
No. 278	Back to the Galaxy (College Park, MD 1992)	93-71543	1-56396-227-6
No. 279	Advanced Accelerator Concepts (Port Jefferson, NY 1992)	93-71773	1-56396-191-1
No. 280	Compton Gamma-Ray Observatory (St. Louis, MO 1992)	93-71830	1-56396-104-0
No. 281	Accelerator Instrumentation Fourth Annual Workshop (Berkeley, CA 1992)	93-072110	1-56396-190-3
No. 282	Quantum 1/f Noise & Other Low Frequency Fluctuations in Electronic Devices (St. Louis, MO 1992)	93-072366	1-56396-252-7
No. 283	Earth and Space Science Information Systems (Pasadena, CA 1992)	93-072360	1-56396-094-X

	Title	L.C. Number	ISBN
No. 284	US-Japan Workshop on Ion Temperature Gradient-Driven Turbulent Transport (Austin, TX 1993)	93-72460	1-56396-221-7
No. 285	Noise in Physical Systems and 1/f Fluctuations (St. Louis, MO 1993)	93-72575	1-56396-270-5
No. 286	Ordering Disorder: Prospect and Retrospect in Condensed Matter Physics: Proceedings of the Indo-U.S. Workshop (Hyderabad, India 1993)	93-072549	1-56396-255-1
No. 287	Production and Neutralization of Negative Ions and Beams: Sixth International Symposium (Upton, NY 1992)	93-72821	1-56396-103-2
No. 288	Laser Ablation: Mechanismas and Applications-II: Second International Conference (Knoxville, TN 1993)	93-73040	1-56396-226-8
No. 289	Radio Frequency Power in Plasmas: Tenth Topical Conference (Boston, MA 1993)	93-72964	1-56396-264-0
No. 290	Laser Spectroscopy: XIth International Conference (Hot Springs, VA 1993)	93-73050	1-56396-262-4
No. 291	Prairie View Summer Science Academy (Prairie View, TX 1992)	93-73081	1-56396-133-4
No. 292	Stability of Particle Motion in Storage Rings (Upton, NY 1992)	93-73534	1-56396-225-X
No. 293	Polarized Ion Sources and Polarized Gas Targets (Madison, WI 1993)	93-74102	1-56396-220-9
No. 294	High-Energy Solar Phenomena: A New Era of Spacecraft Measurements (Waterville Valley, NH 1993)	93-74147	1-56396-291-8
No. 295	The Physics of Electronic and Atomic Collisions: XVIII International Conference (Aarhus, Denmark, 1993)	93-74103	1-56396-290-X
No. 296	The Chaos Paradigm: Developments an Applications in Engineering and Science (Mystic, CT 1993)	93-74146	1-56396-254-3
No. 297	Computational Accelerator Physics (Los Alamos, NM 1993)	93-74205	1-56396-222-5
No. 298	Ultrafast Reaction Dynamics and Solvent Effects (Royaumont, France 1993)	93-074354	1-56396-280-2
No. 299	Dense Z-Pinches: Third International Conference (London, 1993)	93-074569	1-56396-297-7

	Title	L.C. Number	ISBN
No. 300	Discovery of Weak Neutral Currents: The Weak Interaction Before and After (Santa Monica, CA 1993)	94-70515	1-56396-306-X
No. 301	Eleventh Symposium Space Nuclear Power and Propulsion (3 Vols.) (Albuquerque, NM 1994)	92-75162	1-56396-305-1 (set) 156396-301-9 (pbk. set)
No. 302	Lepton and Photon Interactions/ XVI International Symposium (Ithaca, NY 1993)	94-70079	1-56396-106-7
No. 303	Slow Positron Beam Techniques for Solids and Surfaces Fifth International Workshop (Jackson Hole, WY 1992)	94-71036	1-56396-267-5
No. 304	The Second Compton Symposium (College Park, MD 1993)	94-70742	1-56396-261-6
No. 305	Stress-Induced Phenomena in Metallization Second International Workshop (Austin, TX 1993)	94-70650	1-56396-251-9
No. 306	12th NREL Photovoltaic Program Review (Denver, CO 1993)	94-70748	1-56396-315-9
No. 307	Gamma-Ray Bursts Second Workshop (Huntsville, AL 1993)	94-71317	1-56396-336-1
No. 308	The Evolution of X-Ray Binaries (College Park, MD 1993)	94-76853	1-56396-329-9
No. 309	High-Pressure Science and Technology—1993 (Colorado Springs, CO 1993)	93-72821	1-56396-219-5 (set)
No. 310	Analysis of Interplanetary Dust (Houston, TX 1993)	94-71292	1-56396-341-8
No. 311	Physics of High Energy Particles in Toroidal Systems (Irvine, CA 1993)	94-72098	1-56396-364-7
No. 312	Molecules and Grains in Space (Mont Sainte-Odile, France 1993)	94-72615	1-56396-355-8
No. 313	The Soft X-Ray Cosmos ROSAT Science Symposium (College Park, MD 1993)	94-72499	1-56396-327-2
No. 314	Advances in Plasma Physics Thomas H. Stix Symposium (Princeton, NJ 1992)	94-72721	1-56396-372-8
No. 315	Orbit Correction and Analysis in Circular Accelerators (Upton, NY 1993)	94-72257	1-56396-373-6

	Title	L.C. Number	ISBN
No. 316	Thirteenth International Conference on Thermoelectrics (Kansas City, Missouri 1994)	95-75634	1-56396-444-9
No. 317	Fifth Mexican School of Particles and Fields (Guanajuato, Mexico 1992)	94-72720	1-56396-378-7
No. 318	Laser Interaction and Related Plasma Phenomena 11th International Workshop (Monterey, CA 1993)	94-78097	1-56396-324-8
No. 319	Beam Instrumentation Workshop (Santa Fe, NM 1993)	94-78279	1-56396-389-2
No. 320	Basic Space Science (Lagos, Nigeria 1993)	94-79350	1-56396-328-0
No. 321	The First NREL Conference on Thermophotovoltaic Generation of Electricity (Copper Mountain, CO 1994)	94-72792	1-56396-353-1
No. 322	Atomic Processes in Plasmas Ninth APS Topical Conference (San Antonio, TX)	94-72923	1-56396-411-2
No. 323	Atomic Physics 14 Fourteenth International Conference on Atomic Physics (Boulder, CO 1994)	94-73219	1-56396-348-5
No. 324	Twelfth Symposium on Space Nuclear Power and Propulsion (Albuquerque, NM 1995)	94-73603	1-56396-427-9
No. 325	Conference on NASA Centers for Commercial Development of Space (Albuquerque, NM 1995)	94-73604	1-56396-431-7
No. 326	Accelerator Physics at the Superconducting Super Collider (Dallas, TX 1992-1993)	94-73609	1-56396-354-X
No. 327	Nuclei in the Cosmos III Third International Symposium on Nuclear Astrophysics (Assergi, Italy 1994)	95-75492	1-56396-436-8
No. 328	Spectral Line Shapes, Volume 8 12th ICSLS (Toronto, Canada 1994)	94-74309	1-56396-326-4
No. 329	Resonance Ionization Spectroscopy 1994 Seventh International Symposium (Bernkastel-Kues, Germany 1994)	95-75077	1-56396-437-6
No. 330	E.C.C.C. 1 Computational Chemistry F.E.C.S. Conference (Nancy, France 1994)	95-75843	1-56396-457-0
No. 331	Non-Neutral Plasma Physics II (Berkeley, CA 1994)	95-79630	1-56396-441-4

Title	L.C. Number	ISBN
No. 332 X-Ray Lasers 1994 Fourth International Colloquium (Williamsburg, VA 1994)	95-76067	1-56396-375-2
No. 333 Beam Instrumentation Workshop (Vancouver, B. C., Canada 1994)	95-79635	1-56396-352-3
No. 334 Few-Body Problems in Physics (Williamsburg, VA 1994)	95-76481	1-56396-325-6
No. 335 Advanced Accelerator Concepts (Fontana, WI 1994)	95-78225	1-56396-476-7 (set) 1-56396-474-0 (Book) 1-56396-475-9 (CD-Rom)
No. 336 Dark Matter (College Park, MD 1994)	95-76538	1-56396-438-4
No. 337 Pulsed RF Sources for Linear Colliders (Montauk, NY 1994)	95-76814	1-56396-408-2
No. 338 Intersections Between Particle and Nuclear Physics 5th Conference (St. Petersburg, FL 1994)	95-77076	1-56396-335-3
No. 339 Polarization Phenomena in Nuclear Physics Eighth International Symposium (Bloomington, IN 1994)	95-77216	1-56396-482-1
No. 340 Strangeness in Hadronic Matter (Tucson, AZ 1995)	95-77477	1-56396-489-9
No. 341 Volatiles in the Earth and Solar System (Pasadena, CA 1994)	95-77911	1-56396-409-0
No. 342 CAM -94 Physics Meeting (Cacun, Mexico 1994)	95-77851	1-56396-491-0
No. 343 High Energy Spin Physics Eleventh International Symposium (Bloomington, IN 1994)	95-78431	1-56396-374-4
No. 344 Nonlinear Dynamics in Particle Accelerators: Theory and Experiments (Arcidosso, Italy 1994)	95-78135	1-56396-446-5
No. 345 International Conference on Plasma Physics ICPP 1994 (Foz do Iguaçu, Brazil 1994)	95-78438	1-56396-496-1
No. 346 International Conference on Accelerator-Driven Transmutation Technologies and Applications (Las Vegas, NV 1994)	95-78691	1-56396-505-4
No. 347 Atomic Collisions: A Symposium in Honor of Christopher Bottcher (1945-1993) (Oak Ridge, TN 1994)	95-78689	1-56396-322-1

	Title	L.C. Number	ISBN
No. 348	Unveiling the Cosmic Infrared Background (College Park, MD, 1995)	95-83477	1-56396-508-9
No. 349	Workshop on the Tau/Charm Factory (Argonne, IL, 1995)	95-81467	1-56396-523-2
No. 350	International Symposium on Vector Boson Self-Interactions (Los Angeles, CA 1995)	95-79865	1-56396-520-8
No. 351	The Physics of Beams Andrew Sessler Symposium (Los Angeles, CA 1993)	95-80479	1-56396-376-0
No. 352	Physics Potential and Development of $\mu^+\mu^-$ Colliders: Second Workshop (Sausalito, CA 1994)	95-81413	1-56396-506-2
No. 353	13th NREL Photovoltaic Program Review (Lakewood, CO 1995)	95-80662	1-56396-510-0
No. 354	Organic Coatings (Paris, France, 1995)	96-83019	1-56396-535-6
No. 355	Eleventh Topical Conference on Radio Frequency Power in Plasmas (Palm Springs, CA 1995)	95-80867	1-56396-536-4
No. 356	The Future of Accelerator Physics (Austin, TX 1994)	96-83292	1-56396-541-0
No. 357	10th Topical Workshop on Proton-Antiproton Collider Physics (Batavia, IL 1995)	95-83078	1-56396-543-7
No. 358	The Second NREL Conference on Thermophotovoltaic Generation of Electricity	95-83335	1-56396-509-7
No. 359	Workshops and Particles and Fields and Phenomenology of Fundamental Interactions (Puebla, Mexico 1995)	96-85996	1-56396-548-8
No. 360	The Physics of Electronic and Atomic Collisions XIX International Conference (Whistler, Canada, 1995)	95-83671	1-56396-440-6
No. 361	Space Technology and Applications International Forum (Albuquerque, NM 1996)	95-83440	1-56396-568-2
No. 362	Two-Center Effects in Ion-Atom Collisions (Lincoln, NE 1994)	96-83379	1-56396-342-6
No. 363	Phenomena in Ionized Gases XXII ICPIG (Hoboken, NJ, 1995)	96-83294	1-56396-550-X
No. 364	Fast Elementary Processes in Chemical and Biological Systems (Villeneuve d'Ascq, France, 1995)	96-83624	1-56396-564-X

	Title	L.C. Number	ISBN
No. 365	Latin-American School of Physics XXX ELAF Group Theory and Its Applications (México City, México, 1995)	96-83489	1-56396-567-4
No. 366	High Velocity Neutron Stars and Gamma-Ray Bursts (La Jolla, CA 1995)	96-84067	1-56396-593-3
No. 367	Micro Bunches Workshop (Upton, NY, 1995)	96-83482	1-56396-555-0
No. 368	Acoustic Particle Velocity Sensors: Design, Performance and Applications (Mystic, CT, 1995)	96-83548	1-56396-549-6
No. 369	Laser Interaction and Related Plasma Phenomena (Osaka, Japan 1995)	96-85009	1-56396-445-7
No. 370	Shock Compression of Condensed Matter-1995 (Seattle, WA 1995)	96-84595	1-56396-566-6
No. 371	Sixth Quantum 1/f Noise and Other Low Frequency Fluctuations in Electronic Devices Symposium (St. Louis, MO, 1994)	96-84200	1-56396-410-4
No. 372	Beam Dynamics and Technology Issues for + - Colliders 9th Advanced ICFA Beam Dynamics Workshop (Montauk, NY, 1995)	96-84189	1-56396-554-2
No. 373	Stress-Induced Phenomena in Metallization (Palo Alto, CA 1995)	96-84949	1-56396-439-2
No. 374	High Energy Solar Physics (Greenbelt, MD 1995)	96-84513	1-56396-542-9
No. 375	Chaotic, Fractal, and Nonlinear Signal Processing (Mystic, CT 1995)	96-85356	1-56396-443-0
No. 376	Chaos and the Changing Nature of Science and Medicine: An Introduction (Mobile, AL 1995)	96-85220	1-56396-442-2
No. 377	Space Charge Dominated Beams and Applications of High Brightness Beams (Bloomington, IN 1995)	96-85165	1-56396-625-7
No. 378	Surfaces, Vacuum, and Their Applications (Cancun, Mexico 1994)	96-85594	1-56396-418-X
No. 379	Physical Origin of Homochirality in Life (Santa Monica, CA 1995)	96-86631	1-56396-507-0
No. 380	Production and Neutralization of Negative Ions and Beams / Production and Application of Light Negative Ions (Upton, NY 1995)	96-86435	1-56396-565-8
No. 381	Atomic Processes in Plasmas (San Francisco, CA 1996)	96-86304	1-56396-552-6

	Title	L.C. Number	ISBN
No. 382	Solar Wind Eight (Dana Point, CA 1995)	96-86447	1-56396-551-8
No. 383	Workshop on the Earth's Trapped Particle Environment (Taos, NM 1994)	96-86619	1-56396-540-2
No. 384	Gamma-Ray Bursts (Huntsville, AL 1995)	96-79458	1-56396-685-9
No. 385	Robotic Exploration Close to the Sun: Scientific Basis (Marlboro, MA 1996)	96-79560	1-56396-618-2
No. 386	Spectral Line Shapes, Volume 9 13th ICSLS (Firenze, Italy 1996)		1-56396-656-5
No. 387	Space Technology and Applications International Forum (Albuquerque, NM 1997)	96-80254	1-56396-679-4 (Case set) 1-56396-691-3 (Paper set)
No. 388	Resonance Ionization Spectroscopy 1996 Eighth International Symposium (State College, PA 1996)	96-80324	1-56396-611-5
No. 389	X-Ray and Inner-Shell Processes 17th International Conference (Hamburg, Germany 1996)	96-80388	1-56396-563-1
No. 390	Beam Instrumentation Proceedings of the Seventh Workshop (Argonne, IL 1996)	97-70568	1-56396-612-3
No. 391	Computational Accelerator Physics (Williamsburg, VA 1996)	97-70181	1-56396-671-9
No. 392	Applications of Accelerators in Research and Industry: Proceedings of the Fourteenth International Conference (Denton, TX 1996)	97-71846	1-56396-652-2
No. 393	Star Formation Near and Far Seventh Astrophysics Conference (College Park, MD 1996)	97-71978	1-56396-678-6
No. 394	NREL/SNL Photovoltaics Program Review Proceedings of the 14th Conference— A Joint Meeting (Lakewood, CO 1996)	97-72645	1-56396-687-5
No. 395	Nonlinear and Collective Phenomena in Beam Physics (Arcidosso, Italy 1996)	97-72970	1-56396-668-9
No. 396	New Modes of Particle Acceleration— Techniques and Sources (Santa Barbara, CA 1996)	97-72977	1-56396-728-6
No. 397	Future High Energy Colliders (Santa Barbara, CA 1997)	97-73333	1-56396-729-4

Title	L.C. Number	ISBN
No. 398 Advanced Accelerator Colliders Seventh Workshop (Lake Tahoe, CA 1996)	97-72788	1-56396-697-2 (set) 1-56396-727-8 (cloth) 1-56396-726-X (CD-Rom)
No. 399 The Changing Role of Physics Departments (College Park, MD 1996)	97-74866	1-56396-698-0
No. 400 High Energy Physics First Latin Symposium (Yucatan, México 1996)	97-73971	1-56396-686-7
No. 401 Thermophotovoltaic Generation of Electricity Third NREL Conference (Colorado Springs, CO 1997)	97-74374	1-56396-734-0
No. 402 Astrophysical Implications of the Laboratory Study of Presolar Materials (St. Louis, MO 1996)	97-74679	1-56396-664-6
No. 403 Radio Frequency Power in Plasmas 12th Topical Conference (Savannah, GA 1997)	97-74472	1-56396-709-X
No. 404 Future Generations Photovoltaic Technologies First NREL Conference (Denver, CO 1997)	97-74386	1-56396-704-9
No. 405 Beam Stability and Nonlinear Dynamics (Santa Barbara, CA 1996)	97-74676	1-56396-731-6
No. 406 Laser Interaction and Related Plasma Phenomena 13th International Conference (Monterey, CA 1997)	97-76763	1-56396-696-4
No. 407 Deep Inelastic Scattering and QCD 5th International Workshop (Chicago, IL 1997)	97-74677	1-56396-716-2
No. 408 The Ultraviolet Universe at Low and High Redshift (College Park, MD 1997)	97-76762	1-56396-708-1
No. 409 Dense 2-Pinches 4th International Conference (Vancouver, Canada 1997)	97-76959	1-56396-610-7
No. 410 Proceedings of the 4th Compton Symposium (Williamsburg, VA 1997)	97-77179	1-56396-659-X (set)
No. 411 Applied Non-Linear Dynamics Near the Millenium (San Diego, CA 1997)	97-77035	1-56396-736-7